The Physics of Musical Instruments

Second Edition

Neville H. Fletcher Thomas D. Rossing

The Physics of Musical Instruments

Second Edition

With 485 Illustrations

Neville H. Fletcher
Research School of Physical
Sciences and Engineering
Australian National University
Canberra, A.C.T. 0200
Australia

Thomas D. Rossing
Department of Physics
Northern Illinois University
DeKalb, IL 60115
USA

Cover illustration: French horn. © The Viesti Collection, Inc.

Library of Congress Cataloging-in-Publication Data
Fletcher, Neville H. (Neville Horner)
The physics of musical instruments / Neville H. Fletcher : Thomas D. Rossing. — 2nd ed.
p. cm.
Includes bibliographical references (p.) and index.

1. Music — Acoustics and physics. 2. Musical instruments — Construction. I. Rossing, Thomas D., 1929– II. Title.
ML3805.F58 1998
784.19'01'53—dc21 97-35360

ISBN 978-1-4419-3120-7 e-ISBN 978-0-387-21603-4

Printed in the United States of America. (BPI/MVY)

springeronline.com

Preface

When we wrote the first edition of this book, we directed our presentation to the reader with a compelling interest in musical instruments who has "a reasonable grasp of physics and who is not frightened by a little mathematics." We are delighted to find how many such people there are.

The opportunity afforded by the preparation of this second edition has allowed us to bring our discussion up to date by including those new insights that have arisen from the work of many dedicated researchers over the past decade. We have also taken the opportunity to revise our presentation of some aspects of the subject to make it more general and, we hope, more immediately accessible. We have, of course, corrected any errors that have come to our attention, and we express our thanks to those friends who pointed out such defects in the early printings of the first edition.

We hope that this book will continue to serve as a guide, both to those undertaking research in the field and to those who simply have a deep interest in the subject.

June 1997 N.H.F and T.D.R.

Preface to the First Edition

The history of musical instruments is nearly as old as the history of civilization itself, and the aesthetic principles upon which judgments of musical quality are based are intimately connected with the whole culture within which the instruments have evolved. An educated modern Western player or listener can make critical judgments about particular instruments or particular performances but, to be valid, those judgments must be made within the appropriate cultural context.

The compass of our book is much less sweeping than the first paragraph might imply, and indeed our discussion is primarily confined to Western musical instruments in current use, but even here we must take account of centuries of tradition. A musical instrument is designed and built for the playing of music of a particular type and, conversely, music is written to be performed on particular instruments. There is no such thing as an "ideal" instrument, even in concept, and indeed the unbounded possibilities of modern digital sound-synthesis really require the composer or performer to define a whole set of instruments if the result is to have any musical coherence. Thus, for example, the sound and response of a violin are judged against a mental image of a perfect violin built up from experience of violins playing music written for them over the centuries. A new instrument may be richer in sound quality and superior in responsiveness, but if it does not fit that image, then it is not a better violin.

This set of mental criteria has developed, through the interaction of musical instruments makers, performers, composers, and listeners, over several centuries for most musical instruments now in use. The very features of particular instruments that might be considered as acoustic defects have become their subtle distinguishing characteristics, and technical "improvements" that have not preserved those features have not survived. There are, of course, cases in which revolutionary new features have prevailed over tradition, but these have resulted in almost new instrument types—the violin and cello in place of the viols, the Boehm flute in place of its baroque ancestor, and the saxophone in place of the taragato. Fortunately, perhaps, such profound changes are rare, and most instruments of today have evolved quite slowly, with minor tonal or technical improvements reflecting the gradually changing mental image of the ideal instrument of that type.

The role of acoustical science in this context is an interesting one. Centuries of tradition have developed great skill and understanding among the makers of musical instruments, and they are often aware of subtleties that are undetected by modern acoustical instrumentation for lack of precise technical criteria for their recognition. It is difficult, therefore, for a scientist to point the way forward unless the problem or the opportunity has been identified adequately by the performer or the maker. Only rarely do all these skills come together in a single person.

The first and major role of acoustics is therefore to try to understand all the details of sound production by traditional instruments. This is a really major program, and indeed it is only within the past few decades that we have achieved even a reasonable understanding of the basic mechanisms determining tone quality in most instruments. In some cases even major features of the sounding mechanism itself have only recently been unravelled. This is an intellectual exercise of great fascination, and most of our book is devoted to it. Our understanding of a particular area will be reasonably complete only when we know the physical causes of the differences between a fine instrument and one judged to be of mediocre quality. Only then may we hope that science can come to the help of music in moving the design or performance of contemporary instruments closer to the present ideal.

This book is a record of the work of very many people who have studied the physics of musical instruments. Most of them, following a long tradition, have done so as a labor of love, in time snatched from scientific or technical work in a field of more immediate practical importance. The community of those involved is a world-wide and friendly one in which ideas are freely exchanged, so that, while we have tried to give credit to the originators wherever possible, there will undoubtedly be errors of oversight. For these we apologize. We have also had to be selective, and many interesting topics have perforce been omitted. Again the choice is ours, and has been influenced by our own particular interests, though we have tried to give a reasonably balanced treatment of the whole field.

The reader we had in mind in compiling this volume is one with a reasonable grasp of physics and who is not frightened by a little mathematics. There are fine books in plenty about the history of particular musical instruments, lavishly illustrated with photographs and drawings, but there is virtually nothing outside the scientific journal literature that attempts to come to grips with the subject on a quantitative basis. We hope that we have remedied that lack. We have not avoided mathematics where precision is necessary or where hand-waving arguments are inadequate, but at the same time we have not pursued formalism for its own sake. Detailed physical explanation has always been our major objective. We hope that the like-minded reader will enjoy coming to grips with this fascinating subject.

The authors owe a debt of gratitude to many colleagues who have contributed to this book. Special thanks are due to Joanna Daly and Barbara Sullivan, who typed much of the manuscript and especially to Virginia Plemons, who typed most of the final draft and prepared a substantial part of the artwork. Several colleagues assisted in the proofreading, including Rod Korte, Krista McDonald, David Brown, George Jelatis, and Brian Finn. We are grateful to David Peterson, Ted Mansell, and other careful readers who alerted us to errors in the first printing. Thanks are due to our many colleagues for allowing us to reprint figures and data from their publications, and to the musical instrument manufacturers that supplied us with photographs. Most of all, we thank our colleagues in the musical acoustics community for many valuable discussions through the years that led to our writing this book.

December 1988

Neville H. Fletcher
Thomas D. Rossing

Contents

Part I

Vibrating Systems

1

Free and Forced Vibrations of Simple Systems

Mechanical, acoustical, or electrical vibrations are the sources of sound in musical instruments. Some familiar examples are the vibrations of strings (violin, guitar, piano, etc.), bars or rods (xylophone, glockenspiel, chimes, clarinet reed), membranes (drums, banjo), plates or shells (cymbal, gong, bell), air in a tube (organ pipe, brass and woodwind instruments, marimba resonator), and air in an enclosed container (drum, violin, or guitar body).

In most instruments, sound production depends upon the collective behavior of several vibrators, which may be weakly or strongly coupled together. This coupling, along with nonlinear feedback, may cause the instrument as a whole to behave as a complex vibrating system, even though the individual elements are relatively simple vibrators.

In the first eight chapters, we will discuss the physics of mechanical and acoustical oscillators, the way in which they may be coupled together, and the way in which they radiate sound. Since we are not discussing electronic musical instruments, we will not deal with electrical oscillators except as they help us, by analogy, to understand mechanical and acoustical oscillators.

Many objects are capable of vibrating or oscillating. Mechanical vibrations require that the object possess two basic properties: a stiffness or springlike quality to provide a restoring force when displaced and inertia, which causes the resulting motion to overshoot the equilibrium position. From an energy standpoint, oscillators have a means for storing potential energy (spring), a means for storing kinetic energy (mass), and a means by which energy is gradually lost (damper). Vibratory motion involves the alternating transfer of energy between its kinetic and potential forms.

The inertial mass may be either concentrated in one location or distributed throughout the vibrating object. If it is distributed, it is usually the mass per unit length, area, or volume that is important. Vibrations in distributed mass systems may be viewed as standing waves.

The restoring forces depend upon the elasticity or the compressibility of some material. Most vibrating bodies obey Hooke's law; that is, the restoring force is proportional to the displacement from equilibrium, at least for small displacement.

1.1 Simple Harmonic Motion in One Dimension

The simplest kind of periodic motion is that experienced by a point mass moving along a straight line with an acceleration directed toward a fixed point and proportional to the distance from that point. This is called simple harmonic motion, and it can be described by a sinusoidal function of time $t : x(t) = A \sin 2\pi f t$, where the amplitude A describes the maximum extent of the motion, and the frequency f tells us how often it repeats.

The period of the motion is given by

$$T = \frac{1}{f}. \tag{1.1}$$

That is, each T seconds the motion repeats itself.

A simple example of a system that vibrates with simple harmonic motion is the mass–spring system shown in Fig. 1.1. We assume that the amount of stretch x is proportional to the restoring force F (which is true in most springs if they are not stretched too far), and that the mass slides freely without loss of energy. The equation of motion is easily obtained by combining Hooke's law, $F = -Kx$, with Newton's second law, $F = ma = m\ddot{x}$. Thus,

$$m\ddot{x} = -Kx$$

and

$$m\ddot{x} + Kx = 0,$$

where

$$\ddot{x} = \frac{d^2x}{dt^2}.$$

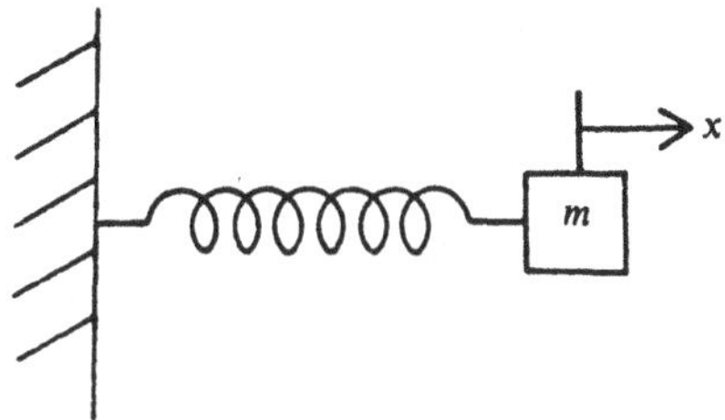

FIGURE 1.1. Simple mass–spring vibrating system.

The constant K is called the spring constant or stiffness of the spring (expressed in newtons per meter). We define a constant $\omega_0 = \sqrt{K/m}$, so that the equation of motion becomes

$$\ddot{x} + \omega_0^2 x = 0. \tag{1.2}$$

This well-known equation has these solutions:

$$x = A\cos(\omega_0 t + \phi) \tag{1.3}$$

or

$$x = B\cos\omega_0 t + C\sin\omega_0 t, \tag{1.4}$$

from which we recognize ω_0 as the natural angular frequency of the system.

The natural frequency f_0 of our simple oscillator is given by the relation $f_0 = (1/2\pi)\sqrt{K/m}$, and the amplitude by $A = \sqrt{B^2 + C^2}$; ϕ is the initial phase of the motion. Differentiation of the displacement x with respect to time gives corresponding expressions for the velocity v and acceleration a:

$$v = \dot{x} = -\omega_0 A\sin(\omega_0 t + \phi), \tag{1.5}$$

and

$$a = \ddot{x} = -\omega_0^2 A\cos(\omega_0 t + \phi). \tag{1.6}$$

The displacement, velocity, and acceleration are shown in Fig. 1.2. Note that the velocity v leads the displacement by $\pi/2$ radians (90°), and the acceleration leads (or lags) by π radians (180°).

Solutions to second-order differential equations have two arbitrary constants. In Eq. (1.3) they are A and ϕ; in Eq. (1.4) they are B and C. Another alternative is to describe the motion in terms of constants x_0 and v_0, the displacement and velocity when $t = 0$. Setting $t = 0$ in Eq. (1.3) gives $x_0 = A\cos\phi$, and setting $t = 0$ in Eq. (1.5) gives $v_0 = -\omega_0 A\sin\phi$.

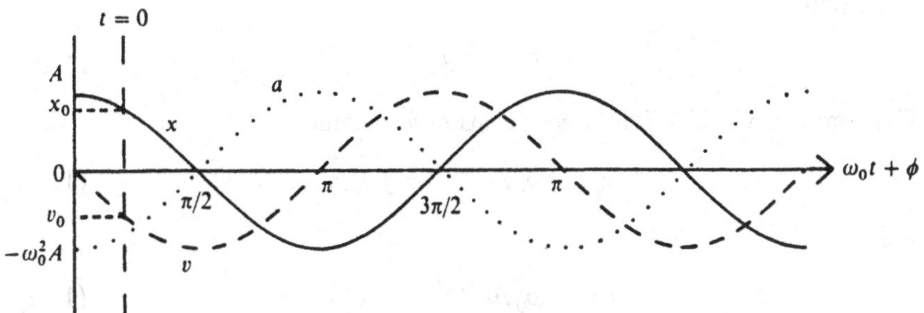

FIGURE 1.2. Relative phase of displacement x, velocity v, and acceleration a of a simple vibrator.

From these we can obtain expressions for A and ϕ in terms of x_0 and v_0:

$$A = \sqrt{x_0^2 + \left(\frac{v_0}{\omega_0}\right)^2},$$

and (1.7)

$$\phi = \tan^{-1}\left(\frac{-v_0}{\omega_0 x_0}\right).$$

Alternatively, we could have set $t = 0$ in Eq. (1.4) and its derivative to obtain $B = x_0$ and $C = v_0/\omega_0$ from which

$$x = x_0 \cos \omega_0 t + \frac{v_0}{\omega_0} \sin \omega_0 t. \tag{1.8}$$

1.2 Complex Amplitudes

Another approach to solving linear differential equations is to use exponential functions and complex variables. In this description of the motion, the amplitude and the phase of an oscillating quantity, such as displacement or velocity, are expressed by a complex number; the differential equation of motion is transformed into a linear algebraic equation. The advantages of this formulation will become more apparent when we consider driven oscillators.

This alternate approach is based on the mathematical identity $e^{\pm j\omega_0 t} = \cos \omega_0 t \pm j \sin \omega_0 t$, where $j = \sqrt{-1}$. In these terms, $\cos \omega_0 t = \mathrm{Re}(e^{\pm j\omega_0 t})$, where Re stands for the "real part of." Equation (1.3) can be written

$$\begin{aligned} x =& A\cos(\omega_0 t + \phi) = \mathrm{Re}[Ae^{j(\omega_0 t+\phi)}] = \mathrm{Re}(Ae^{j\phi}e^{j\omega_0 t}) \\ =& \mathrm{Re}(\tilde{A}e^{j\omega_0 t}). \end{aligned} \tag{1.9}$$

The quantity $\tilde{A} = Ae^{j\phi}$ is called the complex amplitude of the motion and represents the complex displacement at $t = 0$. The complex displacement $\tilde{x}$ is written

$$\tilde{x} = \tilde{A}e^{j\omega_0 t}. \tag{1.10}$$

The complex velocity $\tilde{v}$ and acceleration $\tilde{a}$ become

$$\tilde{v} = j\omega_0 \tilde{A}e^{j\omega_0 t} = j\omega_0 \tilde{x}, \tag{1.11}$$

and

$$\tilde{a} = -\omega_0^2 \tilde{A}e^{j\omega_0 t} = -\omega_0^2 \tilde{x}. \tag{1.12}$$

Each of these complex quantities can be thought of as a rotating vector or phasor rotating in the complex plane with angular velocity ω_0, as shown in

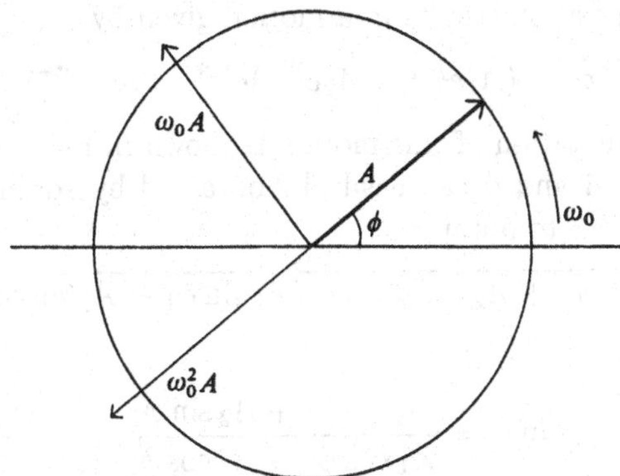

FIGURE 1.3. Phasor representation of the complex displacement, velocity, and acceleration of a linear oscillator.

Fig. 1.3. The real time dependence of each quantity can be obtained from the projection on the real axis of the corresponding complex quantities as they rotate with angular velocity ω_0.

1.3 Superposition of Two Harmonic Motions in One Dimension

Frequently, the motion of a vibrating system can be described by a linear combination of the vibrations induced by two or more separate harmonic excitations. Provided we are dealing with a linear system, the displacement at any time is the sum of the individual displacements resulting from each of the harmonic excitations. This important principle is known as the principle of linear superposition. A linear system is one in which the presence of one vibration does not alter the response of the system to other vibrations, or one in which doubling the excitation doubles the response.

1.3.1 Two Harmonic Motions Having the Same Frequency

One case of interest is the superposition of two harmonic motions having the same frequency. If the two individual displacements are

$$\tilde{x}_1 = A_1 e^{j(\omega t + \phi_1)}$$

and

$$\tilde{x}_2 = A_2 e^{j(\omega t + \phi_2)},$$

their linear superposition results in a motion given by

$$\tilde{x}_1 + \tilde{x}_2 = (A_1 e^{j\phi_1} + A_2 e^{j\phi_2}) e^{j\omega t} = A e^{j(\omega t + \phi)}. \tag{1.13}$$

The phasor representation of this motion is shown in Fig. 1.4.

Expressions for A and ϕ can easily be obtained by adding the phasors $A_1 e^{j\omega\phi_1}$ and $A_2 e^{j\omega\phi_2}$ to obtain

$$A = \sqrt{(A_1 \cos\phi_1 + A_2 \cos\phi_2)^2 + (A_1 \sin\phi_1 + A_2 \sin\phi_2)^2}, \tag{1.14}$$

and

$$\tan\phi = \frac{A_1 \sin\phi_1 + A_2 \sin\phi_2}{A_1 \cos\phi_1 + A_2 \cos\phi_2}. \tag{1.15}$$

What we have really done, of course, is to add the real and imaginary parts of $\tilde{x}_1$ and $\tilde{x}_2$ to obtain the resulting complex displacement $\tilde{x}$. The real displacement is

$$x = \mathrm{Re}(\tilde{x}) = A\cos(\omega t + \phi). \tag{1.16}$$

The linear combination of two simple harmonic vibrations with the same frequency leads to another simple harmonic vibration at this same frequency.

1.3.2 More Than Two Harmonic Motions Having the Same Frequency

The addition of more than two phasors is accomplished by drawing them in a chain, head to tail, to obtain a single phasor that rotates with angular velocity ω. This phasor has an amplitude given by

$$A = \sqrt{(\sum A_n \cos\phi_n)^2 + (\sum A_n \sin\phi_n)^2}, \tag{1.17}$$

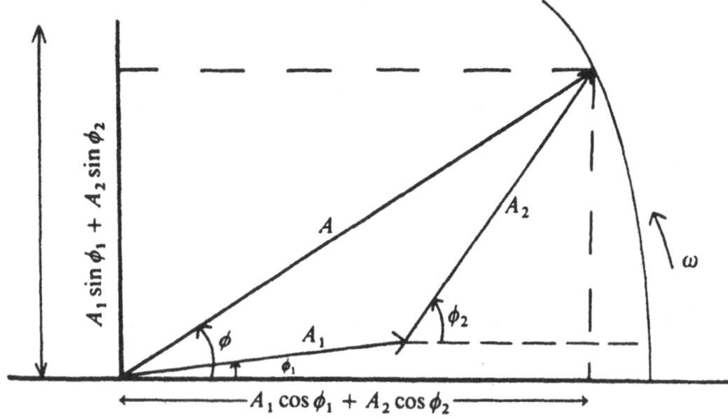

FIGURE 1.4. Phasor representation of two simple harmonic motions having the same frequency.

and a phase angle ϕ obtained from

$$\tan\phi = \frac{\sum A_n \sin\phi_n}{\sum A_n \cos\phi_n}. \tag{1.18}$$

The real displacement is the projection of the resultant phasor on the real axis, and this is equal to the sum of the real parts of all the component phasors:

$$x = A\cos(\omega t + \phi) = \sum A_n \cos(\omega t + \phi_n). \tag{1.19}$$

1.3.3 Two Harmonic Motions with Different Frequencies: Beats

If two simple harmonic motions with frequencies f_1 and f_2 are combined, the resultant expression is

$$\tilde{x} = \tilde{x}_1 + \tilde{x}_2 = A_1 e^{j(\omega_1 t + \phi_1)} + A_2 e^{j(\omega_2 t + \phi_2)}, \tag{1.20}$$

where A, ω, and ϕ express the amplitude, the angular frequency, and the phase of each simple harmonic vibration.

The resulting motion is not simple harmonic, so it cannot be represented by a single phasor or expressed by a simple sine or cosine function. If the ratio of ω_2 to ω_1 (or ω_1 to ω_2) is a rational number, the motion is periodic with an angular frequency given by the largest common divisor of ω_2 and ω_1. Otherwise, the motion is a nonperiodic oscillation that never repeats itself.

The linear superposition of two simple harmonic vibrations with nearly the same frequency leads to periodic amplitude vibrations or beats. If the angular frequency ω_2 is written as

$$\omega_2 = \omega_1 + \Delta\omega, \tag{1.21}$$

the resulting displacement becomes

$$\begin{aligned} \tilde{x} &= A_1 e^{j(\omega_1 t + \phi_1)} + A_2 e^{j(\omega_1 t + \Delta\omega t + \phi_2)} \\ &= [A_1 e^{j\phi_1} + A_2 e^{j(\phi_2 + \Delta\omega t)}] e^{j\omega_1 t}. \end{aligned} \tag{1.22}$$

We can express this in terms of a time-dependent amplitude $A(t)$ and a time-dependent phase $\phi(t)$:

$$\tilde{x} = A(t) e^{j(\omega_1 t + \phi(t))}, \tag{1.23}$$

where

$$A(t) = \sqrt{A_1^2 + A_2^2 + 2A_1A_2\cos(\phi_1 - \phi_2 - \Delta\omega t)}, \tag{1.24}$$

and

$$\tan\phi(t) = \frac{A_1 \sin\phi_1 + A_2 \sin(\phi_2 + \Delta\omega t)}{A_1 \cos\phi_1 + A_2 \cos(\phi_2 + \Delta\omega t)}. \tag{1.25}$$

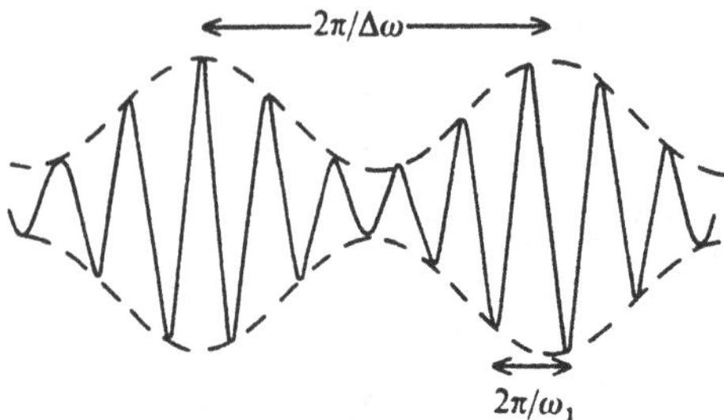

FIGURE 1.5. Waveform resulting from linear superposition of simple harmonic motions with angular frequencies ω_1 and ω_2.

The resulting vibration could be regarded as approximately simple harmonic motion with angular frequency ω_1 and with both amplitude and phase varying slowly at frequency $\Delta\omega/2\pi$. The amplitude varies between the limits $A_1 + A_2$ and $|A_1 - A_2|$.

In the special case where the amplitudes A_1 and A_2 are equal and ϕ_1 and $\phi_2 = 0$, the amplitude equation [Eq. (1.24)] becomes

$$A(t) = A_1\sqrt{2 + 2\cos\Delta\omega_1 t} \tag{1.26}$$

and the phase equation [Eq. (1.25)] becomes

$$\tan\phi(t) = \frac{\sin\Delta\omega_1 t}{1 + \cos\Delta\omega_1 t}. \tag{1.27}$$

Thus, the amplitude varies between $2A_1$ and 0, and the beating becomes very pronounced.

The displacement waveform (the real part of $\tilde{x}$) is illustrated in Fig. 1.5. This waveform resembles the waveform obtained by modulating the amplitude of the vibration at a frequency $\Delta\omega/2\pi$, but they are not the same. Amplitude modulation results from nonlinear behavior in a system, which generates spectral components having frequencies ω_1 and $\omega_1 \pm \Delta\omega$. The spectrum of the waveform in Fig. 1.5. has spectral components ω_1 and $\omega_1 + \Delta\omega$ only.

Audible beats are heard whenever two sounds of nearly the same frequency reach the ear. The perception of combination tones and beats is discussed in Chapter 8 of Rossing (1982) and other introductory texts on musical acoustics.

1.4 Energy

The potential energy E_p of our mass–spring system is equal to the work done in stretching or compressing the spring:

$$E_{\mathrm{p}} = -\int_0^x F\,dx = \int_0^x Kx\,dx = \tfrac{1}{2}Kx^2. \tag{1.28}$$

Using the expression for x in Eq. (1.3) gives

$$E_p = \tfrac{1}{2}KA^2\cos^2(\omega_0 t + \phi). \tag{1.29}$$

The kinetic energy is $E_k = \frac{1}{2}mv^2$, and using the expression for v in Eq. (1.5) gives

$$E_k = \tfrac{1}{2}m\omega_0^2A^2\sin^2(\omega_0 t + \phi) = \tfrac{1}{2}KA^2\sin^2(\omega_0 t + \phi). \tag{1.30}$$

The total energy E is then

$$E = E_p + E_k = \tfrac{1}{2}KA^2 = \tfrac{1}{2}m\omega_0^2A^2 = \tfrac{1}{2}mU^2, \tag{1.31}$$

where U is the maximum velocity. The total energy in our loss-free system is constant and is equal either to the maximum potential energy (at maximum displacement) or the maximum kinetic energy (at the midpoint).

1.5 Damped Oscillations

There are many different mechanisms that can contribute to the damping of an oscillating system. Sliding friction is one example, and viscous drag in a fluid is another. In the latter case, the drag force F_r is proportional to the velocity:

$$F_r = -R\dot{x},$$

where R is the mechanical resistance. The drag force is added to the equation of motion:

$$m\ddot{x} + R\dot{x} + Kx = 0$$

or

$$\ddot{x} + 2\alpha\dot{x} + \omega_0^2x = 0, \tag{1.32}$$

where $\alpha = R/2m$ and $\omega_0^2 = K/m$.

We assume a complex solution $\tilde{x} = \tilde{A}e^{\gamma t}$ and substitute into Eq. (1.32) to obtain

$$(\gamma^2 + 2\alpha\gamma + \omega_0^2)\tilde{A}e^{\gamma t} = 0. \tag{1.33}$$

This requires that $\gamma^2 + 2\alpha\gamma + \omega_0^2 = 0$ or that

$$\gamma = -\alpha \pm \sqrt{\alpha^2 - \omega_0^2} = -\alpha \pm j\sqrt{\omega_0^2 - \alpha^2} = -\alpha \pm j\omega_d, \tag{1.34}$$

where $\omega_d = \sqrt{\omega_0^2 - \alpha^2}$ is the natural angular frequency of the damped oscillator (which is less than that of the same oscillator without damping). The general solution is a sum of terms constructed by using each of the two values of γ:

$$\tilde{x} = e^{-\alpha t}(\tilde{A}_1e^{j\omega_d t} + \tilde{A}_2e^{-j\omega_d t}). \tag{1.35}$$

The real part of this solution, which gives the time history of the displacement, can be written in several different ways as in the loss-free case. The expressions that correspond to Eqs. (1.3) and (1.4) are

$$x = Ae^{-\alpha t}\cos(\omega_{\rm d}t + \phi), \tag{1.36}$$

and

$$x = e^{-\alpha t}(B\cos\omega_{\rm d}t + C\sin\omega_{\rm d}t). \tag{1.37}$$

Setting $t = 0$ in Eq. (1.37) and its derivatives gives the displacement in terms of the initial displacement x_0 and initial velocity v_0:

$$x = e^{-\alpha t}(x_0\cos\omega_{\rm d}t + \frac{v_0 + \alpha x_0}{\omega_{\rm d}}\sin\omega_{\rm d}t). \tag{1.38}$$

Figure 1.6 shows a few cycles of the displacement for different values of α when $v_0 = 0$.

The amplitude of the damped oscillator is given by $x_0 e^{-\alpha t}$, and its motion is not strictly periodic. Nevertheless, the time between zero crossings in the same direction remains constant and equal to $T_{\rm d} = 1/f_{\rm d} = 2\pi/\omega_{\rm d}$, which is defined as the period of the oscillation. The time interval between successive maxima is also $T_{\rm d}$, but the maxima and minima are not exactly halfway between the zeros.

One measure of the damping is the time required for the amplitude to decrease to $1/e$ of its initial value x_0. This time, τ, is called by various names, such as decay time, lifetime, relaxation time, and characteristic time; it is given by

$$\tau = \frac{1}{\alpha} = \frac{2m}{R}. \tag{1.39}$$

When $\alpha \geq \omega_0$, the system is no longer oscillatory. When the mass is displaced, it returns asymptotically to its rest position. For $\alpha = \omega_0$, the

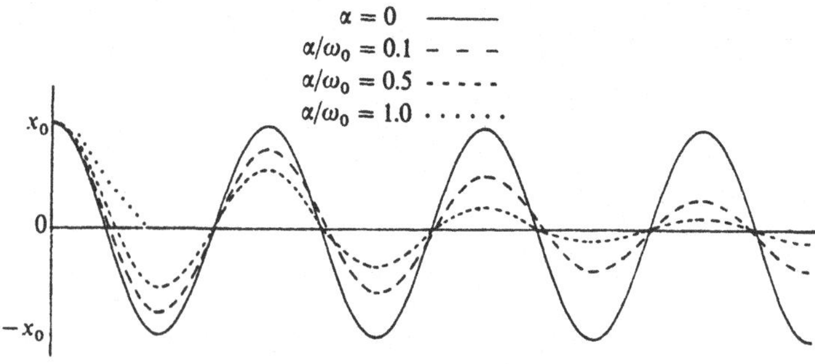

FIGURE 1.6. Displacement of a harmonic oscillator with $v_0 = 0$ for different values of damping. The relaxation time is given by $1/\alpha$. Critical damping occurs when $\alpha = \omega_0$.

system is critically damped, and the displacement is

$$x_c = x_0(1 + \alpha t)e^{-\alpha t}. \tag{1.40}$$

For $\alpha > \omega_0$, the system is overdamped and returns to its rest position even more slowly.

It is quite obvious that the energy of a damped oscillator decreases with time. The rate of energy loss can be found by taking the time derivative of the total energy:

$$\begin{aligned}\frac{d}{dt}(E_{\mathrm{p}} + E_{\mathrm{K}}) &= \frac{d}{dt}\left[\tfrac{1}{2}Kx^2 + \tfrac{1}{2}m\dot{x}^2\right] = Kx\dot{x} + m\dot{x}\ddot{x} \\ &= \dot{x}(Kx + m\ddot{x}) = \dot{x}(-R\dot{x}) = -2\alpha m\dot{x}^2, \end{aligned} \tag{1.41}$$

where use has been made of Eq. (1.32). Equation (1.41) tells us that the rate of energy loss is the friction force $-R\dot{x}$ times the velocity $\dot{x}$.

Often a Q factor or quality factor is used to compare the spring force to the damping force:

$$Q = \frac{Kx_0}{R\omega_0 x_0} = \frac{K}{R\omega_0} = \frac{\omega_0}{2\alpha}. \tag{1.42}$$

1.6 Other Simple Vibrating Systems

Besides the mass–spring system already described, the following are familiar examples of systems that vibrate in simple harmonic motion.

1.6.1 A Spring of Air

A piston of mass m, free to move in a cylinder of area-S and length-L [see Fig. 1.7(a)], vibrates in much the same manner as a mass attached to a spring. The spring constant of the confined air turns out to be $K = \gamma p_{\mathrm{a}} S/L$, so the natural frequency is

$$f_0 = \frac{1}{2\pi}\sqrt{\frac{\gamma p_{\mathrm{a}} S}{mL}}, \tag{1.43}$$

where p_{a} is atmospheric pressure, m is the mass of the piston, and γ is a constant that is 1.4 for air.

1.6.2 Helmholtz Resonator

In the Helmholtz resonator shown in Fig. 1.7(b), the mass of air in the neck serves as the piston and the large volume of air V as the spring. The mass of air in the neck and the spring constant of the confined air are given by

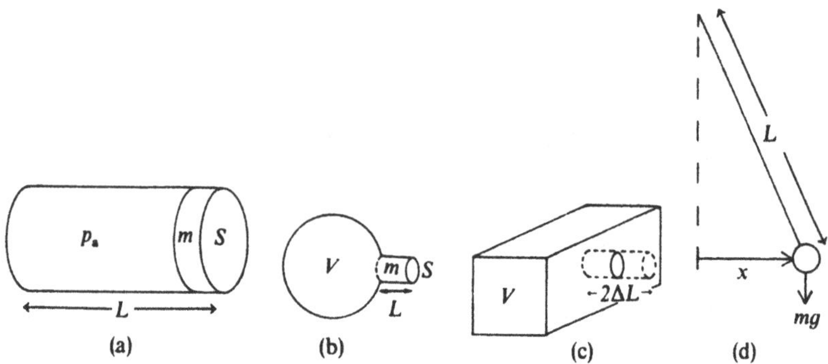

FIGURE 1.7. Simple vibrating systems: (a) piston in a cylinder; (b) Helmholtz resonator with neck of length L; (c) Helmholtz resonator without a neck; and (d) simple pendulum.

the expressions

$$m = \rho SL,$$

and (1.44)

$$K = \rho S^2 c^2 / V,$$

where ρ is the air density and c is the speed of sound.

The natural frequency of vibration is given by

$$f_0 = \frac{1}{2\pi}\sqrt{\frac{K}{m}} = \frac{c}{2\pi}\sqrt{\frac{S}{VL}}. \tag{1.45}$$

Note that the smaller the neck diameter, the lower the natural frequency of vibration, a result which may appear surprising at first glance.

The Helmholtz resonator in Fig. 1.7(c) has no neck to delineate the vibrating mass, but the effective length can be estimated by taking twice the "end correction" of a flanged tube (which is $8/3\pi \cong 0.85$ times the radius a). Thus,

$$m = \rho SL = \rho(\pi a^2)\left(\frac{16a}{3\pi}\right) = 5.33\rho a^3. \tag{1.46}$$

The natural frequency of a neckless Helmholtz resonator with a large face is thus expressed as

$$f_0 = \frac{c}{2\pi}\sqrt{\frac{1.85a}{V}}. \tag{1.47}$$

If the face of the resonator surrounding the hole is not large, the natural frequency will be slightly higher. The Helmholtz resonator is discussed again in Section 8.15 and the end correction in Section 8.3.

1.6.3 Simple Pendulum

A simple pendulum, consisting of a mass m attached to a string of length l [Fig. 1.7(d)], oscillates in simple harmonic motion provided that $x \ll l$. Assuming that the mass of the string is much less than m, the natural frequency is given by

$$f_0 = \frac{1}{2\pi}\sqrt{\frac{g}{l}}, \tag{1.48}$$

where g is the acceleration due to gravity. Note that the frequency does not depend on the mass.

1.6.4 Electrical RLC Circuit

In the electrical circuit, shown in Fig. 1.8, the voltages across the inductor, the resistor, and the capacitor, respectively, should add to zero:

$$L\frac{di}{dt} + Ri + \frac{1}{C}\int i\,dt = 0.$$

Differentiating each term leads to an equation that is analogous to Eq. (1.32) for the simple mechanical oscillator:

$$L\ddot{i} + R\dot{i} + \frac{1}{C}i = 0$$

or (1.49)

$$\ddot{i} + 2\alpha\dot{i} + \omega_0^2 i = 0,$$

where $\alpha = R/2L$ and $\omega_0^2 = 1/LC$

The solution to Eq. (1.49) can be written as

$$i = I_0 e^{-\alpha t}\cos(\omega_d t + \phi), \tag{1.50}$$

which represents a current oscillating at a frequency $\sqrt{\omega_0^2 - \alpha^2}/2\pi$, with an amplitude that decays exponentially. If $a \ll \omega_0$ (small damping), the

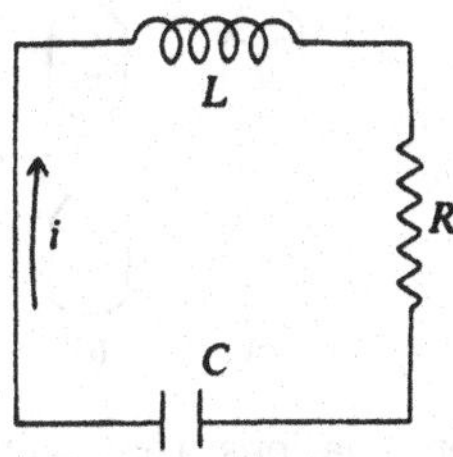

FIGURE 1.8. Simple electrical oscillator with inductance L, resistance R, and capacitance C.

frequency of oscillation is approximately

$$f_0 = \frac{\omega_0}{2\pi} = \frac{1}{2\pi\sqrt{LC}}, \tag{1.51}$$

and the current has a waveform similar to that shown in Fig. 1.6.

1.6.5 Combinations of Springs and Masses

Several combinations of masses and springs are shown in Fig. 1.9, along with their resonance frequencies. Note the effect of combining springs in series and parallel combinations. Two springs with spring constants K_1 and K_2 will have a combined spring constant $K_p = K_1 + K_2$ when connected in parallel but only $K_s = K_1K_2/(K_1 + K_2)$ in series. When $K_1 = K_2$, the parallel and series values become $2K_1$ and $K_1/2$, respectively. The combinations in Fig. 1.9 all have a single degree of freedom. In Section 1.12, we discuss two-mass systems with two degrees of freedom; that is, the two masses move independently.

1.6.6 Longitudinal and Transverse Oscillations of a Mass–Spring System

Consider the vibrating system shown in Fig. 1.10. Each spring has a spring constant K, a relaxed length a_0, and a stretched length a. Thus, each spring exerts a tension $K(a - a_0)$ on the mass when it is in its equilibrium position ($x = 0$). When the mass is displaced a distance x, the net restoring force

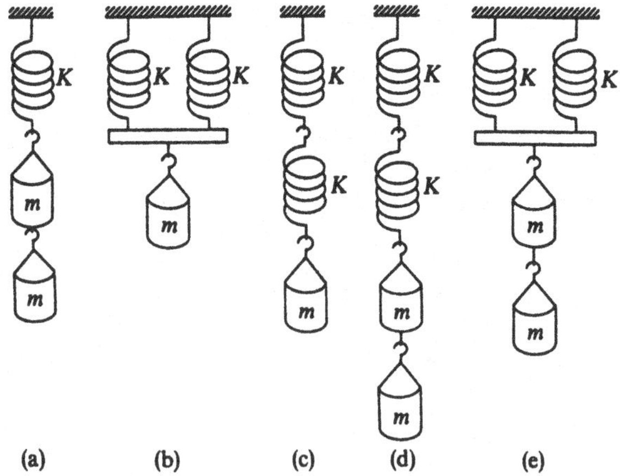

FIGURE 1.9. Mass–spring combinations that vibrate at single frequencies: (a) $f_0 = (1/2\pi)\sqrt{K/2m}$; (b) $f_0 = (1/2\pi)\sqrt{2K/m}$; (c) $f_0 = (1/2\pi)\sqrt{k/2m}$; (d) $f_0 = (1/2\pi)\sqrt{K/4m}$; (e) $f_0 = (1/2\pi)\sqrt{K/m}$.

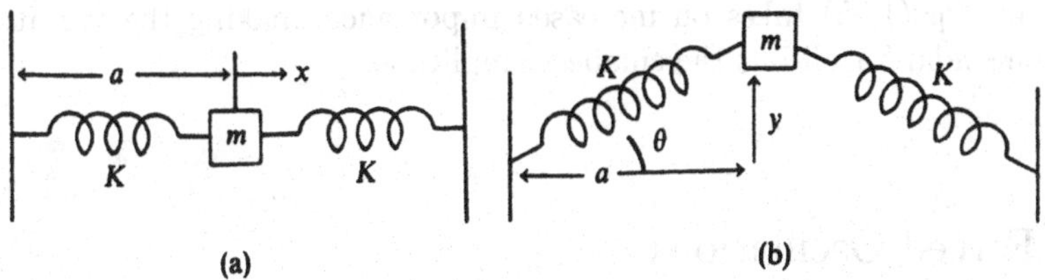

FIGURE 1.10. Longitudinal (a) and transverse (b) oscillations of a mass–spring system.

is the difference between the two tensions:

$$F_x = K(a - x - a_0) - K(a + x - a_0) = -2Kx. \tag{1.52}$$

The natural frequency for longitudinal vibration is thus given by

$$f_l = \frac{1}{2\pi}\sqrt{\frac{2K}{m}}. \tag{1.53}$$

Now, consider transverse vibrations of the same systems, as shown in Fig. 1.10(b). When the mass is displaced a distance y from its equilibrium position, the restoring force is due to the y component of the tension:

$$\begin{aligned} F_y &= -2K(\sqrt{a^2 + y^2} - a_0)\sin\theta = -2K(\sqrt{a^2 + y^2} - a_0)\frac{y}{\sqrt{a^2 + y^2}} \\ &= -2Ky\left(1 - \frac{a_0}{\sqrt{a^2 + y^2}}\right). \end{aligned} \tag{1.54}$$

For small deflection y, the force can be written as

$$\begin{aligned} F_y &= -\,2Ky\left[1 - \frac{a_0}{a}\left(1 + \frac{y^2}{a^2}\right)^{-1/2}\right] \\ &\cong -\,2Ky\left(1 - \frac{a_0}{a}\right) - \frac{Ka_0}{a^3}y^3. \end{aligned} \tag{1.55}$$

When the springs are stretched to several times their relaxed length ($a \gg a_0$), the force is approximately $-2Ky$, and the natural frequency is practically the same as the frequency for longitudinal vibrations given in Eq. (1.53):

$$f_t \cong \frac{1}{2\pi}\sqrt{\frac{2K}{m}}. \tag{1.56}$$

When the springs are stretched only a small amount from their relaxed length ($a \cong a_0$), however, the first term in Eq. (1.55) becomes very small, so the vibration frequency is considerably smaller than that given in Eqs. (1.53) and (1.56). Furthermore, the contribution from the cubic term in Eq. (1.55) takes on increased importance, making the vibration nonsinusoidal for all but the smallest amplitude.

1.7 Forced Oscillations

When a simple oscillator is driven by an external force $f(t)$, as shown in Fig. 1.11, the equation of motion Eq. (1.32) then becomes

$$m\ddot{x} + R\dot{x} + Kx = f(t). \tag{1.57}$$

The driving force $f(t)$ may have harmonic time dependence, it may be impulsive, or it may even be a random function of time. For the case of a sinusoidal driving force $f(t) = F\cos\omega t$ turned on at some time, the solution to Eq. (1.57) consists of two parts: a transient term containing two arbitrary constants, and a steady-state term that depends only on F and ω.

To obtain the steady-state solution, it is advantageous to write the equation of motion in complex form:

$$m\ddot{\tilde{x}} + R\dot{\tilde{x}} + K\tilde{x} = Fe^{j\omega t}. \tag{1.58}$$

Since this equation is linear in x and the right-hand side is a harmonic function with angular frequency ω, in the steady state the left-hand side should be harmonic with the same frequency. Thus, we replace $\tilde{x}$ by $\tilde{A}e^{j\omega t}$ and obtain

$$\tilde{A}e^{j\omega t}(-\omega^2 m + j\omega R + K) = Fe^{j\omega t}. \tag{1.59}$$

The complex displacement is

$$\tilde{x} = \frac{Fe^{j\omega t}}{K - \omega^2 m + j\omega R} = \frac{\tilde{F}/m}{\omega_0^2 - \omega^2 + j\omega 2\alpha}, \tag{1.60}$$

where $\tilde{F} = Fe^{j\omega t}$, $\omega_0^2 = K/m$, and $\alpha = R/2m$.

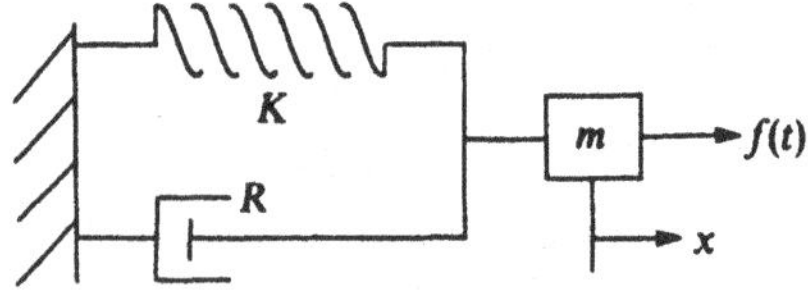

FIGURE 1.11. A damped harmonic oscillator with driving force $f(t)$.

Differentiation of $\tilde{x}$ gives the complex velocity $\tilde{v}$:

$$\tilde{v} = \frac{Fe^{j\omega t}}{R + j(\omega m - K/\omega)} = \frac{\tilde{F}\omega/m}{2\omega\alpha + j(\omega^2 - \omega_0^2)}. \tag{1.61}$$

The mechanical impedance $\tilde{Z}$ is defined as $\tilde{F}/\tilde{v}$:

$$\tilde{Z} = \tilde{F}/\tilde{v} = R + j(\omega m - K/\omega) = R + jX_m, \tag{1.62}$$

where $X_m = \omega m - K/\omega$ is the mechanical reactance. The actual steady-state displacement is given by the real part of Eq. (1.60):

$$x = \mathrm{Re}\frac{\tilde{F}}{j\omega\tilde{Z}} = \frac{F}{\omega Z}\sin(\omega t + \phi). \tag{1.63}$$

A quantity $x_s = F/K = F/m\omega_0^2$ can be defined as the static displacement of the oscillator produced by a constant force of magnitude F. At very low frequency, the displacement amplitude will approach F/K, and the oscillator is said to be stiffness dominated. When $\omega = \omega_d$, the amplitude becomes

$$x_0 = F/2\alpha m\omega_0 = Qx_s. \tag{1.64}$$

In other words, Q becomes a sort of amplification factor, which is the ratio of the displacement amplitude at resonance ($\omega_0 = \omega$) to the static displacement.

There is a direct relation between the damping coefficient α, the decay time τ of Eq. (1.39), and the width of the resonance peak in Fig. 1.12. If we take the absolute value of both sides of Eq. (1.61), then we see that the denominator, which largely determines the shape of the resonance, is $[(\omega^2 - \omega_0^2)^2 + 4\omega^2\alpha^2]^{1/2}$. Provided that we are only concerned with frequencies ω quite close to ω_0, we can write this approximately as $2\omega_0[(\omega - \omega_0)^2 + \alpha^2]^{1/2}$. The magnitude of the denominator thus increases by a factor $2^{1/2}$ relative to its value at $\omega = \omega_0$ when $|\omega - \omega_0| \approx \alpha$. The response decreases by the same factor, which represents a 3 dB decline from the peak value at the resonance. This 3 dB half-width of the resonance curve, measured in radians per second, is thus equal to the damping coefficient α, and also, by Eq. (1.39), to the reciprocal of the decay time τ in seconds. The 3 dB full-width $\Delta\omega$ of the curve is $2\alpha = R/m$, and its relative value $2\Delta\omega/\omega_0$ is equal to Q^{-1}.

At high frequency ($\omega \gg \omega_0$), the displacement falls toward zero. The frequency response of a simple oscillator for different values of α (or Q) is shown in Fig 1.12(a). The magnitude of x is less than x_s for frequencies above $\omega_0\sqrt{2 - \delta^2}$ (where $\delta = 1/Q = 2\alpha/\omega_0$), which, for small values of α, is about $\sqrt{2}\omega_0$. If $\alpha > \omega_0/\sqrt{2}$, $x < x_s$ at all frequencies.

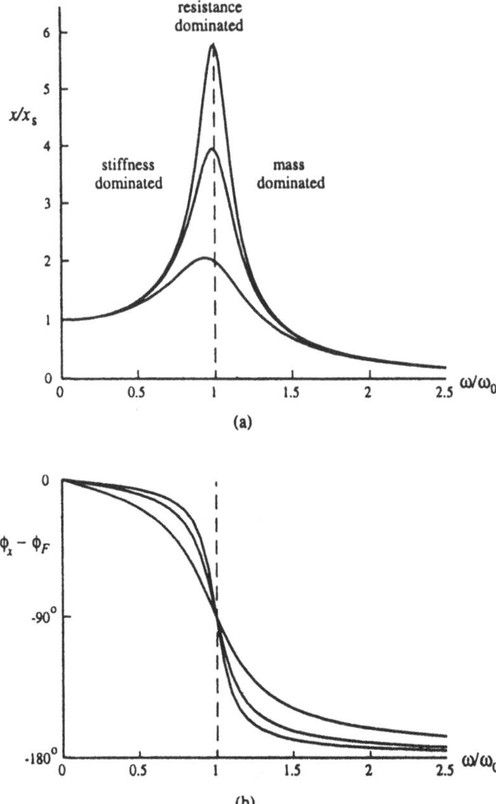

FIGURE 1.12. Frequency dependence of the magnitude x and phase $(\phi_x - \phi_F)$ of the displacement of a linear harmonic oscillator.

The phase angle between the displacement and the driving force is the phase angle of the denominator in Eq. (1.60):

$$\phi_x - \phi_F = \tan^{-1} \frac{2\alpha\omega}{\omega^2 - \omega_0^2}. \tag{1.65}$$

At low frequency ($\omega \cong 0$), $\phi_x - \phi_F = 0$. When $\omega = \omega_0$, $\phi_x - \phi_F = 90°$, and at high frequency ($\omega \gg \omega_0$), $\phi_x - \phi_F \cong 180°$, as shown in Fig. 1.12(b).

There are other convenient ways to represent the frequency response of a simple oscillator. One way is to show how the real and imaginary parts of the mechanical impedance $\tilde{Z}(= \tilde{F}/\tilde{v})$ or the mechanical admittance (mobility) $\tilde{Y} = 1/\tilde{Z}(= \tilde{v}/\tilde{F})$ vary with frequency. At resonance, the real part of the admittance has its maximum value, while that of the impedance remains equal to R at all frequencies. The imaginary parts of both quantities are zero at resonance. Figure 1.13 shows the real and imaginary parts of the

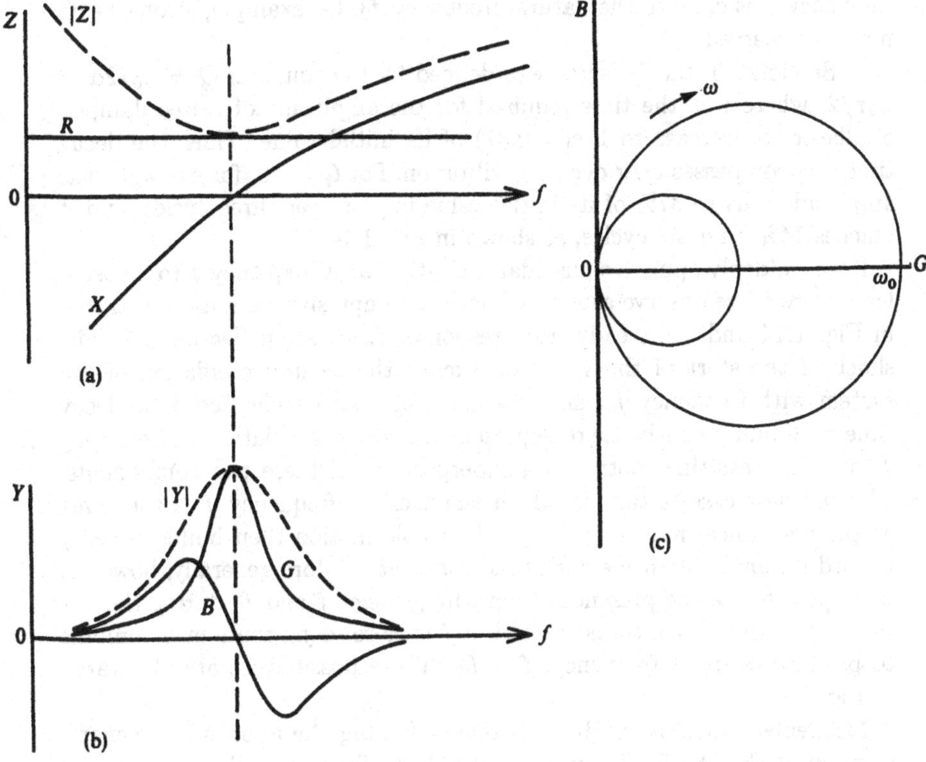

FIGURE 1.13. Real and imaginary parts of the mechanical impedance and admittance for a harmonic oscillator of the same type as in Fig. 1.12: (a) mechanical impedance; (b) mechanical admittance or mobility; (c) Nyquist plot showing the imaginary part of admittance versus the real part, with frequency as a parameter.

mechanical impedance and admittance for an oscillator of the same type as in Fig. 1.12. The graph of imaginary part versus the real part in Fig. 1.13(c) is sometimes called a Nyquist plot.

1.8 Transient Response of an Oscillator

When a driving force is first applied to an oscillator, the motion can be quite complicated. We expect to find periodic motions at the natural frequency f_0 of the oscillator as well as the driving frequency f (or at all its component frequencies if the driving force is not harmonic). If the oscillator is heavily damped, the transient motion decays rapidly, and the oscillator quickly settles into its steady-state motion. If the damping is small, however, the transient behavior may continue for many cycles of oscillation. If the driving frequency f is close to the natural frequency f_0, for example, strong beats may be observed.

In Section 1.5, the Q factor was defined by the equation $Q = \omega_0/2\alpha = \omega_0\tau/2$, where τ is the time required for the amplitude of a free damped oscillator to decrease to $1/e(= 0.37)$ of its initial value. Thus, the decay time τ encompasses Q/π cycles of vibration. For $Q = 10$, for example, the amplitude falls to 37% of its initial value in just over three cycles, and it reaches 14% after six cycles, as shown in Fig. 1.14.

If we suddenly apply a sinusoidal excitation with frequency f to an oscillator at rest, we observe aspects of both the impulsive response illustrated in Fig. 1.14 and the steady-state response discussed in Section 1.7. The shock of the start of the vibration excites the natural oscillation of the system with frequency f_0, and this dies away with a characteristic decay time τ. Simultaneously, there is present the forced oscillation at frequency f, and the resulting motion is a superposition of these two components. The simplest case is that in which the exciting frequency f is the same as the resonance frequency f_0, for the whole motion then builds steadily toward its final amplitude with time constant τ. More generally, however, we expect to see the presence of both frequencies f and f_0 during the duration τ of the attack transient and, if f is close to f_0, these may combine to produce beats at frequency $|f - f_0|$. These possibilities are illustrated in Fig. 1.15.

Mathematically, the problem is one of finding the appropriate general solution of Eq. (1.57). Because Eq. (1.57) is a linear equation, the general solution is a combination of the general solution of the homogeneous equa-

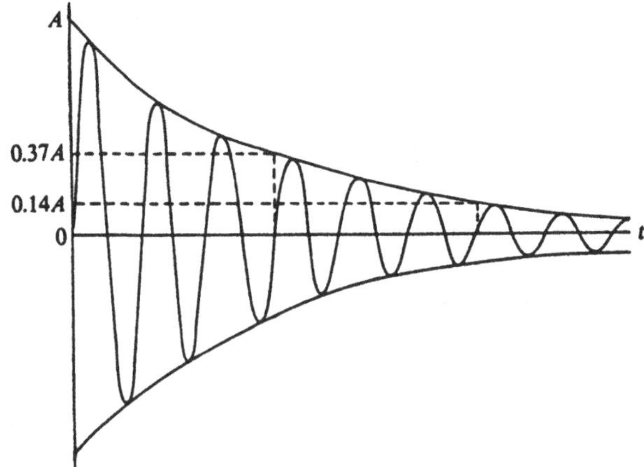

FIGURE 1.14. Response of a damped oscillator ($Q = 10$) to impulsive excitation (by the application of a large force for a very short time, for example). The amplitude falls to 37% of its initial value in time τ, which corresponds to Q/π cycles.

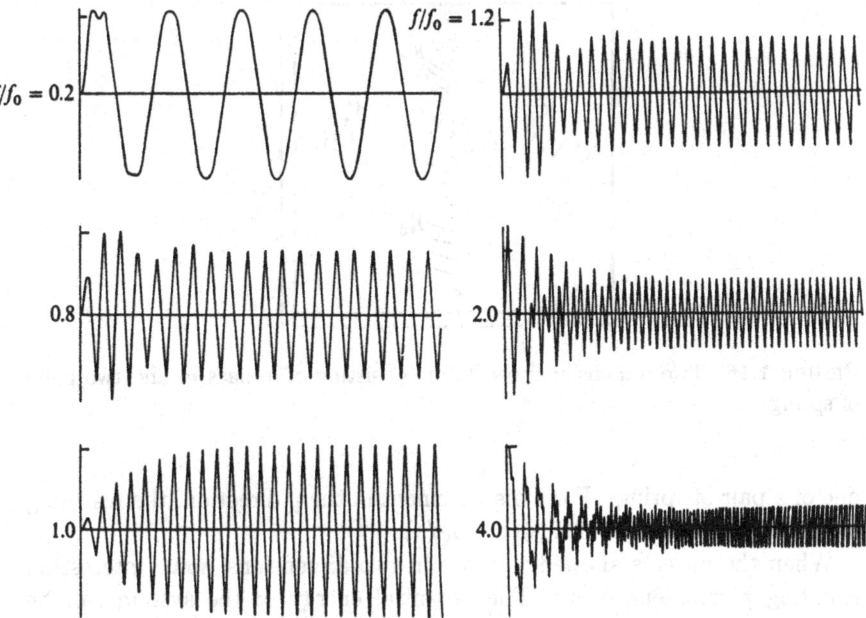

FIGURE 1.15. Response of a simple oscillator to a sinusoidal force applied suddenly. The ratio f/f_0 varies from 0.2 to 4.0, and $Q = 10$ in each case. Note that the scale of amplitude is different in each case (from Fletcher, 1982).

tion, Eq. (1.32), and a particular solution of Eq. (1.57), which we take to be the steady-state solution [Eq. (1.63)].

$$x = Ae^{-\alpha t}\cos(\omega_{\mathrm{d}}t + \phi) + \frac{F}{\omega Z}\sin(\omega t + \phi), \tag{1.66}$$

where A and ϕ are arbitrary constants to be determined by the initial conditions. If the damping is small, ω_{d} can be replaced by ω_0.

When the driving frequency matches the natural frequency ($\omega = \omega_0$), the amplitude builds up exponentially to its final value without beats, as shown in Fig. 1.15(c). Note that irrespective of how the oscillator starts its motion, it eventually settles down to this steady-state motion.

1.9 Two-Dimensional Harmonic Oscillator

An interesting oscillating system is the one shown in Fig. 1.16, which results from adding a second pair of springs to the system in Fig. 1.10. The displacement of the mass m from its equilibrium position is given by coordinates x and y, and both pairs of springs exert restoring forces. For a displacement in the x direction, the restoring force is approximately $F_x = -2K_{\mathrm{A}}x - 2K_{\mathrm{B}}x(1 - b_0/b)$, where b_0 is the unstretched length of

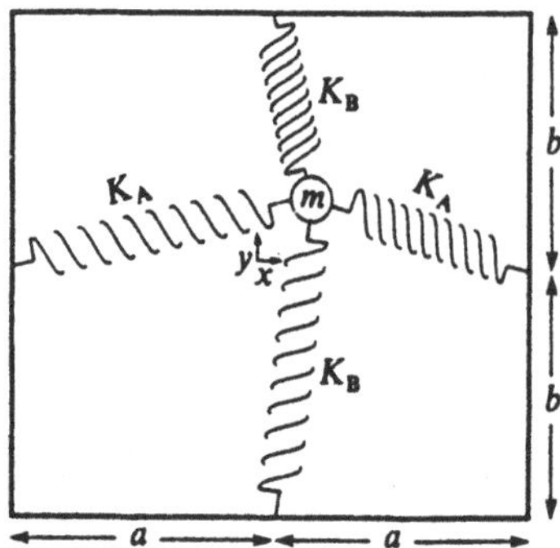

FIGURE 1.16. Two-dimensional oscillator consisting of a mass m and two pairs of springs.

one of a pair of springs. For a displacement in the y direction, the restoring force is $F_y = -2K_\mathrm{B}y - 2K_\mathrm{A}y(1 - a_0/a)$.

When the mass is allowed to move in two dimensions, some interesting coupling phenomena occur. The potential energy of the system can be written as

$$\begin{aligned} E_\mathrm{p} &= \tfrac{1}{2}K_\mathrm{A}[\sqrt{(a+x)^2+y^2}-a_0]^2 + \tfrac{1}{2}K_\mathrm{A}[\sqrt{(a-x)^2+y^2}-a_0]^2 \\ &\quad + \tfrac{1}{2}K_\mathrm{B}[\sqrt{(b+x)^2+y^2}-a_0]^2 + \tfrac{1}{2}K_\mathrm{B}[\sqrt{(b-x)^2+y^2}-a_0]^2. \end{aligned} \tag{1.67}$$

f_x is obtained by differentiating Eq. (1.67). If we retain terms only to third order in x and y, we obtain the expression

$$F_\mathrm{x} \cong -2(K_\mathrm{A}+K_\mathrm{B})x + \frac{2K_\mathrm{B}b_0}{b}x + \left(\frac{2K_\mathrm{A}a_0}{a^3} + \frac{2K_\mathrm{B}b_0}{b^3}\right)xy^2 - \frac{K_\mathrm{B}b_0}{b^3}x^3. \tag{1.68}$$

Note that the third term, which is of third order, couples the x and y motions. For small amplitudes of vibration, however, the x and y motions are independent. Thus, we can solve the independent equations of motion,

$$\ddot{x} + \frac{2K_\mathrm{A} + 2K_\mathrm{B}(1 - b_0/b)}{m}x = 0,$$

and (1.69)

$$\ddot{y} + \frac{2K_\mathrm{A}(1 - a_0/a) + 2K_\mathrm{B}}{m}y = 0,$$

to obtain two independent or normal modes of vibration with natural frequencies:

$$f_1 = \frac{1}{2\pi}\sqrt{\frac{2K_\mathrm{A} + 2K_\mathrm{B}(1 - b_0/b)}{m}},$$

and (1.70)

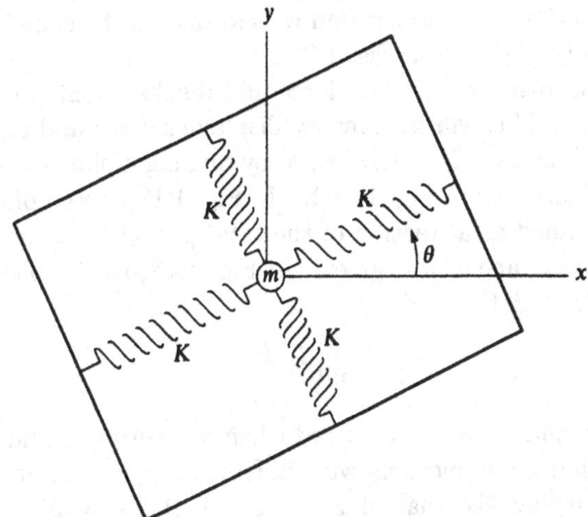

FIGURE 1.17. Two-dimensional oscillator of Fig. 1.16 rotated through an angle θ. The normal modes remain unchanged, but the normal coordinates are no longer x and y.

$$f_2 = \frac{1}{2\pi}\sqrt{\frac{2K_A(1-a_0/a)+2K_B}{m}}.$$

When a_0 and b_0 are much smaller than a and b, f_1 and f_2 differ only slightly. When this oscillating system is set up on an air table and the mass is initially set into motion at 45° to the x and y axes, a slowly changing Lissajous figure is observed as the x and y components of motion change their relative phases.

Since each of the normal modes corresponds to motion along one coordinate only, we call the x and y coordinates the normal coordinates of the motion. In general, the normal coordinates of a two-dimensional oscillator will not be the x and y axes. If the springs were oriented at angles θ to the axes, for example, the normal coordinates (which still lie in the directions of the springs) would be $\psi_1 = x\cos\theta + y\sin\theta$ and $\psi_2 = y\cos\theta - x\sin\theta$ as illustrated in Fig. 1.17.

1.10 Graphical Representations of Vibrations: Lissajous Figures

There are several useful ways to represent a vibrating object with a graphic display device, such as a cathode-ray oscilloscope or an X–Y plotter. Perhaps the most common way is to make a plot of position (or velocity) versus time by incorporating a transducer that gives an electrical output proportional to position (or velocity). With a multiple-trace oscilloscope, the position, velocity, and acceleration waveforms may be combined, giving a display of the type shown in Fig. 1.2.

Another useful display combines force and displacement [Eq. (1.63)] or force and velocity. This can be done by displaying force and displacement as functions of time, as in Fig. 1.18(a), or by making a plot of displacement as a function of force, as in Fig. 1.18(b). In Fig. 1.18(a), the phase angle ϕ would be determined as a fraction of the total period (multiplied by 360° to obtain the phase angle in degrees). In Fig. 1.18(b), the phase angle is obtained from the relationship

$$\phi = \sin^{-1}\frac{B}{A}. \tag{1.71}$$

Note that the display must be centered when measuring A and B.

Two related harmonic motions with different frequencies are often represented in a display like that of Fig. 1.18(b). If $\omega_2 = \omega_1 + \Delta\omega$, as in Eq. (1.19), the display will cycle between a straight line ($\phi = 0, 180°$), a horizontal or vertical ellipse ($\phi = 90°, 270°$), and ellipses of other orientations, as ϕ advances with a frequency $\Delta\omega/2\pi$.

When ω_2 and ω_1 are related by the relationship $m\omega_1 = n\omega_2$, where m and n are integers, stable patterns result. These patterns are called Lissajous figures in honor of Jules Antoine Lissajous. Examples of such figures are shown in Fig. 1.19.

1.11 Normal Modes of Two-Mass Systems

Further understanding of normal modes and normal coordinates of oscillating systems comes from considering the two-mass system in Fig. 1.20(a). The analysis is simplified by letting all three spring constants and both masses be the same. Letting x_1 and x_2 be the displacements of the two masses, we write the equations of motion:

$$m\ddot{x}_1 + Kx_1 + K(x_1 - x_2) = 0,$$

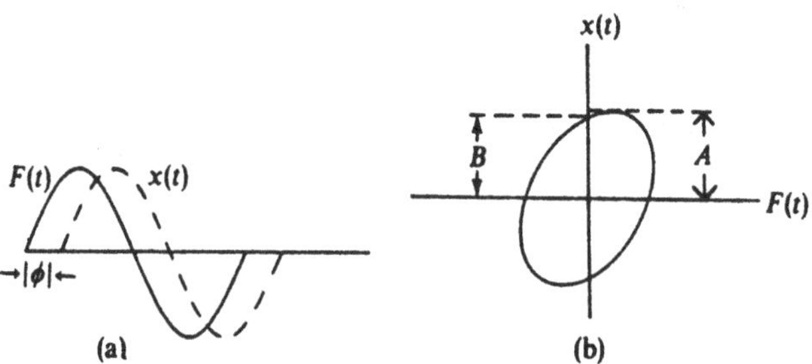

FIGURE 1.18. Two useful displays of (a) force $F(t)$ and (b) displacement $x(t)$ from which the phase angle ϕ can be determined.

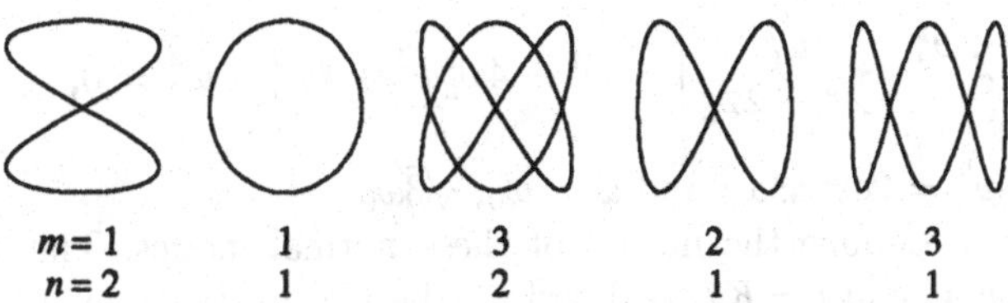

FIGURE 1.19. Lissajous figures for vibrations of frequencies $m\omega$ amd $n\omega$, m and n being small integers.

and (1.72)

$$m\ddot{x}_2 + Kx_2 + K(x_2 - x_1) = 0.$$

In order to find the normal modes, we assume harmonic solutions $x_1 = A_1 \cos \omega t$ and $x_2 = A_2 \cos \omega t$, and substitute them into Eq. (1.72) to obtain

$$-\omega^2 A_1 + \frac{2K}{m} A_1 - \frac{K}{m} A_2 = 0,$$

and

$$-\omega^2 A_2 + \frac{2K}{m} A_2 - \frac{K}{m} A_1 = 0.$$

Letting $K/m = \omega_0^2$ as before, these equations can be written as

$$(\omega^2 - 2\omega_0^2)A_1 + \omega_0^2 A_2 = 0,$$

and (1.73)

$$\omega_0^2 A_1 + (\omega^2 - 2\omega_0^2)A_2 = 0.$$

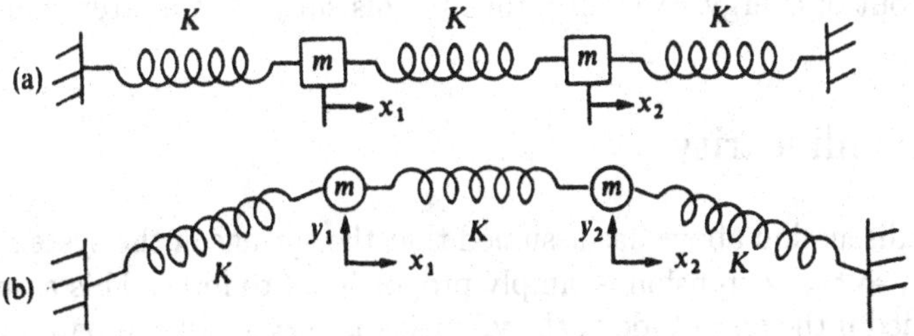

FIGURE 1.20. Oscillating systems consisting of two masses and three springs. In (a) the masses move in a line; in (b) they move in a plane, so transverse oscillations are possible as well as longitudinal oscillations.

The normal mode frequencies are obtained by setting the determinant of the coefficients equal to zero:

$$\begin{vmatrix} \omega^2 - 2\omega_0^2 & \omega_0^2 \\ \omega_0^2 & \omega^2 - 2\omega_0^2 \end{vmatrix} = \omega^4 - 4\omega_0^2\omega^2 + 4\omega_0^4 - \omega_0^4 = 0, \qquad (1.74)$$

from which $\omega^2 = 2\omega_0^2 \pm \omega_0^2$ and $\omega = \omega_0, \sqrt{3}\omega_0$.

It is easy to deduce the nature of these normal modes. The one with angular frequency $\omega_0 (= K/m)$ describes the two masses moving together in the same direction ($x_1 = x_2$), so that the center spring is not stretched. Thus, each mass is acted on by one spring, and the frequency is the same as the one-mass system in Fig. 1.1. The normal mode of higher frequency $\sqrt{3}\omega_0$ consists of the masses moving in opposite directions ($x_1 = -x_2$), so that the center spring is stretched twice as much as either of the end springs.

This result can be obtained in a more formal way by substituting each value of ω into the Eq. (1.73), in turn, and solving for A_1 and A_2:

$$\omega = \omega_0: \quad (\omega_0^2 - 2\omega_0^2)A_1 + \omega_0^2 A_2 = 0 \quad \text{from which} \quad A_1 = A_2,$$

and (1.75)

$$\omega = \sqrt{2}\omega_0: \quad (3\omega_0^2 - 2\omega_0^2)A_1 + \omega_0^2 A_2 = 0 \quad \text{from which} \quad A_1 = -A_2.$$

The normal coordinates are thus written as

$$\psi = x_1 + x_2 \quad \text{and} \quad \psi_2 = x_1 - x_2.$$

Now, consider the two-mass system in Fig. 1.20(b), where each mass is free to move in two directions. We define four coordinates x_1, x_2, y_1, and y_2. By analogy with the one-dimensional oscillator, we can see that there are now four normal modes: two transverse modes (motion in the y directions) and two longitudinal modes (motion in the x directions). Each longitudinal mode will be higher in frequency than the corresponding transverse mode, as in the one-mass system discussed in Section 1.9. Each mode can be described as motion along a normal coordinate. A system given an initial excitation along a single normal coordinate (or vibrating in a single normal mode) would ideally remain in that same normal mode of vibration until it runs out of energy. We will return to this subject in a later chapter.

1.12 Nonlinearity

All the discussion above has assumed that the springs in the system behave linearly, so that extension is simply proportional to force. This assumption is justified if the amplitude of the vibration is very small compared with the characteristic dimensions of the system. For vibrations of large amplitude, however, real springs do not behave in this simple way, whether thay are made from helices of metal or solid pieces of rubber. Nearly all springs become appreciably harder at large extensions or large compressions, so that the force law has a form like

$$F = K(x + \beta x^2), \tag{1.76}$$

where x is the displacement from the equilibrium position, scaled in terms of the natural length of the spring, and the nonlinear coefficient β is positive and very much less than unity. In some systems with peculiar geometry, β may in fact be negative, leading to a softer spring at large amplitudes, and there may also be a further asymmetrical term αx inside the brackets.

As we discuss in detail in Chapter 5, such nonlinearity can lead to many interesting and acoustically important effects. Among them are

1. Dependence of vibration mode frequencies upon amplitude.
2. Generation of overtones that are exact harmonics of the fundamental oscillation ("harmonic distortion").
3. Forced oscillation at submultiples of the driving frequency.
4. Chaotic oscillations.

Appendix

A1.1 Alternative Ways of Expressing Harmonic Motion

The solutions to the equation $\ddot{x} + \omega_0^2 x = 0$ can be written in three ways:

$$x = A\cos(\omega_0 t + \phi), \tag{1.3}$$

$$x = B\cos\omega_0 t + C\sin\omega_0 t, \tag{1.4}$$

and

$$x = \mathrm{Re}(\tilde{A}e^{j\omega_0 t}), \tag{A1.1}$$

where

$$\tilde{A} = Ae^{j\phi} = A\cos\phi + jA\sin\phi. \tag{A1.2}$$

In order to establish a relationship between these constants, we can expand the cosine in Eq. (1.3):

$$x = A\cos\phi\cos\omega_0 t - A\sin\phi\sin\omega_0 t. \tag{A1.3}$$

Comparison with Eq. (1.4) gives the relationships

$$B = A\cos\phi,$$

$$C = -A\sin\phi,$$

and (A1.4)

$$\phi = \tan^{-1}(-C/B).$$

Comparing Eq. (1.4) with Eq. (A1.2), it is clear that

$$\mathrm{Re}(\tilde{A}) = A\cos\phi = B,$$

and

$$\mathrm{Im}(\tilde{A}) = A\sin\phi = -C.$$

Yet a fourth useful form is obtained by writing $x = De^{pt}$ and noting that this will be a solution to the differential equation when $p^2 = -\omega_0^2$ or $p = \pm j\omega_0$. The general solution can then be written as

$$x = \tilde{D}e^{j\omega_0 t} + \tilde{D}^* e^{-j\omega_0 t}. \tag{A1.5}$$

$\tilde{D}$ and $\tilde{D}^*$ are complex conjugates, as are $e^{j\omega_0 t}$ and $e^{-j\omega_0 t}$, of course; thus, the general solution is *real* (since any number added to its complex conjugate is real).

Expanding the exponentials in Eq. (A1.5) gives

$$x = \tilde{D}\cos\omega_0 t + j\tilde{D}\sin\omega_0 t + \tilde{D}^*\cos\omega_0 t - j\tilde{D}^*\sin\omega_0 t.$$

Comparison with Eq. (A1.1) gives

$$\tilde{D} + \tilde{D}^* = 2\mathrm{Re}(\tilde{D}) = A\cos\phi,$$

and

$$j(\tilde{D} - \tilde{D}^*) = -2\mathrm{Im}(\tilde{D}) = -A\sin\phi. \tag{A1.6}$$

To summarize, we have four forms of the solution given by Eqs. (1.3), (1.4), (A1.1), and (A1.5). Each form includes two arbitrary constants. Although in Eqs. (A1.1) and (A1.5) the constants are complex, x is real in each case. In Eq. (A1.5), the real displacement x is obtained by taking the real part of a complex displacement x; in Eq. (A1.1), however, the real displacement is obtained by adding two terms that are complex conjugates.

A1.2 Equivalent Electrical Circuit for a Simple Oscillator

Many mechanical systems are mathematically equivalent to corresponding electrical systems. It is often helpful to represent a mechanical oscillating system by an equivalent electrical circuit, so that electrical network theory can be applied. The simple mechanical oscillator in Fig. 1.11 [and in Fig. A.1(a)], for example, can be represented by the equivalent electrical circuit in Fig. A.1(c). In the two electrical circuits, we identify velocity $\dot{x}$ with current i, displacement x with charge q, and force $f(t)$ with voltage $v(t)$. Mass m is then analogous to inductance L and stiffness to reciprocal capacitance $1/C$; resistance R appears in both circuits.

The mechanical and electrical impedances are

$$Z_{\mathrm{m}} = R + jX_{\mathrm{m}} = R + j(\omega m - K/\omega), \tag{A1.7a}$$

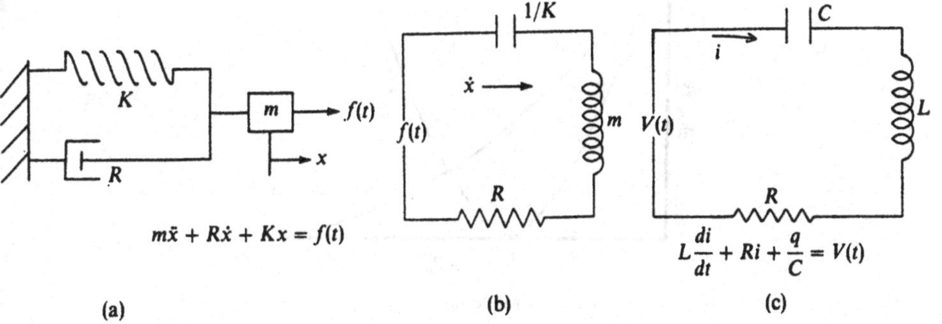

FIGURE A1.1. (a) A simple mechanical oscillator. (b) Its equivalent electrical circuit. (c) Circuit of electrical analogs.

and

$$Z_e = R + jX_e = R + j(\omega L - 1/\omega C). \tag{A1.7b}$$

The resonance frequencies are

$$f_0 = \frac{1}{2\pi}\sqrt{\frac{K}{m}},$$

and

$$f_0 = \frac{1}{2\pi\sqrt{LC}}. \tag{A1.8}$$

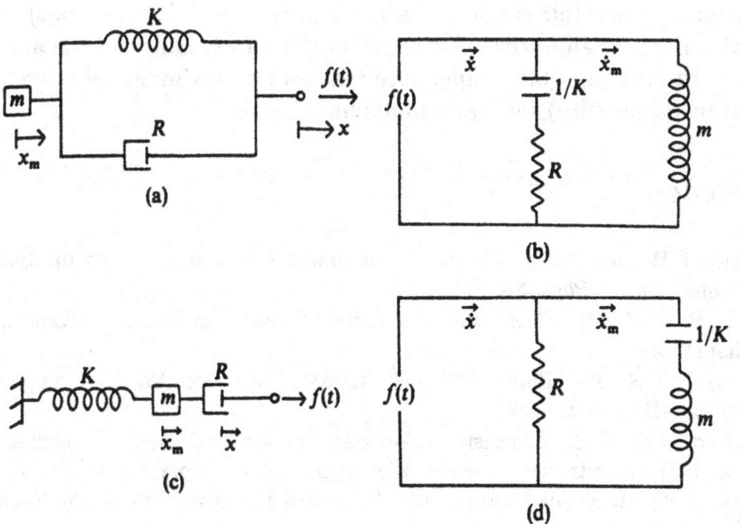

FIGURE A1.2. Mechanical oscillating systems and their equivalent electrical circuits.

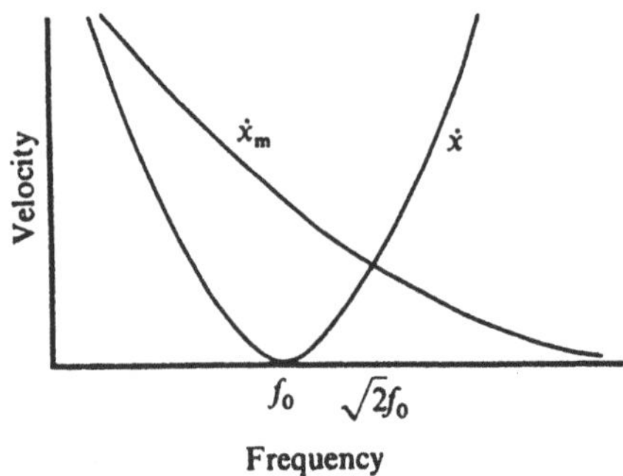

FIGURE A1.3. Frequency response of the oscillating system shown in Fig. A1.2(a) as represented by the equivalent circuit in Fig. A1.2(b).

The oscillator in Fig. A1.1(a) is represented by a series circuit, because all the elements experience the same displacement x. If the force were applied to the end of the spring opposite the mass, as in Fig. A1.2(a), the system would be represented by the parallel circuit shown in Fig. A1.2(b). Similarly, the system in Fig. A1.2(c) has the equivalent circuit shown in Fig. A1.2(d).

The reciprocal of electrical impedance is electrical admittance. In mechanical systems, the reciprocal of impedance is called *mechanical admittance* or *mobility*. Mobility Y is velocity divided by force.

Note that the oscillating system in Fig. A1.2(a) is represented by a circuit [Fig. A1.2(b)] in which the two reactive elements ($1/K$ and m) are in parallel. At the natural frequency ω_0, $\dot{x}$ has a minimum rather than a maximum value (although the velocity of the mass $\dot{x}_m$ does not). This behavior is called an *antiresonance* rather than a resonance. At an antiresonance, the driving point impedance reaches its maximum value and the admittance (mobility) reaches a minimum.

References

Arnold, T.W., and Case, W. (1982). Nonlinear effects in a simple mechanical system. *Am. J. Phys.* **50**, 220.

Beyer, R.T. (1974). "Nonlinear Acoustics" (Naval Sea Systems Command), Chapter 2.

Crawford, F.S. Jr. (1965). "Waves." Berkeley Physics, Vol. 3, Chapter 1. McGraw-Hill, New York.

Fletcher, N.H. (1982). Transient response and the musical characteristics of bowed-string instruments. *Proc. Wollongong Coop. Workshop on the Acoustics of Stringed Instruments* (A. Segal, ed.). University of Wollongong, Australia.

Kinsler, L.E., Frey, A.R., Coppens, A.B., and Sanders, J.V. (1982). "Fundamentals of Acoustics," 3rd ed., Chapter 1. Wiley, New York.

Lee, E.W. (1960). Non-linear vibrations. *Contemp. Phys.* **2**, 143.

Main, I.G. (1978). "Vibrations and Waves in Physics." Cambridge Univ. Press, London and New York.

Morse, P.M. (1948). "Vibration and Sound," 2nd ed., Chapter 2. McGraw-Hill, New York. Reprinted 1976, Acoustical Soc. Am., Woodbury, New York.

Rossing, T.D. (1982). "The Science of Sound," Addison-Wesley, Reading, Massachusetts.

Skudrzyk, E. (1968). "Simple and Complex Vibrating Systems," Chapters 1 and 2. Pennsylvania State University, University Park, Pennsylvania.

2

Continuous Systems in One Dimension: Strings and Bars

In the last chapter, we considered vibrating systems consisting of one or more masses, springs, and dampers. In this chapter, we will focus on systems in which these elements are distributed continuously throughout the system rather than appearing as discrete elements. We begin with a system composed of several discrete elements, then allow the number of elements to grow larger, eventually leading to a continuum.

2.1 Linear Array of Oscillators

The oscillating system with two masses in Fig. 1.20 was shown to have two transverse vibrational modes and two longitudinal modes. In both the longitudinal and transverse pairs, there is a mode of low frequency in which the masses move in the same direction and a mode of higher frequency in which they move in opposite directions.

The normal modes of a three-mass oscillator are shown in Fig. 2.1. The masses are constrained to move in a plane, and so there are six normal modes of vibration, three longitudinal and three transverse. Each longitudinal mode will be higher in frequency than the corresponding transverse mode. If the masses were free to move in three dimensions, there would be $3 \times 3 = 9$ normal modes, three longitudinal and six transverse.

Increasing the number of masses and springs in our linear array increases the number of normal modes. Each new mass adds one longitudinal mode and (provided the masses move in a plane) one transverse mode. The modes of transverse vibration for mass/spring systems with $N = 1$ to 24 masses are shown in Fig. 2.2; note that as the number of masses increases, the system takes on a wavelike appearance. A similar diagram could be drawn for the longitudinal modes.

As the number of masses in our linear system increases, we take less and less notice of the individual elements, and our system begins to resemble a vibrating string with mass distributed uniformly along its length. Presumably, we could describe the vibrations of a vibrating string by writing

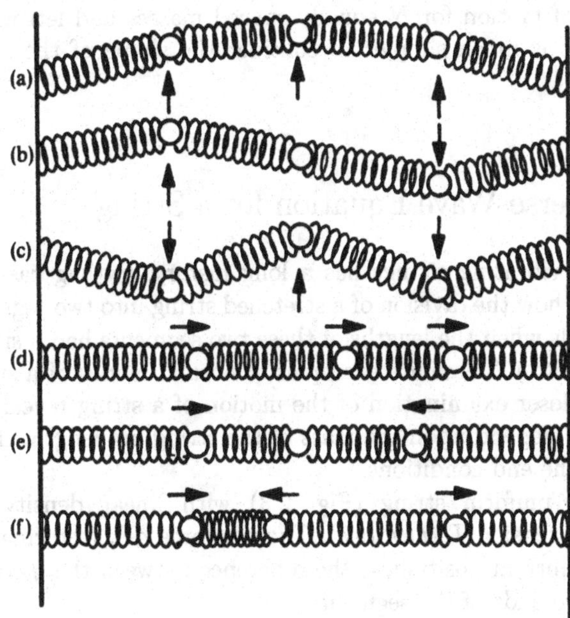

FIGURE 2.1. Normal modes of a three-mass oscillator. Transverse mode (a) has the lowest frequency and longitudinal mode (f) the highest.

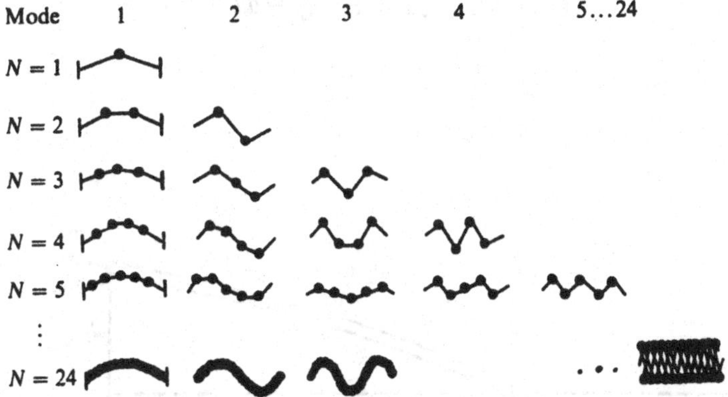

FIGURE 2.2. Modes of transverse vibration for mass/spring systems with different numbers of masses. A system with N masses has N modes.

N equations of motion for N equally spaced masses and letting N go to infinity, but it is much simpler to consider the shape of the string as a whole.

2.2 Transverse Wave Equation for a String

The study of vibrating strings has a long history. Pythagoras is said to have observed how the division of a stretched string into two segments gave pleasing sounds when the lengths of these two segments had a simple ratio (2:1, 3:1, 3:2, etc.). These are examples of normal modes of a string fixed at its ends. Closer examination of the motion of a string reveals that the normal modes depend upon the mass of the string, its length, the tension applied, and the end conditions.

Consider a uniform string (Fig. 2.3) with linear density μ(kg/m) stretched to a tension T (newtons). The net force dF, restoring segments ds to its equilibrium position, is the difference between the y components of T at the two ends of the segment:

$$dF_y = (T \sin\theta)_{x+dx} - (T \sin\theta)_x.$$

Applying the Taylor's series expansion

$$f(x+dx) = f(x) + \frac{\partial f(x)}{\partial x}\,dx + \cdots$$

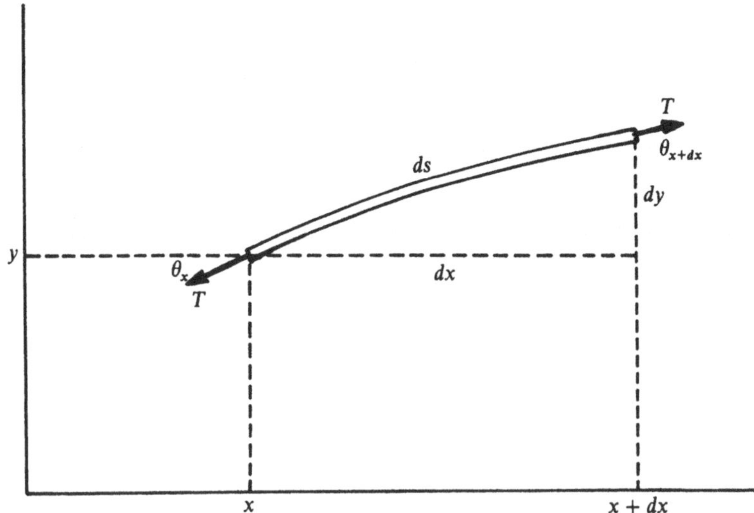

FIGURE 2.3. Segments of a string with tension T.

to $T\sin\theta$ and keeping first-order terms gives

$$dF_y = [(T\sin\theta)_x + \frac{\partial(T\sin\theta)}{\partial x}dx] - (T\sin\theta)_x = \frac{\partial(T\sin\theta)}{\partial x}dx. \quad (2.1)$$

For small displacement y, $\sin\theta$ can be replaced by $\tan\theta$, which is also $\partial y/\partial x$:

$$dF_y = \frac{\partial(T\partial y/\partial x)}{\partial x}dx = T\frac{\partial^2 y}{\partial x^2}dx. \quad (2.2)$$

The mass of the segment ds is μds, so Newton's second law of motion becomes

$$T\frac{\partial^2 y}{\partial x^2}dx = (\mu ds)\frac{\partial^2 y}{\partial t^2}. \quad (2.3)$$

Since dy is small, $ds \cong dx$. Also, we write $c^2 = T/\mu$ and obtain

$$\frac{\partial^2 y}{\partial t^2} = \frac{T}{\mu}\frac{\partial^2 y}{\partial x^2} = c^2\frac{\partial^2 y}{\partial x^2}. \quad (2.4)$$

This is the well-known equation for transverse waves in a vibrating string.

2.3 General Solution of the Wave Equation: Traveling Waves

The general solution of Eq. (2.4) can be written in a form credited to d'Alembert (1717–1783):

$$y = f_1(ct - x) + f_2(ct + x). \quad (2.5)$$

The function $f_1(ct - x)$ represents a wave traveling to the right with a velocity c; similarly, $f_2(ct + x)$ represents a wave traveling to the left with the same velocity. The nature of functions f_1 and f_2 is arbitrary; they could be sinusoidal or they could describe impulsive waves, for example. In fact, the two independent functions f_1 and f_2 can be chosen so that their sum represents any desired initial displacement $y(x,0)$ and velocity $\partial y/\partial t = \dot{y}(x,0)$.

Differentiation of Eq. (2.5) by x and t leads to

$$\partial y/\partial x = -f_1' + f_2',$$

and (2.6)

$$\partial y/\partial t = c(f_1' + f_2'),$$

where f_1' and f_2' are derivatives of the two functions with respect to their arguments.

2.4 Reflection at Fixed and Free Ends

In order to understand wave reflection at the ends of a string, we first consider what happens to a single pulse at fixed and free ends of a string, as indicated in Fig. 2.4. By fixed end, we understand that the string is securely fastened, but free end requires some explanation. We need to maintain tension in the x direction, but we want the string to move freely in the y direction. Thus, we imagine it is fastened to a massless ring that slides up and down on a rod without friction.

1. At a *fixed end*, $y = 0$. Assuming that the string is fixed at $x = 0$, the general solution [Eq. (2.5)] becomes

$$y = 0 = f_1(ct - 0) + f_2(ct + 0),$$

   from which (2.7)

$$f_1(ct) = -f_2(ct).$$

   Thus, an up pulse reflects as a down pulse, as shown in Fig. 2.4(a).
2. At a *free end*, $\partial y/\partial x = 0$ because no transverse force is possible. Thus, from Eq. (2.6),

$$f_1'(ct) = f_2'(ct). \tag{2.8a}$$

   Integration of Eq. (2.8a) gives

$$f_1(ct) = f_2(ct). \tag{2.8b}$$

   An up pulse now reflects as an up pulse, as shown in Fig. 2.4(b).

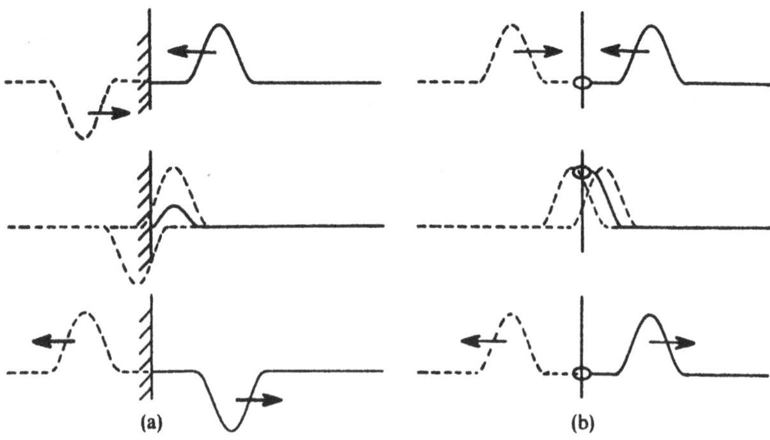

FIGURE 2.4. Reflection of a pulse at a fixed end (a) and at a free end (b). In (a) the appropriate boundary condition can be met by having an imaginary pulse of opposite phase meet the real pulse at $x = 0$. In (b) the imaginary pulse has the same phase.

Of course, many other end conditions are possible. For example, the string may be attached to a string with a different linear density μ, to a spring, or to a mass. A particularly important case is that of an end support that is nearly fixed but yields slightly, such as the bridge of a piano or violin.

2.5 Simple Harmonic Solutions to the Wave Equation

In order to see how simple harmonic motions are propagated along a string, we let the functions f_1 and f_2 in the general solution [Eq. (2.5)] each consist of a sine term and a cosine term

$$\begin{aligned} y(x,t) &= A \sin \frac{\omega}{c}(ct - x) + B \cos \frac{\omega}{c}(ct - x) + C \sin \frac{\omega}{c}(ct + x) \\ &\quad + D \cos \frac{\omega}{c}(ct + x) \\ &= A \sin(\omega t - kx) + B \cos(\omega t - kx) + C \sin(\omega t + kx) \\ &\quad + D \cos(\omega t + kx). \end{aligned} \tag{2.9}$$

where $k = \omega/c = 2\pi/\lambda$ is known as the wave number.

Alternatively, we could have used the complex notation

$$\tilde{y}(x,t) = \tilde{A}e^{j(\omega t - kx)} + \tilde{B}e^{j(\omega t + kx)}, \tag{2.10}$$

where $\tilde{y}$, $\tilde{A}$, and $\tilde{B}$ are complex. In this case, $y(x,t) = \mathrm{Re}\tilde{y}(x,t)$.

2.6 Standing Waves

Consider a string of length L fixed at $x = 0$ and $x = L$. The first condition $y(0,t) = 0$ requires that $A = -C$ and $B = -D$ in Eq. (2.9), so

$$y = A[\sin(\omega t - kx) - \sin(\omega t + kx)] + B[\cos(\omega t - kx) - \cos(\omega t + kx)]. \tag{2.11}$$

Using the sum and difference formulas, $\sin(x \pm y) = \sin x \cos y \pm \cos x \sin y$ and $\cos(x \pm y) = \cos x \cos y \mp \sin x \sin y$

$$\begin{aligned} y &= 2A \sin kx \cos \omega t - 2B \sin kx \sin \omega t \\ &= 2[A \cos \omega t - B \sin \omega t] \sin kx. \end{aligned} \tag{2.12}$$

The second condition $y(L,t) = 0$ requires that $\sin kL = 0$ or $\omega L/c = n\pi$. This restricts ω to values $\omega_n = n\pi c/L$ or $f_n = n(c/2L)$. Thus, the string has normal modes of vibration:

$$y_n(x,t) = (A_n \sin \omega_n t + B_n \cos \omega_n t) \sin \frac{\omega_n x}{c}. \tag{2.13}$$

These modes are harmonic, because each f_n is n times $f_1 = c/2L$.

The general solution of a vibrating string with fixed ends can be written as a sum of the normal modes:

$$y = \sum_n (A_n \sin \omega_n t + B_N \cos \omega_n t) \sin k_n x, \tag{2.14}$$

and the amplitude of the nth mode is $C_n = \sqrt{A_n^2 + B_n^2}$. At any point $y(x,t) = \sum_n y_n(x,t)$.

Alternatively, the general solution could be written as

$$y = \sum_n C_n \sin(\omega_n t + \phi_n) \sin k_n x, \tag{2.15}$$

where C_n is the amplitude of the nth mode and ϕ_m is its phase.

2.7 Energy of a Vibrating String

When a string vibrates in one of its normal modes, the kinetic and potential energies alternately take on their maximum value, which is equal to the total energy, just as in the simple mass–spring system discussed in Section 1.3. Thus, the energy of a mode can be calculated by considering either the kinetic or the potential energy. The maximum kinetic energy of a segment vibrating in its nth mode is

$$dE_n = \frac{\omega_n^2 \mu}{2} (A_n^2 + B_n^2) \sin^2 \frac{n\pi x}{L} \, dx.$$

Integrating over the entire length gives

$$E_n = \frac{\omega_n^2 \mu L}{4} (A_n^2 + B_n^2) = \frac{\omega_n^2 \mu L}{4} C_n^2. \tag{2.16}$$

The potential and kinetic energies of each mode have a time average value that is $E_n/2$. The total energy of the string can be found by summing up the energy in each normal mode:

$$E = \sum_n E_n.$$

2.8 Plucked String: Time and Frequency Analyses

When a string is excited by bowing, plucking, or striking, the resulting vibration can be considered to be a combination of several modes of vibration. For example, if the string is plucked at its center, the resulting vibration will consist of the fundamental plus the odd-numbered harmonics. Fig. 2.5 illustrates how the modes associated with the odd-numbered harmonics, when each is present in the right proportion, add up at one instant in time to give the initial shape of the center-plucked string. Modes 3, 7, 11, etc., must be opposite in phase from modes, 1, 5, and 9 in order to give maximum displacement at the center, as shown at the top. Finding the normal mode spectrum of a string, given its initial displacement, calls for *frequency analysis* or *Fourier analysis*.

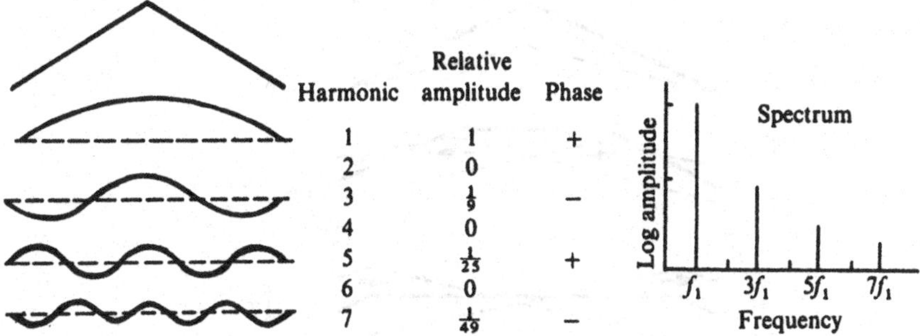

FIGURE 2.5. Frequency analysis of a string plucked at its center. Odd-numbered modes of vibration add up in appropriate amplitude and phase to give the shape of the string.

Since all the modes shown in Fig. 2.5 have different frequencies of vibration, they quickly get out of phase, and the shape of the string changes rapidly after plucking. The shape of the string at each moment can be obtained by adding the normal modes at that particular time, but it is more difficult to do so because each of the modes will be at a different point in its cycle. The resolution of the string motion into two pulses that propagate in opposite directions on the string, which we might call *time analysis*, is illustrated in Fig. 2.6. If the string is plucked at a point other than its center, the spectrum or recipe of the constituent modes is different, of course. For example, if the string is plucked $\frac{1}{5}$ of the distance from one end, the spectrum of mode amplitudes shown in Fig. 2.7 is obtained. Note that the 5th harmonic is missing. Plucking the string $\frac{1}{4}$ of the distance from the end suppresses the 4th harmonic, etc. (In Fig. 2.5, plucking it at $\frac{1}{2}$ the distance eliminated the 2nd harmonic as well as other even-numbered ones.)

A time analysis of the string plucked at $\frac{1}{5}$ of its length is shown in Fig. 2.8. A bend racing back and forth within a parallelogram boundary can be viewed as the resultant of two pulses (dashed lines) traveling in opposite directions. Each of these pulses can be described by one term in d'Alembert's solution [Eq. (2.5)].

Each of the normal modes described in Eq. (2.13) has two coefficients A_n and B_n whose values depend upon the initial excitation of the string. These coefficients can be determined by Fourier analysis. Multiplying each side of Eq. (2.14) and its time derivative by $\sin m\pi x/L$ and integrating from 0 to L gives the following formulae for the Fourier coefficients:

$$A_n = \frac{2}{\omega_n L}\int_0^L \dot{y}(x,0)\sin\frac{n\pi x}{L}\,dx, \tag{2.17}$$

$$B_n = \frac{2}{L}\int_0^L y(x,0)\sin\frac{n\pi x}{L}\,dx. \tag{2.18}$$

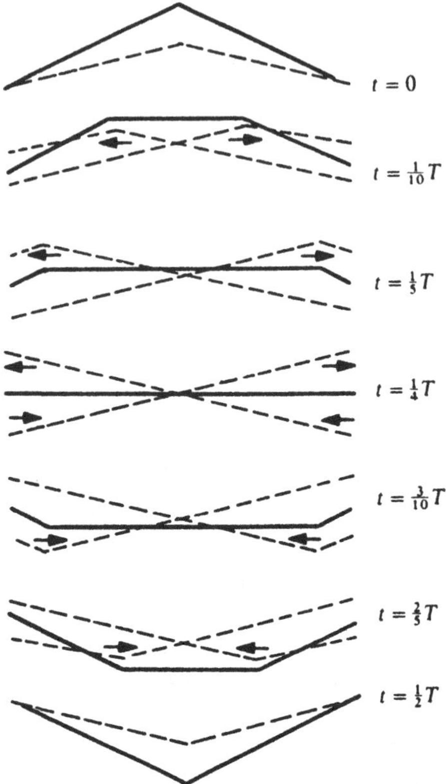

FIGURE 2.6. Time analysis of the motion of a string plucked at its midpoint through one half cycle. Motion can be thought of as due to two pulses traveling in opposite directions.

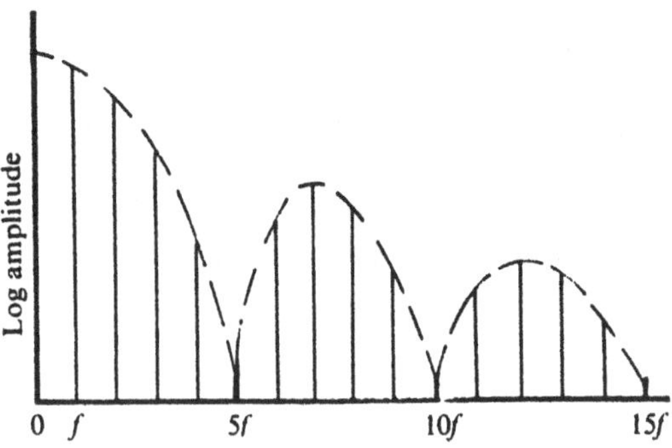

FIGURE 2.7. Spectrum of a string plucked one-fifth of the distance from one end.

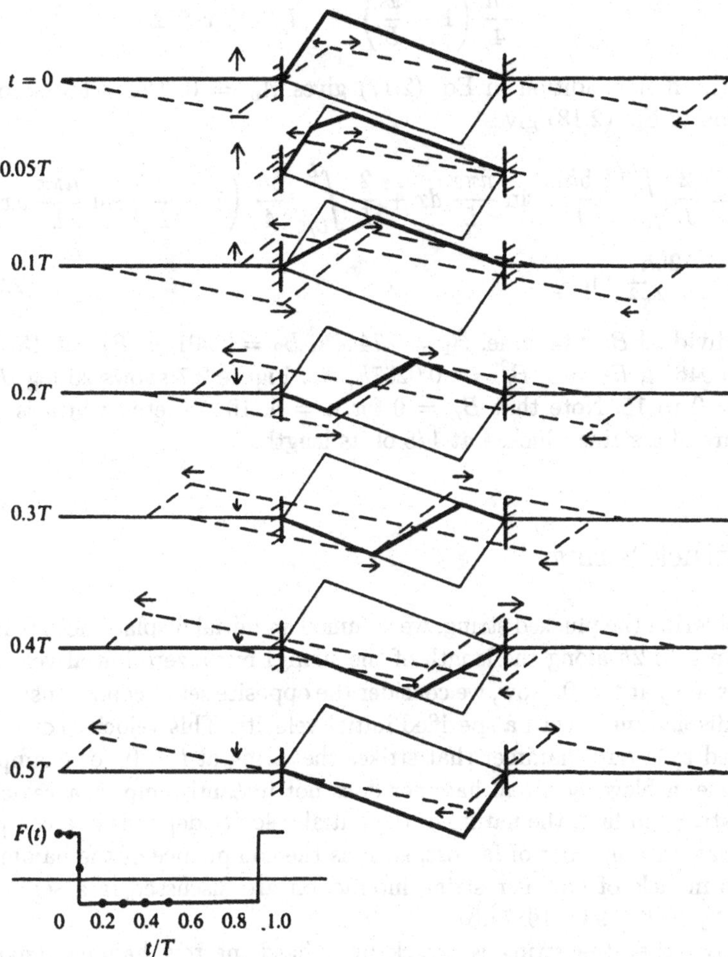

FIGURE 2.8. Time analysis through one half cycle of the motion of a string plucked one-fifth of the distance from one end. The motion can be thought of as due to two pulses [representing the two terms in Eq. (2.5)] moving in opposite directions (dashed curves). The resultant motion consists of two bends, one moving clockwise and the other counterclockwise around a parallelogram. The normal force on the end support, as a function of time, is shown at the bottom.

Using these formulae, we can calculate the Fourier coefficients for the string of length L plucked with amplitude h at one-fifth of its length, as shown in the time analysis in Fig. 2.8. The initial conditions are

$$\dot{y}(x,0) = 0,$$

$$y(x,0) = \frac{5h}{L}x, \qquad 0 \leqq x \leqq L/5, \tag{2.19}$$

$$= \frac{5h}{4}\left(1 - \frac{x}{L}\right), \qquad L/5 \leqq x \leqq L.$$

Using the first condition in Eq. (2.17) gives $A_n = 0$. Using the second condition in Eq. (2.18) gives

$$B_n = \frac{2}{L}\int_0^{L/5} \frac{5h}{L} x \sin\frac{n\pi x}{L}\,dx + \frac{2}{L}\int_{L/5}^{L} \frac{5h}{4}\left(1 - \frac{x}{L}\right)\sin\frac{n\pi x}{L}\,dx$$

$$= \frac{25h}{2n^2\pi^2}\sin\frac{n\pi}{5}. \tag{2.20}$$

The individual B_n's become: $B_1 = 0.7444h$, $B_2 = 0.3011h$, $B_3 = 0.1338h$, $B_4 = 0.0465h$, $B_5 = 0$, $B_6 = -0.0207h$, etc. Figure 2.7 shows 20 log $|B_n|$ for $n = 0$ to 15. Note that $B_n = 0$ for $n = 5$, 10, 15, etc., which is the signature of a string plucked at 1/5 of its length.

2.9 Struck String

In considering the plucked string, we assumed an initial displacement (varying from 0 to $2h$ along the length of the string) but a zero initial velocity (everywhere) at $t = 0$. Now, we consider the opposite set of conditions: zero initial displacement with a specified initial velocity. This velocity could be imparted by a hard hammer that strikes the string at $t = 0$, for example. Of course, a blow by a real hammer does not instantly impart a velocity to the string; in fact, the nature of the initial velocity depends in a complicated way on a number of factors, such as the compliance of the hammer. Various models of hammer–string interaction are discussed in a series of papers by Hall (1986, 1987a,b).

Suppose that the string is struck by a hard, narrow hammer having a velocity V. After a short time t, a portion of the string with length $2ct$ and mass $2\mu ct$ is set into motion. As this mass increases and becomes comparable to the hammer mass M, the hammer is slowed down and would eventually be stopped. With a string of finite length, however, reflected impulses return while the hammer still has appreciable velocity, and these reflected impulses interact with the hammer in a rather complicated way, causing it to be thrown back from the string.

At the point of contact, the string and hammer together satisfy the equation

$$M\frac{\partial^2 y}{\partial t^2} = T\Delta\left(\frac{\partial y}{\partial x}\right), \tag{2.21}$$

while elsewhere the string continues to satisfy Eq. (2.4). The discontinuity in the string slope $\Delta\left(\dfrac{\partial y}{\partial x}\right)$, according to Eq. (2.21), is responsible for the force that slows down the hammer. Equation (2.21) is satisfied at the contact point by a velocity

$$v(t) = Ve^{-t/\tau}, \tag{2.22}$$

where $\tau = Mc/2T$ may be termed the deceleration time (Hall, 1986). The corresponding displacement is

$$y(t) = V\tau(1 - e^{-t/\tau}). \tag{2.23}$$

The displacement at the contact point approaches $VMc/2T$, and the velocity approaches zero. If the string were very long, the displacement and velocity elsewhere on the string could be found by substituting $t - [(x - x_0)/c]$ for t in Eqs. (2.24) and (2.25), as shown in Fig. 2.9.

Only when the string is very long or the hammer is very light does the hammer stop, as in Fig. 2.9. In a string of finite length, reflected pulses return from both ends of the strings and interact with the moving hammer in a fairly complicated way. Eventually, the hammer is thrown clear of the string, and the string vibrates freely in its normal modes.

In general, the harmonic amplitudes in the vibration spectrum of a struck string fall off less rapidly with frequency than those of the plucked strings shown in Figs. 2.5 and 2.7. For a very light hammer whose mass M is much less than the mass of the string M_s, the spectrum dips to zero for harmonic numbers that are multiples of $1/\beta$ (where the string is struck at a fraction β of its length), but otherwise does not fall off with frequency, as shown in

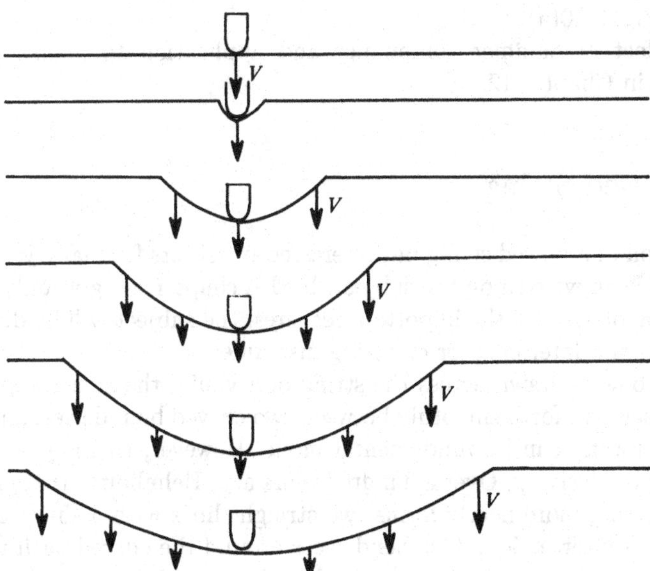

FIGURE 2.9. Displacement and velocity of a long string at successive times after being struck by a hard narrow hammer having a velocity V.

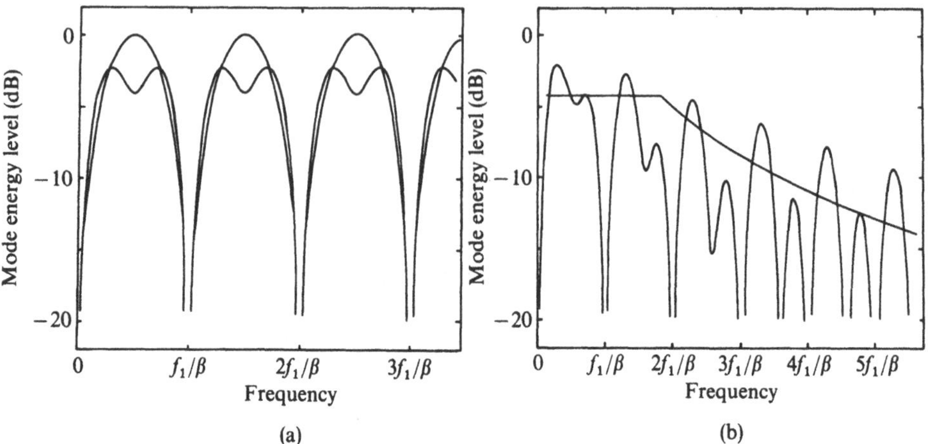

FIGURE 2.10. Spectrum envelopes for a string struck at a fraction β of its length: (a) hammer mass $M \ll$ string mass M_s; (b) $M = 0.4/\beta M_s$ (from Hall, 1986).

Fig. 2.10(a). (The spectrum of sound radiated by a piano, which may be quite different from the vibration spectrum of the string, will be discussed in Chapter 12).

If the hammer mass is small but not negligible compared to the mass of the string, the spectrum envelope falls off as $1/n$ (6 dB/octave) above a mode number given by $n_m = 0.73M_s/M$, as shown in Fig. 2.10(b). Note that for high harmonic (mode) numbers, there are missing modes between those in Fig. 2.10(a).

The effect of hammer compliance and application to pianos will be discussed in Chapter 12.

2.10 Bowed String

The motion of a bowed string has interested physicists for many years, and much has been written on the subject. In this chapter, we give only a brief description of some of the important features; the subject will be discussed more fully in a later chapter on string instruments.

As the bow is drawn across the string of a violin, the string appears to vibrate back and forth smoothly between two curved boundaries, much like a string vibrating in its fundamental mode. However, this appearance of simplicity is deceiving. Over a hundred years ago, Helmholtz (1877) showed that the string more nearly forms two straight lines with a sharp bend at the point of intersection. This bend races around the curved path that we see, making one round trip each period of the vibration.

To observe the string motion, Helmholtz constructed a vibration microscope, consisting of an eyepiece attached to a tuning fork. This was driven in sinusoidal motion parallel to the string, and the eyepiece was focused on a bright-colored spot on the string. When Helmholtz bowed the string, he saw a Lissajous figure (see Section 1.10). The figure was stationary when the tuning fork frequency was an integral fraction of the string frequency. Helmholtz noted that the displacement of the string followed a triangular pattern at whatever point he observed it, as shown in Fig. 2.11. The velocity waveform at each point alternates between two values.

Other early work on the subject was published by Krigar-Menzel and Raps (1891) and by C.V. Raman (1918). More recent experiments by Schelleng (1973), McIntyre, et al. (1981), Lawergren (1980), Kondo and Kubata (1983), and by others have verified these early findings and have greatly added to our understanding of bowed strings. An excellent discussion of the bowed string is given by Cremer (1981).

The motion of a bowed string is shown in Fig. 2.12. A time analysis in Fig. 2.12(A) shows the Helmholtz-type motion of the string; as the bow moves ahead at a constant speed, the bend races around a curved path. Fig. 2.12(B) shows the position of the point of contact at successive times; the letters correspond to the frames in Fig. 2.12(A). Note that there is a single bend in the bowed string, whereas in the plucked string (Fig. 2.8), we had a double bend.

The action of the bow on the string is often described as a stick and slip action. The bow drags the string along until the bend arrives [from (a) in Fig. 2.12(A)] and triggers the slipping action of the string until it is picked up by the bow once again [frame (c)]. From (c) to (i), the string moves at the speed of the bow. The velocity of the bend up and down the string is the usual $\sqrt{T/\mu}$.

The envelope around which the bend races [the dashed curve in Fig. 2.12(A)] is composed of two parabolas with a maximum amplitude that is proportional, within limits, to the bow velocity. It also increases as the string is bowed nearer to one end.

The actual string motion may be a superposition of several Helmholtz-type motions. Also, if the bowing point is at an integral fraction of the length, so that certain harmonics are not excited, the displacement curves

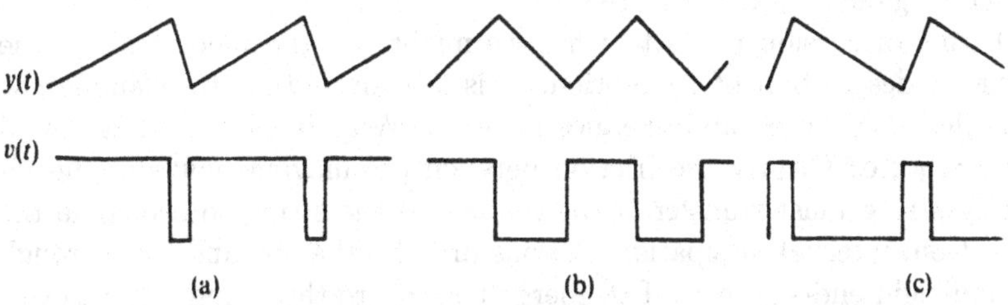

FIGURE 2.11. Displacement and velocity of a bowed string at three positions along its length: (a) at $x = L/4$, (b) at the center, and (c) at $x = 3L/4$.

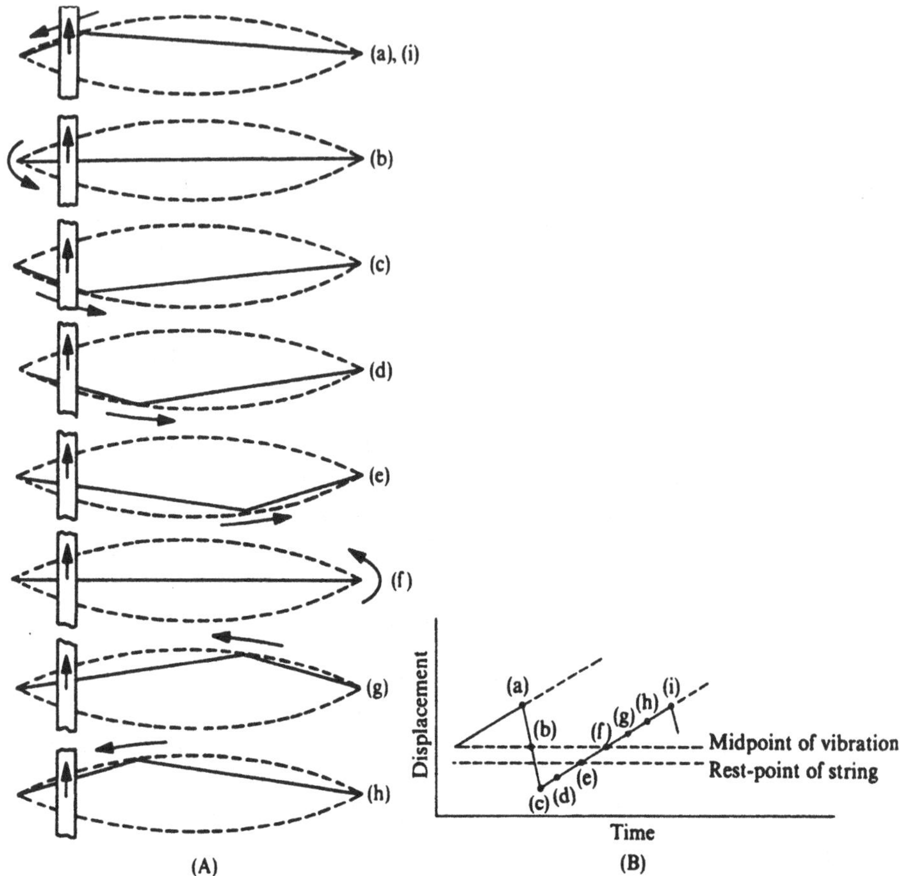

FIGURE 2.12. Motion of a bowed string. (A) Time analysis of the motion, showing the shape of the string at eight successive times during the cycle. (B) Displacement of the bow (dashed line) and the string at the point of contact (solid line) at successive times. The letters correspond to the letters in (A).

take on ripples. These and many other details of bowed string motion are treated elegantly by Cremer (1981).

All this discussion refers to what we might call the kinematics of the string—a description of its motion. It is also important to examine the dynamics—the forces and energies involved. Why, in fact, does a bowed string vibrate? Clearly the friction between the moving bow and the vibrating string must transfer enough power to the string to overcome the losses from internal dissipation, viscous drag by the air, and loss through the non-rigid end-supports. The energy transfer to the string in each cycle of motion is just the frictional force multiplied by the displacement and integrated over the cycle, so that, since the string returns to its initial position once in each cycle, the energy transferred is zero unless the frictional force varies with velocity.

As we know from elementary physics, however, there is a difference between static friction—the maximum tangential force between two surfaces when their relative velocity is zero—and dynamic friction—the tangential force when one surface is sliding over the other. Dynamic friction is generally less than static friction, so that, on an inclined plane for example, once a body begins to slide it continues to do so. The frictional curve actually has the shape shown in Fig. 2.13. Dynamic friction is a nonlinear function of relative velocity, and static friction is undetermined in sign or magnitude, except that its maximum absolute value (which is usually called the static friction) has a definite magnitude for the two surfaces in question and for the applied normal force. The exact shape of the curve depends upon the nature of the two surfaces.

If the frictional behavior is as specified by the curve in Fig. 2.13, then we can see the origin of the energy transfer. Suppose that a moving bow is lowered onto a string that is already vibrating with small amplitude. When the string is moving in the same direction as the bow, energy is transferred to it, while when it is moving in the opposite direction to the bow, the frictional force dissipates string energy. In the first situation, however, the velocity of the string relative to the bow is smaller than in the second, so that the frictional force is larger and the energy transferred to the string in this half-cycle is greater than that lost by the string in the following half-cycle. The amplitude of the vibration will therefore increase.

There is, however, a limit to this increase in vibrational amplitude of the string. At some critical amplitude, the string velocity in the forward direction will exactly equal the bow velocity. Any further increase in amplitude will take the string past the point A in Fig. 2.13 and the frictional force will

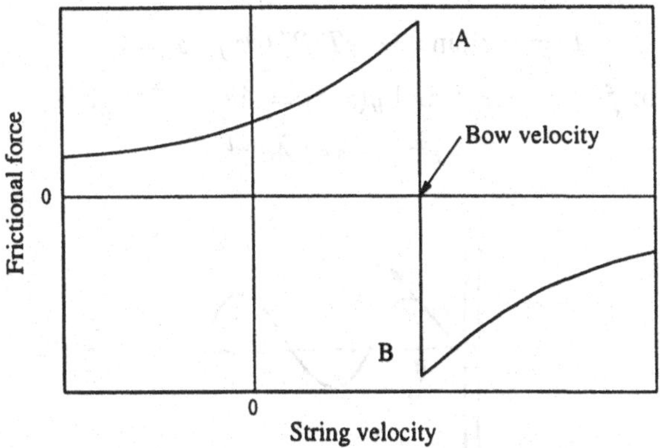

FIGURE 2.13. The dependence of frictional force upon relative velocity for a bow moving across a string.

suddenly change sign to point B, taking energy from the string and clipping the waveform. From this point onwards, the string effectively sticks to the bow for part of each cycle, and the full Helmholtz motion develops.

This is, however, only one way in which the player of a bowed-string instrument may begin a note. Quite often the player may rest the stationary bow on the string and then suddenly begin to move it at a steady speed. Because of the large magnitude of static friction, the string is drawn aside by the bow until the restoring force due to tension exceeds the maximum static friction. The string then slips and the Helmholtz motion begins.

This discussion, of course, simplifies matters greatly. In the first case there is a gradual transition from sinusoidal motion to Helmholtz motion, while in the second case the initial motion resembles that of a plucked string, with two Helmholtz waves proceeding in opposite senses. Frictional forces damp out one of these waves quite quickly, however, and the normal motion is established. The need for this damping out of contrary motion is one of the reasons that violins are fitted with rather lossy strings.

2.11 Driven String: Impedance

One way to excite a string is to apply a transverse force $f(t)$ to one end. We first consider an infinite string with tension T and a transverse force $\tilde{f}(t) = Fe^{j\omega t}$ as shown in Fig. 2.14. Since the string is infinitely long and the force is applied at the left end, the solution consists only of waves moving to the right.

$$y(x,t) = \tilde{A}e^{j(\omega t - kx)},$$

where $\tilde{A}$ is a complex constant giving the amplitude and the phase with respect to the driving force and $k = \omega/c = 2\pi/\lambda$.

Since there is no mass concentrated at $x = 0$, the driving force should balance the transverse component of the tension:

$$\tilde{F} = -T\sin\theta \cong -T(\partial\tilde{y}/\partial x) \quad x = 0. \tag{2.24}$$

Substitution of $\tilde{f}(t) = Fe^{j\omega t}$ and $\tilde{y}(x,t) = \tilde{A}e^{j(\omega t - kx)}$ gives

$$Fe^{j\omega t} = jkT\tilde{A}e^{j\omega t}$$

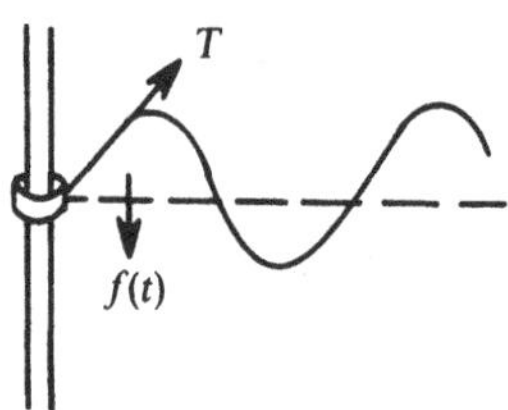

FIGURE 2.14. Forces at $x = 0$ on a string free to move in the y direction.

or

$$\tilde{A} = F/jkT, \tag{2.25}$$

so

$$\tilde{y}(x,t) = \frac{-jF}{kT} e^{j(\omega t - kx)}.$$

The velocity $\tilde{u} = \partial\tilde{y}/\partial t$ becomes

$$\tilde{u}(x,t) = \frac{F\omega}{kT} e^{j(\omega t - kx)} = \frac{Fc}{T} e^{j(\omega t - kx)}. \tag{2.26}$$

We define the mechanical input impedance Z_{in} as the ratio of force to velocity at the driving point, so

$$Z_{\text{in}} = \frac{\tilde{f}(t)}{\tilde{u}(0,t)}. \tag{2.27}$$

In a string of infinite length (or a string terminated so that no reflections occur), $\tilde{Z}_{\text{in}}$ equals *the characteristic impedance* Z_0, which is a real quantity; the input impedance is purely resistive in an infinite string.

$$Z_0 = \frac{T}{c} = \sqrt{T\mu} = \mu c. \tag{2.28}$$

The behavior of a string of finite length is more complicated because $\partial\tilde{y}/\partial x$ at $x = 0$ depends upon the reflected wave as well.

$$\tilde{y}(x,t) = \tilde{A}e^{j(\omega t - kx)} + \tilde{B}e^{j(\omega t + kx)}. \tag{2.29}$$

Assume that the string is fixed at $x = L$ and driven at $x = 0$ as before. Substitution of Eq. (2.29) into Eq. (2.24) gives

$$Fe^{j\omega t} = T(jk\tilde{A} - jk\tilde{B})e^{j\omega t}. \tag{2.30}$$

The boundary condition at $x = L$ gives

$$0 = \tilde{A}e^{-jkL} + \tilde{B}e^{jkL}. \tag{2.31}$$

Solving Eqs. (2.30) and (2.31) together gives

$$\tilde{A} = \frac{Fe^{jkL}}{2jkT\cos kL},$$

and

$$\tilde{B} = \frac{Fe^{-jkL}}{-2jkT\cos kL},$$

from which (2.32)

$$\tilde{y}(x,t) = \frac{F}{kT}\frac{\sin k(L-x)}{\cos kL}e^{j\omega t},$$

and

$$\tilde{u}(x,t) = \frac{j\omega F}{kT}\frac{\sin k(L-x)}{\cos kL}e^{j\omega t}. \tag{2.33}$$

The input impedance at $x = 0$ is

$$\tilde{Z}_{\text{in}} = f(t)/\tilde{u}(x,t) = \frac{-jkT}{\omega} \cot kL = -jZ_0 \cot kL. \tag{2.34}$$

This impedance is purely reactive and varies from 0 ($kL = \pi/2$, $3\pi/2$, etc.) to $\pm j\infty$($kL = 0$, π, etc.). These are the resonances and antiresonances of the string, respectively.

2.12 Motion of the End Supports

In Section 2.6, we considered the string to be terminated by two rigid end supports ($y = 0$ at $x = 0$ and $x = L$). We will now consider what happens when one of the end supports is not completely rigid.

We can generally describe the termination by writing its complex impedance. If the imaginary part of the complex impedance is masslike, the resonances of the string will be raised slightly above those given by Eq. (2.13); if it is springlike, on the other hand, the resonance frequencies will be lowered. The real part of the complex impedance is indicative of the rate of energy transfer from the spring to the support (the bridge and soundboard of a guitar, for example).

Let us consider a string fixed at $x = 0$ and terminated at $x = L$ by a support that can be characterized by a mass m. The transverse force exerted on the mass by the string is $-T(\partial\tilde{y}/\partial x)_{x=L}$. By Newton's second law,

$$-T(\partial\tilde{y}/\partial x)_L = m(\partial^2\tilde{y}/\partial t^2)_L. \tag{2.35}$$

Applying the boundary condition at $x = 0$ to Eq. (2.11) gives

$$0 = \tilde{A}e^{j\omega t} + \tilde{B}e^{j\omega t}, \tag{2.36}$$

so $\tilde{A} = -\tilde{B}$, and the harmonic solution becomes

$$\tilde{y}(x,t) = \tilde{A}(-e^{-kx} + e^{kx})e^{j\omega t} = \tilde{A} \sin kxe^{j\omega t}. \tag{2.37}$$

Substituting Eq. (2.37) into Eq. (2.35) gives

$$-kT\tilde{A} \cos kLe^{j\omega t} = -\omega^2 m\tilde{A} \sin kLe^{j\omega t},$$

and

$$\cot kL = \frac{\omega^2 m}{kT} = \frac{ky^2 m}{T}\frac{km}{\mu} = \frac{m}{M}kL, \tag{2.38}$$

where $M = \mu L$ is the total mass of the string. The transcendental equation, Eq. (2.38), can be solved graphically, as shown in Fig. 2.15, for two values of m/M. As $m \gg M$, the roots approach the values $k = n\pi$ for the string fixed at $x = L$ as well as at $x = 0$. Note that the normal mode

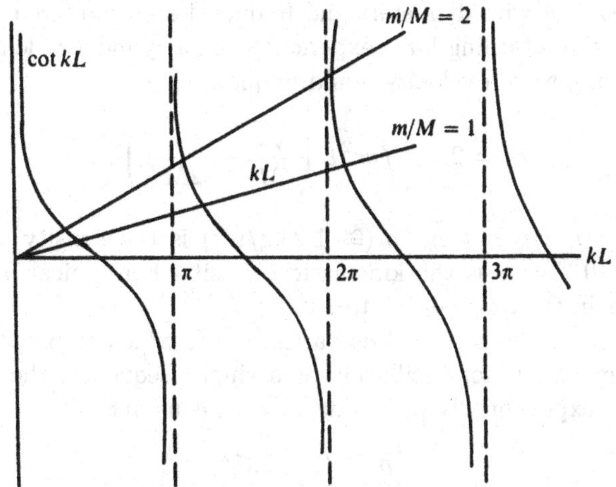

FIGURE 2.15. Graphical solution of cot $kL = (m/M)kL$.

frequencies obtained from Eq. (2.38) are slightly compressed from the harmonic relationship; the frequency of the lowest mode is raised slightly more than the second.

2.13 Damping

Damping of vibrating strings can generally be attributed to three different loss mechanisms: (1) air damping, (2) internal damping, and (3) transfer of energy to other vibrating systems. The damping due to these mechanisms will vary with frequency, and their contributions will be comparable in size in many systems. (Fletcher, 1976, 1977).

2.13.1 Air Damping

A vibrating string is not a good sound radiator. The reason for this is that the string acts as a dipole source, producing a compression in front and a rarefaction behind as it moves; its radius is so small that these effectively cancel each other. This does not mean, however, that the string has little interaction with the air. Viscous flow of air around the moving string may be the major cause of damping of its vibrations under some conditions.

The complex problem of viscous drag on a vibrating string was solved long ago by Stokes (1851), who showed that the force on the string has two components. One is an additional masslike load that lowers the mode frequencies very slightly; the other produces exponential decay of amplitude.

Over a range of wire diameters and frequencies encountered in musical instruments, the retarding force experienced by a cylinder of length L and radius r moving with a velocity v and frequency f is

$$F_{\mathrm{r}} = 2\pi^2 \rho_{\mathrm{a}} f v r^2 L \left(\frac{\sqrt{2}}{M} + \frac{1}{2M^2} \right), \tag{2.39}$$

where $M = (r/2)\sqrt{2\pi f/\eta_{\mathrm{a}}}$, $\rho_{\mathrm{a}}(\cong 1.2\,\mathrm{kg/m^3})$ is the density of air, and $\eta_{\mathrm{a}}(\cong 1.5 \times 10^{-5}\mathrm{m^2/s}$ is the kinematic viscosity. For typical harpsichord strings, M is in the range of 0.3 to 1.0.

Since $F_r \propto v$, the rate of loss varies as v^2, which is proportional to kinetic energy. Thus, for oscillation at a single frequency, the amplitude should decay exponentially with a decay time constant τ_1.

$$\tau_1 = \frac{\rho}{2\pi \rho_{\mathrm{a}} f} \left(\frac{2M^2}{2\sqrt{2}M + 1} \right). \tag{2.40}$$

The decay time is proportional to the wire density, but depends in a more complicated way on wire radius and frequency. $\tau_1 \propto \rho r^2$ at low frequency and $\tau_1 \propto \rho r/\sqrt{f}$ at high frequency.

2.13.2 Internal Damping

String material has so far been characterized by its radius, its density, and its Young's modulus, but more can be said than this. All real materials show an elastic behavior in which, when a stress is applied, an instantaneous strain occurs and then, over some characteristic time τ, the strain increases slightly. This second elongation may be moderately large or extremely small, and the time τ may be anything from less than a millisecond to many seconds. We treat this topic in more detail in Chapter 22, but a brief discussion is appropriate here.

The behavior can be represented by making the Young's modulus for the material complex:

$$E = E_1 + jE_2. \tag{2.41}$$

According to a relaxation formula attributable to Debye, E_2 has a peak at the relaxation frequency $\omega = 1/\tau$. Equation (2.41) can, however, be used in the more general case where many relaxation times contribute, both E_1 and E_2 varying with frequency. This behavior is simple to understand, E_1 being contributed by normal elastic bond distortions and E_2 by relaxation processes such as dislocation motion or the movement of kinks in polymer chains. Typically E_2/E_1 may be less than 10^{-4} in hard crystals, rather larger in metals, and perhaps as large as 10^{-1} in some polymer materials, though in such cases it may also depend on temperature. One elastic constant is really inadequate to describe even isotropic materials, but we shall neglect this added complication here.

By substituting Eq. (2.41) into the equation of motion, the decay time for this internal damping can be found to be

$$\tau_2 = \frac{1}{\pi f} \frac{E_1}{E_2}. \tag{2.42}$$

Clearly, this type of internal damping is a material property independent of string radius, length, or tension. It is generally negligible for solid metal strings but may become the prime damping mechanism for gut or nylon strings. The decay time due to this mechanism is clearly shortest at high frequencies if, as is often the case, E_1 is nearly independent of frequency.

This is, however, not the only internal damping mechanism. In metal strings there is also damping due to thermal conduction and the movement of dislocations, while in compound strings consisting of twisted fibers, or in strings with metal cladding, there will also be internal friction deriving from the relative motion of the component parts.

2.13.3 Energy Loss Through the Supports

In considering energy transfer to a movable support (and through it to other vibrating systems), it is easier to consider admittance than impedance. Mechanical *admittance* (the reciprocal of impedance) is the ratio of velocity to force, and its real part G is called *conductance.*

For a given string mode n, the velocity imparted to the support can be written

$$v_n = \alpha G F_n, \tag{2.43}$$

where F_n is the vertical component of the force and α is a constant. An analysis of the energy loss process again leads to an exponential energy decay with a time constant given by (Fletcher, 1977)

$$\tau_3 = (8\mu L f^2 G)^{-1}. \tag{2.44}$$

When all three mechanisms contribute to damping, the decay time τ is obtained by adding reciprocals:

$$1/\tau = 1/\tau_1 + 1/\tau_2 + 1/\tau_3. \tag{2.45}$$

This relationship is shown schematically in Fig. 2.16 on the assumption that G and E_2 are independent of frequency. The curves show the various contributions to the decay time as functions of frequency on the assumption that we are dealing with a single string whose frequency is raised by reducing its length. Also indicated are the directions in which the various curves would move in response to increases in various string parameters. The curve for the resultant decay time is a smoothed lower envelope to the individual decay time curves.

In most musical instruments, the rate at which energy is transferred from the string to the bridge and soundboard is quite small. For thin metal

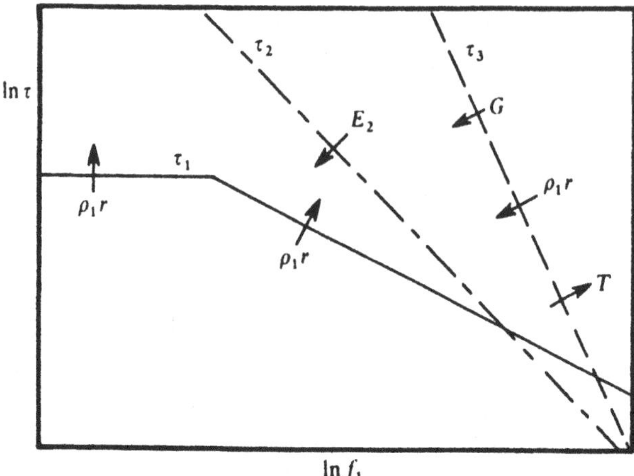

FIGURE 2.16. Schematic behavior of decay times τ_1 caused by various mechanisms as functions of the fundamental frequency f_1 of the string, which is assumed to be varied by changing only the string length. τ_1 is determined by air damping, τ_2 by internal damping, and τ_3 by loss to the support. Arrows indicate the directions in which the curves would be shifted by an increase in the string radius r, the density tension T, and the imaginary part of the Young's modulus E_2, and by the mechanical conductivity G of the bridge (Fletcher, 1976).

strings, the decay time is determined mostly by air viscosity, and so the decay time for the upper partials varies as $1/\sqrt{f}$. For instruments with gut or nylon strings, internal damping becomes dominant for most modes, and the decay time for the upper partials varies as $1/f$. Such strings therefore have a much less brilliant sound than do metal strings. If a finger tip is used to stop the string or if the bridge is so light that end losses predominate, the decay time for the upper partials varies more nearly as $1/f^2$.

2.14 Longitudinal Vibrations of a String or Thin Bar

Longitudinal waves in a string are much less common than transverse waves. Nevertheless, they do occur, and they may give rise to standing waves or longitudinal modes of vibration. Unlike transverse waves, their velocity (and hence their frequency) does not change with tension (except for possible changes in the physical properties of the string). Longitudinal waves in a thin bar travel at the same velocity as do longitudinal waves in a string of the same material.

When a bar is strained, elastic forces are produced. Consider a short segment of length dx of a bar having a cross section area S, as shown in Fig. 2.17, and acted upon by a force $F(x)$. The plane at x moves a distance w to the right while the plane at $x + dx$ moves a distance $w + dw$. The stress is given by F/S and the strain (change in length per unit of original length) by $\partial w/\partial x$. If E is Young's modulus, Hooke's law can be written

$$\frac{F}{S} = E\frac{\partial w}{\partial x}. \tag{2.46}$$

Expanding $F + dF$ in a Taylor's series, and differentiating Eq. (2.46) gives

$$dF = F(x + dx) - F(x) = \frac{\partial F}{\partial x}\,dx = SE\frac{\partial^2 w}{\partial x^2}\,dx. \tag{2.47}$$

The mass segment under consideration is $\rho S\,dx$, and thus, the equation of motion becomes

$$\begin{aligned} \rho S\,dx\frac{\partial^2 w}{\partial t^2} &= SE\frac{\partial^2 w}{\partial x^2}\,dx, \\ \frac{\partial^2 w}{\partial t^2} = \frac{E}{\rho}\frac{\partial^2 w}{\partial x^2} &= c_L^2\frac{\partial^2 w}{\partial x^2}. \end{aligned} \tag{2.48}$$

This is a one-dimensional wave equation for waves with a velocity $c_L = \sqrt{E/\rho}$.

The general solution of Eq. (2.48) has the same form as Eq. (2.5) for transverse waves in a string:

$$w = y_1(c_L t - x) + y_2(c_L t + x), \tag{2.49}$$

where the function $y(x,t)$ specifies the form of the wave disturbance. The normal modes of vibration depend upon the end conditions. If both ends are fixed, or if both ends are free, the mode frequencies are given by

$$f_n = n\frac{c_L}{2L}, \quad n = 1, 2, 3, \ldots, \tag{2.50}$$

and for a bar fixed at one end and free at the other,

$$f_m = m\frac{c_L}{4L}, \quad m = 1, 3, 5, \ldots. \tag{2.51}$$

If the bar (or string) is terminated by a movable support, the modal frequencies are found by methods similar to that described in Section 2.12.

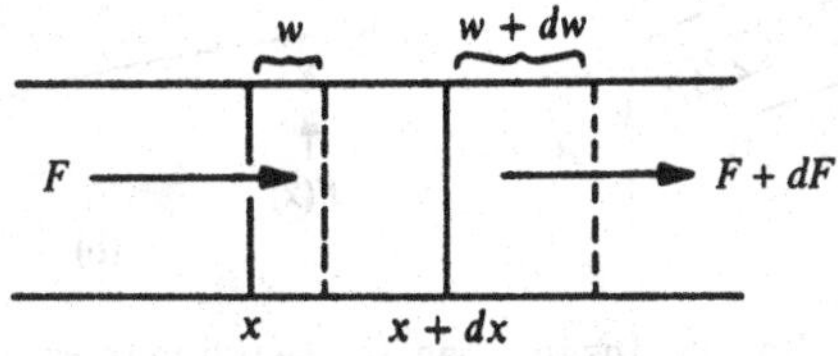

FIGURE 2.17. Forces and strains in a short segment of a bar (or string).

2.15 Bending Waves in a Bar

A bar or rod is capable of transverse vibrations in somewhat the same manner as a string. The dependence of the frequency on tension is more complicated than it is in a string, however. In fact, a bar vibrates quite nicely under zero tension, the elastic forces within the bar supplying the necessary restoring force in this case.

When a bar is bent, the outer part is stretched and the inner part is compressed. Somewhere in between is a neutral axis whose length remains unchanged, as shown in Fig. 2.18. A filament located at a distance z below the neutral axis is compressed by an amount $zd\phi$. The strain is $zd\phi/dx$, and the amount of force required to produce the strain is

$$E\,dS\,z\frac{d\phi}{dx}, \tag{2.52}$$

where dS is the cross sectional area of the filament and E is Young's modulus.

The moment of this force about the center line is

$$dM = [E\,dS\,z(d\phi/dx)]z,$$

and so the total moment to compress all the filaments is

$$M = \int dM = E\frac{d\phi}{dx}\int z^2\,dS. \tag{2.53}$$

It is customary to define a constant K called the *radius of gyration* of the cross section such that

$$K^2 = \frac{1}{S}\int z^2\,dS, \tag{2.54}$$

where $S = \int dS$ is the total cross section. The radii of gyration for a few familiar shapes are shown in Fig. 2.19. The bending moment is thus

$$M = E\frac{d\phi}{dx}SK^2 \cong -ESK^2\frac{\partial^2 y}{\partial x^2}, \tag{2.55}$$

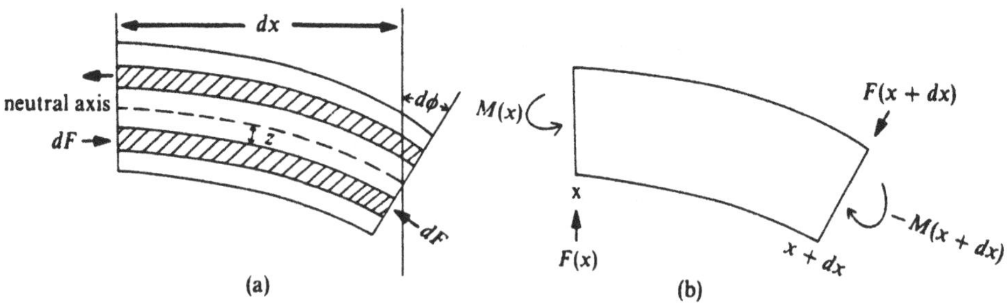

FIGURE 2.18. (a) Bending strains in a bar. (b) Bending moments and shear forces in a bar.

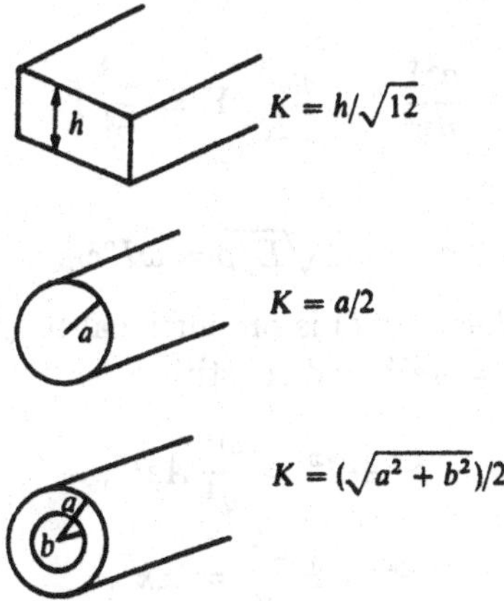

FIGURE 2.19. Radii of gyration for some simple shapes.

since $d\phi \cong -\left(\frac{\partial^2 y}{\partial x^2}\right) dx$ for small $d\phi$.

The bending moment is not the same for every part of the bar. In order to keep the bar in equilibrium, there must be a shearing force F with a moment $F\,dx$, as shown in Fig. 2.18(b).

$$F\,dx = (M + dM) - M = dM,$$

and

$$F = \frac{\partial M}{\partial x} = -ESK^2 \frac{\partial^3 y}{\partial x^3}.$$

But the shearing force F is not constant, either; the net force $dF = (\partial F/\partial x)dx$ produces an acceleration perpendicular to the axis of the bar. The equation of motion is

$$\begin{aligned} \left(\frac{\partial F}{\partial x}\right) dx &= (\rho S\,dx)\frac{\partial^2 y}{\partial t^2}, \\ -ESK^2 \frac{\partial^4 y}{\partial x^4} &= \rho S \frac{\partial^2 y}{\partial t^2}, \\ \frac{\partial^2 y}{\partial t^2} &= -\frac{EK^2}{\rho}\frac{\partial^4 y}{\partial x^4}. \end{aligned} \tag{2.56}$$

This is a fourth-order differential equation. It is not possible to construct a general solution from transverse waves traveling with velocity v, as in the longitudinal case. The velocity of transverse waves is, in fact, quite dependent on frequency; that is, the bar has *dispersion.*

We write the complex displacement as $\tilde{y} = \tilde{Y}(x)e^{j\omega t}$.;

$$\frac{\partial^2 \tilde{y}}{\partial t^2} = -\omega^2 \tilde{Y} e^{j\omega t} \text{ and } \frac{\partial^4 \tilde{y}}{\partial x^4} = \frac{d^4 \tilde{Y}}{dx^4} e^{j\omega t}, \text{ so } -\omega^2 \tilde{Y} = -\frac{EK^2}{\rho}\frac{d^4 \tilde{Y}}{dx^4}$$

or

$$\frac{d^4\tilde{Y}}{dx^4} = \frac{\rho\omega^2}{EK^4}\tilde{Y} = \frac{\omega^4}{v^4}\tilde{Y}$$

where

$$v^2 = \omega K\sqrt{E/\rho} = \omega K c_L.$$

Note that the wave velocity $v(f)$ is proportional to $\sqrt{\omega}$.

We now write $\tilde{Y}(x) = \tilde{A}e^{\gamma x}$ and substitute

$$\gamma^4 A e^{\gamma x} = \frac{\omega^4}{v^4} A e^{\gamma x},$$

$$\gamma^2 = \pm\frac{\omega^2}{v^2} = \pm k^2,$$

or (2.57)

$$\gamma = \pm\frac{\omega}{v} = \pm k \quad \text{or} \quad \pm j\frac{\omega}{v} = \pm jk$$

where we have defined the propagation number $k = \omega/v$. The complete solution is a sum of four terms, each corresponding to one of the roots of Eq. (2.57):

$$\tilde{y}(x,t) = e^{j\omega t}(\tilde{A}e^{kx} + \tilde{B}e^{-kx} + \tilde{C}e^{jkx} + \tilde{D}e^{-jkx}). \tag{2.58}$$

Since $e^{\pm x} = \cosh x \pm \sinh x$ and $e^{\pm jx} = \cos x \pm j\sin x$, another way of writing Eq. (2.58) is

$$y = \cos(\omega t + \phi)\left[A\cosh kx + B\sinh kx + C\cos kx + D\sin kx\right], \tag{2.59}$$

where A, B, C, and D are now real constants (Kinsler et al., 1982).

Since the equation of motion is a fourth-order equation, we have four arbitrary constants. We thus need four boundary conditions (two at each end) to determine them.

2.16 Bars with Fixed and Free Ends

We will consider three different end conditions for a bar: free, supported (hinged), and clamped. For each of these, we can write a pair of boundary conditions. At a free end, there is no torque and no shearing force, so the second and third derivatives are both zero, as given in Fig. 2.20. At a simply supported (or hinged) end, there is no displacement and no torque, so y and its second derivative are zero. At a clamped end, y and its first derivative are zero.

1. Example I: A bar of length L free at both ends. The boundary conditions at $x = 0$ become

$$\frac{\partial^2 y}{\partial x^2} = 0 = k^2(A - C)\cos(\omega t + \phi),$$

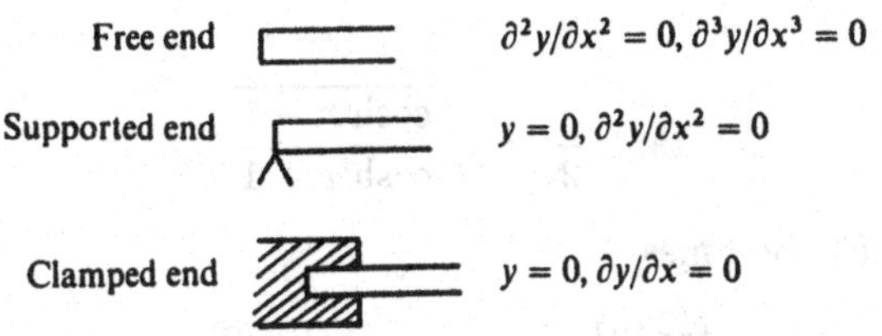

Free end		$\partial^2 y/\partial x^2 = 0, \partial^3 y/\partial x^3 = 0$
Supported end		$y = 0, \partial^2 y/\partial x^2 = 0$
Clamped end		$y = 0, \partial y/\partial x = 0$

FIGURE 2.20. Three different end conditions for a bar.

and

$$\frac{\partial^3 y}{\partial x^3} = 0 = k^3(B - D)\cos(\omega t + \phi),$$

from which $A = C$ and $B = D$, so the general solution [Eq. (2.59)] becomes

$$y(x,t) = \cos(\omega t + \phi)[A(\cosh kx + \cos kx) + B(\sinh kx + \sin kx)]. \quad (2.60)$$

At $x = L$, the boundary conditions become

$$\frac{\partial^2 y}{\partial x^2} = 0 = \cos(\omega t + \phi)k^2[A(\cosh kL - \cos kL) + B(\sinh kL - \sin kL)],$$

$$\frac{\partial^3 y}{\partial x^3} = 0 = \cos(\omega t + \phi)k^3[A(\sinh kL + \sin kL) + B(\cosh kL - \cos kL)].$$

These equations can have a common solution only for certain values of ω. Setting the expressions in brackets equal to zero, and dividing the first by the second gives

$$\frac{\cosh kL - \cos kL}{\sinh kL + \sin kL} = \frac{\sinh kL - \sin kL}{\cosh kL - \cos kL}.$$

Cross multiply and note that $\sin^2 x + \cos^2 x = \cosh^2 x - \sinh^2 x = 1$:

$$\cosh^2 kL - 2\cosh kL \cos kL + \cos^2 kL = \sinh^2 kL - \sin^2 kL,$$

$$2 - 2\cosh kL \cos kL = 0,$$

or

$$\cosh kL = \frac{1}{\cos kL}. \quad (2.61)$$

This equation could be solved by graphing the two functions, but this is not very practical since the hyperbolic cosine increases exponentially. An alternative is to make use of the identities

$$\tan\frac{x}{2} = \sqrt{\frac{1 - \cos x}{1 + \cos x}}$$

and

$$\tanh \frac{x}{2} = \sqrt{\frac{\cosh x - 1}{\cosh x + 1}},$$

so that Eq. (2.61) becomes

$$\tan(kL/2) = \pm \tanh(kL/2). \tag{2.62}$$

A graph of these two functions is shown in Fig. 2.21. The intersections of these curves give roots $\omega L/2v = \pi/4(3.011, 5, \ldots)$. But $v^2 = \omega K\sqrt{E/\rho}$ so $\omega^2 = (v^2\pi^2/4L^2)(3.011^2, 5^2, 7^2, \ldots)$, and the allowed frequencies are given by

$$f_n = \frac{\pi K}{8L^2}\sqrt{\frac{E}{\rho}}[3.011^2, 5^2, 7^2, \ldots, (2n+1)^2]. \tag{2.63}$$

The frequencies and nodal positions for the first four bending vibrational modes of a bar with free ends are given in Table 2.1. Note that the frequencies are not harmonically related as they were for longitudinal modes.

2. Example II: A bar of length of L clamped at $x = 0$ and free at $x = L$. The boundary conditions at $x = 0$ now lead to the result that $A + C = 0 = B + D$. The transcendental equation, which gives the allowed

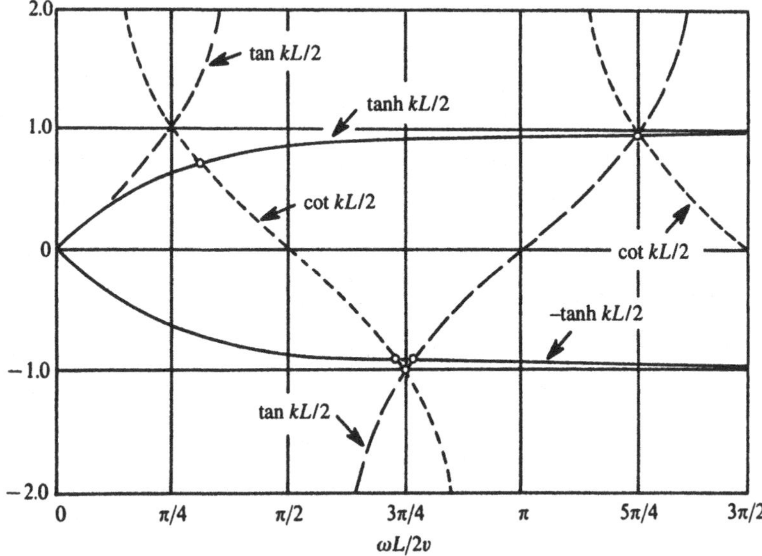

FIGURE 2.21. Curves showing tangent, cotangent, and hyperbolic tangent functions (adapted from Kinsler et al., "Fundamentals of Acoustics," 3rd ed., copyright © 1982, John Wiley & Sons, Inc. Reprinted by permission).

TABLE 2.1. Characteristics of transverse vibrations in a bar with free ends.

Frequency (Hz)	Wavelength (m)	Nodal positions (m from end of 1-m bar)
$f_1 = 3.5607K/L^2\sqrt{E/\rho}$	$1.330L$	0.224, 0.776
$2.756f_1$	$0.800L$	0.132, 0.500, 0.868
$5.404f_1$	$0.572L$	0.094, 0.356, 0.644, 0.906
$8.933f_1$	$0.445L$	0.073, 0.277, 0.500, 0.723, 0.927

values of frequency, is now [see equation 3.55 in Kinsler et al. (1982)]

$$\cot(kL/2) = \pm\tanh(kL/2).$$

Again, we obtain the frequencies by using Fig. 2.21:

$$f_n = \frac{\pi K}{8L^2}\sqrt{\frac{E}{\rho}}[1.194^2, 2.988^2, 5^2, \ldots, (2n-1)^2]. \tag{2.64}$$

These frequencies are in the ratios $f_2 = 6.267f_1$, $f_3 = 17.55f_1$, $f_4 = 35.39f_1$, etc. The lowest frequency has the frequency $f_1 = (0.5598K/L^2)\sqrt{E/\rho}$, which is only about $\frac{1}{6}$ of the lowest frequency of the same bar with two free ends.

3. Example III: A bar of length L with simple supports (hinges) at the ends. The frequencies are given by

$$f_n = \frac{\pi K}{2L^2}\sqrt{\frac{E}{\rho}}m^2 \qquad m = 1, 2, 3, \ldots. \tag{2.65}$$

These frequencies are considerably lower than those given by Eq. (2.63), since the bending wavelengths are longer than the corresponding modes of the free bar, as shown in Fig. 2.22.

2.17 Vibrations of Thick Bars: Rotary Inertia and Shear Deformation

Thus far, we have considered transverse motion of the bar due to the bending moment only. Such a simplified model is often called the Euler–Bernoulli beam. It is essentially correct for a long, thin bar or rod. The Timoshenko beam, a model that considers rotary inertia and shear stress, is preferred in considering thick bars.

As a beam bends, the various elements rotate through some small angle. The rotary inertia is thus equivalent to an increase in mass and results in a slight lowering of vibrational frequencies, especially the higher ones.

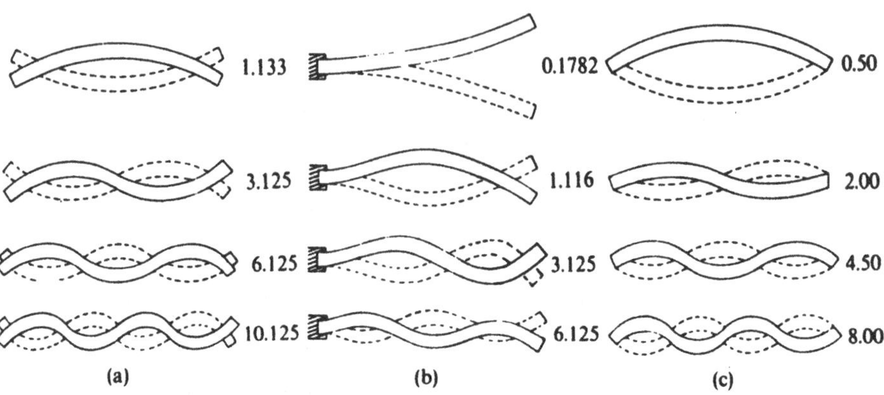

FIGURE 2.22. Bending vibrations of (a) a bar with two free ends, (b) a bar with one clamped end and one free end, and (c) a bar with two supported (hinged) ends. The numbers are relative frequencies; to obtain actual frequencies, multiply by $(\pi K/L^2)\sqrt{E/\rho}$.

Shear forces, which we considered in deriving the equation of motion [Eq. (2.56)], tend to deform the bar; in particular they cause rectangular elements to become parallelograms and thus decrease the transverse deflection slightly. Therefore, the frequencies of the higher modes are decreased slightly in a thick bar as compared with a thin one.

2.18 Vibrations of a Stiff String

In real strings, the restoring force is partly due to the applied tension and partly due to the stiffness of the string (although the former usually dominates). Thus, the equation of motion of a flexible string [Eq. (2.4)] can be modified by adding a term appropriate to bending stiffness:

$$\mu \frac{\partial^2 y}{\partial t^2} = T \frac{\partial^2 y}{\partial x^2} - ESK^2 \frac{\partial^4 y}{\partial x^4}. \tag{2.66}$$

In this equation, μ is mass per unit length, T is tension, E is Young's modulus, S is the cross-sectional areas and K is the radius of gyration, as before.

This equation can be solved easily for the case of a string with pinned ends, for the mode functions are then simply $\sin n\pi x/L$ where L is the string length. The result is

$$f_n = n f_1^\circ (1 + Bn^2)^{1/2} \tag{2.67a}$$

where f_1° is the fundamental frequency of the same string without stiffness and $B = \pi^2 ESK^2/TL^2$. Fletcher (1964) has derived an approximate solution for the case of clamped ends, which for $B \ll 1$ has the form

$$f_n \approx nf_1^\circ(1 + Bn^2)^{1/2}[1 + (2/\pi)B^{1/2} + (4/\pi^2)B]. \tag{2.67b}$$

Clamping the ends thus raises the frequencies a little. Stretching of the partials is actually slightly reduced, though this is not apparent from (2.67b).

String stiffness is of considerable importance in the tuning of piano strings. To minimize beating between the upper strings and the inharmonic overtones of the lower strings, octaves are stretched to ratios that are greater than 2:1. In a 108-cm (42-inch) upright piano, for example, the fourth harmonic of C_4 (middle C) is about 4 cents (0.2%) sharp, but this increases to about 18 cents (1.1%) for C_5 (Kent, 1982). In a large grand piano with long strings, inharmonicity is considerably less, but in small spinets it is substantially greater.

The stiffness of a violin string is of considerable importance when it is excited by bowing. In our discussion of Helmholtz-type motion in Section 2.10, we envisioned a very sharp bend propagating back and forth on an ideal string with great flexibility. On a real string, however, the bend is rounded appreciably by the stiffness of the string (and to a lesser extent by damping of high-frequency components). The rounded Helmholtz bend is sharpened each time it passes the bow, however, and so the resultant motion represents an equilibrium between the rounding and sharpening process. Two effects that depend upon rounding of the Helmholtz corner are noise due to small variations in period (jitter) and the note flattening with increased bow pressure (Cremer, 1981; McIntyre and Woodhouse, 1982).

2.19 Dispersion in Stiff and Loaded Strings: Cutoff Frequency

Waves on an ideal string travel without dispersion; that is, the wave velocity is independent of frequency. Thus, a pulse does not change its shape as it propagates back and forth on an ideal, lossless string. Two sources of dispersion in real strings are stiffness and mass loading. In addition, loss mechanisms in strings are frequency dependent.

In an ideal string, ω and k are related by the simple expression $\omega = ck$, so the wave velocity equals the slope of the line obtained by plotting ω versus k, as in Fig. 2.23(a). In order to draw a graph of ω versus k for a stiff string, we write Eq. (2.67a) as

$$\omega_n = nk\sqrt{\frac{T}{\mu}}(1 + Bn^2)^{1/2} = nk\sqrt{\frac{T}{\mu}}(1 + \alpha k^2)^{1/2}. \tag{2.68}$$

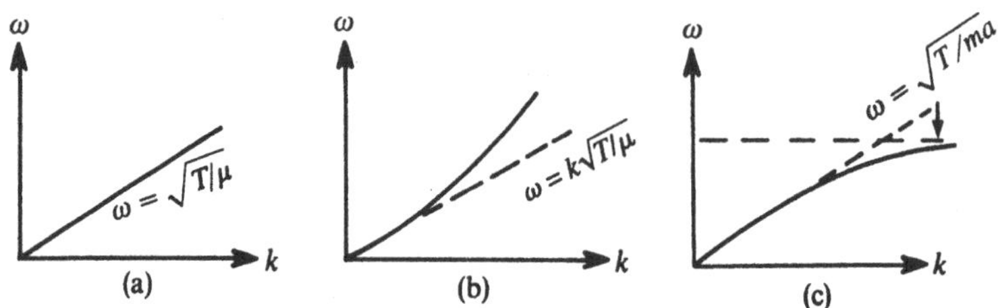

FIGURE 2.23. Graphs of ω versus k for (a) an ideal string, (b) a stiff string, and (c) a string loaded with masses spaced a distance a apart.

The graph of Eq. (2.68) is shown in Fig. 2.23(b). When the graph of ω versus k is a curved line, we observe two different wave velocities: a phase velocity v_ϕ and a group velocity v_g. These are given by

$$v_\phi = \omega/k \quad \text{and} \quad v_g = d\omega/dk. \tag{2.69}$$

The phase velocity is the velocity of a wave crest or a given phase angle, whereas the group velocity is the velocity of the wave envelope of a given amplitude of the wave packet.

For a string loaded with equally spaced masses, dispersion of a different type is observed. The dispersion relationship can be written (see p. 76 in Crawford, 1965) as

$$\omega(k) = 2\sqrt{\frac{T}{ma}}\sin ka, \tag{2.70}$$

where a is the spacing between beads of mass m. The maximum value of ω is $2\sqrt{T/ma}$, which occurs when $k = \pi/a$. The allowed values of k between 0 and π/a equal the number of normal modes of the system, which is also equal to the number of equally spaced masses.

The maximum value of ω divided by 2π is called the cutoff frequency f_c:

$$f_c = \frac{1}{\pi}\sqrt{\frac{T}{ma}}, \tag{2.71}$$

It represents the highest frequency of wave disturbance that can propagate on the loaded string. Note that when $k = \pi/a$, the group velocity $v_g = d\omega/dk = 0$.

2.20 Torsional Vibrations of a Bar

Torsional waves are a third type of wave motion possible in a bar or rod. The equation of motion for torsional waves is derived by equating the net torque acting on an element of the bar to the product of moment of inertia and angular acceleration. Young's modulus is replaced by the shear modulus.

The resulting wave equation is quite similar to the equation for longitudinal waves.

Torsional waves in a bar, like compressional waves (but unlike bending waves), are nondispersive; that is, they have a wave velocity that is independent of frequency

$$c_T = \sqrt{\frac{GK_T}{\rho I}}, \tag{2.72}$$

where GK_T is the torsional stiffness factor that relates a twist to the shearing strain produced, ρI is the polar moment of inertia per unit length, ρ is density, and G is the shear modulus. For a circular rod, $K_T \cong I$, so the velocity is $\sqrt{G/\rho}$; for square and rectangular bars, it is slightly less, as shown in Fig. 2.24.

In many materials the shear modulus G is related to the Young's modulus E and Poisson's ratio ν by the equation

$$G = \frac{E}{2(1+\nu)}. \tag{2.73}$$

In aluminum, for example, $\nu = 0.33$, so $G = 0.376E$, and the ratio of torsional to longitudinal wave velocity in a circular aluminum rod is 0.61.

The torsional modes of vibration of a bar with free ends have frequencies that equal the torsional wave velocity times $n/2L$ in direct analogy to the longitudinal modes. [If one end of the bar is clamped, the frequencies become $m/4L(m = 1, 3, 5, \ldots)$.]

Bowing a violin string excites torsional waves as well as transverse waves. For a steel E string tuned to 660 Hz, the torsional wave speed v_T is about 7.5 times the transverse wave speed c, but for a gut E string, $v_T/c \cong 2$ (Schelleng, 1973). Gillan and Elliott (1989) found values of v_T/c from 2.6 to 7.6 in violin strings and 5.7 for a steel cello string. Furthermore, they found damping factors from 1% to 7.7%, which suggests that torsional damping is dominated by internal damping in the string rather than reflection losses at the bridge and nut. Torsional waves change the effective compliance of the string and affect the mechanics of the bow/string interaction in other ways as well (Cremer, 1981).

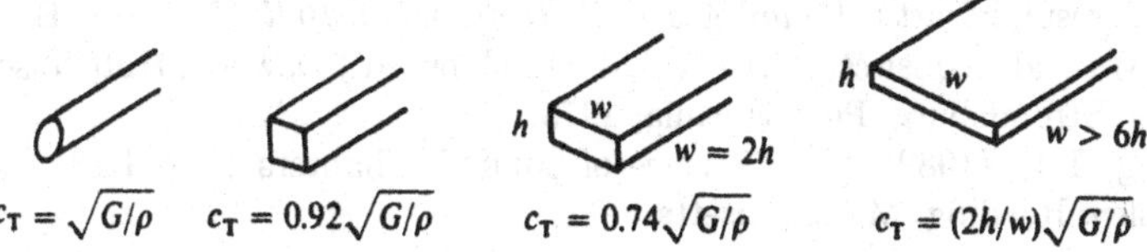

FIGURE 2.24. Torsional wave velocities for bars with different cross sections.

References

Crawford, F.S., Jr. (1965). "Waves," Chapter 2. McGraw-Hill, New York.

Cremer, L. (1981). "Physik der Geige." Hirtzel Verlag, Stuttgart. English translation by J.S. Allen, MIT Press, Cambridge, Massachusetts, 1984.

Fletcher, H. (1964). Normal vibration frequencies of a stiff piano string. *J. Acoust. Soc. Am.* **36**, 203–209.

Fletcher, N.H. (1976). Plucked strings—a review. *Catgut Acoust. Soc. Newsletter*, No. 26, 13–17.

Fletcher, N.H. (1977). Analysis of the design and performance of harpsichords. *Acustica* **37**, 139–147.

Gillan, F.S., and Elliott, S.J. (1989). Measurement of the torsional modes of vibration of strings on instruments of the violin family. *J. Sound Vib.* **130**, 347–351.

Hall, D.E. (1986). Piano string excitation in the case of small hammer mass. *J. Acoust. Soc. Am.* **79**, 141–147.

Hall, D.E. (1987a). Piano string excitation II: General solution for a hard narrow hammer. *J. Acoust. Soc. Am.* **81**, 535–546.

Hall, D.E. (1987b). Piano string excitation III: General solution for a soft narrow hammer. *J. Acoust. Soc. Am.* **81**, 547–555.

Helmholtz, H.L.F. (1877). "On the Sensations of Tone," 4th ed., trans. A.J. Ellis (Dover, New York, 1954).

Kent, E.L. (1982). Influence of irregular patterns in the inharmonicity of piano-tone partials upon piano tuning practice. *Das Musikinstrument* **31**, 1008–1013.

Kinsler, L.E., Frey, A.R., Coppens, A.B., and Sanders, J.V. (1982). "Fundamentals of Acoustics," 3rd ed., Chapters 2 and 3. New York.

Kondo, M., and Kubota, H. (1983). A new identification expression of Helmholtz waves. *SMAC 83.* (A. Askenfelt, S. Felicetti, E. Jansson, and J. Sundberg, eds.), pp. 245–261. Royal Swedish Acad. Music, Stockholm.

Krigar-Menzel, O., and Raps, A. (1981), Aus der Sitzungberichten, *Ann. Phys. Chem.* **44**, 613–641.

Lawergren B. (1980). On the motion of bowed violin strings. *Acustica* **44**, 194–206.

McIntyre, M.E., Schumacher, R.T., and Woodhouse, J. (1981). Aperiodicity of bowed string motion. *Acustica* **49**, 13–32.

McIntyre, M.E., and Woodhouse, J. (1982). The acoustics of stringed musical instruments. *Interdisc. Sci. Rev.* **3**, 157–173. Reprinted in "Musical Acoustics: Selected Reprints" (T.D. Rossing, ed.), Am. Assn. Physics Teach., College Park, Maryland, 1988.

Raman, C.V. (1918). On the mechanical theory of the vibrations of bowed strings and of musical instruments of the violin family, with experimental verification of the results: Part I. *Indian Assoc. Cultivation Sci. Bull.* **15**, 1–158. Excerpted in "Musical Acoustics, Part I" (C.M. Hutchins, ed.) Dowden, Hutchinson, and Ross, Stroudsburg, Pennsylvania, 1975.

Rossing, T.D. (1982). "The Science of Sound," Chapters 10 and 13. Addison-Wesley, Reading, Massachusetts.

Schelleng, J.C. (1973). The bowed string and the player. *J. Acoust. Soc. Am.* **53**, 26–41.

Schelleng, J.C. (1974). The physics of the bowed string. *Scientific Am.* **235**(1), 87–95.

Stokes, G.G. (1851). On the effect of the internal friction of fluids on the motion of pendulums. *Trans. Cambridge Phil. Soc.* **9**, 8. Reprinted in G.G. Stokes, "Mathematical and Physical Papers" vol. 3 Cambridge University Press, Cambridge 1922, pp. 1–140.

Weinreich, G. (1979). The coupled motions of piano strings. *Scientific Am.* **240**(1), 118–27.

3

Two-Dimensional Systems: Membranes, Plates, and Shells

In this chapter, we will consider two-dimensional, continuous vibrating systems, with and without stiffness. An ideal membrane, like an ideal string, has no stiffness of its own, and thus, its oscillations depend upon the restoring force supplied by an externally applied tension. A plate, on the other hand, like a bar, can vibrate with fixed or free ends and with or without external tension.

3.1 Wave Equation for a Rectangular Membrane

The simplest two-dimensional system we can consider is a rectangular membrane with dimensions L_x and L_y, with fixed edges, and with a surface tension T that is constant throughout.

Consider an element with area density σ, as shown in Fig. 3.1. It has been displaced a small distance dz, and the surface tension T acts to restore it to equilibrium. The forces acting on the edges dx have the magnitude $T\,dx$,

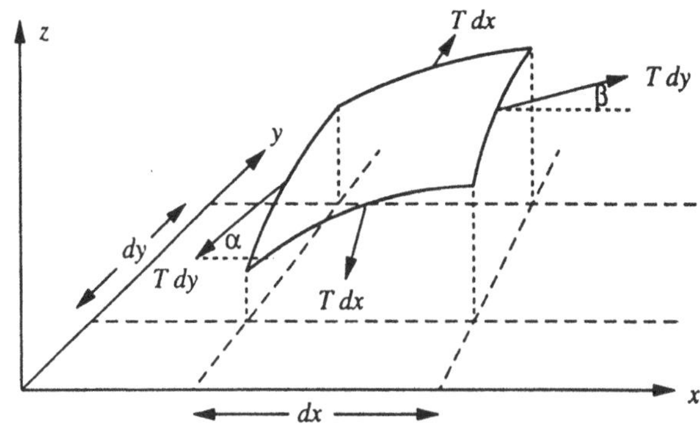

FIGURE 3.1. Forces on a rectangular membrane element.

and their vertical components are $-T \sin\alpha\, dx$ and $-T \sin\beta\, dx$. For small α and β,

$$\sin\alpha \approx \tan\alpha = \left(\frac{\partial z}{\partial y}\right)_{y+dy}$$

and

$$\sin\beta \approx \tan\beta = \left(\frac{\partial z}{\partial y}\right)_{y}.$$

The force F on the element in the z direction is therefore

$$F_y = T\,dx \left[\left(\frac{\partial z}{\partial y}\right)_{y+dy} - \left(\frac{\partial z}{\partial y}\right)_{y}\right] = T\,dx\,\frac{\partial^2 z}{\partial y^2}\,dy.$$

Similarly, the vertical component of the forces acting on the edges dy is

$$F_x = T\,dy\,\frac{\partial^2 z}{\partial y^2}\,dx.$$

The total force in the z direction on element $dx\,dy$ is $F = F_x + F_y$, so the equation of motion is

$$T\,dx\,dy\left(\frac{\partial^2 z}{\partial x^2} + \frac{\partial^2 z}{\partial y^2}\right) = \sigma\,dx\,dy\,\frac{\partial^2 z}{\partial t^2}$$

or

$$\frac{\partial^2 z}{\partial t^2} = \frac{T}{\sigma}\left(\frac{\partial^2 z}{\partial x^2} + \frac{\partial^2 z}{\partial y^2}\right) = c^2\nabla^2 z. \tag{3.1}$$

This is a wave equation for transverse waves with a velocity $c = \sqrt{T/\sigma}$. It is easily solved by writing the deflection $z(x, y, t)$ as a product of three functions, each of a single variable: $z(x, y, t) = X(x)Y(y)\Phi(t)$. The second derivatives are

$$\frac{\partial^2 z}{\partial x^2} = \frac{d^2 X}{dx^2}Y\Phi, \quad \frac{\partial^2 z}{\partial y^2} = \frac{d^2 Y}{dy^2}X\Phi, \quad \text{and} \quad \frac{\partial^2 z}{\partial t^2} = \frac{d^2\Phi}{dt^2}XY,$$

so that the equation becomes

$$\frac{1}{\Phi}\frac{d^2\Phi}{dt^2} = \frac{c^2}{X}\frac{d^2 X}{dx^2} + \frac{c^2}{Y}\frac{d^2 Y}{dy^2}. \tag{3.2}$$

This equation can only be true if each side of the equation is a constant, which we denote as ω^2. This gives two equations:

$$\frac{d^2\Phi}{dt^2} + \omega^2\Phi = 0,$$

with solutions $\Phi(t) = E\sin\omega t + F\cos\omega t$, and

$$\frac{1}{X}\frac{d^2 X}{dx^2} + \frac{\omega^2}{c^2} = -\frac{1}{Y}\frac{d^2 Y}{dy^2}.$$

Again, each side must equal a constant, which we will call k^2. This gives

$$\frac{d^2X}{dx^2} + \left(\frac{\omega^2}{c^2} - k^2\right) X = 0,$$

with solutions $X(x) = A \sin \sqrt{(\omega^2/c^2) - k^2}x + B \cos \sqrt{(\omega^2/c^2) - k^2}x$, and

$$\frac{d^2Y}{dy^2} + k^2Y = 0,$$

with solutions $Y(y) = C \sin ky + D \cos ky$. For a rectangular membrane of dimensions L_x by L_y, fixed at all four sides, the boundary conditions require that $z = 0$ for $x = 0$, $x = L_x$, $y = 0$, and $y = L_y$. From the first condition, we see that $B = 0$; from the second,

$$A \sin \sqrt{\frac{\omega^2}{c^2} - k^2}L_x = 0, \quad \text{so} \quad \sqrt{\frac{\omega^2}{c^2} - k^2}L_x = m\pi, \quad \text{and}$$

$$X(x) = A \sin \frac{m\pi x}{L_x},$$

with $m = 1, 2, \ldots$. From the third, $D = 0$; and from the fourth, $C \sin kL_y = 0$, so $kL_y = n\pi$ and $Y(y) = C \sin(n\pi/L_y)y$, with $n = 1, 2, \ldots$. Therefore,

$$\begin{aligned} z_{mn} &= A \sin \frac{m\pi x}{L_x} \sin \frac{n\pi y}{L_y} (E \sin \omega t + F \cos \omega t) \\ &= \sin \frac{m\pi x}{L_x} \sin \frac{n\pi y}{L_y} (M \sin \omega t + N \cos \omega t), \quad m = 1, 2, \ldots. \end{aligned} \tag{3.3}$$

To determine the modal frequencies, solve $\sqrt{(\omega^2/c^2) - k^2} = m\pi/L_x$ for ω:

$$\omega^2 = \left(\frac{m\pi}{L_x}\right)^2 c^2 + k^2c^2 = \left(\frac{m\pi}{L_x}\right)^2 c^2 + \left(\frac{n\pi}{L_y}\right)^2 c^2,$$

and

$$f_{mn} = \frac{1}{2}\sqrt{\frac{T}{\sigma}}\sqrt{\frac{m^2}{L_x^2} + \frac{n^2}{L_y^2}}, \quad m, n = 1, 2, \ldots. \tag{3.4}$$

Comparison of Eqs. (3.3) and (3.4) with Eqs. (2.13) and (2.14), which describe the modes in a string, suggests that the normal modes of a rectangular membrane might be called two-dimensional string modes. Standing waves in the x direction appear to be independent of standing waves in the y direction. Some of the modes are illustrated in Fig. 3.2.

3.2 Square Membranes: Degeneracy

In a square membrane ($L_x = L_y$), $f_{mn} = f_{nm}$; the mn and nm modes are said to be *degenerate*, since they have the same frequency. Although there

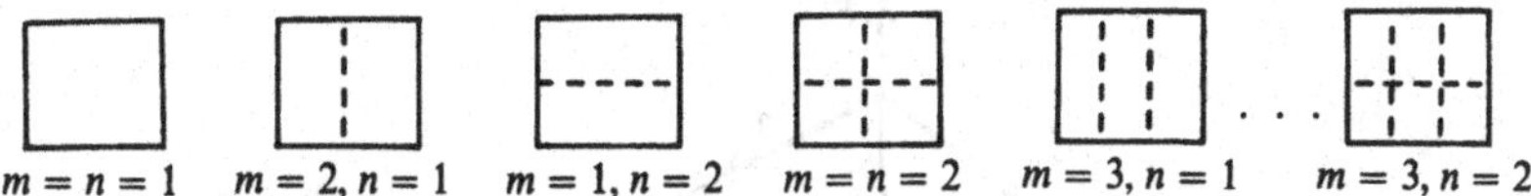

FIGURE 3.2. Some normal modes of a rectangular membrane.

are now fewer allowed frequencies, there are just as many characteristic functions as in the rectangular case. In fact, the membrane can vibrate with simple harmonic motion at a frequency f_{mn} with any of an infinite number of different shapes corresponding to different values of a and b in the equation.

$$z(x, y, t) = (az_{mn} + bz_{mn}) \cos \omega_{mnt} \quad \text{where} \quad a^2 + b^2 = 1. \tag{3.5}$$

Various combinations of z_{13} and z_{31} are shown in Fig. 3.3.

An important difference between a string and a membrane is in the reaction to a force applied at a single point, as shown in Fig. 3.4. A string pulled aside by a force F applied a distance x from one end will deflect a distance h so that $T(h/x)$ and $T[h/(L - x)]$ add up to F. An ideal membrane, on the other hand, cannot support a point force F, and the displacement theoretically becomes infinite no matter how small the force! If a force F is applied to a small circle of radius r at the center of a membrane of radius a, the displacement becomes

$$z = \frac{2F}{T} \ln \frac{a}{r}, \tag{3.6}$$

which goes to infinity as $r \to 0$ (Morse, 1948, p. 176).

3.3 Circular Membranes

For a circular membrane, the wave equation [Eq. (3.1)] should be written in polar coordinates by letting $x = r \cos \phi$ and $y = r \sin \phi$.

$$\frac{\partial^2 z}{\partial t^2} = c^2 \left(\frac{\partial^2 z}{\partial r^2} + \frac{1}{r} \frac{\partial z}{\partial r} + \frac{1}{r^2} \frac{\partial^2 z}{\partial \phi^2} \right). \tag{3.7}$$

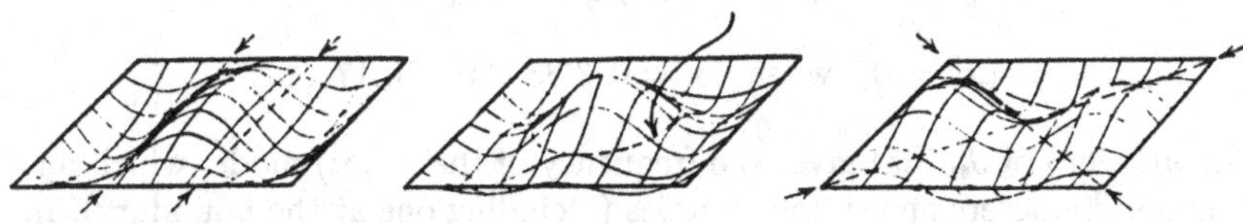

FIGURE 3.3. Degenerate modes of vibration in a square membrane corresponding to different values of a and b in Eq. (3.5). Arrows point to the nodal lines (Morse, 1948).

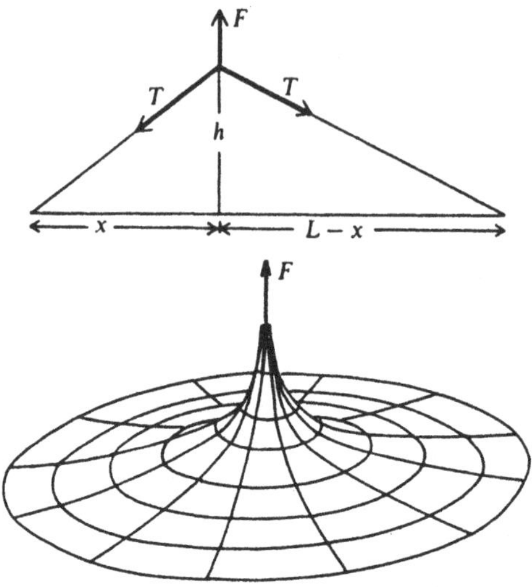

FIGURE 3.4. Reaction of a string and membrane to a force applied at a point.

We write solutions of the form $z(r, \phi, t) = R(r)\Phi(\phi)e^{j\omega t}$ leading to the equations:

$$\frac{d^2R}{dr^2} + \frac{1}{r}\frac{dR}{dr} + \left(\frac{\omega^2}{c^2} - \frac{m^2}{r^2}\right)R = 0,$$

and (3.8)

$$\frac{d^2\Phi}{d\phi^2} + m^2\Phi = 0.$$

The solution to the second equation is $\Phi(\phi) = Ae^{\pm jm\phi}$. The first equation is a form of Bessel's equation $(d^2y/dx^2) + (1/x)(dy/dx) + [1 - (m^2/x^2)]y = 0$ with $y = R$ and $x = kr = \omega r/c$. The solutions are Bessel functions of order m. Each of these functions $J_0(x)$, $J_1(x)$, ..., $J_m(x)$ goes to zero for several values of x as shown in Fig. 3.5.

$$J_0(x) = 0 \quad \text{when} \quad x = 2.405, 5.520, 8.654, \ldots.$$

$$J_1(x) = 0 \quad \text{when} \quad x = 0, 3.83, 7.02, 10.17, \ldots.$$

The nth zero of $J_m(kr)$ gives the frequency of the (m, n) mode, which has m nodal diameters and n nodal circles (including one at the boundary). In the fundamental (0, 1) mode, the entire membrane moves in phase. The first 14 modes of an ideal membrane and their relative frequencies are given in Fig. 3.6.

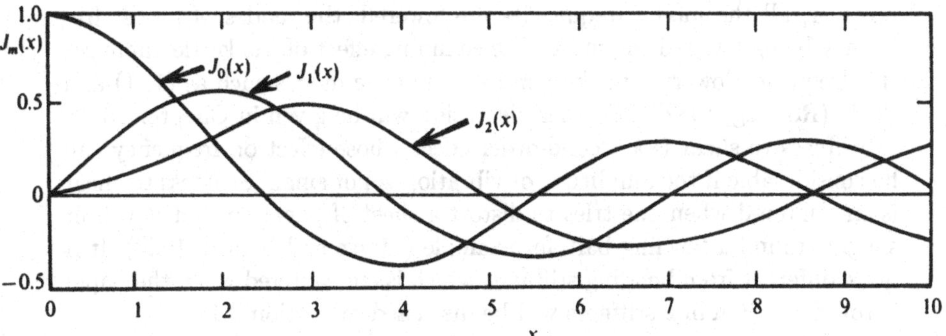

FIGURE 3.5. First three Bessel functions.

3.4 Real Membranes: Stiffness and Air Loading

The normal mode frequencies of real membranes may be quite different from those of an ideal membrane given in Fig. 3.6. The principal effects in the membrane acting to change the mode frequencies are air loading, bending stiffness, and stiffness to shear. In general, air loading lowers the modal frequencies, while the other two effects tend to raise them. In thin membranes, air loading is usually the dominant effect.

The effect on frequency of the air loading depends upon the comparative velocities for waves in the membrane and in air, and also upon whether the air is confined in any way. A confined volume of air (as in a kettledrum, for example) will raise the frequency of the axisymmetric modes, especially the (0, 1) mode. When a membrane vibrates in an unconfined sea of air,

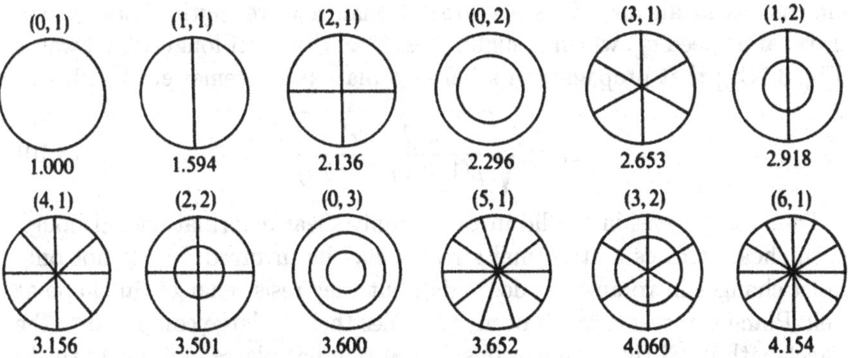

FIGURE 3.6. First 14 modes of an ideal membrane. The mode designation (m, n) is given above each figure and the relative frequency below. To convert these to actual frequencies, multiply by $(2.405/2\pi a)\sqrt{T/\sigma}$, where a is the membrane radius.

however, all the modal frequencies are lowered, the modes of lowest frequency being lowered the most. The confining effect of the kettle enhances this frequency lowering in the non-axisymmetric modes such as (1, 1) and (2, 1) (Rossing, 1982b). Further discussion will be given in Chapter 18.

Stiffness to shear is a second-order effect whose effect on frequency can be considerable if the amplitude of vibration is not small. Stiffness to shear is encountered when one tries to distort a sheet of paper so that it will fit snugly around a bowling ball, for example (Morse and Ingard, 1968). It is quite different from bending stiffness, which is encountered when the paper is rolled up. Bending stiffness will be discussed in Section 3.12.

3.5 Waves in a Thin Plate

A plate may be likened to a two-dimensional bar or a membrane with stiffness. Like a bar, it can transmit compressional waves, shear waves, torsional waves, or bending waves; and it can have three different boundary conditions: free, clamped, or simply supported (hinged).

A plate might be expected to transmit longitudinal (compressional) waves at the same velocity as a bar: $c_L = \sqrt{E/\rho}$. This is not quite the case, however, since the slight lateral expansion that accompanies a longitudinal compression is constrained in the plane of the plate, thus adding a little additional stiffness. The correct expression for the velocity of *longitudinal* waves in an infinite plate is

$$c_L = \sqrt{\frac{E}{\rho(1-\nu^2)}}, \tag{3.9}$$

where ν is Poisson's ratio ($\nu \simeq 0.3$ for most materials).

Actually, pure longitudinal waves [Fig. 3.7(a)] occur only in solids whose dimensions in all directions are greater than a wavelength. These waves travel at a speed c'_L, which is slightly less than the quasi-longitudinal waves [Fig. 3.7(b)] that propagate in a bar or a plate (see Cremer et al., 1973).

$$c'_L = \sqrt{\frac{E(1-\nu)}{\rho(1+\nu)(1-2\nu)}}. \tag{3.10}$$

Transverse waves in a solid involve mainly shear deformations, although both shear stresses and normal stresses may be involved. Solids not only resist changes in volume (as do fluids), but they resist changes in shape as well. Plane transverse waves occur in bodies that are large compared to the wavelength in all three dimensions, but also in flat plates of uniform thickness (see Chapter 2 in Cremer et al., 1973). Transverse waves propagate at the same speed as torsional waves in a circular rod ($c_T = \sqrt{G/\rho}$). The shear modulus G is considerably smaller than Young's modulus E, so transverse and torsional waves propagate at roughly 60% of the speed of longitudinal

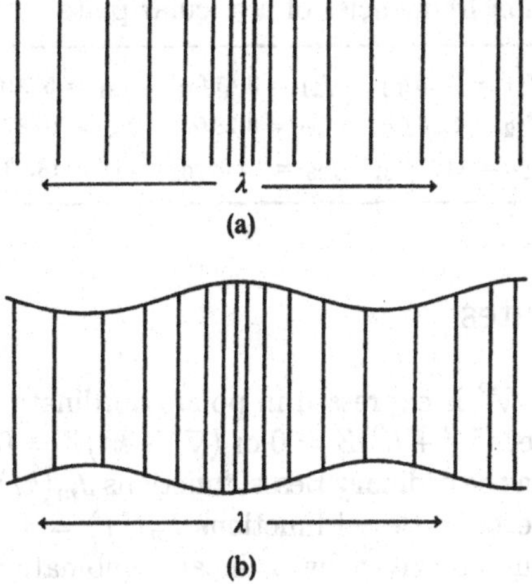

FIGURE 3.7. (a) Pure longitudinal wave in an infinite solid. (b) Quasi-longitudinal wave in a bar or plate.

waves. The radiation of sound in both cases is rather insignificant compared to the case of bending waves, which we now discuss.

The equation of motion for bending or flexural waves in a plate is

$$\frac{\partial^2 z}{\partial t^2} + \frac{Eh^2}{12\rho(1-\nu^2)}\nabla^4 z = 0, \tag{3.11}$$

where ρ is density, ν is Poisson's ratio, E is Young's modulus, and h is the plate thickness. For harmonic solutions, $z = Z(x, y)e^{j\omega t}$:

$$\nabla^4 Z - \frac{12\rho(1-\nu^2)\omega^2}{Eh^2} Z = \nabla^4 Z - k^4 Z = 0, \tag{3.12}$$

where

$$k^2 = \frac{\sqrt{12}\omega}{h}\sqrt{\frac{\rho(1-\nu^2)}{E}} = \frac{\sqrt{12}\omega}{c_{\mathrm{L}}h}.$$

Bending waves in a plate are dispersive; that is, their velocity v depends upon the frequency

$$v(f) = \omega/k = \sqrt{\omega h c_{\mathrm{L}}/\sqrt{12}} = \sqrt{1.8 f h c_{\mathrm{L}}}. \tag{3.13}$$

The frequency of a bending wave is proportional to k^2:

$$f = \omega/2\pi = 0.0459 h c_{\mathrm{L}} k^2. \tag{3.14}$$

The values of k that correspond to the normal modes of vibration depend, of course, on the boundary conditions.

TABLE 3.1. Vibration frequencies of a circular plate with clamped edge.

$f_{01} = 0.4694 c_L h/a^2$	$f_{11} = 2.08 f_{01}$	$f_{21} = 3.41 f_{01}$	$f_{31} = 5.00 f_{01}$	$f_{41} = 6.82 f_{01}$
$f_{02} = 3.89 f_{01}$	$f_{12} = 5.95 f_{01}$	$f_{22} = 8.28 f_{01}$	$f_{32} = 10.87 f_{01}$	$f_{42} = 13.71 f_{01}$
$f_{03} = 8.72 f_{01}$	$f_{13} = 11.75 f_{01}$	$f_{23} = 15.06 f_{01}$	$f_{33} = 18.63 f_{01}$	$f_{43} = 22.47 f_{01}$

3.6 Circular Plates

For a circular plate, ∇^2 is expressed in polar coordinates, and $Z(r, \phi)$ can be a solution of either $(\nabla^2 + k^2)Z = 0$ or $(\nabla^2 - k^2)Z = 0$. Solutions of the first equation contain the ordinary Bessel functions $J_m(kr)$, and solutions to the second, the hyperbolic Bessel functions $I_m(kr) = j^{-m} J_m(jkr)$. Thus, the possible solutions are given by a linear combination of these Bessel functions times an angular function:

$$Z(r, \phi) = \cos(m\phi + \alpha)[AJ_m(kr) + BI_m(kr)]. \tag{3.15}$$

If the plate is clamped at its edge $r = a$, then $Z = 0$ and $\partial Z/\partial r = 0$. The first of these conditions is satisfied if $AJ_m(ka) + BI_m(ka) = 0$, and the second if $AJ'_m(ka) + BI'_m(ka) = 0$.

The allowed values of k are labeled k_{mn}, where m gives the number of nodal diameters and n the number of nodal circles in the corresponding normal mode:

$$\begin{aligned} k_{01} &= 3.189/a, & k_{11} &= 4.612/a, & k_{21} &= 5.904/a, \\ k_{02} &= 6.306/a, & k_{12} &= 7.801/a, & k_{22} &= 9.400/a, \\ k_{03} &= 9.425/a, & k_{13} &= 10.965/a, & k_{23} &= 12.566/a, \end{aligned}$$

$$[k_{mn} \to (2n + m)\pi/2a \quad \text{as} \quad n \to \infty].$$

The corresponding mode frequencies are given in Table 3.1.

A plate with a free edge is more difficult to handle mathematically. The boundary conditions used by Kirchoff lead to a rather complicated expression for k_{mn}, which reduces to $(2n + m)\pi/2r$ for large ka (see Rayleigh, 1894). The (2, 0) mode is now the fundamental mode; the modal frequencies are given in Table 3.2. The mode frequencies for a plate with a simply supported (hinged) edge are given in Table 3.3.

The frequencies in Tables 3.1–3.3 are derived mainly from calculations given by Leissa (1969). Measurements on two large brass plates by

TABLE 3.2. Vibration frequencies of a circular plate with free edge.

—	—	$f_{20} = 0.2413 c_L h/a^2$	$f_{30} = 2.328 f_{20}$	$f_{40} = 4.11 f_{20}$	$f_{50} = 6.30 f_{20}$
$f_{01} = 1.73 f_{20}$	$f_{11} = 3.91 f_{20}$	$f_{21} = 6.71 f_{20}$	$f_{31} = 10.07 f_{20}$	$f_{41} = 13.92 f_{20}$	$f_{51} = 18.24 f_{20}$
$f_{02} = 7.34 f_{20}$	$f_{12} = 11.40 f_{20}$	$f_{22} = 15.97 f_{20}$	$f_{32} = 21.19 f_{20}$	$f_{42} = 27.18 f_{20}$	$f_{52} = 33.31 f_{20}$

TABLE 3.3. Vibration frequencies of a circular plate with a simply supported edge.

$f_{01} = 0.2287c_L h/a^2$	$f_{11} = 2.80 f_{01}$	$f_{21} = 5.15 f_{01}$
$f_{02} = 5.98 f_{01}$	$f_{12} = 9.75 f_{01}$	$f_{22} = 14.09 f_{01}$
$f_{03} = 14.91 f_{01}$	$f_{13} = 20.66 f_{01}$	$f_{23} = 26.99 f_{01}$

Waller (1938) are in good agreement with the data in Table 3.2. Some modes of circular plates are shown in Fig. 3.8.

Chladni (1802) observed that the addition of one nodal circle raised the frequency of a circular plate by about the same amount as adding two nodal diameters, a relationship that Rayleigh (1894) calls Chladni's law. For large values of ka, $ka \simeq (m + 2n)\pi/2$, so that f is proportional to $(m + 2n)^2$. The modal frequencies in a variety of circular plates can be fitted to families of curves: $f_{mn} = c(m + 2n)^p$. In flat plates, $p = 2$, but

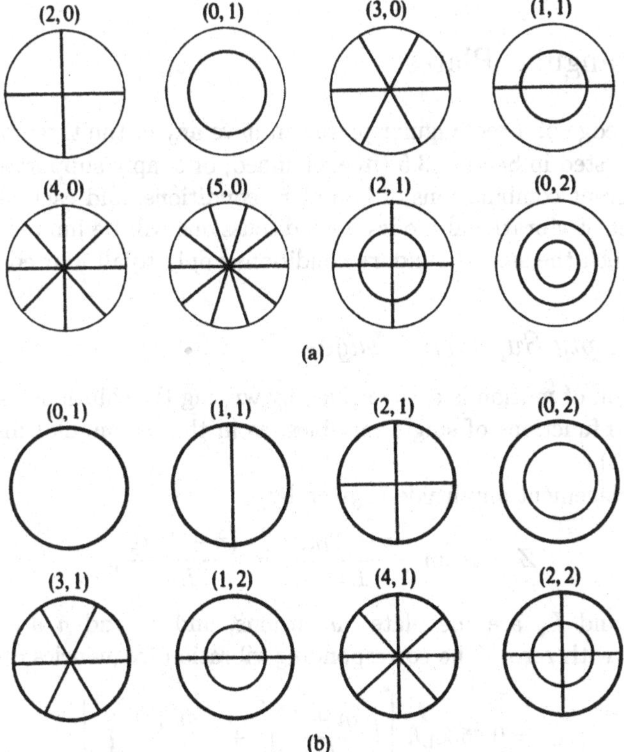

FIGURE 3.8. Vibrational modes of circular plates: (a) free edge and (b) clamped or simply supported edge. The mode number (n, m) gives the number of nodal diameters and circles, respectively.

in nonflat plates (cymbals, bells, etc.), p is generally less than 2 (Rossing, 1982c).

3.7 Elliptical Plates

The frequencies of an elliptical plate of moderately small eccentricity with a clamped edge are given approximately by the formula (Leissa, 1969)

$$f \approx \frac{0.291 c_{\mathrm{L}} h}{a^2} \sqrt{1 + \frac{2}{3}\left(\frac{a}{b}\right)^2 + \left(\frac{a}{b}\right)^4}, \tag{3.16}$$

where a and b are the semimajor and semiminor axes. An elliptical plate with $a/b = 2$ has frequencies 37% greater than a circular plate with the same area.

Waller (1950) shows Chladni patterns and gives relative frequencies for elliptical plates with $a/b = 2$ and $a/b = 5/4$. The nodal patterns resemble those in rectangular plates of similar shape.

3.8 Rectangular Plates

Since each edge of a rectangular plate can have any of the three boundary conditions listed in Section 3.5 (free, clamped, or simply supported), there are 27 different combinations of boundary conditions, and each leads to a different set of vibrational modes. Our discussions will be limited to three cases in which the same boundary conditions apply to all four edges.

3.8.1 Simply Supported Edges

The equation of motion is easily solved by writing the solutions as a product of three functions of single variables, as in the rectangular membrane (Section 3.1).

The displacement amplitude is given by

$$Z = A \sin \frac{(m+1)\pi x}{L_x} \sin \frac{(n+1)\pi y}{L_y}, \tag{3.17}$$

where L_x and L_y are the plate dimensions, and m and n are integers (beginning with zero). The corresponding vibration frequencies are

$$f_{mn} = 0.453 c_{\mathrm{L}} h \left[\left(\frac{m+1}{L_x}\right)^2 + \left(\frac{n+1}{L_{\mathrm{y}}}\right)^2\right]. \tag{3.18}$$

The displacement is similar to that of a rectangular membrane, but the modal frequencies are not. Note that the nodal lines are parallel to the edges, this is not the case for plates with free or clamped edges, as we shall see.

It is convenient to describe a mode in a rectangular plate by $(m,\ n)$, where m and n are the numbers of nodal lines in the y and x directions, respectively (not counting nodes at the edges). To do this, we use $m+1$ and $n+1$ in Eq. (3.17) rather than m and n, as in a rectangular membrane [Eq. (3.4)]. Thus, the fundamental mode is designated (0, 0) rather than (1, 1).

3.8.2 Free Edges

Calculating the modes of a rectangular plate with free edges was described by Rayleigh as a problem "of great difficulty." However, Rayleigh's own methods lead to approximate solutions that are close to measured values, and refinements by Ritz bring them even closer. Results of many subsequent investigations are summarized by Leissa (1969).

The limiting shapes of a rectangle are the square plate and the thin bar. The modes of a thin bar with free ends have frequencies [from Eq. (2.63)]

$$f_n = \frac{0.113h}{L^2}\sqrt{\frac{E}{\rho}}\,[3.0112^2, 5^2, \ldots, (2n+1)^2]. \tag{3.19}$$

The nth mode has $n+1$ nodal lines perpendicular to the axis of the bar. As the bar takes on appreciable width, bending along one axis causes bending in a perpendicular direction. This comes about because the upper part of the bar above the neutral axis (see Fig. 2.17) becomes longer (and thus narrower), while the lower part becomes shorter (and thus wider). We have already seen how Poisson's constant ν is a measure of the lateral contraction that accompanies a longitudinal expansion in a plate (Section 3.5) and how the factor $1-\nu^2$ appears in the expression for both longitudinal and bending wave velocities [Eqs. (3.9) and (3.13)].

Several bending modes in a rectangular plate can be derived from the bending modes of a bar. The $(m,\ 0)$ modes might be expected to have nodal lines parallel to one pair of sides, and the $(0,\ n)$ modes would have nodes parallel to the other pair of sides. Because of the coupling between bending motions in the two directions, however, the modes are not pure bar modes. The nodal lines become curved, and the plate takes on a sort of saddle shape (i.e., concave in one direction but convex in the perpendicular direction). This can be called anticlastic bending, and it is quite evident in the modes of two different rectangular plates shown in Fig. 3.9.

It is interesting to note how the combinations develop in a rectangle as L_x/L_y approaches unity. Fig. 3.10 shows the shapes of two modes that are descendants of the (2, 0) and (0, 2) beam modes in rectangles of varying L_x/L_y. When $L_x \gg L_y$, the (2, 0) and (0, 2) modes appear quite independent. However, as $L_y \to L_x$, the beam modes mix together to form two

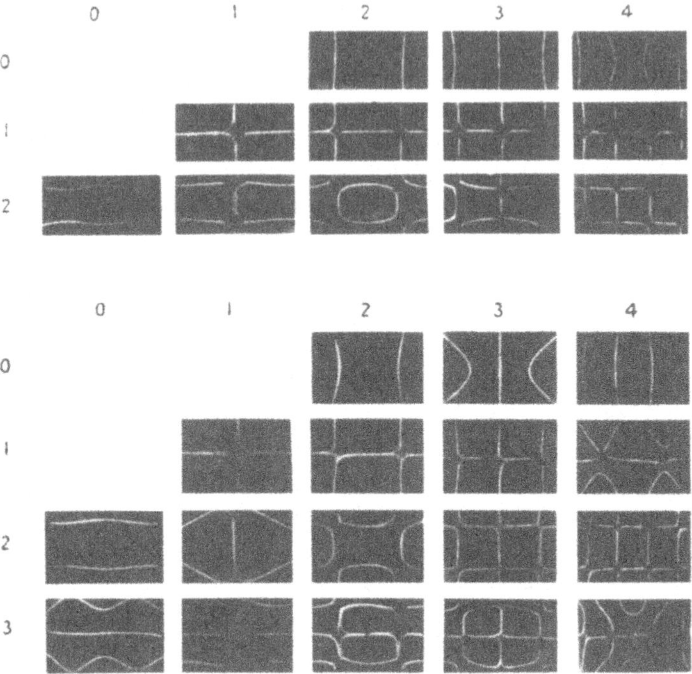

FIGURE 3.9. Chladni patterns showing the vibrational modes of rectangular plates of different shapes: (a) $L_x/L_y = 2$; (b) $L_x/L_y = 3/2$ (Waller, 1949).

new modes. In the square, the mixing is complete, and two combinations are possible depending upon whether the component modes are in phase or out of phase.

Frequencies for the modes that have as their bases the (2, 0) and (0, 2) beam modes are shown in Fig. 3.11. The frequencies have been normalized to L_x, and the normalized frequency of the (2, 0) mode is seen to

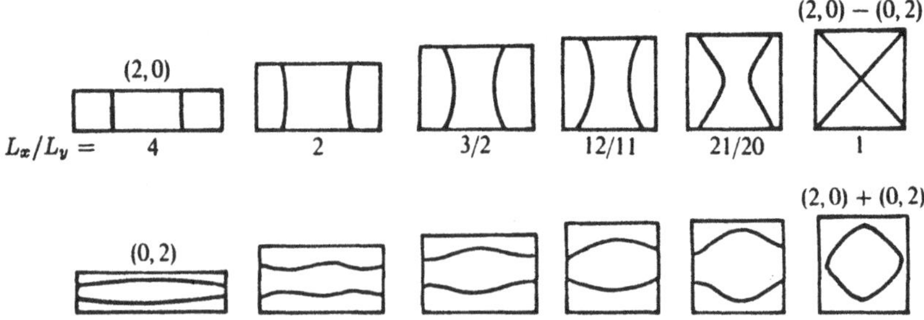

FIGURE 3.10. Mixing of the (2, 0) and (0, 2) modes in rectangular plates with different L_x/L_y ratios (after Waller, 1961).

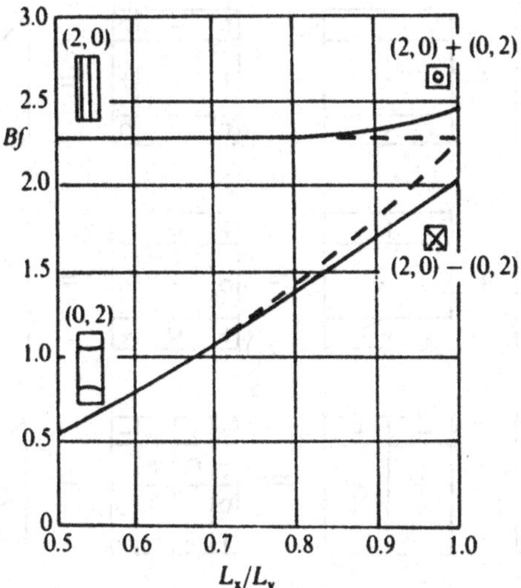

FIGURE 3.11. Normalized frequencies for the (2, 0) and (0, 2) modes (and modes based on combinations of these) in rectangular plates with free edges and varying L_x/L_y ratios (from Warburton, 1954). $B = 2.21(L_x^2/h)\sqrt{\rho(1-\nu^2)/E}$.

be relatively independent of L_y. The dashed curves are obtained from an approximate formula using the Rayleigh method, whereas the solid curves are from a more exact numerical calculation (Warburton, 1954).

3.9 Square Plates

It is obvious from Fig. 3.10 that in plates with $L_x \gg L_y$ (or $L_y \gg L_x$), two normal modes are similar to the (2, 0) and (0, 2) beam modes, with nodal lines nearly parallel to the edges. As the plate becomes more nearly square, these modes are replaced by normal modes that are essentially linear combinations of the beam modes. The nodal patterns of the two modes can be understood from the graphical construction in Fig. 3.12. Zeros denote regions in which the contributions from the (2, 0) and (0, 2) modes cancel each other and lead to nodes. The $(2,0)+(0,2)$ and $(2,1)-(0,2)$ modes are sometimes referred to as the ring mode and x mode, respectively, on account of the shapes of their nodal patterns.

From Fig. 3.11, it is clear that the $(2,0)+(0,2)$ ring mode has a higher frequency than the $(2,0) - (0,2)$ x mode. In the x mode, the bending motions characteristic of the (2, 0) and (0, 2) beam modes aid each other through an elastic interaction that we call Poisson coupling, since its strength depends upon the value of Poisson's constant. In the $(2,0)+(0,2)$ ring mode, however, there is an added stiffness due to the fact that the $(2,0)$ and (0, 2) bending motions oppose each other. Thus, the Poisson coupling

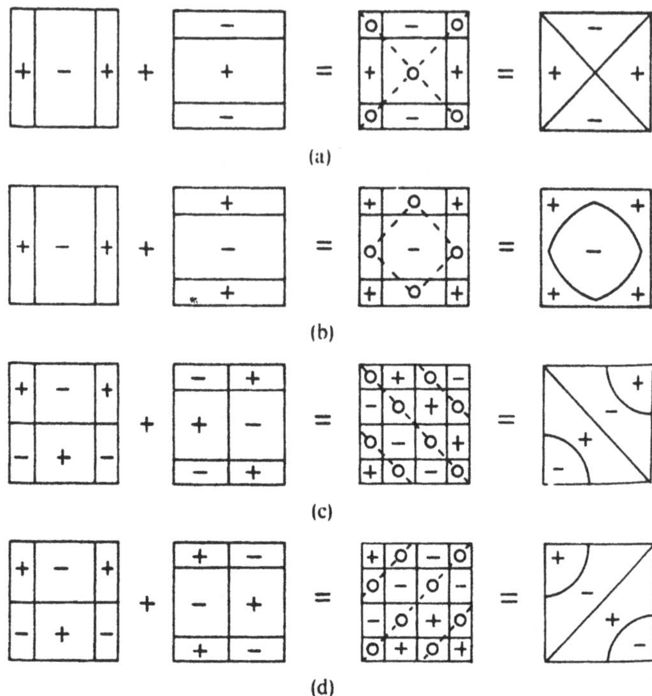

FIGURE 3.12. Graphical construction of combination modes in a square isotropic plate: (a) $(2,0)-(0,2)$, x mode; (b) $(2,0)+(0,2)$, ring mode; (c) $(2,1)-(1,2)$ mode; and (d) $(2,1)+(1,2)$ mode.

splits a modal degeneracy that otherwise would have existed in a square plate. The ratio of the $(2,0)+(0,2)$ and $(2,0)-(0,2)$ mode frequencies is (Warburton, 1954)

$$\frac{f+}{f-} = \sqrt{\frac{1+0.7205\nu}{1-0.7205\nu}}. \tag{3.20}$$

Also shown in Fig. 3.12 are the $(2,1)-(1,2)$ and $(2,1)+(1,2)$ modes, which have the same frequency as the (2, 1) and (1, 2) modes, since Poisson coupling does not aid or oppose either combination. Thus, any of these four modes can be excited depending upon where the driving force is applied. There are, in fact, a large number of degenerate modes, all linear combinations of the (2, 1) and (1,2) modes, which can be excited.

The first 10 modes of an isotropic square plate with free edges are shown in Fig. 3.13. The mode of lowest frequency, the (1, 1) mode, is a twisting mode in which opposite corners move in phase. Its frequency is given by

$$f_{11} = \frac{c_{\mathrm{T}}}{2L_y} = \frac{h}{L_x L_y}\sqrt{\frac{G}{\rho}}\,\frac{h}{L^2}\sqrt{\frac{E}{2\rho(1+\nu)}} = \frac{hc_{\mathrm{L}}}{L^2}\sqrt{\frac{1-\nu}{2}}, \tag{3.21}$$

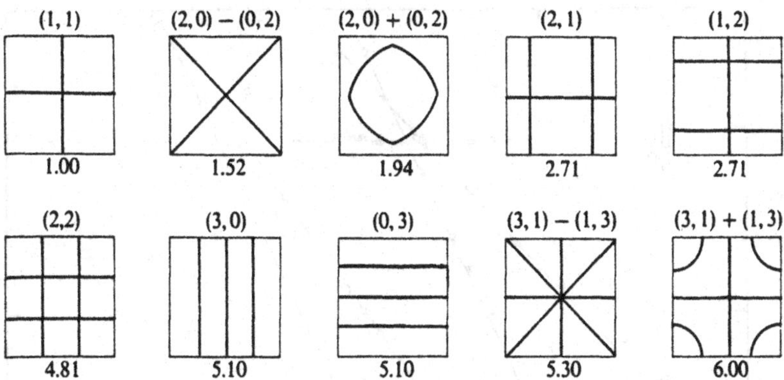

FIGURE 3.13. The first 10 modes of an isotropic square plate with free edges. The modes are designated by m and n, the numbers of nodal lines in the two directions, and the relative frequencies for a plate with $\nu = 0.3$ are given below the figures.

where the torsional wave velocity c_T from Fig. 2.23 is used (subject to the restriction that $L_x > 6h$). In this equation, h is the thickness and G is the shear modulus.

Note that the (2, 1) and (1, 2) modes form a degenerate pair, as do the (3, 0) and (0, 3) modes. However, Poisson coupling removes the degeneracy in the case of the (3, 1)/(1, 3) pair just as it does in the (2, 0)/(0, 2) case. The general rule is that a nondegenerate pair of modes $(m, n \pm n, m)$ exists in a square plate when $m - n = \pm 2, 4, 6, \ldots$.

The modal frequencies in an aluminum plate with a varying length to width ratio are shown in Fig. 3.14. In this case, L_x was kept constant as L_y was varied, so the frequency of the (3, 0) mode, for example, is unchanged. The (1, 1) mode has a slope of 1, as predicted by Eq. (3.19). The (0, 3) bending mode has a slope of 2, as does the (0, 2) mode above and below the region of $L_x = L_y$. The (2, 1) mode, which combines twisting and bending motions, has a slope of about $\frac{4}{3}$.

3.10 Square and Rectangular Plates with Clamped Edges

The first eight modes of a square plate with clamped edges are shown in Fig. 3.15. There is considerable variation in the mode designation by various authors, and so we have used the same designation that was used in Section 3.8.1 for a plate with simply supported edges: m and n are the numbers of nodes in the directions of the y and x axes, respectively, not counting the nodes at the edges. The fundamental (0, 0) mode has a frequency: $f_{00} = 1.654 c_L h/L^2$, where h is the thickness, L is length, and

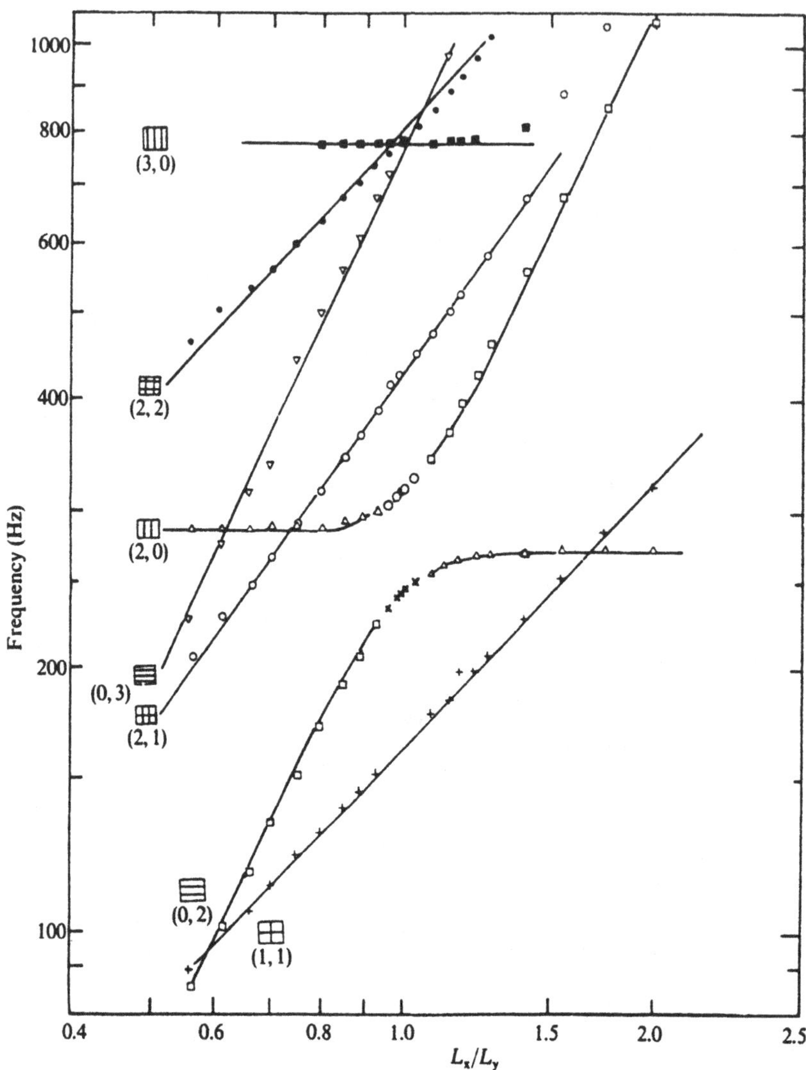

FIGURE 3.14. Modal frequencies of an isotropic aluminum plate (L_x constant). Lines representing the (1, 1) and (2, 2) twisting modes have a slope of 1; lines representing the (0, 2) and (0, 3) bending modes have a slope of 2 (from Caldersmith and Rossing, 1983).

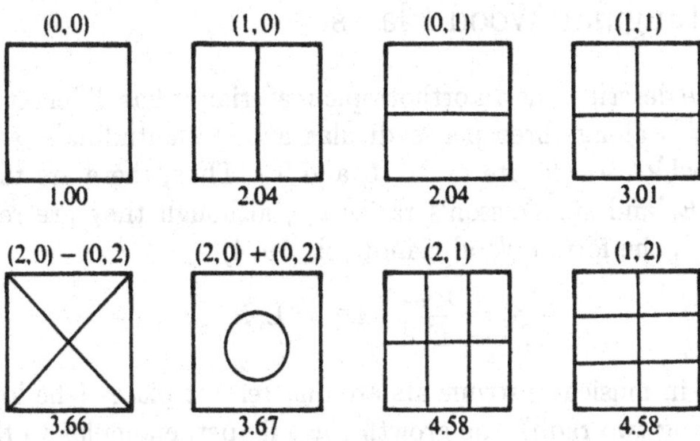

FIGURE 3.15. Modal patterns for the first eight modes of a square plate with clamped edges. Relative frequencies are given below the patterns.

$c_L = [E/\rho(1-\nu^2)]^{1/2}$ is the longitudinal wave velocity (Leissa, 1969). The relative frequencies of the modes are given below the patterns in Fig. 3.15.

Comparing the modes of the square plate with clamped edges to one with free edges, we note that

1. the (1, 1) mode has a frequency nearly 10 times greater than the (1, 1) mode in a free plate;
2. three other modes exist below the (1, 1) mode in the clamped plate;
3. the X mode and ring mode are only about 0.5% different in frequency, and the diameter of the ring node is smaller than it is in a free plate;
4. nondegenerate mode pairs (m, $n \pm n$, m) exist when $m - n = \pm 2, 4, 6, \ldots$, as in the free plates, but the transition from modes characteristic of rectangular plates to those of square plates changes much more abruptly as $L_x \rightarrow L_y$ in clamped plates than in free plates (Warburton, 1954).

Relative frequencies of rectangular plates with clamped edges (from Leissa, 1969) are given in Table 3.4. The actual frequencies can be obtained by multiplying the relative frequencies by $1.654 c_L h/L_y^2$.

TABLE 3.4. Relative vibrational frequencies of rectangular plates with clamped edges.

Mode	$L_x/L_y = 1$	1.5	2	2.5	3	∞
(0,0)	1.00	0.75	0.68	0.66	0.64	0.62
(0,1)	2.04	1.88	1.82	1.79	1.78	1.72
(1,0)	2.04	1.16	1.16	0.88		
(1,1)	3.01	2.27	2.02	1.91	1.86	1.72

3.11 Rectangular Wood Plates

Wood can be described as an orthotropic material; it has different mechanical properties along three perpendicular axes (longitudinal, radial, and tangential, which we denote by L, R, and T). Thus, there are three elastic moduli E_i and six Poisson's ratios ν_{ij}, although they are related by expressions of the form (Wood Handbook, 1974)

$$\frac{\nu_{ij}}{E_i} = \frac{\nu_{ji}}{E_j}, \quad i, j = \mathrm{L, R, T}.$$

Most plates in musical instruments are quarter-cut plates (the log is split or sawed along two radii); the growth rings lie perpendicular to the plate. For a quarter-cut plate, the axes L and R lie in the plane and the axis T in the direction of the thickness. Thus, the constants of interest are E_{L}, E_{R}, ν_{LR}, and $\nu_{\mathrm{RL}} = \nu_{\mathrm{LR}} E_{\mathrm{R}}/E_{\mathrm{L}}$.

To describe the vibrational modes of wood plates generally requires four elastic constants. These may be Young's moduli along (E_x) and across (E_y) the grain, the in-plane shear modulus G, and the larger of the two Poisson ratios ν_{xy}. For a quarter-cut plate in the xy plane, $E_x = E_{\mathrm{L}}$, $E_y = E_{\mathrm{R}}$, $G = G_{\mathrm{LR}}$, $\nu_{xy} = \nu_{\mathrm{LR}}$, and $\nu_{\mathrm{yx}} = \nu_{\mathrm{RL}}$. For other plate orientations, the elastic constants of the plate may be combinations with other elastic constants of the wood (see Fig. 1 in McIntyre and Woodhouse, 1986).

Since $E_{\mathrm{L}} > E_{\mathrm{R}}$ in wood, the modal patterns in Fig. 3.13 that are particularly characteristic of a square plate will not exist in square wood plates. However, the corresponding modal patterns may appear in rectangular wood plates when the ratio of length (along the grain) to width (across the grain) is

$$\frac{L_x}{L_y} = \left(\frac{E_{\mathrm{L}}}{E_{\mathrm{R}}}\right)^{1/4}.$$

For Sitka spruce, for example, $E_{\mathrm{L}}/E_{\mathrm{R}} = 12.8$, $\nu_{\mathrm{RL}} = 0.029$, and $\nu_{\mathrm{LR}} = 0.37$ (Wood Handbook, 1974). Thus $(2, 0 \pm 0, 2)$ combination modes of the type in Fig. 3.15, for example, might be expected in a rectangular plate with $L_x/L_y = 1.9$. (This is somewhat greater than the value 1.5, which has appeared several places in the literature.)

The modal frequencies of a quarter-cut spruce plate are shown in Fig. 3.16. The $(2, 0 \pm 0, 2)$ ring mode and X mode occur at about $L_x/L_y = 2$, as expected. The curve relating the frequency of the (1, 1) mode to L_x/L_y has a slope of one, as in the aluminum plate in Fig. 3.14, and for the (2, 0) mode, the slope is two.

Some modal patterns obtained in the same spruce plate are shown in Fig. 3.17. The (2, 0) and (0, 2) beam modes appear relatively unmodified when L_x/L_y is well above or below the critical ratio of 2.08 where the x mode and ring mode appear. Note that the (2, 0) and (0, 2) modes mix oppositely in phase for L_x/L_y above and below 2.08.

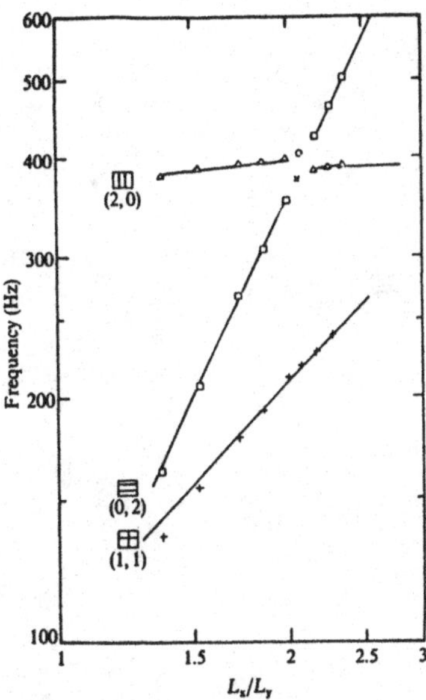

FIGURE 3.16. Modal frequencies of a quarter-cut spruce plate (L_x constant). The (1, 1) twisting mode has a slope of 1; the (0, 2) bending mode has a slope of 2 (Caldersmith and Rossing, 1983).

Modes of vibration closely resembling the x mode and ring mode, along with the (1, 1) twisting mode, have been used by violin makers for centuries to tune the top and back plates of violins before assembling the instruments. If a plate is held with the fingers at a nodal line and tapped at an antinode, the trained ear of a skilled violin maker can ascertain whether the plate has a clear and full ring. In recent years, many violin makers, following the lead of Carleen Hutchins, have used Chladni patterns to test and guide the tuning of these three important modes (see Fig. 3.18).

Many of the formulas in Chapters 2 and 3 for vibrations of bars and plates of isotropic material are easily modified for wood by substituting E_y or E_x for E and $\nu_{yx}\nu_{xy}$ for ν^2. For example, Eqs. (3.18) and (3.19) become

$$f_{mn} = 0.453h \left[c_x \left(\frac{m+1}{L_x} \right)^2 + c_y \left(\frac{n+1}{L_y} \right)^2 \right], \tag{3.18$'$}$$

where

$$c_x = \sqrt{E_x/\rho(1 - \nu_{xy}\nu_{yx})} \quad \text{and} \quad c_y = \sqrt{E_y/\rho(1 - \nu_{xy}\nu_{yx})}$$

and

$$f_n = \frac{0.113h}{L^2} (c_x c_y)^{1/2} [(3.0112)^2, 5^2, \ldots, (2n+1)^2]. \tag{3.19$'$}$$

The torsional stiffness D_{xy} of a wooden plate depends upon E_x, E_y, and G, but it can be approximated by the geometric average $(D_x D_y)^{1/2}$ of

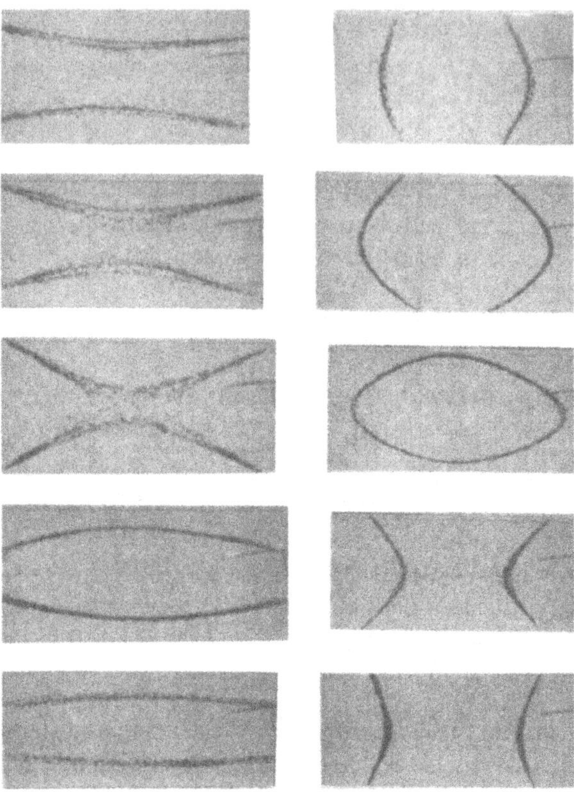

FIGURE 3.17. Modal shapes of the $A(2,0) + B(0,2)$ combination modes in a quarter-cut spruce plate. In (a) and (e), $|A| \ll 1$ (opposite in sign in the two cases). In (f) and (j), $|B| \ll 1$. In (c) and (h), $A/B = \pm 1$ (Caldersmith and Rossing, 1983).

the bending stiffness in the x and y directions (Caldersmith, 1984). Thus, Eq. (3.21) can be written as

$$f_{11} = \frac{h}{L_x L_y}\sqrt{\frac{G}{\rho}} \approx \frac{h(c_y c_x)^{1/2}}{L_y L_x}\sqrt{\frac{1-(\nu_{yx}\nu_{xy})^{1/2}}{2}}. \tag{3.21$'$}$$

The analog to Eq. (2.73) is, approximately,

$$G = \frac{\sqrt{E_x E_y}}{2[1+(\nu_{xy}\nu_{yx})^{1/2}]}. \tag{2.73$'$}$$

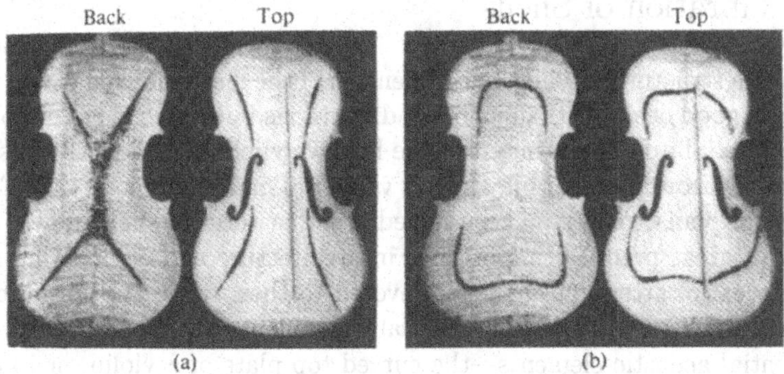

FIGURE 3.18. Chladni patterns showing two modes of vibration in the top and back of a viola (Hutchins, 1977).

3.12 Bending Stiffness in a Membrane

In Chapter 2, we described a stiff string as being slightly barlike and added a term to the equation of motion to represent the bending stiffness. We follow the same approach now by describing a stiff membrane as being slightly platelike, and we add a term to the equation of motion [Eq. (3.1) or (3.7)] to represent the bending stiffness:

$$\frac{\partial^2 z}{\partial t^2} = \frac{T}{\sigma}\nabla^2 z - \frac{h^2 E}{12\rho(1-\nu^2)}\nabla^4 z = c^2\nabla^2 z - S^4\nabla^4 z, \tag{3.22}$$

where T is tension, σ is mass per unit area, h is thickness, E is Young's modulus, ρ is density, ν is Poisson's ratio, and c is the speed of transverse waves in the membrane without stiffness. Assuming a solution $z = AJ_m(kr)\cos m\theta \cos \omega t$ leads to the equation

$$k^2 + \frac{k^4 S^4}{c^2} - \frac{\omega^2}{c^2} = 0. \tag{3.23}$$

Solving for the frequency gives

$$f = \frac{\omega}{2\pi} = \frac{ck}{2\pi}\sqrt{1 + \frac{k^2 S^4}{c^2}}. \tag{3.24}$$

For a typical Mylar membrane used on a kettledrum, $E = 3.5 \times 10^9\,\mathrm{N/m^2}$, $h = 1.9 \times 10^{-4}$ m, a (radius) $= 0.328$ m, $\rho = 1.38 \times 10^3\,\mathrm{kg/m^3}$, and $\sigma = 0.262\,\mathrm{kg/m^2}$, so that $c = 100$ m/s, $S^4 = 8.7 \times 10^{-3}\,\mathrm{m^4/s^2}$. For $k_{11} = 10\,\mathrm{m^{-1}}$, the k^2S^4/c^2 term in Eq. (3.24) is about 10^{-4}, so the frequency of the (1, 1) mode is raised by only about 0.005% by the effect of bending stiffness. In other drums, the frequency change may be as large as 0.1%, but this is still negligible.

3.13 Vibration of Shells

A shell is a structure in which one dimension of the material from which it is made is a good deal less than its other dimensions but which is not simply a plane plate. Simple shells may thus be hollow cylinders, hollow spheres, or sections cut from these, or may have a variety of more complex forms. Even sections of planar plates may be joined together to form shells. Shells are clearly of great practical importance in architecture and engineering and a great deal of attention has been devoted to their static and vibrational properties (Leissa, 1973). Many musical instruments also incorporate shells as essential acoustic elements—the curved top-plate of a violin, or indeed the whole violin body, the quasi-spherical shell of an orchestral cymbal, and the more complex shape of a church bell are all examples. We leave aside the shells that constitute the structure or brass and woodwind instruments, since actual shell vibrations contribute little, if anything, to the instrument sound in these cases.

Love (1888) and Rayleigh (1894) laid the foundations of the study of the vibrations of shells and identified two types of vibrational modes, which Rayleigh called extensional and nonextensional (or flexural), respectively. In an extensional mode, there is a first-order change in the length of a line drawn on the shell surface, and the elastic forces associated with this extension provide a significant restoring force—indeed, for a very thin shell, essentially the only restoring force. In a nonextensional mode, on the other hand, there is no first-order change in the length of lines drawn on the shell surface, and restoring forces are provided by the flexural stiffness of the shell. The frequencies of the purely extensional modes of a thin shell are independent of the shell thickness, since both the mass and the stiffness are proportional to the shell thickness. In a nonextensional mode, on the other hand, the mass is proportional to the shell thickness and the elastic stiffness to the cube of this quantity, as for a plate, so that the frequency is proportional to shell thickness. Modes can be classified in this way only for rather simple shell shapes. For a more general geometry, if the shell has thickness h, then the mode frequencies will have the form

$$f_{m,n} = (A_{m,n} + B_{m,n}h^2)^{1/2}, \tag{3.25}$$

where $A_{m,n}$ and $B_{m,n}$ are constants. For a sufficiently thin shell, the modes of lowest frequency are nonextensional, implying that $A_{m,n} = 0$. This at first sight surprising result, in which the zeroth-order term in (3.25) is set to zero and the second-order term retained, comes about because the elastic energy associated with extensional vibrations is so large, relative to that for nonextensional vibrations, that they are inhibited.

The whole subject of shell vibration is so complex that we can outline only a few of the very simplest cases, and we restrict these to those that are relevant to musical instruments. In particular we should realize that, in general, each vibrational mode involves a combination of longitudinal and transverse motions, though it is only the transverse motions that are acoustically important. A selection of classic papers covering the field, including most of those mentioned below, is given in Kalnins and Dym (1976), while Soedel (1993) provides a modern account. Specifically, we shall discuss just the case of a shallow spherical shell, which is a first approximation to the behavior of the top-plate of a violin-like instrument or to the vibration of a cymbal, and that of a circular cylinder, which appears in music as an orchestral chime or tubular bell. The vibrations of church bells are rather distantly related to those of a cylindrical shell open at one end and closed at the other, but are most usefully discussed from a practical viewpoint, as we take up in Chapter 21.

3.13.1 Shallow Spherical Shells

The behavior of the lowest axisymmetric mode of vibration of a shallow circular cap cut from a spherical shell has been treated by Reissner (1946, 1955). In particular, we are concerned to note the transition in behavior between an initially flat plate and the same plate curved to form a shallow spherical shell. Conceptually this is the transition between the flat top-plate of a guitar and the arched top-plate of a violin, though this is complicated by the anistropic elasticity and fibrous nature of wood.

Suppose we have a flat plate of radius a and thickness h with $h \ll a$. The frequency of the lowest mode depends upon the boundary conditions at the edge of the plate, as discussed in Section 3.6. The mode shape can be expressed as a combination of Bessel functions with argument kr, as in Eq. (3.15), and the lowest mode is specified by a particular value of $ka = \mu_0$, which depends upon the boundary conditions. For a spherical cap of radius a, as illustrated in Fig. 3.19, Reissner (1955) shows that, similarly, the lowest mode is specified by a value of $ka = \mu$ that is constant at about 3 for a shell with free edges but which, in the case of a shell with clamped edges, increases from about 3 when the rise H at the center of the cap is less than or comparable with the thickness h of the shell, to a value of about 6 when $H \gg h$.

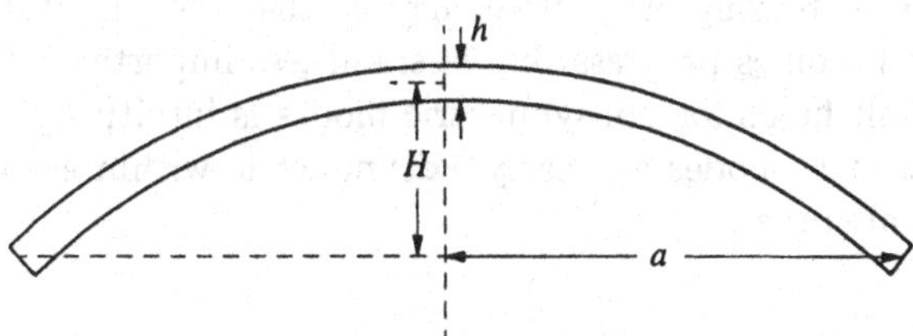

FIGURE 3.19. Geometrical parameters for a spherical-cap shell. The theory applies only to cases in which $h/a \ll 1$ and $H/a \ll 1$.

Reissner's final result for the frequency ω of the lowest axisymmetric mode of the shell, which is actually an extensional mode, is

$$\frac{\omega}{\omega_0} = \left[\frac{1}{1-\nu^2}\left(\frac{\mu}{\mu_0}\right)^4 + \frac{48}{\mu_0^4}\left(\frac{H}{h}\right)^2\right]^{1/2}, \tag{3.26}$$

where ω_0 is the lowest-mode frequency for a flat plate with the same boundary conditions. Detailed calculation shows that the shell mode frequency ω rises steadily from ω_0, when $H/h \ll 1$ and the shell is essentially flat, to an asymptotic behavior with $\omega/\omega_0 \approx 0.68H/h$ for a shell with clamped edges, or $\omega/\omega_0 \approx 0.84H/h$ for a shell with free edges.

If the shell is thin enough that H/h is greater than about 20, and shallow enough that $H/a < 0.25$, then the first term in Eq. (3.26) can be neglected and the second term, along with an explicit expression for ω_0, gives

$$\omega \approx 2\left(\frac{E}{\rho}\right)^{1/2}\frac{H}{a^2}. \tag{3.27}$$

Surprisingly, this result is approximately independent of whether the shell is clamped or free at its circumference, though this is not true in general for the frequencies of higher modes.

We can apply these conclusions, at least in qualitative form, to shells with more complex boundary geometry. The practical importance of the results is that a quite moderate arching of a plate to form a shallow shell increases its stiffness greatly, the mode frequency being doubled if the arch height is a little more than twice the shell thickness. The top plate of a violin is thus stiff enough that it requires no internal bracing, while the flat soundboards of other instruments require strong bracing if they are to support the normal stress component of the string tension. The arch of the violin top-plate similarly raises the frequency of its lowest mode, while planar soundboards require bracing for this purpose.

In musical instruments we are, of course, interested in many vibrational modes, not just the fundamental. Kalnins (1963) investigated this problem for the nonsymmetric vibrations of a spherical-cap shell, and showed that the higher modes are much less affected by shell curvature than is the axisymmetric fundamental. We would expect this result, in the case of higher axisymmetric modes, from the form of Eq. (3.26). Both μ_0 and μ increase steadily with increasing mode frequency, so that the curvature-dependent second term becomes progressively less and less important. The fact that a similar result holds for nonsymmetric modes is intuitively correct, since we can treat such modes as nearly axisymmetric within each cell defined by their nodal curves.

3.13.2 Cylindrical Shells

The vibration of thin cylindrical shells can be approached in rather the same manner, and there has been a good deal of attention paid to the subject. From a musical point of view, however, cylindrical shells occur mainly as the relatively rigid supporting structures for light membranes in drums, or as tubes with length very much greater than the shell radius in orchestral chimes.

In the first case, there are two deformation modes of prime interest. The first is the lowest extensional mode, which consists of a uniform radial displacement of the shell walls. The curvature of the shell makes it very stiff against this sort of deformation, so that it is well able to support the inward tension of the two drumheads. If the radius of the drum is r and the drumhead tension is T, then the tangential compressive force in the shell walls is just rT. The compressive stress in the shell of a small drum with deep and moderately thick walls is therefore not large.

Nonextensional modes of deformation, for the lowest of which the cross-section of the drum shell becomes elliptical, can enter in two ways. The first is static—the drum can be distorted by an external force and is not very strong to resist it. Indeed, the principal stabilizing influence against this sort of distortion is the membrane tension in the drumhead. More importantly from an acoustical point of view, it is usual to strike a drum at a point well away from its center in order to excite higher membrane modes. This applies normal stresses with angular dependence like $\cos n\phi$ to the shell, and these are able to excite nonextensional shell modes with the same angular pattern. It is unlikely that these wall vibrations contribute appreciably to the radiated sound in a normal drum. The situation is different, however, in non-membrane drums in which the shell is excited directly by striking it with wooden beaters. Such a drum might best be considered to be a sort of wooden bell.

In the higher modes of a cylindrical shell, the normal displacement has a form rather like $\cos(k_n z + \beta)\cos m\phi$, where z is the axial coordinate and ϕ is the angular coordinate. The phase constant β depends upon the end conditions. To be more accurate we should include hyperbolic functions in the axial direction, and recognize that the normal displacements are accompanied by tangential displacements in both the z and ϕ directions. The axisymmetric extensional mode for which $m = 0$ has a high frequency and is not generally excited. The nonextensional modes have a multipole structure, with sections of the surface moving in opposite directions, and do not radiate very effectively except for small values of m, as we shall see in Chapter 7.

To a first approximation we could consider a church bell as a short cylinder with the end at $z = 0$ rigidly closed, so that $\beta = \pi/2$. The musically important modes are essentially those associated with the first few allowed values of k and m, omitting $m = 0$. Because of the complex profile of a typical bell, however, we cannot say anything from first principles about the relative frequencies of these modes, or indeed about their precise nodal patterns, and we defer their consideration to Chapter 21.

An orchestral chime or tubular bell, on the other hand, is essentially a long narrow pipe, as also is the common wind-chime. The dimensions are such that this cylindrical shell can best be considered as a form of bar, with a radius of gyration as defined in Fig. 2.18. The mode frequencies for simple transverse vibrations are then given by Eq. (2.63) and vary with the longitudinal mode number n approximately as $(n + \frac{1}{2})^2$. There are, of course, higher modes to be considered, particularly those with $m > 0$ associated with distortions of the tube cross section. There are also corrections to the simple formula (2.63) for the transverse mode frequencies to allow for coupling to distortions of the cross-section, rotary inertia, and other minor effects (Flugge, 1962). The effect of these corrections, broadly, is to slightly lower the frequencies of the higher modes relative to those predicted by the thin-bar formula.

3.14 Driving Point Impedance

In Section 1.7, we discussed the mechanical impedance $\tilde{Z}$ (and its reciprocal, the mobility or mechanical admittance $\tilde{Y}$) of a simple oscillator. In Fig. 1.13(a), the real part of the admittance reaches its maximum value at resonance, while the imaginary part goes through zero. On the Nyquist plot in Fig. 1.13(c), the entire frequency span from $\omega = 0$ to $\omega \to \infty$ represents a complete circle.

In a more complex system, the impedance depends upon the location at which the force $\tilde{F}$ and the velocity $\tilde{v}$ are measured. If the force $\tilde{F}$ is applied at a single point and the velocity $\tilde{v}$ is measured at the same point, the quotient $\tilde{F}(\mathbf{n}_1)/\tilde{v}(\mathbf{n}_1) = \tilde{Z}$ is called the driving-point impedance. Measuring $\tilde{F}$ and $\tilde{v}$ at different locations gives a transfer impedance.

The driving point impedance is often measured by means of an impedance head, which incorporates both a force transducer and accelerometer, as shown in Fig. 3.20. Both transducers employ piezoelectric crystals, and the accelerometer has an inertial mass attached to the crystal, as shown. Since attaching the impedance head adds mass m to the structure, the measured impedance Z_1' is the true driving point impedance Z_1 plus the impedance $j\omega m$ of the added mass m below the force transducer:

$$Z_1' = Z_1 + j\omega m. \tag{3.28}$$

The second term can be minimized by placing the force transducer next to the structure and making its mass as small as possible.

In practice, the impedance head is generally attached to an electromagnetic shaker, which furnishes the driving force or excitation. The output of the accelerometer is integrated to obtain a velocity signal, which is divided by the force to obtain the admittance (or vice versa for impedance). The driving force may be a swept sinusoid, random noise, or pseudorandom noise. When random noise is used, it is necessary to employ a real-time spectrum analyzer to obtain the admittance or impedance of the structure as a function of frequency (often called the frequency response function).

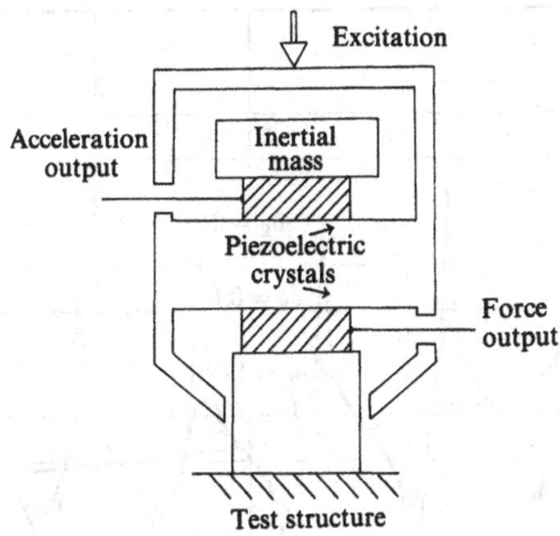

FIGURE 3.20. Impedance head consisting of an accelerometer and force transducer.

3.15.1 Impedance of Infinite Bars and Thin Plates

The input impedance for longitudinal waves in an infinitely long bar is simply μc_L, where μ is the mass per unit length and c_L is the longitudinal wave speed. The input impedance for flexural waves is a little more difficult to calculate, since the speed of flexural waves is frequency dependent (Section 2.15), and also because in flexure, there are exponentially decaying near fields in addition to the propagating waves (Cremer et al., 1973). The impedance is

$$\tilde{Z}(f) = \mu v(f) \frac{1+j}{2}. \tag{3.29}$$

The impedance is complex and increases with $\sqrt{f}$, as does the flexural wave speed $v(f)$, up to a certain limiting frequency.

The input impedance of a thin, isotropic, infinite plate, on the other hand, turns out to be real and independent of the frequency (Cremer et al., 1973):

$$Z = 8\sqrt{B\sigma} = 8\sqrt{\frac{E\sigma}{1-\nu^2}\frac{h^3}{12}}, \tag{3.30}$$

where B is the flexural stiffness, σ is the mass per unit area, h is the thickness, and ν is Poisson's ratio.

3.15.2 Impedance of Finite Bars and Plates

When the driving point impedance or admittance of a finite structure is plotted as a function of frequency, a series of maxima and minima are added to the curves for the corresponding infinite structure. The normalized

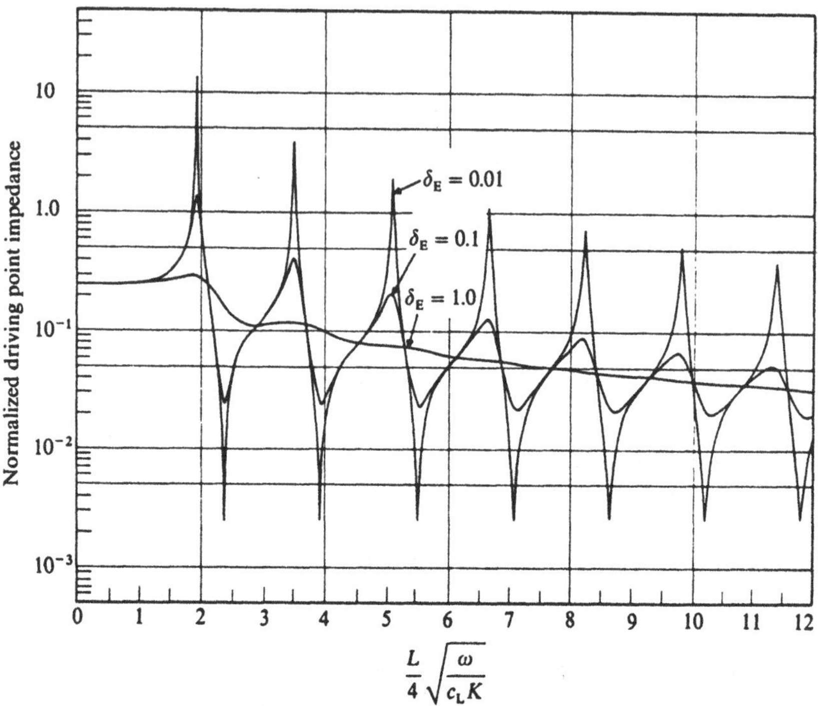

FIGURE 3.21. Normalized driving-point impedance at one end of a bar with free ends. Note that the horizontal axis is proportional to $\sqrt{f}$; the normalized impedance, which includes a factor $1/f$, decreases with frequency even though the actual driving-point impedance increases as $\sqrt{f}$. Three different values of damping are shown (Snowdon, 1965).

driving-point impedance at one end of a bar with free ends is shown in Fig. 3.21. The heavily damped curve ($\delta = 1$) approximates Eq. (3.29) for an infinite bar, while the lightly damped curve ($\delta = 0.01$) has sharp maxima and minima.

The minima in Fig. 3.21 correspond to normal modes of the bar, whereas the maxima occur at frequencies for which the bar vibrates with a nodal line passing through the driving point. Impedance minima (admittance maxima) correspond to resonances, while impedance maxima (admittance minima) correspond to antiresonances.

The driving-point admittance at two different locations on a rectangular plate with simply supported edges is shown in Fig. 3.22. In these graphs, resonances corresponding to normal modes of the plate give rise to maxima on the curves. Note that some normal modes are excited at both driving points but some are not (when a node occurs too near the driving point).

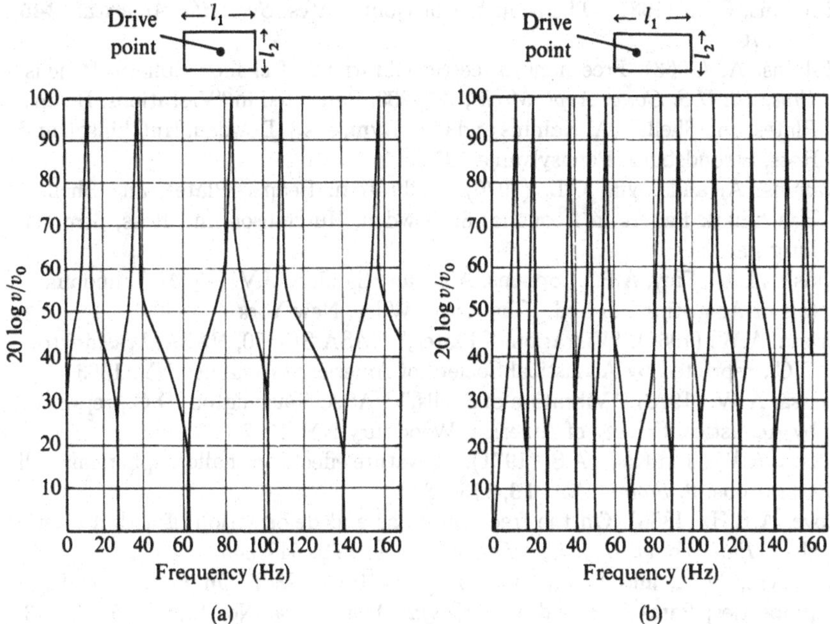

FIGURE 3.22. Driving-point admittance of a rectangular plate with simply-supported edges: (a) driven at the center and (b) driven off center (after Cremer et al., 1973).

References

Caldersmith, G.W. (1984). Vibrations of orthotropic rectangular plates. *Acustica* **56**, 144–152.

Caldersmith, G., and Rossing, T.D. (1983). Ring modes, X-modes and Poisson coupling. *Catgut Acoust. Soc. Newsletter*, No. 39, 12–14.

Caldersmith, G., and Rossing, T.D. (1984). Determination of modal coupling in vibrating rectangular plates. *Applied Acoustics* **17**, 33–44.

Chladni, E.F.F. (1802). "Die Akustik," 2nd ed. Breitkopf u. Härtel, Leipzig.

Cremer, L. (1984). "The Physics of the Violin." Translated by J.S. Allen, M.I.T. Press, Cambridge, Massachusetts.

Cremer, L., Heckl, M., and Ungar, E.E. (1973). "Structure-Borne Sound," Chapter 4. Springer Verlag, Berlin and New York.

Flügge, W. (1962). "Statik und Dynanik der Schalen." Springer-Verlag, Berlin.

French, A.P. (1971). "Vibrations and Waves," p. 181 ff. Norton, New York.

Hearmon, R.F.S. (1961). "Introduction to Applied Anisotropic Elasticity." Oxford Univ. Press, London.

Hutchins, C.M. (1977). Another piece of the free plate tap tone puzzle. *Catgut Acoust. Soc. Newsletter*, No. 28, 22.

Hutchins, C.M. (1981). The acoustics of violin plates. *Scientific American* **245** (4), 170.

Kalnins. A. (1963). Free nonsymmetric vibrations of shallow spherical shells. *Proc. 4th U.S. Cong. Appl. Mech.*, 225–233. Reprinted in "Vibrations: Beams, Plates, and Shells" (A. Kalnins and C.L. Dym, eds.), Dowden, Hutchinson and Ross, Stroudsburg, Pennsylvania 1976.

Kalnins, A. and Dym, C.L. (1976). "Vibration: Beams, Plates, and Shells," Benchmark Papers in Acoustics 8, Dowden, Hutchinson and Ross, Stroudsburg Pa.

Kinsler, L.E., Frey, A.R., Coppens, A.B., and Sanders, J.V. (1982). "Fundamentals of Acoustics," 3rd ed., Chapter 4. Wiley, New York.

Leissa, A.W. (1969). "Vibration of Plates," NASA SP-160, NASA, Washington, D.C., reprinted by Acoustical Society of America, Woodbury NY, 1993.

Leissa, A.W. (1973). "Vibration of Shells," NASA, Washington, D.C., reprinted by Acoustical Society of America, Woodbury NY, 1993.

Leissa, A.W., and Kadi, A.S. (1971). Curvature effects on shallow spherical shell vibrations. *J. Sound Vibr.* **16**, 173–187.

Love, A.E.H. (1888). On the free vibrations and deformation of a thin elastic shell. *Phil. Trans. Roy. Soc. London* Ser.A, **179**, 543–546.

McIntyre, M.E., and Woodhouse, J. (1984/1985/1986). On measuring wood properties, Parts 1, 2, and 3, *J. Catgut Acoust. Soc.* No. 42, 11–25; No. 43, 18–24; and No. 45, 14–23.

Morse, P.M. (1948). "Vibration and Sound," Chapter 5. McGraw-Hill, New York. Reprinted 1976, Acoustical Soc. Am., Woodbury, New York.

Morse, P.M., and Ingard, K.U. (1968). Chapter 5. "Theoretical Acoustics," McGraw-Hill, New York. Reprinted 1986, Princeton Univ. Press, Princeton, New Jersey.

Rayleigh, Lord (1894). "The Theory of Sound," Vol. 1, 2nd ed., Chapters 9 and 10. Macmillan, New York. Reprinted by Dover, New York, 1945. Vol. 1, pp.395–432

Reissner, E. (1946). On vibrations of shallow spherical shells. *J. Appl. Phys.* **17**, 1038–1042.

Reissner, E. (1955). On axi-symmetrical vibrations of shallow spherical shells. *Q. Appl. Math.* **13**, 279–290.

Rossing, T.D. (1982a). "The Science of Sound," Chapters 2 and 13. Addison-Wesley, Reading, Massachusetts.

Rossing, T.D. (1982b). The physics of kettledrums. *Scientific American* **247** (5), 172–178.

Rossing, T.D. (1982c). Chladni's law for vibrating plates. *American J. Physics* **50**, 271–274.

Snowdon, J.C. (1965). Mechanical impedance of free-free beams. *J. Acoust. Soc. Am.* **37**, 240–249.

Soedel, W. (1993). "Vibrations of Plates and Shells," second edition. Marcel Dekker, New York.

Ver, I.L., and Holmer, C.I. (1971). Interaction of sound waves with solid structures. In "Noise and Vibration Control" (L.L. Beranek, ed.). McGraw-Hill, New York.

Waller, M.D. (1938). Vibrations of free circular plates. Part I: Normal modes. *Proc. Phys. Soc.* **50**, 70–76.

Waller, M.D. (1949). Vibrations of free rectangular plates. *Proc. Phys. Soc. London* **B62**, 277–285.

Waller, M.D. (1950). Vibrations of free elliptical plates. *Proc. Phys. Soc. London* **B63**, 451–455.

Waller, M.D. (1961). "Chladni Figures: A Study in Symmetry." Bell, London.

Warburton, G.B. (1954). The vibration of rectangular plates. *Proc. Inst. Mech. Eng.* **A168**, 371–384.

Wood Handbook (1974). Mechanical properties of wood. In "Wood Handbook: Wood as an Engineering Material." U.S. Forest Products Laboratory, Madison, Wisconsin.

4

Coupled Vibrating Systems

4.1 Coupling Between Two Identical Vibrators

In Chapter 1, we considered the free vibrations of a two-mass system with three springs of equal stiffness; we found that there were two normal modes of vibration. Such a system could have been viewed as consisting of two separate mass/spring vibrators coupled together by the center spring (see Fig. 1.20). If the coupling spring were made successively weaker, the two modes would become closer and closer in frequency.

A similar behavior is exhibited by two simple pendulums connected by a spring, as in Fig. 4.1. Each pendulum, vibrating independently at small amplitude, has a frequency given by

$$f_0 = \frac{1}{2\pi}\sqrt{\frac{g}{l}},$$

where l is the length of the pendulum and g is the acceleration of gravity.

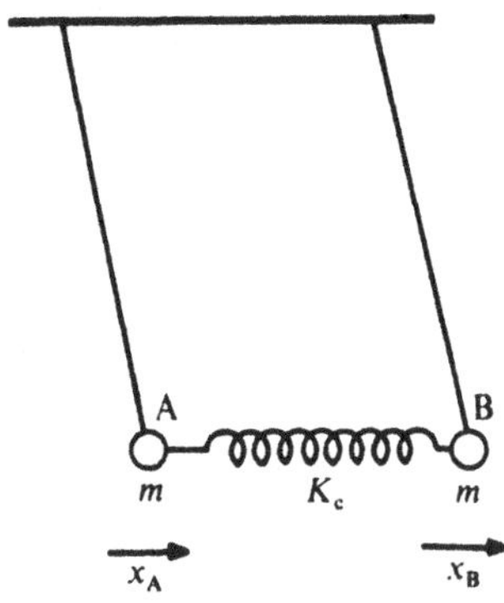

FIGURE 4.1. Two simple pendulums coupled by a spring.

The coupled system has two normal modes of vibration given by

$$f_1 = \frac{1}{2\pi}\sqrt{\frac{g}{l}} \quad \text{and} \quad f_2 = \frac{1}{2\pi}\sqrt{\left(\frac{g}{l} + \frac{2K_c}{m}\right)}, \tag{4.1}$$

where K_c is the spring constant and m is mass. The pendulums move in phase in the mode of lower frequency and in opposite phase in the mode of higher frequency.

The coupling can be expressed in terms of a coupling frequency $\omega_c = \sqrt{K_c/m}$ in which case $\omega_1^2 = \omega_0^2$ and $\omega_2^2 = \omega_0^2 + 2\omega_c^2$. If either pendulum is clamped in place, the other pendulum oscillates at a frequency $\omega^2 = \omega_0^2 + \omega_c^2$. This is not a normal mode frequency, however. If the system is started with pendulum A at its rest position and pendulum B in a displaced position, for example, the resulting motion changes with time. During each swing, pendulum B gives up some of its motion to pendulum A until pendulum B finds itself at rest and pendulum A has all the motion. Then, the process reverses. The exchange of energy between the two pendulums takes place at a rate that depends on the coupling frequency ω_c.

4.2 Normal Modes

Solution of the equations of motion to obtain the normal modes is relatively easy in this case. The equations of motion for small vibrations are

$$m\ddot{x}_A + \frac{mg}{l}x_A + K(x_A - x_B) = 0, \tag{4.2a}$$

and

$$m\ddot{x}_B + \frac{mg}{l}x_B + K(x_B - x_A) = 0. \tag{4.2b}$$

These can be rewritten in terms of ω_0 and ω_c, previously defined:

$$\ddot{x}_A + (\omega_0^2 + \omega_c^2)x_A - \omega_c^2 x_B = 0, \tag{4.3a}$$

and

$$\ddot{x}_B + (\omega_0^2 + \omega_c^2)x_B - \omega_c^2 x_A = 0. \tag{4.3b}$$

If the two equations are added together, we obtain

$$\frac{d^2}{dt^2}(x_A + x_B) + \omega_0^2(x_A + x_B) = 0. \tag{4.4a}$$

If they are subtracted, we obtain

$$\frac{d^2}{dt^2}(x_A - x_B) + (\omega_0^2 + 2\omega_c^2)(x_A - x_B) = 0. \tag{4.4b}$$

These are equations for simple harmonic motion. In the first, the variable is $(x_A + x_B)$, and the frequency is $\omega_0/2\pi$. In the second, the variable is

$(x_A - x_B)$, and the frequency is $\sqrt{\omega_0^2 + 2\omega_c^2}/2\pi$. These represent the two normal modes described previously.

It is sometimes desirable to define normal coordinates q_1 and q_2 along which displacements can take place independently. In this case, $\mathbf{q}_1 = \mathbf{x}_A + \mathbf{x}_B$ and $\mathbf{q}_2 = \mathbf{x}_A - \mathbf{x}_B$; the normal coordinates are rotated 45° from the old, as shown in Fig. 4.2. Thus, the transformation from (x_A, x_B), the coordinates of the individual pendulum, to (q_1, q_2), the normal coordinates, can be found geometrically. In mode 1, the system oscillates along the coordinate q_1 with amplitude Q_1 and angular frequency ω_1, and in mode 2 it oscillates along q_2 with amplitude Q_2 and angular frequency ω_2. [The actual frequencies are given by Eq. (4.1).] Thus, x_A and x_B can be written as

$$x_A = \frac{Q_1}{\sqrt{2}} \cos \omega_1 t + \frac{Q_2}{\sqrt{2}} \cos \omega_2 t, \tag{4.5a}$$

and

$$x_B = \frac{Q_1}{\sqrt{2}} \cos \omega_1 t - \frac{Q_2}{\sqrt{2}} \cos \omega_2 t. \tag{4.5b}$$

Q_1 and Q_2 are determined by the initial displacements.

If the system is given an initial displacement $x_A(0) = A_0$, $x_B(0) = 0$, then $Q_1 = A_0/\sqrt{2}$, $Q_2 = A_0/\sqrt{2}$, and Eqs. (4.5) become

$$x_A = \frac{A_0}{2} (\cos \omega_1 t + \cos \omega_2 t), \tag{4.6a}$$

and

$$x_B = \frac{A_0}{2} (\cos \omega_1 t - \cos \omega_2 t). \tag{4.6b}$$

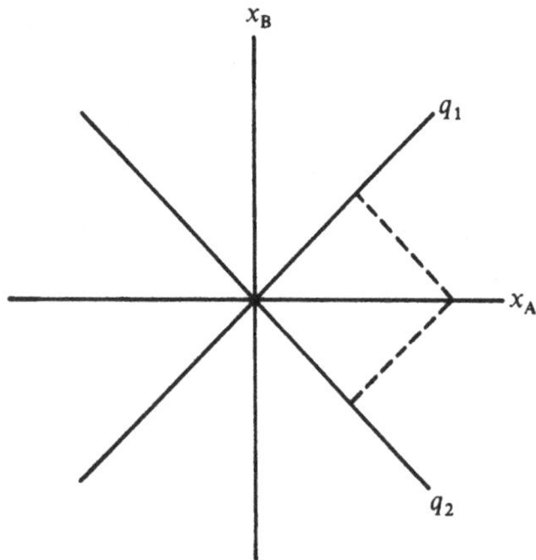

FIGURE 4.2. Relationship of the normal coordinates (q_1, q_2) of two coupled oscillators to the individual coordinates (x_A, x_B).

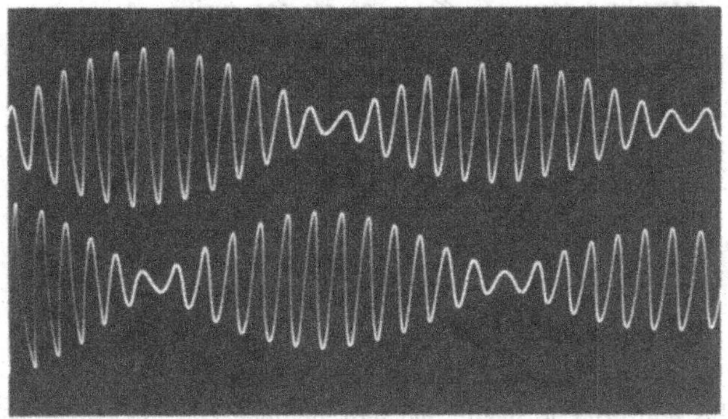

FIGURE 4.3. Motion of two coupled pendulums with flashlight bulbs attached to the bobs. Pendulum B had an initial displacement; pendulum A had none. Note the exchange of kinetic energy at a rate ω_m/π, and also note the effect of damping (from French, 1971).

These equations can be rewritten:

$$x_{\rm A} = A_0 \cos\left(\frac{\omega_2 - \omega_1}{2} t\right) \cos\left(\frac{\omega_2 + \omega_1}{2} t\right), \tag{4.7a}$$

and

$$x_{\rm B} = A_0 \sin\left(\frac{\omega_2 - \omega_1}{2} t\right) \sin\left(\frac{\omega_2 + \omega_1}{2} t\right). \tag{4.7b}$$

These can be interpreted as oscillations at a frequency $\overline{\omega} = (\omega_2 + \omega_1)/2$ with the amplitude modulated at a frequency $\omega_m = (\omega_2 - \omega_1)/2$. Note that the oscillations of the two pendulums at frequency $\overline{\omega}$ are 90° different in phase, and so are the oscillations in the amplitude at frequency ω_m. The photograph of two coupled pendulums in Fig. 4.3 illustrates this.

4.3 Weak and Strong Coupling

In the case we have been discussing, the average frequency and modulation frequency are

$$\overline{f} = \frac{\omega_2 + \omega_1}{4\pi} = \frac{1}{4\pi}\left(\sqrt{\omega_0^2 + 2\omega_c^2} + \omega_0\right),$$

and (4.8)

$$f_m = \frac{\omega_2 - \omega_1}{4\pi} = \frac{1}{4\pi}\left(\sqrt{\omega_0^2 + 2\omega_c^2} - \omega_0\right).$$

In the weak coupling case, $\omega_c \ll \omega_0$, so we can write

$$\overline{f} = \frac{1}{4\pi}\left[\omega_0\left(1+\frac{2\omega_c^2}{\omega_0^2}\right)^{1/2}+\omega_0\right] \simeq \frac{\omega_0}{2\pi}\left(1+\frac{\omega_c^2}{2\omega_0^2}\right), \tag{4.9a}$$

and

$$f_m = \frac{1}{4\pi}\left[\omega_0\left(1+\frac{2\omega_c^2}{\omega_0^2}\right)^{1/2}+\omega_0\right] \simeq \frac{\omega_c^2}{4\pi\omega_0}. \tag{4.9b}$$

When the coupling is weak, we can neglect the energy stored in the coupling spring and characterize the motion as an interchange of energy between pendulum A and pendulum B at a rate given by f_m. The pendulum whose amplitude is increasing is the one that is lagging in phase by 90°, as expected for a vibrator absorbing power from a driving force at resonance. The two pendulums alternate as the driver and the driven. When it comes to rest, the driver suddenly changes phase by 180° so that it can become the driven vibrator.

In the case where $\omega_c = 2\omega_0$ (a strong coupling), $\overline{f} = 2f_0$ and $f_m = f_0$. The modulation frequency is half the average frequency, so the excursions are alternately large and small. The kinetic energy is exchanged rapidly between the two bobs, as shown in Fig. 4.4. The normal mode frequencies in this case are in a 3:1 ratio: $f_1 = f_0$ and $f_2 = 3f_0$. The initial conditions that give the motion in Fig. 4.4 are $x_A(0) = A_0$ and $x_B(0) = 0$; that is, bob B is held at its rest point while bob A is displaced and both are released together.

Note that the motions shown in Fig. 4.4 are described by Eqs. (4.6a) and (4.6b), but they could also be obtained graphically from Fig. 4.2 by marking off appropriate time intervals between $A_0/2$ and $-A_0/2$ along both the q_1 and q_2 axes. The displacements x_A and x_B at each time are found by projecting $q_1(t)$ and $q_2(t)$ on the appropriate axis and adding them.

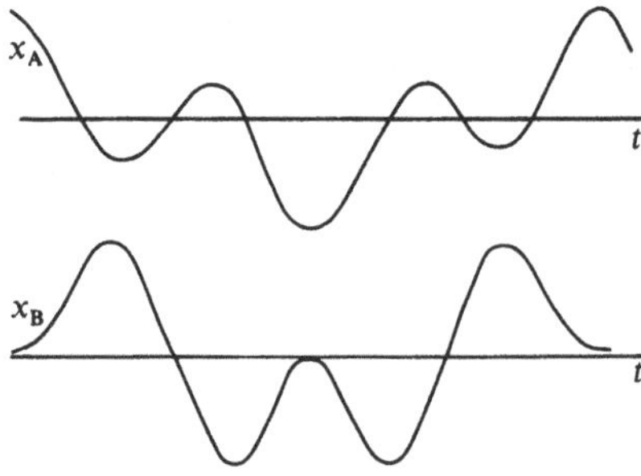

FIGURE 4.4. Motion of two strongly coupled pendulums with $\omega_c = 2\omega_0$.

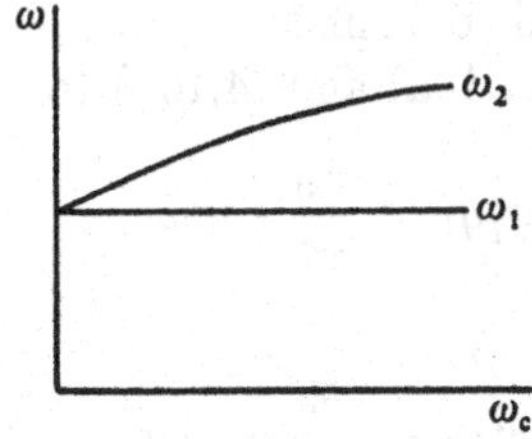

FIGURE 4.5. Dependence of the mode frequencies on the coupling strength for two coupled pendulums.

In the case of two coupled pendulums, the lower frequency ω_1 is independent of the coupling strength; this is not true of all coupled oscillations. What is generally true, however, is that the separation between ω_1 and ω_2 increases with the coupling strength, as shown in Fig. 4.5.

4.4 Forced Vibrations

We have seen how in a two-mass vibrating system, the motion of each mass can be described as a superposition of two normal modes. For free vibrations, the amplitudes and phases are determined by the initial conditions. In a system driven in steady state, on the other hand, the amplitudes and phases depend upon the driving frequency. Our intuition tells us that large amplitudes will occur when the driving frequency is close to one of the normal-mode frequencies.

Consider the two-mass system in Fig. 4.6. The normal-mode angular frequencies are $\omega_1 = \sqrt{K/m} = \omega_0$ and $\omega_2 = \sqrt{(K/m) + (2K_c/m)} = \sqrt{\omega_0^2 + 2\omega_c^2}$. Suppose that a driving force $F_0 \cos \omega t$ is applied to mass A. The equations of motion are

$$\ddot{x}_A + (\omega_0^2 + \omega_c^2)x_A - \omega_c^2 x_B = \frac{F_0}{m} \cos \omega t, \tag{4.10a}$$

and

$$\ddot{x}_B + (\omega_0^2 + \omega_c^2)x_B - \omega_c^2 x_A = 0. \tag{4.10b}$$

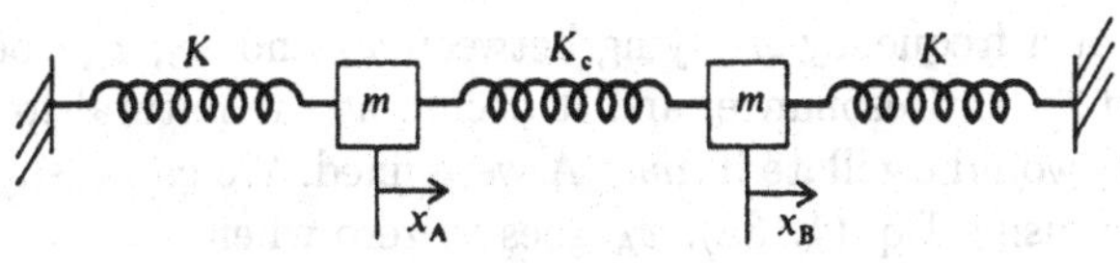

FIGURE 4.6. Two-mass oscillator (compare Fig. 4.1.).

Again, we introduce normal coordinates $q_1 = x_A + x_B$ and $q_2 = x_A - x_B$; we add and subtract Eqs. (4.10a) and (4.10b) to obtain

$$\ddot{q}_1 + \omega_1^2 q_1 = \frac{F_0}{m}\cos\omega t, \tag{4.11a}$$

and

$$\ddot{q}_2 + \omega_2^2 q_2 = \frac{F_0}{m}\cos\omega t, \tag{4.11b}$$

where $\omega_1^2 = \omega_0^2$ and $\omega_2^2 = \omega_0^2 + 2\omega_c^2$. Note that the same driving force appears in both normal-mode equations.

These are the equations of two harmonic oscillators with natural frequencies ω_1 and ω_2 and with no damping. The steady-state solutions are [see Eq. (1.39)]

$$q_1 = \frac{F_0/m}{\omega_1^2 - \omega^2}\cos\omega t, \tag{4.12a}$$

and

$$q_2 = \frac{F_0/m}{\omega_2^2 - \omega^2}\cos\omega t. \tag{4.12b}$$

The displacements of the masses A and B are

$$\begin{aligned} x_A &= \frac{q_1 + q_2}{2} = \frac{F_0}{2m}\frac{\omega_1^2 - \omega^2 + \omega_2^2 - \omega^2}{(\omega_1^2 - 0^2)(\omega_2^2 - \omega^2)}\cos\omega t \\ &= \frac{F_0}{m}\frac{(\omega_1^2 + \omega_c^2) - \omega^2}{(\omega_1^2 - \omega^2)(\omega_2^2 - \omega^2)}\cos\omega t, \end{aligned} \tag{4.13a}$$

and

$$\begin{aligned} x_B &= \frac{q_1 - q_2}{2} = \frac{F_0}{2m}\frac{\omega_2^2 - \omega^2 - \omega_1^2 + \omega^2}{(\omega_1^2 - 0^2)(\omega_2^2 - \omega^2)}\cos\omega t \\ &= \frac{F_0}{m}\frac{\omega_c^2}{(\omega_1^2 - \omega^2)(\omega_2^2 - \omega^2)}\cos\omega t, \end{aligned} \tag{4.13b}$$

The steady-state displacement amplitudes as functions of ω, from Eq. (4.13), are shown in Fig. 4.7(a). Below the first resonance ω_1, the displacements x_A and x_B are in the same direction, whereas above the second resonance ω_2, they are in opposite directions. With no damping, the amplitudes approach infinity at the resonance and the phase jumps abruptly. If damping were included, the phase would vary smoothly as one goes through the resonances.

Note that at a frequency ω_A lying between ω_1 and ω_2, x_A goes to zero. This is called an antiresonance, and it occurs at the natural frequency at which mass B would oscillate if mass A were fixed. We can easily calculate this frequency using Eq. (4.13a). x_A goes to zero when

$$\omega^2 = \omega_1^2 + \omega_c^2 = \omega_A^2 = \omega_2^2 - \omega_c^2. \tag{4.14}$$

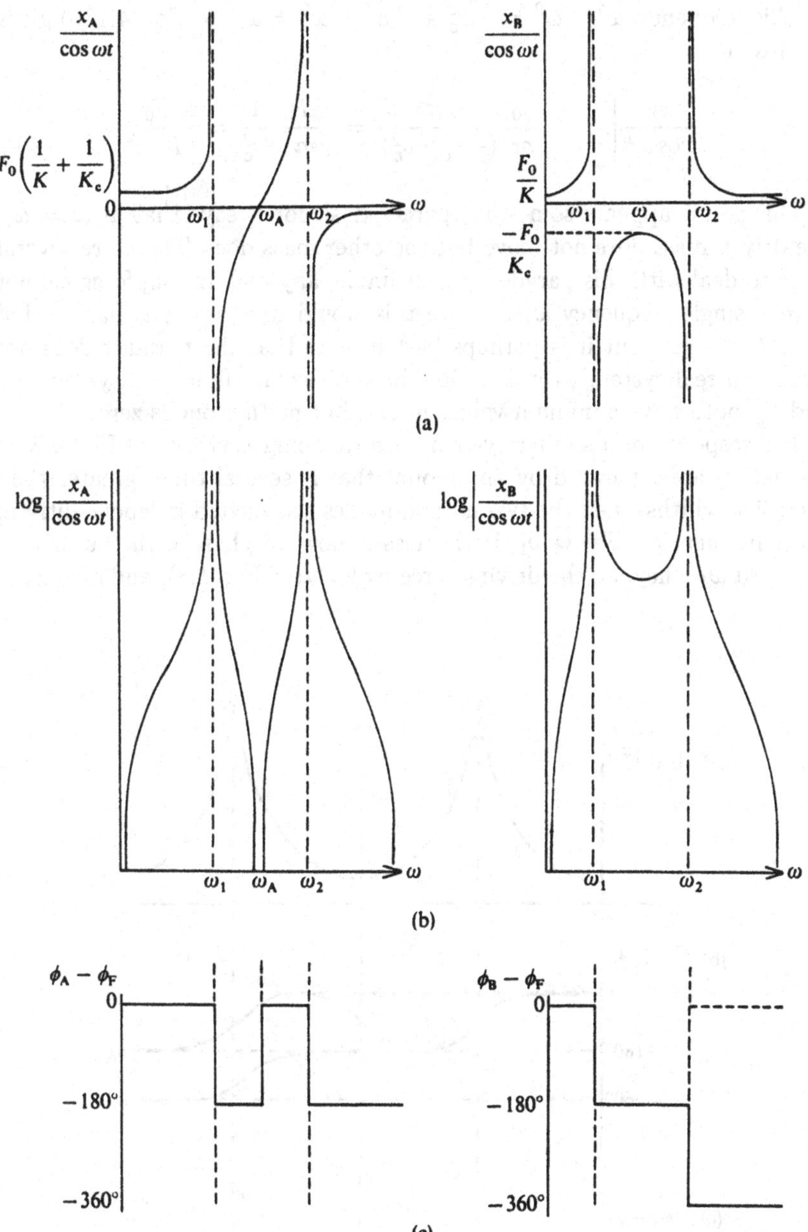

FIGURE 4.7. Frequency response of the two-mass coupled system in Fig. 4.6 to a driving force applied to mass A. (a) Amplitudes of x_A and x_B. (b) Absolute values of amplitude on a logarithmic scale. (c) Phases of masses A and B with respect to the driving force.

At this frequency, $\omega_1^2 - \omega^2 = -\omega_c^2$ and $\omega_2^2 - \omega^2 = \omega_c^2$, so Eq. (4.13b) gives the result

$$\left.\frac{x_B}{\cos\omega t}\right|_{\omega_A} = \frac{F_0}{m}\frac{\omega_c^2}{(-\omega_c^2)(\omega_c^2)} = \frac{-F_0}{m}\frac{1}{\omega_c^2} = \frac{-F_0}{K_c}.$$

This result appears somewhat paradoxical, for it says that at $\omega = \omega_A$ the driven mass does not move but the other mass does. There are several ways to deal with this paradox (for example, any real driving force cannot have a single frequency ω, because this would imply that it has existed since $t = -\infty$), but it is perhaps best to note that the paradox does not exist in a real system with damping, however small. In such a system, x_A and x_B both have minimum values at ω_A, but neither one is zero.

The response of a similar system with damping is shown in Fig. 4.8. If ω_1 and ω_2 are separated by an amount that is several times greater than their line widths, then the two normal modes are excited independently at these frequencies. Below ω_1, both masses move in phase with the driving force. At ω_1, they lag the driving force by 90° (see Fig. 1.8), and above ω_1,

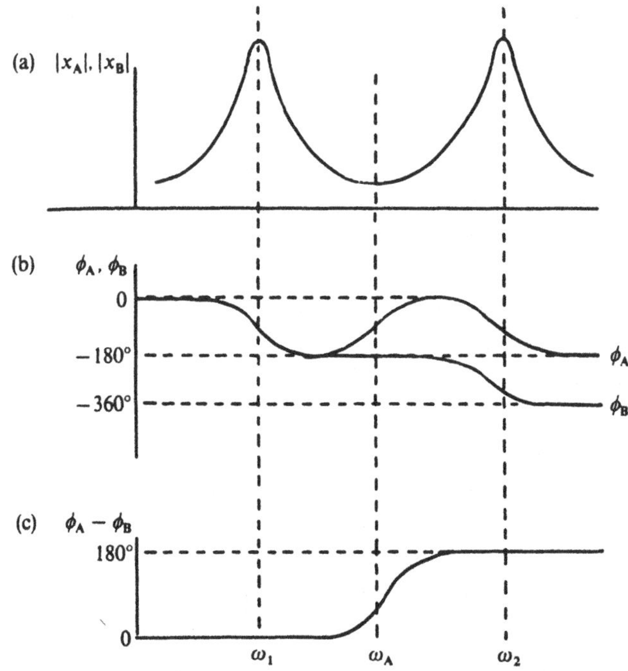

FIGURE 4.8. (a) Amplitude of mass A or mass B as a function of driving frequency when a force is applied to mass A. (b) Phases of masses A and B with respect to the driving force. (c) Phase difference between x_A and x_B.

the phase difference is 180°. Between ω_A and ω_2, mass A again moves in phase with the driving force, while above ω_2 mass B does so.

4.5 Coupled Electrical Circuits

We have already seen in Chapter 1 the usefulness of equivalent electrical circuits in understanding mechanical oscillatory systems. We will now consider some examples of coupled electrical circuits that are the bases of useful equivalent circuits.

The electrical circuit shown in Fig. 4.9 consists of two RLC circuits having natural frequencies ω_a and ω_b when there is no coupling:

$$\omega_a = \frac{1}{\sqrt{L_a C_a}} \quad \text{and} \quad \omega_b = \frac{1}{\sqrt{L_b C_b}}.$$

We consider the effect of coupling through the mutual inductance M existing between the two inductors. The voltage equations in the two loops are

$$L_a \frac{di_a}{dt} + M\frac{di_b}{dt} + R_a i_a + \frac{1}{C_a}\int i_a dt = 0,$$

and (4.15)

$$L_b \frac{di_b}{dt} + M\frac{di_a}{dt} + R_b i_b + \frac{1}{C_b}\int i_b dt = 0.$$

Differentiating and dividing by L_a and L_b gives

$$\ddot{i}_a + \frac{M}{L_a}\ddot{i}_b + \frac{R_a}{L_a}\dot{i}_a + \frac{1}{L_a C_a} i_a = 0,$$

and

$$\ddot{i}_b + \frac{M}{L_b}\ddot{i}_a + \frac{R_b}{L_b}\dot{i}_b + \frac{1}{L_b C_b} i_b = 0.$$

Assuming harmonic solutions $i_a = I_a e^{j\omega t}$ and $i_b = I_b e^{j\omega t}$, replacing $1/L_a C_a$ and $1/L_b C_b$ by ω_a^2 and ω_b^2, and assuming R_a and R_b to be small,

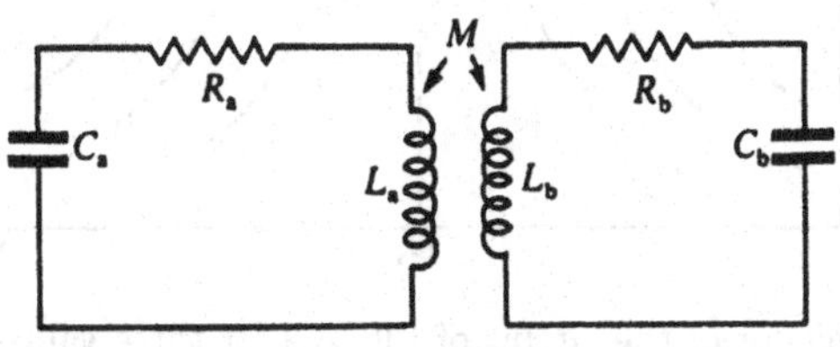

FIGURE 4.9. Two RLC circuits coupled by mutual inductance M.

we obtain

$$(\omega^2 - \omega_{\rm a}^2) I_{\rm a} = -\omega^2 \frac{M}{L_{\rm a}} I_{\rm b},$$

and (4.16)

$$(\omega^2 - \omega_{\rm b}^2) I_{\rm b} = -\omega^2 \frac{M}{L_{\rm b}} I_{\rm a}.$$

Multiplying these two equations together gives

$$(\omega^2 - \omega_{\rm a}^2)(\omega^2 - \omega_{\rm b}^2) = \omega^4 \frac{M^2}{L_{\rm a} L_{\rm b}} = k^2 \omega^4, \tag{4.17}$$

where $k^2 = M^2/L_{\rm a}L_{\rm b}$ is the coupling coefficient. Solving for ω gives the resonance frequencies.

A particularly simple case occurs when $\omega_{\rm a}^2 = \omega_{\rm b}^2$. Then,

$$(\omega^2 - \omega_{\rm a}^2)^2 = k^2 \omega^4,$$
$$\omega^2 - \omega_{\rm a}^2 = \pm k \omega^2,$$

and

$$\omega = \pm \frac{\omega_{\rm a}}{\sqrt{1 \pm k}}.$$

The two positive frequencies are

$$\omega_1 = \frac{\omega_{\rm a}}{\sqrt{1+k}} \quad \text{and} \quad \omega_2 = \frac{\omega_{\rm a}}{\sqrt{1-k}}. \tag{4.18}$$

The current amplitudes for three different values of coupling are shown in Fig. 4.10. With tight coupling, a pronounced dip occurs between the peaks at ω_1 and ω_2.

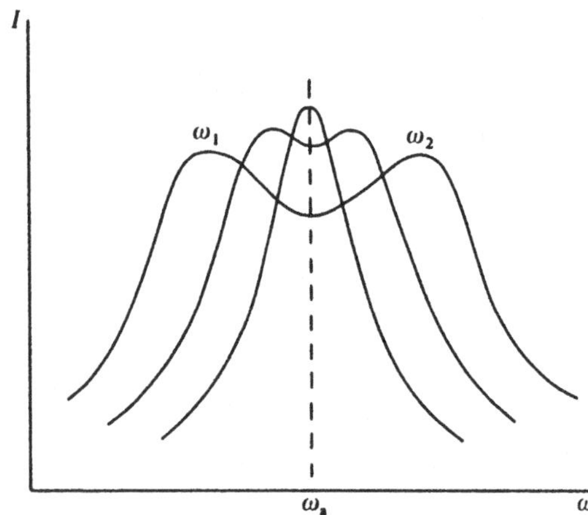

FIGURE 4.10. Behavior of the circuit of Fig. 4.9 for three values of the coupling constant k.

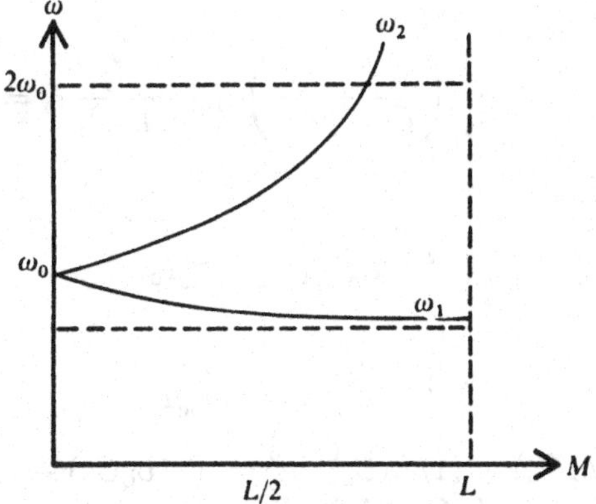

FIGURE 4.11. Variation of modal frequencies with mutual inductance M of identical coupled circuits.

If the two circuits are identical ($L_a = L_b = L$ and $C_a = C_b = C$),

$$\omega_1 = \frac{1}{\sqrt{C(L+M)}} \quad \text{and} \quad \omega_2 = \frac{1}{\sqrt{C(L-M)}}. \tag{4.19}$$

These values are shown in Fig. 4.11.

Next, we consider the circuit shown in Fig. 4.12, which consists of two LC circuits plus a coupling capacitor C_c. The differential equations are

$$L_a \ddot{i}_a + \frac{1}{C_a} i_a + \frac{1}{C_c}(i_a - i_b) = 0$$

and

$$L_b \ddot{i}_b + \frac{1}{C_b} i_b + \frac{1}{C_c}(i_b - i_a) = 0. \tag{4.20}$$

Again, assuming harmonic solutions, $i_a = I_1 e^{j\omega t}$ and $i_b = I_2 e^{j\omega t}$, leads to

$$-\omega^2 I_a + \frac{1}{L_a}\left(\frac{1}{C_a} + \frac{1}{C_c}\right) I_a - \frac{1}{L_a C_c} I_b = 0,$$

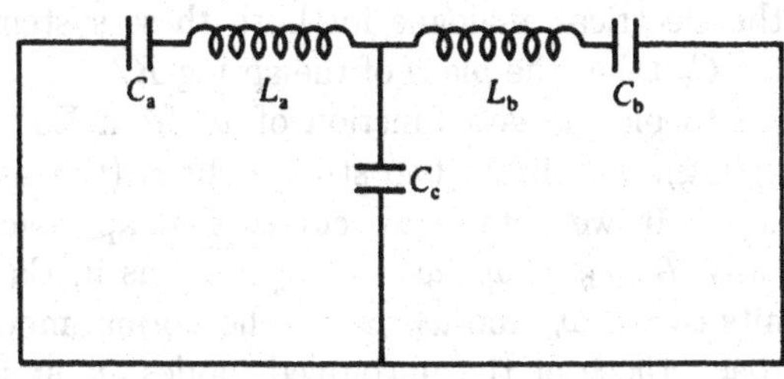

FIGURE 4.12. Two LC circuits coupled by a coupling capacitor C_c.

and

$$-\omega^2 I_b + \frac{1}{L_b}\left(\frac{1}{C_b} + \frac{1}{C_c}\right) I_b - \frac{1}{L_b C_c} I_a = 0,$$

from which

$$(\omega^2 - \omega_a^2) I_a = -\omega_{ac}^2 I_b,$$

and (4.21)

$$(\omega^2 - \omega_b^2) I_b = -\omega_{bc}^2 I_a,$$

where $\omega_a^2 = (1/L_a C_a) + (1/L_a C_c)$, $\omega_b^2 = (1/L_b C_b) + (1/L_b C_c)$, $\omega_{ac}^2 = 1/L_a C_c$, and $\omega_{bc}^2 = 1/L_b C_c$. Multiplying the above equations together gives

$$(\omega^2 - \omega_a^2)(\omega^2 - \omega_b^2) = \frac{1}{L_a L_b C_c^2} = \omega_{ac}^2 \omega_{bc}^2. \tag{4.22}$$

In the case $\omega_a = \omega_b$, we obtain

$$(\omega^2 - \omega_a^2) = \pm \omega_{ac}\omega_{bc} = \pm \omega_c^2,$$

from which $\omega = \sqrt{\omega_a^2 \pm \omega_c^2}$, and we obtain the two frequencies

$$\omega_1 = \sqrt{\omega_a^2 - \omega_c^2} = \omega_a \sqrt{1-k},$$

and (4.23)

$$\omega_2 = \sqrt{\omega_a^2 + \omega_c^2} = \omega_a \sqrt{1+k},$$

In the special case where $L_a = L_b = L$ and $C_a = C_b = C$, we can designate

$$\omega_1 = \sqrt{\omega_a^2 - \omega_c^2} \quad \text{as} \quad \omega_0,$$

so (4.24)

$$\omega_2 = \sqrt{\omega_0^2 + 2\omega_c^2},$$

as in the mechanical systems shown in Figs. 4.1 and 4.6. Thus, the circuit in Fig. 4.12 is the electrical analogue for both these systems, where the coupling capacitor C_c takes the place of the spring K_c.

It is instructive to plot ω as a function of ω_b from Eq. (4.22). When $\omega_c = 0$ (no coupling), we obtain two straight lines (the dashed lines in Fig. 4.13). For $\omega_c > 0$, we obtain two curves that approach the dashed lines asymptotically. At $\omega_b = \omega_a$, $\omega = \sqrt{\omega_a^2 \pm \omega_c^2}$ as in Eq. (4.23). Note that in both limits $\omega_b \ll \omega_a$ and $\omega_b \gg \omega_a$ the normal-mode frequencies ω_1 and ω_2 approach those of the uncoupled modes ω_a and ω_b, and the normal-modes resemble those of the uncoupled LC circuits.

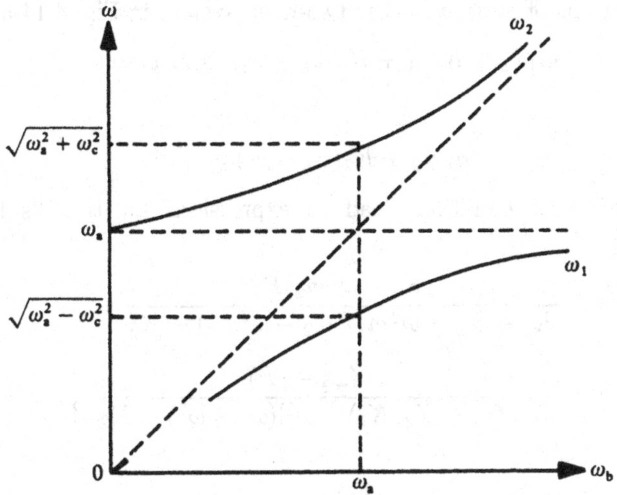

FIGURE 4.13. Normal-mode frequencies of the circuit in Fig. 4.12. The dashed lines indicate the uncoupled case ($k \to 0$ or $C_c \to \infty$).

4.6 Forced Vibration of a Two-Mass System

A coupled system with wide application is the two-mass system in which a sinusoidal driving force is applied to mass m_1, as shown in Fig. 4.14(a). This system is the prototype, for example, of a bass reflex loudspeaker system, a guitar with the ribs fixed, and the dynamic absorber used to damp machine vibrations. The equivalent electrical circuit is also shown in Fig. 4.14(b).

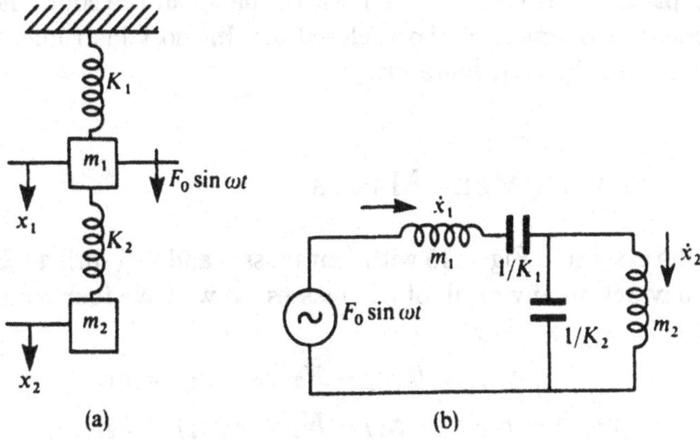

FIGURE 4.14. Two-mass vibrator (a) and its equivalent electrical circuit (b).

The equations of motion for the two-mass system in Fig. 4.14 are

$$m_1\ddot{x}_1 + K_1x_1 + K_2(x_1 - x_2) = F_0 \sin\omega t,$$

and

$$m_2\ddot{x}_2 + K_2(x_2 - x_1) = 0.$$

Solutions of these equations lead to expressions for the displacement amplitudes:

$$\begin{aligned} X_1 &= \frac{(K_2 - \omega^2 m_2)F_0}{(K_1 + K_2 - \omega^2 m_1)(K_2 - \omega^2 m_2) - K_2^2} \\ &= \frac{F_0(\omega_2^2 - \omega^2)}{m_1[\omega_1^2(1 + K_2/K_1) - \omega^2](\omega_2^2 - \omega^2) - K_2\omega_2^2}, \end{aligned} \tag{4.25a}$$

and

$$\begin{aligned} X_2 &= \frac{k_2F_0}{(K_1 + K_2 - \omega^2 m_1)(K_2 - \omega^2 m_2) - K_2^2} \\ &= \frac{F_0\omega_2^2}{m_1[\omega_1^2(1 + K_2/K_1) - \omega^2](\omega_2^2 - \omega^2) - K_2\omega_2^2}. \end{aligned} \tag{4.25b}$$

Like the two-mass system discussed in Section 4.4, this system has two resonances and one antiresonance. X_1 goes to zero at the antiresonance frequency $\omega = \omega_2$, and both X_1 and X_2 approach infinity when the denominator goes to zero.

A dynamic absorber consists of a small mass m_2 and spring K_2 attached to the primary vibrating system and selected so that $\omega_1 = \omega_2$. Then, the amplitude of mass m_1 goes to zero at the original resonance frequency ω_1. Similarly, a bass reflex speaker is often designed so that the resonance frequency ω_2 of the enclosure is the same as that of the loudspeaker cone ω_1. Then, the cone is restrained from moving at its resonance frequency.

In a guitar or violin, m_1 and K_1 represent the mass and spring constant of the top plate, m_2 is the effective mass of the air in the sound hole, and K_2 is the spring constant of the enclosed air. In most instruments, ω_2 is substantially less than ω_1, however.

4.7 Systems with Many Masses

Consider the system in Fig. 4.15 with four masses and five springs. External forces f_i may act on any or all of the masses, so we have four equations of motion:

$$\begin{aligned} m_1\ddot{x}_1 + K_1x_1 - K_2(x_2 - x_1) &= F_1, \\ m_2\ddot{x}_2 + K_2(x_2 - x_1) - K_3(x_3 - x_2) &= F_2, \\ m_3\ddot{x}_3 + K_3(x_3 - x_2) - K_4(x_4 - x_3) &= F_3, \end{aligned}$$

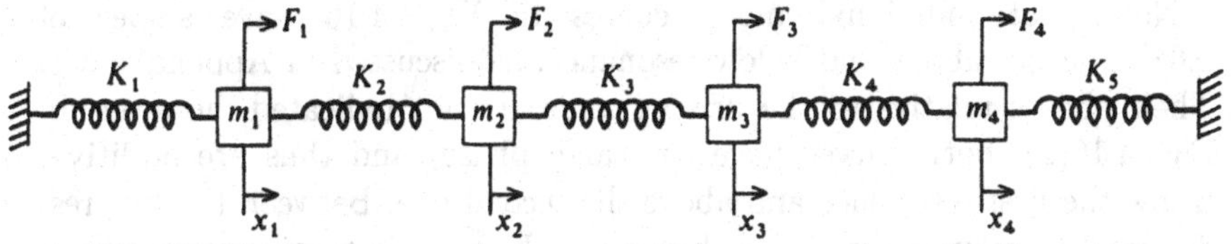

FIGURE 4.15. Four-mass vibrating system.

and

$$m_4\ddot{x}_4 + K_4(x_4 - x_3) + K_5x_4 = F_4. \tag{4.26}$$

Adding the four equations gives

$$\sum_{i=1}^{4} m_i\ddot{x}_i = \sum_{i=1}^{4} F_i - K_1x_1 - K_5x_4. \tag{4.27}$$

Assuming harmonic solutions, $x_i = X_i \sin\omega t$ leads to four algebraic equations, in which the coefficients of X_1 to X_4 form the determinant:

$$\Delta = \begin{vmatrix} d_1 & -K_2 & 0 & 0 \\ -K_2 & d_2 & -K_3 & 0 \\ 0 & 0 & d_3 & -K_4 \\ 0 & 0 & -K_4 & d_4 \end{vmatrix},$$

where $d_1 = K_1+K_2-m_1\omega^2$, $d_2 = K_2+K_3-m_2\omega^2$, $d_3 = K_3+K_4-m_2\omega^2$, and $d_4 = K_4 + K_5 - m_4\omega^2$. The amplitude X_1 is given by

$$X_1 = \frac{1}{\Delta} = \begin{vmatrix} F_1 & -K_2 & 0 & 0 \\ F_2 & d_2 & -K_3 & 0 \\ F_3 & -K_3 & d_3 & -K_4 \\ F_4 & 0 & -K_4 & d_4 \end{vmatrix},$$

and the other three amplitudes are given by analogous expressions (Jacobson and Ayre, 1958).

4.8 Graphical Representation of Frequency Response Functions

There are several ways to represent the frequency response function of a vibrating system (see the appendix to this chapter). One useful representation is a mobility plot illustrated in Fig. 4.16 for a lightly damped system with two degrees of freedom [a linear two-mass system as in Fig. 1.20(a), for example]. If a force F is applied to mass m_1, the driving-point mobility is $\dot{x}_1/F$ and the transfer mobility is $\dot{x}_2/F$.

Note that individual mode curves in Fig. 4.16 have slopes of ±6 dB/octave above and below resonance, as discussed in Appendix A.1. Their phases relative to the driving force are not indicated, however. In Fig. 4.16(a), both curves have the same phase (and thus are additive) below the first resonance and above the second one. Between the two resonances, they are subtractive, however, which leads to an antiresonance where they cross. The opposite situation occurs in the transfer mobility curves in Fig. 4.16(b), where the curves are additive between the resonances but subtractive elsewhere. [This same behavior is noted in Fig. 4.7(b).]

There is a general rule that if two consecutive modes have modal constants with the same sign, then there will be an antiresonance at some frequency between the natural frequencies of the two modes. If they have

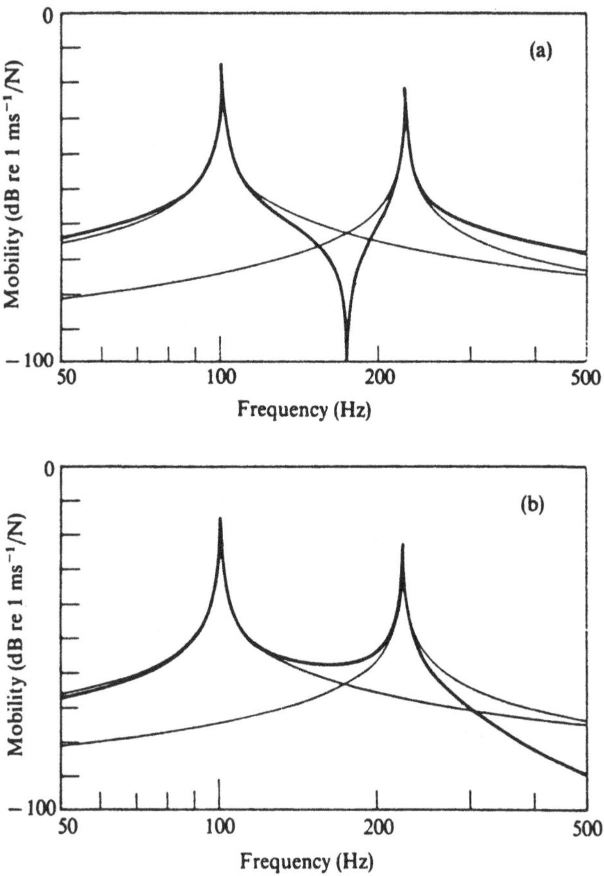

FIGURE 4.16. Mobility plot for a lightly damped system with two degrees of freedom: (a) driving-point mobility and (b) transfer mobility. The lighter curves indicate the contributions from the two normal modes of the system (after Ewins, 1984).

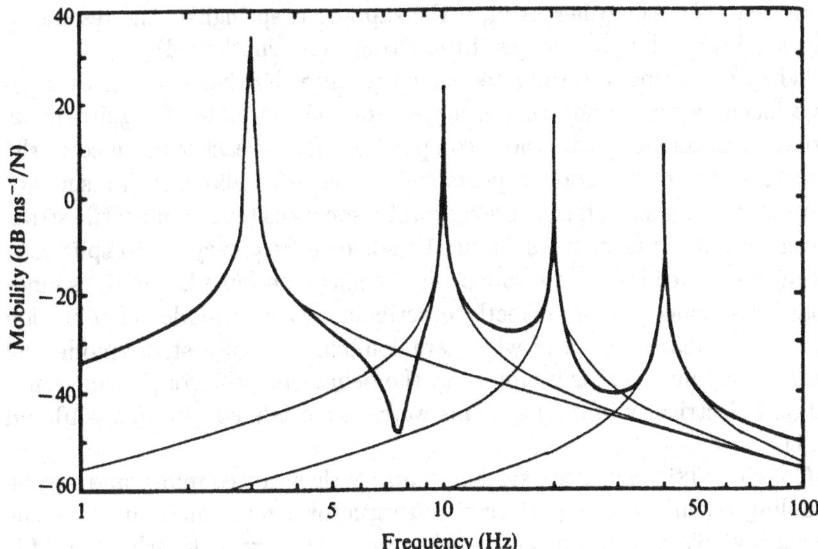

FIGURE 4.17. Example of a transfer mobility curve for a lightly damped structure with four degrees of freedom. The lighter curves indicate the contributions from the four normal modes, which contribute to the frequency response function (after Ewins, 1984).

opposite signs, there will not be an antiresonance, but just a minimum, as in Fig. 4.16(b). The modal constant is the product of two eigenvector elements, one at the drive point and one at the response point. In the case of driving-point mobility, the two points are the same, so the modal constant must be positive. This leads to alternating resonances and antiresonances, as appear in the driving-point admittance (mobility) curves in Fig. 3.22 and also in the driving-point impedance (reciprocal of mobility) curve in Fig. 3.21.

Figure 4.17 shows an example of a transfer mobility curve of a system with four degrees of freedom (four normal modes). The compliance of this structure is positive over part of the frequency range and negative over part of it. The modal constant in this example changes sign from the first to the second mode (resulting in an antiresonance) but not from the second to the third mode for the driving and the observation points selected.

4.9 Vibrating String Coupled to a Soundboard

In Section 2.12, we considered a string terminated by a nonrigid end support (such as the bridge of a string instrument). If the support is masslike, the resonance frequencies will be raised slightly (as if the string were shortened). On the other hand, if the support is springlike, the resonance frequencies will be lowered (as if the string were lengthened).

When the string is terminated at a structure that has resonances of its own (such as the soundboard of a piano or the top plate of a guitar), the situation becomes a little more complicated. Below each resonance of the structure, the termination appears springlike, while above each resonance it appears masslike. Thus, the structural resonances tend to push the string resonances away from the structural resonance frequency, or to split each string resonance into two resonances, one above and one below the soundboard resonance. More correctly described, two new modes of vibration have been created, both of which are combinations of a string mode and a structural mode. This is similar to the behavior of the coupled mechanical and electrical vibrating systems we have considered (see Fig. 4.10 and 4.11).

Gough (1981) discusses systems both with weak coupling and strong coupling when $\omega_n = \omega_\mathrm{B}$ (string and structural resonances at the same frequency). Weak coupling occurs when $m/n^2M < \pi^2/4Q_\mathrm{B}^2$, where m/M is the ratio of the string mass to the effective mass of the structural resonance, n is the number of the resonant mode excited on the string, and Q_B is the Q value of the structural resonance.

In the weak coupling limit, the coupling does not perturb the frequencies of the two normal modes (when the unperturbed frequencies of the string and soundboard coincide). However, the damping of the two modes is modified by the coupling.

In the strong coupling limit $m/n^2M > \pi^2/4Q_\mathrm{B}^2$, however, the coupling splits the resonance frequencies of the normal modes symmetrically about the unperturbed frequencies and both modes now have the same Q value of $2Q_\mathrm{B}$. At the lower frequency, the string and soundboard move in phase, whereas at the higher frequency, they move in opposite phase.

Figure 4.18 shows the frequency splitting as a function of m/n^2M when the string and soundboard mode frequencies coincide ($\omega_n = \omega_\mathrm{B}$). Ω_+ and Ω_- are the mode frequencies in the coupled system, and n is the number of the string mode.

The frequencies of the normal modes, when the resonance frequencies of the string and soundboard are different, are shown in Fig. 4.19 for the weak and strong coupling cases. The mode frequencies are given by the solid curves and the half-widths of the resonances by the dashed curves (Gough, 1981).

4.10 Two Strings Coupled by a Bridge

In most string instruments, several strings (from 4 in a violin to more than 200 in a piano) are supported by the same bridge. This leads to coupling between the strings, which may be strong or weak, depending upon the

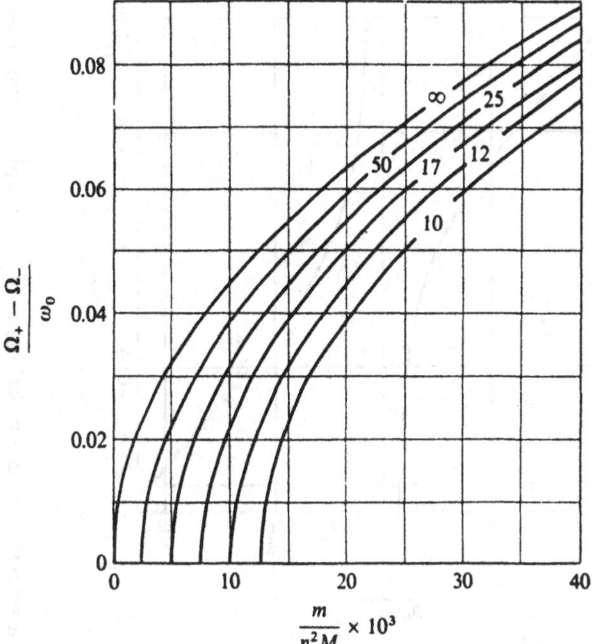

FIGURE 4.18. Normal mode splitting as a function of the ratio of string mass m to soundboard mass M times the mode number n on the string when the resonance frequencies of the string and soundboard mode are the same. Ω_+ and Ω_- are the mode angular frequencies in the coupled system, and Q-values for the soundboard appear on each curve (Gough, 1981).

relative impedances of the bridge and strings. The discussion of coupled piano strings by Weinreich (1977) and the discussion of violin strings by Gough (1981) are especially recommended.

Although an exact description of the interaction of two strings coupled to a common bridge would require the solution of three simultaneous equations describing the three normal modes of the system in the coupling direction, the problem can be simplified by recognizing that the string resonances are generally much sharper than those of the bridge and soundboard. Thus, close to a string resonance, the impedance of the bridge can be considered to be a slowly varying function of frequency.

We consider first the case of two identical strings. When the impedance of the bridge is mainly reactive, the coupling produces a repulsion between the frequencies of the normal modes, as shown in Fig. 4.20(a). [This is quite similar to the behavior of the string–soundboard coupling in Fig. 4.19(b).] When the impedance of the bridge is mainly resistive, however (as it will be at a resonance of the soundboard), the frequencies of the normal modes coalesce over a region close to the crossover frequency, as shown

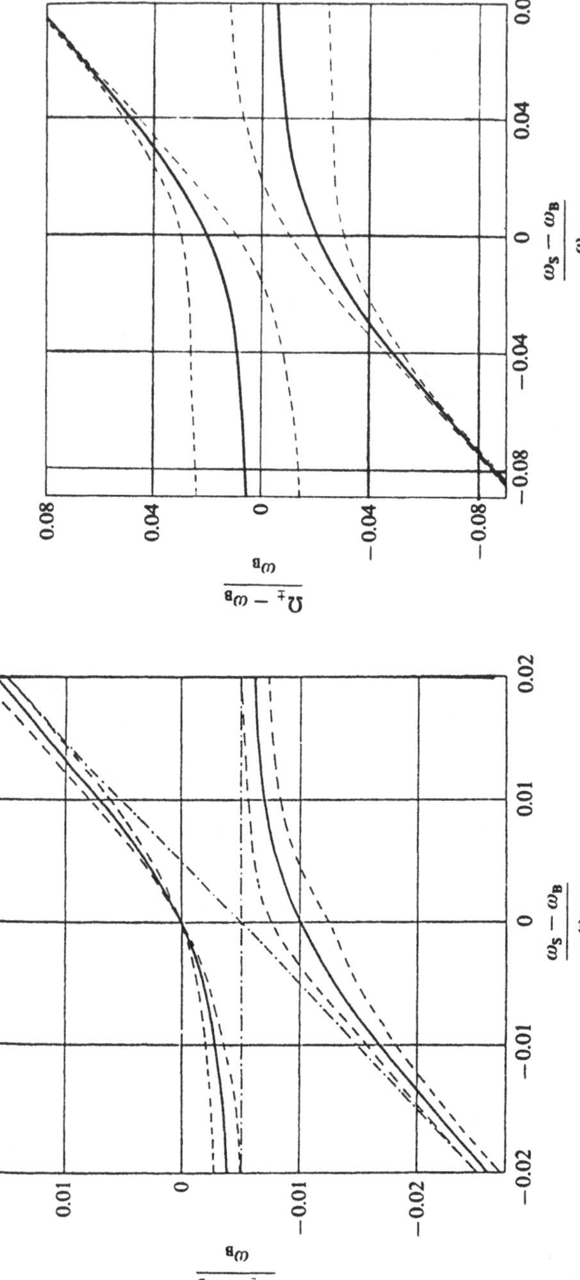

FIGURE 4.19. Normal mode frequencies of a string coupled to a soundboard as a function of their uncoupled frequencies ω_n and ω_B: (a) weak coupling and (b) strong coupling (Gough, 1981).

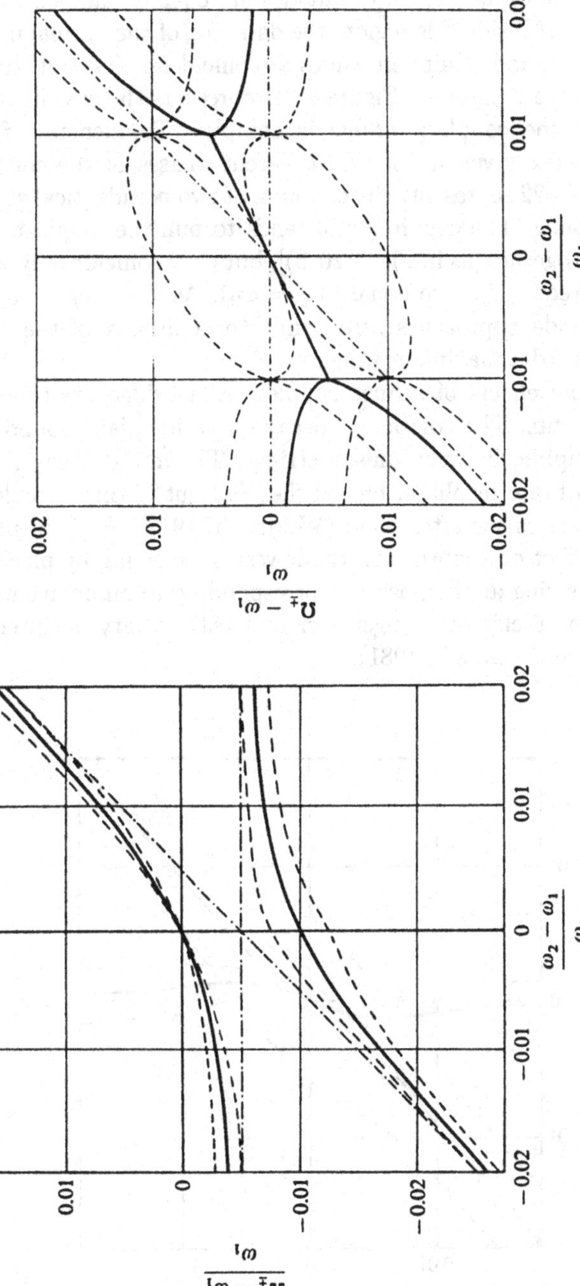

FIGURE 4.20. The normal modes of two identical strings coupled together by a bridge: (a) reactive coupling and (b) resistive coupling. ω_2 and ω_1 are the unperturbed string frequencies; Ω_+ and Ω_- are the frequencies of the coupled system. The solid curves are the frequencies and the dashed curves give the half-widths. The dash-dot lines give the frequencies of the strings without coupling (Gough, 1981).

in Fig. 4.20(b). Outside this region, the modes are equally damped by the bridge impedance, but inside this region, the damping of the normal modes approaches maximum and minimum values at coincidence (Gough, 1981).

Additional cases are of interest. Figure 4.21 represents the case in which the admittance of the coupling bridge is complex. Frequencies of the two normal modes are given for several different phases of the coupling admittance. Figure 4.22 represents the coupling of two nonidentical strings with resistive coupling. The coupling still tends to pull the frequencies of the normal modes together [as in Fig. 4.20(b)], but they coincide only when the unperturbed frequencies are equal ($\omega_1 = \omega_2$). As $\omega_2 - \omega_1 \to 0$, the damping of one mode approaches zero, while the damping of the other mode increases toward a maximum value.

Several interesting effects of strings coupled by a bridge are observed in musical instruments. The compound decay curve for piano sound is a direct result of coupling between unison strings. The initial decay is fast while the strings vibrate in phase, but as they fall out of phase, a slower decay rate characterizes the aftersound (Weinreich, 1977). A violin player can simulate the effect of a vibrato on the lowest open string by placing a finger on the next string at the position corresponding to an octave above the bowed string and rocking the finger back and forth to vary the intensity of the octave harmonic (Gough, 1981).

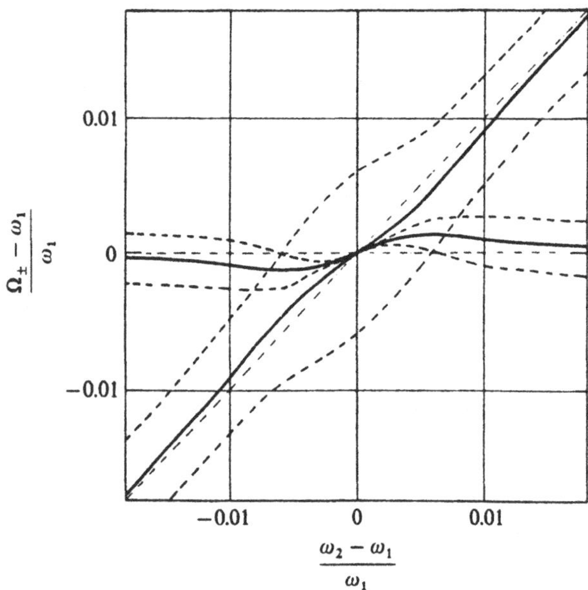

FIGURE 4.21. The normal modes of two nonidentical strings with resistive coupling. Solid curves give the frequencies and dashed curves the line widths (Gough, 1981).

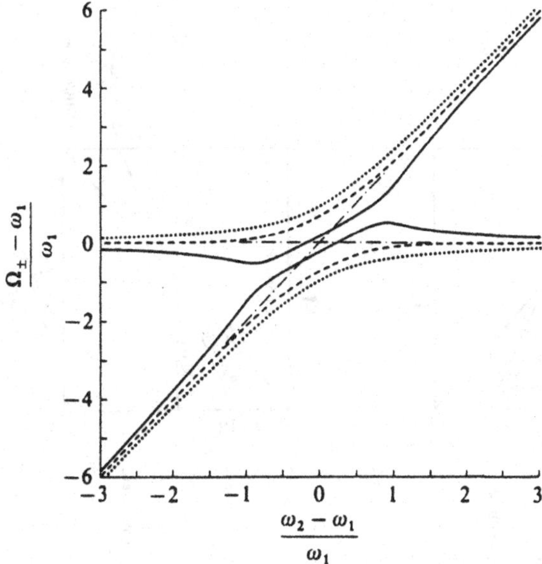

FIGURE 4.22. The normal modes of a two-string system coupled by a bridge with a complex impedance. The dotted curve represents the greatest reactance, the solid curve the least (after Weinreich, 1977).

APPENDIX

A4.1 Structural Dynamics and Frequency Response Functions

To determine the dynamical behavior of a structure in the laboratory, we frequently apply a force $\mathbf{F}$ at some point (x, y, z) and determine the response of the structure at the same point or some other point (x', y', z'). To describe the response, we may measure displacement $\mathbf{r}$, velocity $\mathbf{v}$, or acceleration $\mathbf{a}$. In the simplest case, $\mathbf{F}$, $\mathbf{r}$, $\mathbf{v}$, and $\mathbf{a}$ are in the same direction, so we speak of F, x, v, and a. Some examples of methods used to measure these variables are as follows:

F can be measured with a load cell or force transducer;
a can be measured with an accelerometer;
v can be measured with a phonograph cartridge and stylus or it can be determined by integrating a; probing the near-field sound with a microphone provides a pretty good estimate of v at a point on a surface nearby; and
x can be determined by holographic interferometry or by integrating v or a.

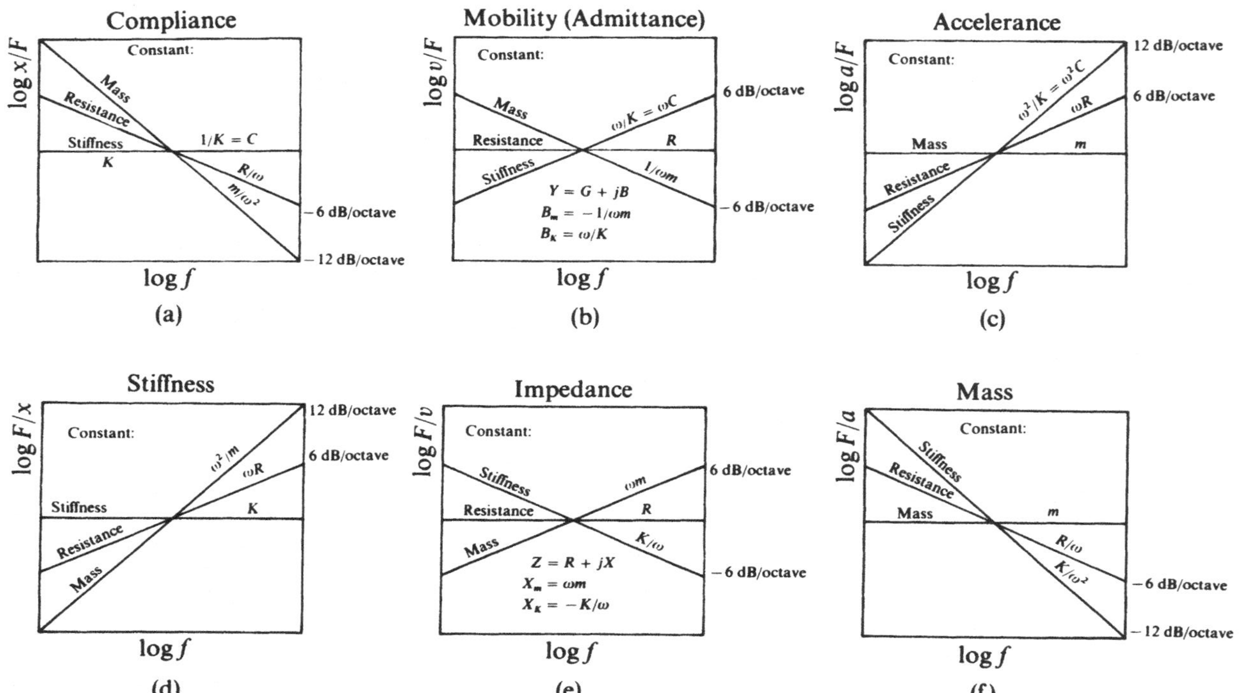

FIGURE A4.1. Six examples of frequency response functions showing (a) compliance, (b) mobility, (c) accelerance, (d) stiffness, (e) impedance, and (f) dynamic mass as functions of frequency. Individual curves are drawn for single elements having constants mass, resistance, and stiffness, respectively.

From these measured variables, we can construct one or more frequency response functions of interest. These include mobility (v/F), accelerance (a/F), compliance (x/F), impedance (F/v), dynamic mass (F/a), and stiffness (F/x). The logarithmic minigraphs in Fig. A.4.1 illustrate how these relate to the static parameters stiffness K, mass m, and resistance R. If $(x, y, z) = (x', y', z')$, we use the prefix driving point (e.g., driving-point mobility); otherwise, we use the prefix transfer.

Combining all three elements (mass, stiffness, and resistance) into an oscillator having a single degree of freedom with light damping leads to the frequency response functions in Fig. A.4.2 (only three of the six functions are shown). Note the ± 6 dB/octave slopes at high and low frequency in the mobility plot, the -12 dB/octave slope in the compliance plot, and the $+12$ dB/octave slope in the accelerance plot, consistent with Fig. A.4.1.

Figures A.4.1 and A.4.2 show logarithms of absolute values of the parameters of interest and therefore are devoid of information about phase. To represent the phase as well as magnitude of the frequency response function of interest, a second graph is often added, as in Fig. A.4.3. At a resonance, the phase changes by 180° (0 to 180° in the compliance plot, 90° to 270° in the mobility plot, or 180° to 360° in the accelerance plot), as indicated in Figs. 1.12 or 4.8.

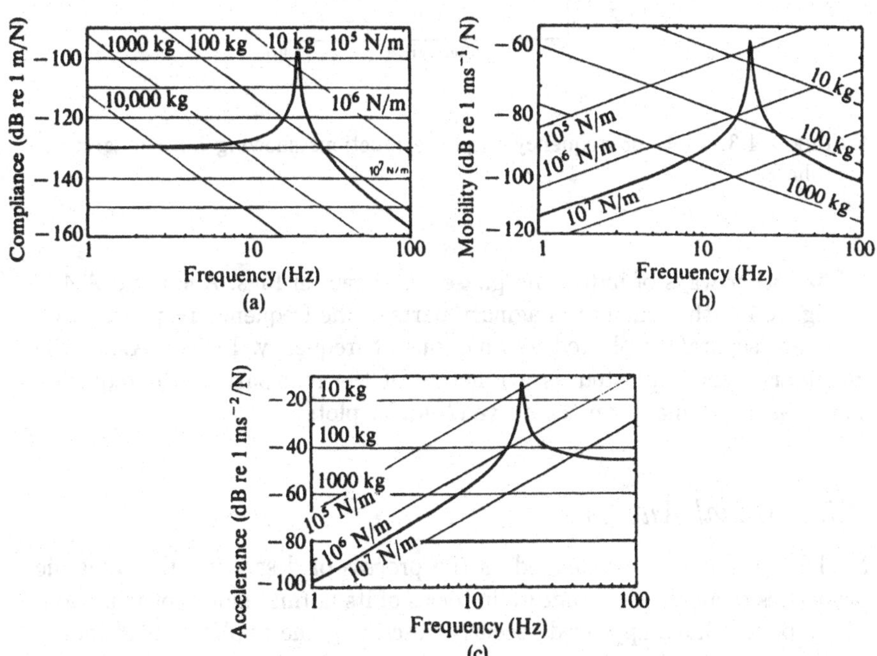

FIGURE A4.2. Three examples of frequency response functions of a lightly damped oscillator with a single degree of freedom (after Ewins, 1984).

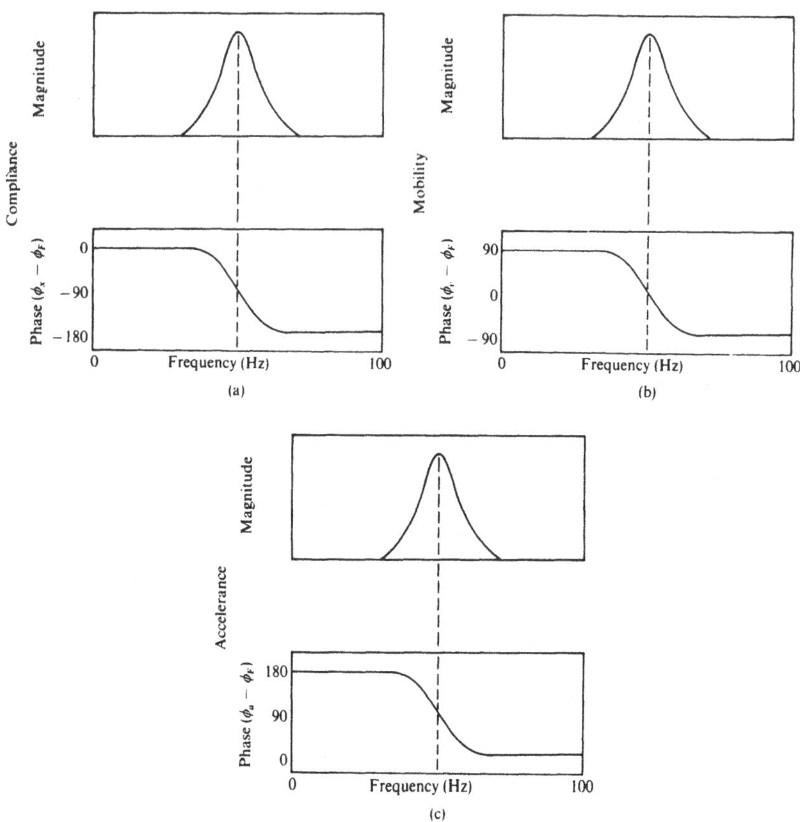

FIGURE A4.3. Plots of frequency response functions showing both magnitude and phase.

Two other ways of indicating phase are shown in Figs. A.4.4 and A.4.5. In Fig. A.4.4, the real and imaginary parts of the frequency response functions are separately plotted as functions of frequency. In Fig. A.4.5, the imaginary part is plotted as a function of the real part, with frequency shown as a parameter on the curve (Nyquist plots).

A4.2 Modal Analysis

Modal analysis may be defined as the process of describing the dynamic properties of an elastic structure in terms of its normal modes of vibration. Many papers have appeared recently describing the application of modal analysis to a wide range of structures from fresh apples to large aircraft. Papers on modal analysis generally deal with either mathematical modal analysis or modal testing.

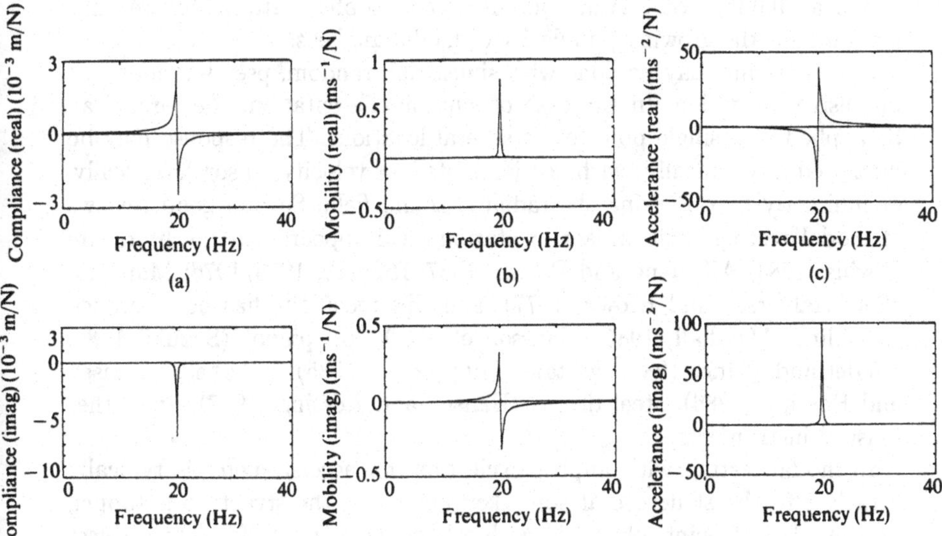

FIGURE A4.4. Real and imaginary parts of several frequency response functions: (a) compliance, (b) mobility, and (c) accelerance.

In mathematical modal analysis, one attempts to uncouple the structural equations of motion by means of some suitable transformation, so that the uncoupled equations can be solved. The frequency response of the structure can then be found by summing the respective modal responses in accordance with their degree of participation in the structural motion.

In experimental modal testing, one excites the structure at one or more points and determines the response at one or more points. From these sets of data, the natural frequencies (eigenfrequencies), mode shapes (eigenfunctions), and damping parameters are determined, often by the use of multidimensional curve-fitting routines on a digital computer. In fact it

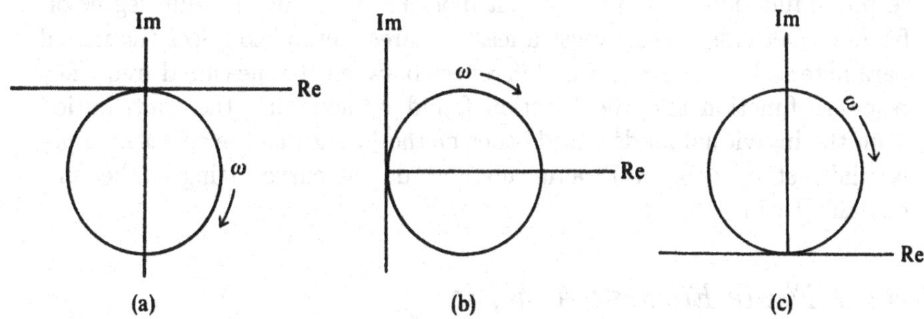

FIGURE A4.5. Nyquist plots show real and imaginary parts of (a) compliance, (b) mobility, and (c) accelerance.

is the availability of digital computers and sophisticated software that accounts for the growing popularity of modal analysis.

Modal testing may be done with sinusoidal, random, pseudorandom, or impulsive excitation. In the case of sinusoidal excitation, the force may be applied at a single point or at several locations. The response may be measured mechanically (with accelerometers or velocity sensors), optically, or indirectly by observing the radiated sound field. Several good reviews of modal testing with impact excitation have appeared in the literature (Ewins, 1984; Allemang and Brown, 1987; Ramsey, 1975/1976; Marshall, 1986; Halvorsen and Brown, 1977), and this technique has been applied to violins (Marshall, 1985; Jansson et al., 1986), pianos (Suzuki, 1986; Kindel and Wang, 1987), guitars (Popp et al., 1985), handbells (Hansen and Rossing, 1986), steel drums (Hansen and Rossing, 1987), and other musical instruments.

In modal testing with impact excitation, an accelerometer is typically attached to the structure at some key point and the structure is tapped at a number of points on a grid with a hammer having a force transducer or load cell. Each force and acceleration waveform is Fourier transformed, and a transfer function $H(\omega)$ is calculated. If a force is applied at i and the response is measured at j, the transfer function $H_{ij}(\omega)$ gives the best estimate of the frequency response function

$$H_{ij}(\omega) = \frac{S_i^*(\omega)S_j(\omega)}{S_i^*(\omega)S_i(\omega)} = \frac{G_{ij}}{G_{ii}},$$

where $S_i^*(\omega)$ is the complex conjugate of the force spectrum $S_i(\omega)$, G_{ii} is the power spectrum of the exciting force, and G_{ij} is the cross spectrum of the force and response.

Several different algorithms may be used to extract the modal parameters from the measured frequency response functions. Single-degree-of-freedom methods include a "peak picking" method, which uses the imaginary (quadrature) component of the response function as the modal coordinate, and a "circle fit" method, which fits the best circle to the data in the Argand plane [whose coordinates are the real and imaginary parts of the response function, as in the Nyquist plot in Fig. A.4.5(a)]. Multidegree-of-freedom methods generally use a least-squares method to select the modal parameters that minimize the differences between the measured frequency response function and the function found by summing the contribution from the individual modes. Still other methods, such as the complex exponential method and the Ibrahim method, do the curve fitting in the time domain (Ewins, 1984).

A4.3 Finite Element Analysis

The finite element method is a powerful numerical analysis method that can be used to calculate the vibrational modes of elastic structures. The method assumes that a structure or system can be modeled by an assemblage of building blocks (called elements) connected only at discrete points (called nodes). A complex structure is often divided into a number of familiar substructures, such as plates, beams, shells, and lumped masses.

This concept of modeling a system as a collection of discrete points was introduced by Courant (1943). His suggestion, that the Rayleigh–Ritz approach using an assumed response function for the system could be applied to triangular elements, became the mathematical basis for finite element analysis. The actual finite element terminology was later introduced by Clough (1960) and others. Clough's method became widely adopted because digital computers were available to perform the complex numerical calculations.

Finite element analysis is essentially an extension of matrix structural analysis methods that have been applied to beams and trusses for some time. This analysis is based on a set of equations of the form

$$(\mathbf{u})(\mathrm{K}) = (\mathbf{F}),$$

where $(\mathbf{u})$ is a displacement vector, $(\mathbf{F})$ is a force vector, and $(\mathbf{K})$ is the stiffness matrix in which a typical element K_{ij} gives the force F_i at the ith node due to a unit displacement u_j at the jth node.

General purpose finite element codes, such as NASTRAN, ANSYS, SAP, and ADINA, include routines for solving the equations of motion in matrix form.

$$(\mathbf{M})(\ddot{u}) + (\mathbf{R})(\dot{u}) + (\mathbf{K})(u) = (\mathbf{F})\cos(\omega t + \gamma),$$

where $(\mathbf{M})$, $(\mathbf{R})$, and $(\mathbf{K})$ are the mass, damping and stiffness matrices. Smaller programs adapted from these large, general purpose systems have made it possible to apply finite element methods to structures of modest size using microcomputers.

To improve the efficiency of finite element calculations, so-called "eigenvalue economizer" routines are often used. These routines reduce the size of the dynamical matrix by condensing it around master nodes (Rieger, 1986). Guidance in the selection of such nodes can often be obtained from the results of modal testing. In this and others ways, modal analysis and finite element analysis have become complementary methods for studying the dynamical behavior of large and small structures, including musical instruments.

References

Allemang, R.J., and Brown, D.L. (1987). Experimental modal analysis. In *"Handbook of Experimental Mechanics"* (A.S. Kabayashi ed.). Prentice-Hall, Englewood Cliffs, New Jersey.

Clough, R.W. (1960). The finite method in plane stress analysis. *Proc. 2nd ASCE Conf. on Electronic Computation*, Pittsburgh, Pennsylvania.

Courant, R. (1943). Variational methods for the solution of problems of equilibrium and vibrations. *Bull. Am. Math. Soc.* **49**, 1.

Ewins, D.J. (1984). "Modal Testing: Theory and Practice." Research Studies Press, Letchworth, England.

French, A.P. (1971) "Vibrations and Waves." W.W. Norton, New York.

Gough, C.E. (1981). The theory of string resonances on musical instruments. *Acustica* **49**, 124–141.

Halvorsen, W.G., and Brown, D.L. (1977). Impulse techniques for structural frequency response testing. *Sound and Vibration* **11** (11), 8–21.

Hansen, U.J., and Rossing, T.D. (1986). Modal analysis of a handbell (abstract). *J. Acoust. Soc. Am.* **79**, S92.

Hansen, U.J., and Rossing, T.D. (1987). Modal analysis of a Caribbean steel drum (abstract). *J. Acoust. Soc. Am.* **82**, S68.

Jacobson, L.S., and Ayre, R.S. (1958). "Engineering Vibrations." McGraw-Hill, New York.

Jansson, E., Bork, I., and Meyer, J. (1986). Investigation into the acoustical properties of the violin. *Acustica* 62, 1–15.

Kindel, J., and Wang, I. (1987). Modal analysis and finite element analysis of a piano soundboard, *Proc. 5th Int'l Conf. on Modal Analysis* (IMAC), 1545–1549.

Marshall, K.D. (1986). Modal analysis of a violin. *J. Acoust. Soc. Am.* **77**, 695–709.

Marshall, K.D. (1987). Modal analysis: A primer on theory and practice. *J. Catgut Acoust. Soc.* 46, 7–17.

Popp, J., Hansen, U., Rossing. T.D., and Strong, W.Y. (1985). Modal analysis of classical and folk guitars (abstract). *J. Acoust. Soc. Am.* **77**, S45.

Ramsey, K.A. (1975/1976). Effective measurements for structural dynamics testing. *Sound and Vibration* **9** (11), 24–34; **10** (4), 18–31.

Rieger, N.F. (1986). The relationship between finite element analysis and modal analysis. *Sound and Vibration* **20** (1), 20–31.

Suzuki, H. (1986). Vibration and sound radiation of a piano soundboard. *J. Acoust. Soc. Am.* **80**, 1573–1582.

Weinreich, G. (1977). Coupled piano strings. *J. Acoust. Soc. Am.* **62**, 1474–1484.

5

Nonlinear Systems

Many of the mechanical elements constituting a musical instrument behave approximately as linear systems. By this we mean that the acoustic output is a linear function of the mechanical input, so that the output obtained from two inputs applied simultaneously is just the sum of the outputs that would be obtained if they were applied separately. For this statement to be true for the instrument as a whole, it must also be true for all its parts, so that deflections must be proportional to applied forces, flows to applied pressures, and so on. Mathematically, this property is reflected in the requirement that the differential equations describing the behavior of the system are also linear, in the sense that the dependent variable occurs only to the first power. An example is the equation for the displacement y of a simple harmonic oscillator under the action of an applied force $F(t)$:

$$m\frac{d^2y}{dt^2} + R\frac{dy}{dt} + Ky = F(t), \tag{5.1}$$

where m, R, and K are, respectively, the mass, damping coefficient, and spring coefficient, all of which are taken to be constants. Then, if $y_1(t)$ is the solution for $F(t) = F_1(t)$, and $y_2(t)$ that for $F(t) = F_2(t)$, the solution for $F = F_1 + F_2$ will be $y = y_1 + y_2$.

A little consideration shows, of course, that this description must be an over-simplification (Beyer, 1974; Hagedorn, 1995). Mass is indeed conserved (apart from relativistic effects), but spring coefficients cannot remain constant when displacements approach the original dimensions of the system, nor can we expect damping behavior to remain unchanged when turbulence or other complicated effects intervene. It is therefore important to know how to treat such nonlinearities mathematically so that we can make use of these techniques when we come to examine the behavior of real musical instruments. Mathematically, the problem is one of solving equations like Eq. (5.1) when at least one of the coefficients m, R, or K depends on the dependent variable y. In interesting practical cases, we will nearly always be concerned with small deviations from linearity, so that these coefficients can be expanded as rapidly convergent power series in y or dy/dt, provided y is small compared with the dimensions of the system.

One sort of problem exemplified by Eq. (5.1) is a percussion instrument or a plucked string instrument in which the forcing function $F(t)$ is external, of limited duration, and given quite explicitly. The behavior of such impulsively excited oscillators or resonators is relatively simple for oscillation amplitudes that are not too large, as we shall see presently, but can embrace a wide range of chaotic regimes at large excitation.

More complex and inherently more essentially nonlinear is the situation encountered in self-excited steady-tone instruments, such as bowed strings or windblown pipes. Here, the forcing function $F(t)$ consists of a steady external part (the bow velocity or the blowing pressure) whose effect on the system is somehow determined by the existing amplitude of the oscillation. Thus, for example, the force between a bow and a string depends upon their relative velocities, while the flow through a reed valve depends upon the pressure difference across it. In this case, Eq. (5.1) is generalized to

$$m\frac{d^2y}{dt^2} + R\frac{dy}{dt} + Ky = F\left(y, \frac{dy}{dt}, t\right), \tag{5.2}$$

where again m, R, and K may be weak functions of y. For such a system, as we shall see presently, the whole behavior depends quite crucially upon the various nonlinearities present in the coefficients and in the function F.

The whole field of nonlinear and chaotic dynamics has grown so strongly in recent years that we can examine only that small portion that is clearly relevant to the behavior of musical instruments. The reader seeking a broader perspective should consult one of the excellent books available, such as that by Moon (1992).

5.1 A General Method of Solution

The systems we shall meet in musical instruments will ultimately prove to be much more complex than described by an equation like Eq. (5.2) since musical oscillators such as strings, plates or air columns generally have infinitely many possible vibrational modes rather than just a single one, but we can learn a great deal as a necessary preliminary by studying the nonlinear oscillator described by Eq. (5.2). In fact, musical instruments generally consist of a nearly linear resonator, the string, plate, or air column, described by the left-hand side of Eq. (5.2), excited by a generator F, which has quite nonlinear behavior. The nonlinearity in F is crucial to the description of the system.

For convenience of notation, we rewrite Eq. (5.2) in the form

$$\ddot{y} + \omega_0^2 y = g(y, \dot{y}, t), \tag{5.3}$$

where a dot signifies d/dt, ω_0 is the resonant frequency,

$$\omega_0 = \left(\frac{K}{m}\right)^{1/2}, \tag{5.4}$$

and

$$g = \frac{F - R\dot{y}}{m} \equiv f - 2\alpha\dot{y}, \tag{5.5}$$

where we have written $f \equiv F/m$ and $2\alpha \equiv R/m$.

If the damping, nonlinearity, and driving force on the system are all small, so that $g \to 0$ in Eq. (5.3), then the solution has the sinusoidal form

$$y(t) = a\sin(\omega_0 t + \phi), \tag{5.6}$$

where a is the amplitude and ϕ is the phase of the oscillation. If g is not identically zero but is small compared with the terms on the left side of Eq. (5.3), then it is reasonable to suppose that the true solution may have a form like

$$y(t) = a(t)\sin[\omega_0 t + \phi(t)], \tag{5.7}$$

where a and ϕ are both slowly varying functions of time. Now from Eq. (5.7), it is clear that

$$\dot{y} = \dot{a}\sin(\omega_0 t + \phi) + a(\omega_0 + \dot{\phi})\cos(\omega_0 t + \phi). \tag{5.8}$$

This is a rather complicated expression, and it is clear that a given behavior of y and $\dot{y}$ as functions of time could be described in several ways depending on how this functional dependence was partitioned between $\dot{a}$ and $\dot{\phi}$. There is a great simplification if we can arrange matters so that

$$\dot{y} = a\omega_0\cos(\omega_0 t + \phi), \tag{5.9}$$

which requires that

$$\dot{a}\sin(\omega_0 t + \phi) + a\dot{\phi}\cos(\omega_0 t + \phi) = 0. \tag{5.10}$$

If we assume Eq. (5.10) to have been satisfied and substitute Eqs. (5.7) and (5.9) into Eq. (5.3), then we find

$$\dot{a}\omega_0\cos(\omega_0 t + \phi) - a\omega_0\dot{\phi}\sin(\omega_0 t + \phi) = g. \tag{5.11}$$

Since Eqs. (5.10) and (5.11) must be satisfied simultaneously, we can solve for $\dot{a}$ and $\dot{\phi}$ to obtain

$$\dot{a} = \frac{g}{\omega_0}\cos(\omega_0 t + \phi), \tag{5.12}$$

and

$$\dot{\phi} = -\frac{g}{a\omega_0}\sin(\omega_0 t + \phi), \tag{5.13}$$

where g is written in terms of y and $\dot{y}$ given by Eqs. (5.7) and (5.9).

The essence of the approximation is now to neglect all terms in Eqs. (5.12) and (5.13) except those that vary slowly in comparison with ω_0. The resulting trends are then denoted by $\langle \dot{a} \rangle$ and $\langle \dot{\phi} \rangle$, respectively. When Eqs. (5.12) and (5.13) are substituted back into Eq. (5.7), we find that the "instantaneous" frequency at time t is just the time derivative of the total phase $\omega t + \phi$ and thus, neglecting rapidly varying components of ϕ, is equal to $\omega_0 + \langle \dot{\phi} \rangle$. The amplitude is similarly

$$a(t) = a(t_0) + \int_{t_0}^{t} \langle \dot{a} \rangle \, dt. \tag{5.14}$$

Within this approximation, it is thus possible to calculate the entire behavior of the system once its initial state is known. We shall use this formalism quite extensively in our later discussion.

5.2 The Nonlinear Oscillator

If this approach is to be convincing, then it is necessary, of course, that it reproduce, with adequate accuracy, the standard results for simple cases. To see that this does happen without undue labor, let us examine a few typical examples.

First, consider the simple damped system with no external forcing. For this case, in Eq. (5.5), $f = 0$ and $g = -2\alpha\dot{y}$. Thus, using Eq. (5.9) to insert g into Eq. (5.12) and neglecting terms at frequency $2\omega_0$, we find $\langle \dot{a} \rangle = -\alpha a$, while similarly from Eq. (5.13), $\langle \dot{\phi} \rangle = 0$. The solution is therefore

$$a(t) \approx a(t_0) e^{-\alpha(t-t_0)} \sin(\omega_0 t + \phi_0), \tag{5.15}$$

which is in adequate agreement with the exact solution, provided $\alpha \ll \omega_0$.

Next, take the case of a damped harmonic oscillator driven at some frequency ω that is close to its resonance. Here,

$$g = f \sin \omega t - 2\alpha\dot{y}, \tag{5.16}$$

so that, from Eqs. (5.12) and (5.13),

$$\langle \dot{a} \rangle = \frac{f}{2\omega_0} \sin[(\omega - \omega_0)t - \phi] - \alpha a, \tag{5.17}$$

and

$$\langle \dot{\phi} \rangle = -\frac{f}{2a\omega_0} \cos[(\omega - \omega_0)t - \phi]. \tag{5.18}$$

The actual motion of the system clearly depends upon the amplitude and phase of its initial state, but the important thing to check is the final steady state as $t \to \infty$. If $\langle \dot{\phi} \rangle$ is to be constant, then from Eq. (5.18) we must have

$$\phi = (\omega - \omega_0)t + \phi_0, \tag{5.19}$$

so that

$$\langle \dot{\phi} \rangle = -\frac{f}{2a\omega_0} \cos \phi_0. \tag{5.20}$$

The requirement that Eqs. (5.19) and (5.20) be consistent gives

$$\cos \phi_0 = -\frac{2a\omega_0(\omega - \omega_0)}{f}, \tag{5.21}$$

so that the oscillator vibrates with frequency $\omega_0 + \langle \dot{\phi} \rangle = \omega$, in synchronism with the external force. The steady amplitude for which $\langle \dot{a} \rangle = 0$ is given by Eq. (5.17) as

$$a = -\frac{f}{2\alpha\omega_0} \sin \phi_0, \tag{5.22}$$

or, using Eq. (5.21),

$$\begin{aligned} |a| &= \frac{f}{2\alpha\omega_0} \left[1 - \frac{4a^2\omega_0^2(\omega - \omega_0)^2}{f^2} \right]^{1/2} \\ &\approx \frac{f}{2\alpha\omega_0} \left[1 - \frac{(\omega - \omega_0)^2}{2\alpha^2} \right], \end{aligned} \tag{5.23}$$

which is the response near the top of a normal resonance curve. The phase difference ϕ_0 between the force and the resulting displacement is also correctly given, being $-90°$ at resonance (as shown in Fig. 1.12).

The method set out in Section 5.1 is called, for obvious reasons, the method of slowly varying parameters (Van der Pol, 1934; Bogoliubov and Mitropolsky, 1961; Morse and Ingard, 1968; Hagedorn, 1995). In nearly all cases, it allows us to calculate the entire time evolution of the system from a given initial state simply by integrating Eqs. (5.17) and (5.18), a procedure that must generally be carried out numerically. The results again agree quite closely with the exact solutions, where these can be obtained, subject only to the restrictions that ω is close to ω_0 and that small terms of higher frequency are neglected in the solution.

The method can also be used in an obvious way to follow the behavior of nonlinear oscillators. A simple example is the case in which the spring parameter K of the oscillator gets progressively stiffer or weaker as the displacement y increases, for example, as $K \to K + \beta m y^2$. Referring to Eqs. (5.3)–(5.5), we find that

$$g = f(t) - 2\alpha\dot{y} - \beta y^2. \tag{5.24}$$

The resulting form of Eq. (5.3) is known as Duffing's equation. Some of its more interesting properties have been discussed by Prosperetti (1976) and by Ueda (1979). This expression [Eq. (5.24)] can be simply inserted in the $\langle \cdots \rangle$ forms of Eqs. (5.12) and (5.13), and these equations integrated to give the behavior.

If $f(t)$ is a sinusoidal excitation of frequency ω, then we can readily calculate the steady response curve of the oscillator. Two cases are shown in Fig. 5.1, that for $\beta = b/m > 0$, which corresponds to a spring that hardens with increasing amplitude, and that for $\beta = b/m < 0$, which corresponds to a softening spring. The overhanging part of the curve represents an unstable situation and the oscillator exhibits amplitude transitions with hysteresis, as shown by the broken lines in Fig. 5.1. These transitions are given by integration of Eqs. (5.17) and (5.18). If $f(t)$ is zero after an initial impulsive supply of energy, for example, then the equations show that, as the amplitude decays toward zero in a more or less exponential man-

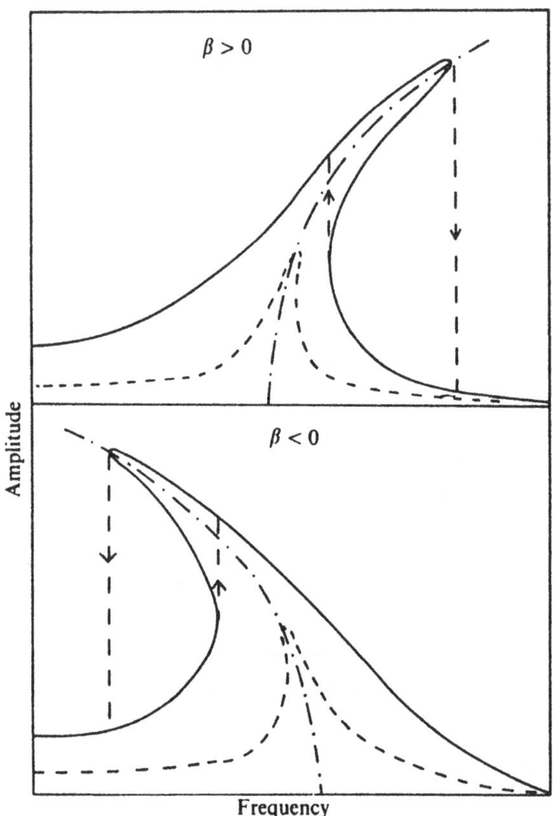

FIGURE 5.1. The frequency response of a nonlinear oscillator of the Duffing type excited by an external sinusoidal force, for the cases of hardening ($\beta > 0$) and softening ($\beta < 0$) behavior. The dotted curves show the response for a smaller exciting force, and the dash-dot curves show the frequency-amplitude relation for free vibration. The vertical broken lines show transitions as the frequency is swept with a constant exciting force.

ner, the frequency glides from its large-amplitude value toward its limiting small-amplitude value along the spine of the curve.

It is not necessary to use the full power of this method if all that we require is an approximate solution for the general trend of the resonance behavior. It is useful to outline how this analysis goes for the Duffing oscillator, since this is a prototype for many simple nonlinear systems. For the case defined by Eq. (5.24), if $f(t) = -f \cos \omega t$, the equation of motion can be written

$$\ddot{y} = -\omega_0^2 y - \beta y^3 - 2\alpha \dot{y}. \tag{5.25}$$

If we substitute $y = a \cos \omega t$ into this equation as a first approximation, and use the fact that $\cos^3 x = \frac{3}{4} \cos x + \frac{1}{4} \cos 3x$, then collection of the terms in $\cos \omega t$ gives a cubic equation in the amplitude a

$$a\omega^2 = \omega_0^2 a + \tfrac{3}{4} \beta a^3 + f. \tag{5.26}$$

This equation can be solved easily enough if we take the damping parameter α to be zero, though the full solution is a good deal more difficult (Morse and Ingard, 1968). If we go one step further, however, and set the external force f equal to zero as well, then we can write down the behavior of the undamped resonance frequency as a function of amplitude. The result is

$$\omega = (\omega_0^2 + \tfrac{3}{4} \beta a^2)^{1/2}, \tag{5.27}$$

which appears as the dot-dash curve in Fig. 5.1. This treatment also gives an approximate expression for the amplitude a_3 of the third harmonic in terms of that of the fundamental,

$$a_3 \approx \frac{\beta a^3}{36\omega^2}. \tag{5.28}$$

While this expression is only a first approximation, it gives a useful estimate.

5.3 The Self-Excited Oscillator

Of particular interest in musical instruments, and indeed in many fields of electronic technology as well, is an oscillator that is arranged so as to modulate some external steady flow of air, electricity, or some other quantity, with a part of the resulting modulated flow then being fed back in an appropriate phase to excite the oscillator. The system equation has the general form

$$\ddot{y} + \omega_0^2 y = g(y, \dot{y}), \tag{5.29}$$

where the form of g depends on the arrangement of the system, and the magnitude of g is related to the magnitude of the external force, which provides energy to drive the oscillator.

The best known case is the Van der Pol (1927) oscillator, for which

$$g = a\dot{y}(1 - y^2). \tag{5.30}$$

Inserting this into Eq. (5.3), we see that the damping of the oscillator is negative for $|y| < 1$ and positive for $|y| > 1$. Oscillations of small amplitude thus tend to grow, while oscillations of large amplitude are damped. There is a stable oscillation regime or "limit cycle" of amplitude $a = 2$ for which the energy losses when $|y| > 1$ just balance the energy gains when $|y| < 1$. The oscillations are fairly closely sinusoidal, provided a is not very much greater than unity but, of course, there is some admixture of higher odd harmonics.

The behavior of a Van der Pol oscillator and indeed of other similar self-excited systems is well encompassed by our formalism, and simple insertion of the appropriate form of $g(y, \dot{y})$ into the $\langle\cdots\rangle$ forms of Eqs. (5.12) and (5.13) provides a prescription from which the development of the system can be calculated. Discussion of specific cases will be left until the underlying physical systems have been introduced, but a few general comments are in order. The most important are the observations that the quiescent state $y = 0$ is always a possible solution, but that this state is unstable to small fluctuations δy, $\delta\dot{y}$, if the part of $g(\delta y, \delta\dot{y})$ in phase with $\delta\dot{y}$ is positive. If this is true, then such fluctuations will lead to growth of the displacement y. The second important set of generalizations inquires whether or not the oscillation settles into a stable limit cycle. This depends on the detail of the nonlinearity and thus upon the physics of the system. Clearly, all musically useful systems do settle to a stable cycle of nonzero amplitude.

5.4 Multimode Systems

All real systems of finite extent have an infinite number of possible vibration modes and, in musically useful resonators, these modes are generally nearly linear in behavior and have well-separated characteristic frequencies, often in nearly harmonic relationship. This does not mean that musical oscillating systems are nearly linear, but rather that such systems usually consist of a nearly linear multimode resonator excited by some nonlinear feedback mechanism.

Suppose that the equation describing wavelike propagation in the oscillator has the form

$$\mathcal{L}\psi - \frac{\partial^2\psi}{\partial t^2} = 0, \tag{5.31}$$

where $\mathcal{L}$ is some linear differential operator typically involving ∇^2 or ∇^4. If we separate variables by writing solutions to this equation as products of spatial functions and a time variation like $\sin \omega t$, then the eigenvalue equation is

$$\mathcal{L}\psi_n + \omega_n^2\psi_n = 0. \tag{5.32}$$

where the eigenfrequencies ω_n are determined from the requirement that the eigenfunctions ψ_n should satisfy appropriate boundary conditions, usually $\psi_n = 0$ or $\nabla\psi_n = 0$, at the surfaces of the resonator.

If we extend Eq. (5.31) to include on the right side a force of unit magnitude (in appropriate units) and frequency ω applied at the point $\mathbf{r}_0$, then this equation becomes

$$\mathcal{L}G_\omega + \omega^2 G_\omega = -\delta(\mathbf{r} - \mathbf{r}_0), \tag{5.33}$$

where $\delta(\mathbf{r} - \mathbf{r}_0)$ is the Dirac delta function, and we have written ψ as $G_\omega(\mathbf{r}, \mathbf{r}_0)$, which is the Green function for the system at frequency ω, taken to obey the same boundary conditions as do the ψ_n. If we assume that G_ω can be expanded as

$$G_\omega(\mathbf{r}, \mathbf{r}_0) = \sum_n a_n \psi_n(\mathbf{r}), \tag{5.34}$$

and substitute this into Eq. (5.33), then, using Eq. (5.32),

$$\sum_n a_n(\omega_n^2 - \omega^2)\psi_n(\mathbf{r}) = \delta(\mathbf{r} - \mathbf{r}_0). \tag{5.35}$$

If we multiply both sides by $\psi_n(\mathbf{r})$ and integrate over the whole volume of the resonator using the usual orthonormality condition

$$\int \psi_n(\mathbf{r})\psi_m(\mathbf{r})d\mathbf{r} = \delta_{mn}, \tag{5.36}$$

then we find an expression for a_n that, substituted back into Eq. (5.34), gives

$$G_\omega(\mathbf{r}, \mathbf{r}_0) = \sum_n \frac{\psi_n(\mathbf{r}_0)\psi_n(\mathbf{r})}{\omega_n^2 - \omega^2}. \tag{5.37}$$

Clearly, this Green function has simple poles at $\omega = \pm\omega_n$, where the ω_n are the resonance frequencies of the system.

Since the resonator system is assumed to be linear, and since $G(\mathbf{r}, \mathbf{r}_0)$ is the response caused by a force of unit magnitude and frequency ω applied at $\mathbf{r}$, we can write the general response Ψ to a set of forces of different frequencies ω and phases θ and distributed over the resonator like the functions $F_\omega(\mathbf{r})$ as the simple sum

$$\Psi(\mathbf{r}, t) = \sum_\omega \left[\int G_\omega(\mathbf{r}, \mathbf{r}_0)F_\omega(\mathbf{r}_0)d\mathbf{r}_0\right] \sin(\omega t + \theta). \tag{5.38}$$

This formulation does not include damping. It could be included as an extension to the linear theory and has the effect of moving the poles of

G_ω slightly off the real axis, but it is more conveniently incorporated along with nonlinear effects at a later stage.

The essential feature of Eq. (5.38) for our present purpose is expressed by Eq. (5.37), which shows that the mode ψ_n with frequency ω_n is excited principally by force components with frequency ω lying close to ω_n. Indeed, if we write the excitation of this nth mode in the form

$$y_n = a_n \sin(\omega t + \phi_n)\psi_n(\mathbf{r}), \tag{5.39}$$

when it is being driven at frequency ω by a force $F_\omega(\mathbf{r}_0)$ applied at the point $\mathbf{r}_0$, then Eqs. (5.37) and (5.38) are equivalent to the sum of the results from the set of differential equations:

$$\ddot{y}_n + \omega_n^2 y_n = \sum_\omega F_\omega \sin(\omega t + \theta_\omega). \tag{5.40}$$

If we retain in Eq. (5.40) only those force components F_ω with frequencies near ω_n, then we can apply the method of slowly varying parameters as developed previously for a simple oscillator. The forcing term on the right side can be written to include the damping term $-2\alpha_n \dot{y}_n$ and, in a general situation, it will contain driving forces with frequencies near ω_n derived from nonlinear terms involving many of the other modes of the system. In general, therefore, Eq. (5.40) has the form

$$\ddot{y}_n + \omega_n^2 y_n = g_n(y_n, y_m, \ldots; \dot{y}_n, \dot{y}_m, \ldots), \tag{5.41}$$

and we omit from g_n, as explicitly evaluated, all terms except those for which the frequency is close to ω_n.

Equation (5.41) is a shorthand way of writing an infinite set of coupled differential equations, one for each of the modes ω_n. In practice, however, the very-high-frequency modes will have small excitation amplitudes, so that the system can be reduced to a finite and indeed relatively small set of N coupled equations describing those modes that are appreciably excited. Each of these equations can be manipulated to give $\langle \dot{a}_n \rangle$ and $\langle \dot{\phi}_n \rangle$, and the resulting $2N$ equations can be easily integrated numerically to define the behavior of the system.

Clearly, the linear physics of a musical system resides in the study of the modes ψ_n and their characteristic frequencies ω_n, while the nonlinear physics involves elucidation of the coupling functions g_n. Both these matters are quite specific to individual systems, so we will not consider particular examples at this stage of our discussion.

As with simple oscillators, so multimode systems can be divided into those that are purely dissipative and can be excited only by an impulsive or time-varying external force, and those that are self-exciting with only a steady external supply of energy. Gongs are typical examples of the former class and can exhibit many interesting phenomena as energy is passed back and forth between the modes by the agency of the nonlinear coupling terms

g_n. Musical instruments producing steady tones, such as winds or bowed strings, belong to the second class.

5.5 Mode Locking in Self-Excited Systems

For many musical resonators, such as stretched strings or air columns in pipes, the normal-mode frequencies are very nearly, but not quite, in integral ratio. Precisely harmonic systems do not exist. Now, if the coupling function g_n in Eq. (5.40) is linear or has no nonlinear terms involving combinations of different modes, then the system becomes essentially uncoupled and each mode takes on an excitation frequency close to its natural frequency ω_n. These frequencies are never precisely in integer ratios, so the resulting total excitation has a nonrepeating waveform.

Sustained tones from real musical instruments do, however, have precisely repeating waveforms, apart from deliberate vibrato effects, and so their individual modes must be somehow locked into precise frequency and phase relationships despite the inharmonicities of the natural resonances. It is important to see how this is accomplished (Fletcher, 1978).

Consider a system for which just two modes are appreciably excited by the feedback mechanism and suppose that their natural frequencies ω_n and ω_m are related approximately, but not exactly, as the ratio of the two small integers n and m:

$$m\omega_n \approx n\omega_m. \tag{5.42}$$

Now, from Eq. (5.42), the leading nonlinear term by which mode m can provide a driving force at nearly the frequency of mode n involves the amplitudes a_n and a_m in the form $a_m^n a_n^{m-1}$. Similarly, mode n can influence mode m in proportion to $a_n^m a_m^{n-1}$. The coefficients of these terms depend upon the Taylor expansion of the nonlinear driving functions g_n and g_m, respectively, and generally decrease sharply as n and m increase. These two functions will be of the same general form but may differ in detail. The directions in which the frequencies of modes m and n are pushed by their $\langle\dot{\phi}\rangle$ terms in Eq. (5.13) will depend upon the combinations of phase angles ϕ_n and ϕ_m involved, and these will vary rapidly if the two modes are not locked together. Once locking occurs, however, this represents a stable situation. This argument can be generalized to the case of more than two interacting modes.

The conditions favoring mode locking are thus that the inharmonicity of the modes not be too great, that the integers n and m linking the modes be as small as possible, that the mode amplitudes be large, and that the nonlinearity of the coupling function be as large as possible. When these conditions are fulfilled, then all modes contributing to an instrumental sound will rapidly settle down to give a phase and frequency locked repetitive waveform. This situation is specially favored when one of the dominant modes excited by the feedback mechanism is the fundamental, for then $m = 1$ and the integer combination is as simple as possible.

Conversely, the conditions favoring nonlocking of modes and thus the production of complex multiphonic effects are great mode inharmonicity (often produced by peculiar venting arrangements in wind instruments), a low excitation level, and preferential excitation of modes other than the fundamental. The complexity of the possible variations on this theme makes it desirable to consider it in the context of particular instruments, though we might mention at this stage the discussions of bowed strings by Popp and Stelter (1990) and by Müller and Lauterborn (1996), and of reed-driven air columns by Keefe and Laden (1991).

5.6 Nonlinear Effects in Strings

Our treatment of strings in Chapter 2 was entirely linear, even in relation to those phenomena such as stiffness that make the mode frequencies of a real string slightly inharmonic. Nonlinear effects are, however, quite apparent in the behavior of strings in musical instruments and have some important practical consequences. The signature of their presence is that the effect under study varies with the amplitude of the initial excitation in a way that is not just a simple scaling of all other amplitudes involved.

We can distinguish two sorts of nonlinearity. One, which we discussed in Section 1.12, arises from a nonlinearity intrinsic to a material within the system. This applies particularly to elastic moduli, which may increase at large strains, though in some materials they may decrease. Because these effects are essentially arbitrary, we shall not include them in our discussion here. More interesting and important are nonlinearities that arise from the geometry of the system itself, simple though it may be. It is to these that we refer in the next two sections.

Many studies of nonlinearity involve strings driven into continuous vibration (extensive references are given by Hanson et al., 1994), but these do not particularly concern us here, except perhaps marginally in the case of bowed-string instruments, although here the constraining effect of the bow is dominant. The more relevant case of nonlinear effects in the free vibration of strings has been studied by Carrier (1945), by Elliott (1980), and by Gough (1984), among others, and a recent review is given by Valette (1995). There are, in fact, several different nonlinear phenomena that are of interest, and we treat them in turn. The first is a simple dependence of the natural frequency of a vibrating string upon its amplitude of vibration, the second is a transfer of energy between the two possible polarizations of string vibration, and the third is the generation of missing modes in a complex vibration. All these have their origin in a nonlinear interaction between transverse vibrations of the string and its tension.

Suppose we have a string of cross-sectional area S, Young's modulus E, and material density ρ stretched with tension T between rigid supports a distance L apart. Then the equation of motion, as we saw in Section 2.2, is

$$\rho S \frac{\partial^2 y}{\partial t^2} = T \frac{\partial^2 y}{\partial x^2}, \tag{5.43}$$

where we have taken the x direction to be along the string and supposed the vibration to be polarized in the (x, y) plane. The normal modes are given by

$$y_n(x,t) = a_n \sin\left(\frac{n\pi x}{L}\right) \cos \omega_n t \tag{5.44}$$

with $\omega_n = (n\pi/L)(T/\rho S)^{1/2}$.

In the course of the vibration $y_n(x,t)$, the string is displaced from its initial straight configuration, and its length increases by an amount

$$\delta L = \int_0^L \left[1 + \left(\frac{dy}{dx}\right)^2\right]^{1/2} dx - L \approx \frac{n\pi^2 a_n^2}{8L^2}(1 - \cos 2\omega_n t). \tag{5.45}$$

The string tension therefore increases by an amount $\delta T = ES\delta L/L$, which involves both a steady and an oscillating term. If the displacement involves several modes n, then their contributions are simply summed, because all cross-terms in Eq. (5.45) vanish during the integration.

The effect of principal musical significance is the quasi-steady increase in tension $\overline{\delta T} = ES\overline{\delta L}/L$, which is proportional to the sum of the squares of the mode excitation amplitudes. This raises the frequency of all modes by the same factor, the fractional frequency shift being $\delta\omega/\omega = ES\overline{\delta T}/TL$. In addition, however, the term in $\cos 2\omega_n t$ in (5.45), when substituted in (5.43), gives a further frequency shift for the mode of frequency ω_n only, the magnitude of which is $\frac{1}{2}\,\delta\omega_n/\omega_n = ES\Delta T_n/TL$, where ΔT_n is the tension increase associated with mode n alone. This second contribution to the frequency shift destroys the harmonicity of the vibration, even for an otherwise ideal string. As the free excitation decays with time, the pitch glides back toward its small-amplitude value. The extent of this glide depends on the length, tension, and Young's modulus of the string, as well as on the square of the excitation amplitude, but the effect is generally an unpleasant "twang" when a rather loose metal string is plucked to large amplitude. The effect can be minimized by using strings with a low Young's modulus, such as nylon or silk, rather than steel, and by designing the instrument so that the strings are as long and as taut as possible, without coming perilously close to their breaking stress.

The oscillating term in the tension cannot excite the mode of double frequency directly in an ideally clamped string, but does contribute a third-harmonic component to the string motion and therefore to the vibration communicated to the end supports. If the string passes at an angle over a bridge that is not completely rigid, as is normally the case in stringed instruments, then the oscillating tension force causes a transverse motion of the bridge at frequency $2\omega_n$ and this can excite mode $2n$. Indeed, if we recognize that the angle at which the string passes over the bridge varies at frequency ω_n, then we see that there is also the possibility of excitation of the mode $3n$ with frequency $3\omega_n$. These effects, which have been demonstrated both theoretically and experimentally (Legge and Fletcher, 1984), are significant only if the string is excited in such a way that these particular modes are initially absent from the vibration.

Another, and more important, effect of the nonlinearly oscillating tension is that it provides a mechanism coupling together the two possible polarization modes of the string vibration (Elliott, 1980; Gough, 1984; Tufillaro, 1989; Hanson et al., 1994; Valette, 1995). Suppose the string is initially vibrating in the (x, y) plane at frequency ω_n. Then, as we have seen, this generates an oscillating tension of frequency $2\omega_n$. When this is substituted back into the analog of Eq. (5.43) for string motion in the (x, z) plane, then there is a forcing term $g(t)$ that varies as $a_n^2(\cos\omega_n t + \cos 3\omega_n t)$. When this is substituted in an equation like Eq. (5.12) for the mode with z polarization and frequency ω_n, we find that there is a steady term generated, so that the mode amplitude grows. This process is called parametric amplification because one of the parameters of the system, in this case the tension, provides the driving force. As the vibration with z polarization grows, it either takes energy from, or feeds energy back to, that with y polarization, depending on their relative phases, so that there is a continuing interaction.

Another way to look at this is to recognize that, in an ideally clamped string, the y and z oscillations are equivalent. In a linear approximation they can exist independently with arbitrary amplitudes, so that the general motion of the string is elliptically polarized. When nonlinear terms are taken into account however, the only completely stable motions are those in which the string vibration is either plane polarized or else circularly polarized. In all other cases the polarization ellipse precesses slowly as energy is transferred from one mode to the other and then back again. In cross section, therefore, the motion of the string has the form shown in Fig. 5.2.

The rate of rotation of the polarization ellipse has been calculated by Elliott (1980), by Gough (1984) and by Valette (1995), for the case of a single mode on the string. The axial ratio and orientation of the ellipse are fixed by the conditions of the initial excitation, which gives amplitudes a and b say, to the y and z motions at $t = 0$. The angular precession velocity Ω of the polarization ellipse is then calculated to be

$$\Omega = A\left(\frac{ES}{T}\right)\left(\frac{ab}{L^2}\right)\omega, \tag{5.46}$$

where ω is the oscillation frequency and A is a numerical factor of order unity, about the exact value of which the two treatments differ. In typical

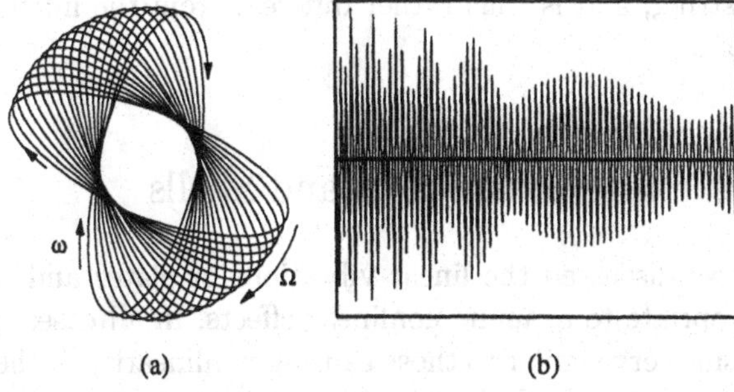

FIGURE 5.2. (a) Cross-sectional view of the precessional motion of a string vibrating at a single frequency ω. The precession rate Ω is proportional to the area of the vibration ellipse. Only part of a precession cycle is shown. (b) Typical time variation of the normal force on the bridge as the string vibration decays, for $a : b = 2 : 1$.

cases the precession frequency $\Omega/2\pi$ is of the order of 1 Hz, though it clearly depends greatly on the system considered, on the mode, and on the amplitude of the string vibration. The precession rate decreases as the vibration amplitude decays. The situation is thus as in Fig. 5.2, which shows in (a) the motion of the string when viewed in a plane normal to its length, and in (b) the calculated normal force on the bridge, in a typical case, as the string vibration decays.

We might expect this behavior to be significant for the sound of plucked-string instruments, since the two polarizations of string vibration will couple differently to the bridge and soundboard and thus to the radiated sound. Harpsichord sound, for example, does indeed show quasi-periodic variations in the radiated amplitude from a single string, but this effect may have a rather simpler origin (Gough, 1981). In any stringed instrument, the string excites the soundboard through a bridge to which it is attached. The mechanical admittance of this bridge, however, is not zero, as for a rigid support, and is different for motions parallel to or perpendicular to the soundboard. This small bridge admittance shifts the frequencies of the two string polarizations by different amounts, so that the relative phases of the normal and tangential motion vary, even in the linear approximation, possibly changing the amplitude of the radiated sound. Whatever the origin in particular cases, however, this effect is an important feature of real plucked-string sound, and is complicated by the fact that different string modes precess at different rates.

When we come to consider bowed-string instruments, there are many nonlinear effects possible that can be discussed in a manner similar to that used in this chapter (Müller and Lauterborn, 1996). Most of the interesting nonlinearity derives, however, from the frictional interaction between the bow and the string, and is thus rather different from the intrinsic effects discussed here.

5.7 Nonlinear Effects in Plates and Shells

In Chapter 3 we discussed the linear vibrations of plates and shells, and it is now appropriate to examine nonlinear effects. In all cases these arise from mechanisms very similar to those causing nonlinearity in the behavior of ideal strings, as discussed above. In the case of the extensional modes of curved shells, tension forces enter directly and are linearly proportional to displacement. In addition, however, and for both the nonextensional modes of shells and the necessarily nonextensional transverse modes of plates, there are always tension forces proportional to the squares and higher powers of the displacements. It is from these that nonlinear effects arise.

The range of possible plate and shell geometries is so large that we cannot survey it here. It is, however, only in the case of the rather thin shells of gongs that significant nonlinear effects can be heard, because only in this case do we encounter adequately large vibration amplitudes. We can understand something of the general behavior by considering just the case of a spherical-cap shell (Grossman et al., 1969, Fletcher, 1985). In the case of a flat circular plate, it can be shown (Fletcher, 1985) that the frequency of the lowest axisymmetric mode for a plate of thickness h varies with amplitude a approximately as

$$\omega \approx \omega_0 \left[1 + 0.16 \left(\frac{a}{h}\right)^2\right], \tag{5.47}$$

where ω_0 is the mode frequency at infinitesimal amplitude. The origin of this frequency shift is just the same as that for the string discussed above. The reason for the dependence on plate thickness is the fact that this quantity determines the relative importance of stiffness and tension forces in plate behavior. The behavior of a gong with a flat central vibrating section is thus rather like that of a string, and the pitch glides downwards as the vibration decays after an initial vigorous excitation. The actual vibration, and thus the radiated sound, contains not only the shifted normal mode frequencies of the plate but also all odd harmonics of those mode frequencies, generated by distortions of each modal waveform—there are no distinct modes corresponding to these added frequencies.

A more interesting case is that of a spherical cap shell of height H and thickness h. The behavior is now more complicated because the vibration is asymmetrical about its rest point. Once again we consider only the lowest axisymmetric mode. The results calculated by Fletcher (1985) are shown in Fig. 5.3 and depend not only on the ratio a/H between the vibration am-

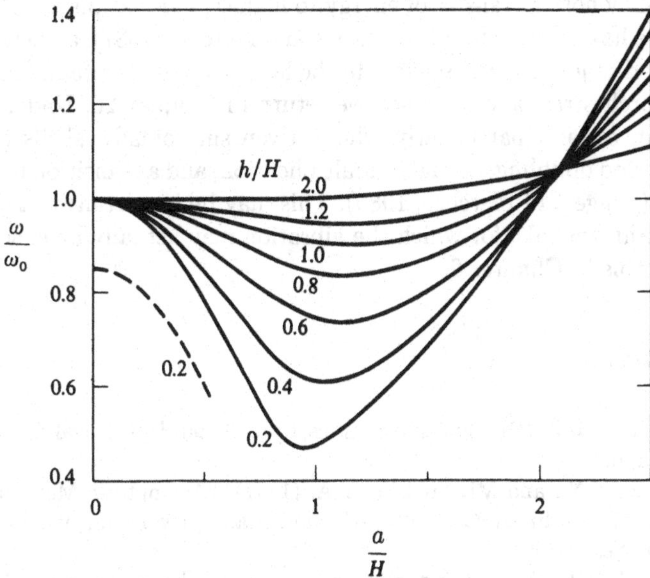

FIGURE 5.3. Calculated frequency ω for the lowest axisymmetric mode of a spherical-cap shell of dome height H and thickness h when vibrating with amplitude a. The mode frequency for infinitesimal vibrations is ω_0. The broken curve shows the limited range of stable vibrations for a moderately thin everted shell (Fletcher, 1985).

plitude and the shell height, but also on the geometrical quantity h/H. If the shell is very thin, so that $h/H \ll 1$, then nonlinear tension effects dominate the behavior. As the vibration amplitude is increased to approach H, the mode frequency falls to about half its small-amplitude value, while further increase in amplitude causes it to rise again. For progressively thicker shells, the nonlinearity is less important because of the dominance of bending stiffness, and very thick shells with $h \gg H$ behave essentially as flat plates.

This behavior in a musical instrument is not generally welcome in Western music, but has an important place in Chinese opera, where contrasting upward and downward gliding gongs are used for dramatic effect. As with other nonlinear systems, each mode is accompanied by harmonics caused by modal distortion, the amplitude of the mth harmonic of a given mode varying initially as the mth power of the amplitude of its fundamental. For spherical-cap gongs, however, this simple relation ceases to hold once the vibration amplitude becomes comparable with the dome height H (Rossing and Fletcher, 1983).

In addition to these pitch-glide effects, plates and shells can exhibit a variety of more complex behaviors when excited to large amplitude. Among these we may note a transfer of energy to higher modes at places where the shell shape has a sharp curvature (Legge and Fletcher, 1987), a phenomenon that is used to great musical effect in the large Chinese tamtam, familiar in Western orchestras, and to which we return in Chapter 20. Under certain circumstances, and particularly when driven sinusoidally, shells can also exhibit period doubling, or higher multiplication, and a transition to chaotic behavior (Legge and Fletcher, 1989). This may be important to the sound of orchestral cymbals, for which the vibration may actually be chaotic. We return to this in Chapter 20.

References

Beyer, R.T. (1974). "Nonlinear Acoustics," pp. 60–90. U.S. Naval Sea Systems Command.

Bogoliubov, N.N., and Mitropolsky, Y.A. (1961). "Asymptotic Methods in the Theory of Non-linear Oscillations." Hindustan, New Delhi, and Gordon & Breach, New York.

Carrier, G.F. (1945). On the nonlinear vibration problem of the elastic string. *Q. Appl. Math.* **3**, 157–165.

Elliott, J.A. (1980). Intrinsic nonlinear effects in vibrating strings. *Am. J. Phys.* **48**, 478–480.

Fletcher, N.H. (1978). Mode locking in nonlinearly excited inharmonic musical oscillators. *J. Acoust. Soc. Am.* **64**, 1566–1569.

Fletcher, N.H. (1985). Nonlinear frequency shifts in quasispherical-cap shells: Pitch glide in Chinese gongs. *J. Acoust. Soc. Am.* **78**, 2069–2073.

Gough, C. (1981). The theory of string resonances on musical instruments. *Acustica* **49**, 124–141.

Gough, C. (1984). The nonlinear free vibration of a damped elastic string. *J. Acoust. Soc. Am.* **75**, 1770–1776.

Grossman, P.L., Koplik, B., and Yu, Y-Y. (1969). Nonlinear vibration of shallow spherical shells. *J. Appl. Mech.* **36**, 451–458.

Hagedorn, P. (1995). Mechanical oscillations. In "Mechanics of Musical Instruments," ed. A. Hirschberg, J. Kergomard and G. Weinreich. Springer-Verlag, Vienna and New York, pp. 7–78.

Hanson, R.J., Anderson, J.M., and Macomber, H.K. (1994). Measurement of nonlinear effects in a driven vibrating wire. *J. Acoust. Soc. Am.* **96**, 1549–1556.

Keefe, D.H., and Laden, B. (1991). Correlation dimension of woodwind multiphonic tones. *J. Acoust. Soc. Am.* **90**, 1754–1765.

Lauterborn, W., and Parlitz, U. (1988). Methods of chaos physics and their application to acoustics. *J. Acoust. Soc. Am.* **84**, 1975–1993.

Legge, K.A., and Fletcher, N.H. (1984). Nonlinear generation of missing modes on a vibrating string. *J. Acoust. Soc. Am.* **76**, 5–12.

Legge, K.A., and Fletcher, N.H. (1987). Non-linear mode coupling in symmetrically kinked bars. *J. Sound Vibr.* **118**, 23–34.

Legge, K.A., and Fletcher, N.H. (1989). Nonlinearity, chaos, and the sound of shallow gongs. *J. Acoust. Soc. Am.* **86**, 2439–2443.

Mettin, R., Parlitz, U., and Lauterborn, W. (1993). Bifurcation structure of the driven Van der Pol oscillator. *Int. J. Bifurc. Chaos* **3**, 1529–1555.

Moon, F.C. (1992). "Chaotic and Fractal Dynamics. An Introduction for Applied Scientists and Engineers." John Wiley, New York.

Morse, P.M., and Ingard, K.U. (1968). "Theoretical Acoustics," pp. 828–882. McGraw-Hill, New York. Reprinted 1986, Princeton Univ. Press, Princeton, New Jersey.

Müller, G., and Lauterborn, W. (1996). The bowed string as a nonlinear dynamical system. *Acustica* **82**, 657–664.

Popp, K., and Stelter, P. (1990). Stick-slip vibrations and chaos. *Phil. Trans. Roy. Soc. Lond.* **A332**, 89–105.

Prosperetti, A. (1976). Subharmonics and ultraharmonics in the forced oscillations of weakly nonlinear systems. *Am. J. Phys.* **44**, 548–554.

Rossing, T.D., and Fletcher, N.H. (1983). Nonlinear vibrations in plates and gongs. *J. Acoust. Soc. Am.* **73**, 345–351.

Tufillaro, N.B. (1989). Nonlinear and chaotic string vibrations. *Am. J. Phys.* **57**, 408–414.

Ueda, Y. (1979). Randomly transitional phenomena in the system governed by Duffing's equation. *J. Statistical Phys.* **20**, 181–196.

Valette, C. (1995) The mechanics of vibrating strings. In "Mechanics of Musical Instruments," ed. A. Hirschberg, J Kergomard and G. Weinreich, Springer-Verlag, Vienna and New York, pp. 115–183.

Van der Pol, B. (1927). Forced oscillations in a circuit with non-linear resistance. *Phil. Mag.* **7**(3), 65–80.

Van der Pol, B. (1934). The nonlinear theory of electric oscillations. *Proc. I.R.E.* **22**, 1051–1086.

Part II

Sound Waves

6

Sound Waves in Air

The sensation we call sound is produced primarily by variations in air pressure that are detected by their mechanical effect on the tympana (ear drums) of our auditory system. Motion of each tympanum is communicated through a linked triplet of small bones to the fluid inside a spiral cavity, the cochlea, where it induces nerve impulses from sensory hair cells in contact with a thin membrane (the basilar membrane). Any discussion of details of the physiology and psychophysics of the hearing process would take us too far afield here. The important point is the dominance of air pressure variation in the mechanism of the hearing process. Direct communication of vibration through the bones of the head to the cochlea is possible, if the vibrating object is in direct contact with the head, and intense vibrations at low frequencies can be felt by nerve transducers in other parts of the body, for example in the case of low organ notes, but this is not part of the primary sense of hearing.

The human sense of hearing extends from about 20 Hz to about 20 kHz, though the sensitivity drops substantially for frequencies below about 100 Hz or above 10 kHz. This frequency response is understandably well matched to human speech, most of the energy of which lies between 100 Hz and 10 kHz, with the information content of vowel sounds concentrated in the range of 300 Hz–3 kHz and the information content of consonants mostly lying above about 1 kHz. Musical sounds have been evolved to stimulate the sense of hearing over its entire range, but again most of the interesting information lies in the range of 100 Hz–3 kHz.

Since the ears respond to pressure only in their immediate vicinity, we devote this and the following chapter to a discussion of the way in which pressure variations—sound waves—propagate through the air and to the way in which vibrating objects couple to the air and excite sound waves.

6.1 Plane Waves

Waves will propagate in any medium that has mass and elasticity, or their equivalents in nonmechanical systems. Solid materials, which have both shear and compressive elasticity, allow the propagation of both shear (transverse) and compressive (longitudinal) waves so that their behavior can be very complicated (Morse and Feshbach, 1953, pp. 142–151). Fluids, and in particular gases such as air, have no elastic resistance to shear, though they do have a viscous resistance, and the only waves that can propagate in them are therefore longitudinal, with the local motion of the air being in the same direction as the propagation direction of the wave itself.

When sound waves are generated by a small source, they spread out in all directions in a nearly spherical fashion. We shall look at spherical waves in detail a little later. It is simplest in the first place to look at a small section of wave at a very large distance from the source where the wave fronts can be treated as planes normal to the direction of propagation. In the obvious mathematical idealization, we take these planes to extend to infinity so that the whole problem has only one space coordinate x measuring distance in the direction of propagation.

Referring to Fig. 6.1, suppose that ξ measures the displacement of the air during passage of a sound wave, so that the element ABCD of thickness dx moves to A′B′C′D′. Taking S to be the area normal to x, the volume of this element then becomes

$$V + dV = S\,dx\left(1 + \frac{\partial \xi}{\partial x}\right). \tag{6.1}$$

Now suppose that p_a is the total pressure of the air. Then the bulk modulus K is defined quite generally by the relation

$$dp_{\mathrm{a}} = -K\frac{dV}{V}. \tag{6.2}$$

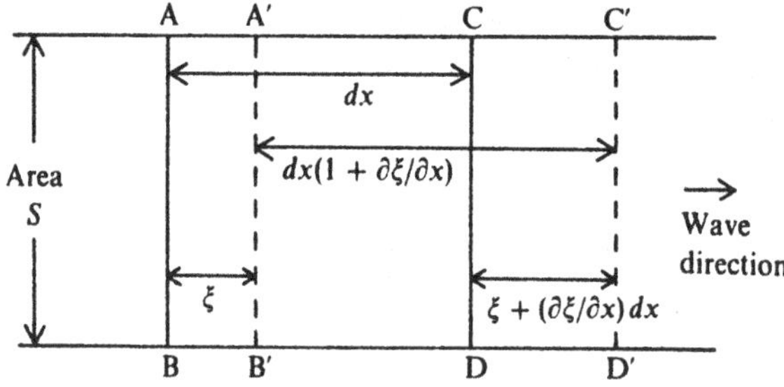

FIGURE 6.1. In passage of a plane wave of displacement ξ, the fluid on plane AB is displaced to A′B′ and that on CD to C′D′.

We can call the small, varying part dp_a of p_a the sound pressure or acoustic pressure and write it simply as p. Comparison of Eq. (6.2) with Eq. (6.1), noting that V is just $S\,dx$, then gives

$$p = -K\frac{\partial \xi}{\partial x}. \tag{6.3}$$

Finally, we note that the motion of the element ABCD must be described by Newton's equations so that, setting the pressure gradient force in the x direction equal to mass times acceleration,

$$-S\left(\frac{\partial p}{\partial x}\,dx\right) = \rho S\,dx\,\frac{\partial^2 \xi}{\partial t^2},$$

or

$$-\frac{\partial p}{\partial x} = \rho\frac{\partial^2 \xi}{\partial t^2}. \tag{6.4}$$

Then, from Eqs. (6.3) and (6.4),

$$\frac{\partial^2 \xi}{\partial t^2} = \frac{K}{\rho}\frac{\partial^2 \xi}{\partial x^2}, \tag{6.5}$$

or, differentiating Eq. (6.5) again with respect to x and Eq. (6.3) twice with respect to t,

$$\frac{\partial^2 p}{\partial t^2} = \frac{K}{\rho}\frac{\partial^2 p}{\partial x^2}. \tag{6.6}$$

Equations (6.5) and (6.6) are two different versions of the one-dimensional wave equation, one referring to the acoustic displacement ξ and the other to the acoustic pressure p. They apply equally well to any fluid if appropriate values are used for the bulk modulus K and density ρ. For the case of wave propagation in air, we need to decide whether the elastic behavior is isothermal, and thus described by the equation

$$p_a V = \text{constant} = nkT, \tag{6.7}$$

where T is the absolute temperature, or whether it is adiabatic, and so described by

$$p_a V^\gamma = \text{constant}, \tag{6.8}$$

where $\gamma = C_p/C_v = 1.4$ is the ratio of the specific heats of air at constant pressure and at constant volume, respectively, and p_a, as before, is the average atmospheric pressure.

Clearly, the temperature tends to rise in those parts of the wave where the air is compressed and to fall where it is expanded. The question is, therefore, whether appreciable thermal conduction can take place between these two sets of regions in the short time available as the peaks and troughs of the wave sweep by. It turns out (Fletcher, 1974) that at ordinary acoustic wavelengths the pressure maxima and minima are so far apart that no appreciable conduction takes place, and the behavior is therefore adiabatic. Only at immensely high frequencies does the free-air propagation tend to become isothermal. For sound waves in pipes or close to solid objects, on the other hand, the behavior also becomes isothermal at very low frequencies—below about 0.1 Hz for a 20 mm tube. Neither of these cases need concern us here.

Taking logarithms of Eq. (6.8) and differentiating, we find, using Eq. (6.2),

$$K = \gamma p_a, \tag{6.9}$$

so that Eq. (6.6) becomes

$$\frac{\partial^2 p}{\partial t^2} = c^2 \frac{\partial^2 p}{\partial x^2}, \tag{6.10}$$

where

$$c^2 = \frac{K}{\rho} = \frac{\gamma p_a}{\rho}, \tag{6.11}$$

and similarly for ξ from Eq. (6.5). As we shall see in a moment, the quantity c is the propagation speed of the sound wave.

It is easy to verify, by differentiation, that possible solutions of the wave equation [Eq. (6.10)] have the form

$$p(x,t) = f_1(x - ct) + f_2(x + ct), \tag{6.12}$$

where f_1 and f_2 are completely general continuous functions of their arguments. We can also see that $f_1(x-ct)$ represents a wave of arbitrary spatial shape $f_1(x - x_0)$ or of arbitrary time behavior $f_1(ct_0 - ct)$ propagating in the $+x$ direction with speed c. Similarly, $f_2(x + ct)$ represents a different wave propagating in the $-x$ direction, also with speed c. In the case of air, or any other nearly ideal gas, Eqs. (6.7) and (6.11) show that

$$c(T) = \left(\frac{T}{T_0}\right)^{1/2} c(T_0), \tag{6.13}$$

where $c(T)$ is the speed of sound at absolute temperature T. There is, however, no variation of c with atmospheric pressure. For air at temperature ΔT degrees Celsius and 50% relative humidity,

$$c \approx 332(1 + 0.00166\,\Delta T)\ \mathrm{m\,s^{-1}}, \tag{6.14}$$

giving $c \approx 343\ \mathrm{m\ s^{-1}}$ at room temperature.

The wave equation [Eq. (6.10)] was discussed in detail in Chapter 2 in relation to waves on a string, and its two-dimensional counterpart in Chapter 3. There is no need to repeat this discussion here except to remind ourselves that it is usual to treat Eq. (6.10) in the frequency domain where the solutions have the form

$$p = Ae^{-jkx}e^{j\omega t} + Be^{jkx}e^{j\omega t}, \tag{6.15}$$

where $k = \omega/c$ and the A and B terms represent waves traveling to the right and the left, respectively. [If we adopt the conventions of quantum mechanics and write time dependence as $\exp(-i\omega t)$, as for example in Morse (1948), then j should be replaced by $-i$.]

If we consider a wave of angular frequency ω traveling in the $+x$ direction, then we can set $B = 0$ and $A = 1$ in Eq. (6.15) and write

$$p = e^{-jkx}e^{j\omega t} \rightarrow \cos(-kx + \omega t), \tag{6.16}$$

where the second form of writing is just the real part of the first. From Eq. (6.5), ξ has a similar form, though with a different amplitude and perhaps a phase factor. We can connect p and ξ through Eq. (6.4), from which

$$jkp = j\rho\omega\frac{\partial\xi}{\partial t}, \tag{6.17}$$

or, if we write u for the acoustic fluid velocity $\partial\xi/\partial t$ and remember that $k = \omega/c$, then

$$p = \rho c u. \tag{6.18}$$

The acoustic pressure and acoustic fluid velocity (or particle velocity) in the propagation direction are therefore in phase in a plane wave.

This circumstance makes it useful to define a quantity z called the wave impedance (or sometimes the specific acoustic impedance):

$$z = \frac{p}{u} = \rho c. \tag{6.19}$$

It is clearly a property of the medium and its units are Pa m^{-1} s or kg $\mathrm{m}^{-2}\ \mathrm{s}^{-1}$, sometimes given the name rayls (after Lord Rayleigh). For air at temperature ΔT°C and standard pressure,

$$\rho c \approx 428(1 - 0.0017\,\Delta\mathrm{T})\ \mathrm{kg\,m^{-2}s^{-1}}. \tag{6.20}$$

In much of our discussion, we will need to treat waves in 3 space dimensions. The generalization of Eq. (6.10) to this case is

$$\frac{\partial p}{\partial t^2} = c^2\nabla^2 p. \tag{6.21}$$

This differential equation can be separated in several coordinate systems to give simple treatments of wave behavior (Morse and Feshbach, 1953, pp. 499–518, 655–666). Among these are rectangular coordinates, leading simply to three equations for plane waves of the form of Eq. (6.10), and spherical polar coordinates, which we consider later in this chapter.

Before leaving this section, however, we should emphasize that we have consistently neglected second-order terms by assuming $p \ll p_a$, so that the resulting wave equation [Eq. (6.10) or Eq. (6.21)] is linear. This is of great assistance in development of the theory and turns out to be an adequate approximation even in very intense sound fields such as exist, for example, inside a trumpet. At even higher intensities, however, and well below the shock-wave limit, nonlinear terms begin to have a detectable effect (Beyer, 1974). It will not be necessary for us to use such extensions of the theory in this book.

6.2 Spherical Waves

When we assume a time dependence $\exp j\omega t$, the wave equation, Eq. (6.21), takes the form

$$\nabla^2 p + k^2 p = 0, \tag{6.22}$$

where $k = \omega/c$. This is known as the Helmholtz equation and is separable, and therefore relatively easily treated, in rectangular, spherical polar, and cylindrical polar coordinates. In spherical coordinates,

$$\nabla^2 p = \frac{1}{r^2}\frac{\partial}{\partial r}\left(r^2 \frac{\partial p}{\partial r}\right) + \frac{1}{r^2 \sin\theta}\frac{\partial}{\partial \theta}\left(\sin\theta \frac{\partial p}{\partial \theta}\right) + \frac{1}{r^2 \sin^2\theta}\frac{\partial^2 p}{\partial \phi^2}, \tag{6.23}$$

and the solution to Eq. (6.22) is the sum of a series of products of radial functions multiplied by spherical harmonics. The intensity pattern in the wave can therefore be very complicated. Of particular interest, however, is the simplest case in which p has no dependence on θ or ϕ but spreads uniformly from a single point at the origin.

For such a simple spherical wave, Eq. (6.23) becomes

$$\nabla^2 p = \frac{1}{r^2}\frac{\partial}{\partial r}\left(r^2 \frac{\partial p}{\partial r}\right), \tag{6.24}$$

and we can simplify matters even further by writing $p = \psi/r$, giving for Eq. (6.22)

$$\frac{\partial^2 \psi}{\partial r^2} + k^2 \psi = 0, \tag{6.25}$$

which is just the one-dimensional wave equation. The general solution for p is therefore a superposition of an outgoing and an incoming wave given by

$$p = \left(\frac{A}{r} e^{-jkr} + \frac{B}{r} e^{jkr}\right) e^{j\omega t}. \tag{6.26}$$

To find the acoustic particle velocity u, we use the equivalent of Eq. (6.4) in the form

$$\rho \frac{\partial u}{\partial t} = -\nabla p = -\frac{\partial p}{\partial r}, \tag{6.27}$$

where the second form of writing is possible since p depends only on r and t. Explicitly then, from Eq. (6.27) for the case of an outgoing wave ($B = 0$),

$$u = \frac{A}{r\rho c}\left(1 + \frac{1}{jkr}\right) e^{-jkr} e^{j\omega t}. \tag{6.28}$$

In the far field, when r is much greater than one wavelength ($kr \gg 1$), u is simply $p/\rho c$ as in the plane wave case, as we clearly expect. Within about one sixth of a wavelength of the origin, however, kr become less than unity. The velocity u then becomes large and shifted in phase relative to p.

The wave impedance for a spherical wave depends on distance from the origin, measured in wavelengths (i.e., on the parameter kr), and has the value

$$z = \frac{p}{u} = \rho c \left(\frac{jkr}{1 + jkr}\right). \tag{6.29}$$

Near the origin, $|z| \ll \rho c$, while $|z| \to \rho c$ for $kr \gg 1$.

The behavior of spherical waves with angular dependence involves more complex mathematics for its solution (Morse, 1948, pp. 314–321), and we will not take it up in detail here, though we will meet the topic again in the next chapter.

6.3 Sound Pressure Level and Intensity

Because there is a factor of about 10^6 between the acoustic pressure at the threshold of audibility and the limit of intolerable overload for the human ear, and because within that range subjective response is more nearly logarithmic than linear (actually, it is more complicated than this—see Rossing, 1982, Chapter 6), it is convenient to do something similar for acoustic pressures. We therefore define the sound pressure level (SPL or L_{p}) for an acoustic pressure p to be

$$L_{\mathrm{p}} = 20 \log_{10}\left(\frac{p}{p_0}\right), \tag{6.30}$$

measured in decibels (dB) above the reference pressure p_0. By convention both p and p_0 are rms values and p_0 is taken to be 20 μPa, which is approximately the threshold of human hearing in its most sensitive range from 1 to 3 kHz. On this SPL scale, 1 Pa is approximately 94 dB and the threshold of pain is about 120 dB. The normal range for music listening is about 40 to 100 dB. It is usual, however, to specify sound pressure levels in the environment after applying a filter (Type-A weighting) to allow for the decreased sensitivity of human hearing at low and high frequencies. The SPL is then given in dB(A).

The sensitivity of normal human hearing is shown in Fig. 6.2, as originally determined by Fletcher and Munson (1933). Each contour passes through

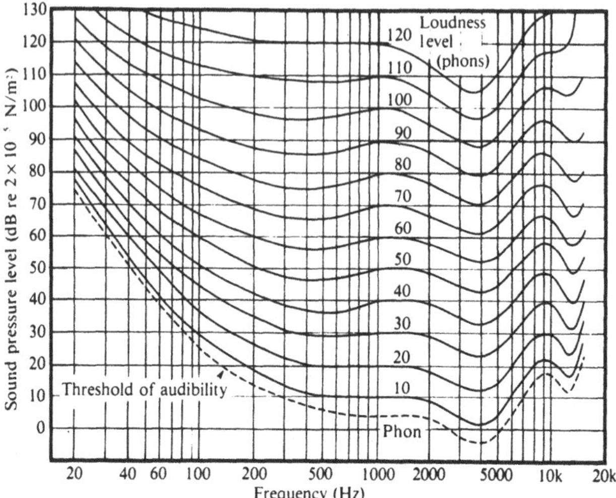

FIGURE 6.2. Equal loudness curves for human hearing. The curve marked MAF (minimum audible field) is the nominal threshold of binaural hearing for pure tones, and the 120-phon curve is the threshold of pain.

points of subjectively equal loudness and is labeled with a loudness level in phons, taken by definition to be the SPL associated with that contour at a frequency of 1 kHz. The normal threshold of hearing for pure tones is a few decibels above the 0 phon contour, and the threshold of pain, or at least discomfort, is at about 120 phon. The A-weighting curve used in specifying environmental (or musical) noise (or sound) levels is approximately the inverse of the 40-phon contour, which corresponds to very quiet sounds.

Because human hearing responds to acoustic pressure at a point, or rather at two points corresponding to two ears that have little acoustic coupling, the sound pressure level is usually the relevant quantity to specify. The SPL, however, depends on the environment and in particular on the reverberant quality of the space in which the sound source and listener are situated. In many cases, it is more useful physically to know the acoustic energy carried through a surface by sound waves. This quantity is called the acoustic intensity I and is measured in watts per square meter. Again, a logarithmic scale is convenient, and we define the intensity level (IL or L_I) to be

$$L_{\rm I} = 10 \log_{10} \left(\frac{I}{I_0} \right) \tag{6.31}$$

in decibels. [The factor is 10 rather than 20 as in Eq. (6.30), since I is proportional to p^2.] The reference intensity I_0 is taken as 10^{-12} W m^{-2}, which makes the SPL and the IL very nearly equal for a plane wave. For a standing wave, of course, the sound pressure level may be large, but the intensity will be small since the intensities in the two waves tend to cancel because of their opposite propagation directions.

To evaluate the intensity of a plane wave, we first calculate the energy density as a sum of kinetic and potential energy contributions and then average over space and time. This energy is transported with the wave speed c, so we just multiply by c to get the results. Without going into details (Kinsler et al., 1982, p. 110), we find the result

$$I = \rho c u^2 = \frac{p^2}{\rho c} = pu, \tag{6.32}$$

where p and u are taken as rms quantities (otherwise, a factor of $\frac{1}{2}$ should be inserted into each result).

Analysis for a spherical wave is more complex (Kinsler et al., 1982, p. 112) because much of the kinetic energy in the velocity field near the origin—specifically the part associated with the term $1/jkr$ in Eq. (6.28)—is not radiated because of its 90° phase shift relative to the acoustic pressure. The radiated intensity is given as in Eq. (6.32) by

$$I = \frac{p^2}{\rho c}, \tag{6.33}$$

where an rms value of p is implied, but the other forms of the result in Eq. (6.32) do not apply.

The total power P radiated in a spherical wave can be calculated by integrating $I(r)$ over a spherical surface of radius r, giving

$$P = \frac{4\pi r^2 p(r)^2}{\rho c}. \tag{6.34}$$

From Eq. (6.26), P is independent of r, as is obviously required. To get some feeling for magnitudes, a source radiating a power of 1 mW as a spherical wave produces an intensity level, or equivalently a sound pressure level, of approximately 79 dB at a distance of 1 m. At a distance of 10 m, assuming no reflections from surrounding walls or other objects, the SPL is 59 dB. These figures correspond to radiation from a typical musical instrument, though clearly a great range is possible. The disparity between this figure and the powers of order 100 W associated with amplifiers and loudspeakers is explained by the facts that the amplifier requires adequate power to avoid overload during transients and the normal operating level is only a few watts, while the loudspeaker itself has an efficiency of only about 1% in converting electrical power to radiated sound.

6.4 Reflection, Diffraction, and Absorption

When a wave encounters any variation in the properties of the medium in which it is propagating, its behavior is disturbed. Gradual changes in the medium extending over many wavelengths lead mostly to a change in the wave speed and propagation direction—the phenomenon of refraction. When the change is more abrupt, as when a sound wave in air strikes a solid object, such as a person or a wall, then the incident wave is generally mostly reflected or scattered and only a small part is transmitted into or through the object. That part of the wave energy transmitted into the object will generally be dissipated by internal losses and multiple reflections unless the object is very thin, like a lightweight wall partition, when it may be reradiated from the opposite surface.

It is worthwhile to examine the behavior of a plane pressure wave Ae^{-jkx} moving from a medium of wave impedance z_1 to one of impedance z_2. In general, we expect there to be a reflected wave Be^{jkx} and a transmitted wave Ce^{-jkx}. The acoustic pressures on either side of the interface must be equal, so that, taking the interface to be at $x = 0$,

$$A + B = C. \tag{6.35}$$

Similarly, the displacement velocities must be the same on either side of the interface, so that, using Eq. (6.19) and noting the sign of k for the various waves,

$$\frac{A - B}{z_1} = \frac{C}{z_2}. \tag{6.36}$$

We can now solve Eqs. (6.35) and (6.36) to find the reflection coefficient:

$$\frac{B}{A} = \frac{z_2 - z_1}{z_2 + z_1}. \tag{6.37}$$

and the transmission coefficient:

$$\frac{C}{A} = \frac{2z_2}{z_2 + z_1}. \tag{6.38}$$

These coefficients refer to pressure amplitudes. If $z_2 = z_1$, then $B = 0$ and $C = A$ as we should expect. If $z_2 > z_1$, then, from Eq. (6.37), the reflected wave is in phase with the incident wave and a pressure maximum is reflected as a maximum. If $z_2 < z_1$, then there is a phase change of 180° between the reflected wave and the incident wave and a pressure maximum is reflected as a minimum. If $z_2 \gg z_1$ or $z_2 \ll z_1$, then reflection is nearly total. The fact that, from Eq. (6.38), the transmitted wave will have a pressure amplitude nearly twice that of the incident wave if $z_2 \gg z_1$ is not a paradox, as we see below, since this wave carries a very small energy.

Perhaps even more illuminating than Eqs. (6.37) and (6.38) are the corresponding coefficients expressed in terms of intensities, using Eq. (6.32). If the incident intensity is $I_0 = A^2/z_1$, then the reflected intensity I_r is given by

$$\frac{I_r}{I_0} = \left(\frac{z_2 - z_1}{z_2 + z_1}\right)^2, \tag{6.39}$$

and the transmitted intensity I_t by

$$\frac{I_t}{I_0} = \frac{4z_2z_1}{(z_2 + z_1)^2}. \tag{6.40}$$

Clearly, the transmitted intensity is nearly zero if there is a large acoustic mismatch between the two media and either $z_2 \gg z_1$ or $z_2 \ll z_1$.

These results can be generalized to the case of oblique incidence of a plane wave on a plane boundary (Kinsler et al., 1982, pp. 131–133), and we then encounter the phenomenon of refraction, familiar from optics, with the reciprocal of the velocity of sound c_i in each medium taking the place of its optical refractive index.

All these results can be extended in a straightforward way to include cases where the wave impedances z_i are complex quantities $(r_i + jx_i)$ rather than real. In particular, the results [Eqs. (6.39) and (6.40)] carry over directly to this more general situation, the reflection and transmission coefficients generally depending upon the frequency of the wave.

If the surface of the object is flat, on the scale of a sound wavelength, and its extent is large compared with the wavelength, then the familiar rules of geometrical optics are an adequate approximation for the treatment of reflections. It is only for large areas, such as the walls or ceilings of concert halls, that this is of more than qualitative use in understanding behavior (Beranek, 1962; Rossing, 1982; Meyer, 1978).

At the other extreme, an object that is small compared with the wavelength of the sound wave involved will scatter the wave almost equally in all directions, the fractional intensity scattered being proportional to the sixth power of the size of the object. When the size of the object ranges from, for example, one-tenth of a wavelength up to 10 wavelengths, then scattering behavior is very complex, even for simply shaped objects (Morse, 1948, pp. 346–356; Morse and Ingard, 1968, pp. 400–449).

There is similar complexity in the "sound shadows" cast by objects. Objects that are very large compared with the sound wavelength create well-defined shadows, but this situation is rarely encountered in other than architectural acoustics. More usually, objects will be comparable in size to the wavelength involved, and diffraction around the edges into the shadow zone will blur its edges or even eliminate the shadow entirely at distances a few times the diameter of the object. Again, the discussion is complex even for a simple plane edge (Morse and Ingard, 1968, pp. 449–458). For the purposes of this book, a qualitative appreciation of the behavior will be adequate.

Even in an unbounded uniform medium, such as air, a sound wave is attenuated as it propagates, because of losses of various kinds (Kinsler et al., 1982, Chapter 7). Principal among the mechanisms responsible are viscosity, thermal conduction, and energy interchange between molecules with differing external excitation. If we write

$$k \to \frac{\omega}{c} - j\alpha, \tag{6.41}$$

so that the wave amplitude decays as $e^{-\alpha x}$ for a plane wave or as $(1/r)e^{-\alpha r}$ for a spherical wave, then α, or rather the quantity 8686α corresponding to attenuation in decibels per kilometer, is available in standard tables (Evans and Bass, 1986). In normal room air with relative humidity greater than about 50%, most of the attenuation is caused by molecular energy exchange with water vapor. The behavior is not simple, but for a frequency f, in hertz, it has roughly the form

$$\begin{aligned} a &\approx 4\times 10^{-7} f, && 100\,\text{Hz} < f < 1\,\text{kHz}, \\ a &\approx 1\times 10^{-10} f^2, && 2\,\text{kHz} < f < 100\,\text{kHz}. \end{aligned} \tag{6.42}$$

If the relative humidity is very low, say less than 20%, then α is increased by as much as a factor 10 over most of this frequency range. For completely dry air, α is increased by nearly a factor 30 below 100 Hz but is decreased by a factor of about 4 between 10 kHz and 100 kHz.

For many musical applications in small rooms, this absorption can be neglected, but this is not true of large concert halls, since the absorption at 10 kHz amounts to about 0.1 dB m^{-1}. This gives a noticeable reduction in brightness of the sound at the back of the hall.

More important, in most halls, is the absorption of sound upon reflection from the walls, ceilings, furnishings, and audience. If an impulsive sound is made in a hall, then the sound pressure level decays nearly linearly, corresponding to an exponential decay in sound pressure. The time T_{60} for the level to decay by 60 dB is known as the reverberation time. Details of the behavior are complicated, but to a first approximation T_{60} is given by the Sabine equation:

$$T_{60} = \frac{0.161V}{\sum_i \alpha_i S_i}, \tag{6.43}$$

where V is the hall volume (in cubic meters) and the S_i are areas (in square meters) of surface with absorption coefficients α_i. Satisfactory reverberation time is important for good music listening but depends on building size and musical style. For a small hall for chamber music, T_{60} may be as small as 0.7 s, for a large concert hall, it may be 1.7 to 2 s, and for a large cathedral suitable for nineteenth century organ music, as long as 10 s (Meyer, 1978; Beranek, 1962, pp. 555–69).

In musical instruments, we will often be concerned with sound waves confined in tubes or boxes or moving close to other surfaces. In such cases, there is attenuation caused by viscous forces in the shearing motion near the surface and by heat conduction from the wave to the solid. Both these effects are confined to a thin boundary layer next to the surface, the thickness of which decreases as the frequency increases. The viscous and thermal boundary layers have slightly different thicknesses, δ_{v} and δ_{t}, given by (Benade, 1968)

$$\delta_{\text{v}} = \left(\frac{\eta}{2\pi\rho}\right)^{1/2} f^{-1/2} \approx 1.6\times 10^{-3} f^{-1/2},$$

and (6.44)

$$\delta_t = \left(\frac{\kappa}{2\pi\rho C_p}\right)^{1/2} f^{-1/2} \approx 1.9 \times 10^{-3} f^{-1/2},$$

where ρ is the density, η the viscosity, κ the thermal conductivity, and C_p the specific heat of air (per unit volume). For sound waves in musical instruments, these boundary layers are thus about 0.1 mm thick. We shall return to consider these surface losses in more detail in Chapter 8.

6.5 Normal Modes in Cavities

While many problems involve the propagation of acoustic waves in nearly unconfined space, many other practical problems involve the acoustics of enclosed spaces such as rooms, tanks, and other cavities. We defer discussion of wave propagation in pipes and horns—which generally have at least one open end—to Chapter 8, and consider here only completely enclosed volumes.

The mathematical principles involved are straightforward—we simply need to solve the wave propagation equation (6.21) with appropriate boundary conditions on the walls of the enclosure. To be more explicit, the three-dimensional wave equation can be written in vector notation in the form

$$\frac{\partial^2 p}{\partial t^2} = c^2 \nabla^2 p \tag{6.45}$$

and we need to express the vector differential operator ∇ in coordinates related to the geometry of the enclosure so that the boundary conditions are simple. If the enclosure is rectangular, for example, then the wave equation (6.45) becomes

$$\frac{\partial^2 p}{\partial t^2} = c^2 \left(\frac{\partial^2 p}{\partial x^2} + \frac{\partial^2 p}{\partial y^2} + \frac{\partial^2 p}{\partial z^2}\right) \tag{6.46}$$

and we can write $p(x, y, z) = p_x(x)\, p_y(y)\, p_z(z)$. We return to this case in a moment.

The boundary conditions at the boundaries of the enclosure depend upon the physical nature of the walls. Suppose that the walls present an acoustic wave impedance z_w, then at the walls we must have

$$p = z_w u = \frac{j z_w}{\rho\omega} \frac{\partial p}{\partial n}, \tag{6.47}$$

where u is the acoustic particle velocity, and the second form of writing comes from Eq. (6.4), with n being the coordinate normal to the surface of the wall. It is easy enough to deal with this general boundary condition if the walls are either simply springy or simply slack and heavy, so that z_{w} is purely imaginary. More generally the walls will have resistive losses so that z_{w} will be a complex quantity. The simplest case is that in which the walls are completely rigid, for then $z_{\mathrm{w}} = \infty$ and Eq. (6.47) reduces to

$$\frac{\partial p}{\partial n} = 0. \tag{6.48}$$

It is usually an adequate approximation for real enclosures such as rooms or the interior of musical instruments to adopt Eq. (6.48) as a boundary condition when determining the mode frequencies and to make allowance for any real part in z_{w} by adding damping to the modes.

To illustrate some aspects of the mode problem, suppose that we have a rectangular enclosure with sides of lengths a, b, and c. Solutions of Eq. (6.46) that satisfy the boundary condition (6.48) on all the walls have the form

$$p(x, y, z, t) = A \cos\left(\frac{l\pi x}{a}\right) \cos\left(\frac{m\pi y}{b}\right) \cos\left(\frac{n\pi z}{c}\right) \sin \omega t, \tag{6.49}$$

where l, m, n are integers (including the possibility of zero) and the frequency ω satisfies

$$\omega = \pi c \left(\frac{l^2}{a^2} + \frac{m^2}{b^2} + \frac{n^2}{c^2}\right)^{1/2}. \tag{6.50}$$

The frequencies given by this equation are the mode frequencies for the enclosure. Clearly their distribution depends upon the relative values of a, b, and c, and thus on the shape of the enclosure. Figure 6.3 shows such

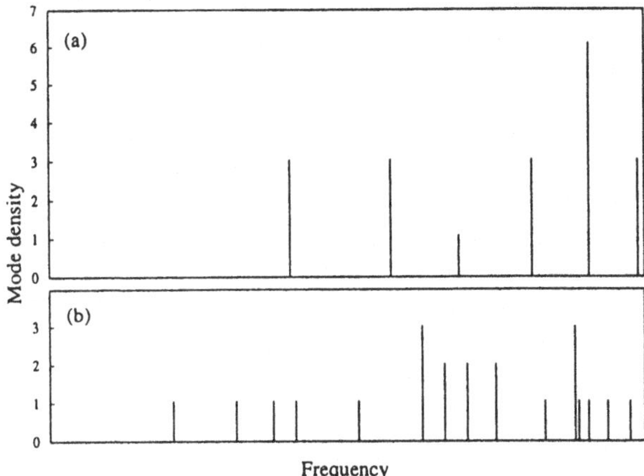

FIGURE 6.3. Distribution of mode frequencies for two rectangular rooms of equal volume and dimension ratios (a) 2:2:2 and (b) 1:2:3. Where mode frequencies are coincident, the relevant line has been lengthened proportionally.

mode distributions for two representative rooms with equal volume but with dimensions in the ratio 2:2:2 and 1:2:3, respectively. It is one of the important parameters of good architectural acoustics to arrange the room shape so that the frequencies of the lower modes of the space are reasonably evenly distributed—any concentration of modes near some particular frequency can give a tonal character to the reverberation that is generally detrimental to music listening. Clearly the 2:2:2 room has a "peaky" response with many coincident resonances, while the 1:2:3 room has a much more even spread and is likely to be more satisfactory musically.

We can adopt the same approach to calculate the mode frequencies for other enclosures with simple geometry, in particular for a spherical cavity and for a cavity in the form of a circular cylinder. For the spherical case the wave equation can be separated in spherical polar coordinates (r, θ, ϕ), and the mode functions for the pressure can be written in terms of spherical Bessel functions and spherical harmonics. The cylindrical case can be separated in cylindrical polar coordinates (r, ϕ, z) and the pressure can be expressed in terms of ordinary Bessel functions and trigonometric (circular) functions. We shall not go through the detailed mathematics here, but simply remark that the resulting mode-frequency distribution appears irregular, as for a rectangular enclosure, but can be calculated in a straightforward way once we are familiar with the properties of Bessel functions.

When we consider the air modes inside a cavity with a much more complex shape, such as a violin body, we can no longer find a coordinate system in which the wave equation can be separated as in Eq. (6.46). Jansson (1977), however, has shown that the distribution of modes in such a cavity can be estimated reliably using perturbation theory. In this approach we start with the known mode frequencies for, say, a rectangular enclosure of appropriate dimensions, and then calculate the changes in these mode frequencies brought about by perturbing the shape into a better match to the real violin.

References

Benade, A.H. (1968). On the propagation of sound waves in a cylindrical conduit. *J. Acoust. Soc. Am.* **44**, 616–623.

Beranek, L.L. (1954). "Acoustics," Chapter 3. McGraw-Hill, New York. Reprinted 1986, Acoustical Society Am., Woodbury, New York.

Beranek, L.L. (1962). "Music, Acoustics and Architecture." Wiley, New York.

Beyer, R.T. (1974). "Nonlinear Acoustics." Naval Sea Systems Command, U.S. Navy.

Evans, L.B., and Bass, H.E. (1986). Absorption and velocity of sound in still air. In "Handbook of Chemistry and Physics," 67th ed., pp. E45–E48. CRC Press, Boca Raton, Florida.

Fletcher, H., and Munson, W.A. (1933). Loudness, its definition, measurement and calculation. *J. Acoust. Soc. Am.* **5**, 82–108.

Fletcher, N.H. (1974). Adiabatic assumption for wave propagation. *Am. J. Phys.* **42**, 487–489.

Jansson, E.V. (1977). Acoustical properties of complex cavities. Prediction and measurement of resonance properties of violin-shaped and guitar-shaped cavities. *Acustica* **37**, 211–221.

Kinsler, L.E., Frey A.R., Coppens, A.B., and Sanders, J.V. (1982). "Fundamentals of Acoustics," 3rd ed., Wiley, New York.

Meyer, J. (1978). "Acoustics and the Performance of Music." Verlag Das Musikinstrument, Frankfurt am Main.

Morse, P.M. (1948). "Vibration and Sound." McGraw-Hill, New York. Reprinted 1981, Acoustical Society Am., Woodbury, New York.

Morse, P.M., and Feshbach, H. (1953). "Methods of Theoretical Physics," 2 vols. McGraw-Hill, New York.

Morse, P.M., and Ingard, K.U. (1968). "Theoretical Acoustics." McGraw-Hill, New York; reprinted 1986, Princeton Univ. Press, Princeton, New Jersey.

Rossing, T.D. (1982). "The Science of Sound," Chapter 23. Addison-Wesley, Reading, Massachusetts.

7

Sound Radiation

In Chapter 6, we discussed some of the basic properties of sound waves and made a brief examination of the way sound waves are influenced by simple structures, such as tubes and cavities. In the present chapter, we take up the inverse problem and look at the way in which vibrating structures can generate sound waves. This is one of the most basic aspects of the acoustics of musical instruments—it is all very well to understand the way in which a solid body vibrates, but unless that vibration leads to a radiated sound wave, we do not have a musical instrument. We might, of course, simply take the fact of sound radiation for granted, and this is often done. This neglects, however, a great deal of interesting and important physics and keeps us from understanding much of the subtlety of musical instrument behavior.

Our plan, therefore, will be to look briefly at the properties of some of the simplest types of sources—monopoles, dipoles, and higher multipoles—to see the behavior we might expect. We then look at radiation from vibrating cylinders, since vibrating strings are so common in musical instruments, and then go on to the much more complicated problem of radiation from the motion of reasonably large and more-or-less flat bodies. These are, of course, the essential sound-radiating elements of all stringed and percussion instruments. The radiation of sound from the vibrating air columns of wind instruments presents a related but rather different set of problems that we defer for discussion in Chapter 8.

7.1 Simple Multipole Sources

The simplest possible source is the point source, which is the limit of a pulsating sphere as its radius tends to zero. Suppose the sphere has a small radius a and that the pulsating flow has a frequency ω and amplitude

$$Q = 4\pi a^2 v(a), \tag{7.1}$$

where $v(a)$ is the radial velocity amplitude at the surface. This object clearly generates a spherical wave, and, from our discussion in Chapter 6 and specifically from Eqs. (6.26) and (6.28), the pressure and velocity amplitudes in such a wave at radius r are given by

$$p(r) = (A/r)e^{-jkr}, \tag{7.2}$$

and

$$v(r) = \frac{A}{\rho c r}\left(1 - \frac{j}{kr}\right)e^{-jkr}. \tag{7.3}$$

Matching Eq. (7.3) to Eq. (7.1) on the surface of the sphere and assuming $ka \ll 1$, we find a value for A from which Eq. (7.2) gives

$$p(r) = \frac{j\omega\rho}{4\pi r}Qe^{-jkr}. \tag{7.4}$$

This result does not depend on the sphere radius a, provided $ka \ll 1$, and so it is the pressure wave generated by a point source of strength Q. Such a source is also called a monopole source. Its radiated power P is simply $\frac{1}{2}p^2/\rho c$ integrated over a spherical surface, whence

$$P = \frac{\omega^2\rho Q^2}{8\pi c}. \tag{7.5}$$

Note that for a given source strength Q the radiated power increases as the square of the frequency.

The next type of source to be considered is the dipole, which consists of two simple monopole sources of strengths $\pm Q$ separated by a distance dz, in the limit $dz \to 0$. A physical example of such a source is the limit of a small sphere oscillating to and fro. Referring to Fig. 7.1, if r_+ and r_- are, respectively, the distances from the positive and negative sources in the dipole to the observation point (r, θ), then the acoustic pressure at this

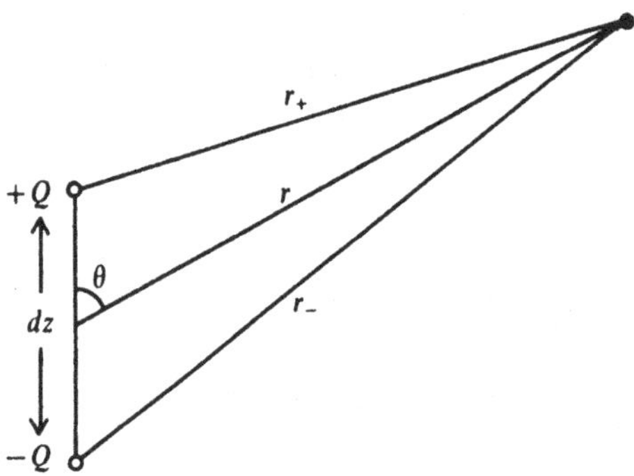

FIGURE 7.1. A dipole source. In the limit $dz \to 0$, $Q \to \infty$, the dipole moment is $\mu = Q\,dz$.

point is, from Eq. (7.4),

$$p = \left(\frac{j\omega\rho}{4\pi}\right)\left(\frac{e^{-jkr_+}}{r_+} - \frac{e^{-jkr_-}}{r_-}\right)Q. \tag{7.6}$$

If we regard r_+ as a function of z so that $r_+ = r - \frac{1}{2}d\cos\theta$, and similarly for r_-, then as $d \to dz$ the difference in brackets in Eq. (7.6) can be replaced by a differential, so that

$$\begin{aligned} p &= \frac{j\omega\rho}{4\pi}\frac{\partial}{\partial z}\left(-\frac{2e^{-jkr_+}}{r_+}\right)Q\,dz \\ &= \frac{\omega^2\rho}{4\pi cr}\left(1 + \frac{1}{jkr}\right)e^{-jkr}\mu\cos\theta, \end{aligned} \tag{7.7}$$

where we have let $Q \to \infty$ as $dz \to 0$, so that the dipole moment $\mu = Q\,dz$ remains finite. The velocity field can be found by taking the gradient of p as usual. In the far field, where $kr \gg 1$, Eq. (7.7) can clearly be simplified by neglecting $1/jkr$ relative to 1. The radiated power P is simply

$$P = \frac{1}{2}\int\int\frac{p^2}{\rho c}r^2\sin\theta\,d\theta\,d\phi = \frac{\omega^4\rho\mu^2}{24\pi c^3}. \tag{7.8}$$

The power radiated by a dipole is thus a very strong function of frequency, and dipole sources are very inefficient radiators at low frequencies. From Eq. (7.7), the radiation is concentrated along the axis of the dipole.

The process of differentiating the field of a monopole to obtain that of a dipole, as expressed in Eq. (7.7), can be thought of as equivalent to reflecting the monopole source, and its radiation field, in a pseudomirror, which changes the sign of the source. The limit operation as we go from Eq. (7.6) to Eq. (7.7) is equivalent to moving the source closer and closer to the mirror plane.

The next step in complication is to reflect the dipole in another mirror plane, as shown in Fig. 7.2, to produce a quadrupole. This can be done in two ways to produce either an axial (longitudinal) quadrupole, in which the four simple poles lie on a straight line, or a plane (lateral) quadrupole,

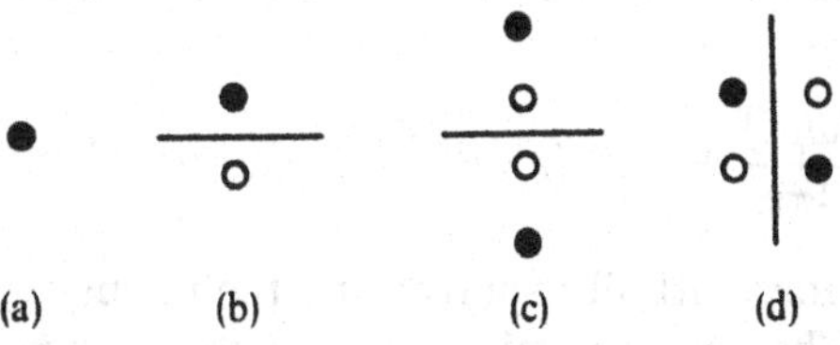

FIGURE 7.2. Generation of two possible configurations of a quadrupole by successive reflection of a monopole, with a sign change on each reflection: (a) monopole, (b) dipole, (c) axial quadrupole, and (d) planar quadrupole.

in which they lie at the corners of a square. The sign of the reflection is always chosen so that the quadrupole has no dipole moment. A physical example of a quadrupole source is a sphere, vibrating so that it becomes alternately a prolate and an oblate spheroid.

The pressure field for a quadrupole can be found by differentiating that for a dipole, Eq. (7.7), either with respect to z for an axial quadrupole or x for a plane quadrupole. To find the far-field radiation terms, the differentiation can be confined to the exponential factor. If the monopole source strength is Q, then we clearly have the following sequence of results for the far-field pressure generated by monopole, dipole, and quadrupole sources, respectively:

$$p_m = jk\left(\frac{\rho c Q}{4\pi r}\right)e^{-jkr}, \tag{7.9}$$

$$p_d = -k^2\left(\frac{\rho c Q\,dz}{4\pi r}\right)\cos(r,z)e^{-jkr}, \tag{7.10}$$

and

$$p_q = -jk^3\left(\frac{\rho c Q\,dz\,dx}{4\pi r}\right)\cos(r,z)\cos(r,x)e^{-jkr}. \tag{7.11}$$

The notation is obvious and, for an axial quadrupole, x is replaced by z in Eq. (7.11). The important points to note are the increasingly complex angular behavior and more steeply decreasing radiation efficiency at low frequencies as we proceed through the series.

7.2 Pairs of Point Sources

To guide our later discussion, it is now helpful to examine briefly the radiation behavior of combinations of several point sources whose separation is not necessarily small compared with the sound wavelength. First, let us treat the case of two monopoles of strength Q separated by a distance d as shown in Fig. 7.3. The sources can be of either the same or opposite sign, and we seek the pressure p at a large distance $r \gg d$ in direction θ. For $r \sim d$, the expression is complicated, as in Eq. (7.6), but for $r \gg d$, we have

$$p \approx \left(\frac{j\omega\rho Q}{4\pi r}\right)e^{-jkr}\left(e^{1/2jkd\cos\theta} \pm e^{-1/2jkd\cos\theta}\right), \tag{7.12}$$

where the plus sign goes with like sources and the minus sign with sources of opposite phases. The absolute value of the square of p is

$$|p^2| = \left(\frac{\omega\rho Q}{2\pi r}\right)^2\begin{Bmatrix}\cos^2\\ \sin^2\end{Bmatrix}\left(\frac{1}{2}kd\cos\theta\right), \tag{7.13}$$

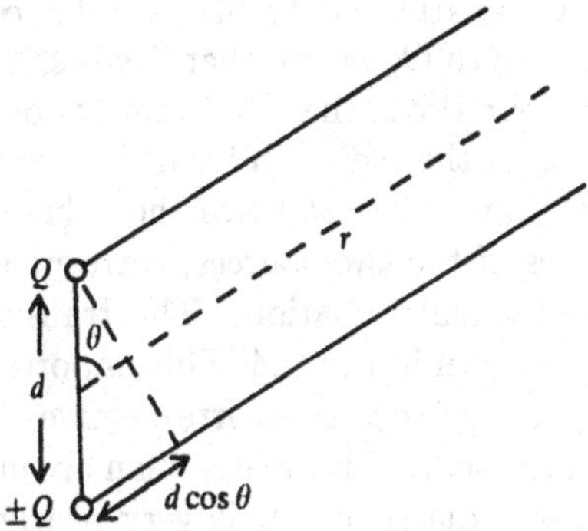

FIGURE 7.3. Radiation from a pair of separated monopoles.

where $\cos^2$ goes with the plus and $\sin^2$ with the minus sign. The total radiated power is then

$$P = \frac{1}{2}\int\int\left(\frac{|p^2|}{\rho c}\right) r^2 \sin\theta \, d\theta \, d\phi = \frac{\omega^2 \rho Q^2}{4\pi c}\left[1 \pm \frac{\sin kd}{kd}\right]. \tag{7.14}$$

The results in Eqs. (7.13) and (7.14) contain a great deal of information. From Eq. (7.13), the angular variation of the acoustic intensity $p^2/\rho c$ is very complex if kd is not small relative to unity, and there are many values of θ for which the intensity vanishes, whether the two sources are in phase or out of phase. From Eq. (7.14), however, the behavior of the total radiated

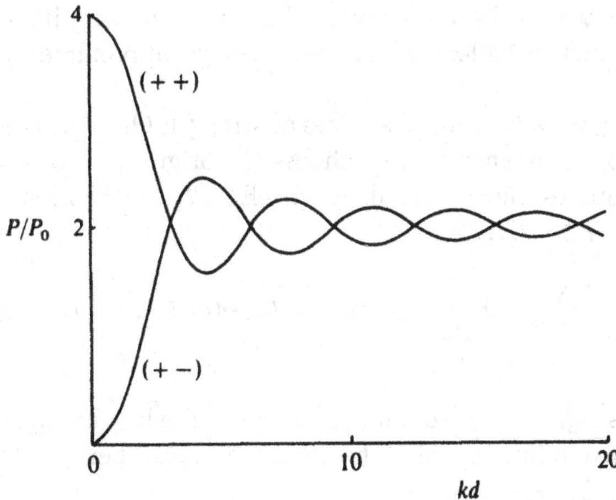

FIGURE 7.4. Total power P radiated from a pair of monopoles of the same (++) or opposite (+−) phase and separation d, as functions of the frequency parameter kd. The power radiated from an isolated monopole is P_0.

power is much simpler. Comparing Eq. (7.14) with the result in Eq. (7.5) for a monopole source of strength Q, we see that if $kd \ll 1$, then the radiated power is either zero or four times that for a single source, corresponding to coherent superposition of the radiation from the two monopoles. On the other hand, if $kd \gg 1$, then P is just twice the value for a single source, irrespective of the phases of the two sources, corresponding to incoherent superposition of the individual radiations. The transition between these two forms of behavior is shown in Fig. 7.4. This important general result is true of the radiation from any two sources, irrespective of exact similarity. If the separation between the sources is greater than about half a wavelength ($kd > 3$), then the total radiated power is very nearly equal to the sum of the powers radiated by the two sources treated independently. A more detailed treatment along the same lines as that given above (Junger and Feit, 1986, Chapter 2) gives information about the angular variation of acoustic intensity and of the acoustic pressure in the near-field region where $r \sim d$, this pressure generally being higher than given by the far-field approximation [Eq. (7.12)] if used for the near field. These details need not concern us here.

7.3 Arrays of Point Sources

In some fields of acoustics, for example in sonar, we need to be concerned with arrays of point sources all related in phase. Something rather similar may apply to radiation from the open finger holes of a woodwind instrument. More usually, we will be concerned with radiation from an extended vibrating source, such as a drumhead or a piano soundboard, which is divided by nodal lines into areas vibrating in antiphase with their neighbors. It is instructive to look at the point-source approximations to these systems.

Suppose we have a line of $2N$ sources of strength Q, each separated from its neighbors by a distance d. If we choose the origin to be at the midpoint of the line of sources, then, by analogy with Eq. (7.12), the acoustic pressure p at a distance $r \gg 2Nd$ is

$$p_{\pm} \approx \left(\frac{j\omega\rho Q}{4\pi r}\right) e^{-jkr} \sum_{n=1}^{N} (\pm 1)^n [e^{(n-1/2)jkd\cos\theta} \pm e^{-(n-1/2)jkd\cos\theta}], \tag{7.15}$$

where the plus sign applies to a line of sources all with the same phase and the minus sign to a line in which the phase alternates between 0 and π. We rewrite Eq. (7.15) as

$$p_{+} \approx \left(\frac{j\omega\rho Q}{4\pi r}\right) e^{-jkr} \sum_{n=1}^{N} 2\cos[(n - \frac{1}{2})kd\cos\theta]$$

$$p_- \approx \left(\frac{j\omega\rho Q}{4\pi r}\right) e^{-jkr} \sum_{n=1}^{N} (-1)^n 2j \sin[(n - \frac{1}{2})kd\cos\theta]. \qquad (7.16)$$

These two series are readily summed, using the expanded form of the trigonometric functions shown in Eq. (7.15), to give

$$p_+ \approx \left(\frac{j\omega\rho Q}{4\pi r}\right) e^{-jkr} \left[\frac{\sin(Nkd\cos\theta)}{\sin(\frac{1}{2}kd\cos\theta)}\right], \qquad (7.17)$$

and

$$p_- \approx \left(\frac{j\omega\rho Q}{4\pi r}\right) e^{-jkr} \left[\frac{(-1)^N \sin(Nkd\cos\theta)}{\cos(\frac{1}{2}kd\cos\theta)}\right]. \qquad (7.18)$$

These results indicate a rather complex radiation pattern in the θ coordinate, but the important features are immediately clear. Let us look first at the case where the sources are all in phase, giving radiated acoustic pressure p_+. The term in square brackets in Eq. (7.17) is always large when $\theta = 90°$, the zero in the numerator being balanced by one in the denominator, and has at that angle the value $2N$. The radiation intensity at $\theta = 90°$ is thus equivalent to that from a source of strength $2NQ$, and, from the form of the bracket, the width of this radiation lobe in radians is approximately $2\pi/Nkd = \lambda/Nd$, where λ is the acoustic wavelength. If $kd < 2\pi$ or $\lambda > d$, then this is the only large maximum of the term in square brackets, which is of order unity at all other angles. The radiation pattern of the array at low frequencies is thus as shown in Fig. 7.5(a).

If the frequency is higher, so that $\lambda < d$ or $kd > 2\pi$, then the denominator in the square brackets of Eq. (7.17) can vanish at other angles θ^* for which

$$\theta^* = \cos^{-1}\left(\frac{2n\pi}{kd}\right), \qquad (7.19)$$

and this will have solutions for one or more values of the integer n. Each zero in the denominator is again balanced by a zero in the numerator, and the bracket again has the value $2N$. The radiation pattern now has the form shown in Fig 7.5(b), with more lobes added at higher frequencies.

The total power radiated by the in-phase array can be found by integrating the intensity $p_+^2/\rho c$ over the surface of a large sphere surrounding the source. The behavior is broadly similar to that shown for the in-phase source pair in Fig. 7.4, except that the high-frequency power approaches NP_0 and the low-frequency power approaches N^2P_0 when the wavelength is greater than the length of the entire array, P_0 being the power radiated by a simple source of strength Q.

The antiphase array, with radiated pressure p_-, behaves rather differently. If $\lambda > 2d$ so that $kd < \pi$, then there are no zeros in the denominator and the term in square brackets is of order unity at all angles θ. The radiated pressure is thus only comparable with that from a single source of

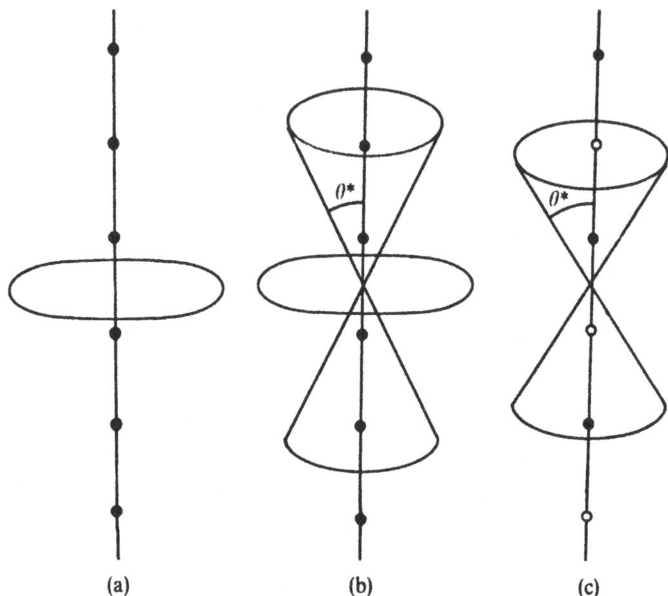

FIGURE 7.5. Radiation intensity patterns for (a) a linear co-phase array of sources of separation d for $kd < 2\pi$ or $\lambda > d$; (b) a co-phase array for $kd > 2\pi$, $\lambda < d$; and (c) an antiphase array for $kd > \pi$, $\lambda < 2d$. There is very little radiation from an antiphase array with $kd < \pi$, $\lambda > 2d$.

strength Q and the radiation process is very inefficient. Most of the flow of the medium is simply a set of closed loops from one source to its antiphase neighbors and, though the acoustic pressure is actually large at distances less than about d from the source line, very little of this escapes as radiation.

At frequencies sufficiently high that $\lambda < 2d$ or $kd > \pi$, a frequency that is half that for an in-phase source, zeros occur in the denominator of Eq. (7.18) for angles θ^* given by

$$\theta^* = \cos^{-1}\left[\frac{(2n-1)\pi}{kd}\right] \tag{7.20}$$

for one or more positive integers n. The radiation pattern then has the form shown in Fig. 7.5(c), and the total radiated power becomes approximately NP_0.

We can extend these methods to the practically important case of a square-grid array of antiphase sources as a prototype for the radiation to be expected from the complex vibrations of plates or diaphragms. It is not necessary to go through this analysis in detail, however, since the important results can be appreciated from the linear case treated above. If the centers of the antiphase regions, represented in the model by point sources, are less than about half of the free-air acoustic wavelength apart, then the radiation efficiency is low (assuming the numbers of co-phase and antiphase sources to be equal) and about equivalent to that of a single simple source. Except in the case of bells, this is nearly always the situation for the free vibrational modes of plates, shells, and diaphragms in musical instruments, because the speed of transverse mechanical waves in the plate is generally much less than the speed of sound in air. However, these modes can be driven at higher frequencies by externally applied periodic forces, and the radiation condition can then be satisfied, as we discuss further in Section 7.8.

7.4 Radiation from a Spherical Source

Although a spherical source pulsating in radius would not at first sight seem to be a good model for any musical instrument, it turns out that the radiation from a source with a pulsating volume, such as a closed drum, is very little dependent on its shape provided its linear dimensions are small in comparison to the relevant sound wavelength in air. We will, however, look at radiation from a pulsating sphere rather more generally than this.

Suppose that the sphere has radius a and pulsates with surface velocity $u\exp(j\omega t)$. If we set this equal to the radial velocity at distance a from a simple source, as given by Eq. (7.3), then from Eq. (7.2) the acoustic pressure at distance r becomes

$$p(r) = \frac{j\rho cka^2u}{r}\left(\frac{1-jka}{1+k^2a^2}\right)e^{-jk(r-a)}. \tag{7.21}$$

There are obvious simplifications if $ka \ll 1$. The radiated power P is then

$$P = \frac{2\pi\rho ck^2a^4u^2}{1+k^2a^2} \tag{7.22}$$

$$\rightarrow 2\pi\rho ca^2u^2 = \frac{\rho cQ^2}{8\pi a^2} \quad \text{if} \quad ka \gg 1, \tag{7.23}$$

$$\rightarrow 2\pi\rho c(ka)^2a^2u^2 = \left(\frac{\rho cQ^2}{8\pi a^2}\right)(ka)^2 \quad \text{if} \quad ka \ll 1, \tag{7.24}$$

where Q is the volume flow amplitude of the source. For a given surface velocity u, the radiated power per unit area increases as $(ka)^2$ while $ka \ll 1$ but then saturates for $ka \gg 1$. Most musical instruments in this approximation operate in the region $ka < 1$, so that, other things being equal, there is usually an advantage in increasing the size of the flexible radiating enclosure.

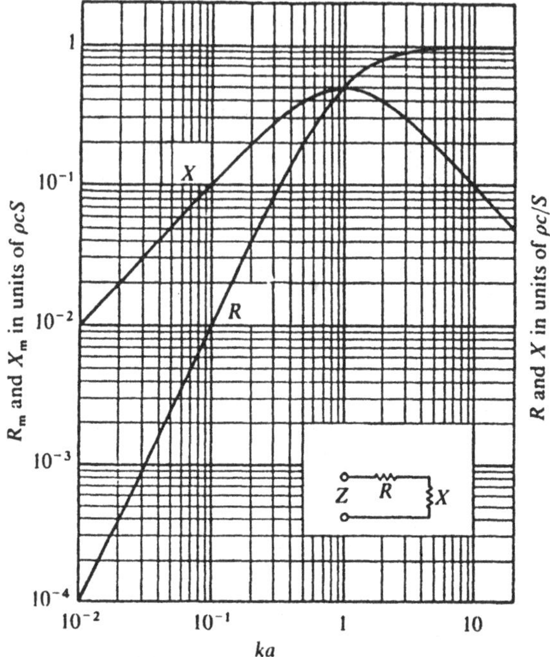

FIGURE 7.6. Real and imaginary parts of the mechanical load $R_m + jX_m$ on the surface of a sphere of radius a pulsating with frequency ω and velocity amplitude u. R_m and X_m are given in units of ρcS, where $S = 4\pi a^2$. R and X are corresponding acoustic quantities in units of $\rho c/S$ (after Beranek, 1954).

It is also useful to know the mechanical load on the spherical surface. From Eq. (7.21), this is simply

$$F = 4\pi a^2 p(a) = \rho cS \left(\frac{k^2a^2 + jka}{1 + k^2a^2} \right) u = (R_{\rm m} + jX_{\rm m})u, \qquad (7.25)$$

where $S = 4\pi a^2$ is the area of the sphere. The real part $R_{\rm m}u$ of this force, which increases as $(ka)^2$ until it saturates at ρcSu, represents the dissipative load of the radiation. The imaginary part $X_{\rm m}u$, which increases as ka for $ka < 1$ and decreases as $(ka)^{-1}$ above $ka = 1$, represents the mass load of the co-moving air. Equation (7.25) is plotted in Fig. 7.6. The load on pulsating volume sources of all shapes behaves very similarly. Also shown in Fig. 7.6 are the acoustic quantities R and X, to be discussed in Chapter 8. From the definition Eq. (8.3), these are just $1/S^2$ times their mechanical counterparts.

7.5 Line Sources

The only common line sources in musical acoustics are transversely vibrating strings. As a first approach, it is convenient to idealize such a string as a cylinder of infinite length and radius a, vibrating with angular frequency ω. Such a source has a dipole character, so that, from the discussion of Section 7.1, we expect it to be an inefficient radiator at low frequencies.

A detailed discussion of this problem is given by Morse (1948, pp. 299–300). All that we need here are the final results for intensity I and total radiated power per unit length P for $ka \ll 1$. These are

$$I \approx \left(\frac{\pi \rho \omega^3 a^4 u^2}{4c^2 r} \right) \cos^2 \phi, \tag{7.26}$$

and

$$P \approx \frac{\rho \pi^2 \omega^3 a^4 u^2}{4c^2}, \tag{7.27}$$

where $u \exp j\omega t$ is the vibration velocity of the string. The waves are, of course, cylindrical. Clearly, there is a very strong dependence on both ω and a and, in fact, the directly radiated acoustic power is almost negligibly small for the string diameters and frequencies commonly met in musical instruments.

For the vibration of a string in its fundamental mode, the infinite cylinder approximation is reasonable, but for higher modes we must recognize that adjoining sections of string vibrate in antiphase relation. Since the transverse wave velocity on the string is significantly less than c, these string sections are separated by less than half the sound wavelength in air, so there is an additional cause for cancellation of the radiated sound intensity, as discussed in Section 7.3.

It is therefore reasonable to neglect the contribution of direct radiation from vibrating strings to the sound of musical instruments. It is only when a vibrating cylinder has a quite large radius, as in the cylinders of tubular bells, that direct radiation becomes significant.

7.6 Radiation from a Plane Source in a Baffle

Few, if any, musical radiators consist of some sort of moving part set in an infinite plane baffle, but we examine the behavior of this system because it is the only case for which a simple general result emerges. Fortunately, it also happens that replacement of the plane baffle by an enclosure of finite size does not have a really major effect on the results, though the changes are significant.

Referring to Fig. 7.7, suppose that the area S on an otherwise rigid plane baffle is vibrating with a velocity distribution $u(\mathbf{r}')$ and frequency ω normal to the plane, all points being either in phase or in antiphase. The small element of area dS at $\mathbf{r}'$ then constitutes a simple source of volume strength $u(\mathbf{r}')dS$, which is doubled to twice this value by the presence of the plane, which restricts its radiation to the half-space of solid angle 2π. The pressure dp produced by this element at a large distance $\mathbf{r}$ is

$$dp(\mathbf{r}) = \frac{j\omega\rho}{2\pi r} e^{-jk|\mathbf{r}-\mathbf{r}'|} u(\mathbf{r}')dS. \tag{7.28}$$

If we take $\mathbf{r}$ to be in the direction (θ, ϕ) and $\mathbf{r}'$ in the direction $(\pi/2, \phi')$, then we can integrate Eq. (7.28) over the whole surface of the plane, remembering that $u = 0$ outside S, to give

$$p(r, \theta, \phi) = \frac{j\omega\rho}{2\pi r} e^{-jkr} \int\!\!\int_S e^{jkr' \sin\theta \cos(\phi-\phi')} u(\mathbf{r}') r'\, d\phi'\, dr'. \tag{7.29}$$

The integral in Eq. (7.29) has the form of a spatial Fourier transform of the velocity distribution $u(\mathbf{r}')$. This is our general result, due in the first place to Lord Rayleigh.

It is now simply a matter of algebra to apply Eq. (7.29) to situations of interest. These include a rigid circular piston and a flexible circular piston (Morse, 1948, pp. 326–335) and both square and circular vibrators excited in patterns with nodal lines (Skudrzyk, 1968, pp. 373–429; Junger and Feit, 1986, Chapter 5). There is not space here to review this work in

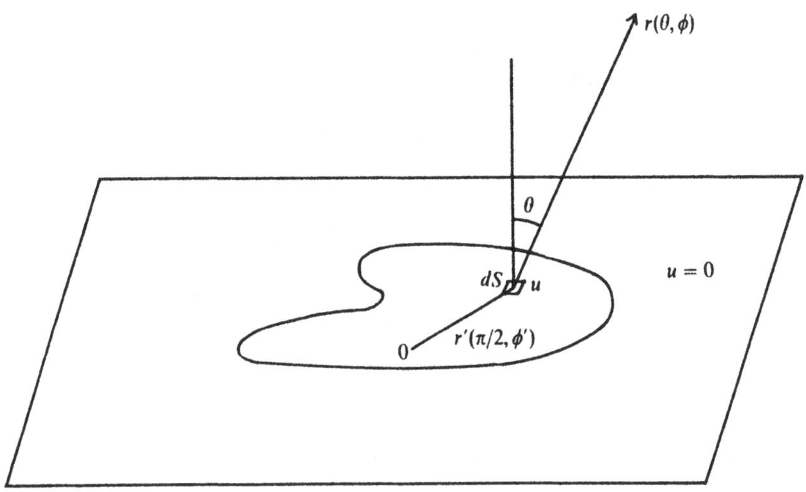

FIGURE 7.7. A vibrating plane source set in an infinite plane baffle. Radiation pressure is evaluated at a point at a large distance r in the direction shown.

detail, but we shall select particular examples and relate the conclusion to the simplified treatments given in the earlier sections of this chapter.

The integral in Eq. (7.29) can be performed quite straightforwardly for the case of a circular piston of radius a with u constant across its surface. The result for the far field (Morse, 1948, pp. 327–328) is

$$p \approx \tfrac{1}{2} j\omega\rho u a^2 \left(\frac{e^{-jkr}}{r}\right)\left[\frac{2J_1(ka\sin\theta)}{ka\sin\theta}\right], \tag{7.30}$$

where J_1 is a Bessel function of order one. The factor in square brackets is nearly unity for all θ if $ka \ll 1$, so the radiation pattern in the half-space $0 \le \theta < \pi/2$ is isotropic at low frequencies. For higher frequencies, the bracket is unity for $\theta = 0$ and falls to zero when the argument of the Bessel function is about 3.83, that is for

$$\theta^* = \sin^{-1}\left(\frac{3.83}{ka}\right). \tag{7.31}$$

The angular width $2\theta^*$ of the primary radiated beam thus decreases nearly linearly with frequency once $ka > 4$. There are some side lobes, but the first of these is already at -18 dB relative to the response for $\theta = 0$, so they are relatively minor.

The force F acting on the piston (Morse, 1948, pp. 332–333; Olson, 1957, pp. 92–93) is

$$F = (R_{\mathrm{m}} + jX_{\mathrm{m}})u = \rho c S u(A + jB), \tag{7.32}$$

where

$$\begin{aligned} A = 1 - \frac{J_1(2ka)}{ka} &= \frac{(ka)^2}{2} - \frac{(ka)^4}{2^2\cdot 3} + \frac{(ka)^6}{2^2\cdot 3^2\cdot 4} - \cdots \\ &\to \tfrac{1}{2}(ka)^2 \quad \text{for} \quad ka \ll 1, \\ &\to 1 \quad \text{for} \quad ka \gg 1, \end{aligned} \tag{7.33}$$

and

$$\begin{aligned} B = \frac{H_1(2ka)}{ka} &= \frac{1}{\pi k^2 a^2}\left[\frac{(2ka)^3}{3} - \frac{(2ka)^5}{3^2\cdot 5} + \frac{(2ka)^7}{3^2\cdot 5^2\cdot 7} - \cdots\right] \\ &\to 8ka/3\pi \quad \text{for} \quad ka \ll 1, \\ &\to 2/\pi ka \quad \text{for} \quad ka \gg 1, \end{aligned} \tag{7.34}$$

where H_1 is a Struve function of order 1. These functions, which apply also to a pipe with an infinite baffle, are shown later in Fig. 8.7. For the moment, we simply note the close agreement between their asymptotic forms and the same quantities for a pulsating sphere of radius a as given in Eq. (7.25) and Fig. 7.6.

When we consider the radiation from a vibrating circular membrane or plate of the type discussed in Chapter 3, we realize that a rigid piston is not a good model for the motion. Indeed, for a membrane, the first mode has the form of a Bessel function $J_0(\alpha_0 r)$, and higher modes are of the form $J_n(\alpha_m r)\cos n\phi$, with the α_m determined by the condition that the functions vanish at the clamped edge. All except the fundamental J_0 mode have either nodal lines or nodal circles or both, and there is a good deal of cross flow and hence low radiation efficiency. All the axisymmetric modes $J_0(\alpha_m r)$, however, have a nonzero volume displacement and hence some monopole radiation component. The J_n modes with $n \neq 0$ have no monopole component, and their radiation is therefore much less efficient. No explicit tabulation of this behavior is readily available, but Morse (1948, pp. 329–332) details a related case for free-edge modes for which the radial slope of the displacement is required to vanish at the boundary—a condition appropriate to a flexible piston closing a flanged circular pipe.

Detailed discussion of radiation from flexible plane vibrators with nodal lines is algebraically complex but reflects the behavior we found for antiphase arrays of point sources, provided that allowance is made for the fact that a finite vibrator may have a net volume flow and hence a monopole radiation contribution. As has already been remarked, antiphase point-source arrays with spacing less than half a wavelength of sound in air are inefficient radiators, and their source strength is of the order of that of a single one of their component sources. Exactly the same result is found for continuous vibrators, with the effective source arising from noncancelling elements at the center and around the edges of the vibrator (Skudrzyk, 1968, pp. 419–429).

Only for the free vibrations of thick metal plates do we reach a situation in which the transverse wave speed in the plate exceeds the sound speed in air, so that high intensity radiated beams can be produced, as shown in Fig. 7.5(c). At 1 kHz, this requires a steel plate about 10 mm in thickness (Skudrzyk, 1968, p. 378)—a situation often encountered in heavy machinery but scarcely applicable to musical instruments.

There is an interesting and important consequence of these conclusions that is investigated in some detail by Skudrzyk (1968, pp. 390–398). If a localized force drives an elastic plate, such as a piano soundboard, then it excites all vibrational modes to amplitudes that are dependent on the frequency separation between the exciting force frequency ω and the resonance frequency ω_n of the mode in question. If $\omega \gg \omega_n$, then the amplitude of mode n will be small, but, because its nodal lines are far apart, it will radiate efficiently at frequency ω. Conversely, a mode with $\omega_n \approx \omega$ may be strongly excited, but, because of the small distance between its nodal lines, it may radiate very poorly. The total power radiated by the forced plate must be found by summing the contributions from all the efficiently radiating modes as well as the smaller contributions from higher modes.

This effect acts to smooth the frequency response of a large planar forced vibrator.

7.7 Unbaffled Radiators

Few if any musical instruments involve a large plane baffle, even one that extends for about a wavelength around the vibrating plate or membrane. In instruments like the timpani, the baffle is folded around so that it encloses one side of the membrane and converts it to a one-sided resonator; the body of the violin serves a somewhat similar function, though there is considerable vibration of the back as well. In instruments like the piano, both sides of the soundboard are able to radiate, but the case provides a measure of separation between them. Only in the case of cymbals, gongs, and bells is there no baffle at all.

For a half-enclosed radiator, like the membrane of the timpani, the enclosed air does, of course, have an effect on the vibration of the membrane. From the point of view of radiation, however, the source is still one-sided, and the major difference from the baffled case arises from the fact that radiation is into a whole space of solid angle 4π rather than into a half-space of solid angle 2π. For high frequencies, such that $ka > 4$, this effect is not large, for the energy of the radiation is concentrated into a broad beam along the direction of the axis ($\theta = 0$), and little of it passes into the half-plane $\theta > \pi/2$ anyway. For low frequencies, $ka < 1$, however, the radiation tends to become isotropic and to fill nearly uniformly the whole 4π solid angle. The vibrating membrane thus experiences a radiation resistance at low frequencies that is only half that for the baffled case, so that the power radiated is reduced by a factor 2, or 3 dB, for a given membrane velocity amplitude. More than this, since the radiation is into 4π rather than 2π, the intensity in any given direction is reduced by a further factor of 2, or 3 dB. Thus, for a given radiator velocity amplitude, the absence of a large baffle leaves the radiated intensity and power unchanged at high frequencies ($ka > 4$) but reduces the radiated intensity in the forward half-space by 6 dB and the total radiated power by 3 dB at low frequencies ($ka < 1$).

These phenomena can have a significant effect on the fullness of tone quality of an instrument in its low-frequency range, and it is usual to increase the bass intensity by providing a reflecting wall close behind the player. This recovers 3 dB of relative intensity in the bass, and a further 3 dB, relative to anechoic conditions, can be recovered from close floor reflections. Of course, these effects are subjectively assumed when listening to normal playing—it is only in anechoic conditions that the loss of intensity at low frequencies becomes noticeable.

Instruments such as the piano have a case structure that goes some way toward separating the two sides of the soundboard acoustically. This is, of course, desirable since they vibrate in antiphase. We have seen, however, that antiphase source distributions cancel each other only if their separation is less than about half a wavelength. A semiquantitative application of this principle gives a distance of about 2 m between the top and bottom of the soundboard of a piano and hence suggests that such cancellations should not occur above about 70 Hz. The different geometries of the lid and the floor in any case ensure that cancellation is only partial. Cancellations do, of course, occur between neighboring antiphase regions on the same side of the soundboard.

Finally, let us look briefly at radiation from cymbals and bells. For a nearly planar radiator, such as a cymbal or tam-tam, cancellation of radiation from opposite surfaces may clearly be very significant. An unbaffled plane piston radiator with both sides exposed acts as a dipole source at low frequencies and, to a good approximation, the mechanical radiation resistance presented to each side is (Olson, 1957, pp. 98–99)

$$\begin{aligned} R_{\mathrm{m}} &\approx 3 \times 10^{-2} \rho c S (ka)^4, \quad ka < 2 \\ &\approx \rho c S, \quad ka > 3. \end{aligned} \tag{7.35}$$

Much of the low-frequency radiation is therefore suppressed. The initial amplitude of the lower modes is, however, usually high, so that some of their sound is heard. Much of the effect of such a gong depends on the shimmer of high-frequency modes, which have comparable radiation efficiency despite near cancellation of radiation from neighboring antiphase areas. Details of the residual radiation from noncancelling areas around the edge of such a gong are discussed by Skudrzyk (1968, pp. 419–429). This overall cancellation has another effect, of course, and that is to reduce the radiation damping of the oscillation and prolong the decay of the sound.

Radiation from curved shells, such as found in bells, is a very complex subject (Junger and Feit, 1986, Chapter 7). There is significant cancellation of radiation from neighboring antiphase regions, but their different geometrical environments inhibit cancellation between the interior and the exterior of the shell. Indeed, modes that would seem at first to be of high order, such as the quadrupole mode associated with the circular to elliptical distortion of a bell, can lead to a change in interior volume and therefore an inside-to-outside dipole source. Further discussion is best left to Chapter 21.

7.8 Radiation from Large Plates

It is often important in practical situations to have some appreciation of the radiation properties of extended vibrating objects. The general subject is difficult, but we can attain some insights by considering simple situations. More details can be found in the texts by Junger and Feit (1986) and by Cremer et al., (1988). Suppose, for example, that we have an infinite plane plate upon which is propagating a plane bending wave of angular frequency ω. The speed $v_p(\omega)$ of this wave is determined by the thickness and elastic properties of the plate, and also by the frequency ω, since a stiff plate is a dispersive medium for transverse waves, as described by Eq. (3.13). If we suppose the plate to lie in the plane $z = 0$ and the plate wave to propagate in the x direction, then we can represent the displacement velocity of the plate surface by the equation

$$u(x, 0) = u_p e^{-jk_p x} e^{j\omega t}, \tag{7.36}$$

where the wave number on the surface of the plate is given by

$$k_p = \omega / v_p. \tag{7.37}$$

In the air above the plate there is no variation of physical quantities in the y direction, and the acoustic pressure $p(x, z)$ and acoustic particle velocity $u(x, z)$ both satisfy the wave equation in the form

$$\frac{\partial^2 p}{\partial x^2} + \frac{\partial^2 p}{\partial z^2} = \frac{1}{c^2}\frac{\partial^2 p}{\partial t^2}, \tag{7.38}$$

which has solutions of the form

$$p(x, z) = \rho c u(x, z) = A e^{-j(k_x x + k_z z)}, \tag{7.39}$$

where k_x and k_z are the components of the wave vector $\mathbf{k}$ in directions respectively parallel to and normal to the plane of the plate and there is a time variation $e^{j\omega t}$ implied. Because the acoustic particle velocity in the air must match the normal velocity of the plate in the plane $z = 0$, we must have

$$k_x = k_p; \qquad A = u_p \tag{7.40}$$

so that, if we substitute this into Eq. (7.39) and thence into Eq. (7.38), we find that

$$k_z^2 = \frac{\omega^2}{c^2} - k_p^2 = \omega^2 \left(\frac{1}{c^2} - \frac{1}{v_p^2} \right). \tag{7.41}$$

Equation (7.41) has very important implications. If the velocity v_p of the wave on the plate is less than the velocity c of sound in air, or equivalently if the wavelength of the wave on the plate is smaller than the wavelength in air, then k_z is imaginary. The acoustic disturbance is then exponentially attenuated in the z direction rather than propagating as a wave. The whole motion of the air is thus confined to the immediate vicinity of the plate surface and there is no acoustic radiation. If, however, $v_p > c$, then an acoustic wave is radiated in a direction making an angle θ with the surface of the plate, given by

$$\tan\theta = \frac{k_z}{k_x} = \left(\frac{v_p^2}{c^2} - 1 \right)^{1/2}. \tag{7.42}$$

The physical reason for this behavior is clear from Fig. 7.8, which illustrates the relation between the plate and air waves. If $v_p > c$, the wavelength λ_p of the plate wave is greater than the wavelength λ of the air wave, and the diagram fits together with the propagation direction θ given by Eq. (7.42). If $v_p < c$, however, then $\lambda_p < \lambda$, and there is no way that the acoustic wave can be matched to the plate wave.

Since the wave speed v_p on the plate increases with increasing frequency, there is a particular critical frequency for any plate above which radiation can occur. Since at this frequency the wave speed on the plate is equal to the sound speed in air, it is usually called the "coincidence frequency." If we have a complex wave propagating on a plate, this means that only the high-frequency components of the wave will be radiated, and their radiation directions will be spread out in angle rather like white light passing through a prism or reflected from a diffraction grating. It is clear that there is a close analogy between this behavior and that already noted for two simple antiphase sources in Section 7.2 and for linear arrays of antiphase sources as discussed in Section 7.3.

If, instead of a propagating wave on the plate, we have a standing wave, then the situation is rather similar. The standing wave has a spatial variation in the x direction given by

$$\cos k_p x = \tfrac{1}{2}\left(e^{jk_p x} + e^{-jk_p x}\right), \tag{7.43}$$

which is just two similar waves propagating in opposite directions. Each behaves as described above, and the resulting radiation pattern, if $v_p > c$, is a superposition of two plane waves at angles $\pm\theta$ to the plane. These interfere to give a plane wave propagating in the z direction but with an amplitude variation $\cos k_p x$ in the x direction.

We can extend this superposition principle to the case in which we have several waves propagating in different directions on the plate. As a special

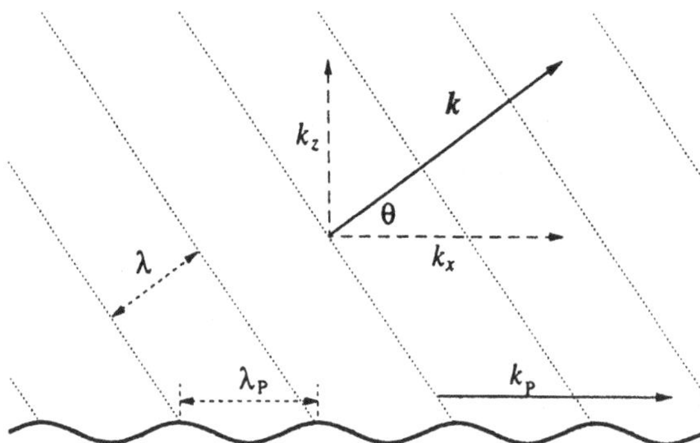

FIGURE 7.8. Relation between a plate wave with propagation number k_p and the acoustic wave with propagation vector k that it radiates, for the case $\lambda_p > \lambda$.

case, we might consider standing waves in the x and y directions, dividing the whole plane into vibrating squares with adjacent squares being in antiphase relation. Once again there is no radiation below the critical frequency, while above that frequency the angular radiation pattern is very complex.

An infinite plane vibrating surface is, of course, something that is not met in practice—all real plates have boundaries, and these modify the idealized behavior discussed above. In an infinite plate, each vibrating region is surrounded by regions of opposite phase, but this arrangement terminates when we reach a boundary. For a plate that is several times larger in all its dimensions than the critical wavelength, there is a transition from good radiation efficiency above the calculated critical frequency to poor radiation efficiency below, but this transition is not completely sharp and there is still some radiation at lower frequencies. As the frequency is decreased below the critical frequency, lower and lower normal modes are progressively excited until the plate is vibrating essentially in its fundamental mode, giving approximately a finite-dipole source, the radiated intensity from which varies as ω^4 for constant panel velocity amplitude, compared with the frequency-independent radiation characteristic above the critical frequency.

The same sort of argument can be applied to radiation from the flexural modes of a long cylinder, which behaves rather like a one-dimensional plate as far as its axial direction is concerned. We then find the same sort of critical frequency relationship for an infinitely long cylinder, and a modified version of this for a cylinder of finite length. These results are of significance for sound radiation from tubular bells.

References

Beranek, L.L. (1954). "Acoustics." McGraw-Hill, New York; reprinted 1986, Acoustical Soc. Am., Woodbury, New York.

Cremer, L., Heckl, M., and Ungar, E.E. (1988). "Structure-Borne Sound," 2nd ed., Springer-Verlag, New York.

Junger, M.C., and Feit, D. (1986). "Sound, Structures, and Their Interaction," 2nd ed., MIT Press, Cambridge, Massachusetts.

Morse, P.M. (1948). "Vibration and Sound." McGraw-Hill, New York; reprinted 1981, Acoustical Soc. Am., Woodbury, New York.

Olson, H.F. (1957). "Acoustical Engineering." Van Nostrand-Reinhold, Princeton, New Jersey.

Skudrzyk, E. (1968). "Simple and Complex Vibratory Systems." Penn. State Univ. Press, University Park, Pennsylvania.

8

Pipes, Horns and Cavities

The wave propagation phenomena in fluids that we have examined in previous chapters have referred to waves in infinite or semi-infinite spaces generated by the vibrational motion of some small object or surface in that space. We now turn to the very different problem of studying the sound field inside the tube of a wind instrument. Ultimately, we shall join together the two discussions by considering the sound radiated from the open end or finger holes of the instrument, but for the moment our concern is with the internal field. We begin with the very simplest cases and then add complications until we have a reasonably complete representation of an actual instrument.

Pipes, cavities, and apertures are, however, also important in instruments other than those of the wind family, and we will later be concerned with sound fields inside both stringed and percussion instruments. We therefore take the opportunity to introduce the general methods of electric network analogs, which are very powerful yet very simple, for the analysis of such systems.

8.1 Infinite Cylindrical Pipes

The simplest possible system of enclosure is an infinite cylindrical pipe or tube with its axis parallel to the direction of propagation of a plane wave in the medium (Morse and Ingard, 1968). If the walls of the pipe are rigid, perfectly smooth, and thermally insulating, then the presence of the tube wall has no effect on wave propagation. A pressure wave propagating in the x direction has the form

$$p(x,t) = p\exp[j(-kx + \omega t)], \tag{8.1}$$

and the resultant acoustic volume flow is, as we saw in Chapter 6,

$$U(x,t) = \left(\frac{Sp}{\rho c}\right)\exp[j(-kx + \omega t)], \tag{8.2}$$

where ω is the angular frequency, k is the angular wave number $k = 2\pi/\lambda = \omega/c$, and S is the cross-sectional area of the pipe. As usual, ρ is the density of and c the velocity of sound in air. The acoustic impedance of the pipe at any point x is defined to be

$$Z_0(x) = \frac{p(x,t)}{U(x,t)} = \frac{\rho c}{S}. \tag{8.3}$$

To treat this problem in more detail, we must solve the wave equation directly in cylindrical polar coordinates (r, ϕ, x). If a is the radius of the pipe and its surface is again taken to be perfectly rigid, then the boundary condition is

$$\frac{\partial p}{\partial r} = 0 \quad \text{at} \quad r = a, \tag{8.4}$$

which implies that there is no net force and therefore no flow normal to the wall. The wave equation in cylindrical coordinates is

$$\frac{1}{r}\frac{\partial}{\partial r}\left(r\frac{\partial p}{\partial r}\right) + \frac{1}{r^2}\frac{\partial^2 p}{\partial \phi^2} + \frac{\partial^2 p}{\partial x^2} = \frac{1}{c^2}\frac{\partial^2 p}{\partial t^2}, \tag{8.5}$$

and this has solutions of the form

$$p_{mn}(r,\phi,x) = p^{\cos}_{\sin}(m\phi) J_m\left(\frac{\pi q_{mn} r}{a}\right)\exp[j(-k_{mn}x + \omega t)], \tag{8.6}$$

where J_m is a Bessel function and q_{mn} is defined by the boundary condition [Eq. (8.4)], so that the derivative $J'_m(\pi q_{mn})$ is zero. The (m, n) mode thus has an (r, ϕ) pattern for the acoustic pressure p with n nodal circles and m nodal diameters, both m and n running through the integers from zero. In the full three-dimensional picture, these become nodal cylinders parallel to the axis and nodal planes through the axis, respectively.

In Fig. 8.1, the pressure and flow velocity patterns for the lowest three modes of the pipe, omitting the simple plane-wave mode, are shown. The pressure patterns have nodal lines as already observed, and there are similar nodal diameters in the transverse flow patterns. Nodal circles for pressure occur for modes of the type $(0, n)$, which have n such nodal circles within the boundary. A general mode (m, n) has both nodal lines and circles in the pressure.

The propagation wave vector k_{mn} for mode (m, n) is obtained by substituting Eq. (8.6) into Eq. (8.5), whence

$$k^2_{mn} = \left(\frac{\omega}{c}\right)^2 - \left(\frac{\pi q_{mn}}{a}\right)^2. \tag{8.7}$$

Thus, while the plane-wave mode with $m = n = 0$ will always propagate with $k = k_{00} = \omega/c$, this is not necessarily true for higher modes. In order for a higher mode (m, n) to propagate, the frequency must exceed the

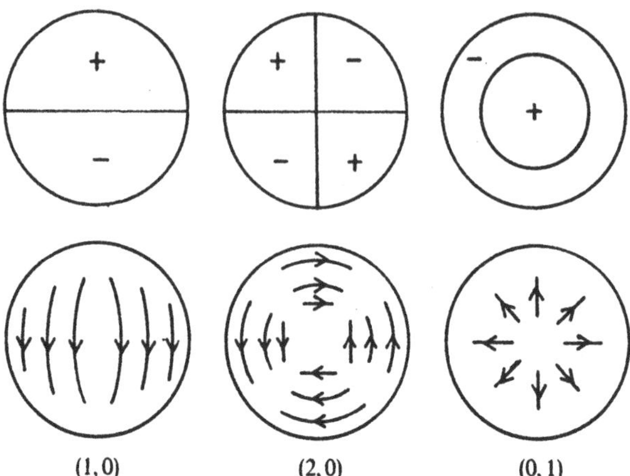

FIGURE 8.1. Pressure and transverse flow patterns for the lowest three transverse modes of a cylindrical pipe. The plane-wave mode is not shown.

critical value

$$\omega_c = \frac{\pi q_{mn} c}{a}. \tag{8.8}$$

For frequencies less than ω_c, k_{mn} is imaginary and Eq. (8.6) shows that the mode is attenuated exponentially with distance. The attenuation is quite rapid for modes well below cutoff, and the amplitude falls by a factor e, or about 10 dB, within a distance less than the pipe radius.

The first higher mode to propagate is the antisymmetric (1, 0) mode, which has a single nodal plane, above a cutoff frequency $\omega_c = 1.84c/a$. Next is the (2, 0) mode, with two nodal planes, for $\omega > 3.05c/a$, and then the lowest nonplanar axial mode (0, 1), for $\omega > 3.80c/a$. Propagating higher modes are thus possible only when the pipe is greater in diameter than about two-thirds of the free-space acoustic wavelength. The nonpropagating higher modes are necessary to explain certain features of the acoustic flow near wall irregularities, such as finger holes or mouthpieces. Indeed, it is possible to match any disturbance distributed over an opening or a vibrating surface in a duct with an appropriate linear combination of duct modes. The plane-wave component of this combination will always propagate along the duct away from the disturbance, but this will not be true for modes with q_{mn} values that are too large. The propagating wave will thus be a low-pass filtered version of the disturbance, while the nonpropagating modes will simply modify the flow in the near neighborhood of the source.

It is helpful to sketch the three-dimensional acoustic flow streamlines associated with a few of these modes for both propagating and nonpropagating cases. This can be done from the form of the pressure pattern given

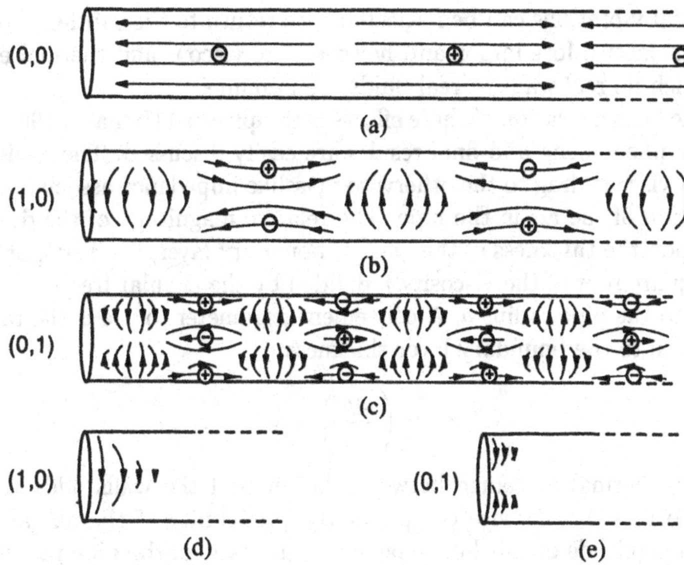

FIGURE 8.2. Acoustic flow patterns and pressure maxima and minima for higher modes in a cylindrical duct. (a)–(c) are modes propagating to the right at a frequency a little above cut-off; (d) and (e) are evanescent modes below cutoff.

by Eq. (8.6) together with the relation

$$u = \frac{j}{\rho\omega} \nabla \mathrm{p} \tag{8.9}$$

for the flow velocity u in a mode excited at frequency ω. Figure 8.2 shows this for the (1, 0) and (0, 1) modes. In the case of the propagating modes, the flow pattern itself moves down the pipe with the characteristic phase velocity of the mode—nearly the normal sound velocity c, except very close to cutoff when the phase velocity is higher than c.

It is not important to go into detail about the impedance behavior of these higher modes, since this depends greatly upon the geometry with which they are driven, the net acoustic flow along the pipe axis being zero except for the plane (0, 0) mode. The impedance is always a real function multiplied by ω/k_{mn}, so it is real for ω above cutoff, becomes infinite at cutoff, and is imaginary below cutoff.

8.2 Wall Losses

So far in our discussion, we have assumed a rigid wall without introducing any other disturbance. In a practical case this can never be achieved, though in musical instruments the walls are at least rigid enough that their mechanical vibrations can be neglected—we return to the subtleties of this statement later. More important, however, are viscous and thermal effects from which no real walls or real fluids are immune.

Detailed consideration of these effects is complicated (Benade, 1968), but the basic phenomena and final results are easily discussed. The walls contribute a viscous drag to the otherwise masslike impedance associated with acceleration of the air in the pipe. The relative magnitude of the drag depends upon the thickness of the viscous boundary layer, itself proportional to the square root of the viscosity η divided by the angular frequency ω, in relation to the pipe radius a. A convenient parameter to use is the ratio of pipe radius to the boundary layer thickness:

$$r_{\mathrm{v}} = \left(\frac{\omega\rho}{\eta}\right)^{1/2} a. \tag{8.10}$$

Similarly, thermal exchange between the air and the walls adds a lossy resistance to the otherwise compliant compressibility of the air, and the relative magnitude of this loss depends on the ratio of the pipe radius a to the thermal boundary layer thickness, as expressed by the parameter

$$r_{\mathrm{t}} = \left(\frac{\omega\rho C_{\mathrm{p}}}{\kappa}\right)^{1/2} a, \tag{8.11}$$

where C_{p} is the specific heat of air at constant pressure and κ is its thermal conductivity. The ratio $(r_t/r_v)^2 = C_p\eta/\kappa$ is the Prandtl number. Near 300 K (27°C), we can insert numerical values to give (Benade, 1968)

$$r_{\mathrm{v}} \approx 632.8af^{1/2}(1 - 0.0029\,\Delta T), \tag{8.12}$$

and

$$r_{\mathrm{t}} \approx 532.8af^{1/2}(1 - 0.0031\,\Delta T), \tag{8.13}$$

where a is the tube radius in meters, f is the frequency in hertz, and ΔT is the temperature deviation from 300 K.

It is clear that the effect of these loss terms will be to change the characteristic impedance Z_0 of the pipe from its ideal real value $\rho c/S$ to a complex quantity. This, in turn, will make the wave number k complex and lead to attenuation of the propagating wave as it passes along the pipe.

The real and imaginary parts of the characteristic impedance Z_0, as fractions of its ideal value $\rho c/S$, are shown in Figs. 8.3 and 8.4, both as functions of r_{v}. The correction to Z_0 begins to be appreciable for $r_{\mathrm{v}} < 10$, while for $r_{\mathrm{v}} < 1$ the real and imaginary parts of Z_0 are nearly equal and vary as r_{v}^{-1}.

It is convenient to rewrite the wave vector k as the complex number $\omega/v - j\alpha$, where α is now the attenuation coefficient per unit length of path and v is the phase velocity. We can then most usefully plot the phase velocity v, measured in units of the free-air sound velocity c, and the attenuation coefficient α, divided by f, both as functions of r_{v}. This is done in Figs. 8.5

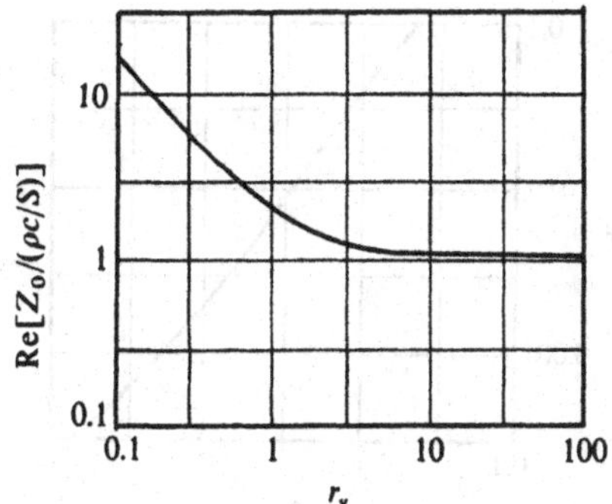

FIGURE 8.3. Real part of the characteristic impedance Z_0, in units of $\rho c/S$, as a function of the parameter r_v (after Benade, 1968).

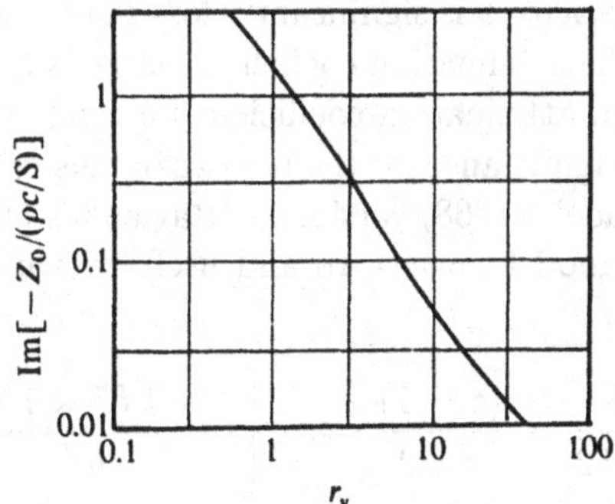

FIGURE 8.4. Imaginary part of the characteristic impedance Z_0, in units of $\rho c/S$, as a function of the parameter r_v (after Benade, 1968).

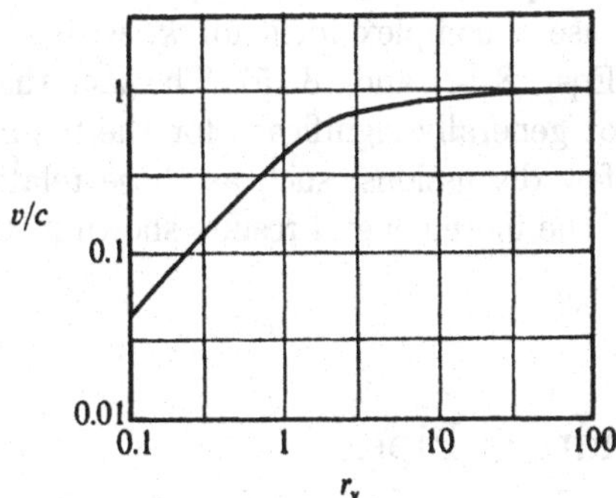

FIGURE 8.5. The phase velocity v, relative to the free-air sound velocity c, as a function of the parameter r_v (after Benade, 1968).

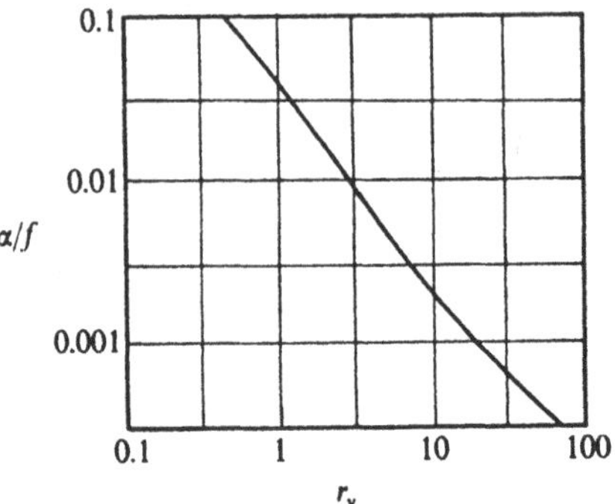

FIGURE 8.6. The attenuation coefficient α in $(\text{meters})^{-1}$ at frequency f, plotted as α/f, as a function of the parameter r_v (after Benade, 1968).

and 8.6. The phase velocity v is significantly less than c for pipes so narrow that $r_v < 10$, while the attenuation coefficient exceeds λ^{-1} if $r_v < 6$. Since the phase velocity and attenuation coefficient for relatively wide tubes are both of fundamental significance for the physics of musical instruments, it is useful to restate Benade's (1968) versions of Rayleigh's (1894) approximate formulas, which are good for $r_v > 10$ and useful down to about $r_v = 3$. They are

$$v \approx c\left[1 - \frac{1}{r_v\sqrt{2}} - \frac{(\gamma - 1)}{r_t\sqrt{2}}\right] \approx c\left[1 - \frac{1.65 \times 10^{-3}}{af^{1/2}}\right], \tag{8.14}$$

and

$$\alpha \approx \frac{\omega}{c}\left[\frac{1}{r_v\sqrt{2}} + \frac{(\gamma - 1)}{r_t\sqrt{2}}\right] \approx \frac{3 \times 10^{-5} f^{1/2}}{a}, \tag{8.15}$$

where α is given in $(\text{meters})^{-1}$ if a is in meters. Here, γ is the ratio of specific heats C_p/C_v, which for air is approximately 1.40.

In most of the more practical discussions that follow, we will find it adequate simply to use a complex form for k, with real and imaginary parts derived from Eqs. (8.14) and (8.15). The fact that Z_0 has a small imaginary part is not generally significant for the main pipes of musical instruments. For a few discussions, such as those related to the smaller tubes of finger holes, the more general results shown in the figures may be necessary.

8.3 Finite Cylindrical Pipes

All of the pipes with which we deal in musical instruments are obviously of finite length, so we must allow for the reflection of a wave from the remote end, whether it is open or closed. Because we are concerned with pipes as closely coupled driven systems, rather than as passive resonators, we shall proceed by calculating the input impedance for a finite length of pipe terminated by a finite load impedance Z_L, rather than examining doubly open or closed pipes in isolation. The terminating impedance Z_L will generally represent an open or a closed end, but it is not restricted to these cases. The development here is essentially the same as that set out in Chapter 2 for a string stretched between nonrigid bridges but, since the results are central to our discussion of pipes and horns, we start again from the beginning.

Suppose the pipe extends from $x = 0$ to $x = L$, and that it is terminated at $x = L$ by the impedance Z_L. The pressure in the pipe is a superposition of two waves, moving to the right and left, respectively, with amplitudes A and B, taken as complex quantities so that they can include a phase factor. Thus, at the point x,

$$p(x,t) = [Ae^{-jkx} + Be^{jkx}]e^{j\omega t}. \tag{8.16}$$

The acoustic particle velocity is similarly a superposition of the particle velocities associated with these two waves and, multiplying by pipe cross section S, the acoustic flow becomes, from Eq. (8.3),

$$U(x,t) = \left(\frac{S}{\rho c}\right)[Ae^{-jkx} - Be^{jkx}]e^{j\omega t}. \tag{8.17}$$

At the remote end $x = L$, pressure and flow are related as required by the terminating impedance Z_L, so that

$$\frac{p(L,t)}{U(L,t)} = Z_L, \tag{8.18}$$

and this equation is enough to determine the complex ratio B/A. If we write for the characteristic impedance of the pipe

$$Z_0 = \rho c/S \tag{8.19}$$

as in Eq. (8.3), then

$$\frac{B}{A} = e^{-2jkL}\left[\frac{(Z_L - Z_0)}{(Z_L + Z_0)}\right], \tag{8.20}$$

and the power reflected from Z_L has a ratio to incident power of

$$R = \left|\frac{B}{A}\right|^2 = \left|\frac{Z_L - Z_0}{Z_L + Z_0}\right|^2. \tag{8.21}$$

Clearly, there is no reflection if $Z_L = Z_0$ and complete reflection if $Z_L = 0$ or ∞. Since Z_0 is real for a lossless tube, there is also perfect reflection if Z_L is purely imaginary; however, if Z_L has a real part that is nonzero, then there will always be some reflection loss.

The quantity in which we are interested now is the input impedance Z_{IN} at the point $x = 0$. From Eqs. (8.16)–(8.19), this is

$$Z_{\mathrm{IN}} = Z_0 \left[\frac{A + B}{A - B} \right], \tag{8.22}$$

or from Eq. (8.20),

$$Z_{\mathrm{IN}} = Z_0 \left[\frac{Z_{\mathrm{L}} \cos kL + jZ_0 \sin kL}{jZ_{\mathrm{L}} \sin kL + Z_0 \cos kL} \right]. \tag{8.23}$$

Two important idealized cases are readily derived. The first corresponds to a pipe rigidly stopped at $x = L$ so that $Z_{\mathrm{L}} = \infty$. For such a pipe,

$$Z_{\mathrm{IN}}^{\mathrm{stopped}} = -jZ_0 \cot kL. \tag{8.24}$$

For the converse case of an ideally open pipe with $Z_{\mathrm{L}} = 0$, which is not physically realizable exactly, as we see below,

$$Z_{\mathrm{IN}}^{\mathrm{open}} = jZ_0 \tan kL. \tag{8.25}$$

The familiar resonance frequencies for open and stopped pipes arise from applying the idealized condition that the input end at $x = 0$ is also open, so that resonances occur if $Z_{\mathrm{IN}} = 0$. For a stopped pipe, this requires that $\cot kL = 0$, giving

$$\omega^{\mathrm{stopped}} = \frac{(2n-1)\pi c}{2L}, \tag{8.26}$$

corresponding to an odd number of quarter wavelengths in the pipe length, while for an ideally open pipe, $\tan kL = 0$, giving

$$\omega^{\mathrm{open}} = \frac{n\pi c}{L}, \tag{8.27}$$

corresponding to an even number of quarter wavelengths, or any number of half wavelengths, in the pipe length.

While Eq. (8.24) applies quite correctly to a physically stopped pipe, the treatment of a physically open pipe is more difficult since, while $Z_{\mathrm{L}} \ll Z_0$, it is not a sufficient approximation to set it to zero. It is relatively straightforward to calculate the radiation load Z_{L} on a pipe that terminates in a plane flange of size much larger than a wavelength (and therefore effectively infinite). The formal treatment of Rayleigh (1894) (Morse, 1948, Olson, 1957) makes the assumption that the wavefront exactly at the open end of the pipe is quite planar, normally a very good approximation, and gives the result

$$Z^{\mathrm{flanged}} = R + jX, \tag{8.28}$$

where, as discussed for Eqs. (7.32)–(7.34),

$$R = Z_0 \left[\frac{(ka)^2}{2} - \frac{(ka)^4}{2^2 \cdot 3} + \frac{(ka)^6}{2^2 \cdot 3^2 \cdot 4} - \cdots \right], \tag{8.29}$$

$$X = \frac{Z_0}{\pi k^2 a^2}\left[\frac{(2ka)^3}{3} - \frac{(2ka)^5}{3^2 \cdot 5} + \frac{(2ka)^7}{3^2 \cdot 5^2 \cdot 7} - \cdots\right], \tag{8.30}$$

and a is the radius of the pipe.

The behavior of R and X as functions of frequency, or more usefully as functions of the dimensionless quantity ka, is shown in Fig. 8.7. If $ka \ll 1$, then $|Z^{\text{flanged}}| \ll Z_0$ and most of the wave energy is reflected from the open end. If $ka > 2$, however, then $Z^{\text{flanged}} \approx Z_0$ and most of the wave energy is transmitted out of the end of the pipe into the surrounding air.

In musical instruments, the fundamental, at least, has $ka \ll 1$, though this is not necessarily true for all the prominent partials in the sound. It is therefore useful to examine the behavior of the pipe in this low-frequency limit. From Eqs. (8.29) and (8.30), $X \gg R$ if $ka \ll 1$, so that

$$Z^{\text{flanged}} \approx jZ_0 k\left(\frac{8a}{3\pi}\right). \tag{8.31}$$

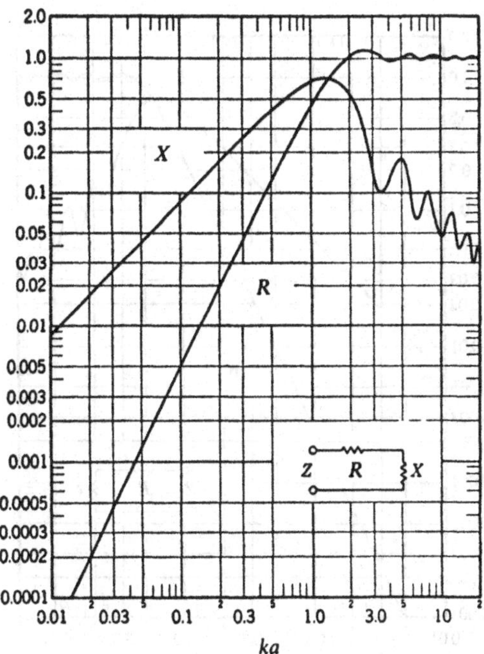

FIGURE 8.7. The acoustic resistance R and the acoustic reactance X, both in units of $\rho c/\pi a^2$, for a circular piston (or open pipe) of radius a set in an infinite plane baffle, as functions of the frequency parameter ka (after Beranek, 1954).

By comparison with Eq. (8.25), since $ka \ll 1$, this is just the impedance of an ideally open short pipe of length

$$\Delta^{\text{flanged}} = \frac{8a}{3\pi} \approx 0.85a. \tag{8.32}$$

It is thus a good approximation in this frequency range to replace the real flanged pipe by an ideally open pipe of length $L + \Delta^{\text{flanged}}$, and to neglect the radiation loss. From Fig. 8.7, it is clear that the end correction Δ^{flanged}, which is proportional to X/ka, decreases slightly as $ka \to 1$ and continues to decrease more rapidly as ka increases past this value.

A real pipe, of course, is not generally flanged, and we need to know the behavior of Z_L in this case. The calculation (Levine and Schwinger, 1948) is very difficult, but the result, as shown in Fig. 8.8, is very similar to that for a flanged pipe. The main difference is that, for $ka \ll 1$, R is reduced by about a factor 0.5 and X by a factor 0.7 because the wave outside the pipe has freedom to expand into a solid angle of nearly 4π rather than just 2π. The calculated end correction at low frequencies is now

$$\Delta^{\text{open}} \approx 0.61a. \tag{8.33}$$

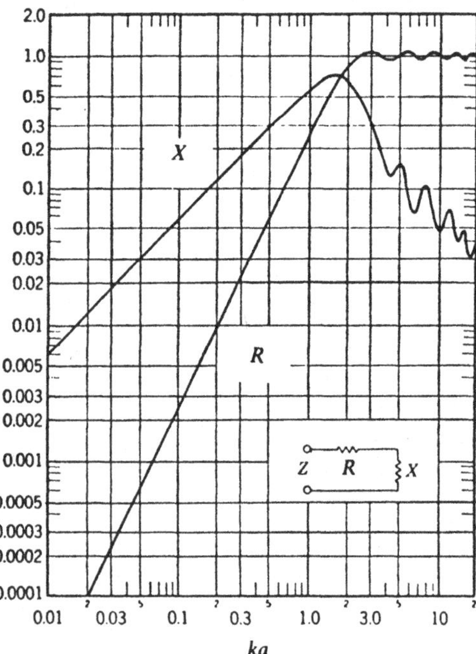

FIGURE 8.8. The acoustic resistance R and the acoustic reactance X, both in units of $\rho c/\pi a^2$ for the open end of a circular cylindrical pipe of radius a, as functions of the frequency parameter ka (after Beranek, 1954).

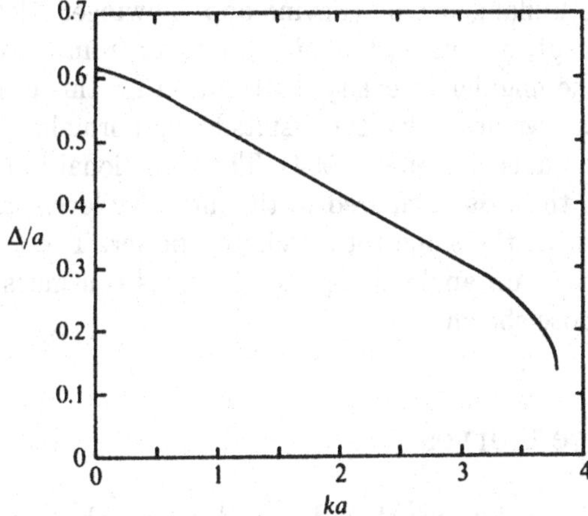

FIGURE 8.9. The calculated end correction Δ for a cylindrical pipe of radius a, plotted as Δ/a, as a function of the frequency parameter ka, (after Levine and Schwinger, 1948).

The calculated variation of this end correction with the frequency parameter ka is shown in Fig. 8.9.

8.4 Radiation from a Pipe

One of our later interests, of course, will be to calculate the sound radiation from musical wind instruments and, as part of this task, it is helpful to know the transformation function between the spectrum of sound energy within the pipe and the total radiated sound energy. This transformation is simply proportional to the behavior of R as a function of frequency, so that, to a good approximation, it rises as (frequency)^2, that is, 6 dB per octave, below the reflection cutoff frequency, defined so that $ka = 2$. Above this frequency, the transformation is independent of frequency. This remark refers, of course, to the total radiated power and neglects directional effects that tend to concentrate the higher frequencies at angles close to the pipe axis.

It is useful to summarize these directional effects here, since they are derived in the course of calculation of the radiation impedance Z_L. The flanged case is simplest (Rayleigh, 1894; Morse, 1948) and gives a radiated intensity at angle θ to the pipe axis proportional to

$$\left[\frac{2J_1(ka\sin\theta)}{ka\sin\theta}\right]^2. \tag{8.34}$$

The result for an unflanged pipe (Levine and Schwinger, 1948) is qualitatively similar except, of course, that θ can extend from 0 to 180° instead of just to 90°. The angular intensity distribution for this case is shown in Fig. 8.10 for several values of ka, the results being normalized to the power radiated along the axis (Beranek, 1954). The directional index (DI) is the intensity level on the axis compared to the intensity level produced by an isotropic source with the same total radiated power. The trend toward a narrower primary beam angle along the pipe axis continues for values of ka larger than those shown.

8.5 Impedance Curves

Finally, in this discussion, we should consider the behavior of pipes with physically realistic wall losses. Provided the pipe is not unreasonably narrow, say $r_v > 10$, then Figs. 8.3 and 8.4 show that we can neglect the small change in the characteristic impedance Z_0 and simply allow the possibility that k is complex for propagation in the pipe. This new k is written $(\omega/v - j\alpha)$ with v given by Eq. (8.14) and α given by Eq. (8.15). This can be simply inserted into Eq. (8.23), along with the appropriate expression for Z_L, to deduce the behavior of the input impedance of a real pipe. The result for an ideally open pipe ($Z_L = 0$) of length L is

$$Z_{IN} = Z_0 \left[\frac{\tanh \alpha L + j \tan(\omega L/v)}{1 + j \tanh \alpha L \tan(\omega L/v)} \right]. \tag{8.35}$$

This expression has maxima and minima at the maxima and minima, respectively, of $\tan(\omega L/v)$. The value of Z_{IN} at the maxima is $Z_0 \coth \alpha L$, and at the minima it is $Z_0 \tanh \alpha L$. By Eq. (8.15), α increases with frequency as $\omega^{1/2}$, so these extrema decrease in prominence at higher frequencies, and Z_{IN} converges toward Z_0. For a pipe stopped at the far end, the factor in square brackets in Eq. (8.35) should simply be inverted.

For narrow pipes the lower resonances are dominated by this wall-loss mechanism, but for wider open pipes radiation losses from the end become more important, particularly at high frequencies. To illustrate some features of the behavior, we show in Fig. 8.11 calculated impedance curves for two pipes each 1 m long and with diameters, respectively, 2 cm and 10 cm. The low-frequency resonances are sharper for the wide pipe than for the narrow pipe because of the reduced relative effect of wall damping, but the high-frequency resonances of the wide pipe are washed out by the effects of radiation damping. We can see that all the impedance maxima and minima have frequencies that are nearly harmonically related, that is, as the ratio of two small integers. In fact, because the end correction decreases with increasing frequency, the frequencies of these extrema are all slightly stretched, and this effect is more pronounced for the wide than for the narrow pipe.

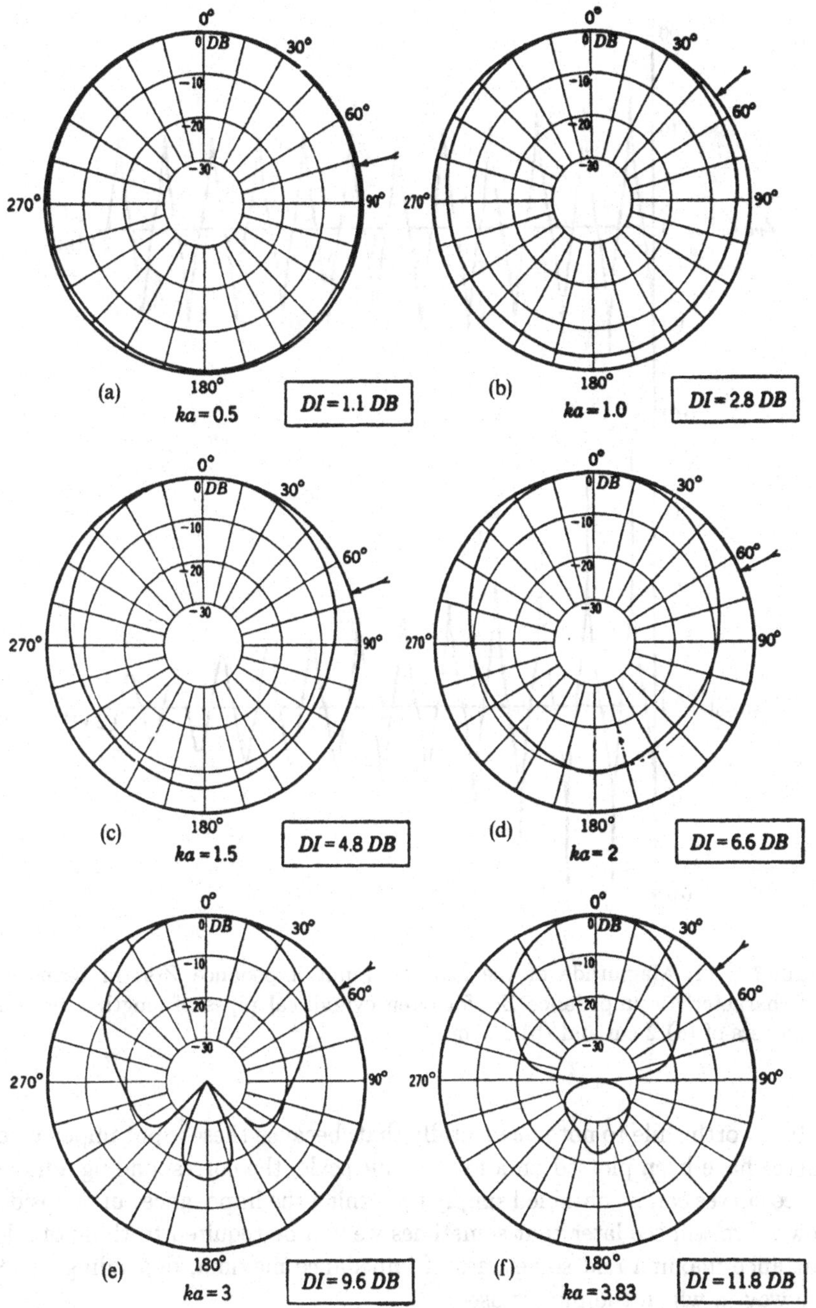

FIGURE 8.10. The directional patterns calculated by Levine and Schwinger for radiation from an unbaffled circular pipe of radius a. The radial scale is in each case 40 dB and the directional index has the calculated value shown (after Beranek, 1954).

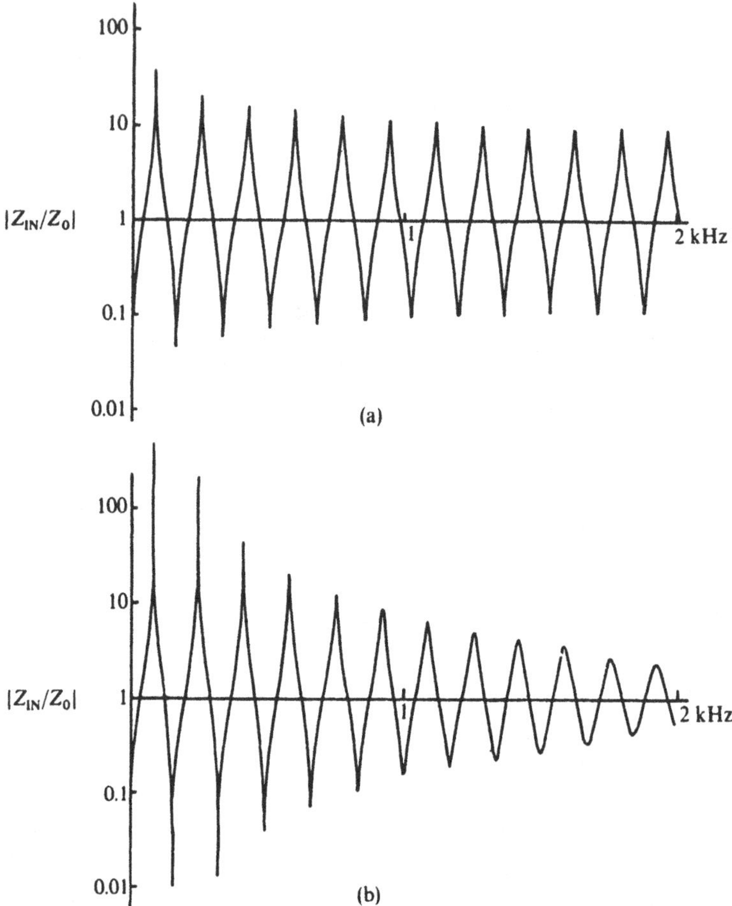

FIGURE 8.11. Magnitude of the acoustic input impedance Z_{IN}, in terms of the characteristic impedance Z_0, for open cylindrical pipes of length 1 m and diameters of (a) 2 cm and (b) 10 cm.

It is worthwhile to note incidentally that, because these input impedance curves have been plotted on a logarithmic scale, the corresponding admittance curves can be obtained simply by turning the impedance curve upside down. We will see later that sometimes we will be required to think of admittance maxima and sometimes of impedance maxima, depending upon the way in which the pipe is used.

When we come to consider musical instruments in detail, we will find that several of them rely upon cylindrical pipes as their sound generators. The most obvious of these is the pipe organ, in which most of the pipes are cylindrical (a few are conical). Tone quality of air-jet-driven pipes is varied by the use of closed and open tubes, by differences in relative diameters, and by differences in the sort of termination at the open end—some are simple open ends, some have slots, some have bells, and some have narrow chimneys. These variations can all be treated on the basis of the above discussion supplemented by a separate consideration of the form of Z_L produced by the termination.

8.6 Horns

Following this introductory discussion of cylindrical pipes, we are now ready to begin a treatment of sound propagation in horns, a horn being defined quite generally as a closed-sided conduit, the length of which is usually large compared with its lateral dimensions. In fact, we shall only treat explicitly horns that are straight and have circular cross section, but much of the discussion is really more general than this. A mathematically detailed discussion of the topic, with copious references, has been given by Campos (1984).

Formulation of the wave propagation problem in an infinitely long horn simply requires solution of the wave equation

$$\nabla^2 p = \frac{1}{c^2}\frac{\partial^2 p}{\partial t^2}, \tag{8.36}$$

subject to the condition that $\mathbf{n}\cdot\nabla p = 0$ on the boundaries, $\mathbf{n}$ being a unit vector normal to the boundary at the point considered. More simply, we suppose the wave to have a frequency ω so that Eq. (8.36) reduces to the Helmholtz equation

$$\nabla^2 p + k^2 p = 0, \tag{8.37}$$

where $k = \omega/c$. Solution of this equation is simple provided that we can choose a coordinate system in which one coordinate surface coincides with the walls of the horn and in which Eq. (8.37) is separable. Unfortunately, the Helmholtz equation is separable only in coordinates that are confocal quadric surfaces or their degenerate forms (Morse and Feshbach, 1953). There are 11 varieties of these coordinate systems, but only a few of them are reasonable candidates for horns. These are rectangular coordinates (a pipe of rectangular cross section), circular cylinder coordinates, elliptic cylinder coordinates, spherical coordinates (a conical horn), parabolic coordinates, and oblate spheroidal coordinates, as shown in Fig. 8.12. Of these, we have already dealt with the circular cylinder case, and the rectangular and elliptic cylinder versions differ from it only in cross-sectional geometry and hence in their higher modes. The parabolic horn is not musically practical since it cannot be made to join smoothly onto a mouthpiece, so we are left with the conical horn and the horn derived from oblate spheroidal coordinates, which will prove to be of only passing interest.

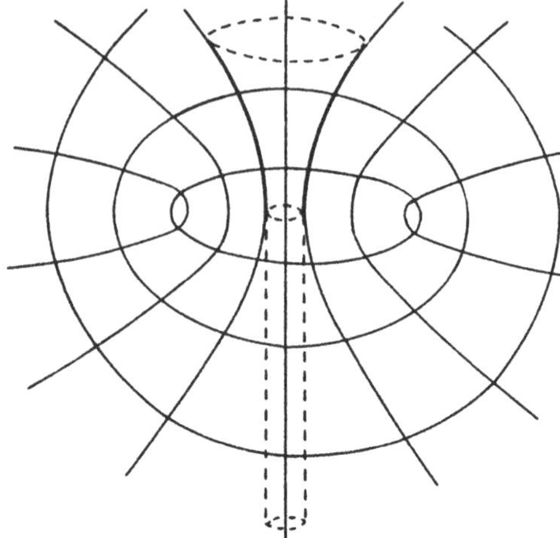

FIGURE 8.12. The oblate spheroidal coordinate system in which the wave equation is separable. If the hyperboloid of revolution (shown in heavy outline) is taken as the horn, then the oblate spheroidal surfaces orthogonal to this and lying within it are the wave fronts. Note that such a hyperboloid horn can be smoothly joined to a cylindrical pipe of appropriate diameter, as shown.

We deal with the oblate spheroidal case first, because it illustrates some of the difficulties we will have to face later. The hornlike family of surfaces consists of hyperboloids of revolution of one sheet, as shown in Fig. 8.12. At large distances, these approach conical shapes, but near the origin they become almost cylindrical. Indeed, one could join a simple cylinder parallel to the axis in the lower half plane to a hyperboloid horn in the upper half plane without any discontinuity in slope of the walls. The important thing to notice, however, is the shape of the wavefronts as shown by the orthogonal set of coordinate surfaces. These are clearly curved and indeed they are oblately spheroidal, being nearly plane near the origin and nearly spherical at large distances. Waves can propagate in this way as a single mode, like the plane waves in a cylinder. Such behavior is possible only for separable coordinate systems. For nonseparable systems that we may try to separate approximately, there will always be an admixture of higher modes. Horn systems resembling a cylinder joined to a narrow-angle hyperboloid horn as described above are in fact used in many brass instruments, though not because of any consideration of separability of the wave equation. Indeed, once the length of the horn is made finite, we produce an unresolvable inseparability near the open end so that there is no real practical design assistance derived from near separability inside the horn.

Rather than setting out the exact solution for a hyperboloid or a conical horn in detail, let us now go straight to the approximate solution for propagation in an infinite horn of rather general shape. We assume that we have some good approximation to the shapes of the wavefronts—something more or less spherical and, since the wave fronts must be orthogonal to the horn walls, centered approximately at the apex of the cone that is locally tangent to the horn walls, as shown in Fig. 8.13. This description will be exact for a conical horn, but only an approximation for other shapes. If $S(x)$ is the area of this wavefront in the horn at position x, defined by its intersection with the axis, then, during an acoustic displacement ξ, the fractional change in the volume of air in the horn at position x is $(1/S)\partial(S\xi)/\partial x$. This contrasts with the simpler expression $\partial\xi/\partial x$ for a plane wave in unconfined space. Proceeding now as for the plane-wave case, we find a wave equation of the form

$$\frac{1}{S}\frac{\partial}{\partial x}\left(S\frac{\partial p}{\partial x}\right) = \frac{1}{c^2}\frac{\partial^2 p}{\partial t^2}, \tag{8.38}$$

which is known as the Webster equation (Webster, 1919; Eisner, 1967), although its origins date back to the time of Bernoulli. Actually, in Webster's case, the curvature of the wave fronts was neglected so that x was taken as the geometrical distance along the horn axis and S as the geometrical cross section at position x. This plane-wave approximation is good for horns that are not rapidly flaring, but breaks down for a horn with large flare. Various simple modifications to the Webster equation have been proposed (Weibel, 1955; Keefe et al., 1993), all of which improve its approximation by replacing the plane-wave surfaces by curved surfaces, chosen so as to meet the horn surface and the axis normally. All essentially ignore the transverse flow that is necessitated by the fact that an elementary volume element between successive wavefronts is thicker on the axis than at its edges, and simply assume constant pressure throughout the element.

Here, as well, we have assumed that p is constant across the wave front in the horn, which is equivalent to assuming separability. This is not a bad approximation for horns that do not flare too rapidly, but we must not expect too much of it in extreme cases. In this spirit, we now make the transformation

$$p = \psi S^{-1/2} \tag{8.39}$$

in the reasonable expectation that, with the even spreading of wave energy across the wavefront, ψ should be essentially constant in magnitude, independent of x. If we also assume that p varies with angular frequency ω and write S in terms of a local equivalent radius a so that

$$S = \pi a^2, \tag{8.40}$$

then Eq. (8.38) becomes

$$\frac{\partial^2 \psi}{\partial x^2} + \left(k^2 - \frac{1}{a}\frac{\partial^2 a}{\partial x^2}\right)\psi = 0, \tag{8.41}$$

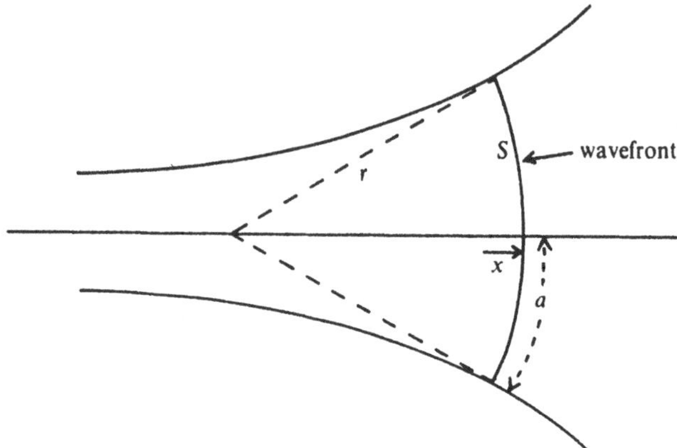

FIGURE 8.13. In a horn, the wavefront has approximately the form of a spherical cap of area S and effective radius r based upon the local tangent cone and cutting the axis at a point with coordinate x.

where $k = \omega/c$. This form of the equation, noted by Stevenson (1951) and later by Benade and Jansson (1974), serves as a good basis for discussion of the behavior of horns. In a formal sense, it is the exact analog of the Schrödinger equation in quantum mechanics, with the "horn function"

$$F \equiv \frac{1}{a}\frac{d^2a}{dx^2} = \frac{1}{2S}\frac{\partial^2 S}{\partial x^2} - \frac{1}{4S^2}\left(\frac{\partial S}{\partial x}\right)^2 \tag{8.42}$$

taking the place of the potential.

The important thing to notice about Eq. (8.41) is that the wave function ψ, and hence the original pressure wave p, is propagating or nonpropagating according as $k^2 \gtrless F$, which is just what we would expect for a quantum particle of normalized energy k^2 meeting a potential barrier. The frequency $\omega = kc$ for which we have equality is called the cutoff frequency at this part of the horn. A visual estimate of the magnitude of F at a given position x can be made, as illustrated in Fig. 8.14, by observing that a is essentially the transverse radius of curvature R_{T} of the horn at point x while $(d^2a/dx^2)^{-1}$ is close to the external longitudinal radius of curvature R_{L}, provided that the wall slope da/dx is small. Thus,

$$F \approx \frac{1}{R_{\mathrm{L}}R_{\mathrm{T}}}. \tag{8.43}$$

Of course, this is no longer a good approximation when the wall slope, or local cone angle, is large, and we must then use the expression

$$F = \frac{1}{a}\frac{d^2a}{dx^2}, \tag{8.44}$$

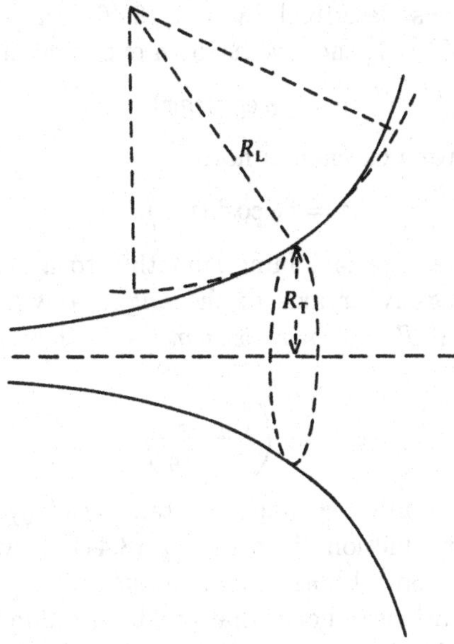

FIGURE 8.14. The geometry of a horn at any one place is characterized by the external longitudinal radius of curvature R_L and the internal transverse radius of curvature R_T.

with a interpreted as the equivalent internal radius measured along the wavefront as discussed previously.

Of particular theoretical simplicity is the class of horns called Salmon horns (Salmon, 1946a, b), for which the horn function F, and therefore the cutoff frequency ω_0, is constant along the whole length of the horn (Morse, 1948). Clearly, from Eq. (8.44), this implies

$$a = Ae^{mx} + Be^{-mx}, \tag{8.45}$$

where $F = m^2$ and m is called the flare constant. It is more convenient to rewrite Eq. (8.45) as

$$a = a_0[\cosh(mx) + T\sinh(mx)], \tag{8.46}$$

where T is an alternative parameter. The pressure wave in the horn then has the form

$$p = \left(\frac{p_0}{a}\right) e^{j\omega t} e^{-j\sqrt{k^2-m^2}x} \tag{8.47}$$

and is nonpropagating if $k < m$. These expressions should strictly all be interpreted in terms of curved wavefront coordinates, as in Fig. 8.13, but it is usual to neglect this refinement and simply use the plane-wave approximation.

The family of horns described by Eq. (8.46) has several important degenerate forms. If $T = 1$, then we have an exponential horn:

$$a = a_0 \exp(mx). \tag{8.48}$$

If $T = 0$, then we have a catenoidal horn:

$$a = a_0 \cosh(mx), \tag{8.49}$$

which has the nice feature of joining smoothly to a cylindrical pipe extending along the negative x axis to the origin, as was the case for the hyperboloidal horn. If $T = 1/mx_0$ and $m \to 0$, then we have a conical horn:

$$a = a_0 \left(1 + \frac{x}{x_0}\right), \tag{8.50}$$

with its vertex at $-x_0$ and a semiangle of $\tan^{-1}(a_0/x_0)$. Consideration of the value of the horn function given by Eq. (8.44) shows that $F = 0$ for this case, so that the conical horn has no cutoff.

Many of the applications of horns that are discussed in textbooks involve situations in which the diameter of the open end of the horn is so large that there is no appreciable reflection. The horn then acts as an efficient impedance transformer between a small diaphragm piston in the throat and a free spherical wave outside the mouth. Exponential and catenoidal horns have near-unity efficiency, as defined by Morse (1948), above their cutoff frequencies, while the efficiency of a conical horn never becomes zero but rises gradually with increasing frequency until it reaches unity. We shall not discuss these situations further—those interested should consult Morse (1948) or Olson (1957).

8.7 Finite Conical Horns

It is useful to quote results analogous to Eq. (8.23) for the throat impedance of a truncated conical or exponential horn terminated by a mouth impedance Z_L. Typically this might be the radiation impedance at an open mouth, for which we can use the expressions in Section 8.3, though these require some modification in careful work because of the curvature of the wavefronts (Fletcher and Thwaites, 1988). Another case of some interest is that in which the remote end of the horn is stopped rigidly so that $Z_L = \infty$.

For a conical horn with a throat of area S_1 located at position x_1, a mouth of area S_2 at position x_2, and length $L = x_2 - x_1$, we find (Olson, 1957)

$$Z_{IN} = \frac{\rho c}{S_1} \left\{ \frac{jZ_L[\sin(kL - \theta_2)/\sin\theta_2] + (\rho c/S_2)\sin kL}{Z_L[\sin(kL + \theta_1 - \theta_2)/\sin\theta_1 \sin\theta_2] - (j\rho c/S_2)[\sin(kL + \theta_1)/\sin\theta_1]} \right\}, \tag{8.51}$$

where $\theta_1 = \tan^{-1} kx_1$ and $\theta_2 = \tan^{-1} kx_2$, both x_1 and x_2 being measured along the axis from the position of the conical apex. Similarly, for an exponential horn of the form of Eq. (8.48) and length L,

$$Z_{\text{IN}} = \frac{\rho c}{S_1} \left[\frac{Z_{\text{L}} \cos(bL + \theta) + j(\rho c/S_2) \sin bL}{jZ_{\text{L}} \sin bL + (\rho c/S_2) \cos(bL - \theta)} \right], \tag{8.52}$$

where $b^2 = k^2 - m^2$ and $\theta = \tan^{-1}(m/b)$. It is not simple to allow for wall effects in these expressions, since the imaginary part of k varies with position in the horn. For a horn with a wide mouth and not too narrow a throat, radiation effects may dominate so that k can be taken as real. This is, however, not a valid approximation in musical instruments, which use long horns of quite small diameter. We shall see that more complex calculations are necessary in such cases.

The expression [Eq. (8.51)] for the input impedance of a conical horn, measured at the end that is at a distance x_1 from the apex, deserves some further discussion. In the first place, we should note that it is applicable for the impedance at either the wide or the narrow end of a conical pipe. For a flaring cone, $x_2 > x_1$ and $L > 0$, while for a tapering cone, $x_2 < x_1$ and $L < 0$.

In the second place, we should examine several special cases of open and stopped cones, making the approximation that $Z_{\text{L}} = 0$ at an open end and $Z_{\text{L}} = \infty$ at a closed end. For a cone of length L with an ideally open end $Z_{\text{L}} = 0$, Eq. (8.51) gives, for either the large or the small end of a cone, the formal result

$$Z_{\text{IN}} = j\left(\frac{\rho c}{S_1}\right) \frac{\sin kL \sin \theta}{\sin(kL + \theta_1)}, \tag{8.53}$$

This does not imply that the input impedance is the same from both ends, since, as noted above, the sign of L and the magnitude of θ_1 are different in the two cases.

Zeros in Z_{IN} occur at frequencies for which $\sin kL = 0$, so that these frequencies are the same in each case and exactly the same as those for a cylindrical pipe with the same length L. To allow for the finite reactance associated with the radiation impedance Z_{L}, it is approximately correct, for a narrow cone, to simply add an appropriate end correction equal to 0.6 times the open end radius to the geometrical length L, as discussed in relation to Eq. (8.33).

The infinities in Z_{IN} occur, however, at frequencies that differ between the two cases and are not simply midway between those of the zeros, as was the case with a cylindrical pipe. Rather, the condition for an infinity in Z_{IN} is, from Eq. (8.53),

$$\sin(kL + \theta_1) = 0, \tag{8.54}$$

or equivalently,

$$kL = n\pi - \tan^{-1} kx_1. \tag{8.55}$$

For a cylinder, $x_1 \to \infty$ so that $\tan^{-1} kx_1 = \pi/2$ and $kL = (n - \frac{1}{2})\pi$, as we already know. For a cone measured at its narrow end, L is positive and, since $\tan^{-1}\theta < \pi/2$ for all θ, the frequencies of the maxima in Z_{IN} are higher than those for a cylinder of the same length. The converse is true for a tapering cone. If the cone is nearly complete, so that $kx_1 \ll 1$, then $\tan^{-1} kx_1 \approx kx_1$ and, since $L = x_2 - x_1$, Eq. (8.55) becomes $kx_2 \approx n\pi$, so that the frequencies of the maxima in impedance approach those of an open cylinder of length $x_2/2$. Figure 8.15 shows an input impedance curve for an incomplete cone calculated for the narrow end, and with a radiation impedance as in Fig. 8.8 assumed at the open mouth.

Cones that are stopped at the remote end are of less musical interest. If $Z_{\mathrm{L}} = \infty$, then, from Eq. (8.51),

$$Z_{\mathrm{IN}} = j\left(\frac{\rho c}{S_1}\right)\frac{\sin(kL - \theta_2)\sin\theta_1}{\sin(kL + \theta_1 - \theta_2)}. \tag{8.56}$$

The zeros are given by $(kL - \theta_2) = n\pi$, which, writing $L = x_2 - x_1$, becomes

$$(kx_2 - \tan^{-1} kx_2) - kx_1 = n\pi. \tag{8.57}$$

Thus, if we are considering a tapering cone and the distance x_2 from the stopped end to the imaginary apex is small enough that kx_2 is rather less than unity, the bracketed terms nearly cancel and the cone behaves approximately as though it is of length x_1 and complete to its vertex. No such simplification occurs for a flaring cone or for the infinities in the impedance, for which the condition is that $kL + \theta_1 - \theta_2 = n\pi$.

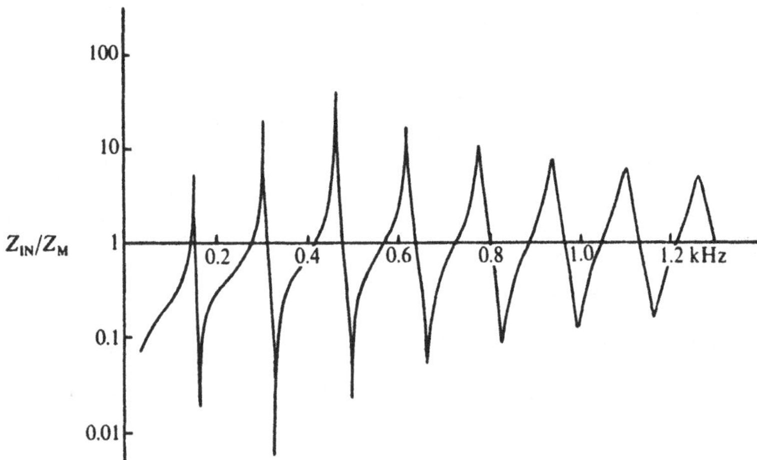

FIGURE 8.15. Magnitude of the input impedance Z_{IN} of a conical horn of length 1 m, throat diameter 1 cm, and mouth diameter 10 cm, relative to the impedance $Z_M = (\rho c/S_1)$, as a function of frequency.

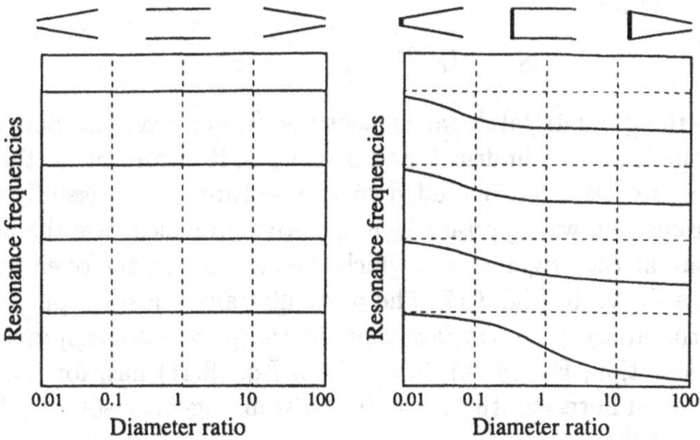

FIGURE 8.16. The frequencies of the first four resonances of open and closed conical horns, as a function of the ratio of the end diameters (Data from Ayers et al. 1985).

The simple behavior of the modes of finite conical horns, as discussed above, is illustrated in Fig. 8.16. The left part of the figure shows the behavior with shape of the modes of a doubly open conical frustrum horn, the frequencies of which are constant. In contrast, the right part of the figures shows the mode frequencies for a conical frustrum horn with one end closed rigidly. The doubly open horns are relevant to the behavior of flute-like instruments and organ flue pipes, while the singly closed horns are representative of the resonators of reed or lip-driven instruments, the closed end being identified with the reed or lip driver.

An extensive discussion of the conical horn in musical acoustics has been given by Ayers et al. (1985). Using straightforward physics, this paper treats both conical frustra and complete cones and points out a number of misconceptions, or at least erroneous expositions, in the standard physics literature. Benade (1988) also discusses conical horns and shows how their properties can be represented by a combination of a cylindrical duct, a transformer, and several terminating impedances. For some purposes this simplified approach is illuminating. Finally, as we discuss in Section 8.15, we can also treat finite horns, combinations of horns and pipes, and other arrangements by the methods of network analog theory.

8.8 Bessel Horns

One particular family of horns that is worthy of attention because of its formal simplicity and that provides a good approximation to musically useful horns (Benade, 1959; Benade and Jansson, 1974) is the Bessel horn family, defined by

$$S = Bx^{-2\varepsilon} \quad \text{or} \quad a = bx^{-\varepsilon}, \tag{8.58}$$

where x is the geometrical distance measured from a reference point $x = 0$. If $\varepsilon = 0$, the horn is cylindrical, and if $\varepsilon = -1$, it is conical, so that these two degenerate cases are included in the picture. More usefully for our present discussion, we suppose ε to be positive, in which case the horn has a rapid flare at the origin $x = 0$, which thus represents the open mouth of the horn as shown in Fig. 8.17. The particular analytical simplicity of the Bessel horns arises from the fact that, in the plane-wave approximation, the wave equation, Eq. (8.38), in the form Eq. (8.41) has, for the case of an ideally open horn mouth at $x = 0$, the standing-wave solution (Jahnke and Emde, 1938)

$$\psi = x^{1/2} J_{\varepsilon+1/2}(kx), \tag{8.59}$$

where J is a Bessel function, hence the name of the horn family. From Eq. (8.39), the pressure standing wave has the form

$$p(x) = Ax^{\varepsilon+1/2} J_{\varepsilon+1/2}(kx). \tag{8.60}$$

The existence of this analytical solution is a great help for semiquantitative discussion of the behavior of this family of horns, whose shape can be made to vary very considerably by choice of the parameter ε. If we are considering a horn composed of segments of different Bessel horns joined end to end, then Eq. (8.60) must be supplemented in each segment by a similar term involving the Neumann function $N_{\varepsilon+1/2}(kx)$. We can then calculate the behavior, in the plane-wave approximation, of composite horns having sections of Bessel, exponential, conical, and cylindrical geometry joined end to end (Pyle, 1975). We will not follow such a course in detail but, instead,

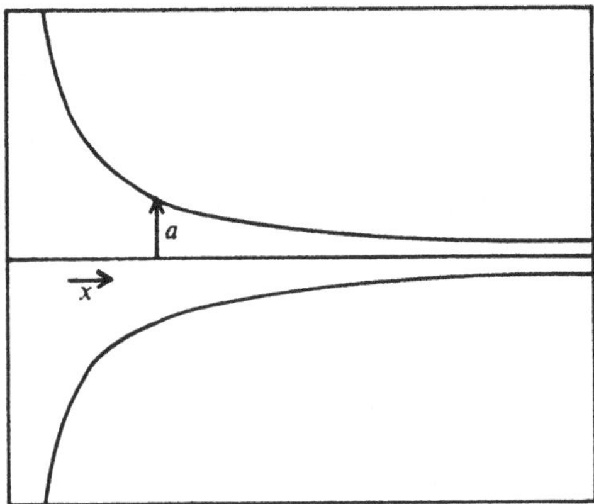

FIGURE 8.17. The form of a Bessel horn with parameter $\varepsilon > 0$.

we will examine briefly the behavior of waves near the mouth of a Bessel horn to show some of the complications involved.

As we saw in Eqs. (8.41)–(8.44), the propagation of a wave in a flaring horn is governed by the value of the horn function F at the point concerned. If F is greater than k^2, the wave is attenuated exponentially with distance and a reflection occurs rather than propagation. For a horn mouth of Bessel type, F is easily calculated in the plane-wave approximation, in which wave coordinate x is replaced by the axial geometrical coordinate, and has the form shown in Fig. 8.18(a). F goes to infinity at the mouth of the horn, so that waves of all frequencies are reflected, and the reflection point for waves of low frequency (small k) is further inside the mouth than for those of high frequency.

Close to the open mouth, however, the plane-wave approximation is clearly inadequate, and the spherical approximation gives a much better picture (Benade and Jansson, 1974). The function F in this approximation is shown in Fig. 8.18(b). Its form is similar to the plane-wave version, but F never becomes infinite, so that there is an upper cutoff frequency above which waves can propagate freely from the mouth without reflection.

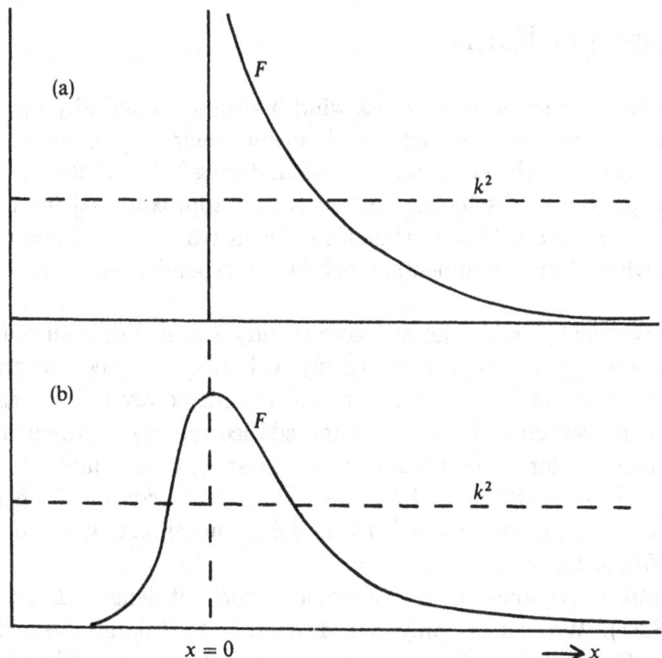

FIGURE 8.18. The horn function F for a Bessel horn calculated on the basis of (a) the plane-wave approximation and (b) the spherical-wave approximation. When $k^2 < F$, the wave is attenuated instead of being propagated (after Benade and Jansson, 1974).

An infinite F, as in the plane-wave approximation, would in fact confine all sound energy within the horn because of the infinite barrier height; the more realistic curve given by the spherical approximation allows some wave energy to leak out at any frequency by tunneling through the barrier.

When we consider the losses in a real horn in detail, not only must we supplement the standing-wave solution [Eq. (8.60)] with extra terms in $N_{\varepsilon+1/2}(kx)$, which combine to represent the small fraction of energy contained in propagating waves lost through the mouth of the horn, but we must also take account of wall losses by adding a small imaginary part $-j\alpha$ to k. For a horn more complicated in profile than a simple cylinder, α depends on the local horn radius and therefore varies from place to place along the horn. The calculations are then quite involved (Kergomard, 1981). Fortunately, we can ignore these complications in our present discussion though they must be taken into account in any really accurate computations.

Because the horns of real musical instruments do not conform exactly to any of the standard types we have considered, we will not go into further detailed discussion of them at this stage, but defer this until we come to describe the instruments themselves in a later chapter.

8.9 Compound Horns

As we see in Chapter 14, most brass wind instruments actually have horn profiles that are nearly cylindrical for about half their length, starting from the mouthpiece, and then expand to an open flaring bell. In modern instruments, the profile of this flaring section is well approximated by a Bessel horn of the form of Eq. (8.58), while for older instruments and some of the more mellow modern instruments, much of the expanding section is nearly conical.

It is not worthwhile to model such compound horns in detail, since real instruments do not conform precisely to any such oversimplified prescription. The complications of mode tuning are illustrated, however, by consideration of the frequencies of the input impedance maxima—which are the sounding modes—for a compound horn consisting of a cylindrical and a conical section smoothly joined together. Part of the complication is produced by the acoustical mismatch at the joint, but similar mode behavior would be found for other profiles.

The input impedance Z_c for a conical horn of length L_1 is given by Eq. (8.51). We can simplify our discussion by taking the radiation impedance Z_L at the open mouth to be zero, giving, as another form of Eq. (8.53),

$$Z_c = \frac{j\rho c}{S_1}\left(\cot kL_1 + \frac{1}{kx_1}\right)^{-1}, \tag{8.61}$$

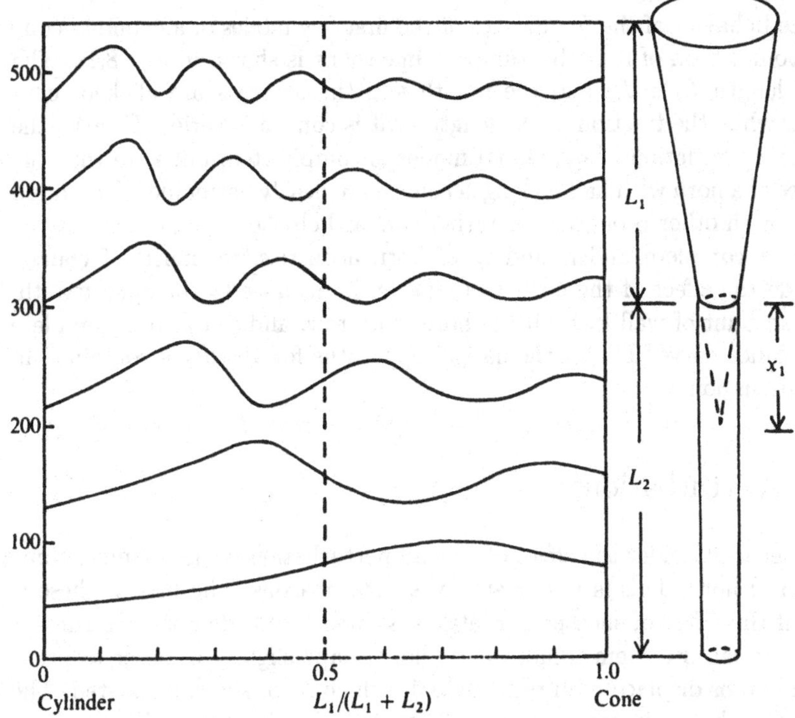

FIGURE 8.19. Frequencies of the input impedance maxima for a compound horn, of the shape shown, as a function of the fraction of horn length that is conical.

where S_1 is the throat area and x_1 is the distance from the throat to the vertex of the cone, as shown in Fig. 8.19. We now take Z_c as the terminating impedance for the cylindrical section of length L_2 and matching area S_1. From Eq. (8.23), the input impedance to the compound horn then becomes infinite if

$$jZ_c \sin kL_2 + \left(\frac{\rho c}{S_1}\right) \cos kL_2 = 0, \tag{8.62}$$

and, with Eq. (8.61), this immediately leads to the condition

$$\tan kL_2 - \cot kL_1 - \left(\frac{1}{kx_1}\right) = 0. \tag{8.63}$$

Solution of this equation for $k = \omega/c$ then gives the frequencies of the impedance maxima. The same calculation for a cylinder joined to an exponential horn would have used Eq. (8.52) and led to the formally similar result

$$\tan kL_2 - \left(\frac{b}{k}\right) \cot bL_1 - \left(\frac{m}{k}\right) = 0, \tag{8.64}$$

where $b = (k^2 - m^2)^{1/2}$.

The behavior of the frequencies of the first few modes of a cylinder–cone compound horn of roughly trumpet dimensions is shown in Fig. 8.19. The total length $L_1 + L_2$ and the mouth and throat areas are all kept constant, while the fraction of the length that is conical is varied. Clearly, the frequency variation of individual modes is complicated, but note that the modes of a horn with half its length conical are nearly harmonic. Compound horns with other profiles show rather similar behavior.

For a complete understanding of horn acoustics, we must, of course, include the effect of the nonzero radiation impedance at the open mouth, take account of wall losses if the horn is narrow, and calculate a complete impedance curve. The mathematical apparatus for all this is contained in our discussion.

8.10 Perturbations

Now let us consider the effect of a small perturbation in the shape of some idealized horn. This is important for several reasons. The first of these is that if the effect of such perturbations is understood, then the instrument designer can use them to adjust the horn shape slightly in order to properly align or displace horn resonances in which he or she is interested. The second is that in instruments with finger holes in the side of the horn perturbations are unavoidable, and it is important to understand and control their effects.

Suppose that we know a standing-wave solution $p_0(x,t)$ for a horn of profile $S_0(x)$ that corresponds to a mode of frequency ω_0, as described by Eq. (8.38), with appropriate terminating impedances at each end of the horn. Now, let the bore of the horn be altered by a small amount so that the new cross section becomes

$$S(x,t) = S_0(x,t) + \delta S(x,t). \tag{8.65}$$

This perturbation will change the resonance frequency ω_0 to a new value ω, which we write

$$\omega = \omega_0 + \delta\omega, \tag{8.66}$$

and the pressure distribution will become

$$p(x,t) = \beta p_0(x,t) + p_1(x,t), \tag{8.66}$$

where $\beta \approx 1$ and p_1 is functionally orthogonal to p_0. If we substitute Eqs. (8.65), (8.66), and (8.67) into Eq. (8.38) and also use the unperturbed version of Eq. (8.38) with S_0, ω_0, and p_0, then we can collect all the terms of first order in the perturbations to give

$$\frac{1}{S_0}\frac{d\delta S}{dx}\frac{dp_0}{dx} - \frac{\delta S}{S_0^2}\frac{dS_0}{dx}\frac{dp_0}{dx} = -\frac{2\omega_0\delta\omega}{c^2}p_0 + \text{terms in } p_1. \tag{8.67}$$

Now, we multiply by S_0p_0 and integrate over the whole length of the horn. The terms in p_1 vanish since they are orthogonal to p_0, and we find

$$\delta\omega = -\frac{c^2}{2\omega_0 N}\int S_0 \frac{d}{dx}\left(\frac{\delta S}{S_0}\right) p_0 \frac{dp_0}{dx}\,dx, \tag{8.68}$$

where

$$N = \int S_0 p_0^2\,dx. \tag{8.70}$$

Clearly, these two equations allow us to calculate the shift $\delta\omega$ in the resonance frequency produced by the bore perturbation $\delta S(x)$.

It is easiest to evaluate the effect of such perturbations by considering what happens when the bore is enlarged by a small amount Δ at a position x_0. To do this, we write

$$\delta S(y) = \Delta\delta(x - x_0), \tag{8.71}$$

where $\delta(x - x_0)$ is a Dirac delta function. Substituting this into Eq. (8.69) we find, since $d\delta(x - x_0)/dx$ under an integral yields the negative of the derivative of the integrand at x_0, that

$$\delta\omega = \frac{c^2\Delta}{2\omega_0 N}\left[\frac{d}{dx}\left(p_0\frac{dp_0}{dx}\right)\right]_{x=x_0}. \tag{8.72}$$

To see what this means, suppose that in some region of the horn the pressure pattern has a spatial variation like $\sin kx$. Then the bracket in Eq. (8.72) behaves like $\cos 2kx_0$. Thus, when the perturbation is near a maximum in the pressure variation ($\sin kx_0 \approx 1$), $\cos 2kx_0 \approx -1$ and $\delta\omega$ has its maximum negative value. Conversely, if $\sin kx_0 \approx 0$, then $\cos 2kx_0 \approx +1$, so that near a maximum in the velocity $\delta\omega$ has its maximum positive value. Both these remarks assume that Δ is positive, so that the bore is being enlarged by the perturbation. Opposite conclusions apply for a constriction in the bore. Since different modes have their pressure maxima at different places in the horn, it is possible to change the relative frequencies of selected modes and so effect musically desirable changes in the behavior of the horn.

A few examples make this point clear. Suppose we consider the modes of a cylindrical pipe open at the end $x = 0$ and closed at the other end, $x = L$. The nth normal mode then has a pressure pattern like $\sin[(2n-1)\pi x/2L]$. If the pipe diameter is enlarged near the open end ($x = 0$), then, by Eq. (8.72), $\delta\omega$ is positive for all n, and the frequencies of all modes are raised. Conversely, if the diameter is enlarged near the closed end, the frequencies of all modes are lowered.

More interesting is the case in which the bore is enlarged at a point one-third of the length away from the closed end, that is, at $x = 2L/3$. The bracket in Eq. (8.72) then behaves like $\cos(2\pi/3) = -0.5$ for the first mode, $n = 1$, and like $\cos 2\pi = +1$ for the second mode, $n = 2$. Thus, the frequency of the first mode is lowered while that of the second mode is raised.

8.11 Numerical Calculations

With the ready availability of computers, it is often practically convenient, though generally less instructive, to calculate the behavior of horns numerically. This can be done conveniently only in the plane-wave approximation or the local spherical-wave approximation, so that its accurate use is limited to horns with small flare, but this is adequate for many wind instruments.

The bare bones of the procedure have already been set out, but its use in this way has not been made explicit. The basis of the method is the recognition that an arbitrary horn can always be represented to very good accuracy by a succession of small conical sections joined end to end. In the limit in which the length of the sections becomes infinitesimal, the representation is exact. Now, in the spherical-wave approximation, the input impedance of a section of conical horn of length L is related to its terminating impedance Z_{L} by Eq. (8.51). Indeed, if the length L is small so that the cross section is nearly constant, the propagation constant k in Eq. (8.51) can be made complex to allow for wall losses according to Eqs. (8.14), (8.15), and (8.35). To make a numerical calculation for an arbitrary horn therefore, we simply start from the open end, with Z_{L} the radiation impedance shown in Fig. 8.8. We then use Eq. (8.51) successively for short distances back along the horn until the input throat is reached. A modification of the program readily allows the pressure distribution along the horn to be calculated at the same time.

For the plane-wave version of this calculation, which trades off a simpler calculation at each step against an increased number of steps, we can approximate the horn by a series of very short cylindrical sections. Equation (8.23) replaces the more complicated conical form of Eq. (8.51), but each cylindrical section must be made very short in order to give a reasonable approximation for a flaring horn.

Calculations have been made of the input impedance of a variety of musical horns, using essentially these methods, by Plitnik and Strong (1979), Caussé et al. (1984), Dudley and Strong (1990), and Amir et al. (1993). The calculation is robust and generally gives very good agreement with experiment for the relatively small flare angles of woodwind instruments, though the bells of brass instruments present greater problems.

8.12 Curved Horns

Only a few musical wind instruments have straight horns; for very many of them it is necessary to bend the horn in order to bring its length within a reasonable compass for playing and transport. This bending is significant in instruments such as the bassoon and saxophone, and extreme in the case of brass instruments. It is therefore important to know what effect, if any, it has on the acoustic properties of the horn.

To an obvious first approximation we should expect to measure the effective length of the horn along a curved axis passing through the centroid of its cross-section at every point. As expected, this is a good approximation for cases in which the bend radius R is large compared with the tube radius r, as defined in Fig. 8.20, but for more extreme bends a careful analysis is required. The problem has been considered by Nederveen (1969) and by Keefe and Benade (1983), both of whom come to similar conclusions. If we define a parameter $B = r/R$ to measure the severity of the bend, then the analysis of Keefe and Benade shows that the wave velocity in the duct is increased above the normal sound velocity c by an amount δv and the characteristic impedance of the duct is decreased from its normal value $Z_0 = \rho c/\pi r^2$ by an amount δZ, these quantities being given in terms of the bend parameter B by

$$\frac{\delta v}{c} = -\frac{\delta Z}{Z_0} = \left(\frac{2I}{\pi B}\right)^{1/2} - 1, \tag{8.73}$$

where

$$I = \int_0^{\pi/2} \cos\theta \ln\left(\frac{1 + B\cos\theta}{1 - B\cos\theta}\right) d\theta. \tag{8.74}$$

Figure 8.20 shows this relation in graphical form for values of the bend

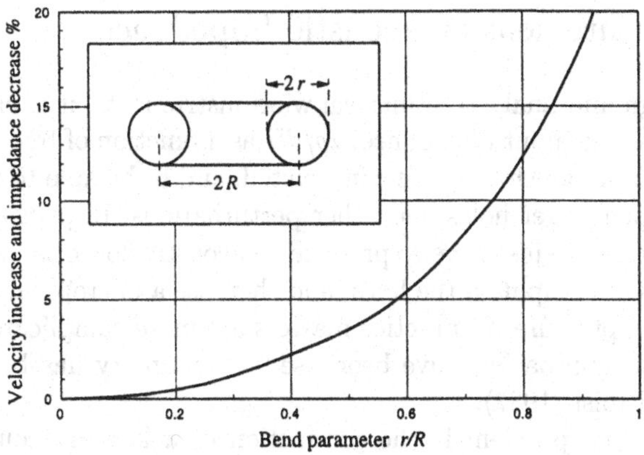

FIGURE 8.20. Increase δv in wave velocity and decrease $-\delta Z$ in duct impedance in a curved circular pipe, as a function of curvature parameter $B = r/R$, calculated from the theories of Nederveen (1969) and of Keefe and Benade (1983). Experimental values are significantly less than those calculated and are not the same for the two quantities.

parameter between zero, a straight tube, and almost 1, which is the geometrical limit. The practical limit for bent tubing in a brass instrument is about $R \approx 0.8$.

The analysis leading to these equations, or the similar equations of Nederveen, is not rigorous, so that it is clearly necessary to check the conclusions against experiment. This was done by Keefe and Benade (1983) for tubing with about the maximum curvature found in brass instruments ($B = 0.728$). For this tubing the theory predicts $\delta v/c = -\delta Z/Z_0 = 8.9\%$. The measured values were significantly smaller than this, however, with $\delta v/c \approx 4.7\%$ and $\delta Z/Z_0 \approx -6.3\%$. The discrepancy between these two values, and between both of them and the theory, is puzzling. Keefe and Benade ascribe it to the neglect of viscosity in the calculation, but it is not clear that there is any shear except at the walls, so it may be some other aspect of the approximations made in the derivation that is to blame. Despite this, the theory does give at least a semiquantitative account of the effect of bending.

From this discussion it is clear that bending a tube shortens its acoustic length and lowers its impedance, and that the corrections may be large enough to require consideration in applications such as design of the valve tubing in trumpets and the larger bends in tubas and saxophones. As pointed out by Nederveen, insertion of a curved section into a cylindrical bore creates least acoustic missmatch if the cross-section of the curved part is reduced slightly from that of the bore, to match the impedances. In the case of a conical bore, the length of the curved section should also be increased slightly, with corresponding decrease in cone angle, to allow for the increased phase velocity.

8.13 Measurement of Acoustic Impedance

In the design and analysis of musical wind instruments, it is clearly necessary to know the input impedance, dp/dU,as a function of frequency, of the basic horn from which they are constructed, and to be able to measure the effects of open finger holes and other perturbations. In principle this is a simple matter—we just have to produce an acoustic flow of known constant amplitude at the input to the horn, and then use a microphone to measure the resulting pressure. In practice matters are more complicated, and several different approaches have been used. A summary has been given by Benade and Ibisi (1987).

The principal problem is the production of a known acoustic flow at the input. To do this we need either a flow source with an impedance that is very much higher than that of the horn being measured, or else some servo-controlled mechanism to maintain a constant flow in the face of changing horn impedance. The earliest approach, introduced by Kent, working in the laboratories of Conn Inc. in the 1950s, was to produce a large acoustic pressure in a cavity driven by a servo-controlled loudspeaker and oscillator, and feed this to the horn under measurement through a capillary tube of high acoustic impedance. Benade also used a similar approach but with a tube full of cotton wool. The capillary method was later refined by Backus (1974), who used an annular capillary, consisting of a rod centered within a narrow tube, to reduce the frequency dependence of the source resistance. Coltman (1968) used a more direct method of servo-control by sensing the motion of the loudspeaker diaphragm using a coil moving in a magnetic field, while Merhaut (1968) similarly used a light diaphragm driven by an enclosed speaker and monitored by electrostatic means as in a condenser microphone. More recently, Benade and Ibisi (1987) have used a solid piezoelectric element with resonance frequency around 5 kHz to generate the flow, this having a sufficiently high impedance that its motion, and therefore the resulting acoustic flow, can be calculated directly from its own resonance curve without the need for any servo-control.

Several independent and rather different methods have also been devised. Fransson and Jansson (1975) used a corona discharge between two closely spaced electrodes, with the steady voltage supplemented by an audio frequency, to generate a volume-flow source. Wolfe et al. (1995a,b) used a broad-band loudspeaker noise source in a cavity and coupled this to the measurement point with a tapered horn, the throat of which was filled with cotton wool to increase resistance and damp out major resonances. Their innovation was to shape the electronic noise input to the loudspeaker automatically in such a way that measurement of the input impedance of a long narrow tube gave the required constant value. The equipment then functioned extremely rapidly to yield a real-time graphical display of input impedance as the horn was perturbed.

Pratt et al. (1977, 1979), on the other hand, turned their backs on the constant-flow approach and used a hot-wire anemometer to measure the flow directly. A rather similar but less direct method was adopted by Chung and Blaser (1980), who used two spaced microphones in a cylindrical tube connecting a noise source to the input of the instrument horn, and processed the resulting measurements to give the acoustic impedance.

8.14 The Time Domain

Nearly all of our discussion has been carried on in the frequency domain—we have examined the propagation in a horn of sinusoidal waves of steady frequency ω. While this is generally the most convenient framework in which to study the physics of musical instruments, it is sometimes helpful to revert to the time domain and examine the buildup and propagation of pressure disturbances along the horn and their reflection from its open end. This is clearly a good way to treat the initial transients of musical sounds, and the time-domain method can also be used for steady tones.

Formally, treatment of a problem in the time domain or the frequency domain must give identical results, but in practice we are forced to make approximations in our analysis in order to get a reasonable answer, and the nature of these approximations can be quite different in the two cases, so that one can usually be employed more easily than the other (Schumacher, 1981; McIntyre et al., 1983; Ayers, 1996).

In Sections 8.3 and 8.7, we gave explicit expressions for the input impedance $Z(\omega)$ of horns of various profiles, while Figs. 8.11 and 8.15 illustrated the behavior of $Z(\omega)$ relative to the characteristic impedance $Z_0 = \rho c/S_1$ of an infinite cylindrical tube having the same area S_1 as the throat of the horn. In general, $Z(\omega)$ displays a long series of more or less sharp impedance peaks at frequencies ω_n that are, for useful musical instruments, related moderately closely to some harmonic series $n\omega_0$.

If now we seek the pressure waveform $p(t)$ observed in the mouth of the horn when the acoustic flow into it is $U(t)$, we can proceed in one of two ways. In the frequency domain, we can express the flow $U(t)$ as a Fourier integral, yielding the Fourier components $U(\omega)$, and then use the definition of the input impedance $Z(\omega)$ to write

$$p(t) = \int_{-\infty}^{\infty} Z(\omega)U(\omega)e^{j\omega t}d\omega. \tag{8.75}$$

Alternatively, we can define an impulse response function, or Green's function, $G(t-t')$, which gives the pressure response at time t to a unit impulse of flow at time t', and then write directly the convolution integral

$$p(t) = \int_{-\infty}^{t} G(t-t')U(t')dt' \equiv G(t) * U(t). \tag{8.76}$$

As we remarked before, these results are formally equivalent, and, indeed, the impulse response function $G(t)$ is simply the Fourier transform of the input impedance function $Z(\omega)$.

As Schumacher (1981) has pointed out, the problem with using Eq. (8.76) as a computational formula from which to derive $p(t)$ arises from the fact that $G(t-t')$ has a considerable extension in time; an acoustic flow pulse injected into the horn reflects from its two ends for some tens of periods before its amplitude is reduced enough for it to become negligible. This follows at once from the sharply peaked nature of $Z(\omega)$, which is an equivalent feature in the frequency domain.

Schumacher proposed a useful way of avoiding this computational problem, and we derive this in a way formalized by Ayers (1996). Suppose the flow enters the instrument from an infinite pipe of matching cross-section S. The plane-wave reflection function $r(\omega)$ at the input to the instrument can be expressed as

$$r(\omega) = \frac{Z(\omega) - Z_0}{Z(\omega) + Z_0}, \tag{8.77}$$

where $Z_0 = \rho c/S$ is the characteristic impedance at the entry to the bore. We can rearrange this equation to the form

$$Z(\omega) = Z_0 + Z_0 r(\omega) + r(\omega)Z(\omega) \tag{8.78}$$

and take the inverse Fourier transform to get

$$G(t) = Z_0\delta(t) + Z_0\delta(t) * r(t) + r(t) * Z(t), \tag{8.79}$$

where $\delta(t)$ is the Dirac delta function and $r(t)$ is the Fourier transform of $r(\omega)$. Convolving this with $U(t)$ and using Eq. (8.76) then gives

$$\begin{aligned} p(t) &= Z_0 U(t) + Z_0 r(t) U(t) + r(t) * p(t) \\ &= Z_0 U(t) + \int_0^\infty r(t')[Z_0 U(t-t') + p(t-t')]dt'. \end{aligned} \tag{8.80}$$

It turns out that $r(t)$ is nearly zero for t less than the wave-return transit time τ along the simple cylindrical part of the bore, and it also has a much smaller extension in time than the original impulse function $G(t)$. The reason for this is obvious, since waves returning to the mouthpiece are not reflected but simply absorbed by the matching input termination. It is therefore relatively straightforward to use Eq. (8.80) as an integral equation from which to calculate numerically the transient and steady-state behavior of a horn-loaded acoustic generator.

As an example, let us apply Eqs. (8.79) and (8.80) to the case of a uniform cylindrical tube open at its far end. Neglecting radiation corrections at the open mouth, we have, from Eq. (8.25),

$$Z = jZ_0 \tan kL, \tag{8.81}$$

where L is the length of the tube and $k = \omega/c$. Then, from Eq. (8.77),

$$r(\omega) = -e^{-2j\omega L/c}. \tag{8.82}$$

Taking the Fourier transform,

$$r(t) = -\int e^{j\omega t} e^{-2j\omega L/c} d\omega = -\delta(t-\tau), \tag{8.83}$$

where

$$\tau = 2L/c. \tag{8.84}$$

Substituting into Eq. (8.80), we find

$$p(t) = Z_0 U(t) - Z_0 U(t-\tau) - p(t-\tau), \tag{8.85}$$

and, applying Eq. (8.80) again to $p(t-\tau)$,

$$p(t) = p(t-2\tau) + Z_0[U(t) - 2U(t-\tau) + U(t-2\tau)]. \tag{8.86}$$

For a lip-driven or reed-driven instrument, $Z_0 U(t)$ is always much smaller than $p(t)$, since the excitation mechanism is pressure controlled. We can therefore neglect the U terms to give

$$p(t) = p(t - 2\tau), \tag{8.87}$$

so that the pipe acts as a quarter-wave resonator with frequency $1/2\tau = c/4L$.

The behavior of a nearly complete conical horn is rather more complex and really requires a different derivation, but the pressure wave reflects to the throat with a time delay τ given by Eq. (8.84) as before. When we consider Eq. (8.85), however, we find from Eqs. (6.26) and (6.27) that the terms Z_0U behave like $1/r^2$ near the origin and so dominate the terms in p, which behave like $1/r$. We can therefore ignore the p terms in Eq. (8.85) and conclude that, in the steady state,

$$U(t) = U(t - \tau), \tag{8.88}$$

so that the horn acts as a half-wave resonator with frequency $1/\tau = c/2L$.

A detailed discussion of the impulse response of a conical horn is complicated by the fact that there is a potential divergence for waves travelling toward the apex. This is removed, in part, when the apex is truncated, as it always is in wind instruments, but there are still analytical problems that have not yet been clearly resolved. If we take the approach by way of a Fourier transform of the input impedance, for example, then we can be left with exponentially growing non-causal functions. The general situation has been discussed by Ayers et al. (1985), and subsequently by Gilbert et al. (1990) and in a series of papers by Agulló et al. (1992), the most recent of which is referenced. This difficulty with the time domain is perhaps surprising, since it could well be argued that natural phenomena occur in the time domain, and it is the frequency domain that is artificial, in the sense that we never really have pure sinusoidal signals of infinite duration.

For a more general type of horn, such as is found in brass wind instruments, there is usually an initial cylindrical section which then flares to conical or Bessel form near the mouth. We expect the reflection behavior to be intermediate between that of a cylinder and a cone of the same length, but the reflected pulse will be considerably distorted by dispersion effects. Details can be found either by direct measurement or by taking the Fourier transform of the input impedance.

Direct measurement of impulse response presents some difficulties, since its definition implies that, after the initial flow impulse, the input should be rigidly blocked. Apart from the experimental difficulty involved with this, it means that the impulse response has a long duration, as we have already remarked. For many purposes it is more useful to measure the modified impulse response in which, after the initial flow impulse, the input is connected to a reflectionless termination. Such an artifice is possible in the case of a horn with an initial cylindrical section, but cannot be simply accomplished when the input section has appreciable flare.

Two approaches have been used for this measurement. One, devised by Agulló et al. (1995), involves inserting between the acoustic driver and the instrument input a section of matching cylindrical tubing sufficiently long that the measurement is over before a reflection returns from the driver end. A second approach, devised by Keefe (1996), involves first determining the reflection impedance of the measuring probe, including the driver, the measuring microphone, and any associated mount, by calibration using the reflection from a closed cylindrical pipe. This information is then used to deconvolve the measured reflection in the real instrument so as to exhibit what its form would be if the input termination were truly nonreflecting. Measurements of this type on real musical instruments are particularly useful in identifying unwanted reflections from irregularities in the bore.

8.15 Network Analogs

While the geometry of musical wind instruments is often very complicated, they can generally be considered to comprise various cavities, tubes, and horns connected together in a fairly simple way. This is true even when we consider the vocal tract of the player as part of the system. It is therefore helpful to have a set of reasonably standard procedures for calculating the acoustic behavior of such acoustic systems.

The basic approach is similar to electrical network theory, with acoustic pressure p taking the place of electric potential and acoustic volume flow U replacing electric current, both considered to be quantities oscillating in time with angular frequency ω. Just as in the electrical case, there is a distinction in approach between network analysis at low frequencies, when the wavelength involved is much larger than the system dimensions and we can use simple lumped components, and at high frequencies when we must use waveguides and transmission lines. Most musical instrument problems belong to the second category, but it is helpful to consider first some examples of low-frequency networks.

Suppose we have a short pipe of length l and cross-section S, and that p is the alternating pressure difference between its ends, considered as an acoustic quantity with time variation ω. The air in the pipe behaves as a simple mass of magnitude ρlS and the force acting on it is pS. Its velocity is U/S, where U is the acoustic volume flow through the pipe, and the relation between force and acceleration then gives

$$p = \left(\frac{\rho l}{S}\right)\frac{dU}{dt} = j\omega\left(\frac{\rho l}{S}\right)U. \tag{8.89}$$

The analogy between this equation and the electrical relation between the voltage V across an inductance L and the current i through it

$$V = L\frac{di}{dt} = j\omega Li \tag{8.90}$$

is immediately apparent. The quantity

$$Z_{\text{pipe}} = j\omega(\rho l/S) \tag{8.91}$$

is an acoustic impedance, as defined in Eq.(8.3), and $\rho l/S$ is called an acoustic inertance.

In just the same way, the pressure inside a cavity of volume V is related to the acoustic current U flowing into the cavity by the equation

$$p = \frac{\gamma p_{\text{a}}}{V} \int U \, dt, \tag{8.92}$$

where p_{a} is the steady atmospheric pressure and γ is the ratio of specific heats for air, so that γp_{a} is the isothermal bulk elastic modulus for air, as discussed in Section 6.1. Since the integral is the total flow into the cavity and is analogous to the electric charge, the quantity $V/\gamma p_{\text{a}}$ is analogous to the electric capacitance, and is generally referred to as the acoustic compliance. A more convenient expression for it, from Eq. (6.11), is $V/\rho c^2$, and the acoustic impedance of the cavity is then

$$Z_{\text{cav}} = -j\rho c^2/V\omega. \tag{8.93}$$

The third acoustic impedance that we need is the analog of an electrical resistance, and generally arises from viscous and thermal losses in air subject to motion and compression. In musical instruments it is generally associated with losses to pipe walls, as discussed in Section 8.2, though we will see below that it is also associated with radiation losses.

Finally we need two terminating elements, representing a rigid stopper and a simple opening, respectively. A rigid stopper simply prevents any acoustic flow, and so is equivalent to an open circuit. A opening may similarly be represented to a first approximation by a short circuit, but to be more accurate we need to ascribe to it the radiation impedance $Z_{\text{rad}} = R + jX$ as discussed in Section 8.3 and given explicitly in Eqs. (8.29) and (8.30). For the open end of an unbaffled pipe of radius a and area $S = \pi a^2$,

$$Z_{\text{rad}} \approx 0.16 \frac{\rho\omega^2}{c} + 0.6 j\omega \frac{\rho a}{S} \tag{8.94}$$

provided $\omega < c/a$.

We have already met the Helmholtz resonator in Section 1.6.2 as an example of an "air spring," and it is now appropriate to examine it in a little more detail as our first example of a network problem. The resonator consists simply of a cavity of volume V vented by a pipe of length l and cross-section S, as shown in Fig. 8.21(a). The network describing the system is shown in Fig. 8.21(b)—note that we have included the radiation impedance at the open end and assumed the resonator to be excited by an external sound wave, represented by a pressure generator p. Setting the driving pressure in the circuit equal to the sum of the pressure drops across

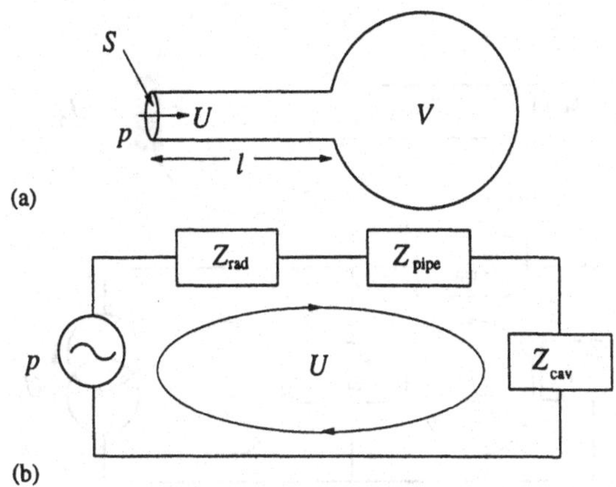

FIGURE 8.21. (a) A simple Helmholtz resonator driven by an external sound field; and (b) the analog network describing its behavior.

the individual circuit elements gives

$$p = (Z_{\text{rad}} + Z_{\text{pipe}} + Z_{\text{cav}})U. \tag{8.95}$$

For a very careful calculation we might want to include as well the effects of viscous and thermal losses to the pipe and cavity walls, but we omit this here. Using the explicit expressions given above, we now find

$$U \approx \left\{ 0.16 \frac{\rho\omega^2}{c} + j \left[\frac{\rho\omega(l + 0.6a)}{S} - \frac{\rho c^2}{V\omega} \right] \right\}^{-1} p. \tag{8.96}$$

This is just a simple damped-resonator equation, with maximum acoustic flow and maximum acoustic pressure in the cavity occurring at the resonance frequency

$$\omega^* \approx c \left[\frac{S}{V(l + 0.6a)} \right]^{1/2}. \tag{8.97}$$

We can use the detailed expression (8.96) to plot out the frequency response of the resonator.

A situation much more closely related to musical instruments is shown in Fig. 8.22(a). Here, instead of being influenced by an external sound wave, the cavity is excited by the vibration of a small piston constituting a part of its wall. This is analogous to the case of a stringed instrument or a cavity percussion instrument in which the wall vibrates, and the cavity is tuned by an aperture or a short pipe. The associated network is shown in Fig. 8.22(b), and consists now of two loops with a flow generator in one of them. We can therefore write two equations

$$Z_{\text{cav}}(U_0 - U) = p_0,$$

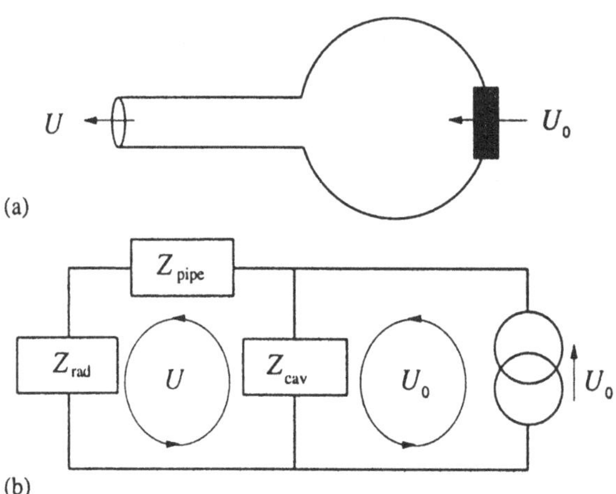

FIGURE 8.22. (a) A Helmholtz resonator driven by the vibration of one of its internal walls; and (b) the analog network describing its behavior.

$$Z_{\text{cav}}(U - U_0) + (Z_{\text{pipe}} + Z_{\text{rad}})U = 0, \tag{8.98}$$

where p_0 is the acoustic pressure in the cavity. The second of these equations gives the acoustic current U through the open neck of the resonator as

$$U = \frac{Z_{\text{cav}} U_0}{Z_{\text{rad}} + Z_{\text{pipe}} + Z_{\text{cav}}}, \tag{8.99}$$

and the first equation of (8.98) can then be used to evaluate the pressure in the cavity. The response, both in terms of flow and internal pressure, has a maximum at the resonance frequency ω^* given by Eq. (8.97). If we want to determine the acoustic power P radiated through the neck of the resonator, then the result is simply

$$P = \tfrac{1}{2} R U^2, \tag{8.100}$$

where R is the resistive part of the radiation impedance.

When the length of the pipes or cavities involved in the instrument we are analyzing becomes comparable with wavelength of the sound involved—indeed when it is much more than one tenth of the wavelength—it is necessary to adopt a more sophisticated approach, for we begin to have appreciable phase shifts between one end and the other. We shall not take this analysis very far, since most problems can be solved using the input impedance methods outlined in Sections 8.7–8.11, but the present approach is more general. The essential thing is to recognize that components such as pipes have two ends, and that the acoustic flows, as well as the pressures, may be different at those ends, as illustrated in Fig. 8.23(a). The pipe or horn in question should therefore be represented by a four-terminal component Z_{ij} as shown in Fig. 8.23(b). The equations connecting these

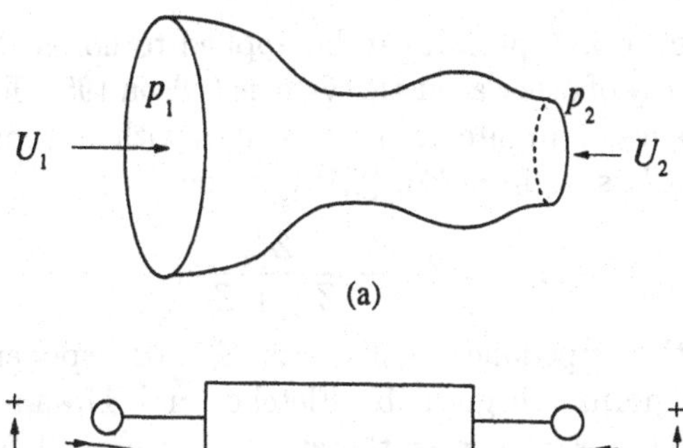

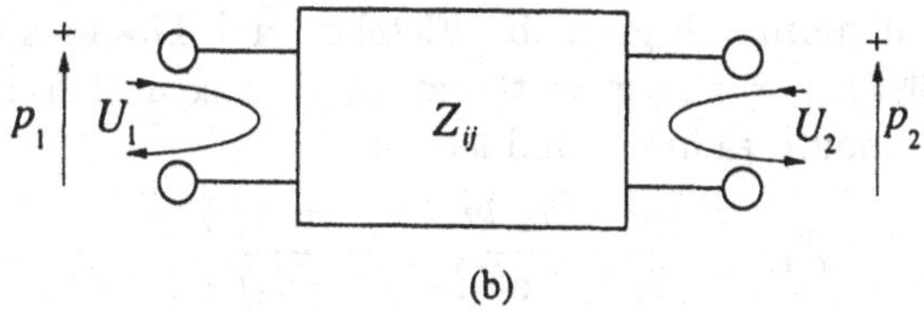

FIGURE 8.23. At high frequencies, the acoustic currents at the two ends of a horn or pipe may differ, as in (a). It must therefore be represented by a four-terminal element as in (b).

pressures and flows, and thereby defining the coefficients Z_{ij}, are

$$p_1 = Z_{11}U_1 + Z_{12}U_2,$$
$$p_2 = Z_{21}U_1 + Z_{22}U_2. \tag{8.101}$$

The impedance coefficients Z_{ij} are generally complex numbers, if resistive losses are included, and satisfy the reciprocity relationship $Z_{21} = Z_{12}$. In the case of a cylindrical tube, of course, symmetry also dictates that $Z_{22} = Z_{11}$, but this is not true in general, and particularly not of horns.

For a simple pipe of length l and cross-section S, we can use Eqs. (8.16) and (8.17) to show that

$$Z_{11} = Z_{22} = -jZ_0 \cot kl,$$
$$Z_{21} = Z_{12} = -jZ_0 \operatorname{cosec} kl, \tag{8.102}$$

where $Z_0 = \rho c/S$ is the characteristic impedance of the pipe and $k = \omega/c$ is real only if wall losses are neglected.

We can easily calculate the input impedance for ideally open or stopped pipes from Eq. (8.101) by setting $p_2 = 0$ or $U_2 = 0$, respectively. After a little algebra we find

$$Z_{\text{IN}}^{\text{open}} = jZ_0 \tan kl \rightarrow \frac{j\omega\rho l}{S},$$
$$Z_{\text{IN}}^{\text{stopped}} = -jZ_0 \cot kl \rightarrow -\frac{j\rho c^2}{\omega l S}. \tag{8.103}$$

The final form of writing is, in each case, the low-frequency result for $kl \ll 1$. These agree exactly with Eqs. (8.24) and (8.25), as indeed they must.

This general network approach can be applied to horns of finite length and to a large variety of other acoustic systems (Olson 1957, Fletcher 1992). The input impedance of a finite horn terminated with an acoustic load Z_L, for instance, is easily seen from Eq. (8.101) to be

$$Z_{IN} = Z_{11} - \frac{Z_{12}^2}{Z_{22} + Z_L}. \tag{8.104}$$

Expressions for the impedance coefficients Z_{ij} for exponential, conical, and paraboloidal horns are given by Fletcher and Thwaites (1988) and by Fletcher (1992). For a horn in the shape of a conical frustrum, with throat radius a, mouth radius b, and length l,

$$Z_{11} = \frac{j\rho c}{S_1} \left[\frac{\sin(kl - \theta_2)\sin\theta_1}{\sin(kl + \theta_1 - \theta_2)} \right], \tag{8.105}$$

$$Z_{22} = -\frac{j\rho c}{S_2} \left[\frac{\sin(kl + \theta_1)\sin\theta_2}{\sin(kl + \theta_1 - \theta_2)} \right], \tag{8.106}$$

$$Z_{12} = Z_{21} = -\frac{j\rho c}{(S_1 S_2)^{1/2}} \left[\frac{\sin\theta_1 \sin\theta_2}{\sin(kl + \theta_1 - \theta_2)} \right], \tag{8.107}$$

where x_1 and x_2 are measured from the apex of the cone to the throat and mouth, respectively, and

$$\theta_1 = \tan^{-1} kx_1, \qquad \theta_2 = \tan^{-1} kx_2. \tag{8.108}$$

Armed with these results, we can now set up network analogs for any system in which we are interested—horns, tubes, cavities, etc.—interconnected in arbitrary ways and excited either externally or internally. Since precise numerical values are assigned to all the quantities involved, we can then calculate frequency response, radiated power, and other quantities of concern. In the case of a Helmholtz resonator, for example, this calculation will reveal not just the fundamental "Helmholtz" resonance, but also higher resonances.

There are, however, limitations to this approach, of which we should be aware. It is essentially a one-dimensional treatment, in which plane waves propagate through the system. This is an adequate approximation provided the sound wavelength is small compared with the transverse dimensions of the ducts and cavities involved. When this assumption is no longer valid, however, we must undertake a proper two- or three-dimensional treatment which essentially involves solving the wave equation in detail.

References

Agulló, J., Barjau, A., and Martínez, J. (1992). On the time-domain description of conical bores. *J. Acoust. Soc. Am.* **91**, 1099–1105.

Agulló, J., Cardona, S., and Keefe, D.H. (1995). Time-domain deconvolution to measure reflection functions for discontinuities in waveguides. *J. Acoust. Soc. Am.* **97**, 1950–1957.

Amir, N., Rosenhouse, G., and Shimony, U. (1993). Input impedance of musical horns and the 'horn function.' *Appl. Acoust.* **38**, 15–35.

Ayers, R.D. (1996). Impulse responses for feedback to the driver of a musical wind instrument. *J. Acoust. Soc. Am.* **100**, 1190–1198.

Ayers, R.D., Eliason, L.J., and Mahgerefteh, D. (1985). The conical bore in musical acoustics. *Am. J. Phys.* **53**, 528–537.

Backus, J. (1974). Input impedance curves for the reed woodwind instruments. *J. Acoust. Soc. Am.* **56**, 1266–1279.

Benade, A.H. (1959). On woodwind instrument bores. *J. Acoust. Soc. Am.* **31**, 137–146.

Benade, A.H. (1968). On the propagation of sound waves in a cylindrical conduit. *J. Acoust. Soc. Am.* **44**, 616–623.

Benade, A.H. (1988). Equivalent circuits for conical waveguides. *J. Acoust. Soc. Am.* **83**, 1764–1769.

Benade, A.H., and Ibisi, M.I. (1987). Survey of impedance methods and a new piezo-disk-driven impedance head for air columns. *J. Acoust. Soc. Am.* **81**, 1152–1167.

Benade, A.H., and Jansson, E.V. (1974). On plane and spherical waves in horns with nonuniform flare. *Acustica* **31**, 80–98.

Beranek, L.L. (1954). "Acoustics," pp. 91–115. McGraw-Hill, New York; reprinted 1986, Acoustical Society Am., Woodbury, New York.

Campos, L.M.B.C. (1984). Some general properties of the exact acoustic fields in horns and baffles. *J. Sound Vibr.* **95**, 177–201.

Caussé, R., Kergomard, J., and Lurton, X. (1984). Input impedance of brass musical instruments—comparison between experiment and numerical models. *J. Acoust. Soc. Am.* **75**, 241–254.

Chung, J.V., and Blaser, D.A. (1980). Transfer function method of measuring in-duct acoustic properties: I. Theory; II. Experiment. *J. Acoust. Soc. Am.* **68**, 907–913 and 914–921.

Coltman, J. (1968). Sounding mechanism of the flute and organ pipe. *J. Acoust. Soc. Am.* **44**, 983–992.

Dudley, J.D., and Strong, W.J. (1990). A computer study of the effects of harmonicity in a brass wind instrument. *Appl. Acoust.* **30**, 116–132.

Eisner, E. (1967). Complete solutions of the "Webster" horn equation. *J. Acoust. Soc. Am.* **41**, 1126–1146.

Fletcher, N.H. (1992). "Acoustic Systems in Biology." Oxford University Press, New York. Chap. 10 and App. B.

Fletcher, N.H., and Thwaites, S. (1988). Response of obliquely truncated simple horns: Idealized models for vertebrate pinnae. *Acustica* **65**, 194–204.

Fransson, P.J., and Jansson, E.V. (1975). The STL-Ionophone: Transducer properties and construction. *J. Acoust. Soc. Am.* **58**, 910–915.

Gilbert, J., Kergomard, J., and Polack, J.D. (1990). On the reflection functions associated with discontinuities in conical bores. *J. Acoust. Soc. Am.* **87**, 1773–1780.

Jahnke, E., and Emde, F. (1938). "Tables of Functions," p. 146. Teubner, Leipzig, Reprinted 1945, Dover, New York.

Keefe, D.H. (1996). Wind-instrument reflection function measurements in the time domain. *J. Acoust. Soc. Am.* **99**, 2370–2381.

Keefe, D.H., Barjau, A., and Agulló, J. (1993). Theory of wave propagation in axisymmetric horns, *Proc. Stockholm Mus. Acoust. Conf., SMAC93*. Royal Swedish Academy of Music, Stockholm, Pub. 79. pp. 496–500.

Keefe, D.H., and Benade, A.H. (1983). Wave propagation in strongly curved ducts. *J. Acoust. Soc. Am.* **74**, 320–332.

Kergomard, J. (1981). Ondes quasi-stationnaires dans les pavilions avec partis visco-thermiques aux parois: Calcul de l'impedance. *Acustica* **48**, 31–43.

Levine, H., and Schwinger, J. (1948). On the radiation of sound from an unflanged pipe. *Phys. Rev.* **73**, 383–406.

McIntyre, M.E., Schumacher, R.T., and Woodhouse, J. (1983). On the oscillation of musical instruments. *J. Acoust. Soc. Am.* **74**, 1325–1345.

Merhaut, J. (1968). Method of measuring the acoustic impedance. *J. Acoust. Soc. Am.* **45**, 331(A).

Morse, P.M. (1948). "Vibration and Sound," 2nd ed pp. 233–288 and 326–338. McGraw-Hill, New York; reprinted 1976, Acoustical Society of Am., Woodbury, New York.

Morse, P.M., and Feshbach, H. (1953). "Methods of Mathematical Physics," Vol. 1, pp. 494–523, 655–666. McGraw-Hill, New York.

Morse, P.M., and Ingard, K.U. (1968). "Theoretical Acoustics," pp. 467–553. McGraw-Hill, New York. Reprinted 1986, Princeton Univ. Press, Princeton, New Jersey.

Nederveen, C.J. (1969). "Acoustical Aspects of Woodwind Instruments." Frits Knuf, Amsterdam, p. 60. (Reprinted by Northern Illinois University Press, DeKalb, 1998.)

Olson, H.F. (1957). "Acoustical Engineering," Van Nostrand-Reinhold, Princeton, New Jersey, pp. 88–123.

Plitnik, G.R., and Strong, W.J. (1979). Numerical method for calculating input impedance of an oboe. *J. Acoust. Soc. Am.* **65**, 816–825.

Pratt, R.L., Elliott, S.J., and Bowsher, J.M. (1977). The measurement of the acoustic impedance of brass instruments. *Acustica* **38**, 236–246.

Pratt, R.L., Elliott, S.J., and Bowsher, J.M. (1979). Comments on acoustic impedance measurements using sine excitation and known volume velocity. *J. Acoust. Soc. Am.* **66**, 905.

Pyle, R.W. (1975). Effective length of horns. *J. Acoust. Soc. Am.* **57**, 1309–1317.

Rayleigh, Lord (1894). "The Theory of Sound," 2 vols. Macmillan, London. Reprinted 1945. Dover, New York.

Rossing, T.D. (1994). Musical instruments. In "Encyclopedia of Applied Physics", VCH Publishers, New York. Vol. 11, pp. 129–171.

Salmon, V. (1946a). Generalized plane wave horn theory. *J. Acoust. Soc. Am.* **17**, 199–211.

Salmon, V. (1946b). A new family of horns. *J. Acoust. Soc. Am.* **17**, 212–218.

Schumacher, R.T. (1981). Ab initio calculations of the oscillations of a clarinet. *Acustica* **48**, 71–85.

Stevenson, A.F. (1951). Exact and approximate equations for wave propagation in acoustic horns. *J. Appl. Phys.* **22**, 1461–1463.

Webster, A.G. (1919). Acoustical impedance, and the theory of horns and of the phonograph. *Proc. Nat. Acad. Sci. (US)* **5**, 275–282.

Weibel, E.S. (1955). On Webster's horn equation. *J. Acoust. Soc. Am.* **27**, 726–727.

Wolfe, J., Smith, J., Brielbeck, G., and Stocker, F. (1995a). A system for real time measurement of acoustic transfer functions. *Acoustics Australia* **23**(1), 19–20.

Wolfe, J., Smith, J., Brielbeck, G., and Stocker, F. (1995b). Making a good musical instrument. *Australian and New Zealand Physicist* **32**, 10–14.

Part III

String Instruments

9

Guitars and Lutes

The modern six-string guitar is a descendant of the sixteenth-century Spanish vihuela, which has its roots in antiquity. Although Boccherini and other composers of the eighteenth century included the guitar in some of their chamber music, the establishment of the guitar as a concert instrument took place largely in the nineteenth century. Fernando Sor (1778–1839) was the first of a long line of Spanish virtuosos and composers for the guitar.

The Spanish luthier Antonio de Torres (1817–1892) contributed much to the development of the modern classical guitar when he enlarged the body and introduced a fan-shaped pattern of braces to the top plate. Francisco Tarrega (1852–1909), perhaps the greatest of all nineteenth century players, introduced the apoyando stroke and generally extended the expressive capabilities of the guitar. Excellent accounts of the historical development of the guitar are given by Jahnel (1981) and by Turnbull (1974).

9.1 Design and Construction of Guitars

The modern guitar, shown in Fig. 9.1, has 6 strings, about 65 cm in length, tuned to E_2, A_2, D_3, G_3, B_3, and E_4 (f = 82, 110, 147, 196, 247, and 330 Hz). The top is usually cut from spruce or redwood, planed to a thickness of about 2.5 mm. The back, also about 2.5 mm thick, is usually a hardwood, such as rosewood, mahogany, or maple. Both the top and back plates are braced, the bracing of the top plate being one of the critical design parameters.

Acoustic guitars generally fall into one of four families of design: classical, flamenco, flat top (or folk), and arch top. Classical and flamenco guitars have nylon strings; flat top and arch top guitars have steel strings. Steel string guitars usually have a steel rod embedded inside the neck, and their sound-boards are provided with crossed bracing. Several designs for bracing are shown in Fig. 9.2.

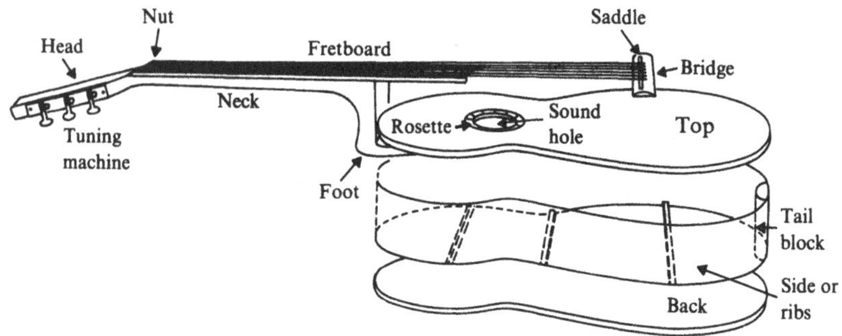

FIGURE 9.1. An exploded view of a guitar, showing its construction (Rossing, 1982a).

9.2 The Guitar as a System of Coupled Vibrators

The guitar can be considered to be a system of coupled vibrators. The plucked strings radiate only a small amount of sound directly, but they excite the bridge and top plate, which in turn transfer energy to the air cavity, ribs, and back plate. Sound is radiated efficiently by the vibrating plates and through the sound hole.

Figure 9.3 is a simple schematic of a guitar. At low frequency, the top plate transmits energy to the back via both the ribs and the air cavity; the bridge essentially acts as part of the top plate. At high frequency, however, most of the sound is radiated by the top plate, and the mechanical properties of the bridge may become significant.

Most of these elements have already been discussed separately. Vibrations of a plucked string were discussed in Section 2.8; plates were discussed in Chapter 3; coupled systems were discussed in Chapter 4; and sound ra-

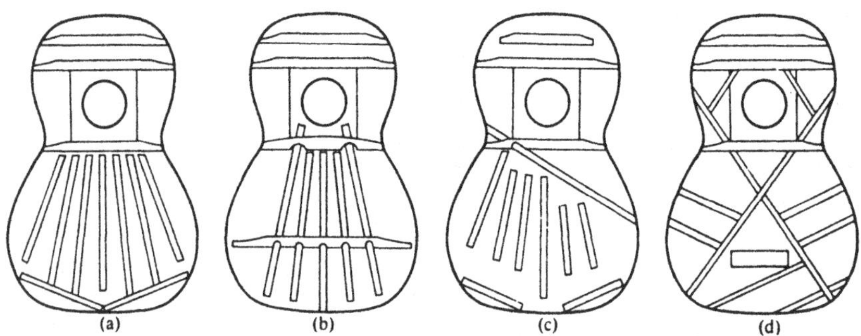

FIGURE 9.2. Various designs for bracing a guitar soundboard: (a) traditional (Torres) fan bracing; (b) Bouchet (France); (c) Ramirez (Spain); (d) crossed bracing (Rossing, 1982a).

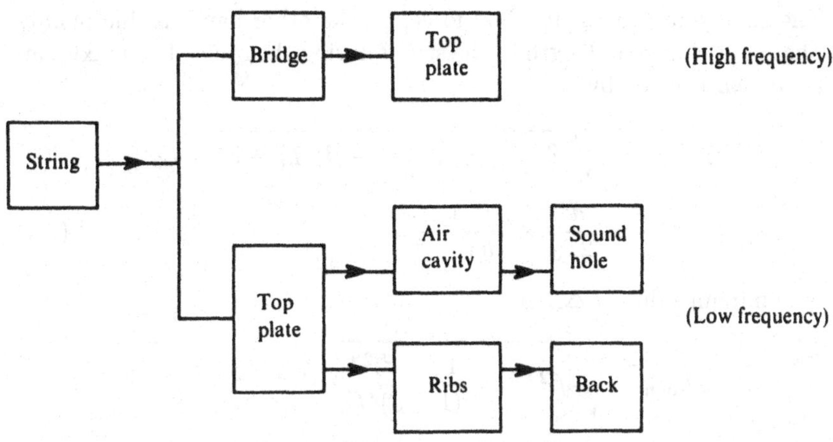

FIGURE 9.3. Simple schematic of a guitar. At low frequency, sound is radiated by the top and back plates and the soundhole. At high frequency, most of the sound is radiated by the top plate.

diation was discussed in Chapter 7. We now proceed to examine them in more detail and apply them to the guitar.

9.3 Force Exerted by the String

The force exerted by a plucked string on the bridge can be estimated by reference to Fig. 2.8. To a first approximation, the force normal to the top plate will be $T \sin\theta$, and the force parallel to the top plate $T\cos\theta$, where T is the string tension and θ is the angle between the string and the plate.

The tension T changes during the cycle, however, as the length of the string changes (see Fig. 2.8). If the string has a cross-sectional area A and an elastic (Young's) modulus E, we can write

$$T = T_0 + \Delta T = T_0 + \frac{EA}{L_0}\Delta L. \tag{9.1}$$

For small θ, the transverse and longitudinal forces become

$$\begin{aligned} F_{\mathrm{T}} &= (T_0 + \Delta T)\sin\theta, \\ F_{\mathrm{L}} &= (T_0 + \Delta T)\cos\theta \simeq T_0 + \Delta T = T_0 + \frac{EA}{L_0}\Delta L. \end{aligned} \tag{9.2}$$

The change in the transverse force during a cycle is primarily due to the change in the direction or slope. If the string is displaced through a distance d at a point βL_0 from the bridge and then released,

$$\sin\theta_1 \simeq \frac{d}{\beta L_0} \quad \text{and} \quad \sin\theta_2 = \frac{d}{(1-\beta)L_0}. \tag{9.3}$$

The change in the longitudinal force, on the other hand, is due mainly to the slight change in length of the string during the cycle. The maximum value of ΔL is given by

$$\begin{aligned}\Delta L_{\max} &= \sqrt{\beta^2 L_0^2 + d^2} + \sqrt{(1-\beta)^2 L_0^2 + d^2} - L_0 \\ &\simeq \frac{d^2}{2L_0} \times \frac{1}{\beta(1-\beta)}. \end{aligned} \tag{9.4}$$

The minimum value of ΔL is

$$\begin{aligned}\Delta L_{\min} &= \sqrt{4\beta^2 L_0^2 + \frac{(1-2\beta)^2 L_0^2}{(1-\beta)^2 L_0^2} d^2} \\ &\quad + \sqrt{(1-2\beta)^2 L_0^2 + \frac{(1-2\beta)^2 L_0^2}{(1-\beta)^2 L_0^2} d^2} - L_0 \\ &\simeq \frac{d^2}{2L_0} \times \frac{1-2\beta}{2\beta(1-\beta)^2}. \end{aligned} \tag{9.5}$$

For $\beta = 1/5$, $\Delta L_{\max} = 3.13 d^2/L_0$, $\Delta L_{\min} = 0.94 d^2/L_0$, $\sin\theta_1 = 5d/L_0$, and $\sin\theta_2 = (5/4)d/L_0$. The force waveforms are shown in Fig. 9.4.

For a typical high-E nylon string, $L_0 = 65$ cm, $A = 0.36$ mm^2, $E = 5\times 10^9$ N/m^2, and $T_0 = 82$ N. For a deflection $d = 3$ mm, $T_0 d/L_0 = 0.38$ N, and $EAd^2/L_0^2 = 0.038$ N, so the maximum transverse force is roughly 16 times greater than the maximum increase in longitudinal force; more important, the amplitude of the transverse force pulses is about 40 times greater than the longitudinal pulses and they couple more efficiently to the top plate. However, the longitudinal force pulses are proportional to d^2 compared to d for the transverse pulses, so the difference diminishes with increasing amplitude.

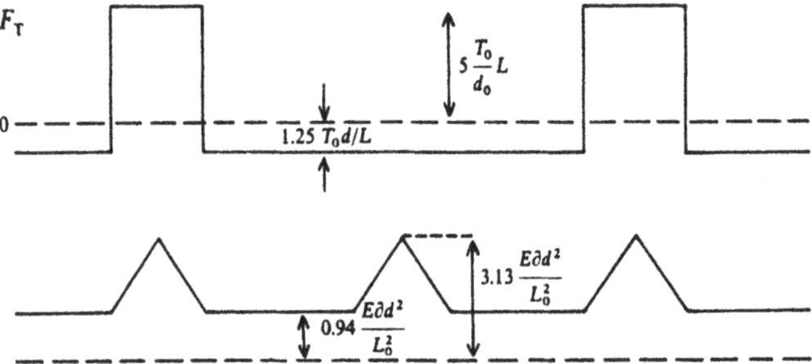

FIGURE 9.4. Waveform of the force on a guitar bridge when a string is plucked with a displacement $d = \frac{1}{5}$ of the distance from the bridge to the nut (similar waveforms were calculated by Houtsma et al., 1975).

The elastic (Young's) modulus for steel is about 40 times greater than for nylon, and static string tensions are about 50% greater, so the longitudinal and transverse force amplitudes will be more nearly equal.

Note that both F_T and F_L increase in amplitude as β is made smaller; in fact for $\beta \ll 1$, both are proportional to $1/\beta$, so plucking nearer the bridge not only leads to greater forces on the bridge but also to an emphasis on the higher harmonics. A larger plucking force is required to achieve the same deflection d, however.

The waveforms and spectra of the transverse bridge force for an ideal flexible string plucked at its center ($\beta = \frac{1}{2}$), one-fifth ($\beta = \frac{1}{5}$), and one-twentieth ($\beta = \frac{1}{20}$) of its length from the bridge are shown in Fig. 9.5. Also shown above the waveforms are the string shapes at successive intervals during the vibration period (compare Fig. 2.8). Note that each βth harmonic is missing from the spectra. Fourier analysis of the transverse force gives the following expression for the amplitude F_n of nth harmonic (Fletcher, 1976a):

$$F_n = \frac{2dT_0}{n\pi L_0} \frac{1}{\beta(1-\beta)} \sin \beta n\pi. \tag{9.6}$$

Well below the first missing harmonic, $\sin \beta n\pi \simeq \beta n\pi$, so

$$F_n \simeq \frac{2dT_0}{L_0}, \quad n\beta \ll 1. \tag{9.7}$$

The discussion thus far has assumed an ideal flexible string, rigid end supports, and a plectrum that is ideally sharp and hard. In the case of a soft broad plectrum, the initial shape of the string before release is two straight segments joined by a smooth curve rather than a sharp angle. If the width of the soft plectrum is δ, then modes with wavelengths shorter than about 2δ are excited very little. This is equivalent to a high-frequency cutoff in the string spectrum at a mode number $n \simeq L/\delta$ (Fletcher, 1976a).

Some effects of string stiffness were discussed in Sections 2.18 and 2.19 and of nonrigid end supports in Section 2.12. If one of the supports is not rigid, nonlinear mode coupling causes the missing harmonics to be excited, typically with a time constant of 0.1 s (Legge and Fletcher, 1984).

It is well known by guitar players that guitar strings, especially the lower strings that are wrapped, become "dead" after a short period of playing. The aging problem is especially critical for steel strings. This problem has been studied theoretically and experimentally in steel strings (Allen, 1976) and in nylon strings (Hanson, 1985), and the conclusion seems to be that the main cause is increased damping due to foreign material that quickly becomes embedded between the windings. Boiling the strings in a cleaning solution often rejuvenates them.

The tension needed to bring a string to the right pitch depends, of course, on the mass of the string, which in turn depends upon its diameter. Players select strings of different gauges, and thus string tensions vary from one

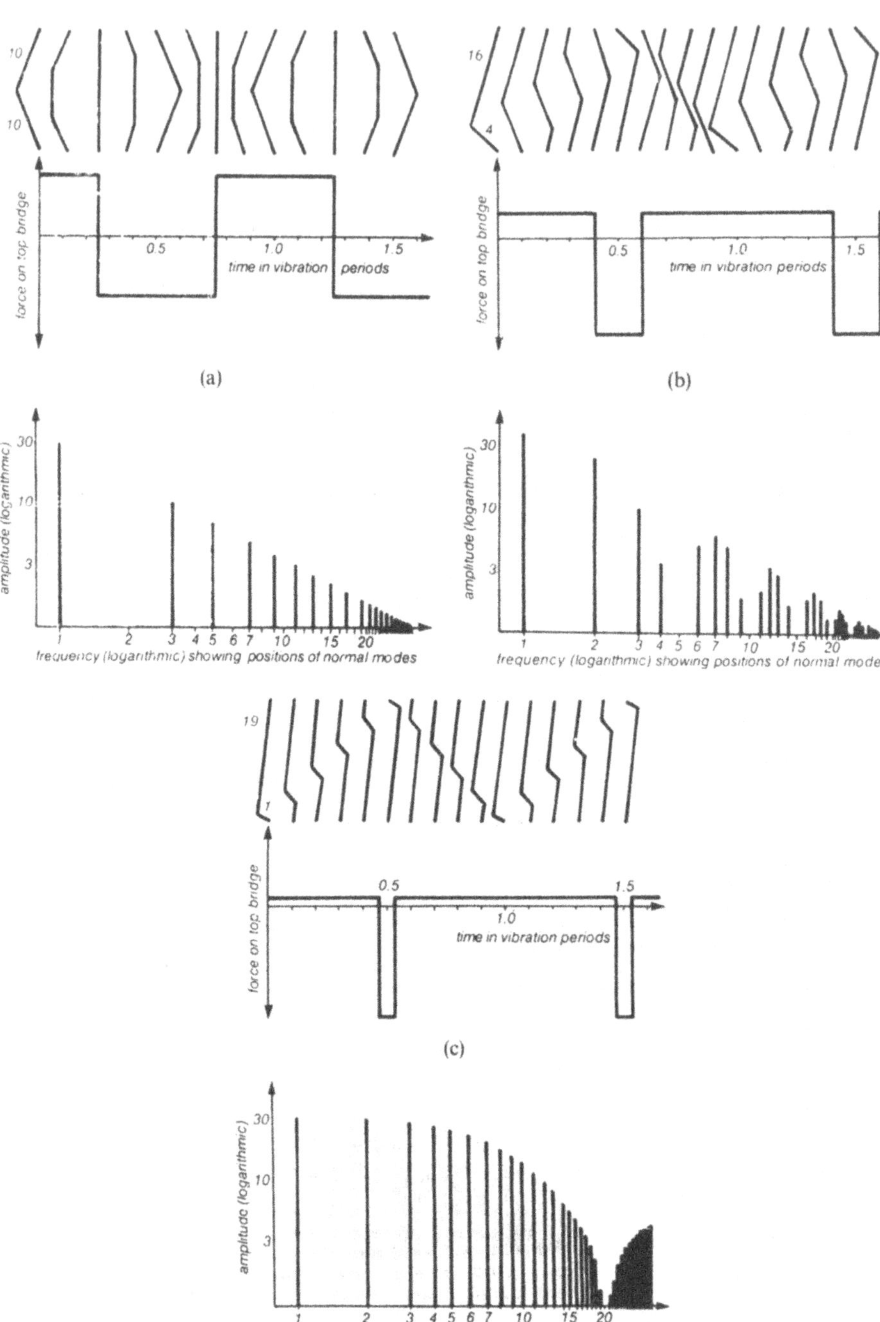

FIGURE 9.5. Waveforms and spectra of the transverse bridge force for a string plucked at its center (a), at $\frac{1}{5}$ of its length (b), and at $\frac{1}{20}$ of its length (c) from the bridge. Also shown above the waveforms are the string shapes at successive intervals during the vibration period (Fletcher, 1976b).

instrument to another. Nylon strings typically require tensions of 50 to 80 N, whereas steel strings require tensions of 100 to 180 N. There appears to be some advantage in selecting string gauges in such a way that the tensions in all six strings will be nearly the same (Houtsma, 1975).

9.4 Modes of Vibration of Component Parts

A guitar top plate vibrates in many modes; those of low frequency bear considerable resemblance to the modes of a rectangular plate described in Chapter 3. The mode shape and frequencies change quite markedly when the braces are added, however, and, in addition, they are totally different if the plate is tested with its edge free, clamped, or simply supported (hinged). Relatively few studies of guitar top plate modes, especially with a free edge, have been reported.

Figure 9.6 shows the observed mode shapes and frequencies of a guitar plate without braces (with a free edge), and Fig. 9.7 shows the modes calculated for a plate with traditional fan bracing (also with a free edge), which are reported to be in good agreement with observed mode frequencies and shapes (Richardson and Roberts, 1985). Mode shapes and frequencies for the first five modes in a classical guitar sans back are shown in Fig. 9.8. These are also in reasonably good agreement with the modes calculated by Richardson and Roberts (1985) for a clamped edge, although the actual boundary condition probably is somewhere between clamped and hinged.

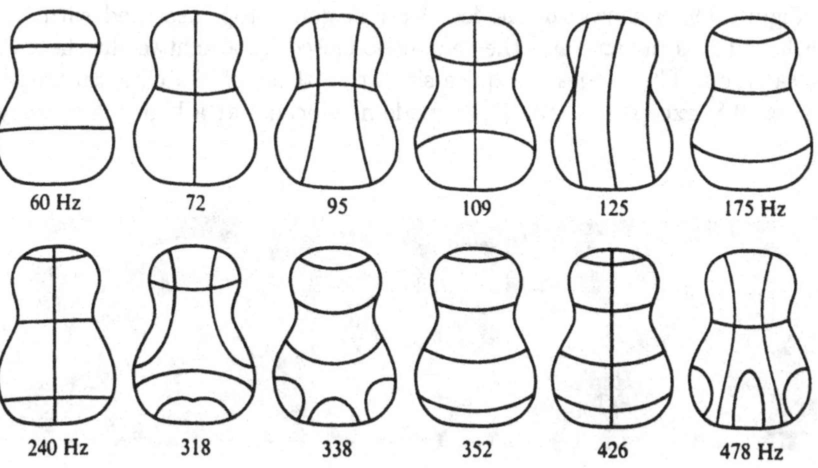

FIGURE 9.6. Vibration modes of a guitar plate blank (without braces) with a free edge (adapted from Rossing, 1982b).

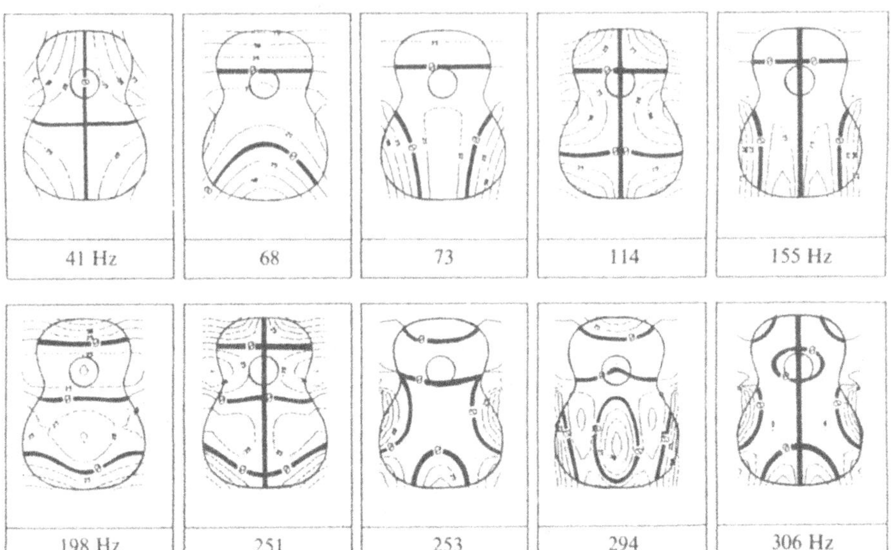

FIGURE 9.7. Vibration modes of a classical guitar top plate with traditional fan bracing (adapted from Richardson and Roberts, 1985).

Obviously, the observed modal shapes and frequencies of the top plate depend upon the exact boundary conditions and acoustic environment during testing. A very convenient and readily reproducible arrangement is to immobilize the back and ribs of the guitar (in sand, for example) and to close the soundhole; a number of guitars have been tested in this way by various investigators.

Figure 9.9(a) shows the modes of a folk guitar top measured with the back and ribs in sand and the soundhole closed by a lightweight sheet of balsa wood. The modes are quite similar to those of the classical guitar in Fig. 9.8, except that the $(1,0)$ mode now occurs at a higher frequency

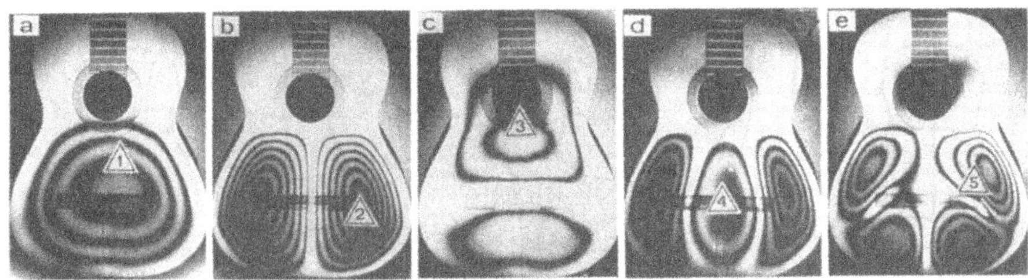

FIGURE 9.8. Vibration modes of a classical guitar top plate glued to fixed ribs but without the back (Jansson, 1971).

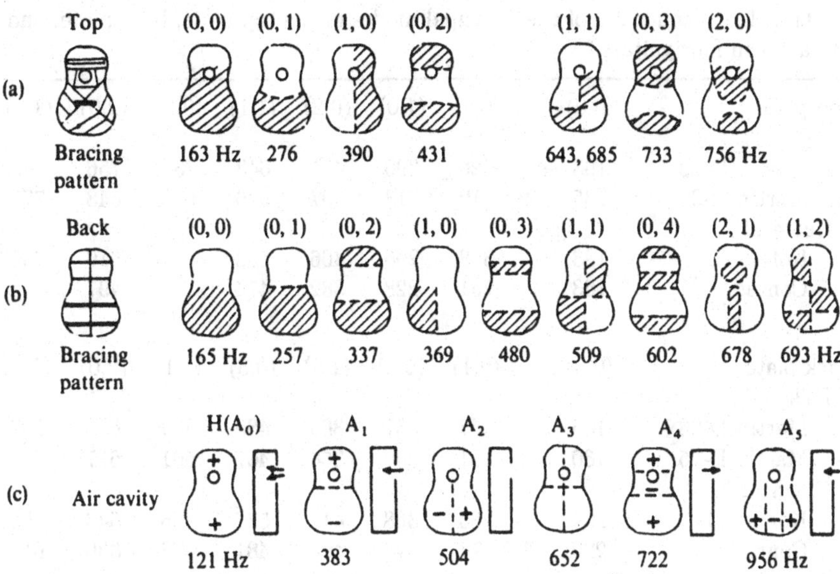

FIGURE 9.9. (a) Modes of a folk guitar top (Martin D-28) with the back and ribs in sand. (b) Modes of the back with the top and ribs in sand. (c) Modes of the air cavity with the guitar body in sand. Modal designations are given above the figures and modal frequencies below (Rossing et al., 1985).

than the (0, 1) mode, and the (2, 0) mode has moved up in frequency and changed its shape because of the crossed bracing.

Generally, the back plate of a guitar is rather simply braced with a center strip and three (most classical guitars) or four (folk guitars) cross braces, as shown in Fig. 9.9(b). Some vibrational modes of the back are shown in Fig. 9.9(b).

Also shown in Fig. 9.9 are the modes of the air cavity of a folk guitar. These were measured with the top, back, and ribs immobilized in sand but with the soundhole open. The lowest mode is the Helmholtz resonance, whose frequency is determined by the cavity volume and the soundhole diameter. There is also a small dependence upon the cavity shape and the soundhole placement, but these are usually not variables in guitar design. The term "Helmholtz resonance" is sometimes applied to the lowest resonance of the guitar (around 90–100 Hz), but this resonance involves considerable motion of the top and back plates and so is not a simple Helmholtz cavity resonance. Higher air modes resemble the standing waves in a rectangular box.

Frequencies of the principal modes of the top plate, back plate, and air cavity in two folk guitars and two classical guitars are given in Table 9.1. The main difference is in the relative frequencies of the (1, 0) and (0, 1) modes in the top plates. In fan-braced classical guitars, the (0, 1)

TABLE 9.1. Frequencies of the principal modes of the top plate, back plate, and air cavity in four guitars.*

Top plate	(0,0)	(0,1)	(1,0)	(0,2)	(1,1)	(0,3)	(2,0)	(1,2)
Folk								
Martin D-28	163	326	390	431	643	733	756	
Martin D-35	135	219	313	397	576	626	648	777
Classical								
Kohno 30	183	388	296	466	558		616	660
Conrad	163	261	228	382	474		497	
Back plate	(0,0)	(0,1)	(0,2)	(1,0)	(0,3)	(1,1)	(2,0)	(1,2)
Folk								
Martin D-28	165	257	337	369	480	509	678	693
Martin D-35	160	231	306	354	467	501	677	
Classical								
Kohno 30	204	285	368	417	537	566	646	856
Conrad	229	277	344	495	481	573	830	611
Air cavity	A_0 (Helmholtz)	A_1 (0,1)	A_2 (1,0)	A_3 (1,1)	A_4 (0,2)	A_5 (2,0)		
Folk								
Martin D-28	121	383	504	652	722	956		
Martin D-35	118	392	512	666	730	975		
Classical								
Kohno 30	118	396	560	674	780			
Conrad	127	391	558	711	772	1033		

*From Rossing et al. (1985).

mode occurs at a higher frequency than the $(1,0)$ mode, while in the cross-braced top plate of the folk guitars and in the back plates of both types, the reverse is generally true. In the Martin D-28, the fundamental modes of the top plate and back plate are tuned to almost the same frequency.

9.5 Coupling of the Top Plate to the Air Cavity: Two-Oscillator Model

If we ignore, for the moment, motion of the back plate and ribs, the guitar can be viewed as a two-mass vibrating system of the type discussed in Chapter 4, particularly Section 4.6. The two-mass model and its electrical equivalent circuit are shown in Fig. 9.10. The vibrating strings apply a force $F(t)$ to the top plate, whose mass and stiffness are represented

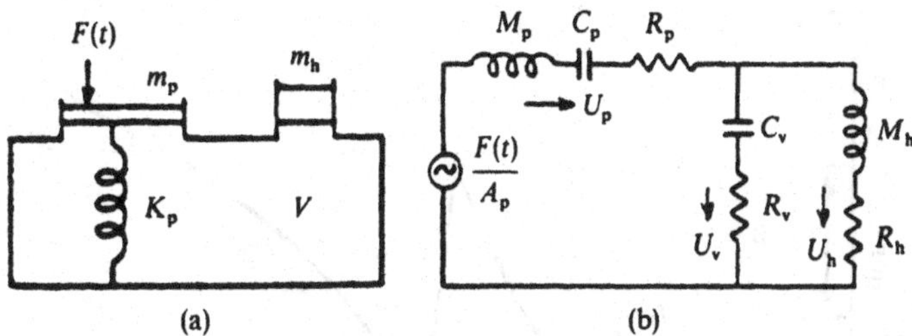

FIGURE 9.10. (a) Two-mass model representing the motion of a guitar with a rigid back plate and ribs. (b) Equivalent electrical circuit for the two-mass model. The equivalent currents are volume velocities (Rossing et al., 1985).

by m_p and K_p. A second piston of mass m_h represents the mass of air in the soundhole, and the volume V of enclosed air acts as the second spring.

The equivalent electrical circuit in Fig. 9.10(b), one of several possible choices for representing Fig. 9.10(a), is an acoustical impedance representation (Beranek, 1954). The equivalent voltage is the force applied to the top plate divided by the effective top plate area, and the equivalent currents are volume velocities (in m^3/s). The following symbols are used:

$M_p = m_p/A_p^2$ is the inertance (mass/area^2) of the top plate (kg/m^4),
$M_h = m_h/A_h^2$ is the inertance of air in the soundhole (kg/m^4),
$C_p = A^2/K_p$ is the compliance of the top plate (N/m^5),
$C_v = V/\rho c^2$ is the compliance of the enclosed air (N/m^5),
U_p is the volume velocity of the top plate (m^3/s),
U_h is the volume velocity of air in the soundhole (m^3/s),
U_v is the volume velocity of air into the cavity (m^3/s),
R_p is the loss (mechanical and radiative) in the top plate,
R_h is the due to radiation from the soundhole, and
R_v is the loss in the enclosure.

The two-mass model predicts two resonances with an antiresonance between them. At the lower resonance, air flows out of the soundhole in phase with the inward moving top plate. In the equivalent circuit in Fig. 9.9(b), this corresponds to U_p and U_v being essentially in phase (they would be exactly in phase if $R_h = R_v = 0$). At the upper resonance, U_p and U_h are essentially opposite in phase; that is, air moves into the soundhole when the top plate moves inward. The antiresonance represents the Helmholtz resonance of the enclosure; U_v and U_h are equal and opposite, and thus U_p is a minimum. This behavior, which is dominant in a guitar at low frequency, is analogous to that of a loudspeaker in a bass reflex enclosure (Caldersmith, 1978).

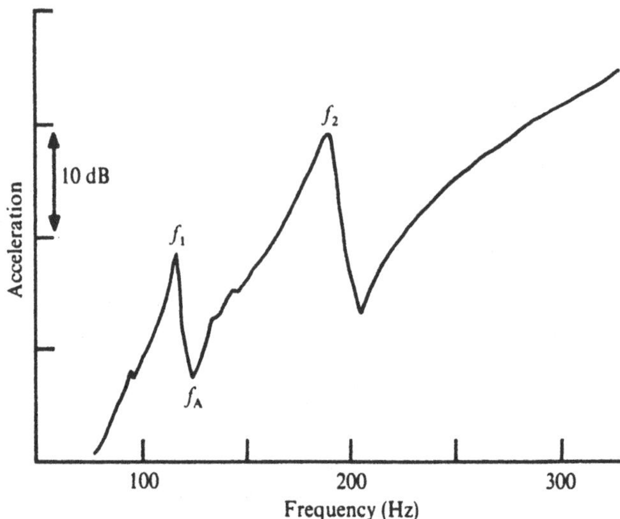

FIGURE 9.11. Low-frequency response curve for a Martin D-28 folk guitar with its back plate and ribs immobilized in sand. The bridge was driven on its treble side by a sinusoidal force of constant amplitude, and the acceleration was recorded at the driving point.

In Fig. 9.11, the response curve for a folk guitar, with its back and ribs immobilized, illustrates the two-mass model. The two resonances occur at f_1 and f_2 and the antiresonance at f_A. The two resonances f_1 and f_2 will span the lowest top plate mode f_p and the Helmholtz resonance f_A; that is, f_A and f_p will lie between f_1 and f_2. In fact, it can be shown that $f_1^2 + f_2^2 = f_A^2 + f_p^2$ (Ross and Rossing, 1979; Christensen and Vistisen, 1980). If $f_p > f_A$ (as it is in most guitars), f_A will lie closer to f_1 than to f_2 (Meyer, 1974).

9.6 Coupling to the Back Plate: Three-Oscillator Model

A three-oscillator model that includes the motion of the back is shown in Fig. 9.12 along with its electrical equivalent circuit. The ribs are immobilized, so that the coupling of the top plate to the back plate is via the enclosed air. Additional circuit elements are the mass M_b, compliance C_b, loss R_b, and volume velocity U_b of the back plate.

The response curve for the three-mass model has a third resonance and a second antiresonance, as shown in Fig. 9.13. In addition f_1 has been moved to a slightly lower frequency, and f_2 may be moved either upward (for $f_b < f_p$) or downward (for $f_b > f_p$), depending upon the resonance

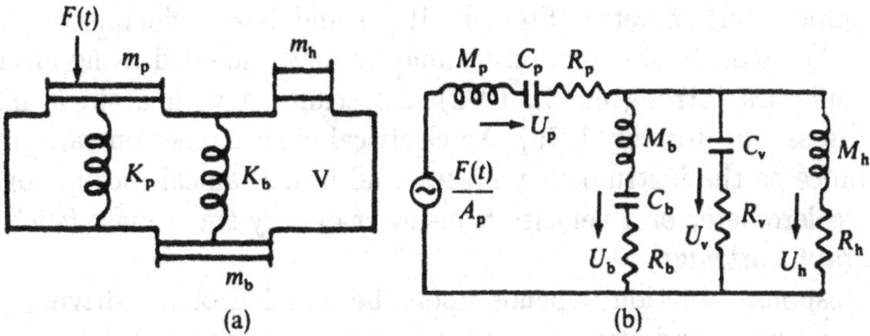

FIGURE 9.12. (a) Three-mass model of a guitar representing coupled motion of the top plate, back plate, and enclosed air. (b) Equivalent electrical circuit for three-mass model. Symbols are similar to those used in Fig. 9.10 (Rossing et al., 1985).

frequencies f_p and f_b of the top and back alone (Christensen, 1982). In most guitars, $f_b > f_p$, so both f_1 and f_2 are shifted downward by interaction with the flexible back (Meyer, 1974).

The three-mass model predicts that $f_1^2 + f_2^2 = f_A^2 + f_B^2$. This relationship has been verified by experimental measurements in several guitars with the ribs immobilized (Rossing, et al., 1985).

9.7 Resonances of a Guitar Body

The frequency response of a guitar is characterized by a series of resonances and antiresonances. In order to determine the vibrational configuration of the instrument at each of its major resonances, it is usually driven sinusoidally at one or more points, and its motion observed optically, acoustically, electrically, or mechanically. Optical sensing techniques include

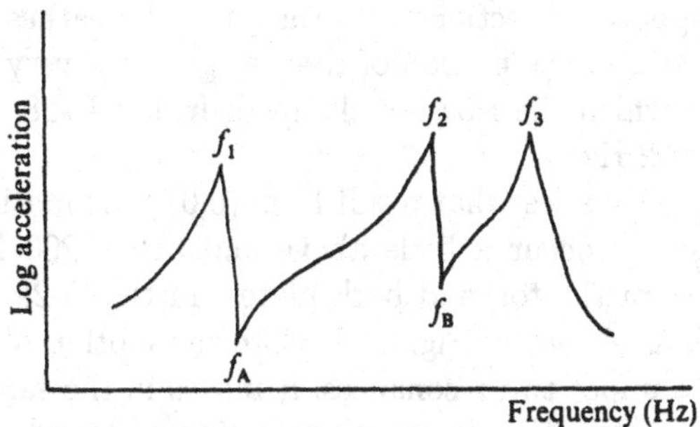

FIGURE 9.13. Frequency response curve predicted by the three-mass model. A third resonance and a second antiresonance have been added to the response curve of the two-mass model.

holographic interferometry (Stetson, 1981) and laser velocimetry (Boullosa, 1981). Acoustical detection techniques have included using an array of microphones (Strong et al., 1982) and scanning with a single microphone (Ross and Rossing, 1979). An electrical pickup relies on variation in capacitance as the instrument vibrates, and a mechanical pickup consists of an accelerometer or a velocity transducer of very small mass (such as a phonograph cartridge).

The response function depends upon the location of the driving point and sensing point and also on how the guitar is supported. A driving point on the bridge is usually selected, but for at least one prominent resonance there is a nodal line near the bridge, and thus it may be overlooked when the drive point or the sensing point lies on the bridge. It is difficult to overemphasize the importance of clearly describing the driving and sensing points and the method of support when reporting experimental results. Suspending the guitar by rubber bands has proved to be quite a satisfactory test configuration.

The configuration of a guitar at one of its resonances is often called a mode of vibration, but it is not necessarily a normal mode or eigenmode of the system. A resonance may result from exciting two or more normal modes. Only when the spacing of the normal modes is large compared with their natural widths does the vibration pattern at a resonance closely resemble that of a normal mode of vibration (Arnold and Weinreich, 1982).

When a guitar is driven at the bridge, the lowest resonance is usually a barlike bending mode. In the Martin D-28 folk guitar that we tested, it occurred at 55 Hz, and probably has little or no musical importance because it lies well below the lowest string frequency. Barlike bending modes at higher frequencies should not be overlooked, however.

Most guitars have three strong resonances in the 100–200 Hz range due to coupling between the (0,0) top and back modes and the A_0 (Helmholtz) air mode. In addition to the coupling via air motion discussed in Section 9.6, the top and back plates in a free guitar are coupled through the motion of the ribs. At the lowest of the three resonances, however, the top and back plates move in opposite directions (i.e., the guitar "breathes" in and out of the soundhole), and so the motion of the free guitar is very little different from the case in which the ribs are clamped. In the D-28, this breathing mode occurs at 102 Hz.

The other two resonances that result from (0,0)-type motion of the component parts usually occur a little above and below 200 Hz, depending upon the stiffness of the top and back plates. In the D-28, they occur at 193 and 204 Hz, as shown in Fig. 9.14. Note the motion of the air in the soundhole; in the upper two resonances, it moves in the same direction as the top plate, thus resulting in strong radiation of sound. At the middle resonance, the lower ribs and tail block move opposite to the main part of the top and back plates. Thus, clamping the ribs lowers this resonance considerably (from 193 to 169 Hz in the D-28; compare Fig. 9.13).

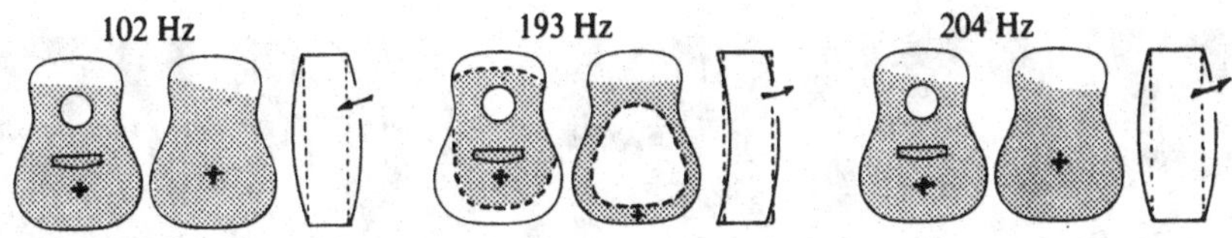

FIGURE 9.14. Vibrational motion of a freely supported Martin D-28 folk guitar at three strong resonances in the low-frequency range.

The (1, 0) modes in the top plate, back plate, and air cavity generally combine to give at least one strong resonance around 300 Hz in a classical guitar but closer to 400 Hz in a cross-braced folk guitar. In the D-28, this coupling is quite strong at 377 Hz. Motion of the (0, 1) type also leads to fairly strong resonances around 400 Hz in most guitars (Fig. 9.15).

Above 400 Hz, the coupling between top and back plate modes appears to be relatively weak, and so the observed resonances are due to resonances in one or the other of the two plates. A fairly prominent (2, 0) top plate resonance is usually observed in classical guitars around 550 Hz, but this mode is much less prominent in folk guitars. Vibrational configurations of a classical guitar top plate at its principal resonance are shown in Fig. 9.16.

9.8 Response to String Forces

In Section 9.3, we considered the parallel and perpendicular forces a string exerts on the bridge when it is plucked. The parallel force at twice the fundamental string frequency was found to be small at ordinary playing amplitudes but increases quadratically, so it can become a factor in loud playing. This force exerts a torque whose magnitude depends upon the bridge height and which could be a factor if it occurred at a frequency near a resonance of the (0, 1), (0, 2), or similar mode having a node near the bridge.

There are, of course, an infinite number of planes in which the string can vibrate; we consider the directions parallel and perpendicular to the bridge.

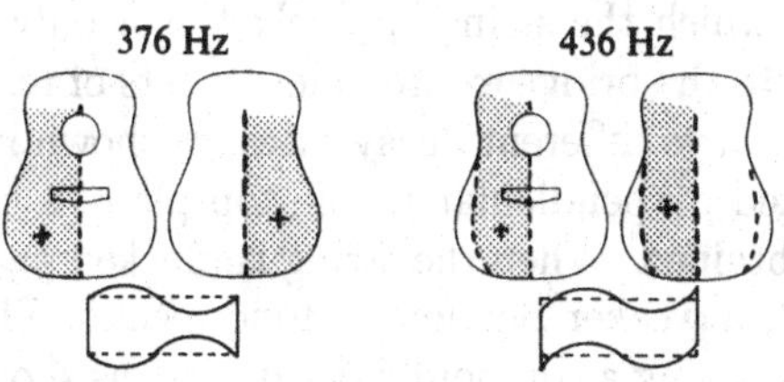

FIGURE 9.15. Vibrational configurations of a Martin D-28 guitar at two resonances resulting from "see-saw" motion of the (1,0) type.

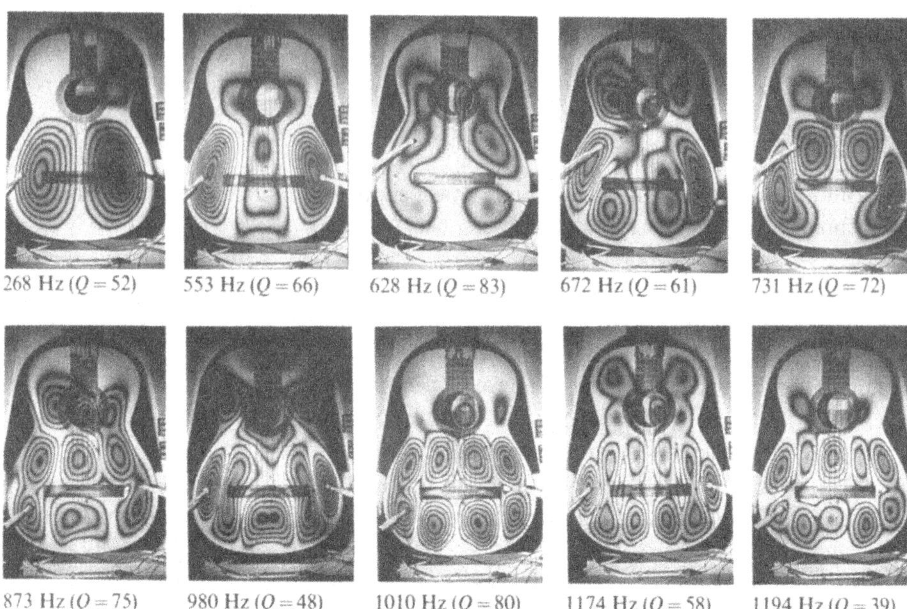

FIGURE 9.16. Time-averaged holographic interferograms of top-plate modes of a guitar (Guitar BR11). The resonant frequencies and Q values of each mode are shown below the interferograms (Richardson and Roberts, 1985).

The force parallel to the bridge encourages rocking motion, and thus it can easily excite resonances of the $(1,0)$ type. A perpendicular force anywhere on the bridge can excite the fundamental $(0,0)$ resonances, and if it is applied at the treble or bass sides, it can excite the $(1,0)$ resonances as well.

The effect of various string forces can be better understood by referring to the holographic interferograms in Fig. 9.17. These show the distortion of the top plate that results from static perpendicular and parallel bridge forces and also a torque caused by twisting one of the center strings.

Presumably, a player can greatly alter the tone of a guitar by adjusting the angle through which the string is plucked. Not only do forces parallel and perpendicular to the bridge excite different sets of resonances, but they result in tones that have different decay rates, as shown in Fig. 9.18. When the string is plucked perpendicular to the top plate, a strong but rapidly decaying tone is obtained. When the string is plucked parallel to the plate, on the other hand, a weaker but longer tone results. Thus, a guitar tone can be regarded as having a compound decay rate, as shown in Fig. 9.18(c). The spectra of the initial and final parts of the tone vary substantially, as do the decay rates.

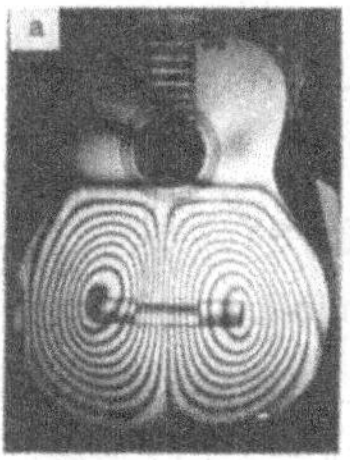

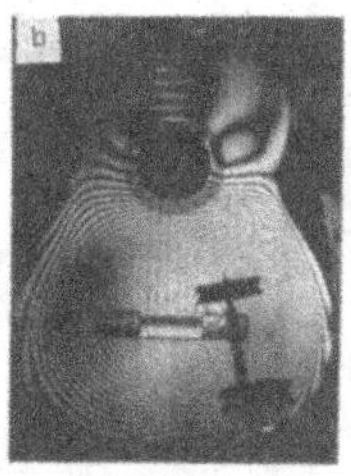

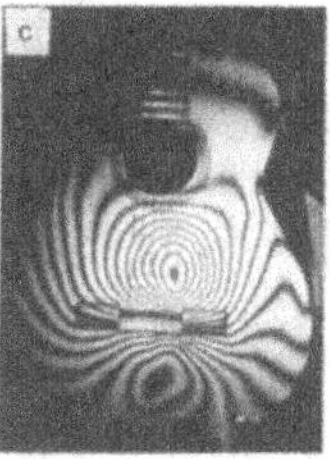

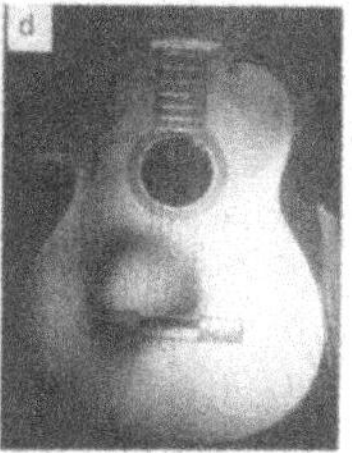

FIGURE 9.17. Holographic interferograms showing top plate distortions for (a) a static force at 1.0 N applied to the sixth string parallel to the bridge, (b) a force of 0.5 N applied to the first string perpendicular to the bridge, (c) a longitudinal force of 2.0 N applied to the first string, and (d) a torque caused by twisting the third string one full turn (Jansson, 1982).

Classical guitarists use primarily two strokes, called *apoyando* and *tirando* (sometimes called the *rest* and *free* strokes). The fingernail acts as sort of a ramp, converting some of the horizontal motion of the finger into vertical motion of the string, as shown in Fig. 9.19. Although the apoyando stroke tends to induce slightly more vertical string motion, there is little difference between the two strokes in this regard. The player can change the balance between horizontal and vertical string motion by varying the angle of the fingertip, however (Taylor, 1978).

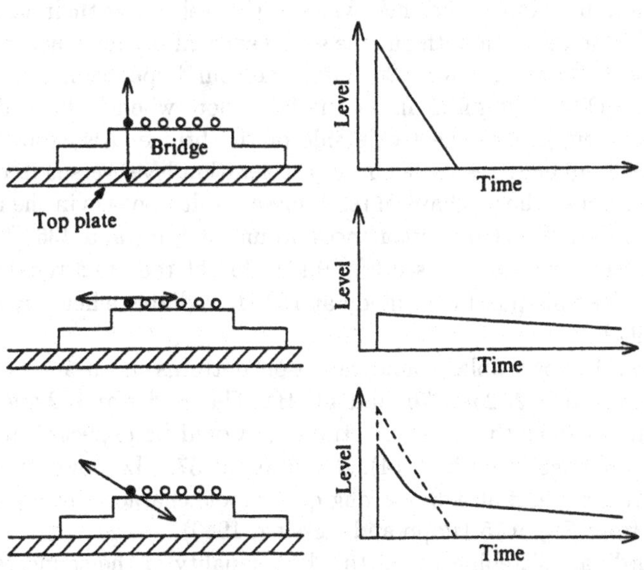

FIGURE 9.18. Decay rates of guitar tone for different plucking directions (Jansson, 1983).

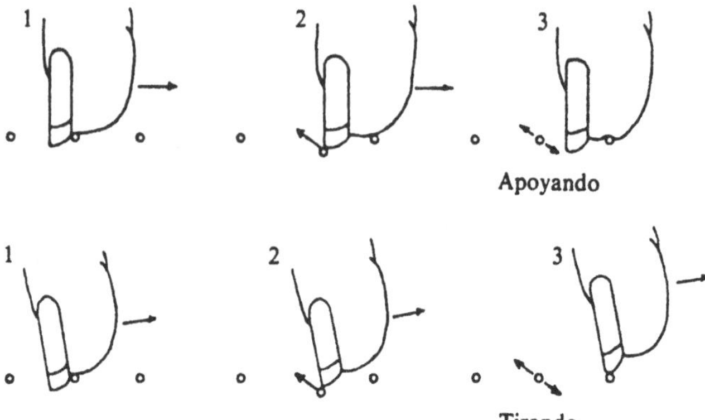

FIGURE 9.19. Finger motion and resulting string motion of apoyando and tirando strokes. In the apoyando stroke, the finger comes to rest on an adjacent string; in the tirando stroke, it rises enough to clear it (after Taylor, 1978).

9.9 Sound Radiation

Sound radiation from a guitar, like any musical instrument, varies with direction and frequency. Even with sinusoidal excitation at a single point (such as the bridge), the radiated sound field is complicated, because several different modes of vibration with different patterns of radiation will be excited at the same time. As pointed out in Section 7.6, some low-frequency modes will radiate quite efficiently when driven well above their natural frequency. This tends to smooth out the sound spectrum, even when measured in a single direction. Figure 9.20 shows the sound spectrum 1 m in front of a Martin D-28 folk guitar in an anechoic room when a sinusoidal force of 0.15 N is applied to the treble side of the bridge. Also shown is the mechanical frequency response curve (acceleration level versus frequency). Note that most of the mechanical resonances result in peaks in the radiated sound, but that the strong resonances around 376 Hz and 436 Hz [which represent "seesaw" motion (see Fig. 9.15)] do not radiate strongly in this direction. The "air pumping" mode at 102 Hz radiates efficiently through the soundhole.

Figure 9.21 shows polar sound radiation patterns in an anechoic room for the modes at 102, 204, 376, and 436 Hz. The modes at 102 and 204 Hz radiate quite efficiently in all directions, as would be expected in view of their mode shapes (see Fig. 9.14). Radiation at 376 Hz, however, shows a dipole character, and at 436 a strong quadrupole character is apparent, as expected from Fig. 9.15 (Popp and Rossing, 1986).

In an ordinary listening room, the directionality of the sound radiation is obscured by reflections from the walls, ceiling, and other surfaces. Furthermore, the sound spectrum for plucked notes will be quite different from

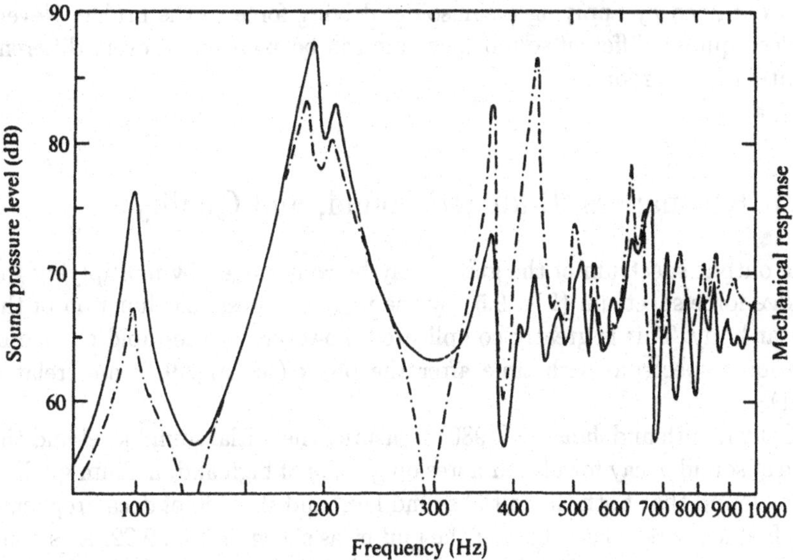

FIGURE 9.20. Mechanical frequency response and sound spectrum 1 m in front of a Martin D-28 folk guitar driven by a sinusoidal force of 0.15 N applied to the treble side of the bridge. Solid curve, sound spectrum; dashed curves, acceleration level at the driving point.

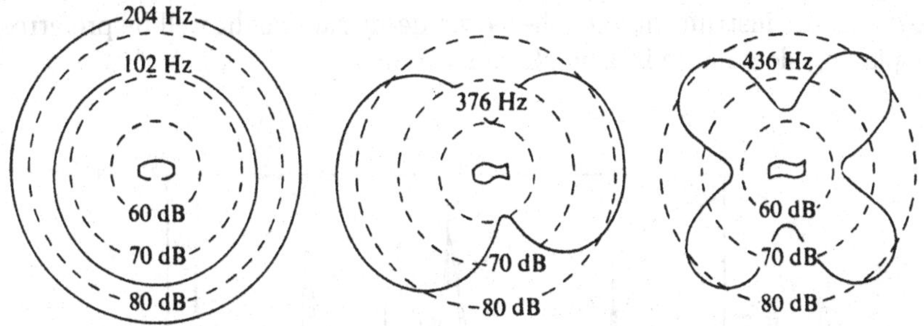

FIGURE 9.21. Sound radiation patterns at four resonance frequencies in a Martin D-28 folk guitar (compare with Figs. 9.14 and 9.15, which show the corresponding vibrational configurations) (from Popp and Rossing, 1986).

that obtained by applying a sinusoidal driving force to the bridge. Nevertheless, quite a different sound spectrum can be expected at every different location in the room.

9.10 Resonances, Radiated Sound, and Quality

The output spectrum of the guitar may be constructed by multiplying the bridge force spectrum (Fig. 9.5) by the frequency response function of the guitar body. This is greatly complicated, however, by the rapid change in the force spectrum with time after the pluck (see Fig. 9.18 and related text).

Caldersmith and Jansson (1980) measured the initial sound level and the rate of sound decay for played notes on guitars of high and medium quality. They found that both the initial sound level and the rate of decay replicate the frequency response curve of the guitar, as shown in Fig. 9.22. At strong resonances, however, the initial levels are slightly lower, and the levels decay faster than predicted by the frequency response curves.

The results of these experiments suggested one criterion for quality in a guitar. The high-quality guitar had both a higher initial sound level and a faster decay rate than the guitar of medium quality, especially for the sixth string, and it would probably be the instrument of choice in a concert hall. However, the instrument with the slower decay rate might well be preferred for playing slow music in a small, quiet room.

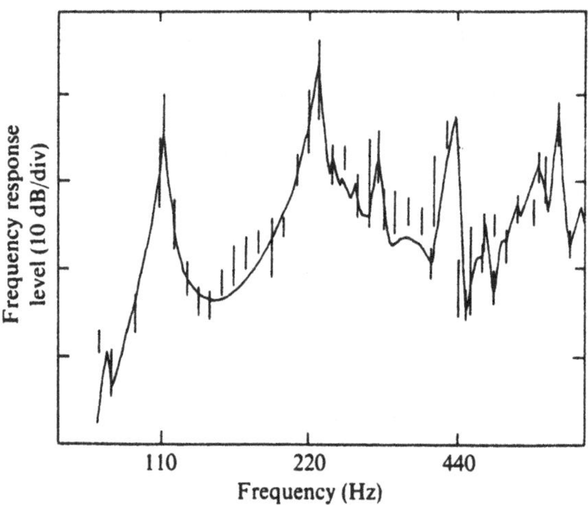

FIGURE 9.22. Comparison of the sound level of the fundamentals of played notes (bars) to the guitar frequency response function (solid curve) with its level adjusted for a good fit. A graph of the rate of sound decay (dB/s) versus frequency similarly follows the frequency response curve (Caldersmith and Jansson, 1980).

Some extensive listening tests were conducted at the Physikalisch-Technische Bundesanstalt in Braunschweig, Germany, to try to correlate quality in guitars to their measured frequency response (Meyer, 1983a). Some of the features that showed the greatest correlation with high quality were

1. the peak level of the third resonance (around 400 Hz),
2. the amount by which this resonance stands above the resonance curve level,
3. the sharpness (Q value) of this resonance,
4. the average level of one-third-octave bands in the range 80–125 Hz,
5. the average level of one-third-octave bands in the range 250–400 Hz,
6. the average level of one-third-octave bands in the range 315–500 Hz,
7. the average level of one-third-octave bands in the range 80–1000 Hz, and
8. the peak level of the second resonance (around 200 Hz).

Negative correlations were found with

1. the sharpness (Q) of first resonance (around 100 Hz),
2. the average level of third octaves in the range 160–250 Hz,
3. the maximum of the third octave levels above 1250 Hz, and
4. splitting of the second resonance into two peaks.

It is interesting to note the importance of the $(0, 1)$-type resonance around 400 Hz in determining guitar quality. Meyer's experiments indicate that the resonance peak should be tall and sharp, thus boosting the sound level of a rather narrow band of frequencies. However, in most guitars the bridge lies close to the nodal line for the $(0, 1)$ mode, and so it does not drive this mode very efficiently.

In a second paper, Meyer (1983b) explores the dependence of this resonance, as well as other quality features, on the constructional details of the guitar, such as number and spacing of struts, addition of transverse braces, size and shape of the bridge, etc. For example, he finds using fewer struts, varying their spacing, adding transverse bracing, and reducing the size of the bridge to have desirable effects.

Although most classical guitars are symmetrical around their center plane, a number of luthiers (e.g., Hauser in Germany and Ramirez in Spain) have had considerable success by introducing varying degrees of asymmetry into their designs. Most asymmetrical designs use shorter but thicker struts on the treble side, thus making the plate stiffer. Three such top plate designs are shown in Fig. 9.23.

The very asymmetric design in Fig. 9.23(c) was proposed by Kasha (1974) and developed by luthiers R. Schneider, G. Eban, and others. It has a split asymmetric bridge (outlined by the dashed line) and closely spaced struts of varying length. A waist bar (WB) bridges the two long struts and the soundhole liner.

Australian luthier Greg Smallman, who builds guitars for John Williams, has enjoyed considerable success by using lightweight top plates supported

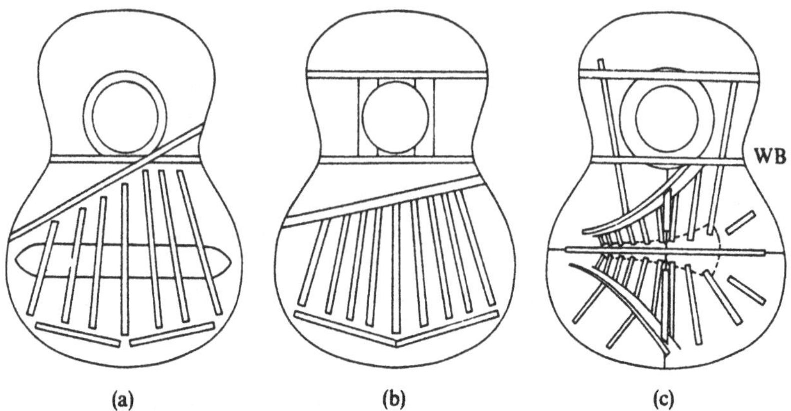

FIGURE 9.23. Examples of asymmetric top plates: (a) Hauser (Germany); (b) Ramirez (Spain); (c) Eban (United States).

by a lattice of braces whose thicknesses are tapered away from the bridge in all directions, as shown in Fig. 9.24. Smallman generally uses carbon-fiber-epoxy struts (typically 3 mm wide and 8 mm high at their tallest point) in order to achieve high stiffness-to-mass ratio and hence high resonance frequencies or "lightness" (Caldersmith and Williams, 1986).

9.11 A Family of Scaled Guitars

Members of guitar ensembles (trios, quartets) generally play instruments of similar design, but Australian physicist/luthier Graham Caldersmith has created a new family of guitars especially designed for ensemble perfor-

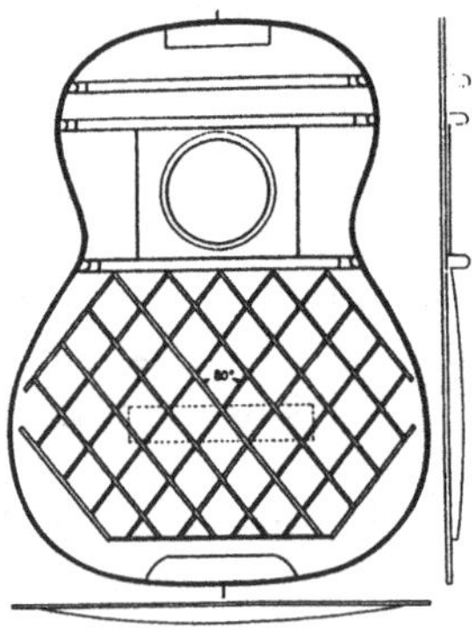

FIGURE 9.24. Lattice bracing of a guitar top plate used by Australian luthier Greg Smallman. Struts are typically of carbon-fiber-epoxy, thickest at the bridge and tapering away from the bridge in all directions (Caldersmith and Williams, 1986).

mance. (Actually he has created two such families: one of classical guitars and one of steel-string folk guitars.) His classical guitar family, including a treble guitar, a baritone guitar, and a bass guitar in addition to the conventional guitar—which becomes the tenor of the family—has been played and recorded extensively by the Australian quartet Guitar Trek (Caldersmith, 1989, 1995a,b).

Caldersmith's guitar families include carefully scaled instruments, whose tunings and resonances are translated up and down by musical fourths and fifths, in much the same way as the new family of violins developed by the Catgut Acoustical Society under the leadership of Carleen Hutchins (see Section 10.15). Caldersmith's bass guitar is a four-string instrument tuned the same as the string bass and the electric bass (E_1, A_1, D_2, G_2), an octave below the four lowest strings of the standard guitar. The baritone is a six-string instrument tuned a musical fifth below the standard, while the treble is tuned a musical fourth above the standard. This configuration facilitates arrangement of existing orchestral, string quartet, and keyboard works as well as new compositions. Transposition of orchestral works usually employs the full family but, in playing string quartet music, the baritone plays the cello part, the standard (tenor) the viola part, and two trebles play the violin parts (Caldersmith, 1995a).

Resonances of three members of the guitar family are appropriately scaled according to pitch. The first treble guitars were made with either thinner strings of standard lengths or with shorter strings of standard gauge. Caldersmith now recommends "slacker" trebles with lower-pitched modes, some of which employ lattice bracing of the Smallman type (Caldersmith, 1995b). Caldersmith's steel-string baritone has X-type bracing, but the X is shallower than the standard Dreadnought pattern of Fig. 9.2(d), and thus stiffens the top less along the grain. The symmetrical chevron brace behind the shallow X allows efficient excitation of the (2, 0) mode, not generally the case in folk guitars. String length is 71 cm, compared to 65 cm in the standard (tenor) guitar.

9.12 Use of Synthetic Materials

Traditionally guitars have top plates of spruce or redwood, with backs and ribs of rosewood or some comparable hardwood. Partly because traditional woods are sometimes in short supply, luthiers have experimented with a variety of other woods, such as cedar, pine, mahogany, ash, elder, and maple. Bowls of fiberglass, used to replace the wooden back and sides of guitars, were developed and patented by the Kaman company in 1966; their Ovation guitars have become very popular, partly because of their great durability.

One of the first successful attempts to build a guitar mostly of synthetic materials was described by Haines et al. (1975). The body of this instrument, built to the dimensions of a Martin folk guitar, used composite sandwich plates with graphite-epoxy facings around a cardboard core. In listening tests, the guitar of synthetic material was judged equal to the wood standard for playing scales, but inferior for playing chords. In France, Charles Besnainou and his colleagues have constructed lutes, violins, violas, cellos, double basses, and harpsichords, as well as guitars, using synthetic materials (Besnainou, 1995). A graphite-epoxy guitar specifically designed to be played outside in inclement weather is the "Rainsong" guitar developed by John Decker and his associates (Decker, 1995).

9.13 Electric Guitars

Although it is possible to attach a contact microphone or some other type of pickup to an acoustic guitar, the electric guitar has developed as a distinctly different instrument. Most electric guitars employ electromagnetic pickups, although piezoelectric pickups are also used.

Electric guitars may have either a solid wood body or a hollow body. Vibrations of the body are relatively unimportant, and since the strings transfer relatively little energy to the body, electric guitars are characterized by a long sustain time. The solid-body electric guitar is less susceptible to acoustic feedback (from loudspeaker to guitar) than an amplified acoustic guitar or a hollow-body electric guitar.

An electromagnetic pickup consists of a coil with a permanent magnet. The vibrating steel strings cause changes in the magnetic flux through the coil, thus inducing electrical signals in the coil. Most pickups provide a separate magnet pole piece for each string. The coil windings surround all six pole pieces, or else six individual coils are connected in a series. The pole pieces, which are adjustable in height, are usually set to be about 1.5 mm below the vibrating string.

Because of the low level of the string-induced signal, guitar pickups are especially susceptible to picking up stray 60-Hz hum from AC power lines. This led to the development to "humbucking" pickups, which combine the outputs of two coils wound in opposite directions so that stray pickup will be largely cancelled.

Most electric guitars have two or three pickups mounted at various positions along the strings that favor various harmonics. The front pickup (nearest the fingerboard) provides the strongest fundamental, whereas the rear pickup (nearest the bridge) is more sensitive to higher harmonics. Switches or individual gain controls allow the guitarist to mix the signals from the pickups as desired.

Most piezoelectric pickups employ a ceramic material, such as lead zirconate titanate (PZT). Piezoelectric pickups may be in direct contact with the string (at the bridge saddle, for example) or they may be placed in contact with the bridge or soundboard of an acoustic guitar.

An important type of electric guitar is the bass guitar or electric bass, widely used in rock and jazz bands. The electric bass generally has four strings tuned in fourths (E_1, A_1, D_2, G_2) and a longer fretboard (90 cm) than the normal electric guitar.

9.14 Frets and Compensation

Spacing the frets on the fretboard presents some interesting design problems. Semitone intervals on the scale of equal temperament correspond to frequency ratios of 1.05946. This is very near the ratio 18:17, which has led to the well-known rule of eighteen. This rule states that each fret should be placed $\frac{1}{18}$ of the remaining distance to the bridge, as shown in Fig. 9.25. Obviously, the fret spacing x decreases as one moves down the fingerboard.

Since the ratio $\frac{18}{17}$ equals 1.05882 rather than 1.05946 (an error of about 0.07%), each semitone interval will be slightly flat if the rule of eighteen is used to locate the frets. By the time the twelfth fret is reached, the octave will be 12 cents ($\frac{12}{100}$ semitone) flat, which is noticeable to the ear. Thus, for best tuning, the more exact figure 17.817 should be used in place of 18; in other words, each fret should be placed 0.05613 of the remaining distance to the bridge.

Another problem in guitar design is the fact that pressing down a string against a fret increases the tension slightly. This effect is much greater in steel strings than nylon, since a much greater force is required to produce the same elongation. Fretted notes will tend to be sharp compared to open ones. The greater the clearance between strings and frets, the greater this sharpening effect will be.

To compensate for this change in tension in fingering fretted notes, the actual distance from the nut to the saddle is made slightly greater than the scale length used to determine the fret spacings. This small extra length is called the string compensation, and it usually ranges from 1 to 5 mm on acoustic guitars but may be greater on an electric bass. Bass strings require more compensation than treble strings, and steel strings require considerably more than nylon strings. A guitar with a high action (larger clearance between strings and frets) requires more compensation than one

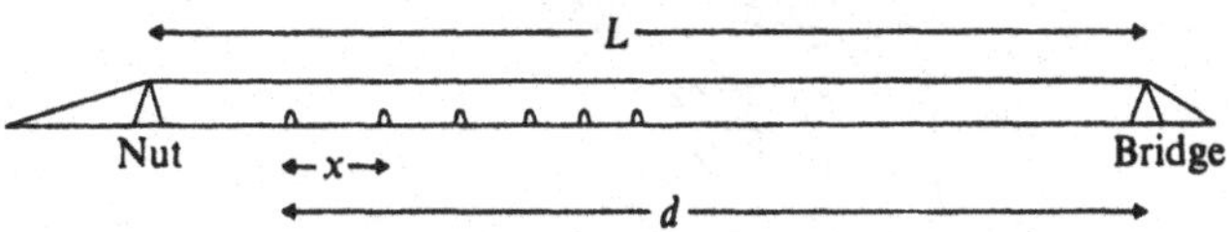

FIGURE 9.25. Fret placement. According to the "rule of eighteen", each fret is placed 1/18 of the remaining distance d towards the bridge or saddle. Greater accuracy is obtained by using 17.817 rather than 18.

with a lower action. Some electric guitars have bridges that allow easy adjustment of compensation for each individual string.

9.15 Lutes

The lute, which probably originated in the Near East, became the most popular instrument throughout much of Europe in the sixteenth and seventeenth centuries. (The name lute appears to have come from the Arabic phrase "al-oud," which means "made of wood.") Many different designs and variations on the basic design have existed through the ages. The long lute, having a neck longer than the body, dates back to at least 2000 B.C. and has modern descendents in several different countries (e.g., the tar of Turkey and Iran, the sitar and vina of India, the bouzouki of Greece, the tambura of India and Yugoslavia, and the ruan of China). The short lute, which dates from about 800 B.C., is the ancestor of the European lute as well as many other plucked string instruments around the world.

Instruments of the sixteenth century generally had 11 strings in 6 courses (all but the uppermost consisting of 2 unison strings), which might be tuned to A_2, D_3, G_3, B_3, E_4, and A_4, although the tuning was often changed to fit the music being played. Sometimes the lower 3 courses were tuned in octaves. In the seventeenth century, an increasing number of bass courses were added; these usually ran alongside the fingerboard, so that they were unalterable in pitch during playing. Lundberg (1988) describes a family of Italian sixteenth/seventeenth century lutes as follows:

Small octave: four courses, string length 30 cm;
Descant: seven courses, string length 44 cm;
Alto: seven courses, string length 58 cm;
Tenor: seven courses, string length 67 cm;
Bass: seven courses, string length 78 cm;

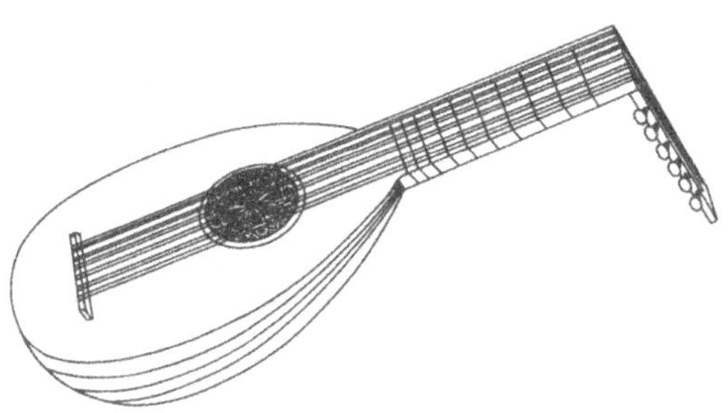

FIGURE 9.26. A sixteenth-century lute.

Octave bass: seven courses, string length 94 cm.

The pear-shaped body of the lute is fabricated by gluing together a number (from 9 up to as many as 37) of thin wooden ribs. The table or soundboard is usually fabricated from spruce, 2.5 to 3 mm thick, although other woods, such as cedar and cypress, have also been used. The table is braced by transverse bars (typically seven) above, below, and at the soundhole (see Jahnel, 1981). A sixteenth century lute is shown in Fig. 9.26.

Only a few studies on the acoustical behavior of lutes have been reported. Firth (1977) measured the input admittance (driving point mobility) at the treble end of the bridge and the radiated sound level 1 m away, which are shown in Fig. 9.27.

Firth associates the peak at 132 Hz with the "Helmholtz air mode" and the peaks at 304, 395, and 602 Hz with resonances in the top plate. Figure 9.28 illustrates 5 such resonances and also shows how the positions of the nodal lines are related to the locations of the bars. The modes at 515 and 652 Hz are not excited to any extent by a force applied to the bridge because they have nodes very close to the bridge.

9.16 Other Plucked String Instruments

A sampling of plucked string instruments is shown in Fig. 9.29. The ud from Iraq has a common ancestry with the European lute. The mandolin is a descendant of the mandola, a small lute of the seventeenth century. Citterns were popular in Europe during the sixteenth and seventeenth centuries.

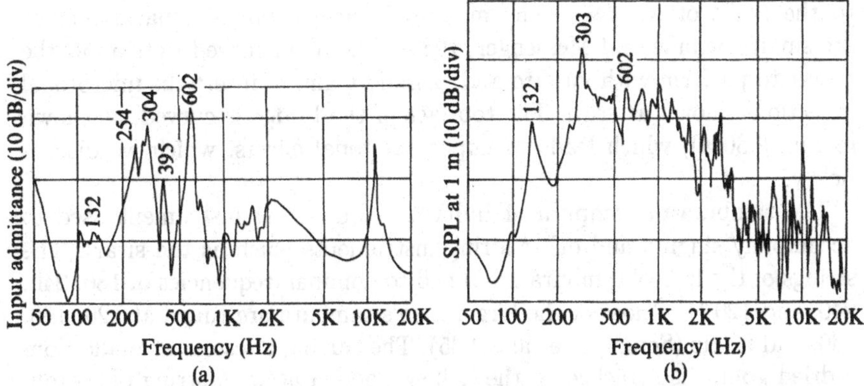

FIGURE 9.27. (a) Mechanical input admittance (mobility) at the treble end of a lute bridge and (b) sound pressure level 1 m from top plate (belly) (Firth, 1977).

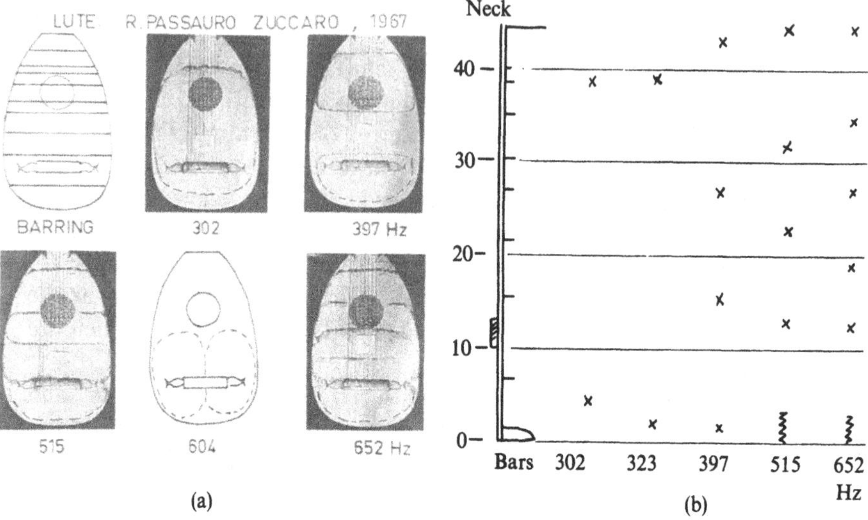

FIGURE 9.28. (a) Barring pattern and nodal patterns in the top plate of a lute at five resonances and (b) locations of nodes compared to the bridge and the bars (Firth, 1977).

The Chinese p'i-p'a has four strings tuned to A_2, D_3, E_3, and A_3. The top plate of wu-t'ung wood carries the frets, and the back is carved from a solid piece of red sandalwood or maple. The bamboo bridge lies near the end of the top plate [see Fig. 9.29(b)]. Major resonances in the top plate occur about 450, 550, and 650 Hz (Feng, 1984).

The sitar is the principal string instrument of North India. The main strings are tuned in fourths, fifths, and octaves to approximately $F_3^{\#}$, $C_2^{\#}$, $G_2^{\#}$, $G_3^{\#}$, $C_3^{\#}$, $C_4^{\#}$, and $C_5^{\#}$. There are also 11 sympathetic strings tuned to the notes of the raga. The measured and calculated inharmonicities are small (Benade and Messenger, 1982). The high, curved frets allow the player to perform with vibrato and glissando, and to insert the microtonal inflections known as sruti. The top face of the bridge is curved, as shown in Fig. 9.30(a), which leads to important tonal effects, which we discuss below.

The tambura or tampura of India is a four-string instrument used to accompany singing and other string instruments (such as the sitar). The strings of the ladies' tambura are tuned to nominal frequencies of 180, 240, 240, and 120 Hz; those of the larger men's tambura are tuned at 120, 160, 160, and 80 Hz (Sengupta et al., 1985). The tumba, a resonator made from a dried gourd, is attached to the hollow wooden neck. A string of cotton, called the jwari or jurali (life giver), is placed between the string and the bridge, as shown in Fig. 9.29(b), giving a slightly different constraint from

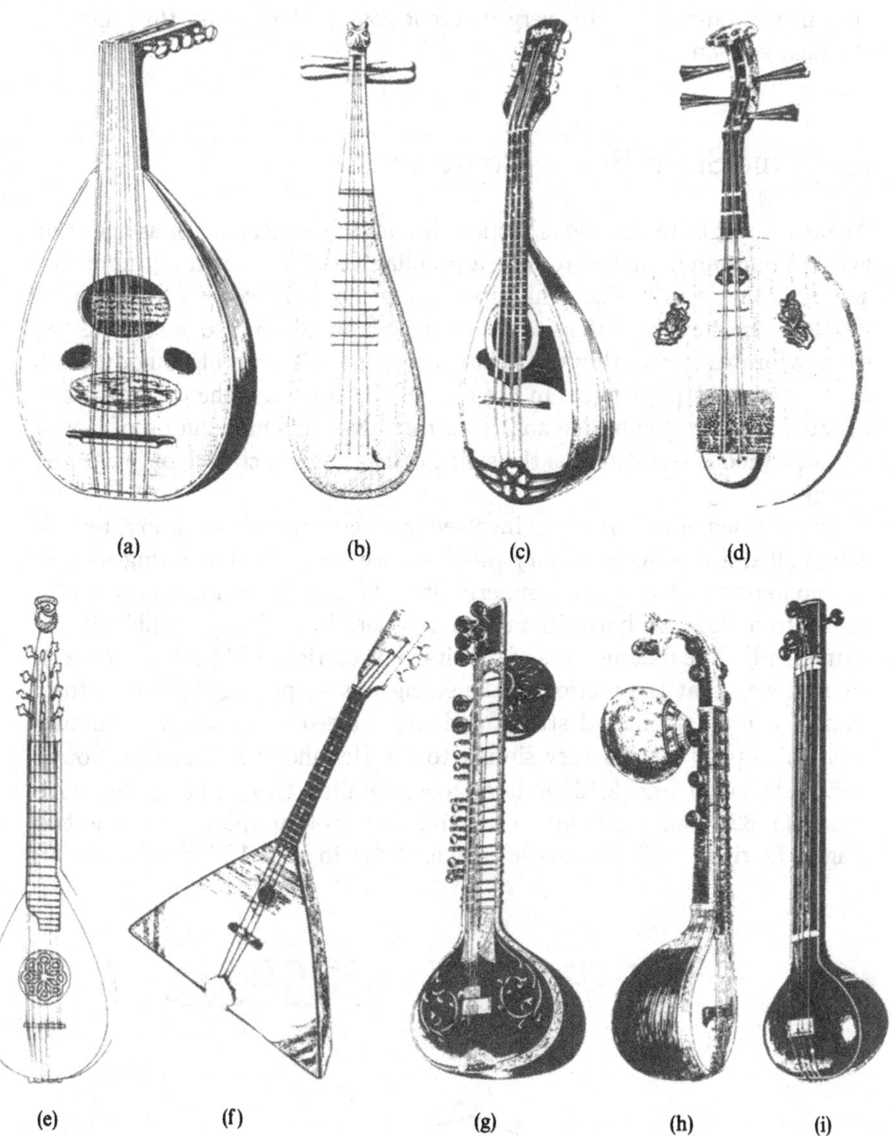

FIGURE 9.29. Plucked string instruments: (a) ud (Iraq), (b) p'i-p'a (China), (c) mandolin (Italy), (d) gekkin (Japan), (e) cittern (seventeenth-century Europe), (f) balalaika (Russia), (g) sitar (North India), (h) vina (South India), and (i) tampura (India).

that in the tambura and a very different sound. We discuss this, also, in the next section.

9.17 One-Sided Bridge Constraints

As mentioned in the previous section, Indian instruments such as the sitar and tambura are designed to have a peculiar one-sided constraint upon the string at the fixed bridge. This situation is shown in exaggerated form in Fig. 9.30. In the case of the sitar, the bridge is simply curved, so that as the string vibrates it smoothly wraps and unwraps from the bridge, modulating the vibrating string length. In the case of the tambura, the cotton thread inserted between the bridge and the string gives this modulation of length a discontinuous character, as the string comes in contact with or leaves the bridge.

The mathematical problems involved in discussing this situation, technically called a moving-boundary problem, are complex. The string motion can no longer be treated as a superposition of normal modes, and it is necessary to model the dynamics in detail, either algebraically, graphically, or numerically. The first analysis of the sitar was carried out by Raman (1922), who showed that the motion of the string was surprisingly different from that of a normal plucked string, and that indeed it consisted essentially of a single-peak motion very similar to the Helmholtz motion of a bowed string shown in Fig. 2.12. A later more-detailed treatment by Burridge et al., (1982) confirmed this result and showed that the motion has two stages. During the first transient stage, the string has 12 identifiable con-

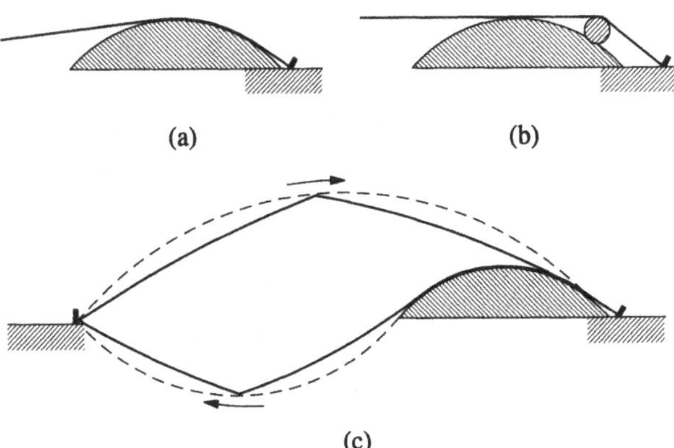

FIGURE 9.30. Exaggerated views of (a) a sitar bridge, (b) a tambura bridge, and (c) the quasi-Helmholtz second-stage motion of a plucked sitar string, as calculated by Burridge et al. (1982).

figurations, some of which expand in time-duration at the expense of others until the second Helmholtz-like motion is achieved. In the Helmholtz-like motion, the upward and downward displacements of the string are slightly different, because of wrapping along the bridge, and the string segments are slightly curved, rather than straight, as shown in Fig. 9.30(c). Both stages of the motion are only quasi-periodic, there is a slight pitch-glide as the amplitude decays, and the bridge constraint removes energy from the string so that its vibration time is not very long. These effects give the sitar its characteristic sound.

In a later publication, Valette (1995) has made a numerical analysis of the behavior of the tambura string. As one might expect, the motion is similar, except that the waveform changes are more abrupt than for the sitar. Valette also established that there is an important role for string stiffness, and the consequent dispersion of wave propagation speeds as discussed in Section 2.18, in producing the characteristic buzzing tambura sound. High-frequency waves, associated with abrupt changes in string slope, propagate ahead of waves of lower frequency, and can be reinforced during the reflection stage at the effectively two-point bridge.

References

Allen, J.B. (1976). On the aging of steel guitar strings. *Catgut Acoust. Soc. Newsletter*, No. 26, 27–29.

Arnold, E.B., and Weinreich, G. (1982). Acoustical spectroscopy of violins. *J. Acoust. Soc. Am.* **72**, 1739–1746.

Benade, A.H., and Messenger, W.G. (1982). Sitar spectrum properties. *J. Acoust. Soc. Am.* **71**, S83 (abstract).

Beranek, L.L. (1954). "Acoustics," Reprinted by Acoust. Soc. Am., Woodbury, New York, 1986. Section 3.4. McGraw-Hill, New York.

Besnainou, C. (1995). From wood mechanical measurements to composite materials for musical instruments: New technology for instrument makers. *MRS Bulletin* **20**(3), 34–36.

Boullosa, R.R. (1981). The use of transient excitation for guitar frequency response testing. *Catgut Acoust. Soc. Newsletter*, No. 36, 17.

Burridge, R., Kappraff, J., and Morshedi, C. (1982). The sitar string: A vibrating string with a one-sided inelastic constraint. *SIAM J. Appl. Math.* **42**, 1231–1251.

Caldersmith, G. (1978). Guitar as a reflex enclosure. *J. Acoust. Soc. Am.* **63**, 1566–1575.

Caldersmith, G. (1989). Towards a classic guitar family. *American Lutherie* No. 18, 20–25.

Caldersmith, G. (1995a). The guitar family, continued. *American Lutherie* No. 41, 10–16.

Caldersmith, G. (1995b). Designing a guitar family. *Appl. Acoust.* **46**, 3–17.

Caldersmith, G.W., and Jansson, E.V. (1980). Frequency response and played tones of guitars. Quarterly Report STL-QPSR 4/1980, Department of Speech Technology and Music Acoustics, Royal Institute of Technology (KTH), Stockholm. pp. 50–61.

Caldersmith, G. and Williams, J. (1986). Meet Greg Smallman. *American Lutherie* No. 8, 30–34.

Christensen, O. (1982). Qualitative models for low frequency guitar function. *J. Guitar Acoustics*, No. 6, 10–25.

Christensen, O., and Vistisen, R.B. (1980). Simple model for low-frequency guitar function. *J. Acoust. Soc. Am.* **68**, 758–766.

Decker, J.A. (1995). Graphite-epoxy acoustic guitar technology. *MRS Bulletin* **20**(3), 37–39.

Feng, S.-Y. (1984). Some acoustical measurements on the Chinese musical instrument P'i-P'a. *J. Acoust. Soc. Am.* **75**, 599–602.

Firth, I. (1977). Some measurements on the lute. *Catgut Acoust. Soc. Newsletter*, No. 27, 12.

Fletcher, N.H. (1976a). Plucked strings—a review. *Catgut Acoust. Soc. Newsletter*, No. 26, 13–17.

Fletcher, N.H. (1976b). "Physics and Music." Heinemann Educational Australia, Richmond, Victoria.

Haines, D.W., Hutchins, C.M., and Thompson, D.A. (1975). A violin and a guitar with graphite-epoxy composite soundboards. *Catgut Acoust. Soc. Newsletter* No. 23, 25–28.

Hanson, R. (1985). Comparison of new and "dead" nylon strings. *J. Acoust. Soc. Am.* **78**, S34 (abstract).

Houtsma, A.J.M. (1975). Fret positions and string parameters for fretted string instruments. *J. Acoust. Soc. Am.* **58**, S131 (abstract).

Houtsma, A.J.M., Boland, R.P., and Adler, N. (1975). A force transformation model for the bridge of acoustic lute-type instruments. *J. Acoust. Soc. Am.* **58**, S131 (abstract).

Houtsma, A.J.M., and Burns, E.M. (1982). Temporal and spectral characteristics of tambura tones. *J. Acoust. Soc. Am.* **71**, S83 (abstract).

Jahnel, F. (1981). "Manual of Guitar Technology." Verlag Das Musikinstrument, Franfurt am Main.

Jansson, E.V. (1971). A study of acoustical and hologram interferometric measurements on the top plate vibrations of a guitar. *Acustica* **25**, 95–100.

Jansson, E.V. (1982). Fundamentals of the guitar tone. *J. Guitar Acoustics*, No. 6, 26–41.

Jansson, E.V. (1983). Acoustics for the guitar player. In "Function, Construction, and Quality of the Guitar" (E.V. Jansson ed.). Royal Swedish Academy of Music, Stockholm. pp. 7–26.

Kasha, M. (1974). Physics and the perfect sound. In "Brittanica Yearbook of Science and the Future." Encyclopedia Brittanica, Chicago.

Legge, K.A., and Fletcher, N.H. (1984). Nonlinear generation of missing modes on a vibrating string. *J. Acoust. Soc. Am.* **76**, 5–12.

Lundberg, R. (1987). Historical lute construction: The Erlangen lectures. *Am. Lutherie* **12**, 32–47.

Meyer, J. (1974). Die abstimmung der grundresonanzen von guitarren. *Das Musikinstrument* **23**, 179–86; English translation in *J. Guitar Acoustics*, No. 5, 19 (1982).

Meyer, J. (1983a). Quality aspects of the guitar tone. In "Function, Construction, and Quality of the Guitar" (E.V. Jansson, ed.). Royal Swedish Academy of Music, Stockholm. pp. 51–75.

Meyer, J. (1983b). The function of the guitar body and its dependence upon constructional details (E.V. Jansson, ed.). Royal Swedish Academy of Music, Stockholm, pp. 77–108.

Popp, J., and Rossing, T.D. (1986). Sound radiation from classical and folk guitars. *International Symposium on Musical Acoustics, West Hartford, Connecticut, July* 20–23.

Raman, C.V. (1922). On some Indian stringed instruments. *Proc. Indian Assoc. Adv. Sci.* **7**, 29–33.

Richardson, B.E., and Roberts, G.W. (1985). The adjustment of mode frequencies in guitars: A study by means of holographic interferometry and finite element analysis. *Proc. SMAC 83*. Royal Swedish Academy of Music, Stockholm. pp. 285–302.

Ross, R.E., and Rossing, T.D. (1979). Plate vibrations and resonances of classical and folk guitars. *J. Acoust. Soc. Am.* **65**, 72; also Ross, R.E. (1979), "The acoustics of the guitar: An analysis of the effect of bracing stiffness on resonance placement," unpublished M.S. thesis, Northern Illinois University.

Rossing, T.D. (1981). Physics of guitars: An introduction. *J. Guitar Acoustics*, No. 4, 45–67.

Rossing, T.D. (1982a). "The Science of Sound." Addison-Wesley, Reading, Massachusetts. Chap. 10.

Rossing, T.D. (1982b). Plate vibrations and applications to guitars. *J. Guitar Acoustics*, No. 6, 65–73.

Rossing, T.D. (1983/1984). An introduction to guitar acoustics. *Guild of American Luthiers Quarterly* **11** (4), 12–18 and **12** (1), 20–29.

Rossing, T.D., Popp, J., and Polstein, D. (1985). Acoustical response of guitars. *Proc. SMAC 83*. Royal Swedish Academy of Music, Stockholm. pp. 311–332.

Sengupta, R., Bannerjee, B.M., Sengupta, S., and Nag, D. (1985). Tonal qualities of the Indian tampura. *Proc. SMAC 83*. Royal Swedish Academy of Music, Stockholm. pp. 333–342.

Stetson, K.A. (1981). On modal coupling in string instrument bodies. *J. Guitar Acoustics*, No. 3, 23–31.

Strong, W.Y., Beyer, T.B., Bowen, D.J., Williams, E.G., and Maynard, J.D. (1982). Studying a guitar's radiation properties with nearfield holography. *J. Guitar Acoustics*, No. 6, 50–59.

Taylor, J. (1978). "Tone Production on the Classical Guitar." Musical New Services, Ltd., London.

Turnbull, H. (1974). "The Guitar from the Renaissance to the Present Day." Batsford, London.

Valette, C. (1995). The mechanics of vibrating strings. In "Mechanics of musical Instruments," Ed. A. Hirschberg, J. Kergomard and G. Weinreich. Springer-Verlag, Vienna and New York. pp. 115–183.

10

Bowed String Instruments

Bowed string instruments have held a special place in music for many years. They form the backbone of the symphony orchestra, and they are widely used as solo instruments and in chamber music as well. They are instruments of great beauty and versatility. Unlike many other musical instruments, bowed string instruments have been the objects of considerable scientific study. Even so, their acoustical behavior is just beginning to be understood.

The violinist draws the bow across the violin strings in order to set them into vibration. The vibrating strings exert forces on the bridge that set the body into vibration, and the body radiates sound. Hidden by this simple description of violin mechanics, however, are the many subtleties that distinguish highly prized instruments from those of lesser quality.

10.1 A Brief History

No one knows who invented the violin, since it developed gradually from the various bowed string instruments used in Europe during the Middle Ages, such as the rebec, the gigue, the lyra, and the vielle.

During the sixteenth century, two families of viols developed: the *viola da gamba* or "leg viol" and the *viola da braccio* or "arm viol." These instruments, which normally had six strings tuned in fourths (except for a major third separating the third and fourth strings), developed in different sizes from treble to bass. They were given a waist to make them easier to bow, and gut frets were attached to their fingerboards. They have remained popular to this day, especially for playing music from their period and accompanying singing.

The instruments in the violin family were developed in Italy during the sixteenth and seventeenth centuries and reached a peak in the eighteenth century in the hands of masters such as Antonio Stradivari (1644–1737) and Giuseppe Guarneri del Gesù (1698–1744) of Cremona. The surviving instruments of these masters continue to be regarded as some of the finest ever produced, although nearly all of them have been altered to produce the more powerful sound required for large concert halls. The alterations include a slightly longer neck set back at a greater angle, longer strings at a higher tension, a higher bridge, and a heavier bass bar.

The viola, tuned a perfect fifth below the violin, is the alto member of the violin family. It is a distinctly different instrument, however, as is the violoncello or cello, the baritone member of the family (its name means little bass viol). The double bass or contrabass, which is tuned in fourths, has mainly developed from the bass viol.

The all important bow was given its present form by Francois Tourte (1747–1835). Two important characteristics of his bows are the inward curving stick of pernambuco wood and the frog with a metal ferrule to keep the bow hair evenly spread. A modern violin bow is 74 to 75 cm long with about 65 cm of free hair. The cello and double bass bows are shorter and heavier than the violin bow.

10.2 Research on Violin Acoustics

Hutchins (1983) has traced the history of violin research from Pythagoras (sixth century B.C.) to the present. Scientific understanding of vibrating strings began with Galileo Galilei (1564–1642) and Marin Mersenne (1588–1648). In his *Harmonie Universelle*, Mersenne discussed stringed instruments and indicated that he could hear at least four overtones in the sound of a vibrating string. The stick and slip action of the bow on the string appears to have been first recognized by Jean-Marie Duhamel (1797–1872).

Significant studies on the violin itself were carried out in the first half of the nineteenth century by Felix Savart (1791–1841), some of them in collaboration with the renowned violin maker Jean Baptiste Vuillaume (1798–1875). Savart constructed experimental instruments for his studies, and collaborated with Vuillaume in developing the large, 12-foot octobasse, whose three thick strings are stopped with the aid of levers and pedals. Savart studied the vibrations of top and back plates on Stradivarius and Guarnerius violins loaned to him by Vuillaume, and by careful experiments he correctly deduced the function of the soundpost.

Herman von Helmholtz (1821–1894) contributed to our understanding of bowed string instruments by both his physical and psychoacoustical experiments. By listening through spherical resonators of graduated sizes (which have come to be known as Helmholtz resonators), he identified some of the various partials in the tones of violins as well as other musical instruments. Using a vibration microscope proposed by Lissajous (1822–1880), Helmholtz observed the stick-slip motion of a violin string during bowing, from which he deduced the sawtooth waveform of the string displacement, now commonly referred to as Helmholtz motion.

A large number of different types of Helmholtz motion were observed by Krigar-Menzel and Raps (1891) using an ingenious optical recording method. These were explained in a comprehensive theoretical treatment of bowed string motion by Raman (1888–1970). Raman used a mechanical bowing machine in his own experiments, which showed that the minimum bowing force needed to maintain stable motion of the string varies directly as the speed of the bow and inversely as the square of the distance from the bridge.

During the decade 1930–1940, significant work was done in Germany by Erwin Meyer, Hermann Backhaus, and Hermann Meinel. After World War II, this tradition was continued by Werner Lottermoser, Frieder Eggers, and Jürgen Meyer. Particularly noteworthy has been the work of Lothar Cremer (1905-1990) plus his students and colleagues. This work is presented in Cremer's book *The Physics of the Violin*, now published in English (1984) as well as in German (1981).

The pioneer researcher in violin acoustics in the United States was Frederick Saunders (1875–1963). Saunders made acoustical comparisons for many violins, old and new, and he also developed a mechanical bowing machine. Saunders, along with Carleen Hutchins, John Schelleng, and Robert Fryxell, founded the Catgut Acoustical Society, an organization that still promotes research in violin acoustics. Among the accomplishments of this organization is the development of the violin octet, a carefully scaled ensemble of eight new violin family instruments (Hutchins, 1967).

Research on violin acoustics is excellently documented in the collections of reprinted papers edited by Hutchins (1995/96) and by Hutchins and Benade (1995), which cover most aspects of published work up to the end of 1993.

10.3 Construction of the Violin

The essential parts of a violin are shown in the exploded view in Fig. 10.1. The four strings of steel, gut, or nylon (wound with silver, aluminum, or steel) are tuned to G_3, D_4, A_4, and E_5. The top plate is generally carved from Norway spruce *(Picea abies* or *Picea excelsis)* and the back, ribs, and neck from curly maple *(Acer platanoides)*. The fingerboard is made of ebony and the bow stick from pernambuco *(Caesalpinia echinata)*. Running longitudinally under the top is a bass bar, and a sound post is positioned near the treble foot of the bridge.

Exact dimensions of violins vary with different makers, but the violin body is usually about 35 cm long and 16 to 20 cm wide in the upper and lower bouts, respectively. Ribs are typically 30 to 32 mm high. A finished top plate varies in thickness from 2.0 to 3.5 mm, a back plate from 2.0 to 6.0 mm. The maximum height of the arch in both plates is about 15 mm.

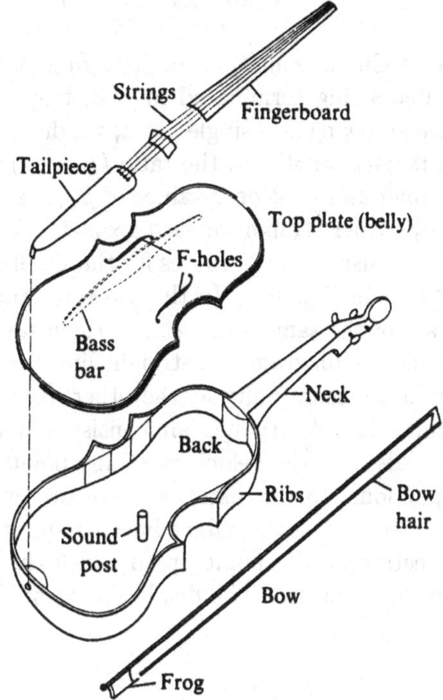

FIGURE 10.1. Parts of a violin (from Rossing, 1982).

Traditionally, violin makers have relied on the sound of "tap tones" and the "feel" of the wood under bending to guide them in carving the plates to their final thickness. More recently, however, many makers have followed the lead of Carleen Hutchins in relying to a large extent on Chladni patterns of three important plate modes to guide them (Hutchins, 1981). A small channel or groove is cut out along the edge of the plate and inlaid with thin strips of wood called purfling. This allows the plates to vibrate more as if they were hinged (simply supported) rather than clamped at the edge.

The total tension in the four strings of a violin is typically about 220 N (50 lb), and this results in a downward force on the bridge of about 90 N. This down-bearing is supported, in part, by the bass bar and the soundpost at opposite sides of the bridge.

10.4 Motion of Bowed Strings

A brief description of the motion of a bowed string was given in Section 2.10. Analysis of the motion showed that one or more bends race around a curved envelope with a maximum amplitude that depends on the bow speed and the bowing position. We now examine the motion in a little more detail.

Raman's model of Helmholtz motion in the string assumed the string to be an ideal, flexible string terminated in real, frequency-independent mechanical resistances. Bowing at a single point, x_b, divided the string into two straight segments with lengths in the ratio $(L - x_b)/x_b = p$. Raman considered both rational and irrational values of p. (Irrational values of p can be represented by rather straightforward geometrical constructions.)

Raman's velocity and displacement curves for the simplest case, with one discontinuity, are shown in Fig. 10.2. In this case, the positive and negative velocity waves are of the same form, and at the instant at which they are coincident, the velocity diagram is a straight line passing through one end of the string with a discontinuity at the other. As this discontinuity moves along the string, the velocity diagram consists of parallel lines passing through its two ends, and the velocities at any point before and after its passage are proportional to the distances from the two ends. The displacement diagrams consist of two straight lines passing through the ends of the string and meeting at the point up to which the discontinuity in the velocity diagram has traveled. The displacement of the string can be

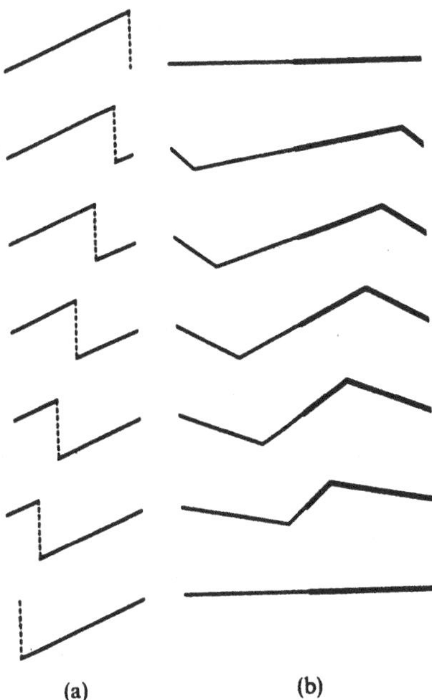

FIGURE 10.2. Helmholtz motion with a single discontinuity in a bowed string: (a) velocity diagram and (b) displacement diagram (Raman, 1918).

written as

$$y(x,t) = \sum_{n=1}^{\infty} a_n \sin \frac{n\pi x}{L} \sin n\omega t. \tag{10.1}$$

Raman's velocity and displacement curves for the case with two discontinuities are shown in Fig. 10.3. The point at which the two discontinuities coincide has been arbitrarily chosen to be the center of the string, which causes the string to vibrate in two segments at twice the fundamental frequency. If the discontinuities cross elsewhere, however (and thus cross twice as often), the frequency of vibration will be that of the fundamental. Note that displacements are measured from the equilibrium configuration of the string under the steady frictional force exerted by the bow (a triangle with its apex at the bowing position), not from the equilibrium position with the bow removed.

Helmholtz (1877) observed that the nth partial and its overtones are absent if the string is bowed at a rational fraction m/n of its length. [In Eq. (10.1), $y_n = 0$ when $x_b = mL/n$, where m is an integer.] This results in deviations from the idealized displacement curve, which Helmholtz described as ripples.

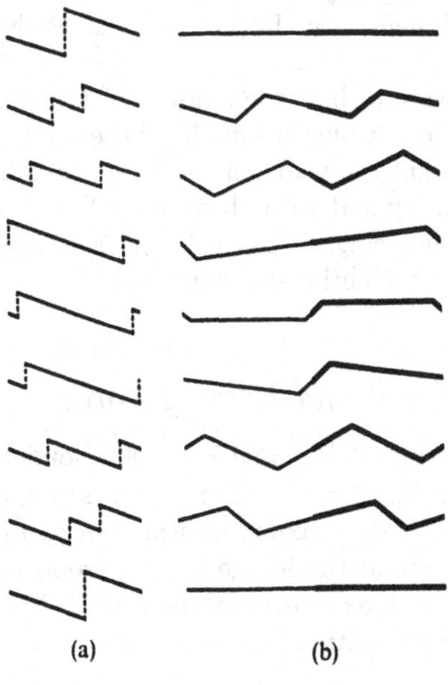

FIGURE 10.3. Helmholtz motion with two discontinuities in a bowed string: (a) velocity diagram and (b) displacement diagram (Raman, 1918).

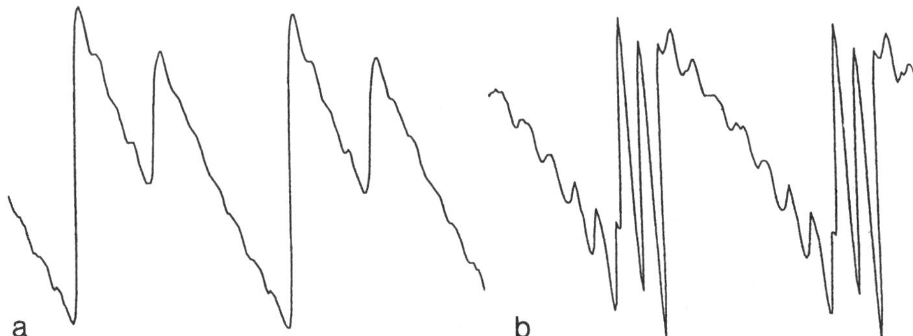

FIGURE 10.4. (a) Bridge force waveform in double-slip motion and (b) Bridge force waveform in multiple-flyback motion (Woodhouse, 1995).

In practice, two periodic regimes besides the Helmholtz motion are sometimes encountered. If the bow force is insufficiently high, "double-slip" motion can occur. A second slip occurs near the middle of the sticking period, and a double sawtooth wave occurs, as shown in Fig. 10.4(a). If the second slip is equal to the first, the note may sound an octave higher, but more commonly the fundamental frequency remains the same although the tone quality changes significantly (string players sometimes describe this as "surface sound"). The second regime is called "multiple-flyback" motion, in which the single flyback of the Helmholtz motion has been replaced by a cluster of them with alternating signs, as shown in Fig. 10.4(b) (Woodhouse, 1995).

Players are well aware of how the amount of bow hair in contact with the string affects bowing. A physical model of the bowed string by Pitteroff (1994) takes into account the width of the bow, the angular motion of the string, bow-hair elasticity and string bending stiffness. The frictional force for the edge of the bow facing the nut is lower than that of the edge facing the bridge throughout the entire sticking time.

10.4.1 Bowing Speed and Bowing Force

String players know that it is possible to play louder by bowing faster and nearer to the bridge. Cremer (1984) points out that under normal playing conditions, both the maximum displacement y_{m} of the string and the peak transverse force at the bridge F_{m} are proportional to $v_{\mathrm{b}}/x_{\mathrm{b}}$, and he substantiates this with experimental data by H. Müller and E. Völker. The relationships can be written

$$y_{\mathrm{m}} = \frac{1}{8f}\frac{v_{\mathrm{b}}}{x_{\mathrm{b}}} \quad \text{and} \quad F_{\mathrm{m}} = \mu c \frac{v_{\mathrm{b}}}{x_{\mathrm{b}}}. \tag{10.2}$$

In these expressions, v_b is the bow speed, x_b is the distance from the bridge to the bowing point, f is the frequency, μ is the mass per unit length, c is the wave speed, and the quantity μc is the characteristic impedance of the string.

For each position of the bow there is a minimum and a maximum bowing force (musicians often refer to it as "bowing pressure") at which the bend can trigger the beginning and end of slippage between the bow and string, as shown in Fig. 10.5. The closer to the bridge the instrument is bowed, the less leeway the player has between minimum and maximum bowing force. Bowing close to the bridge (sul ponticello) requires considerable bowing force and the steady hand of an experienced player. According to Cremer and Lazarus (1968), the minimum bowing force is proportional to v_b/x_b, but Cremer (1984) expresses second thoughts about the validity of this simple relationship. In fact, it appears as if it might be more nearly proportional to v_b/x_b^2.

Askenfelt (1986) used a bow equipped with a resistance wire and strain gauges to record the bowing speed and bowing force employed by two professional violinists under a variety of playing conditions. The bow force was found to vary between 0.5 and 1.5 N; the lowest bow force that still produced a steady tone was 0.1 N. The bow speed varied between 0.1 and 1 m/s; the lowest speed that still produced a steady tone was about 0.04 m/s. A range of 38 dB in bow velocity produced a change in the top plate vibration level of about 30 dB. This discrepancy suggests that the bow to bridge distance changed between the highest and lowest bow force.

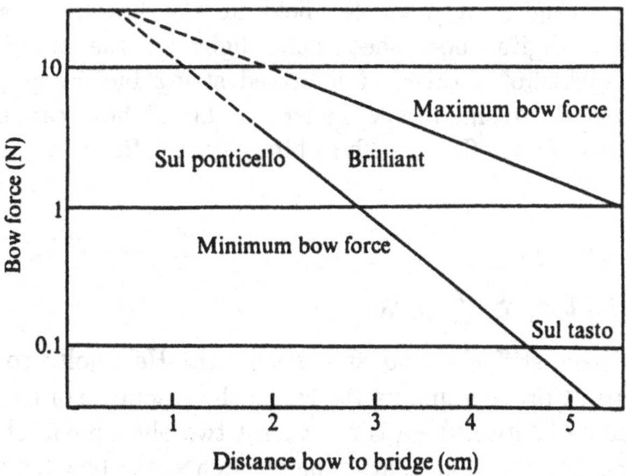

FIGURE 10.5. Range of bowing force for different bow-to-bridge distances for a cello bowed at 20 cm/s (after Schelleng, 1973).

Although most studies of bowing speed and bowing force have considered only steady-state behavior, Pickering (1986a) studied the effect of bowing parameters on the starting characteristics of violin tones. The bow was rapidly accelerated to a constant speed, while the amplitudes of both the fundamental and second harmonic components of string motion were observed as functions of time. With a relatively small bow force of 0.2 N (20-gram load), the second harmonic develops more rapidly than the fundamental. The optimum response occurs when the bow force is increased to 0.6 N. With a bow force of 0.8 N, periodic breakdown of the Helmholtz motion leads to "squawking." Pickering also found that bowing further from the bridge allows greater tolerance to changes in bow force, as would be expected from Fig. 10.5.

10.4.2 Computer Simulation and the Digital Bow

A violin bow can be thought of as a system that senses the instantaneous velocity of the string at the bowing point and exerts a force on the string which is some function of that velocity. Experimental studies on the motion of bowed strings are complicated by the difficulty in measuring and controlling the friction between a string and a rosined bow. In order to achieve constant and reproducible bowing conditions, it is desirable to provide the force on the string electronically. Such a device is called a "digital bow" (Weinreich and Caussé 1991).

A digital bow uses a photoelectric velocity sensor to determine the string's velocity at an imagined "bowing point." A computer determines the corresponding force from a previously programmed frictional characteristic, and the appropriate force is applied by passing a current through the string in a magnetic field at the bowing point. Experiments with a digital bow shed some light on the stability of the Helmholtz "stick-slip" motion of a bowed string but leave some questions unanswered. At high bow speeds, a digital bow can be used to demonstrate nonlinear effects, such as bifurcation (Müller and Lauterborn, 1996).

10.4.3 Effect of Stiffness

Real strings have stiffness, and as a result, the Helmholtz corner is not perfectly sharp. Corner rounding (to which the reactance of the string terminations also contributes) leads to at least two phenomena of interest to string players. One is a slight flattening of pitch as the bow force increases. The other is a type of noise called jitter that is caused by random variations in the period of the string. Variations up to 30 cents between the shortest and longest cycles have been found. The amount of jitter is roughly proportional to the amount of corner rounding (McIntyre and Woodhouse, 1978).

When a rounded Helmholtz bend passes the bow, it is sharpened by the nonlinear friction to an extent that depends upon the bow force. Thus, the actual motion of the string represents an equilibrium between corner rounding by the string and its terminations and corner sharpening by the bow. Furthermore, the amount of corner sharpening is not the same at the two bends, being greater during the transition from slipping to sticking (McIntyre and Woodhouse, 1978).

Another consequence of corner sharpening by the bow is the generation of ripples. When the shape of the bend is changed at the bow, a reflected wave is generated. These secondary waves reflect back and forth between the two bends (Cremer, 1984).

10.4.4 Longitudinal and Torsional Motion

Anyone who has listened to a violin played by an unskilled player will recognize the beginner's squeak that results from bowing with a heavy bow force at a slight forward angle along the string. Longitudinal vibrations of the string give rise to sounds that are easily noticed, since their frequencies are not harmonically related to the main transverse vibrations and they lie in a frequency range (1–5 kHz) of high sensitivity for hearing. Lee and Rafferty (1983) observed longitudinal modes with frequencies of 1350 Hz and 2700 Hz for the G and D strings, respectively, of a violin. Schumacher (1975) argues that longitudinal motion of the bow hair at the point of contact can differ from the nominal bow velocity by as much as 12%.

Torsional vibrations in bars and strings were discussed in Section 2.19. Torsional waves travel at a speed given by

$$c_{\mathrm{T}} = \sqrt{\frac{GK_{\mathrm{T}}}{\rho I}}, \tag{2.72}$$

where G is the shear modulus, K_{T} is the torsional stiffness factor, ρ is the density, and I is the polar moment of inertia. For a homogeneous circular string, $K_{\mathrm{T}} \cong I$, so

$$c_{\mathrm{T}} = \sqrt{\frac{G}{\rho}}. \tag{2.72$'$}$$

In many materials the shear modulus G is related to Young's modulus E and Poisson's ratio ν by the equation

$$G = \frac{E}{2(1+\nu)}, \tag{2.73}$$

so the ratio of torsional and longitudinal wave speeds becomes

$$\frac{c_{\mathrm{T}}}{c_{\mathrm{L}}} = \sqrt{\frac{G}{\rho}} \Big/ \sqrt{\frac{E}{\rho}} = \frac{1}{\sqrt{2(1+\nu)}}. \tag{10.3}$$

For steel strings, $\nu = 0.28$, so this ratio is 0.63.

Of greater interest to the violin player, however, is the ratio of c_{T} to c, the transverse wave speed. This varies from string to string, since c is a function of tension. For a steel E string, $c_{\mathrm{T}}/c \cong 7.7$, whereas for a gut E string, $c_{\mathrm{T}}/c \cong 2$, since gut has a much lower shear modulus G (Schelleng, 1973). For wound strings, the ratio will generally be somewhere between these values.

Torsional waves are rather strongly excited by bowing. If we assume that the outside of the string moves at the speed of the bow v_{b}, then the center of the string moves at a speed

$$v_{\mathrm{s}} = v_{\mathrm{b}} - \omega a, \tag{10.4}$$

where ω is the angular velocity of the string and a is its radius. Thus, the transverse speed of the string's center of mass fluctuates above and below the bow speed due to torsional oscillations.

Cornu (1896) observed these torsional oscillations by attaching a tiny mirror parallel to the string and perpendicular to the bowing direction. More recently, Kondo et al. (1986) have studied torsional motion by observing helical patterns on the string. The striking photograph in Fig. 10.6 of a string with several helical scratches was made with an anamorphic camera, which exaggerates transverse motion.

Damping of torsional vibrations tends to be much higher than damping of transverse vibrations, and thus energy converted from transverse to torsional waves is subject to this higher damping.

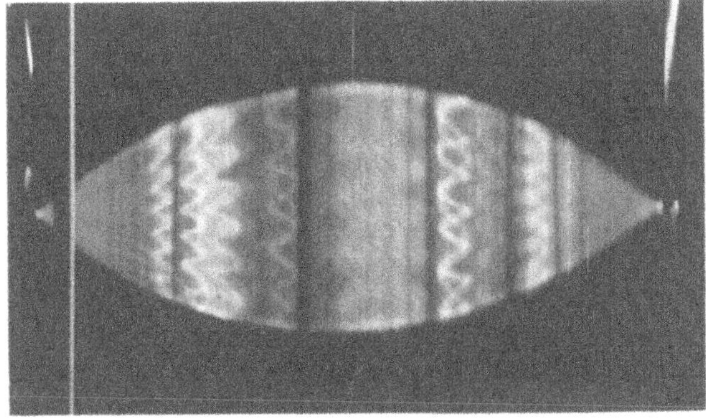

FIGURE 10.6. Lissajous pattern of a steel string showing torsional motion and Helmholtz motion (Kondo et al., 1986).

10.4.5 Finite Width of the Bow

Players are well aware of how much the amount of bow hair in contact with the string affects bowing. A physical model of the bowed string by Pitteroff (1994) takes into account the width of the bow, the angular motion of the string, bow-hair elasticity, and string bending stiffness. The frictional force for the edge of the bow facing the nut is lower than that for the edge facing the bridge throughout the entire sticking time.

Pitteroff and Woodhouse (1998) conclude that bending stiffness is relatively less significant in a finite-width model of bowing than it is in line-bow models. On the other hand, considering the elasticity of bow hairs becomes important in finite-width models. This is because the assumption of rigid bow hair leads to singular forces at both edges of the bow, which are removed when the bow-hair compliance is allowed for. A Young's modulus of about 7 GPa was found in both static and dynamic measurements, although the material properties of horse hair vary from the fresh growth at the base to the older hair at the tip of the tail (Pitteroff and Woodhouse 1998).

Players constantly adjust the angle at which the bow is tilted; a stroke might start with the bow strongly tilted to "ease" the attack, after which the angle of tilt is reduced. Gentle tilting alters the distribution of the bow force while keeping the full width of the bow hairs in contact with the string. Strong tilting, on the other hand, means that only a fraction of the bow width is in contact with the string. Strong tilting is thus associated with a small bow force, as a large bow force would again bring the full width of the bow (and even the bow stick) into contact with the string (Pitteroff and Woodhouse 1998).

10.4.6 Physical Properties of Commercial Strings

The physical properties of some widely used violin strings have been carefully measured by Pickering (1985, 1986b). Among the parameters tabulated are the minimum and maximum diameters, the mass per unit length, the torsional stiffness, the normal playing tension, the elasticity, and the inharmonicity of the overtones.

The nominal tension varies from 34.8 to 84.0 N. Apparently, modern violinists do not heed Leopold Mozart's advice to strive for equal tension in all four strings. The tension range among D strings alone is 34.8 to 61.7 N, whereas E strings tend to run uniformly high (72.3 to 81.0 N), whether solid or wound.

What Pickering calls elasticity is a measure of how much the tension (and hence the pitch) changes for a given change in length; presumably, it is proportional to the elastic modulus E. The elasticity of a steel string is about three times greater than a synthetic string and as much as seven times greater than gut. Not only are steel strings more difficult to tune, but their tension changes appreciably when pressed against the fingerboard.

The frequency of a newly fitted string drops after its initial tuning, and continues to do so for a considerable period of time before its tuning becomes stable. Steel strings tend to stabilize in a few minutes, synthetic strings require about 8 h, while gut strings require as much as 48 h (Pickering, 1986b).

With the aid of a scanning electron microscope, Firth (1987) has studied the construction details of violin strings having a gut core, nylon overwrap, and an outer wrap of silver or aluminum. These details are compared to the inharmonicity and to the preferences of players. Surprisingly, the strings with the lowest inharmonicity ranked low in player preference, but that is probably because of other factors. For one thing, the maximum bow force (Fig. 10.5) is greater with heavier, stiffer strings, which allows the skilled player to put in more energy. Also, greater stiffness tends to weaken the higher overtones, so the player can use greater bow force without producing too sharp a timbre.

Reduction of inharmonicity is most easily accomplished by reducing the core diameter, but it may also be accomplished by using a core of low elastic (Young's) modulus and by using a stranded and twisted cable as the core.

10.4.7 Rosin

While horse hair, like any hair, has small scales covering its surface, so that friction is different for the two directions of motion along its length, this frictional force is rather low on clean hair. The action of the bow therefore depends almost entirely upon the application of rosin and upon its frictional properties.

Violin rosin is a natural gum obtained from conifers such as larch that produce turpentine. The principal ingredients are abetic and pimaric acid. Rosin is quite hard and dry at normal room temperature, but it becomes sticky when warned only slightly (a fact well known to baseball pitchers). When powdered, small particles adhere firmly to the bowhair because of ion exchange. Rubbing the bow on a solid cake of rosin causes fragments of various sizes to adhere to the bow, but the larger ones leave during the first few bow strokes. Small particles trapped between hair and string are warmed by friction and provide the adhesion between bow and string (Pickering, 1984).

Another valuable property of rosin is that as it is warmed, its static coefficient of friction rises rapidly, while its dynamic, or sliding, coefficient drops slightly, as shown in Fig. 10.7. The rosin acts almost like a lubricant during the brief "slip" period. Infrared photos show that the temperature of the string may be raised by 25 to 30 degrees at the point of bowing, and this temperature rise appears to take place in a few tens of milliseconds (Pickering, 1991).

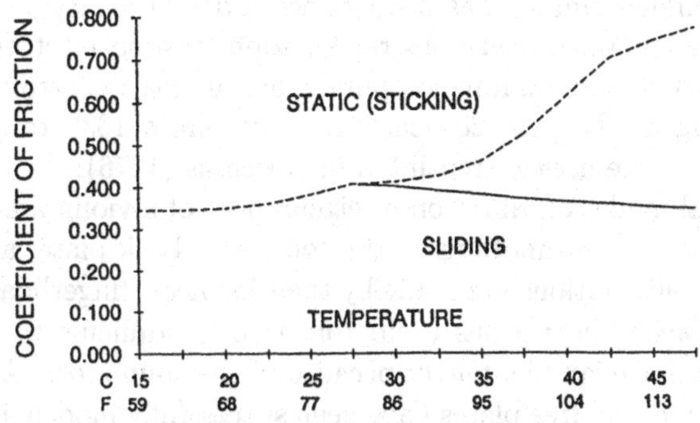

FIGURE 10.7. Frictional coefficient of violin rosin, measured by sliding a polished granite block down a heated polished aluminum bar (Pickering, 1991).

10.4.8 Low-pitched Tones with High Bowing Force

The usual result of using a very high bow force is a raucous sound that a player normally avoids. It is possible, however, with high bow force and very careful bow control to produce sustained periodic sounds with a definite pitch much lower than that obtained with normal bowing. Tones are approximately a musical 3rd, 7th, octave, 9th, and 12th below the free vibration fundamental frequency of the string (Hanson et al., 1994). These low tones, which have been used by concert violinists in concert, are quite different from the small flattening effect described in Section 10.4.3.

In the Helmholtz model for normal bowing, the traveling kink, which has a period the same as that of a freely vibrating string, serves as a synchronizing signal to ensure that the period for the slip-stick process is the same as the natural vibrational period of the string. If the bow force is great enough, however, the maximum transverse frictional force exerted by the bow on the string is sufficient to prevent the Helmholtz kink from triggering the release of the string from the bow hair, and the bow continues in the sticking part of the cycle until some other small signal triggers the slipping process. If careful bow control is exercised, it is possible for a periodic motion with longer than normal period to occur as a result of regular triggering by the same part of the waveform for each cycle. This can result from the combined effects of multiple bow-nut and bow-bridge reflections of transverse waves and reflections of torsional waves (Hanson et al., 1994; Guettler, 1994).

10.5 Violin Body Vibrations

Probably the most important determinant of the sound quality and playability of a string instrument is the vibrational behavior of its body. Although violin body vibrations have been studied for 150 years or more, most of our understanding has been gained during the last 50 or 60 years. The development of optical holography, sophisticated electronic measuring instruments, and digital computers has greatly contributed to our understanding in the past 20 years. Research from 1930 to 1973 is well documented by the articles reprinted in Hutchins (1976).

The normal modes of vibration or eigenmodes of a violin are determined mainly by the coupled motions of the top plate, back plate, and enclosed air. Smaller contributions are made by the ribs, neck, fingerboard, etc. The coupling between the various oscillators is more difficult to model than in the guitar, for example, partly because of the soundpost. Although the vibrational modes of free plates have been successfully modeled by finite element methods (Rodgers, 1986; Rubin and Farrar, 1987; Richardson, et al., 1987), most of our knowledge about eigenmodes of complete instruments is based on experimental studies.

Mode frequencies appear to differ rather widely in different instruments, as can be seen in Fig. 10.8, which shows the mechanical admittances (mobilities) measured at the bridge in six different violins. The most consistent mode in these six instruments is the mode termed the "air resonance" or "f-hole resonance," which occurs around 260–300 Hz. Several prominent resonances (peaks) and antiresonances (dips) occur in the 400- to 800-Hz octave.

Although the frequencies of the prominent peaks are quite different in the admittance curves of Fig. 10.8, the overall curves are quite similar, and that is probably why all six instruments sound like violins. We will discuss this in a later section.

Different investigators have used different systems of nomenclature for the various modes observed in violins. We will follow, for the most part, that employed by Erik Jansson and his co-workers at the Royal Institute of Technology (KTH) in Stockholm.[1] They identify modes according to the primary vibrating element as follows:

Air modes (A_0, A_1, A_2, ...): substantial motion of the enclosed air;
Top modes (T_1, T_2, T_3, ...): motion primarily of the top plate;
Body modes (C_1, C_2, C_3, ...): modes in which the top and back plate move similarly.

The first bending mode C_1 is similar to the first bending mode of a bar, having one nodal line near the bridge and one halfway up the neck. The

[1]A different system of nomenclature is used by Carleen Hutchins and her co-workers. Their air modes (A_0, A_1, ...) appear to be the same as the Swedish ones, and body modes ($B{-}1$, B_0, B_1^-, B_1^+) seem to correspond to modes C_1, N, T_1, and C_3, respectively, in the Swedish system. Hutchins (1993) suggests that the "main wood resonance," the strongest bowed tone in the frequency range 440–570 Hz (Saunders, 1937), results from a combination of the A_1 and B_1 modes, as does the "wood prime" resonance (an octave lower), in all probability.

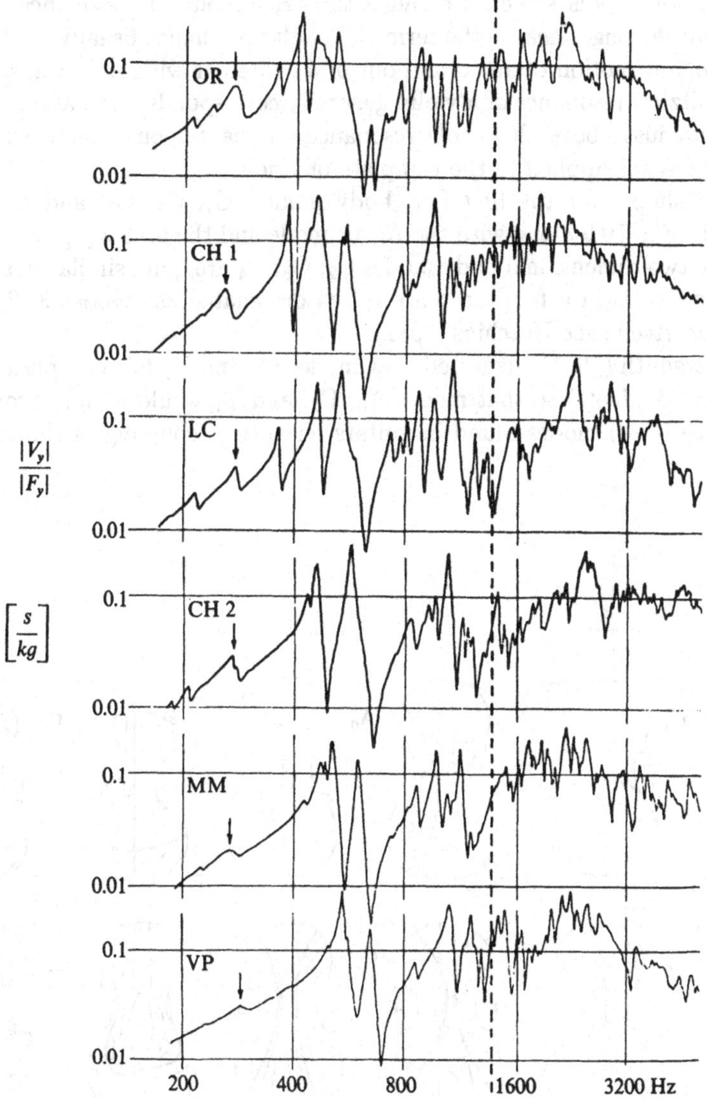

FIGURE 10.8. Mechanical admittances of six violins as measured at the bridges (Beldie, 1974, as presented by Cremer, 1984).

second bending mode N features the motion of the neck and fingerboard. Neither of these modes radiates sounds very efficiently, but they probably contribute to the "feel" of the instrument, and they may couple to some radiative modes having nearly the same frequencies.

The lowest mode of acoustical importance is the main air resonance or f-hole resonance A_0, in which a substantial amount of air moves in and out of the f-holes. It is sometimes called the "Helmholtz air resonance," but this is misleading, because the term Helmholtz resonance usually describes the resonant motion of air in and out of a container with rigid walls. The Helmholtz air resonance of a violin generally corresponds to an admittance minimum just above the f-hole resonance on the response curve when a driving force is applied to the top plate or bridge.

Mode shapes for the first four body modes, C_1, C_2, C_3, and C_4, are shown in Fig. 10.9 along with the A_0 air mode and the first top plate mode T_1. The two-dimensional body modes C_2, C_3, C_4 are quite similar in shape to three modes in a free top plate (often designated as "modes 3, 2, and 5," respectively; see Hutchins, 1981).

Caldersmith (1981) suggested that modes C_2 and T_1 form a "plate fundamental doublet," so that modes A_0, C_2, and T_1 would be analogous to the three (0, 0) modes found in guitars because of coupling of the lowest

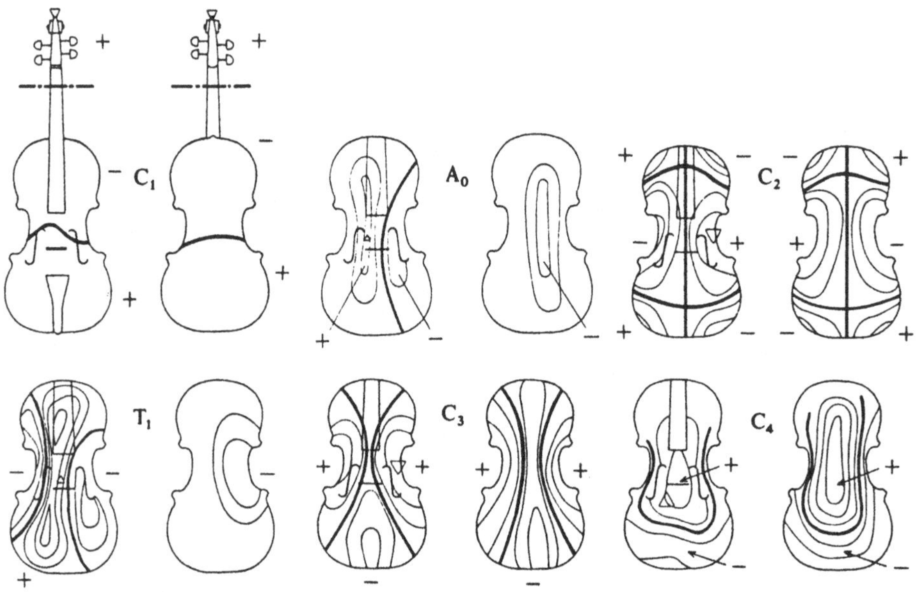

FIGURE 10.9. Modal shapes for six modes in a violin. C_1 (185 Hz): one-dimensional bending [B−1]; A_0 (275 Hz): air flows in and out of the f-holes [A_0]; C_2 (405 Hz): two-dimensional flexure; T_1 (460 Hz): mainly motion of the top plates (sometimes strong motion of the back also) [B_1^-]; C_3 (530 Hz): strong two-dimensional motion of both plates [B_1^+]; C_4 (700 Hz): two-dimensional motion of both plates. Top plate and back plate are shown for each mode. The heavy lines are nodal lines; the direction of motion (viewed from the top) is indicated by + or −. The drive point is indicated by a small triangle (after Alonso Moral and Jansson, 1982a). Mode designations used by Hutchins and co-workers appear in brackets.

modes of the top plate, back plate, and air cavity (see Chapter 9). However, from the mode shapes in Fig. 10.9, it appears more as if T_1 and C_3 form a sort of doublet pair. In a violin, the top and back plates are strongly coupled by the soundpost as well as the ribs, so it is not surprising that there are few modes in which the entire top plate or back plate moves in the same direction.

The modes designated A_0, T_1, C_3, and C_4 in Fig. 10.9 appear to be the most important low-frequency modes in a violin. They are identified on the admittance curve of a Guarneri violin in Fig. 10.10. Further evidence for the important role of modes T_1, C_3, and C_4 came from examining 24 violins entered in a Scandinavian violin maker's competition. Alonso Moral and Jansson (1982b) found that the violins judged highest in quality tended to have uniformly high levels of admittance for these modes (when driven on the bass bar side). Another criterion of quality was a rapid increase in admittance from 1.4 and 3 kHz. Admittance curves for the best violin, driven on both the bass bar and soundpost sides, are shown in Fig. 10.11.

10.5.1 Experimental Methods for Modal Analysis

Quite a number of different experimental methods have been used to study violin body vibrations. The agreement between the results of different investigators using different methods on different violins is not always good.

Early work by Backhaus and Meinel, as well as investigations by Eggers during the 1950s (all reported in Hutchins, 1976), made use of capacitive sensors to detect the plate displacements as a function of position. Menzel and Hutchins (1970) used an optical proximity detector. Luke (1971) and Lee (1975) scanned the plates with optical sensors. Luke applied a sinusoidal force to the bridge, while Lee excited his violin by bowing the open strings.

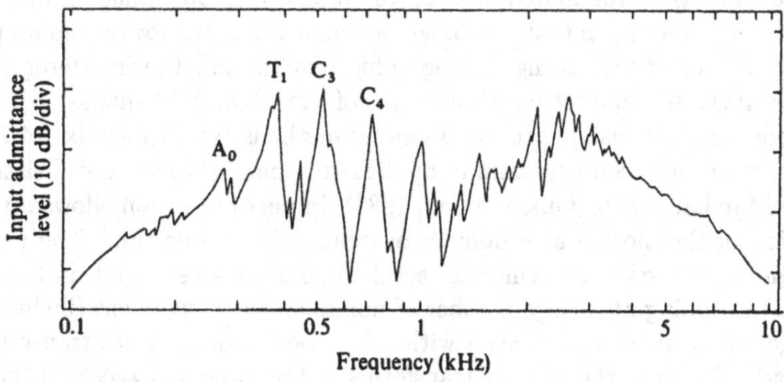

FIGURE 10.10. Input admittance (driving point mobility) of a Guarneri violin driven on the bass bar side (Alonso Moral and Jansson, 1982b).

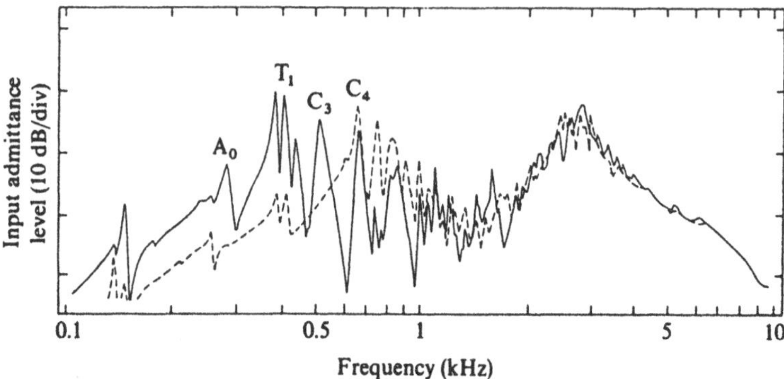

FIGURE 10.11. Input admittance (driving point mobility) for the best violin in a violin makers competition. Solid curve is driven on the bass bar side and dashed curve on the soundpost side. Note the uniform high level of the T_1, C_3, and C_4 peaks (for the bass bar drive) and the steep rise from 1.4 to 3 kHz (Alonso Moral and Jansson, 1982b).

Several optical interferometric techniques, employing laser light sources, provide very accurate records of violin body motion. One is time-average holographic interferometry (Powell and Stetson, 1965), which was used to study violin body resonances by Reinecke and Cremer (1970) and by Jansson et al. (1970). This method records the location of nodal lines with great accuracy and also provides a sort of contour map of displacement amplitude in the antinodal regions. It has the disadvantage of requiring that the instrument be fixed in position, generally by clamping at two or more points, which may affect its modal shapes.

Electronic TV holography offers a convenient way to view modal shapes in real time and also to record them (Stetson, 1990; Jansson et al., 1994). In this technique, a CCD (charge-coupled device) camera is used to record images of the laser-illuminated object. An on-line electronic image processor creates interferograms that can be viewed on a video monitor, avoiding the inconvenience of processing photographic film to view the interferogram. Figure 10.12 shows electronic holograms of the A0 and A1 modes.

Modal analysis using structural testing methods (see Appendix A4.2 in Chapter 4) has been successfully applied to violins (Müller and Geissler, 1983; Marshall, 1985; Jansson et al., 1986). Impact excitation allows investigation of the motion at a number of frequencies at one time. The usual technique consists of attaching an accelerometer at a key point on the violin and tapping at a large number of points on the instrument (including bridge, neck, finger-board, etc.) with a hammer having a force transducer or load cell. From the Fourier transforms of the force and acceleration at each pair of locations, it is possible to compute the modal parameters of interest, including modal frequencies, damping, and mode shapes.

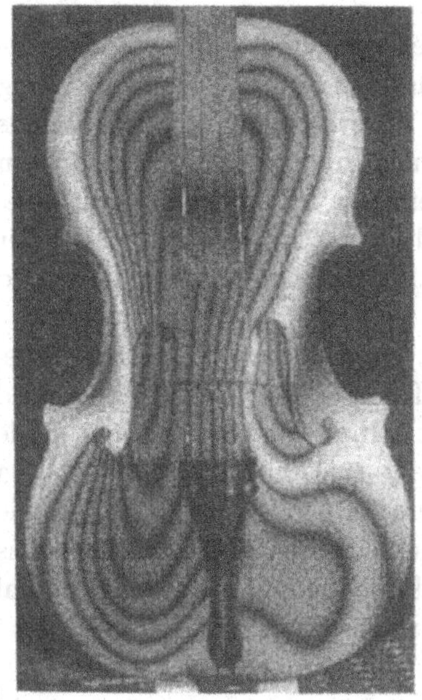

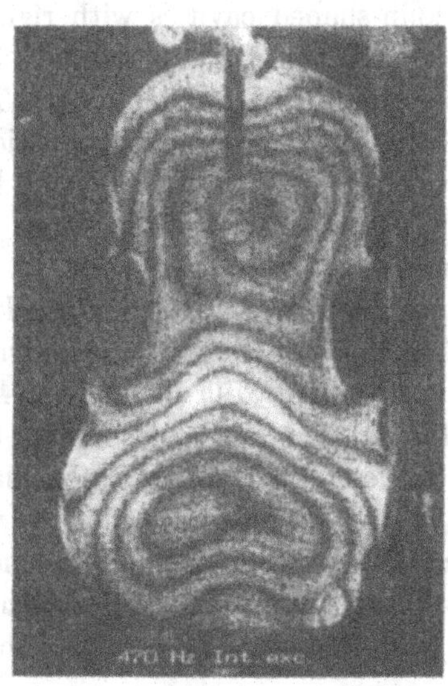

FIGURE 10.12. Interferograms of two air modes in violins using electronic TV holography. (a) A_0 mode excited by sound from a loudspeaker (from Saldner et al., 1996); (b) A_1 mode excited by applying a sound pressure internally (Roberts and Rossing, 1997).

Modal analysis with impact excitation is a fast and convenient method for determining the vibrational modes of an elastic structure. The modal shapes obtained by this method appear to be in fair agreement with those obtained by more precise holographic methods; the locations of antinodes are in better agreement than the locations of nodal lines (Jansson et al., 1986).

Other methods for modal analysis include the time-honored method of Chladni patterns, scanning the near-field sound next to the vibrating surfaces with a microphone, and scanning the vibrating surfaces with a very small accelerometer. Chladni patterns are quite effective for mapping the modes of free plates, but they are somewhat impractical for finished violins.

Jansson et al. (1986) compared three different systems for applying sinusoidal excitation and also three different arrangements for supporting the instrument during testing (hung on rubber bands, loosely clamped at the neck, clamped at the neck and supported at the tailpiece end). Sinusoidal drive was compared to impact excitation.

It is well to point out that each of the experimental methods described in this section has its own advantages and disadvantages. Applying two or more methods to the same instrument and comparing the data is probably the most effective strategy for modal analysis of violin body vibrations.

10.5.2 Violin Cavity Modes

The modes of a violin air cavity have been measured both by constructing violin-shaped cavities with rigid walls (Jansson 1977) and by using sand bags to hold the top and back plates of a violin stationary (Hutchins 1990; Roberts and Rossing 1997). The mode frequencies have also been calculated using equivalent acoustical networks (Shaw 1990), finite element methods (Roberts 1986), and boundary element methods (Bissinger 1996). The results of these various investigations are in reasonably good agreement.

Figure 10.13 shows the first 7 modes of a violin air cavity. The first mode, the (0,0) mode or Helmholtz resonance, is found at 284 Hz. This is followed by a longitudinal (0,1) air mode at 499 Hz and a transverse (1,0) air mode at 1077 Hz. The (0,2) longitudinal air mode at 1190 Hz is followed by a (1,1) mode at 1340 Hz, a longitudinal mode at 1646 Hz, and a (1,2) mode at 1887 Hz. The mode designation (m, n) used in Fig. 10.13, in which m is the number of transverse nodal planes and n is the number of longitudinal nodal planes, is more descriptive than the use of $A_1, A_2, \ldots$, which also risks confusion with the modes of the whole violin.

10.5.3 Free Plates and Assembled Violins

Violin makers have traditionally attached great significance to the vibrations of free top and back plates as determined by listening to tap tones and, more recently, by visual observation of the mode shapes. The vibrational modes of the finished violin, however, are quite different from those observed in the free plates from which they are constructed.

Holographic interferograms of a well-tuned top plate and back plate are shown in Fig. 10.14. More recently, Carleen Hutchins and several other luthiers have recommended tuning the X-mode (mode #2) and the ring mode (mode #5) to the same frequencies (preferably an octave apart) in

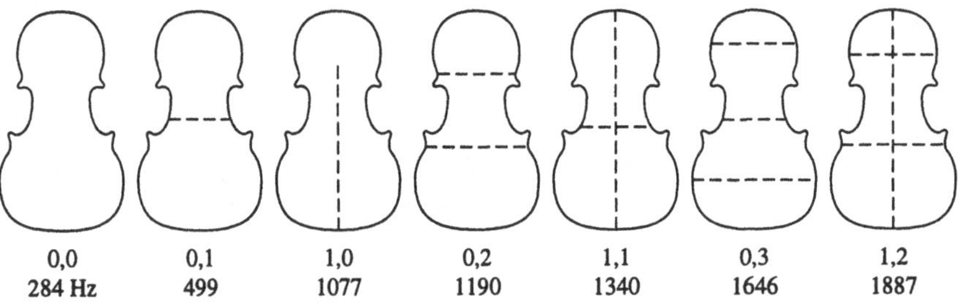

FIGURE 10.13. Modes of a violin air cavity. Mode frequencies are from Roberts and Rossing (1997).

both the top plate and the back plate, referred to as double octave tuning. Approximate suggested frequencies are given by Hutchins (1987):

X-mode(#2)	O-mode(#5)	Type of instrument
165	330	Violin for soft-bowing players and students
170	340	Violin for chamber music and orchestral players
180–185	360–370	Violin for soloists of exceptional bowing technique
115–125	230–250	Violas from $16\frac{1}{4}$ to $17\frac{1}{4}$ in. in length
60–65	120–130	Cellos

In a cello, mode #1 can be tuned an octave below mode #2 in both plates, but in violins and violas the interval should be less than an octave in the back plate.

An interesting study on the relationship between the free plate mode frequencies and those of an assembled violin was made by Alonso Moral (1984). Three top plates, three back plates, and three sets of ribs, with different stiffness, were used to assemble twelve different violins for testing.

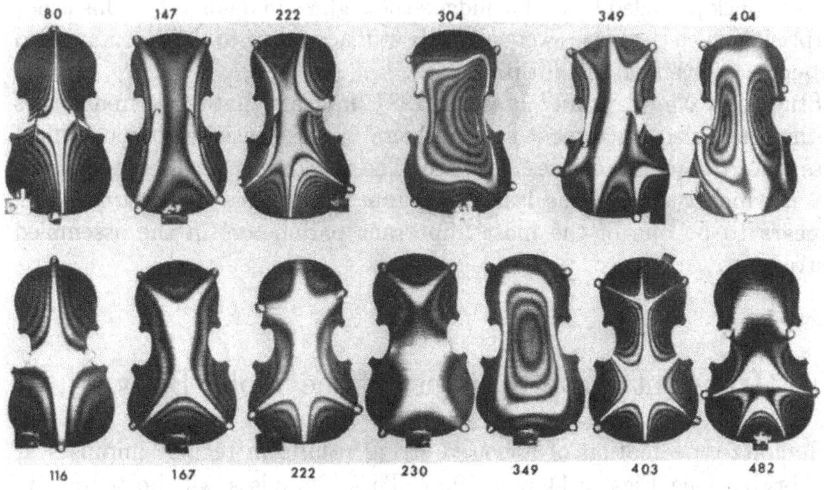

FIGURE 10.14. Time-average holographic interferograms of a free violin top plate and back plate (Hutchins et al., 1971).

The plates and ribs were rated as pliant (P), normal (N), or resistive (R). The masses of the top plates varied from 67 to 93 g, and those of the back plates from 115 to 147 g. The mode frequencies (modes #2 and #5) in the top plates were 189 Hz ± 11% and 388 Hz ± 11%; in the back plates, they were 194 Hz ± 11% and 385 Hz ± 10%. The frequency ratios of the two modes were close to 2 in each plate, and the plates were not varnished.

Alonso Moral found that the A_0 and T_1 modes were most strongly influenced by the properties of the top plate, the C_4 mode was most strongly influenced by the back plate, and the C_3 mode was equally influenced by both the top and back plates. More specifically:

1. The level of the A_0 resonance is very sensitive to the level of the free top plate resonance, but the frequency of the A_0 resonance is insensitive to the top and back plate frequencies.
2. The frequency and level of the T_1 resonance are very sensitive to the frequency and level of the top plate resonance.
3. The frequency shift in the C_3 resonance is about 20% of the shifts in frequencies of the top and/or back plate. The level increase is about 50% of the level increase in the top plate resonance, but decreases by about 40% of the level increase in the back plate.
4. The frequency shift in the C_4 resonance is about 40% of the shifts in frequency in the top and/or back plates, but the level is more sensitive to a shift in the level of the back plate resonance.

Alonso Moral found that combinations of the pliant top with a pliant or normal back plate led to violins judged the highest in quality, but this is not surprising since the plates were all fairly stiff according to the recommended frequencies given earlier in this section.

Studies by Niewczyk and Jansson (1987) indicate that mode frequencies in the assembled instrument are considerably less sensitive to plate thickness than the modes of the free plate. The thickness near the edge of the plate, which is one of the least important parameters in the free plate, appears to be one of the most important parameters in the assembled instrument.

10.6 Transient Wave Response of the Violin Body

Helmholtz-type motion of a bowed string results in regular impulses at the bridge (see Figs. 2.11 and 10.2). These impulses will be mainly in a direction parallel to the top plate but, depending upon the manner of bowing, they may also have a substantial component perpendicular to the plate as well. The transient wave response of the violin body to these impulses is important to understanding the transient sound radiation by the violin.

By means of double-pulse holographic interferometer, Molin et al. (1990, 1991) have recorded the propagating wave field in a violin top and back plates during the first half millisecond or so after an impulse is applied to the bridge. Figure 10.15 shows interferograms of the top and back plates for an impulse parallel to the top plate, and Fig. 10.16 shows interferograms of the top plate for an impulse perpendicular to the top plate. In the case of the parallel excitation (Fig. 10.15), waves of opposite phase propagate from the feet of the bridge in the top plate and from the soundpost in the back plate. In the perpendicular case, the waves propagating from the bridge feet have the same phase, leading to a different pattern when they meet, as seen in Fig. 10.16(c,d).

10.7 Soundpost and Bass Bar

Although a violin, viewed from the outside, appears to be symmetrical, its vibrations are quite asymmetrical due to the soundpost and the bassbar, as observed by Savart (1840) and many other later investigators. The strings are bowed mainly parallel to the top plate, thus inducing a rocking motion of the bridge. Below the first mechanical bridge resonance (about 2500 Hz), the downward forces exerted on the top plate by the bridge feet are nearly

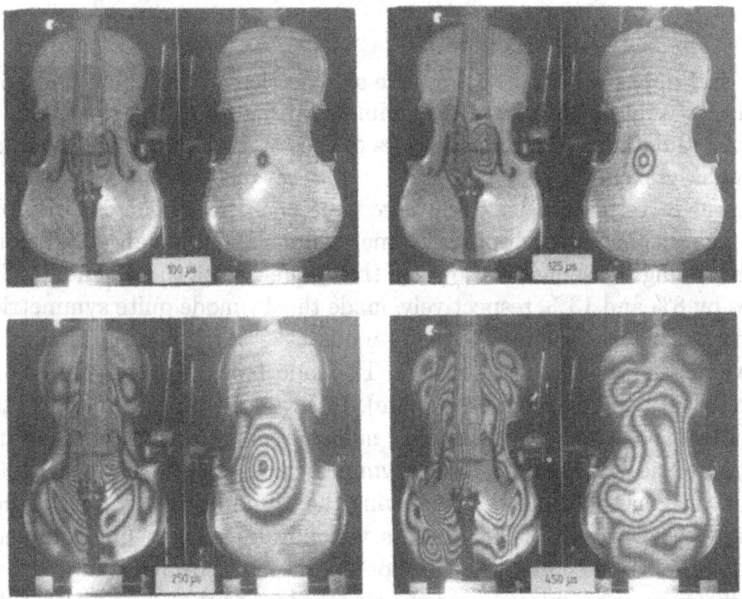

FIGURE 10.15. Interferograms of the top and back plates of a violin at 100 μs, 125 μs, 250 μs, and 450 μs after application of a bridge impulse parallel to the top plate. Note the wave propagation in the top plate is outward from both bridge feet and in the back plate it is outward from the soundpost (Molin et al., 1990).

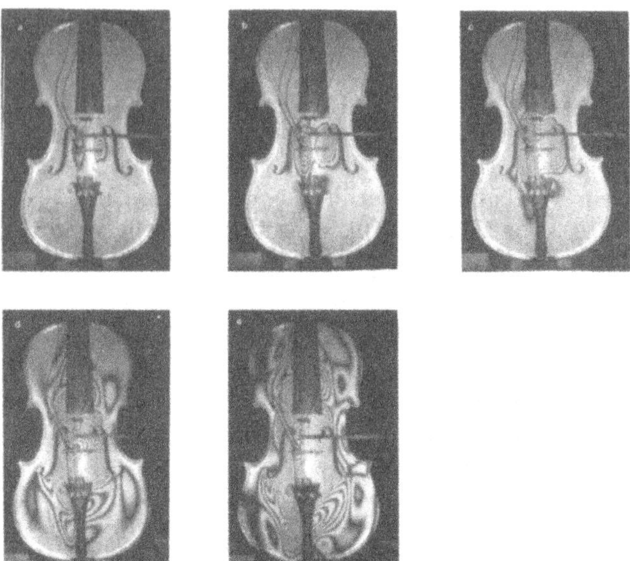

FIGURE 10.16. Interferograms of a violin top plate at 70 μs, 120 μs, 150 μs, 230 μs, and 300 μs after application of a bridge impulse perpendicular to the top plate. Note that the waves propagating from the two bridge feet are in phase. (Molin et al., 1991).

equal and opposite. If the violin were symmetrical, these equal and opposite forces would excite mainly asymmetrical modes of the top plate; the strongly radiating symmetrical modes of the top and back plates would be only weakly excited.

Several investigators have compared the acoustical behavior of violins with and without a soundpost. In one violin, Saldner et al. (1996) found that removing the soundpost lowered the frequencies of the A_0 and T_1 (B_1^-) modes by 8% and 13%, respectively, made the T_1 mode quite symmetrical, raised the frequency of the C_3 (B_1^+) mode by 2%, left the A_1 internal air mode unchanged, and introduced a T_2 mode (consisting largely of antisymmetrical bending of the top plate). In a violin with a soundpost, they concluded, the strongly radiating T_1 mode appears to be a combination of the symmetrical T_1 mode and the asymmetrical T_2 mode. Bissinger (1995), on the other hand, found that removing the soundpost left the frequencies of the T_1 (B_1^-) and C_3 (B_1^+) modes virtually unchanged but completely changed the mode shapes. He also noted a dramatic decrease in the radiation efficiency, as calculated from the observed mode shapes using boundary element methods, in the 500–800 Hz range.

Performers and violinmakers know that moving the soundpost a small distance can cause rather dramatic changes in the sound of a violin. Saldner et al. (1996) compared several modes of a violin with the soundpost in its normal position to those of the same violin with the soundpost relocated about 10 mm closer to the centerline of the violin. With the soundpost nearer the centerline, the T_1 mode was raised about 9% in frequency and the bridge mobility (input admittance) level was increased about 15 dB at the frequency of the C_3 mode, although its frequency was not changed.

Some studies indicate longitudinal vibration and "rocking" motion of the soundpost. Modal analysis results by Marshall (1985) show modes in which the two ends of the soundpost move in opposite directions, although the frequencies are far below the first longitudinal mode of the soundpost itself. Fang and Rodgers (1992) have observed strong rocking motion of the soundpost at about 8000 Hz.

The main function of the bassbar is to stiffen the top plate, both statically and dynamically. The bassbar helps to distribute the downward force of the bridge due to string tension, and it also raises the frequency of vibrational modes of the top plate. Most old violins (made before about 1800) have had the bassbars replaced with stiffer ones in order to accommodate the greater string tension being used today.

Changes in top plate modes have been tracked through several stages of construction (Jansson, et al., 1970; Schleske, 1996). These studies show that the top plate modes change rather little with the installation of the bassbar.

10.8 The Bridge

The primary role of the bridge is to transform the motion of the vibrating strings into periodic driving forces applied by its two feet to the top plate of the instrument. However, generations of luthiers have discovered that shaping the bridge is a convenient and effective way to alter the frequency response curve of the instrument. As Savart (1840) described it:

> If we take a piece of wood cut like a bridge and glue it on to a violin, the instrument will have almost no sound; it begins to improve if feet are formed on the bridge; if we make two lateral slots, the quality of sound takes on increasing value as we cut the bridge completely to the usual form. It is astonishing that by feeling our way we have arrived at the shape currently used, which seems to be the best of all that could be adopted.

Careful experiments by Minnaert and Vlam (1937) confirmed early observations by Raman and others that the bridge has a large number of vibration modes, and that these vibrations are not confined to the plane of the bridge. By attaching mirrors to both the broad and narrow sides of the bridge and observing the deflection of light beams during bowing, they were able to identify separately longitudinal, flexural, and torsional vibrations.

Bladier (1960) and Steinkopf (1963) studied the frequency response functions of cello bridges. Bladier concluded that the bridge acts as an acoustical lever or amplifier with a gain of about 2 (6 dB) between 66 and 660 Hz, decreasing to 1 or below above 660 Hz. Steinkopf modeled the cello bridge as a three-port device, having one input port (the vibrating string) and two output ports (the feet). He observed a major resonance at 1000–1500 Hz in various bridges, and showed that below this resonance the bridge acts like a stiff spring with a reactance varying as $1/f$.

Reinecke (1973) attached piezoelectric force sensors at the string notch and under both feet. In addition, he measured velocity at the upper edge of the bridge with a capacitive sensor, so that he was able to determine both the input impedance and the force transfer functions. He observed resonances around 1000 and 2000 Hz in cello bridges and around 3000 and 6000 Hz in violin bridges standing on a rigid support. The upper curve in Fig. 10.17 shows the input impedance (F_t/v_t), and the lower curve the force transformation functions from the string notch to the two feet (F_s/F_t and F_b/F_t).

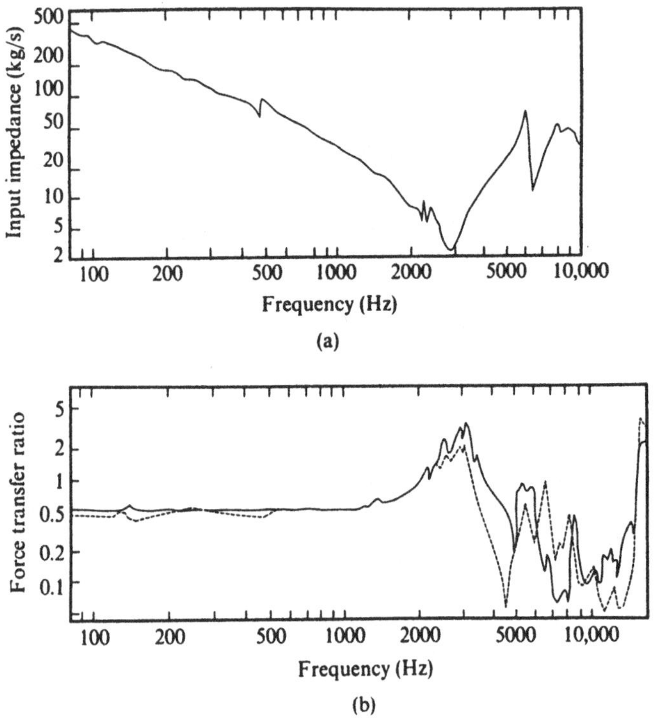

FIGURE 10.17. Frequency response of a violin bridge. (a) Input impedance and (b) force transfer from input (string notch) to outputs (bridge feet); solid curve is soundpost foot and dashed curve is bass bar foot (after Reinecke, 1973).

Note that the force transfer functions have a constant value 0.5 up to about 1000 Hz; equal and opposite forces appear at the two feet. At higher frequencies, the forces are uneven; at some frequencies, the force on the soundpost foot predominates; at some, it is the bass bar foot. Furthermore, Fig. 10.17 shows that the mobility (ratio of velocity to force) is different on the two sides. At any rate, Helmholtz's statement that the bass bar foot alone sets the top plate into motion does not seem to be correct. At the bridge resonances, the impedance reaches a minimum and the force transfer functions have maximum values.

Holographic interferometry provides an effective way to determine the vibrational modes of a bridge. The motions of a violin and a cello bridge at their two main resonances are shown in Fig. 10.18. The motion of the violin bridge at its first resonance is mainly rotational, while in the cello it is mainly bending of its slender legs. Cremer (1984) picturesquely compares their legs to those of a Dachshund and a Great Dane, respectively. Note that the second resonance of the violin mode is excited by a vertical force, and that is why it is less prominent in Fig. 10.17.

In order to darken the sound of a string instrument, the player attaches a mute to the bridge. The main effect of this additional mass is to shift the bridge resonances to lower frequency. Conversely, the resonance frequencies may be raised by additional elasticity in the form of tiny wedges pushed into the slits (Müller, 1979). The effect on the force transfer function is illustrated in Fig. 10.19.

The change in the first resonance frequency with added mass is found to follow a simple mass/spring oscillator model. Hacklinger (1978) determined the stiffness in one violin bridge to be 166,000 N/m and from the resonance frequency, calculated an effective mass of 0.52 g. Adding mutes with masses of 0.32 g and 0.86 g lowered the main resonance from 2850 Hz to 2240 and

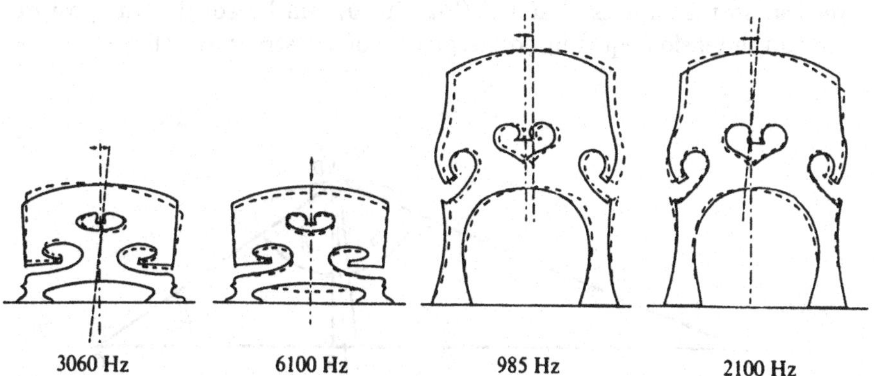

FIGURE 10.18. First two vibrational modes of a violin bridge (left) and a cello bridge (right) (after Reinecke, 1973).

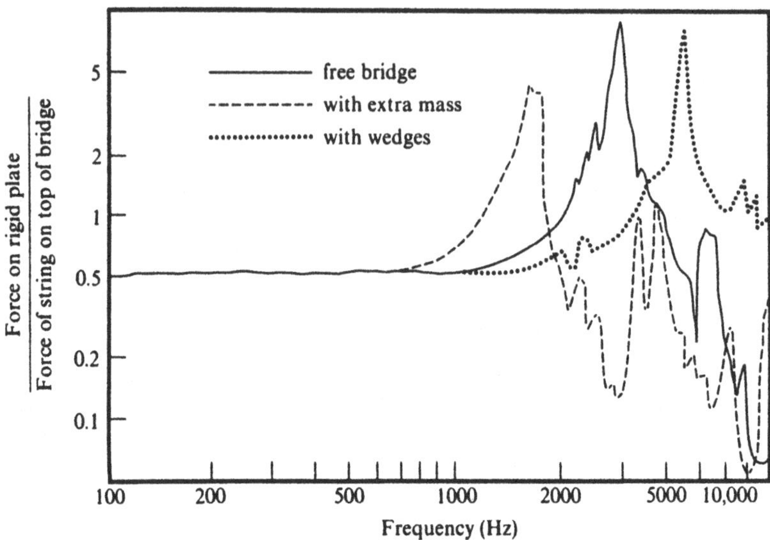

FIGURE 10.19. Force transfer function of the bridge with additional mass (1.5 g) and additional stiffness (data of Reinecke cited in Müller, 1979).

1745 Hz, respectively. Other bridges ranged in stiffness from 160,000 to 950,000 N/m.

For aesthetic reasons bridges are generally carved quite symmetrically. However, Hacklinger (1979) suggests that the treble side should probably be made thicker (stiffer) than the bass side. Because of greater tension in the higher strings, the static force on the soundpost foot is roughly twice that on the bass bar foot. Furthermore, the E string vibrates nearly parallel to the top plate, whereas the G string makes roughly a 50° angle.

Hacklinger (1980) also recommends tilting the bridge slightly away from vertical so that angles α and β (Fig. 10.20) will be equal. This posture makes the tension equal in both segments of the string as well.

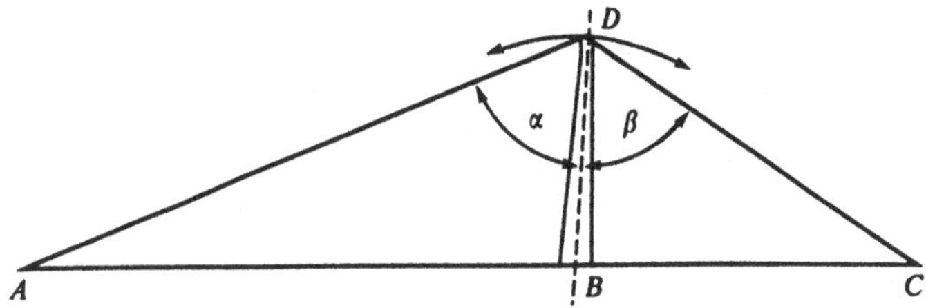

FIGURE 10.20. Tilting the bridge slightly so that $\alpha = \beta$ will result in equal tension in both string segments (Hacklinger, 1980).

10.9 Sound Radiation

A complete description of the sound radiated by a musical instrument should include information about the radiated intensity as a function of both frequency and direction. Ordinarily, measurements should be made in an anechoic environment, although there are ways to determine sound intensity in a non-anechoic environment by using two or more microphones. Sound spectra of violins and other musical instruments in the literature often do not adequately describe the location of the environment(s) of the microphones or the manner and level of excitation.

A graph of sound level versus frequency for a sinusoidally driven violin is shown in Fig. 10.21. The sound level is averaged over 6 microphones 1 m from the violin, so it provides a reasonably good indication of the total radiated power. The power level appears to fall off at a rate of about 9 dB/octave.

Another way to measure the sound power of a source is to measure the sound pressure level in a reverberant room. Narrow bands of noise, rather than sine functions, are used as drive signals in order to average out the resonances of the room. This method was used to obtain the power spectrum in Fig. 10.22(a); the bridge input admittance for the same instrument is shown in Fig. 10.22(b). The drive force was kept constant. Note that the air resonance (just below 300 Hz) is more prominent in the radiation curve than in the admittance curve.

The directional characteristics (in the plane of the bridge) of two violins driven sinusoidally at the bridge are shown in Fig. 10.23. Note the rather dramatic changes that can occur when the frequency is changed by a small amount near a resonance. At higher frequencies, the sound radiation field becomes more directional.

Weinreich (1985) has proposed the term *radiativity* to characterize the ratio of radiated sound field to the force of the vibrating string on the bridge. In order to express the radiativity in terms of a reasonable number of parameters, the string force can be described by its components parallel

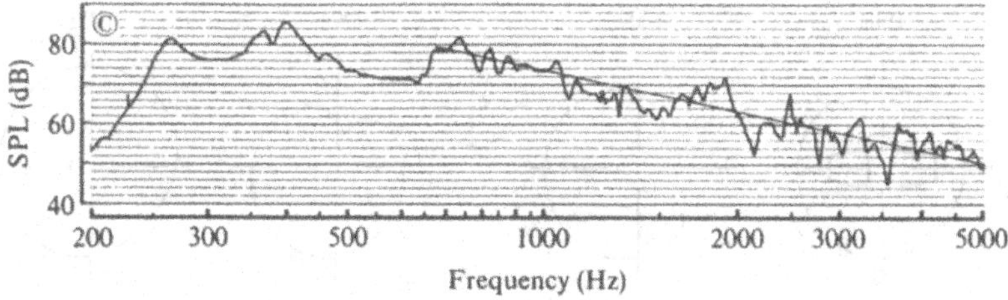

FIGURE 10.21. Sound pressure level averaged over six microphones 1 m away from a sinusoidally driven violin. Straight line has a slope of -9 dB/octave (Jansson et al., 1986).

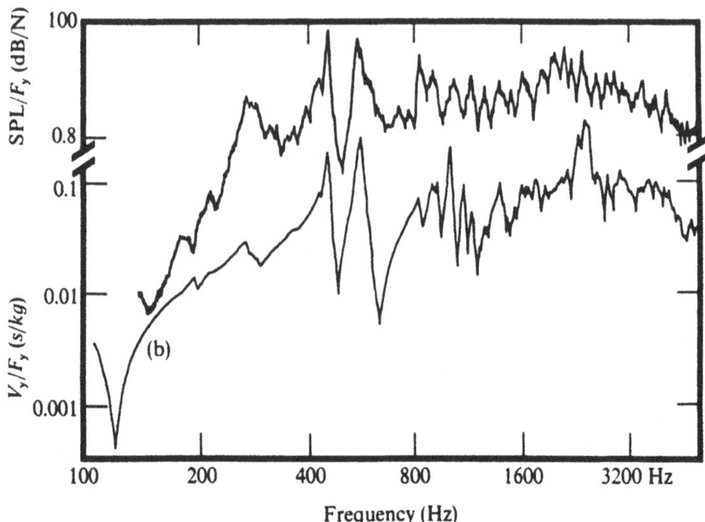

FIGURE 10.22. (a) Sound pressure level of a violin in a reverberant room and (b) bridge input admittance for the same violin on a logarithmic scale (Beldie, 1974).

and perpendicular to the violin top, and the sound field can be expanded in spherical harmonics, giving rise to multipole terms in the resulting series (see Section 7.1). Thus the "monopole radiativity" expresses the radiative monopole moment developed per unit force applied to the bridge at the bridge frequency in question (the two components of force being specified separately).

Thus far, we have discussed the motion of bowed strings, the way in which the bridge transmits the string force to the top plate, and the motion of various parts of the body. The final link in the chain is a discussion of how the moving parts radiate sound. Throughout the discussions, we have described various parts of the violin by transfer functions, which re-

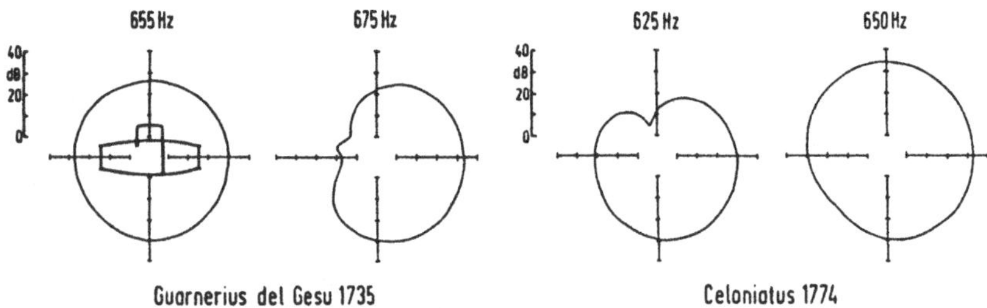

FIGURE 10.23. Directional characteristics of two violins driven sinusoidally at the bridge (Meyer, 1975).

late output to input. Transfer functions found throughout the literature have been expressions of mobility or admittance (velocity divided by force) or of force transfer (output force divided by input force) as functions of frequency. Another useful transfer function might be radiated sound power divided by bridge force (or top plate velocity at the bridge). This would not give us information about the directivity of the sound radiation, however.

10.9.1 Multipole Expansion of the Sound Field

In Chapter 7, it was shown that the sound field radiated by a complex source can be approximated by combining the fields of two or more simple sources (monopoles, dipoles, quadrupoles), appropriately phased. Mathematically, this amounts to a series expansion of the sound pressure in functions having the proper spatial dependence.

The monopole radiativity of two very simple violin models is shown in Fig. 10.24. The first model has no f-holes and its body has a single resonance. At very low frequencies, the radiativity is determined by the (static) stiffness of the wood plus the enclosed air. The radiativity increases as the resonance is approached, and above the resonance frequency it changes sign. The second model adds a single f-hole. At low frequency, air is "pumped" in and out of the f-hole, so that there is no net volume change, hence no monopole radiativity. As the air resonance is approached, the volume motion of the air becomes greater than that of the shell (assuming the air resonance is lower in frequency than the wood resonance), and the monopole radiativity has the opposite sign from the one due to motion of the wood plates alone.

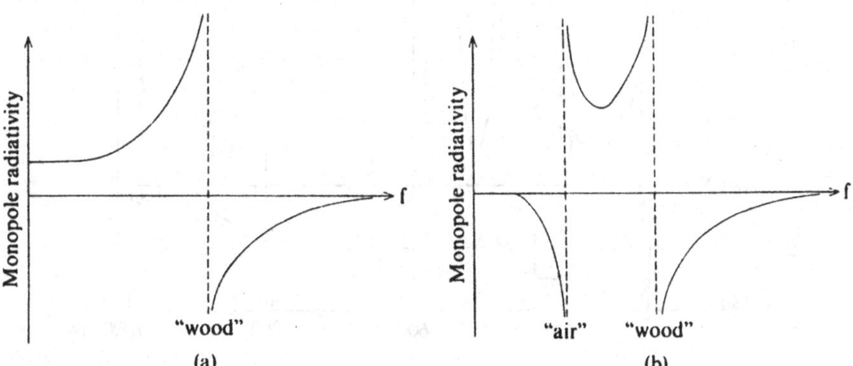

FIGURE 10.24. Monopole radiativity of two very simple violin models. (a) Model with no f-holes has only a single body resonance. (b) Model with an f-hole adds an air resonance (Weinreich, 1985).

It is interesting to note that the dipole radiativity, though small, remains finite as the frequency goes to zero, since the volume displaced by the moving plate is equal and opposite to that pumped in and out of the f-hole. If the model had two f-holes, symmetrically located with respect to the anti-node of the wood vibration, the dipole moment would go to zero but the (linear) quadrupole moment would remain finite. This is of little practical value, however, because dipole and quadrupole radiation are very weak at low frequency.

The monopole radiativity of a violin of medium quality is shown in Fig. 10.25. The solid curves give the magnitude and phase of the radiativity when a sinusoidal force is applied parallel to the violin top, and the dotted curve gives the magnitude of the radiativity for a force perpendicular to the top. Note that the phase advances 180° at each resonance and retreats 180° at each antiresonance (compare Fig. 4.7); ignore jumps of 360°. These graphs are based on a definition of the monopole moment as a volume velocity, and hence the in-phase motion of the plate and air between the air and wood resonances (300 to 450 Hz) is characterized by a phase of −90° rather than 0. Above 450 Hz is a cluster of resonances that collectively act as the "main body resonance," above which the motion of the top plate is governed by inertia rather than elasticity (Weinreich, 1985). The narrow resonances of the strings (294, 392, 440, 588, 660, 784 Hz, etc.) cause no net phase change.

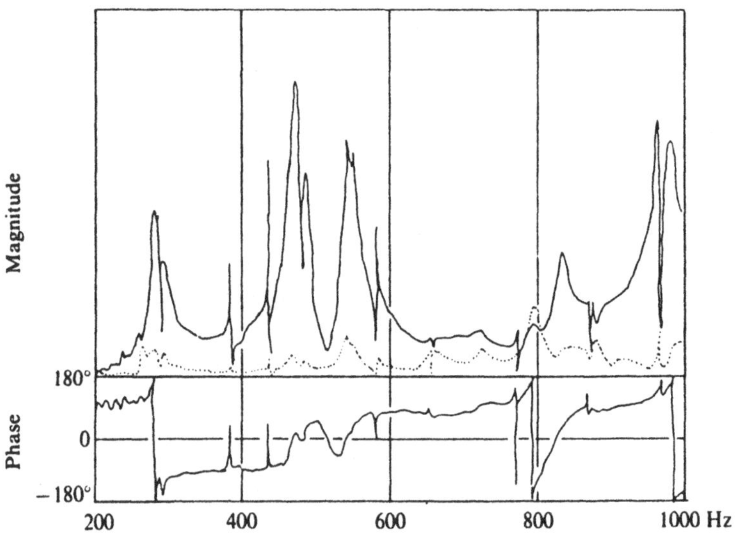

FIGURE 10.25. Monopole radiativity of a violin at low frequency. Solid curve is for force parallel to the top plate; dotted curve is for a perpendicular force (Weinreich, 1985).

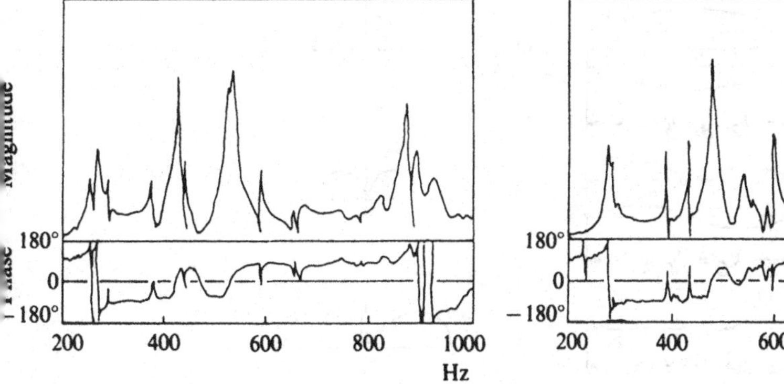

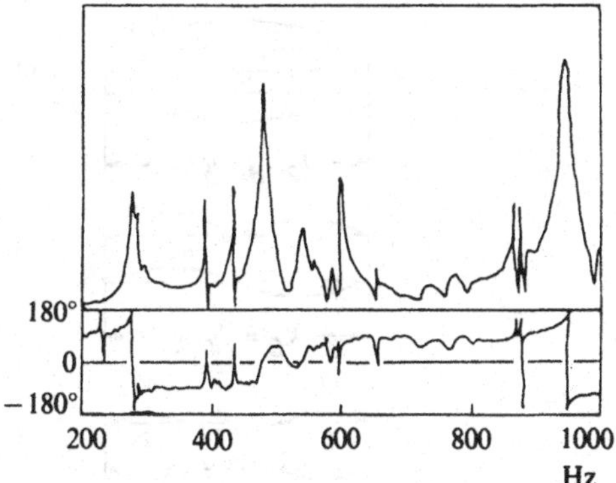

FIGURE 10.26. Monopole radiativity curves of violins by Vuillaume (left) and C. Hutchins (right). Scale is the same as in Fig. 10.25 (Weinreich, 1985).

Monopole radiativity curves of violins by Vuillaume (1862) and by Hutchins are shown in Fig. 10.26. They are quite similar to the one in Fig. 10.25.

10.9.2 Critical Frequency

Since the speed of bending waves in a plate increases with the square root of frequency (see Chapter 3), it will catch up to the speed of sound (in air) at some frequency that depends upon the elastic properties of the plate. This is called the critical frequency or coincidence frequency. The dramatic increase in radiation by a vibrating plate that occurs above the critical frequency was discussed in Chapter 7.

In an infinite plate, flexural waves above the critical frequency radiate at an angle θ given by

$$\theta = \sin^{-1}\left[\frac{c}{v(f)}\right], \tag{10.5}$$

where c is the speed of sound (in air) and $v(f)$ is the speed of bending waves in the plate. This angle θ is similar to the Mach angle for shock waves radiated by a supersonic airplane. Thus, the radiation by the supersonic flexural wave is nearly parallel to the plate when $v(f)$ is only slightly greater than c, but it occurs at greater angles for $v(f) \gg c$.

The radiation behavior of finite plates is somewhat more complicated. The directional characteristics of a symmetrically oscillating two-plate system at various $c/v(f)$ ratios is shown in Fig. 10.27. The number n denotes the number of half bending waves on the plate, so the frequency is proportional to n^2. The $n = 4$ case, where $\lambda_B = \lambda_L$, represents the critical frequency; radiation parallel to the plate ($\theta = 0$) is maximum in this case.

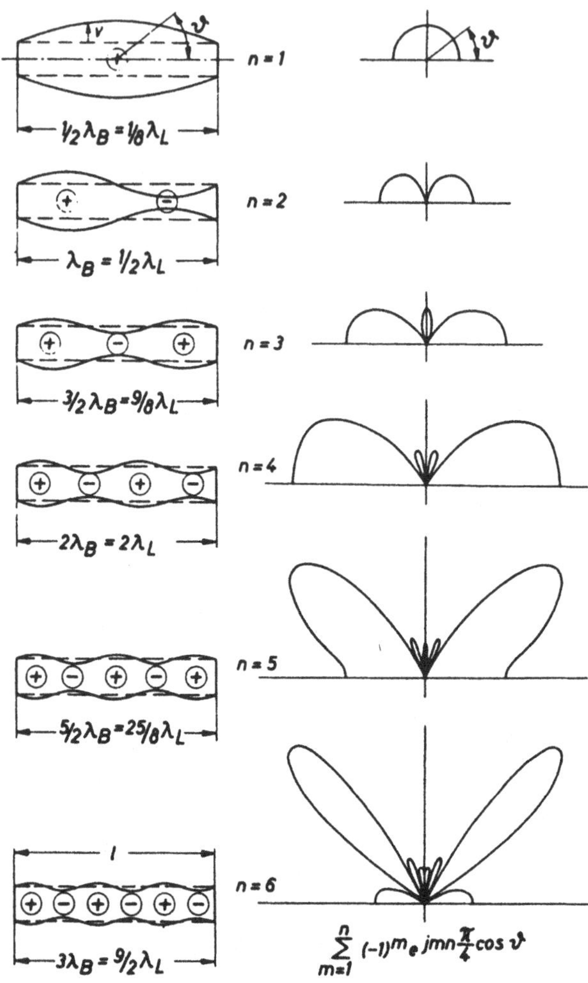

FIGURE 10.27. Directional characteristics of a symmetrically oscillating two-plate system with varying numbers of half bending waves n. Frequency is proportional to n^2; $n = 4$ represents the critical frequency (Cremer, 1984).

As n increases, the radiation pattern approaches that of an infinite plate with maxima at θ, as given in Eq. (10.5).

The critical frequency is difficult to calculate for violin plates. By comparison to a flat plate with approximately the dimensions of a violin, Cremer (1984) estimates critical frequencies of 4870 Hz along the grain and 18,420 Hz across the grain for a violin plate 2.5 mm thick. In a 4.4-mm-thick cello plate, however, the critical frequency for bending waves along the grain would be only 2800 Hz.

10.9.3 Sound Radiation from Bowed Violins

In most studies of sound radiation, a sinusoidal force has been applied to the bridge. When the violin is bowed, however, the complex driving force (whose waveform approximates a sawtooth wave) is rich in harmonic partials. Quite a different power spectrum is then obtained. The fundamental is generally the strongest partial; between the fundamental and the bridge resonance (2500–3000 Hz), the partials decrease about 6 dB/octave because of the sawtooth character of the bow excitation. Above 3000 Hz, the spectrum has a slope of roughly -15 dB/octave because of the sawtooth excitation (-6 dB/octave) combined with the sinusoidal frequency response curve (-9 dB/octave; see Fig. 10.21).

Saunders (1937) used total intensity curves to compare different violins. To record these curves, the instruments were played as loudly as possible without allowing the tone to break, and the sound intensity level was recorded. Total intensity curves for five Stradivarius violins are shown in Fig. 10.28. Hutchins refers to curves employing the hand-bowing technique,

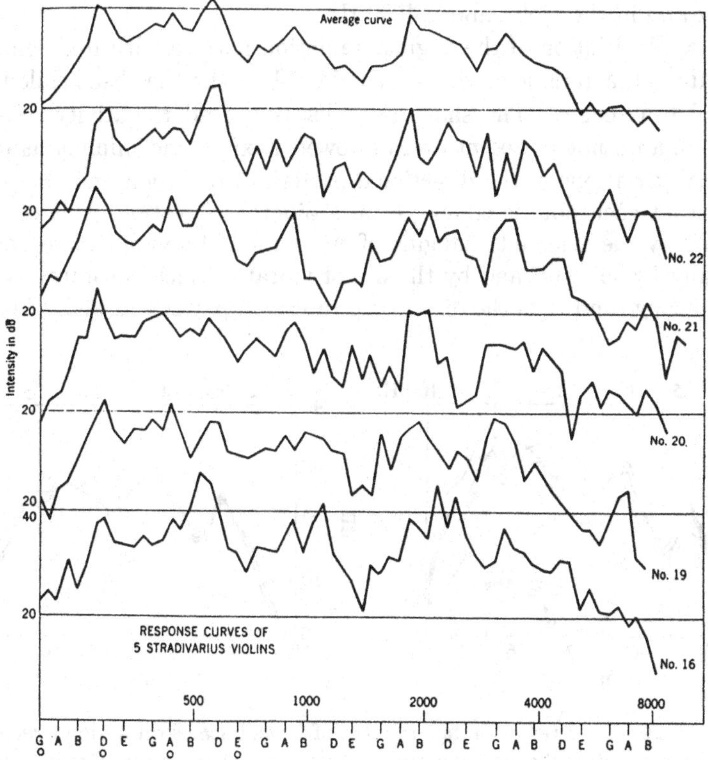

FIGURE 10.28. Total intensity curves for five Stradivarius violins bowed one note at a time (Saunders, 1937).

in which each note is bowed loudly with fast vigorous bow strokes, as Saunders loudness curves.

Another useful technique for analyzing the sound of bowed instruments is to compute long-term-average spectra (LTAS). Scales played on the violin are tape recorded and analyzed by means of 23 filters whose center frequencies and bandwidths approximate the critical bands of hearing (alternatively, the spectrum analysis may be done on a digital computer programmed to simulate the 23 filters). If the directional characteristics of the radiated sound are desired, the recording should be done in an anechoic room; if the total sound power is desired, a reverberation chamber is preferred. Both environments are illustrated in Fig. 10.29.

10.9.4 Frequency Dependence of Directivity

Directional radiation patterns observed by Meyer (1972) for the cello (vertical and horizontal planes) and the violin (horizontal plane) are shown in Fig. 10.30. The shaded areas show the directions in which the radiated sound is within 3 dB of its maximum value averaged over that frequency range. There is some evidence of a critical frequency effect in the cello above 2000 Hz and in the violin above 4500 Hz.

The rapid variation of the angular radiation pattern with frequency, an important characteristic of violin tone, describes what has been called "directional tone color." The sharp and closely spaced radiativity maxima and minima do not represent overall power maxima and minima as much as radical variations in the directional pattern of radiation and create the illusion that each note is coming from a different direction. This effect is enhanced by the constant changing of orientation of a violin while bowing (especially by soloists) and by the use of vibrato, which constantly varies the frequency (and thus the directional radiation pattern as well). This ef-

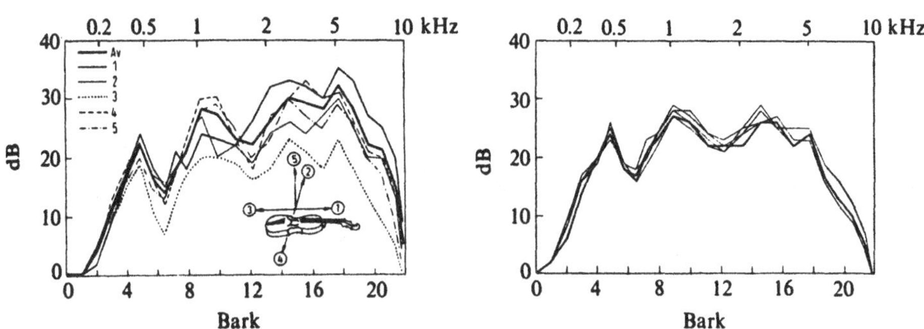

FIGURE 10.29. Long-term-average spectra of bowed scales on a violin recorded at five different microphones in an anechoic room (left) and a reverberation room (right). The sound field in the reverberation room is relatively independent of microphone position. "Bark" corresponds to the number of the critical band (Jansson, 1976).

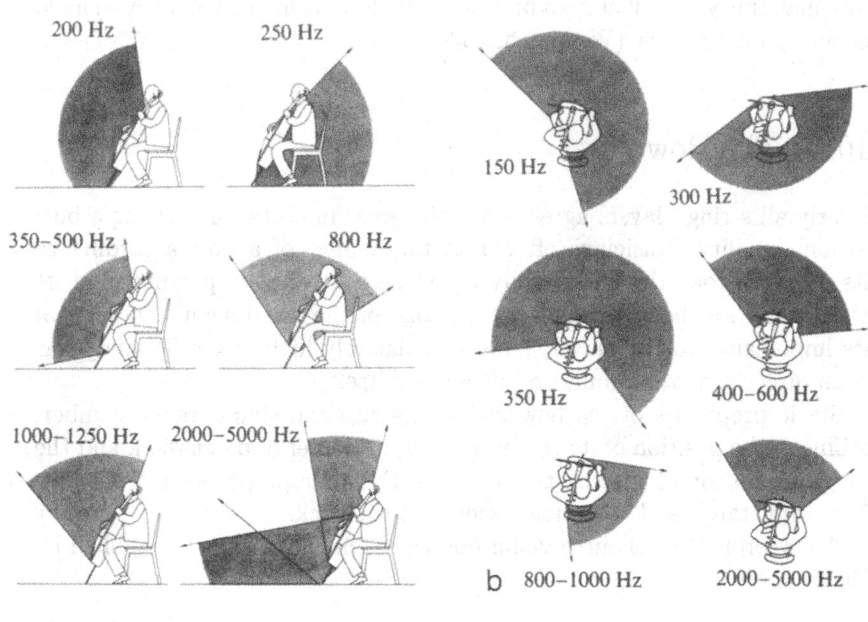

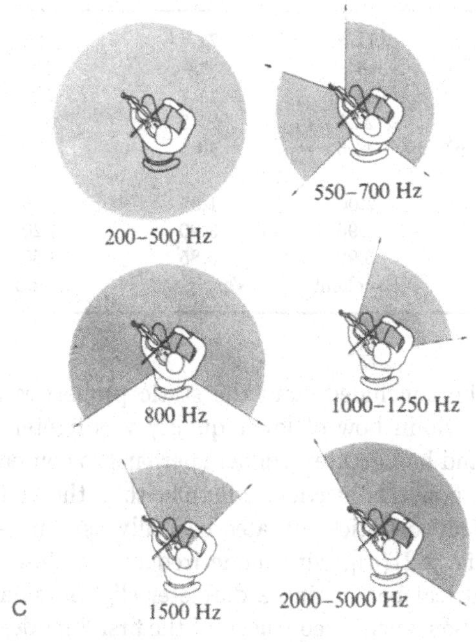

FIGURE 10.30. Principal radiation directions for the cello and violin. (a) cello (vertical plane); (b) cello (horizontal plane); (c) violin (horizontal plane). Shaded areas are the 3 dB limits (Meyer, 1972).

fect makes it very difficult to produce a realistic violin sound with a single ordinary loudspeaker (Weinreich, 1994).

10.10 The Bow

Nearly all string players agree about the great importance of using a bow of high quality. Musicians often rate the quality of a bow according to its playing properties (how easily it can be controlled in playing) and its tonal qualities (how it influences the tone of the instrument). In spite of its importance to the player, however, relatively little scientific study has been directed at the properties of the bow itself.

Static properties of the bow include its size and shape, mass, camber, stiffness, the position of its center of mass, its center of percussion, and the longitudinal compliance of its bow hairs. Dynamic properties include bow hair admittance and vibrational modes of the stick.

A few properties of three violin bows and one viola bow are supplied by Reder (1970):

Property	Violin bow #1	Violin bow #2	Violin bow #3	Viola bow
Stick length (cm)	72.8	72.8	72.8	71.9
Mass (g)	64	58	53	67.5
Camber (mm)	17.7	17.7	20	18.6
Center of mass (cm)	24.8	25.2	23.95	25.0
Center of percussion (cm)	44.5	44.1	45.0	43.7
Deflection (mm)				
0.2-kg load	2.00	1.95	2.10	1.50
0.4-kg load	3.95	3.90	4.20	3.00
0.6-kg load	5.90	5.85	6.50	4.80
Playing quality	Excellent	Very good	Good	Excellent

If anything can be concluded from the static properties listed in this table, it is that the violin bow of lower quality was lighter and less stiff than the best bow and had greater camber (perhaps to compensate for less stiffness). The viola bow was heavier and stiffer than the violin bows.

A freely suspended bow stick vibrates basically as a free-free bar [see Section 2.16 and Fig. 2.21(a)], with mode frequencies close to those of a homogeneous cylindrical bar having a diameter slightly smaller than the thinnest part of the bow stick. Frequencies of the first 8 modes are typically 60, 160, 300, 500, 750, 1000, 1300, and 1700 Hz; damping ratios are typically in the range of 0.2–0.6% of critical damping, corresponding to Q values of 250–80 (Askenfelt, 1992). Pernambuco wood has by far the lowest damping of hardwoods used in musical instruments (with the possible exception of

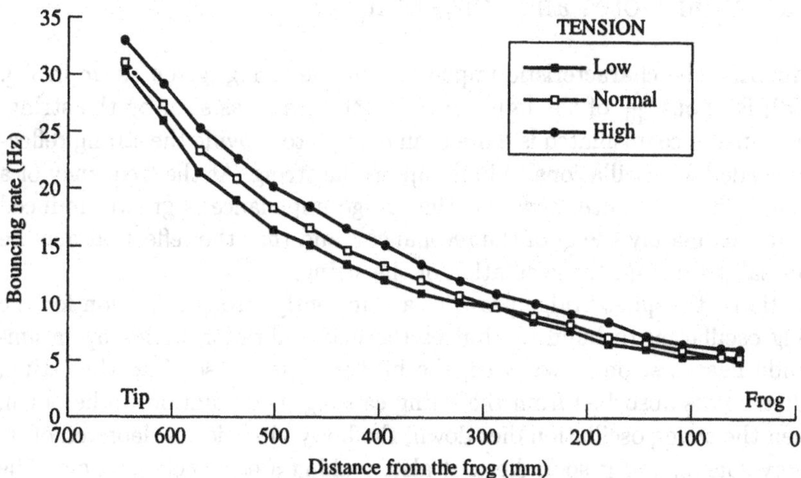

FIGURE 10.31. Measured frequencies of the bouncing bow vs. position of support (frog to tip) for three different bow hair tensions (Askenfelt, 1992).

ebony), and this is probably the reason that it is the preferred wood for violin bows.

Transverse vibrational modes of the assembled bow are only 1 to 7% lower than those of the bow stick alone, but damping ratios are roughly twice as great. An additional transverse mode is found at 60–75 Hz (depending on tension) in the assembled bow, and this is identified as the lowest transverse mode of the bow hair. Bows of high quality generally have a low damping ratio at low frequency, increasing rather markedly with frequency (Askenfelt, 1992). Bissinger (1995) reports similar mode shapes and frequencies in bows, and he compares these to the "bow bounce" frequencies observed when the bow is pivoted at the frog, similar to what a player does in playing spiccato of ricochet. Bow bounce frequencies depend on where the bow contacts the string, of course, but they show surprisingly little dependence on bow hair tension, as shown in Fig. 10.31.

Rapid bouncing bow patterns, such as sautille and ricochet, played at about the middle of the bow, easily generate repeated notes at a rate of about 10 Hz, whereas slower bowing patterns such as spiccato bounce the bow at a frequency much lower than the bouncing bow resonance frequency (Askenfelt, 1992).

The first longitudinal resonance of the bow hair lies between 1500 and 2000 Hz for a 62-cm length, giving a longitudinal wave speed of 1900 to 2500 m/s. From this, a characteristic impedance of 30 kg/s and a tension of 0.3 N on each hair or 60 N for 200 hairs are estimated (Schumacher, 1975). In some preliminary experiments, stick modes were found to have an average spacing of about 300 Hz up to 2000 Hz.

10.11 Wolf Notes and Playability

Ordinarily, the characteristic impedance of the string [given as μc in Eq. (2.28)] is about $\frac{1}{10}$ of the impedance of the bridge, as seen by the string. This impedance mismatch is more than enough to provide the strong reflection needed for oscillations to build up on the string. At the frequency of a strong body resonance, however, the bridge impedance is greatly reduced (by approximately the Q of the resonance), and thus the reflection may be too weak to sustain the oscillation of the string.

Initially, the quiet bridge provides a sufficiently strong reflection for the string oscillation to build up, but as the body vibration builds up in amplitude near a strong resonance, the bridge appears "soft" to the string, and energy is absorbed from the string causing its oscillation to die down. When the string oscillation dies down, the body vibration is deprived of its energy source, and it soon dies out also, making a new cycle possible. The net result is that the string and the body vibrations alternately grow and diminish in amplitude, in somewhat the same fashion as the coupled pendula described in Chapter 4. When this occurs, the player finds it difficult, if not impossible, to maintain a steady sound by bowing. Rather the tone varies strongly and harshly at the exchange frequency, which is typically about 5 Hz. The note at which this unstable situation occurs is called the wolf tone. The alternating amplitudes of the string and body are evident in Fig. 10.32.

There are two reasons why the viola and especially the cello are more susceptible to wolf tone trouble than the violin. One is that the string/body impedance mismatch is less in the larger instruments; that is, their dimensions have not increased in proportion to $1/f$. As Schelleng (1963) described it, "The chronic susceptibility of the cello to this trouble is the price paid for the convenience of a small instrument, small compared to one completely scaled."

The second reason has to do with the bridge height. When the much taller bridge of the cello (see Fig. 10.18) rocks back and forth in response to the vibration of the top plate, it acts as a longer lever in reducing the impedance seen by the string. The reduction in impedance increases as the square of the lever length, which is 2.4 times greater in the cello. Both factors considered, Cremer (1984) estimates that it is 3.4 times easier to generate wolf tones in the cello.

There have been a number of suggestions as to how to suppress wolf tones. Damping the top plate is probably not a good solution, because it would reduce the Q of other resonances as well as the one that causes the wolf tone. One possibility is to couple yet another oscillator with the same resonance frequency but with greater damping than the body resonance. A convenient way to do this is to attach an appropriate mass to the extension of a string between the bridge and tailpiece. A material with high internal damping, such as Plasticine or modeling clay, can be used for the required

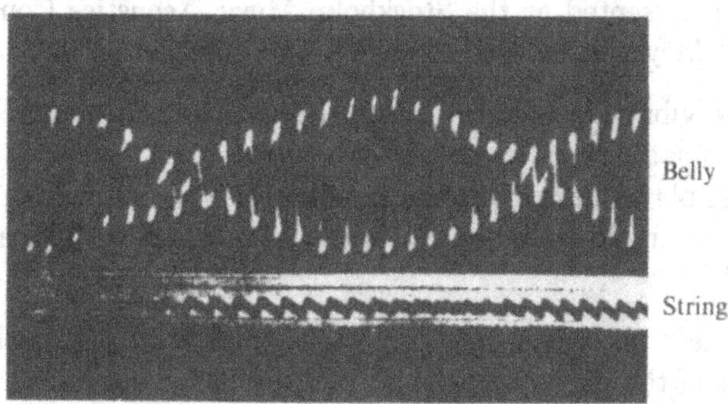

FIGURE 10.32. Oscillograms of the body and string showing alternate growth of amplitude characteristic of a wolf tone (Raman, 1916).

mass. Other dampers attach directly on the body or mount inside the instrument (Güth, 1978). An experienced player can usually suppress a wolf tone by increasing the bowing force or changing the point of bowing, but "an ounce of prevention is worth a pound of cure."

A rather complete discussion of the wolf note phenomenon, as well as an analysis of the playability of violins is given by Woodhouse (1993,1995), who re-examines Schelleng's discussion of the wolf note. For a wolf note to occur, it is necessary that the bow force be below the minimum (see Section 10.4) in such a way that the double-slip mechanism alternates with the normal Helmholtz motion, the transition being triggered by body vibrations.

10.12 Tonal Quality of Violins

A question musical acousticians are often asked by musicians as well as laypersons is "Do you think 'they' will ever discover the secrets of Stradivari?" There is little doubt that instruments crafted by the old Italian masters are highly treasured. Although they have undergone substantial modifications to bring them up to modern pitch standards, most surviving instruments of these masters still have excellent tonal and playing qualities.

The sound quality of a musical instrument depends upon a number of factors. Although most of them are acoustical, relating to the way the instrument vibrates and radiates sound, there are nonacoustical factors that can have a psychological effect on the player as well. If a violin appears to have been made by a master craftsman, it will probably be played accordingly. This is especially true if the player knows of the maker and his reputation. Every string instrument has its own unique playing characteristics, and if you are given the opportunity to play (or to own) a Stradivarius violin, you will no doubt take great care to adapt your playing technique to bring out the best in the instrument.

In a paper presented at the Stockholm Music Acoustics Conference in 1983, Jürgen Meyer attempted to answer two questions:

(1) Does the vibration behavior of old Italian violins differ from that of other violins?
(2) Is it possible to imagine sound qualities of a violin making it appear superior to an old Italian violin, without its losing the typical sound of a violin?

To answer these questions, it is important to try to distinguish between the influence of the player on the instrument and the acoustical properties of the instrument itself.

The main parameters at the player's disposal are related to bowing: bow speed, bow force, and bowing position on the string (and the way in which these relate to each other, especially during the attack). Additional parameters have to do with holding the instrument, correct intonation, and vibrato technique. Although these playing parameters profoundly affect the overall tonal quality, Meyer found that their effect on the sound spectrum is small, being mainly restricted to the slope of the envelope toward high frequencies. Thus, violins need not be bowed by machines in order to compare their playing spectra.

The frequencies of the air resonance (f_0) and the first body resonance (f_1) in 100 violins are compared in Fig. 10.33. Included in this group are 6

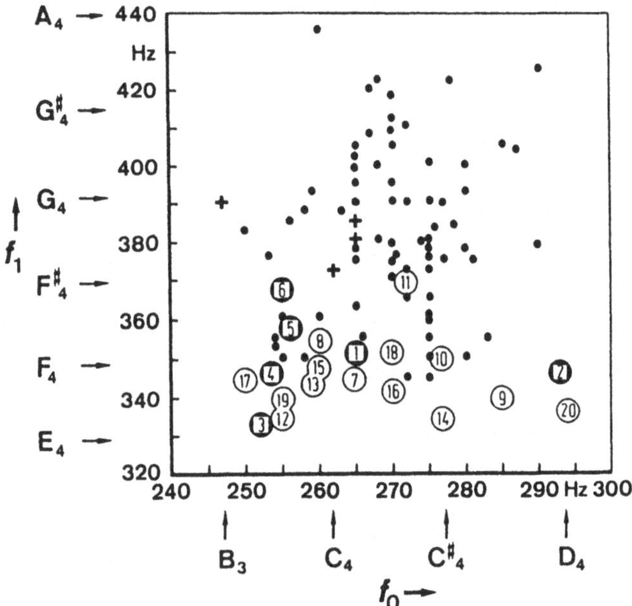

FIGURE 10.33. Tuning of the two lowest resonances of different violins: 1–6 are Stradivarius violins; 7–20 are other old Italian masters (Meyer, 1985).

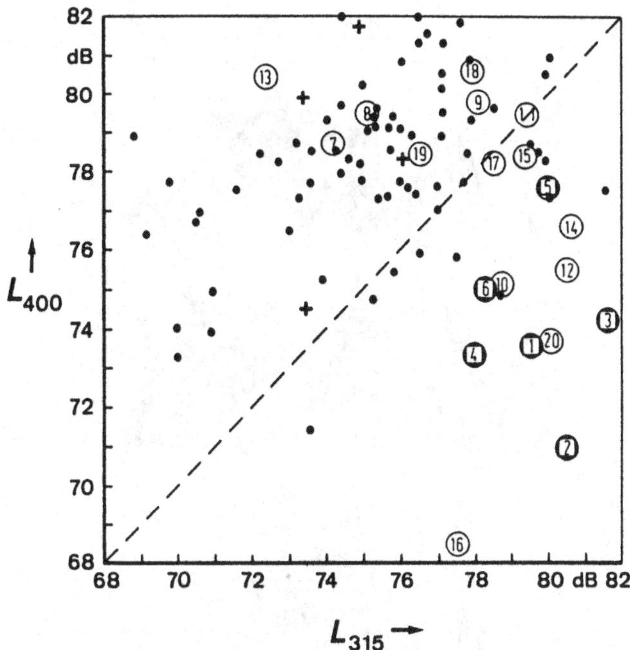

FIGURE 10.34. Averaged level of the response curves of different violins. L_{315}: level of the 315-Hz, third-octave band; L_{400}: level of the 400-Hz, third-octave band, (Meyer, 1985).

violins by Stradivari and 14 by other old Italian masters. One characteristic that stands out is the low tuning of the plate resonance (T_1 in Fig. 10.9) in the old Italian instruments.

Analysis with one-third octave filters provides information about energy distribution. Figure 10.34 shows the levels in two adjacent third-octave bands having center frequencies of 315 and 400 Hz. Each point below the dashed line represents a violin whose level in the 315-Hz band is greater than its level in the 400-Hz band, a characteristic of 13 old Italian instruments (including all 6 Strads), but of very few others.

Response curves of three groups of violins are compared in Fig. 10.35. The old Italian violins are characterized by a broad maximum around 2500 Hz, followed by a marked roll-off toward high frequency. This maximum, which occurs in the range of greatest hearing sensitivity, is quite similar to the singer's formant found in the spectra of most opera singers. (Several other remarkable similarities between the violin and the singing voice were noted by Rakowski, 1985, as well.)

More recently, Dünnwald (1985) has found that two parameters L and N could be used to characterize the sound of old Italian violins. L is the relative sound level at the air resonance frequency, and N is the percentage of tones from an instrument in which the strongest partial lies within the

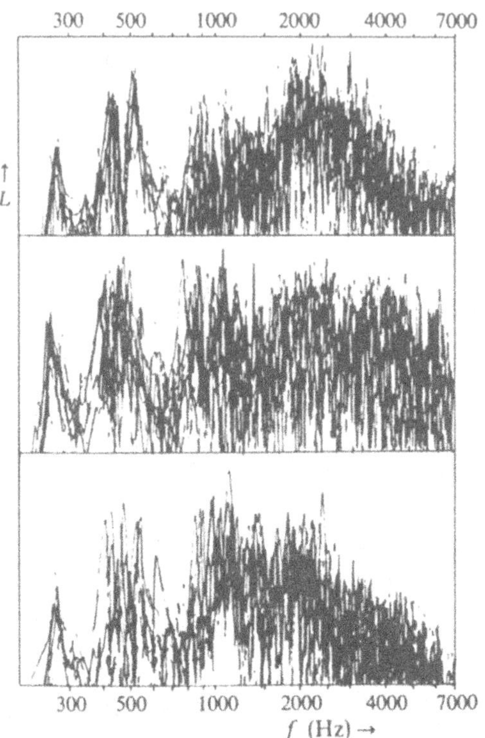

FIGURE 10.35. Response curves of three groups of violins. Top: ten old Italian violins; center: master violins; bottom: factory-made violins (Dünnwald, 1983).

frequency range 1300–2500 Hz. All except one of the old Italian violins lie in one corner of the graph of L versus N in Fig. 10.36.

Hutchins (1993) has found that the frequency spacing Δ between the A_1 cavity (air) mode and the $B_1(C_3)$ body mode is critical to the overall tone and playing qualities of a violin. A violin with a delta of 60–80 Hz is suitable for soloists, one with 40–60 Hz for orchestra players, with 20–40 Hz for chamber music players; a violin with delta below 20 Hz is easy to play but lacks power.

Jansson (1997) finds that violins of high quality tend to have strong peaks in their frequency response curves at about 550 Hz (C_3 mode) and 2500 Hz ("bridge hill").

10.12.1 The Reciprocal Bow

A useful device for evaluating tonal quality is the "reciprocal bow," which uses the principle of reciprocity to hear the sound of a violin without the use of a violinist. A loudspeaker at a typical listening point is supplied with an electrical signal corresponding to the sawtooth motion of a bowed

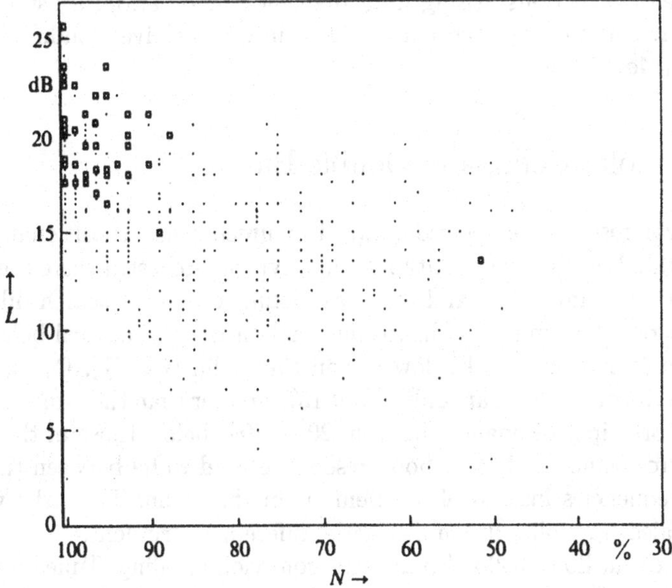

FIGURE 10.36. Quality parameters L and N for 350 violins. (□ = old Italian violins; • = others) (Dünnwald, 1985).

string; the resulting bridge velocity, measured with a phonograph pickup, is heard through headphones or another loudspeaker, thus making audible the sound that the same violin would produce when bowed in an identical way. The electrical signal can also be generated by bowing a violin string on a "silent" violin (without a soundboard). It is relatively easy to compare violins by this method (Weinreich, 1994).

10.12.2 Vibrato

Vibrato is produced on string instruments by a periodic rolling motion of a finger on the string, which produces a frequency modulation. The vibrato frequency is generally between 5 Hz and 8 Hz, about the same as found in vocal vibrato (Small, 1937). The frequency deviation is usually about 25 cents, although it can exceed 35 cents (Meyer 1992). Because of violin body resonances, a frequency deviation generates amplitude modulation in each partial, which may be at the frequency of the finger motion or at twice that frequency, depending upon the frequency of the resonance in relation to the bowed tone. This amplitude modulation may produce sound level changes of 3 to 15 dB.

Vibrato is generally considered to add warmth to the tone of string instruments. Because of the way in which vibrato is produced on the violin, it is generally used only on stopped notes. Blending of strings is generally improved if players synchronize their vibrato. As a result of the time sequence of reflections arising from frequency modulation, a listener in a room may actually experience an increase in the perceived loudness due to vibrato (Meyer 1992).

10.13 Viola, Cello, and Double Bass

Acoustical research on bowed string instruments has traditionally been concentrated on the violin. Although in a typical orchestra there are about one-third as many violas and cellos as violins, one must search diligently to find 1 or 2% as much published material on their acoustical behavior.

The viola is tuned a fifth lower than the violin (C_3, G_3, D_4, and A_4). However, its dimensions are only about 15% greater than those of the violin, and the principal resonances lie from 20 to 40% below those of the violin. The air resonance and main body resonance tend to lie between the open string frequencies instead of on them, as in the violin. Thus, the viola is not a scaled-up violin; it is a distinctly different instrument.

The violoncello or cello also departs from violin scaling. Tuned a twelfth below the violin and an octave below the viola (C_2, G_2, D_3, and A_3), its dimensions are about double those of the violin except for the rib height, which is about four times that of the violin. Its lowest air resonance is close to the open second string frequency, as it is in the violin. The cello modes most resembling the prominent T_1 and C_3 modes in the violin lie near the third and fourth strings (Bynum and Rossing, 1997). String tunings and modal frequencies in the violin, viola, cello, and double bass are compared in Fig. 10.37.

Shapes and frequencies of the A_0, T_1, C_3, and C_4 modes, as observed by electronic TV holography (Bynum and Rossing 1997), are shown in Fig. 10.38. The modes are given the same labels as the corresponding modes in a violin, although that gives the C_4 mode a lower frequency than the C_3 mode. Table 10.1 compares mode frequencies in the cello to the corresponding modes in a violin.

The (0,0) cello air cavity mode or Helmholtz mode was observed at 104 Hz. The modes characterized by longitudinal motion of the internal air occurred at 226 Hz (0,1), 547 Hz (0,2), 742 Hz (0,3), and 1016 Hz (0,4). The modes characterized by transverse air motion in the lower bout were observed at 496 Hz (1,0) and 893 Hz (2,0). The (1,1) mode, characterized by transverse air motion in both bouts (but stronger in the upper bout) was observed at 609 Hz. Other modes are the (1,2) at 820 Hz, the (1,3) at 980 Hz, and the (1,4) at 1129 Hz. With the exception of the (0,0) mode, the frequencies of these modes are about 45% of those of the corresponding violin air cavity modes, which is not surprising since the dimensions of a cello are nearly twice those of a violin. To more nearly match the scaling of the body modes, the (0,0) mode is lowered to 37% of its violin counterpart by appropriate choice of rin height and f-hole area.

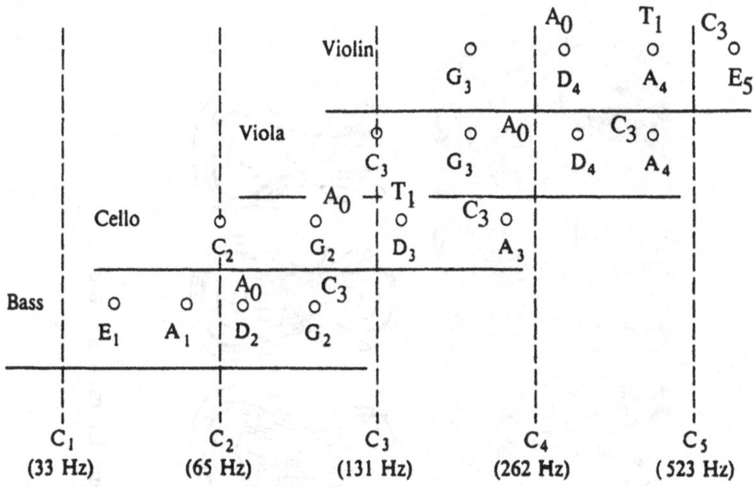

FIGURE 10.37. String tunings and vibrational modes of the violin, viola, cello, and double bass. The open strings are denoted by o.

The double bass (also known as the contrabass or bass viol) still retains features of the viol family: sloping shoulders, flat back, and tuning in fourths rather than fifths (E_1, A_1 D_2, G_2), and many instruments have five instead of four strings. Some players hold the bow in the manner of viol playing, with the hand underneath the stick.

The input admittance (mobility) of a high-quality double bass is shown in Fig. 10.39. The air resonance around 60 Hz and the main body resonance (T_1) at 98 Hz are easily identified. Smaller peaks nearby are probably from modes similar to the C_2, C_3, and C_4 modes in violins. Between 120 and 400 Hz is a series of body resonances, and the bridge resonance lies around 400 Hz. The high-frequency response reaches its maximum around 600 Hz and then rolls off fairly rapidly.

The double bass plays down to E_1 (41 Hz). Notes near the bottom of its range will have two or more partials in each critical band, which will tend to give roughness to its sound. The series of peaks between 120 and 220 Hz in the admittance curve in Fig. 10.39, by acting as a sort of comb filter that de-emphasizes every other partial, may alleviate this roughness somewhat; this may be one of the features that contributes to the high quality of this instrument (Askenfelt, 1982).

10.14 Viols

Viols were popular instruments during the sixteenth and seventeenth centuries, dropped from favor in the eighteenth century, and have enjoyed a

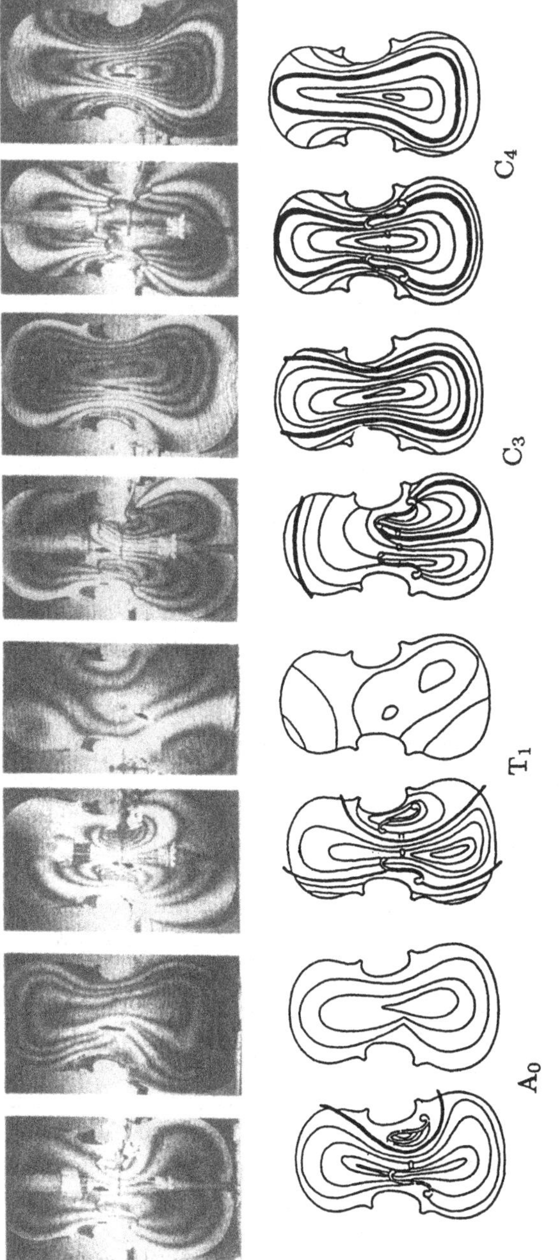

FIGURE 10.38. Shapes and frequencies of the A_0, T_1, C_3 and C_4 modes in a cello (Bynum and Rossing 1997).

TABLE 10.1. Mode frequencies of a cello compared to those of a violin (Bynum and Rossing 1997).

Mode	Freq. (Hz)	Ratio to violin	Comments
C_1	57	0.33	
A_0	102	0.36	
T_1	144	0.32	Stronger in top than back
C_2	170	0.42	Torsional motion
C_4	195	0.29	Ring-shaped node
A_1	203	0.43	Longitudinal air motion
C_3	219	0.39	
A_3	277	0.25	Transverse air motion
A_2	302	0.37	Longitudinal air motion

considerable renaissance in the twentieth century. The three most common members of the viol family are the treble, tenor, and bass. Each has six strings, tuned in fourths except for the major third at the center:

Bass :	D_2	G_2	C_3	E_3	A_3	D_4
Tenor :	A_2	D_3	G_3	B_3	E_4	A_4
Treble :	D_3	G_3	C_4	E_4	A_4	D_5

Toward the end of the seventeenth century, a descant viol or *pardessus de viole*, tuned a fourth above the treble viol (G_3 C_4 F_4 A_4 D_5 G_5), was added. In addition to these rather standard instruments, the viol family has had several other members: the double-bass viol, division (small bass) viol, lyra viol ("viola bastarda"), viola d'amore, and baryton (the latter two with sympathetic strings).

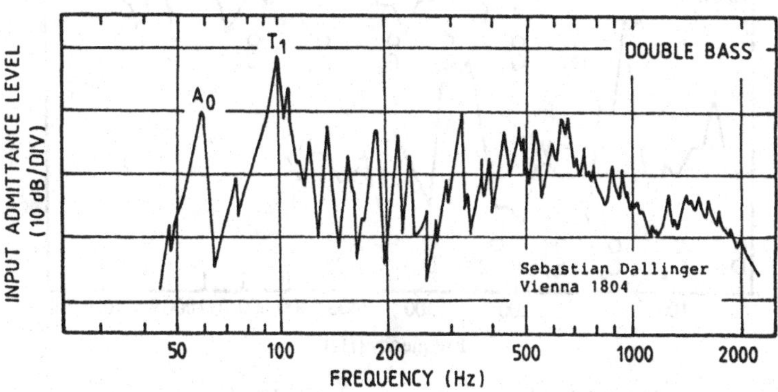

FIGURE 10.39. Input admittance curve for a high-quality double bass (Askenfelt, 1982).

A viol differs from a violin primarily in that its back is flat and braced by three cross bars, the middle one supporting the sound post. The top plate has a shallower arch than that of the violin, and it has c-holes rather than f-holes. The ribs are considerably higher than those of the violin. Bowed loudness curves for a treble and a bass viol are shown in Fig. 10.40 (compare with Fig. 10.28).

The treble viol appears to have been the first musical instrument to be studied by holographic interferometry (or at least to be reported in the literature). Figure 10.41 shows time-average holographic interferograms of the top plate of the treble viol shown in Fig. 10.40. The top plate was held in a jig, an artificial sound post was pushed against it, and it was excited at the back of the bass bar opposite the place where the bass bridge foot would rest.

10.15 A New Violin Family

In 1958, composer Henry Brant suggested to violin researchers Frederick Saunders and Carleen Hutchins that they design and construct a family of scaled violin-type instruments to cover the entire range of orchestral music. Saunders and Hutchins accepted the challenge and, with the help of John Schelleng and other members of the Catgut Acoustical Society, created a new family of eight violins. Several sets of these instruments now exist, and

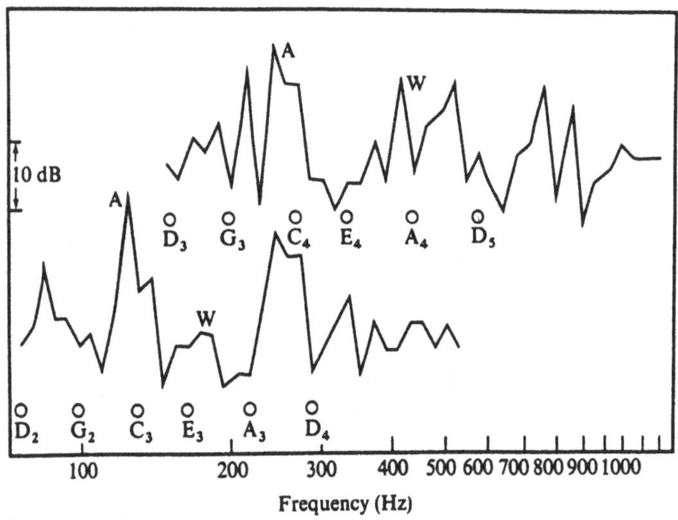

FIGURE 10.40. Bowed loudness curves for a treble viol and bass viol. A and W denote the air resonance and main body resonance (cf. modes A_0 and T_1 in Fig. 10.12). Frequencies of the open strings are noted (Ågren, 1968).

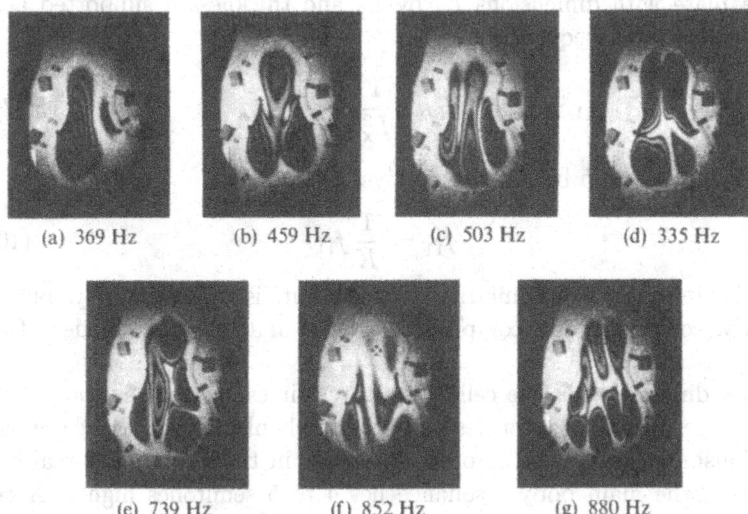

FIGURE 10.41. Holographic interferograms of a treble viol top plate. The first two plate modes probably resemble the modal shapes at the air and main body resonances (244 and 440 Hz) in the complete instrument, since the edge is attached to a jig and a soundpost presses against it (Ågren and Stetson, 1972).

they have been heard in a number of concerts. The reaction of performers and listeners has ranged from great enthusiasm to polite acceptance.

Design of the new family of violins was guided by a theoretical scaling relationship called the general law of similarity. This law states that if all linear dimensions of an instrument are increased by a factor K, keeping all materials the same, the natural vibrational modes decrease in frequency by a factor $1/K$:

$$K = \frac{L_1}{L_2} = \frac{f_2}{f_1}. \tag{10.6}$$

The truth of this law is easy to show for two simple vibrators: a Helmholtz resonator and a flat plate. The natural frequency of a Helmholtz resonator is

$$f = \frac{c}{2\pi}\sqrt{\frac{A}{Vl}}. \tag{1.28}$$

If all linear dimensions are scaled up by a factor K, we obtain

$$f_0' = \frac{c}{2\pi}\sqrt{\frac{K^2A}{(K^3V)(Kl)}} = \frac{c}{2\pi K}\sqrt{\frac{A}{Vl}} = \frac{1}{K}f_0. \tag{10.7}$$

For a plate with dimensions L_x by L_y and thickness h supported at its edges, the lowest frequency is

$$f_{11} = 0.0459 V_L h \left[\frac{1}{L_x^2} + \frac{1}{L_y^2} \right]. \tag{3.17}$$

Scaling $L_x L_y$ and h by a factor K gives

$$f'_{11} = \frac{1}{K} f_{11}. \tag{10.8}$$

This simple law of similarity, it turns out, is approximately, but not exactly, correct for the complicated shapes and vibration modes of real string instruments.

If all dimensions of the cello were three times as great as those of the violin, the natural frequencies of its body should have exactly the same relationship to those of the open strings as in the violin. In a real cello, however, the main body resonance lies 4 to 5 semitones higher. A cello three times as long as the violin (and with $3^3 = 27$ times the weight) would be considered quite impractical for the role it fulfills. The similarity requirements are even more difficult to fulfill in the viola (Cremer, 1984).

From a musical point of view, there would be both advantages and disadvantages to having physical similarity between instruments. Similarity

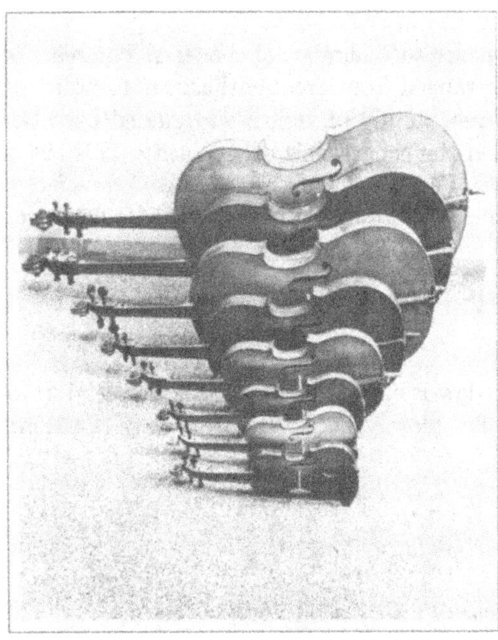

FIGURE 10.42. The eight instruments in the new violin family (photograph by John Castronova).

TABLE 10.2. Tunings, measurements, and scalings of the new violin family compared to conventional bowed string instruments (underlined).

Instrument			Length (cm)			Relative scaling factors		
Name	Tuning	Hz	Overall	Body	String	Body length	Resonance placement	String tuning
Treble	G D A E	392 587 880 1319	48	28.6	26	.75	.50	.50
Soprano	C G D A	262 392 587 880	54–55	31.2	30	.89	.67	.67
Mezzo	G D A E	196 294 440 659	62–63	38.2	32.7	1.07	1.00	1.00
Violin	G D A E	196 294 440 659	59–60	35.5	32.7	1.00	1.00	1.00
Viola	C G D A	132 196 294 440	70–71	43	37–38	1.17	1.33	1.50
Alto	C G D A	132 196 294 440	82–83	50.2	43	1.44	1.50	1.50
Tenor	G D A E	98.0 147 220 330	107	65.4	60.8	1.82	2.00	2.00
Cello	C G D A	65.4 98.0 147 220	124	75–76	68–69	2.13	2.67	3.00
Baritone	C G D A	65.4 98.0 147 220	142	86.4	72	2.42	3.00	3.00
Small bass	A D G C	55.0 73.4 98.0 131	171	104.2	92	2.92	4.00	4.00
Bass	E A D G	41.2 55.0 73.4 98.0	178–198	109 122	104–117	3.09–3.43	4.00	6.00
Contrabass	E A D G	41.2 55.0 73.4 98.0	213–214	130.0	110	3.60	6.00	6.00

[a] From Hutchins, 1980.

could improve balance in small ensembles, but the difference in timbre between dissimilar instruments might be missed.

Getting back to the new violin story: by 1962, the Catgut group had constructed a set of six scaled violins for testing. In 1965, the full set of eight was ready for its first concert performance. Since then, they have appeared in a number of successful concerts in Europe and North America.

The eight instruments in the new violin family, shown in Fig. 10.42, range in length from 48 cm (treble) to 214 cm (contrabass). Table 10.2 compares their tuning, measurements, resonance frequencies, and scaling factors to those of conventional bowed string instruments (underlined).

References

Ågren, C.-H. (1968). A second look at the viol family. *Catgut Acoust. Soc. Newsletter*, No. 10, 6–11.

Ågren, C.-H., and Stetson, K.A. (1972). Measuring the resonances of treble viol plates by hologram interferometry and designing an improved instrument. *J. Acoust. Soc. Am.* **51**, 1971–1983.

Alonso Moral, J. (1984). Form properties of free top plates, of free back plates, and of ribs of properties of assembled violins. Report STL-QPSR 1/1984, pp. 1–29. Speech Transmission Laboratory, Royal Institute of Technology (KTH), Stockholm.

Alonso Moral, J., and Jansson, E. (1982a). Eigenmodes, input admittance, and the function of the violin. *Acustica* **50**, 329–337.

Alonso Moral, J., and Jansson, E. (1982b). Input admittance, eigenmodes, and quality of violins. Report STL-QPSR 2–3/1982, pp. 60–75. Speech Transmission Laboratory, Royal Institute of Technology (KTH), Stockholm.

Askenfelt, A. (1982). Eigenmodes and tone quality of the double bass. *Catgut Acoust. Soc. Newsletter*, No. 38, 34–39.

Askenfelt, A. (1986). Measurement of bow motion and bow force in violin playing. *J. Acoust. Soc. Am.* **80**, 1007–1015.

Askenfelt, A. (1992). Observations on the dynamic properties of violin bows. Report STL-APSR 4/1992, pp. 43–48. Speech transmission Laboratory, Royal Institute of Technology (KTH), Stockholm.

Beldie, I.P. (1974). Vibration and sound radiation of the violin at low frequencies. *Catgut Acoust. Soc. Newsletter*, No. 22, 13–14.

Bissinger, G. (1995). Bounce test, modal analysis, and the playing qualities of the violin bow. *Catgut Acoust. Soc. J.* **2**(8), 17–22.

Bissinger, G. (1996). Acoustic normal modes below 4 kHz for a rigid, closed, violin-shaped cavity. *J. Acoust. Soc. Am.* **100**, 1835–1840.

Bladier, B. (1960). Sur le chevalet du violoncelle. Translated by R.B. Lindsay, in "Musical Acoustics, Part I. Violin Family Components' (C.M. Hutchins, ed.), pp. 296–298. Dowden, Hutchinson, and Ross, Stroudsburg, Pennsylvania, 1975.

Bynum, E., and Rossing, T.D. (1997). Holographic studies of cello vibrations. *Proc. ISMA97*, Edinburgh.

Caldersmith, G. (1981). Plate fundamental coupling and its musical importance. *J. Catgut Acoust. Soc.* **36**, 21–27.

Cremer, L. (1984). "The Physics of the Violin." Translated by J.S. Allen, MIT Press, Cambridge, Massachusetts. "Physik der Geige." S. Hirzel Verlag, Stuttgart, 1981.

Cremer, L., and Lazarus, H. (1968). Der Einflusz des Bogendruckes beim Anstreichen einer Saite. *Proc. ICA, Tokyo.*

Dünnwald, H. (1983). Auswertung von Geigenfrequenzgängen. *Proc. 11th ICA, Paris*, **4**, 373–376.

Dünnwald, H. (1985). Ein Verfahren zur objektiven Bestimmung der Klangualität von Violinen. *Acustica* **58**, 162–169.

Fang, N. J.-J., and Rodgers, O. E. (1992). Violin soundpost elastic vibration. *J. Catgut Acoust. Soc.* **2**(1), 39–40.

Firth, I. (1987). Construction and performance of quality commercial violin strings. *J. Catgut Acoust. Soc.* **47**, 17–20.

Guettler, K. (1994). Wave analysis of a string bowed to anomalous low frequencies. *J. Catgut Acoust. Soc.* **2**(6), 8–14.

Güth, W. (1978). "Gesichtspunkte beider Konstruktion eines Resonanz-Wolfdämpfers fürs Cello. *Acustica* **41**, 177–182.

Hacklinger, M. (1978). Violin timbre and bridge frequency response. *Acustica* **39**, 323–330.

Hacklinger, M. (1979). Violin adjustment—Strings and bridges. *Catgut Acoust. Soc. Newsletter*, No. 31, 17–19.

Hacklinger, M. (1980). Note on bridge inclination. *Catgut Acoust. Soc. Newsletter*, No. 33, 18.

Hanson, R. J., Schneider, A. J., and Halgedahl, F. W. (1994). Anomalous low-pitched tones from a bowed violin string. *Catgut Acoust. Soc. J.* **2**(6), 1–17.

Helmholtz, H.L.F. (1877). "On the Sensations of Tone," 4th ed. Translated by A.J. Ellis, Dover, New York, 1954.

Hutchins, C.M. (1967). Founding a family of fiddles. *Phys. Today* **20**, 23–27.

Hutchins, C.M. (editor) (1975–6). "Musical Acoustics, Parts I and II." Dowden, Hutchinson and Ross, Stroudsburg, Pennsylvania.

Hutchins, C.M. (1980). The new violin family. In "Sound Generation in Winds, Strings, Computers," pp. 182–203. Royal Swedish Academy of Music, Stockholm.

Hutchins, C.M. (1981). The acoustics of violin plates. *Scientific American* **245** (4), 170–186.

Hutchins, C.M. (1983). A history of violin research. *J. Acoust. Soc. Am.* **73**, 1421–1440.

Hutchins, C.M. (1987). Some notes on free plate tuning frequencies for violins, violas and cellos. *J. Catgut Acoust. Soc.* **47**, 39–41.

Hutchins, C.M. (1990). A study of the cavity resonances of a violin and their effects on its tone and playing qualities. *J. Acoust. Soc. Am.* **87**, 392–397.

Hutchins, C.M. (1993). Mode tuning for the violin maker. *J. Catgut Acoust. Soc.* **2**(4), 5–9.

Hutchins, C.M., and Benade, V. (editors) (1997). "Research Papers in Violin Acoustics 1975–1993," 2 vols. Acoustical Society of America, Woodbury, New York.

Hutchins, C.M., Stetson, K.A., and Taylor, P.A. (1971). Clarification of "free plate tap tones" by holographic interferometry. *J. Catgut Acoust. Soc.* **16**, 15–23.

Jansson, E. (1976). Long-term-average spectra applied to analysis of music. Part III: A simple method for surveyable analysis of complex sound sources by means of a reverberation chamber. *Acustica* **34**, 275–280.

Jansson, E.V. (1977). Acoustical properties of complex cavities. Prediction and measurements of resonance properties of violin-shaped and guitar-shaped cavities. *Acustica* **37**, 211–221.

Jansson, E. V. (1997). Admittance measurements of 25 high quality violins. *Acustica* **83**, 337–341.

Jansson, E., Bork, I., and Meyer, J. (1986). Investigation into the acoustical properties of the violin. *Acustica* **62**, 1–15.

Jansson, E. V., Molin, N.-E., and Saldner, H. O. (1994). On eigenmodes of the violin—Electronic holography and admittance measurements. *J. Acoust. Soc. Am.* **95**, 1100–1105.

Jansson, E., Molin, N.-E., and Sundin, H. (1970), Resonances of a violin studied by hologram interferometry and acoustical methods, *Phys. Scripta* **2**, 243–256.

Kondo, M., Kubota, H., and Sakakibara, H. (1986). Measurement of torsional motion of bowed strings with a helix pattern on its surface. Paper K3–8, *Proc. 12th Int'l. Congress on Acoustics, Toronto.*

Krigar-Menzel, O., and Raps, A. (1891). Aus der Sitzungberichten. *Ann. Phys. Chem.* **44**, 613–641.

Lee, A.R., and Rafferty, M.P. (1983). Longitudinal vibrations in violin strings. *J. Acoust. Soc. Am.* **73**, 1361–1365.

Lee, R.M. (1975). An investigation of two violins using a computer graphic display. *Acustica* **32**, 78–88.

Luke, J.C. (1971). Measurement and analysis of body vibrations of a violin. *J. Acoust. Soc. Am.* **49**, 1264–1274.

Marshall, K.D. (1985). Modal analysis of a violin. *J. Acoust. Soc. Am.* **77**, 695–709.

McIntyre, M.E., and Woodhouse, J. (1978). The acoustics of stringed musical instruments. *Interdisciplinary Science Reviews* **3**, 157–173.

Menzel, R.E., and Hutchins, C.M. (1970). The optical proximity detector in violin testing. *Catgut Acoust. Soc. Newsletter*, No. 13, 30–35.

Meyer, J. (1972). Directivity of bowed stringed instruments and its effect on orchestral sound in concert halls. *J. Acoust. Soc. Am.* **51**, 1994–2009.

Meyer, J. (1975). Akustische Untersuchungen zur Klangqualität von Geigen. *Instrumentenbau* **29** (2), 2–8.

Meyer, J. (1985). The tonal quality of violins. *Proc. SMAC* 83. Royal Swedish Academy of Music, Stockholm.

Meyer, J. (1992) Zur klanglichen Wirkung des Streicher-Vibratos. *Acustica* **76**, 283–291.

Minnaert, M., and Vlam, C.C. (1937). The vibrations of the violin bridge. *Physica* **4**, 361–372.

Molin, N.-E., Wåhlin, A. O., and Jansson, E. V. (1990). Transient wave response of the violin body. *J. Acoust. Soc. Am.* **88**, 2479–2481.

Molin, N.-E., Wåhlin, A. O., and Jansson, E. V. (1991). Transient wave response of the violin body revisited. *J. Acoust. Soc. Am.* **90**, 2192–2195.

Müller, H.A. (1979). The function of the violin bridge. *Catgut Acoust. Soc. Newsletter*, No. 31, 19–22.

Müller, H.A., and Geissler, P. (1983). Modal analysis applied to instruments of the violin family. *Proc. SMAC* 83 Royal Swedish Academy of Music, Stockholm.

Müller, G., and Lauterborn, W. (1996). The bowed string as a nonlinear dynamical system. *Acustica* **82**, 657–664.

Niewczyk, B., and Jansson, E. (1987). Experiments with violin plates. Report STL-QPSR 4/1987, pp. 25–42. Speech Transmission Laboratory, Royal Inst. of Tech. (KTH), Stockholm.

Pickering, N. (1984). A study of bow hair and rosin. *J. Violin Soc. Am.* **7**(1), 46–72.

Pickering, N. (1991). "The Bowed String." Amereon, Ltd., Mattituck, NY.

Pickering, N.C. (1985). Physical properties of violin strings. *J. Catgut Acoust. Soc.* **44**, 6–8.

Pickering, N.C. (1986a). Transient response of certain violin strings. *J. Catgut Acoust. Soc.* **45**, 24–26.

Pickering, N.C. (1986b). Elasticity of violin strings. *J. Catgut Acoust. Soc.* **46**, 2–3.

Pitteroff, R. (1994). Modelling of the bowed string taking into account the width of the bow. *Proc. SMAC 93*, Ed. A. Friberg, J. Ewarsson, E. Jansson, and J. Sundberg (Royal Swedish Acad. Music, Stockholm).

Pitteroff, R., and Woodhouse, J. (1998). Mechanics of the contact area between bow and string. Part I: Reflection and transmission behaviour. *Acustica* **84**, 543–562.

Powell, R.L., and Stetson, K.A. (1965). Interferometric vibration analysis by wavefront reconstruction. *J. Opt. Soc. Am.* **55**, 1593–1598.

Rakowski, A. (1985). Bowed instruments—Close relatives of the singing voice. *Proc. SMAC* 83. Royal Swedish Academy of Music, Stockholm.

Raman, C.V. (1916). On the wolf-note in bowed string instruments. *Phil. Mag.* **32**, 391–395.

Raman, C.V. (1918). On the mechanical theory of the vibrations of bowed strings and of musical instruments of the violin family, with experimental verification of the results. Bull. 15, The Indian Association for the Cultivation of Science.

Reder, O. (1970). The search for the perfect bow. *Catgut Acoust. Soc. Newsletter*, No. 13, 21–23.

Reinecke, W. (1973). "Übertragungseigenschaften des Streichinstrumentenstegs. *Catgut Acoust. Soc. Newsletter*, No. 19, 26–34.

Reinecke, W., and Cremer, L. (1970). Application of holographic interferometry to vibrations of the bodies of string instruments. *J. Acoust. Soc. Am.* **48**, 988–992.

Richardson, B.E., Roberts, G.W., and Walker, G. P. (1987). Numerical modelling of two violin plates. *J. Catgut Acoust. Soc.* **47**, 12–16.

Roberts, G.W. (1986). Vibrations of shells and their relevance to musical instruments. PhD dissertation, University College, Cardiff.

Roberts, M., and Rossing, T. D. (1997). Vibrational modes of two violins. *133rd meeting, Acoust. Soc. Am.*, Penn State Univ., June 15–20, 1997.

Roberts, M., and Rossing, T.D. (1997). Normal modes of vibration in violins. *Proc. ISMA97*, Edinburgh.

Rodgers, O.E. (1986). Initial results on finite element analysis of violin backs. *J. Catgut Acoust. Soc.* **46**, 18–23.

Rossing, T.D. (1982). "The Science of Sound." Addison-Wesley, Reading, Massachusetts. Chap. 10.

Rubin, C., and Farrar, D.F., Jr. (1987). Finite element modeling of violin plate vibrational characteristics. *J. Catgut Acoust. Soc.* **47**, 8–11.

Sacconi, S.F. (1979). "The 'Secrets' of Stradivari." Libreris del Convegno, Cremona.

Saldner, H. O., Molin, N.-E., and Jansson, E. V. (1996). Vibration modes of the violin forced via the bridge and action of the soundpost. *J. Acoust. Soc. Am.* **100**, 1168–1177.

Saunders, F.A. (1937). The mechanical action of violins, *J. Acoust. Soc. Am.* **9**, 91–98.

Savart, F. (1840). Des instruments de musique. Translated by D.A. Fletcher, in Hutchins (1976), pp. 15–18.

Schelleng, J.C. (1963). The violin as a circuit. *J. Acoust. Soc. Am.* **35**, 326–338.

Schelleng, J.C. (1973). The bowed string and the player. *J. Acoust. Soc. Am.* **53**, 26–41.

Schleske, M. (1996). Eigenmodes of vibration in the working process of a violin. *Catgut Acoust. Soc. J.* **3**(1), 2–8.

Schumacher, R.T. (1975). Some aspects of the bow. *Catgut Acoust. Soc. Newsletter*, No. 24, 5–8.

Shaw, E.A.G. (1990). Cavity resonance in the violin: Network representation and the effect of damped and undamped rib holes. *J. Acoust. Soc. Am.* **87**, 398–410.

Steinkopf, G. (1963). Unpublished thesis. Techn. Univ. Berlin, described by Cremer (1984).

Small, A.M. (1937). An objective analysis of violin performance. *Univ. Iowa Studies in the Psychology of Music* **4**, 172–231.

Stetson, K.A. (1990). Theory and applications of electronic holography. *Proc. Soc. Experimental Mechanics Conf. On Hologram Interferometry and Speckle Metrology*, 294–300.

Weinreich, G. (1985). Violin radiativity: Concepts and measurements. *Proc. SMAC* 83. Royal Swedish Academy of Music, Stockholm. pp. 99–109.

Weinreich, G. (1994). Radiativity revisited: Theory and experiment ten years later. *Proc. SMAC 93*, Ed. A. Friberg, J. Ewarsson, E. Jansson, and J. Sundberg. Royal Swedish Acad. Music, Stockholm.

Weinreich, G., and Caussé, R. (1991). Elementary stability considerations for bowed-string motion. *J. Acoust. Soc. Am.* **89**, 887–895.

Woodhouse, J. (1993). On the playability of violins. Part I: Reflection functions; and Part II: Minimum bow force and transients. *Acustica* **78**, 125–136 and 137–153.

Woodhouse, J. (1995). Self-sustained musical oscillators. In "Mechanics of Musical Instruments," ed. A. Hirschberg, J. Kergomard, and G. Weinreich. Springer-Verlag, Wien.

11

Harps, Harpsichords, Clavichords, and Dulcimers

In Chapter 9, we discussed plucked-string instruments of the guitar and lute families, having a relatively small number of strings, the sounding length of which can be varied by pressing them with a finger against a set of fixed frets. These instruments are further characterized by being light in weight and having a bulbous hollow sound box and a long neck and fingerboard.

The distinction between these instruments and those of the family we are now to describe is not at all sharp, and many instruments of transitional design have been developed (Baines, 1966), though most of them are now obsolete. The shapes of these instruments varied widely, and they had different combinations of free and fretted strings in profusion. Various classifications have been devised, and the history of many types has been documented by Marcuse (1975, pp. 177–234). For our present discussion, it is convenient to ignore the fretted instruments and consider only those with one open string per note. The hand-plucked instruments then divide conveniently into those with a sounding board parallel to the strings—the zither, the psaltery, and the koto are examples—and those with the soundboard nearly at right angles to the plane of the strings, which belong to the harp family proper. Again, this is only a classification of convenience, and intermediate types do exist.

It is interesting to note that, on the basis of this classification, the harpsichord is really a mechanized psaltery or box zither, rather than a mechanized harp. Some of this former class of instruments, occasionally the psaltery but particularly the dulcimer, were played by striking the strings with hammers instead of plucking them, and so are the ancestors of the clavichord and particularly of the pianoforte.

11.1 Traditional Instruments

There is not space here to give a description of the various kinds of zithers that have survived in the reviving tradition of folk music, and we consider only a few examples. Figure 11.1 shows a typical psaltery from the Middle

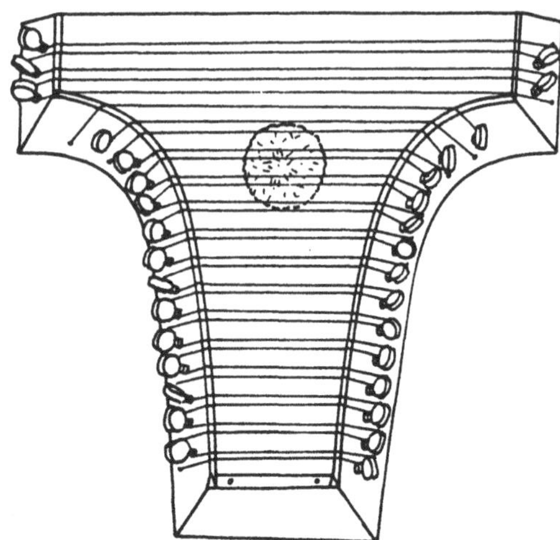

FIGURE 11.1. In plucked-string instruments of the zither family, such as the psaltery shown here, the strings run parallel to the soundboard. There is one string per note and frets are absent.

East. There is one string per note, and these run parallel to the soundboard, which forms the top of an enclosure vented through a perforated rose. The strings are supported on two bridges along the edges of the soundboard, and are plucked by the fingers. Psalteries of this kind differ considerably in size and detailed design and do not appear to have been the subject of acoustical studies.

11.1.1 The Dulcimer

The dulcimer, which is a version of the same instrument played with two wooden hammers, appears also to have originated in the Middle East, though it has become part of the musical culture of many Western countries. The Appalachian dulcimer of the United States, however, is plucked rather than hammered, and so is more properly classified as zither. A large chromatic dulcimer, developed by Pantaleon Hebenstreit of Liepzig around 1690, is sometimes considered to be the forerunner of the pianoforte. Its modern equivalent in terms of size and complexity is the Hungarian cimbalom, a large instrument with a steel frame and multi-ply pin blocks.

Dulcimers are made in a wide variety of pattern, size and compass, with two, three, or even four strings per note. A typical example, shown in Fig. 11.2, has strings running sideways across a trapezoidal soundboard that forms the top of a vented box. A major innovation, however, is that each pair of treble strings is divided in two by a bridge resting on the soundboard, in such a way that the length ratio is 3 : 2, allowing two notes per string,

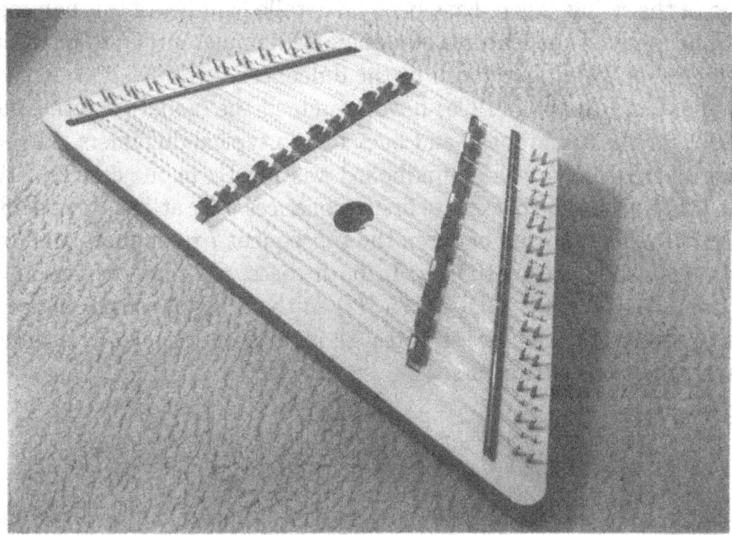

FIGURE 11.2. A typical 12/11 hammered dulcimer by Gillian Alcock. Note the left-hand bridge resting on the soundboard and dividing the 12 treble strings in a harmonic ratio 3 : 2. The 11 bass strings rest on the right-hand bridge.

separated by a musical fifth. Because the bridge is not ideally rigid, there is energy transfer between the two parts of each string, this latter transfer being enhanced by the harmonic relation between their frequencies. The common frequency, which is that of the third mode of the longer part of the string and the second mode of the shorter part, is split by the interaction into two closely separated modes. In one of these modes the two strings are in-phase, giving a large energy transfer to the bridge and soundboard and a consequent loud sound and short decay time. In the other mode, the two vibrations are antiphase, so that there is little energy transfer and a prolonged quiet sound (Peterson, 1995). There is a similar interaction between the two strings of a pair, which are tuned in unison. This coupling behavior was first identified in the piano by Weinreich (1977) and was discussed in Section 4.10. The bass strings on a typical dulcimer rest on another bridge, as shown in the example of Fig. 11.2, and only the longer string section is played. The compass of a typical 12/11 dulcimer with 12 treble and 11 bass strings is two and a half octaves from G_3 to D_6, with a few missing chromatic notes, but fully chromatic instruments with larger compass are also built.

The hammer used to excite the dulcimer is typically a curved wooden striker with a mass between 5 and 10 g, though sometimes hammers faced with leather are used. The contact time for a wooden hammer is typically about 2 ms (Peterson, 1995), which is comparable with the fundamental period of the string. There have been no detailed studies of this hammer impact, but many of the features of pianoforte hammer strike, to be discussed in the next chapter, apply also to the dulcimer.

Canfield and Rossing (1995) have examined the modal patterns and frequencies for the soundboard and back of two typical dulcimers. The lowest resonance frequency of the soundboard was 145 Hz in one case and 177 Hz in the other, while the backplate first resonances were at 143 Hz and 283 Hz, respectively. The spacing between the subsequent 7 resonances was roughly constant at about 50 to 100 Hz. In both cases, the range of the dulcimer was from D_3 (147 Hz) to D_6 (1175 Hz) so that the lowest resonance came close to reinforcing the lowest notes, though there was clearly a difference between the two instruments.

11.1.2 The Koto

The Japanese koto or soh is a close relative of the Chinese ch'in and the Vietnamese dan tranh. As shown in Fig. 11.3, it consists of a shallow rectangular box about 180 cm long, 25 cm wide, and 7 cm high, with a curved arched top plate 3 to 4 cm thick and a flat ribbed base plate with two soundholes. It is made of paulownia (kiri) wood, which has density and mechanical properties rather similar to those of the spruce traditionally used for the top plates of violins. The koto has 13 tightly twisted silk strings which pass over fixed bridges at each end of the instrument and over high individual movable bridges, one for each string, distributed along the body of the instrument. These bridges are set so that the right-hand sections of

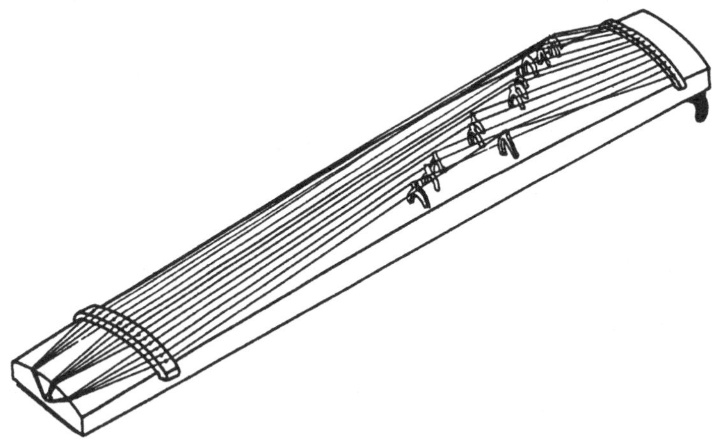

FIGURE 11.3. The Japanese koto or soh is a long, arched zither with 13 silk strings passing over movable bridges. It is played with thimblelike plectra on the fingers of the right hand, while the left hand presses on the nonsounding part of the string to interpolate semitones and produce vibrato effects.

the string, viewed from the player's side, produce a pentatonic scale over about two octaves. The usual scale is G_3 A_3 $B^\flat_3$ D_4 $E^\flat_4$ G_4 A_4 $B^\flat_4$ D_5 $E^\flat_5$ G_5 A_5, but others are also used (Ando, 1989). The player plucks the strings with fingernail-like plectra and, by pressing on the nonsounding part of the string, can produce intermediate semitones, sliding transitions between notes, and vibrato effects.

Ando (1986) examined the vibrations of the body of the koto and found modes, as shown in Fig. 11.4, that have relatively simple transverse nodal lines. The first three modes, which are nearly the same for the top plate alone and for the complete instrument, appear to be simple barlike vibrations of the type discussed in Chapter 2, except that the frequencies have a nearly linear rather than a nearly quadratic increase with mode number. The reason for this is not clear. We might designate these modes (1, 0), (2, 0), and (3, 0), respectively, as discussed in Chapter 3, indicating no mode excitation in the transverse dimension of the plate. The fourth mode at 404 Hz appears anomalous, but could be of the form (1, 1), the elastic anisotropy of the timber giving a smaller rigidity and hence a lower frequency for transverse modes than would be expected from simple dimensional considerations. The expected longitudinal nodal lines may be concealed at the sides of the instrument. The other modes indicated might then be of the form (n, 1). It is interesting to note that the dimensions of the instrument and the existence of soundholes at both ends lead us to expect air resonances at integral multiples of a fundamental near 100 Hz, for which the cavity is half a wavelength long. Close similarity to the plate mode frequencies leads to significant interaction.

The mode near 400 Hz, which has a nearly pure monopole breathing character like the A_0 mode of a violin, as shown in Fig. 10.12, is particularly efficient in radiation and shows up as a sharp peak in the sound spectrum

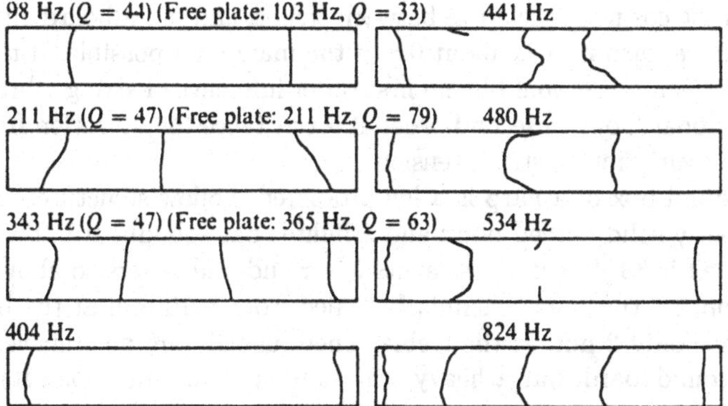

FIGURE 11.4. Nodal curves for the first few modes of the top plate of a koto, with their frequencies in hertz (Ando, 1986).

when the body is excited by a sharp tap. The effects of this resonance, the fact that plate resonances extend right down to the fundamental of the lowest note ($G_3 = 196$ Hz), and the damping effect on the high-frequency string vibrations of their twisted silk construction combine to give the koto a mellow harplike sound. The player has some control over sound quality, as in all manually plucked instruments, by varying the position of the plucking point as discussed in Chapters 2 and 10.

The radiation pattern of the koto should follow the general principles discussed in Chapter 7. The 400 Hz mode has an approximately monopole character, at least as far as body vibrations are concerned, and should radiate efficiently. The other top-plate modes can be approximated by a row of line sources of opposite phases. The separation between these antiphase regions is somewhat less than half the sound wavelength in air at the respective resonance frequencies so that there will be a good deal of radiation cancelling, except for the unbalanced zones at the ends of the instrument. This behavior will be complicated, however, by the extra radiation from the soundholes in the base of the instrument, reflected from the usually hard floor on which it is played. No quantitative studies on these matters seem to have been published.

11.2 The Harp

The harp has a long tradition in western Europe and remains the national stringed instrument in countries such as Wales, Ireland, and Scotland. It is characterized, as we noted, by having a soundboard that is more-or-less perpendicular to, rather than parallel to, the plane of the strings. As we see from Fig. 11.5, however, the angle between individual strings and soundboard is in the range 30–40°, rather than being 90°. A little thought shows that this is necessary if the vibrations of the strings are to have a direct influence on the soundboard to which they are attached. This effect varies as the cosine of the angle between strings and soundboard and so has a value for a harp that is about 0.8 of the maximum possible. If the angle were 90°, then there would be no first-order influence of string vibration on the soundboard, only a second-order effect at twice the vibration frequency caused by variation in string tension.

The sound box of a harp is a long, tapered, hollow structure, generally with a fairly solid curved body and a lighter flat soundboard. The soundboard itself is 30–40 cm across at its lower end and tapers to about 10 cm at the top. Its thickness is similarly scaled from 8–10 mm at the bass end to perhaps only 2 mm at the treble. There usually are no transverse bars on this soundboard, but a heavy longitudinal strut runs along its center, and it is to this that the strings are fixed.

The strings themselves are generally of gut, though some of the bass strings may be of silk overwrapped with metal wire to give greater density.

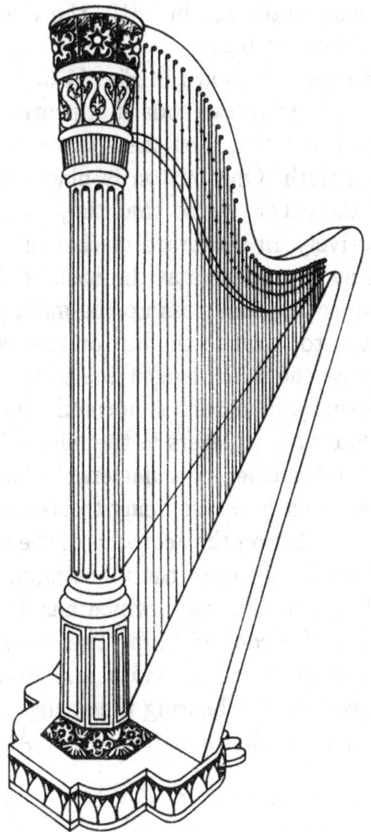

FIGURE 11.5. In both small Welsh, Irish, and Scottish harps and in modern concert harps, the plane of the soundboard is normal to the plane of the strings, and individual strings meet the soundboard at an angle of about 30°. The concert harp has pedal levers to alter the pitch of all strings with the same note name by one or two semitones.

Their number ranges from about 30 in small portable harps to 44 in a modern concert harp. String lengths increase toward the bass, as allowed by the shape of the frame, but, rather than doubling with each octave as required by strictly proportional scaling, the increase is by a factor typically between 1.7 and 1.8 (Firth, 1986). The strings are normally tuned to a major diatonic scale and cover a compass of $4\frac{1}{2}$ to 6 octaves. The top open string on most modern concert harps is $E_7^{\flat}$ (2489 Hz).

While many simple harps have only open strings and are tuned to different scales by use of the tuning pegs at the beginning of a performance, some of them incorporate small brass blades that can be rotated to touch individual strings and, thus, by shortening their vibrating length, to raise their pitch by a semitone. This simple mechanism was further developed by Sebastian Erard who patented, in 1819, the double action now used universally in modern concert harps.

The essence of this action is shown in Fig. 11.6. Each string passes over two brass buttons, each bearing two rods that protrude on the two sides of the string. If a button is rotated, then these two rods grasp the string and shorten its vibrating length. One button is placed about $\frac{1}{18}$ of the way along the string and the other $\frac{2}{18}$, so that they raise the pitch by one or two semitones, respectively. In the Erard mechanism, the rotation of these buttons is controlled by seven levers at the base of the harp, one for each note of the scale. Connecting rods pass up the main pillar of the harp and thence along the curved top arm to the button mechanism. Each lever has three positions that leave the string length unaltered or shorten it by one or two semitones, respectively. The harp is normally tuned with open strings in the key of C♭, so that with all levers in the midposition the key becomes C. Each note can then be sharpened or flattened by one semitone from this midposition, the notes of each octave being treated in the same way.

Very little has been written on the acoustics of the harp, but Firth (1977) has made a detailed examination of the vibrational modes of the soundboard of a Scottish harp, the clarsach, which has 32 strings and is rather more than half the size of a concert harp. The soundboard is 93 cm long and tapers from 30 cm to 10 cm in width and from 8 mm to 2 mm in thickness. Before the central bar bearing the strings is attached, the nodal lines have the general form shown in Fig. 11.7. At low frequencies, the

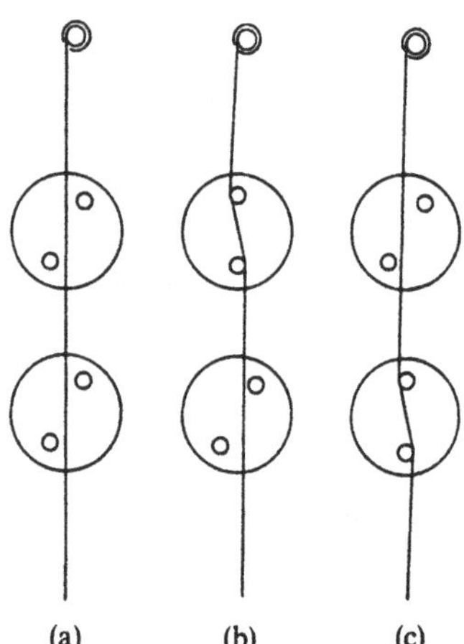

FIGURE 11.6. In the double-action harp of Erard, each string passes between two buttons bearing pegs. When the buttons are rotated separately by the mechanism, the length of the string is reduced by either $\frac{1}{18}$ or $\frac{2}{18}$, thus raising the pitch by one or two semitones. The open strings (a) are tuned in the key C♭ so that (b) corresponds to C♮ and (c) to C♯.

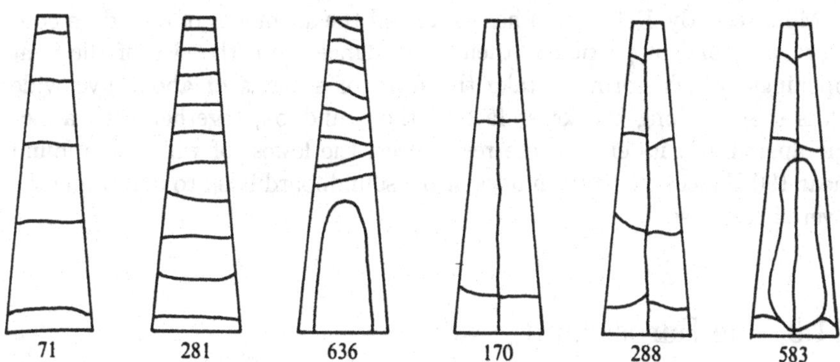

FIGURE 11.7. Nodal patterns for representative low-, medium-, and high-frequency vibration modes of the free soundboard of a small harp. The three patterns at the left are for the soundboard alone, and the three patterns at the right for the soundboard with its central bar (after Firth, 1977).

vibrations are like those of a simple beam, except that the nodal lines are more widely spaced at the broad thick end of the soundboard than at the narrow thin end. At high frequencies, a transverse nodal ring develops at the broad end of the soundboard and the transverse nodal lines are restricted to the narrow end. To a reasonable approximation, the mode frequencies in hertz are given by

$$f_n \approx 6n^{1.7}, \quad n = 1, 2, 3, \ldots. \tag{11.1}$$

When the central bar is added, the mode shapes remain similar to those of the unbarred soundboard except that the ring node develops for a much smaller value of n. The modal frequencies are all greatly raised, as we would expect, and the frequency relation [Eq. (11.1)] is changed to the surprisingly linear form

$$f_n \approx 103n, \tag{11.2}$$

at least for $n \leq 6$. This is reminiscent of the mode-frequency relation noted above for the body of the koto, though the geometry is clearly different.

Firth did not conduct a similar examination of mode shapes for the assembled sound box, but rather conducted a survey of mechanical admittance of the completed structure at positions along the central bar corresponding to string attachment points. He found some correlation between the behavior of sound output for particular strings and mechanical admittance at their point of attachment at their vibration frequency. Particular tonal effects were noted for strings attached at points where the admittance at their characteristic frequency was particularly high or low. Clearly, the design of a really fine instrument must avoid allowing individual strings to have very idiosyncratic behavior, though absolute uniformity is neither expected nor desired.

More recently, Bell (1997) has examined the air modes of a modern concert harp, and in particular their dependance upon the size of the vent openings, which normally take the form of a series of about five wide slots spaced along the lower face of the soundbox, together with a single aperture in its end. These resonances, the lowest of which was found near 190 Hz, couple to the modes of the soundboard itself to determine the overall response.

11.3 The Harpsichord

The harpsichord, or cembalo, and its rather different companion the clavichord, have a long history back to at least the year 1400. The harpsichord was the mainstay of chamber music throughout the baroque and early classical periods until about 1800, when it was supplanted by the more expressive and much louder pianoforte. Fortunately, many excellent instruments have survived from the eighteenth century—the golden age of harpsichord building—and careful studies by Hubbard (1967) and others have led to a modern revival in their construction and playing (Zuckermann, 1969). The history and national traditions of harpsichord construction have been documented by Hubbard (1967) and, more briefly, by Kottick (1987), who also gives practical information on harpsichord maintenance.

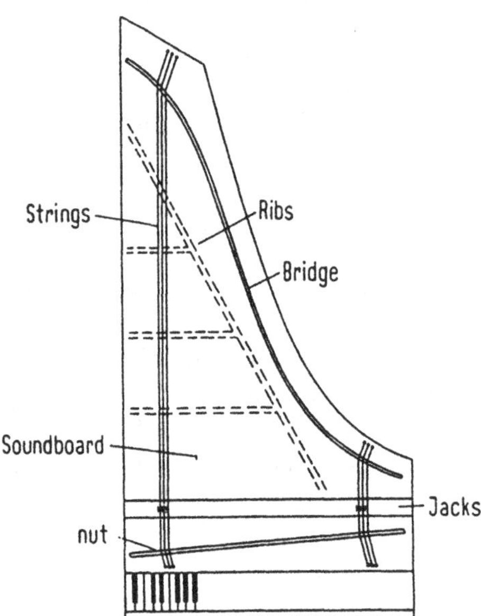

FIGURE 11.8. The structure of a typical one-manual harpsichord with a single choir of strings (Fletcher, 1977).

The general construction of a harpsichord is shown in Fig. 11.8. The keyboard spans about $4\frac{1}{2}$ octaves, from A_1 to F_6, though larger and smaller compasses have been built. The instrument is more or less triangular, with the oblique side generally being curved. Strings run from front to back. In some very small harpsichords (spinets), the case may be angled in from both ends, with the strings running parallel to the longer angled side. The strings are fixed to pegs set in the top of the frame along the bent side, pass at an angle around metal pins set in a curved wooden bridge glued to the soundboard, cross another set of pins in a fixed bridge (the nut) set rigidly on a pin block, and then wrap around individual tuning pegs as in a piano. In the upper half of the compass, the string length doubles on the octave descending the scale, but in the bass this doubling is reduced and the strings of the lowest octave are nearly equal in length. The outer frame of the harpsichord is of light but strong timber, cross-braced from one side to the other. The soundboard is of light timber, generally spruce, as in violin top plates, with the grain running parallel to the strings, and is stiffened by light ribs in a pattern such as shown. Traditional harpsichords are always in the form of a closed box with a solid base. Some open, pianolike instruments have been built in recent years, but they are not historically accurate and are not generally approved.

The action of a harpsichord is shown in Fig. 11.9. Each key pivots on a pin near its midpoint and raises a jack from which protrudes a short quill plectrum. This deflects the string and then bends to let it slip past and vibrate freely. On the return movement, when the key is released, the plectrum, which is mounted on a short spring-loaded lever, slips back under the string and a felt damping pad, also mounted on the jack, stops the sound. The jacks are prevented from leaping out of their slots in fast playing by a wooden cover, lined with thick felt, which also serves to return the jack rapidly in staccato or repetitive playing.

When a harpsichord has only one set of strings, these are now tuned to normal piano pitch, $A_4 = 440$ Hz, but perhaps to A 415 or even A 390 if playing with wind instruments of the period. This pitch is known as "8-foot" by analogy with the pipe organ, a much older keyboard instrument, for which the longest manual pipe at this pitch ($C_2 = 65$ Hz) is about 8 feet (2.4 m) long. The size or scale of a harpsichord is generally specified by giving the length of the C_5 string in the middle of the length-proportional range of the 8-ft string choir. A typical scale is 25 to 35 cm. Some harpsichords have two separate sets of strings, both at 8-ft pitch, plucked by separate jacks at different distances from the end to give variety in harmonic content, as discussed in Chapter 2. Larger harpsichords have a set of shorter strings, passing over a separate bridge on the soundboard and anchored to a separate heavy hitch-pin rail glued to the underside of the soundboard. These strings are tuned to produce notes an octave above the written pitch and are referred to as a 4-ft set, again by analogy with the organ. Some harpsichords have a still further set of strings of the same

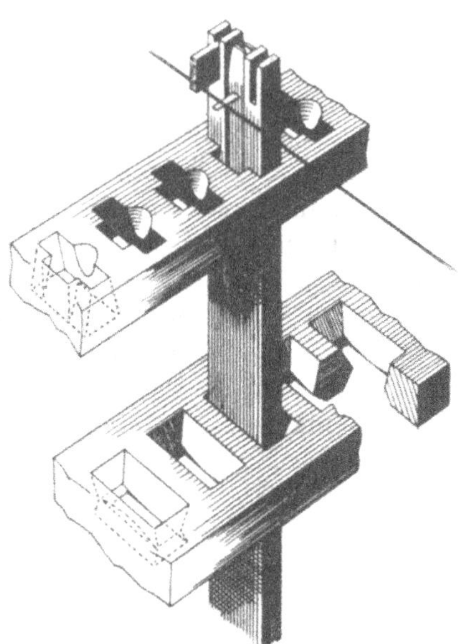

FIGURE 11.9. The mechanism of a harpsichord. The key raises the jack and the plectrum plucks the string. When the key is released, the spring-loaded tongue deflects to allow the plectrum to slip back under the string, while the felt damper pad terminates the string vibration. The stop rail can be moved slightly sideways to disengage all the jacks for a particular choir of strings (Hubbard, 1967).

length as the 8-ft set but of thicker wire, tuned as a 16-ft choir an octave below the written notes. The jacks for each set of strings are moved by the same keys but run in different sets of guide slots. A small motion of the rail holding a set of slots can be made to move the jacks just far enough that their plectra fail to touch the strings, thus silencing them. In this way, a considerable variety of basic harpsichord sounds can be produced. Further variations in tone color are achieved by using plectra of leather or other soft materials instead of quill, or even by moving a set of felt pads against the strings to damp the sound rapidly and give a lutelike effect.

Still larger harpsichords have another complete manual with its own keys, jacks, and strings, but using the same soundboard, to give further variety and allow quick changes in sound quality. Generally, the upper manual is softer than the lower. An ingenious mechanism allows the key action of the two manuals to be coupled together for the fullest sound effect.

The eighteenth-century harpsichord was thus a fully developed and flexible instrument—some even had pedal keyboards like the organ—and served for much solo music as well as being an essential component of the baroque orchestra and the normal accompaniment, along with cello or bassoon, for most vocal and chamber music. It had, and still has, a rich and clear sound, a reasonable volume level, and a precise and flexible action. Its shortcomings, to the ears of musical romantics, lay in its inability to produce an expressive crescendo or accent, and in its lack of volume level adequate to balance the new instrumental forces of the romantic symphony orchestra.

The string courses of a harpsichord are always single, not doubled or tripled as in a piano, though more than one string course may be engaged at once. In nearly all cases, historical instruments used steel or soft iron strings in the treble and brass strings for the lowest octave or so. Wire diameters were quite small, ranging from about 0.2 mm at the top of an 8-ft course to 0.7 mm in the extreme bass. String stiffness effects, which we shall find to be important for the piano, are therefore nearly negligible. The strings of a 4-ft choir were even thinner, while a 16-ft choir had considerably thicker strings in order to preserve a reasonable tension despite their short length. Details are given by Hubbard (1967) and by Thomas (1971).

11.4 Harpsichord Design Considerations

It is instructive to examine the effects of various available design parameters on the acoustic behavior of a harpsichord (Kellner, 1976; Fletcher, 1977) since similar principles apply to other stringed keyboard instruments. In particular, we consider the soundboard, the strings, and the plucking action.

The soundboard, as already noted, is a roughly triangular plate, about a meter in longest dimension, stiffened by light ribs. Its function is to receive energy from the vibrating strings and, because its coupling to the air is very much better than that of a string, as we saw in Chapter 7, to radiate the sound. As with most vibrators, however, we expect that most of the energy will be dissipated in internal losses and only a few percent will be radiated. The fundamental frequency range for a harpsichord extends down to about 50 Hz and the strings produce components to well over 10 kHz, so that a wide frequency response is desired.

Little quantitative information is available in the literature on harpsichord soundboards. Typically, they are about 3 mm thick so that, from the results of Chapter 3, we expect a wave impedance $\sigma_p v_p$, where σ_p is the mass of the plate per unit area and $v_p(f)$ is the frequency-dependent wave velocity, of about $30f^{1/2}$ at frequency f Hz [see Eq. (3.13)]. Typically, this wave impedance is between 30 and 300 N m^{-1} s over the frequency range of interest, so that when resonances with Q values around 10 are allowed for, the impedance presented at the bridge may range from about 3 to 1000 N m^{-1} s. This is comfortably higher than the string impedance μc_s, where μ is the mass of the string per unit length and c_s is the transverse wave velocity along it, which is of order 0.1 N m^{-1} s for the thicker bass strings and correspondingly less in the treble. The impedance mismatch is large, as we require if the string is not to be rapidly damped and critically influenced by soundboard mechanics (Gough, 1981), but not so large that little energy is transferred.

While detailed information on soundboard properties is not available, Weyer (1976) has measured the response of harpsichord soundboards to an impulsive excitation on the bridge, this response being detected in the radiant sound field. As shown in Fig. 11.10, which is a one-third-octave smoothed response, there is essentially no radiation below about 30 Hz, the response goes through a broad maximum around 200 Hz, and radiation extends up to about 15 kHz, though its level is then −40 dB relative to that at 200 Hz. This response was for the soundboard of a large Flemish harpsichord of total length 260 cm. A smaller Italian harpsichord of overall length 202 cm had a more sharply peaked response beginning at about 50 Hz and not extending much above 5 kHz at the −40 dB level.

This behavior is much what we would expect from our discussion in Chapter 7. A sharp tap excites all modes except those having a nodal line passing through the impact point. The lowest soundboard mode for the Flemish harpsichord apparently lies at about 40 Hz, which agrees reasonably with estimates. This mode has a pure breathing character and radiates efficiently. Whether it is aided by a cavity resonance as in the violin, we do not know. The next few modes, having few nodal lines and probably some residual monopole character as well, also radiate efficiently, but above about 1 kHz the cancellation of radiation from neighboring zones becomes increasingly severe and the response falls. The frequency of the fundamental soundboard mode is important in determining the fullness of the sound of the instrument and explains the preference for a long soundboard whenever possible.

Turning now to the strings, it is important to determine the rule by which string diameter should be increased from treble to bass. Certainly, it is necessary to increase the diameter of the relatively under-length strings in the extreme bass so that their tension is not undesirably small, but there is more to it than this simple mechanical consideration, for string gauge also influences loudness and sound decay time.

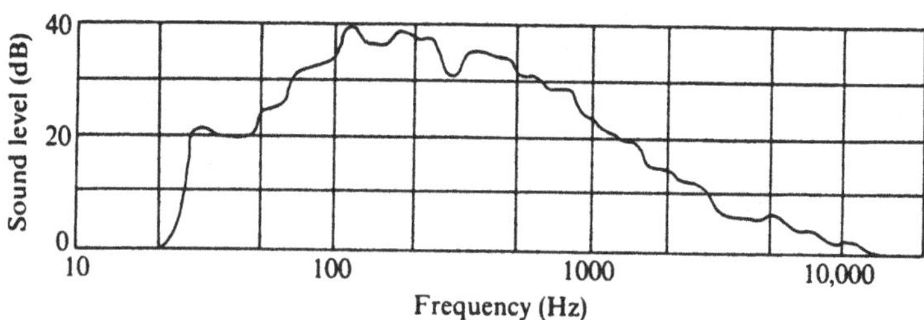

FIGURE 11.10. The radiated sound spectrum of the soundboard of a large harpsichord excited by a sharp tap on the bridge. Resolution is one-third octave (after Weyer, 1976).

For satisfactory playing, it is necessary that the force exerted on the keys be constant over the keyboard compass, and this is essentially equivalent to constant plucking force F, since variation in the pivot position of the keys is limited. If the string diameter is d, its length L, and its tension T and if $\beta \ll 1$ is the position of the plucking point, as discussed in Section 9.3, then the total energy given to the string is

$$E = \tfrac{1}{2} Fa \approx \frac{F^2 \beta L}{2T}, \tag{11.3}$$

where a is the deflection at the plucking point. If we write the tension T in terms of the stress s and string diameter d, then Eq. (11.3) becomes

$$E \approx \frac{2F^2 \beta L}{\pi s d^2}. \tag{11.4}$$

In the upper part of the harpsichord compass, where the string length is inversely proportional to frequency, the stress s must be constant to give correct tuning. Indeed, the strings are usually made long, with s close to the breaking stress, in order to minimize the nonlinear twanging effects discussed in Section 5.6 (Lieber, 1975). To make the stored energy constant, assuming a constant value of β, suggests a rule of the form

$$d \propto L^{1/2} \propto f^{-1/2} \tag{11.5}$$

over this part of the compass. In the extreme bass, where L is constant, s must vary as f^2 and the rule then becomes

$$d \propto f^{-1}. \tag{11.6}$$

A diameter scaling according to Eqs. (11.5) and (11.6) gives a string tension that is constant in the lowest part of the compass and decreases as f^{-1} in the upper proportional-length range.

The rate at which energy is transferred from the string to the soundboard depends on the real part G of the soundboard admittance and on the oscillating force exerted by the string on the bridge. This force has a pulselike form with a magnitude $(1-\beta)F$ for a fraction β of the period and $-\beta F$ for the remainder. Assuming $\beta \ll 1$, this force is essentially a train of impulses each of integrated magnitude βF, so that the rate of transfer of energy to the soundboard has the form

$$P \propto \beta^2 F^2 G, \tag{11.7}$$

though, since G is a function of frequency, this should really be expressed as a sum over the normal mode frequencies of the string. Some small fraction of this energy is radiated as sound, and we expect this to happen most efficiently in the frequency range from about 40 Hz to 1 kHz, as shown in Fig. 11.10.

If the power flow P from string to soundboard represented the only energy loss from the vibrating string, then we could determine the characteristic decay time τ_1 from the relation

$$\tau_1 = \frac{E}{P} \propto \frac{L}{sd^2\beta G}. \tag{11.8}$$

This would tell us that to achieve a sustained tone we should strive for a large string length L, a small plucking ratio β, and a small soundboard conductance G. These same criteria would give a tone with a rich harmonic development but a small amount of radiated power. If β and G were constant over the whole compass—admittedly not achievable in the case of G—then the decay time τ_1 would also be constant.

There is a measure of truth in these considerations, but in reality there is another important loss mechanism caused by viscous flow of air around the moving wire (Fletcher, 1977), as already discussed in Chapter 2. The behavior is rather complicated, but the damping from this mechanism generally exceeds that from losses to the soundboard in the case of a harpsichord. For strings in the proportional-length region of the compass and with $d \propto L^{1/2}$, we find a decay time τ_2 varying as L, or equivalently as f^{-1}. When these two loss mechanisms are combined, we find a resultant decay time τ given by

$$\tau^{-1} = \tau_1^{-1} + \tau_2^{-1}. \tag{11.9}$$

Even this is an oversimplification, since each note has a broad harmonic structure and different overtones behave differently. However, our discussion does give a first approximation to what is really happening.

11.5 Harpsichord Characteristics

The study of particular harpsichords, as measured by Hubbard (1967), Fletcher (1977), and others, confirms many of the principles conjectured in the previous section. Large harpsichords do have a full sound in the bass, the key force is nearly constant over the keyboard, and string courses plucked close to the end (very small β) have a softer, thinner sound, though the decay time is not noticeably longer.

Examination of string diameters shows a rule moderately close to the $f^{-1/2}$ that we predicted, though actually there are only about six different gauges of wire used on a typical instrument. In fact, the strings of the top octave and a half are uniform and significantly thicker than would have been given by this rule.

When we measure the A-weighted sound power radiated by a typical harpsichord, we find a level of about 68 ± 5 dB(A) at a distance of 2 m for notes over the whole compass. For the decay time of the sound to inaudibility, a level drop in this case of about 60 dB, we find a smooth decrease almost exactly as $f^{-1/2}$ from about 20 s for the lowest notes to 5 s for the highest. These times correspond to about 7τ, so that τ itself ranges from 3 s to 0.7 s, in quite good agreement with what would be expected from damping primarily by air viscosity.

The attack transients of harpsichord sounds can also be studied (Trendelenburg et al., 1940; Weyer, 1976, 1976/1977) and related both to the mechanics of the string-plucking action and to the finite time required for transfer of energy to the soundboard. This time is typically of order 0.01 s, but, despite its short duration, it contributes significantly to the characterization of the sound for listeners.

Finally, we note that the plucking ratio β for virtually all harpsichords decreases as we descend toward the bass. For a typical set of strings, β might be nearly 0.5 in the extreme treble but only 0.1 on the lowest note of the compass, though string choirs with other tone colors might have β values only half these. It is not hard to see the reasons behind this. In the first place, the decrease in β in the bass increases the harmonic development in this range and shifts the energy balance toward higher partials, which are both more efficiently radiated and more clearly heard. Second, music of the period often had an important and decisively moving contrapuntal bass line, and this shift toward greater harmonic development in the bass made this line crisper and more audible. We shall find in Chapter 17 that exactly the same principle is used in organ building for the same reasons.

11.6 The Clavichord

The clavichord has an even longer history than the harpsichord, dating back to the twelfth century. A surviving instrument dated 1534 has virtually the same form as the eighteenth-century instruments for which much keyboard music was written. Although more expressive than the harpsichord, the clavichord suffered from having a very small sound power and was almost totally eclipsed by the piano by the end of the eighteenth century. It has, however, been revived, and excellent reproductions of historical instruments are now available.

A typical clavichord is a rectangular box about 1 m long, 30 cm across, and 10 cm deep, as shown in Fig. 11.11. Some 40 strings, arranged as 20 pairs, run in the long direction of the box, are anchored to pins at one end and to tuning pegs at the other, and pass across a curved bridge mounted on a rather small soundboard near one end. The keyboard typically has about 45 keys but achieves a compass of four octaves, from C_2 to C_6, by virtue of omitting some little-used notes in the bass octave.

When a key is depressed, its other end, which bears an upright brass blade or tangent, strikes against a pair of strings and remains held against them by the force of the player's finger. This action not only sets the string into vibration but also defines the vibrating length as being the length of string from the tangent to the bridge. The other end of the string is damped by a pad of felt near the hitch pin and so does not vibrate appreciably. With

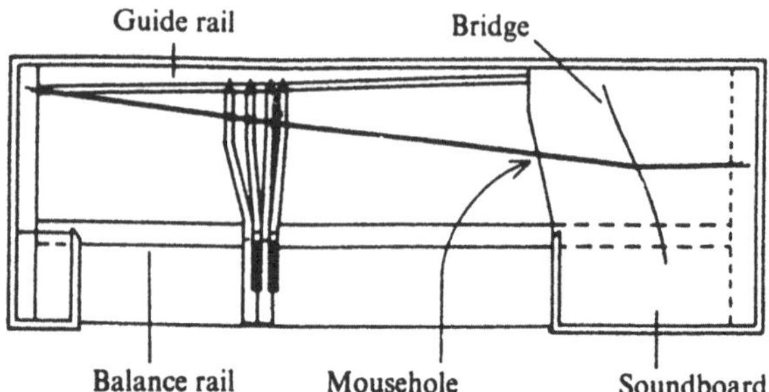

FIGURE 11.11. The typical fretted clavichord. The strings shown are the 10th pair, and there are 20 pairs in all. The overall length of the instrument is 1 m (Thwaites and Fletcher, 1981).

this mechanism, it is possible to use a single pair of strings to produce two or even three different notes, provided they are not required simultaneously. This arrangement is known as fretting and serves to reduce the size of the instrument. Some larger clavichords were built to be fret free.

Clavichords have only one set of strings and these are at 8-ft pitch. The scale, or length of the C_5 string, is similar to that of a small harpsichord, for example, 20 to 25 cm, but the bass strings are much shorter than in a harpsichord. Wire diameters are similar to but rather thicker than those of a harpsichord, from about 0.6 mm at C_2 to 0.3 mm at C_6. The string material is usually brass.

The acoustical design and performance of a typical clavichord has been analyzed in detail by Thwaites and Fletcher (1981). In our discussion, we follow the same course as we did for the harpsichord, but first we need to say a little about the way in which the string is excited.

Measurements show that the tangent hits the string at a speed of about 0.5 m s^{-1} and comes to rest in about 10 ms. There is generally no bounce of the string against the tangent and, when there is, the sound is unsatisfactory. The tangent does, however, vibrate a little under the influence of subsequent string vibrations.

From the viewpoint of the vibrating section of the string, we can simply regard one of its end points as being smoothly displaced, over a time of about 10 ms, through a distance of 1 to 2 mm. This is most easily treated by considering the initial state to be that of the string in its displaced position with an appropriate velocity distribution along its length, corresponding to the traveling wavefront of the end displacement. More important than the form of the string motion itself, however, is the spectral distribution of the transverse force on the bridge. It turns out that, if the tangent comes to rest over several periods of the fundamental mode of the string vibration, as is usually the case, then the bridge force has a spectrum that falls smoothly at about 8 dB/octave. If the deceleration of the tangent is more rapid, then the first few harmonics are weaker than would be expected from this rule, or, equivalently, all but the first few harmonics are stronger.

The total energy given to the string depends on the force of the key stroke, so that the player has some control over dynamics, which is an expressive advantage. While the tangent remains in contact with the string, the player can also execute a pitch vibrato by varying the force exerted on the key and thus the tension of the string.

Turning now to the soundboard, we see from Fig. 11.11 that it forms the top of a cavity in the side of which is a hole—the aptly named mousehole. We therefore expect a cavity resonance at about 160 Hz in the absence of soundboard motion. In fact, the cavity is almost divided in two by the balance rail on which the keys are pivoted, and this leads to another prominent cavity mode at a higher frequency. Both these cavity modes should couple to the plate modes of the soundboard, in much the same way as do the modes of a violin or guitar, as discussed in Chapters 9 and 10. Thwaites and Fletcher (1981) have analyzed these modes in detail and have shown that they account for the first few peaks of the admittance spectrum of the soundboard as measured at the bridge. This spectrum, measured at several positions along the bridge, is shown in Fig. 11.12. It is interesting to note the way in which individual admittance peaks may disappear when the measurement is made at a position close to a node line for the mode concerned. The mean level of the admittance is about 3×10^{-2} m s^{-1} N^{-1}, corresponding to an impedance of 30 N m^{-1} s, which is at the lower end of the range estimated for the harpsichord. This is appropriate since the clavichord has a rather thinner soundboard than the harpsichord and it has no braces.

Because the lower strings of the clavichord are much shorter than those of the harpsichord, the air damping is less, but the greater admittance of the soundboard means that energy transfer through the bridge is greater, and indeed seems to be dominant. The total energy involved is, however, less than for the harpsichord, and the small soundboard is an inefficient radiator. The radiated intensity or sound pressure level at a distance of 1 m is about 48 ± 3 dB(A) over most of the compass, but falls to about 40 dB(A) in the extreme bass.

The behavior of the sound decay of the clavichord is more complex than that of the harpsichord because each note has two strings. This effect has been discussed by Weinreich (1977, 1979) and by Gough (1981). Briefly, as discussed in Section 4.10, coupling between the two strings, through the bridge, can split their motion into a symmetric mode in which both strings vibrate in-phase and an antisymmetric mode in which their motion is antiphase. The in-phase mode exerts a large force on the bridge, produces a loud sound, and decays rather rapidly by energy loss to the soundboard. Indeed, since the force is twice that produced by a single string, the energy transfer is increased by a factor of 4 while the stored energy is only twice that of a single string. The antiphase motion, on the other hand, exerts a

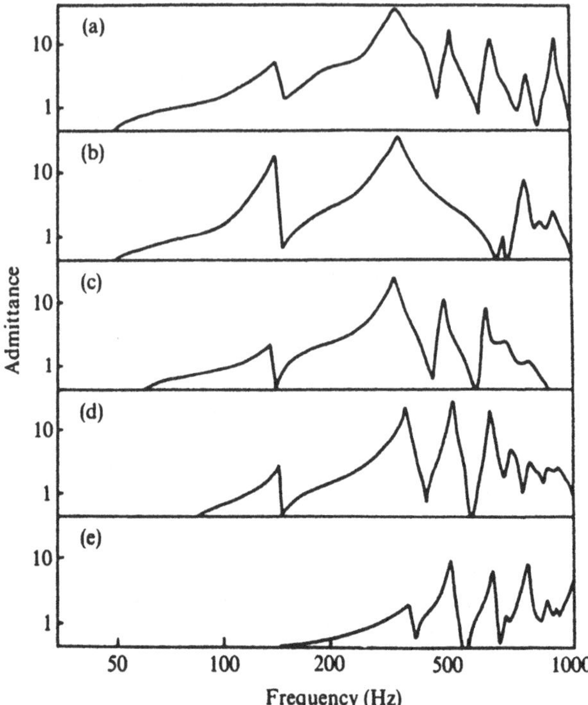

FIGURE 11.12. Mechanical admittance of the soundboard of a clavichord, in dB relative to 10^{-2} m s^{-1} N^{-1}, measured at five different positions along the bridge from the bass-string end at the top to the treble at the bottom of the figure (Thwaites and Fletcher, 1981).

very small force on the bridge, is rather low in sound power, and decays slowly. This behavior can be observed in the clavichord. The overall decay time to inaudibility is about 4 s for notes from C_2 to C_5, but falls to less than 1 s through the top octave to C_6 in the instrument studied.

The clavichord is thus primarily an instrument for private solo playing of music specially written for it, rather than for ensemble playing or public performance.

References

Ando, M. (1989). Koto scales and tuning. *J. Acoust. Soc. Jpn.* **E10**, 279–287.

Ando, Y. (1986). Acoustics of sohs (kotos). *Proc.* 12 *Int. Congress on Acoustics, Toronto*, Vol. 3, paper K1–5.

Baines, A. (1966). "European and American Musical Instruments," Figs. 159–406. Viking Press, New York.

Bell, A.J. (1997). The Helmholtz resonance and higher air modes of the harp soundbox. *Catgut Acoust. Soc. J.* **3**(3), 2–8.

Canfield, G., and Rossing, T.D. (1995). Modes of vibration in hammered dulcimers. *Proc. Int. Symp. Mus. Acoust.* Dourdan. ISMA, Paris. pp. 512–517.

Firth, I.M. (1977). On the acoustics of the harp. *Acustica* **37**, 148–154.

Firth, I.M. (1986). Acoustics of the Irish, Highland and Baroque harps. *Proc.* 12 *Int. Congresss on Acoustics, Toronto*, Vol. 3, paper K3–10.

Fletcher, N.H. (1977). Analysis of the design and performance of harpsichords. *Acustica* **37**, 139–147.

Gough, C.E. (1981). The theory of string resonances on musical instruments. *Acustica* **49**, 124–141.

Hubbard, F. (1967). "Three Centuries of Harpsichord Making." Harvard Univ. Press, Cambridge, Massachusetts.

Kellner, H.A. (1976). Theoretical physics, the harpsichord, and its construction—a physicist's annotations. *Das Musikinstrument* **2**, 187–194.

Kottick, E.L. (1987). "The harpsichord Owner's Guide." Univ. North Carolina Press, Chapel Hill, N.C.

Lieber, E. (1975). Moderne Theorien die Physik der schwingenden Saite und ihre Bedeutung für die musikalische Akustik. *Acustica* **33**, 324–335.

Marcuse, S. (1975). "A Survey of Musical Instruments." Harper and Row, New York.

Peterson, D.R. (1995). Hammer/string interaction and string modes in the hammered dulcimer. *Proc. Int. Symp. Mus. Acoust.* Dourdan. IRCAM, Paris. pp. 533–538.

Thomas, M. (1971). String gauges of old Italian harpsichords. *Galpin Society Journal* **24**, 69–78.

Thwaites, S., and Fletcher, N.H. (1981). Some notes on the clavichord. *J. Acoust. Soc. Am.* **69**, 1476–1483.

Trendelenburg, F., Thienhaus, E., and Franz, E. (1940). Zur Klangwirkung von Klavichord, Cembalo und Flügel. *Akust. Zeits.* **5**, 309–323.

Weinreich, G. (1977). Coupled piano strings. *J. Acoust. Soc. Am.* **62**, 1474–1485.

Weinreich, G. (1979). The coupled motion of piano strings. *Sci. Amer.* **240** (1), 118–127.

Weyer, R.D. (1976). Time-frequency-structures in the attack transients of piano and harpsichord sounds-I. *Acustica* **35**, 232–252.

Weyer, R.D. (1976/1977). Time-varying amplitude-frequency-structures in the attack transients of piano and harpsichord sounds-II. *Acustica* **36**, 241–258.

Zuckermann, W.J. (1969). "The Modern Harpsichord." October House, New York.

12

The Piano

The piano has become the most versatile and popular of all musical instruments. It has a playing range of more than seven octaves (A_0 to C_8), and a wide dynamic range as well. It is widely used as a solo instrument, but it is also used to accompany other solo instruments and especially singing.

The modern piano is a direct descendant of the harpsichord. In 1709, Bartolomeo Cristofori of Florence substituted hammers in place of jacks and rendered the harpsichord capable of gradations in tone. He called his new instrument the "gravicembalo col piano et forte," which later became pianoforte and eventually was shortened to piano. The oldest existing Cristofori piano, dating from 1720, is in the Metropolitan Museum of Art in New York; it can be heard on a recording (Conklin, 1990).

Cristofori's invention was taken up by the German harpsichord and organ builder Gottfried Silbermann, along with his contemporaries Johannes Zumpe and Andreas Stein. Adverse criticism of their early instruments by J.S. Bach may have led to improvements in the action and then to eventual acceptance by Bach and his contemporaries. Zumpe later moved to England, where he teamed up with John Broadwood in developing a number of improvements to the pianoforte. Stein invented an escapement mechanism, and along with his daughter Nanette Streicher, he developed the Viennese type of pianoforte played by Mozart. Pierre Erard is credited with inventing the agraffe and also the double repetition action.

The upright pianoforte was developed by John Hawkins of Philadelphia and Robert Wornum of London about the middle of the nineteenth century. Hawkins also introduced the iron frame that allowed the use of higher tension and thicker wire. Improvements to strings and soundboards, as well as the piano action, continued well into the present century.

Out of this 280-year history, two distinctly different musical instruments have evolved. One is the modern grand piano, which is built in various sizes from the small baby grand (generally 5 to 6 ft in length) to the magnificent concert grand piano (generally 8 to 9 ft in length). The other instrument is the upright piano, which also exists in different sizes (from about 3 to

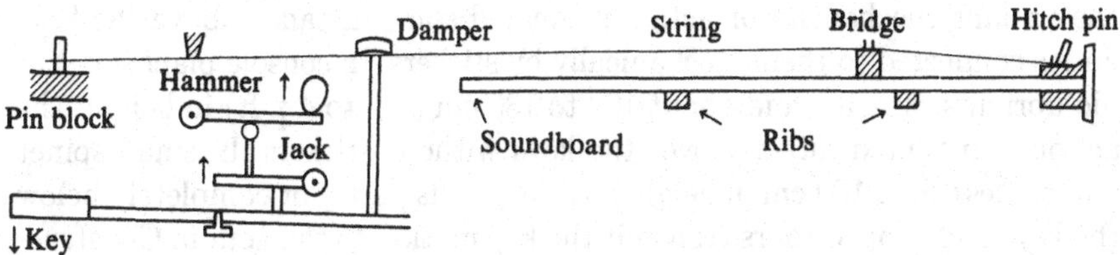

FIGURE 12.1. A simplified diagram of the piano. When a key is depressed, the damper is raised, and the hammer is thrown against the string. Vibrations of the string are transmitted to the soundboard by the bridge (from Rossing, 1982).

5 ft in height). Most pianos have 88 keys, although a few special concert grands have as many as 97.

12.1 General Design of Pianos

The main parts of a piano are the keyboard, the action, the strings, the sound board, and the frame. A simplified diagram of a piano is shown in Fig. 12.1. The strings extend from the pin block across the bridge to the hitch-pin rail at the far end. When a key is depressed, the damper is raised, and the hammer is thrown against the string, setting it into vibration. Vibrations of the string are transmitted to the soundboard by the bridge.

A typical concert grand piano has 243 strings, varying in length from about 2 m at the bass end to about 5 cm at the treble end. Included are 8 single strings wrapped with 1 or 2 layers of wire, 5 pairs of strings also wrapped, 7 sets of 3 wrapped strings, and 68 sets of 3 unwrapped strings. Smaller pianos may have fewer strings, but they play the same number of notes: 88. A small grand piano with 226 strings is shown in Fig. 12.2. Note that the bass strings overlap the middle strings, which allows them to act nearer the center of the soundboard. The acoustical advantages of wrapped strings and multiple strings for most notes will be discussed in Section 12.3.

The soundboard is nearly always made of spruce, up to 1 cm thick, with its grain running the length of the piano. Ribs on the underside of the soundboard stiffen it in the cross-grain direction. The soundboard is the main source of radiated sound, just as is the top plate of a violin or cello.

To obtain the desired loudness, piano strings are held at high tensions that may exceed 1000 N (215 lb). The total force of all the strings in a concert grand is over 20 tons! In order to withstand this force and maintain stability of tuning, pianos have sturdy frames of cast iron.

The basic construction of a large vertical or upright piano is shown in Fig. 12.3. Note that the hammers move horizontally to strike the strings. Upright pianos vary considerably in design, depending upon their size. In full-size upright pianos, which stand 130 to 150 cm ($4\frac{1}{4}$ to 5 ft) in height, the striking mechanism or action is located some distance above the keys and is connected to them mechanically by stickers. In console pianos or studio uprights, which stand about 100 to 130 cm ($3\frac{1}{4}$ to $4\frac{1}{4}$ ft) in height, the action is mounted directly over the keys without stickers. In small spinet pianos (less than 100 cm in height), the action is partly or completely below the keys and drop stickers transmit the key motion to the action. Construction of a console piano and a spinet are shown in Fig. 12.4 (compare to the upright in Fig. 12.3).

12.2 Piano Action

In principle, the piano action transmits the energy from the player's fingers to the appropriate hammer and thence to the string. In practice, this is quite a complicated process, and a lot of ingenuity has gone into the development of modern piano actions.

12.2.1 Grand Piano Action

A typical grand piano action is shown in Fig. 12.5. When a key is pressed down, the capstan sets the whippen into rotation, causing the jack to push against the hammer knuckle or roller, starting the hammer on its journey toward the string. When the key is halfway down, the back end of the key engages the damper lever and begins to raise the damper from the string. As the key continues, the jack bumps into the jack regulator or letoff button, and rotates away from the knuckle. The hammer continues to rotate freely, strikes the string, and rebounds. But, since the jack has been rotated out of the way (tripped), the knuckle lands on the repetition lever instead. The downward-moving hammer rotates the repetition lever, compressing the repetition lever spring. The downward motion of the hammer continues until it catches on the backcheck, which prevents it from bouncing back to the string a second time.

When the key is released slightly, the backcheck releases the hammer tail, allowing the repetition lever to support the hammer just enough for the jack to slip back under the knuckle (as soon as the jack clears the jack regulator). Because the jack clears the regulator when the key is less than halfway up, the action is ready to repeat a note without the need for the key to return to its rest position. This is a desirable feature of the modern grand piano action. The cycle of the action is illustrated in Fig. 12.6.

12.2.2 Upright Piano Action

The action in an upright piano is different from that of a grand piano in several ways. First of all, the hammers and dampers move horizontally, and

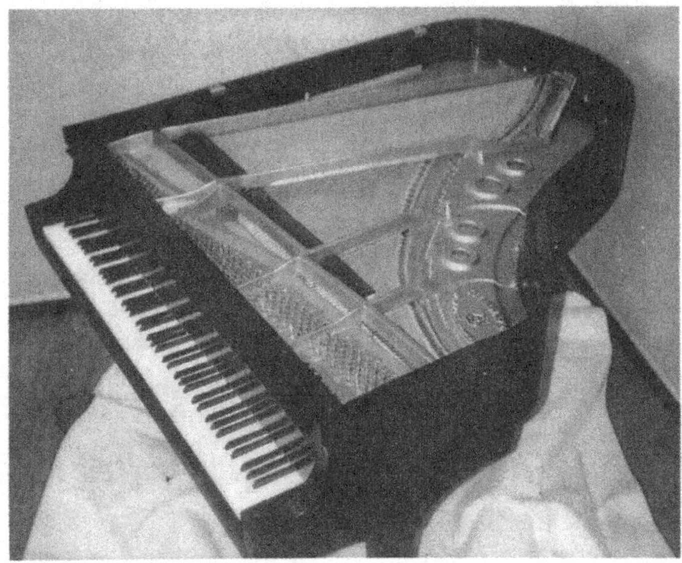

FIGURE 12.2. Top view of a small grand piano showing the cast-iron frame, the overlapping strings, hammers, and dampers. (Courtesy of Steinway & Sons)

FIGURE 12.3. Basic construction of a large upright piano. Note the diagonal orientation of the bass strings. The action is above the keys, and the cast-iron frame can be seen both above and below the keys (Reblitz, 1976).

(a)

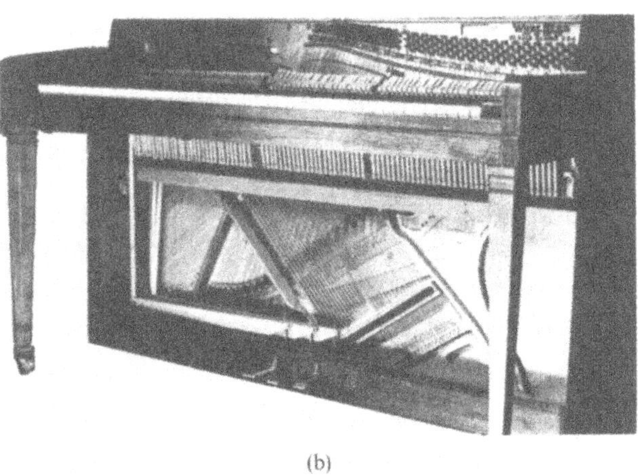

(b)

FIGURE 12.4. (a) Console piano with action mounted above the keyboard. (b) Spinet with action partly below the keyboard (Reblitz, 1976).

thus depend upon spring action to return them to their original positions. Second, the upright has no repetition lever, so the key must be released nearly to its rest position before a note can be repeated. Third, stickers transmit key motion to the whippens when the action is above the keys (in a full-size upright piano) or below the keys (in a spinet). These arrangements are shown in Fig. 12.7.

When a key is depressed, it raises the sticker [(a) and (c)] and the whippen. The whippen pushes the jack against the hammer butt, moving the hammer toward the string. When the key is about halfway down, the spoon

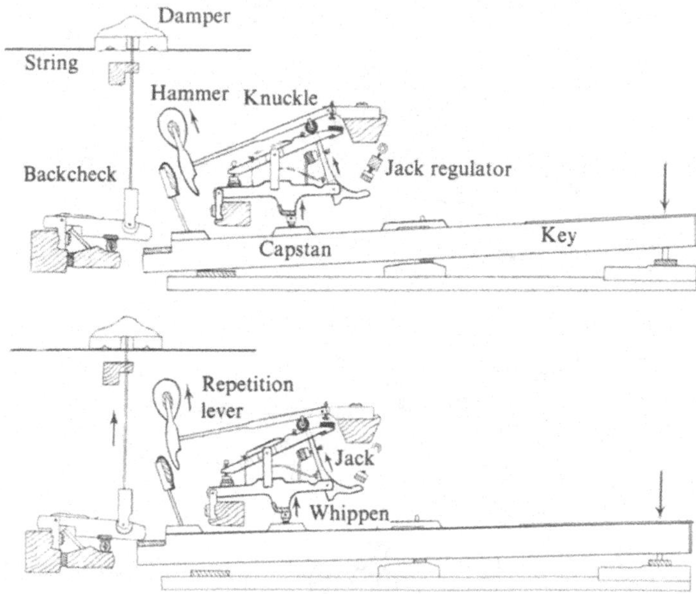

FIGURE 12.5. Typical grand piano action (after Reblitz, 1976).

raises the damper from the string. When the hammer has almost reached the string, the jack heel encounters the regulating button, causing the jack to pivot out from under the hammer butt. The hammer continues on to strike the string and then rebounds to be caught by the backcheck and held until the key is released.

12.2.3 Key Dynamics

Figure 12.8 is a simplified diagram of the key and hammer. When the key is pressed down with force F at point P, the force is transmitted to the hammer by the action of two levers, as shown. During the acceleration phase, the key touch point moves at a velocity $v_{\mathrm{p}}(t)$ while the hammer has a velocity $v_{\mathrm{h}}(t)$. For Steinway and Bechstein grand pianos, $v_{\mathrm{h}}/v_{\mathrm{p}} = 5.5$ during about 70% of the touch, after which the hammer swings free (Dijksterhuis, 1965).

In the following calculation, we use the parameters given by Dijksterhuis (1965). The static force necessary to move the key is about 0.44 N (equivalent to raising a mass of 45 g). To this must be added the force necessary to accelerate the key, the hammer (m_{h}), and the other parts of the action (m_{a}). The key has a moment of inertia of about 1.7×10^{-3} kg m^2. The equivalent mass m_{a} created by the moment of inertia of the action is about 0.0166 kg.

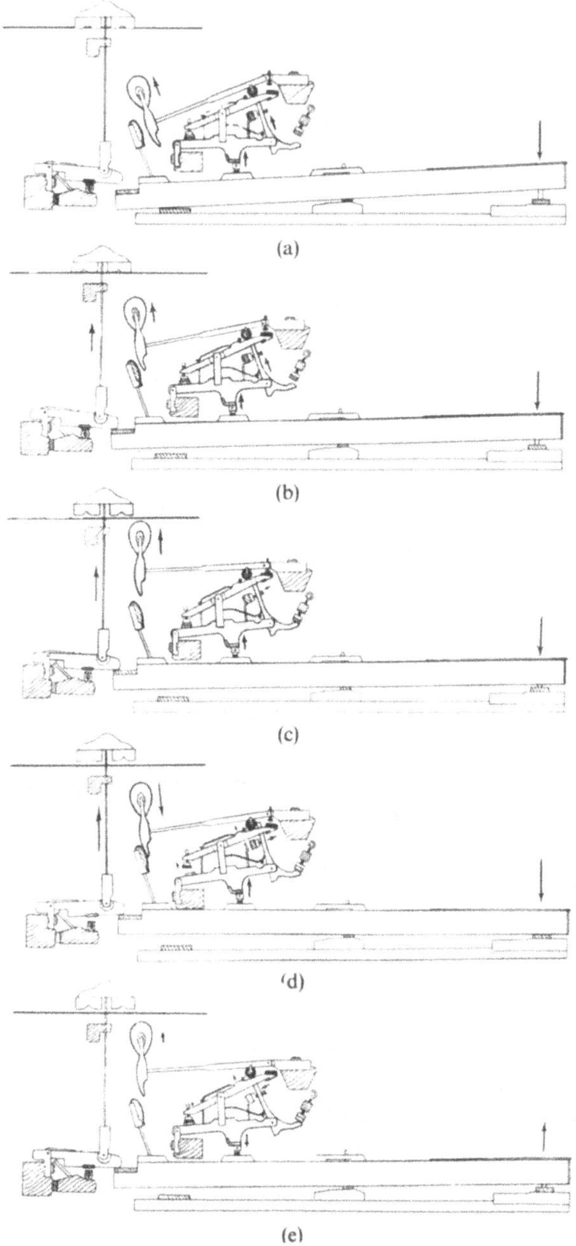

FIGURE 12.6. Operation of the grand piano action. (a) Key is pressed down, raising whippen, jack, and hammer. (b) Key begins to lift the damper. (c) Jack engages the regulator, rotates away from hammer knuckle (roller). (d) Hammer rebounds from string and is caught by the backcheck; roller depresses the repetition lever. (e) When key is released a little, backcheck releases hammer, allowing repetition level to lift hammer until jack returns under knuckle. Key is ready for another cycle (after Reblitz, 1976).

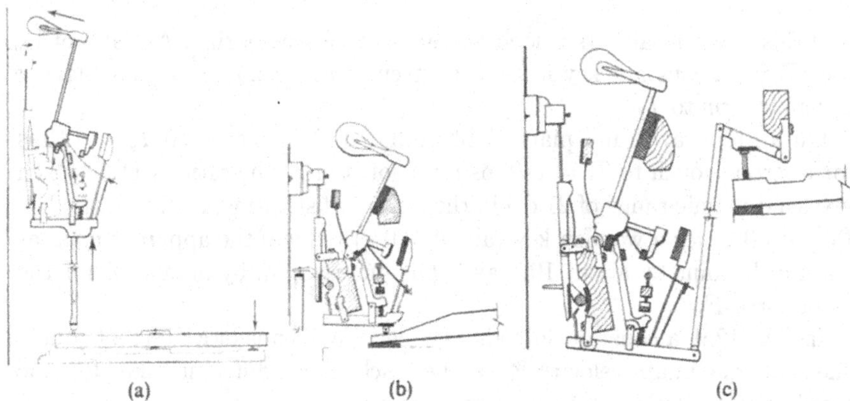

FIGURE 12.7. Upright piano actions: (a) full-size upright (action above the keys), (b) console (directly connected to keys), and (c) spinet (drop action located partly below keys) (after Reblitz, 1976).

The equation of motion for the system is then

$$F - F_s = m_k(0.018a_p) + m_a(0.72a_p) + m_h(5.5a_p). \tag{12.1}$$

where a_p is the acceleration at the touch point. Substituting numerical values for m_k, m_a, and m_h gives

$$a_p = \frac{dv_p}{dt} = 3.3(F - 0.44). \tag{12.2}$$

If the applied force F is kept constant, the time T_s for the key to travel the distance s to the stop is $T_s = (2s/a_p)^{1/2}$. For $s = 9$ mm, this becomes

$$T_s = 0.074(F - 0.44)^{-1/2}. \tag{12.3}$$

When the key strikes the stop, the hammer has a velocity V_0 given by

$$V_0 = 5.5a_pT_s = 1.34(F - 0.44)^{1/2}, \tag{12.4}$$

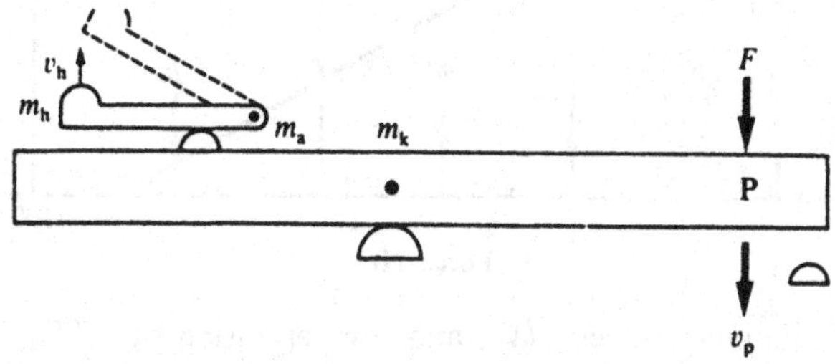

FIGURE 12.8. Simplified diagram of key and hammer.

and this is essentially its velocity when it strikes the string. A graph of V_0 and T_s for a constant key force F is given in Fig. 12.9 for various playing dynamics (pp to ff).

On a Steinway grand piano, Dijksterhuis (1965) measured $T_s = 12$ ms for a strong touch to $T_s = 140$ ms for a soft touch; the ratio is 11.5, which gives a dynamic range of 21 dB in the force. Substituting a light plastic key (about 50 g) for a wooden key (about 130 g) reduces the apparent mass at the touch point by about 10% and thus increases V_0 by about 5% for the same force F.

In Fig. 12.9, a constant force is assumed. By means of a different touch, the same maximum velocity V_0 can be reached at a different time after the touch. Dijksterhuis (1965) compares three cases:

Case 1: Constant force F_1.

Case 2: Force increases with time as $F_2\left(1 - \cos\dfrac{\pi t}{T_2}\right)$.

Case 3: Force decreases with time as $F_3\left(1 - \cos\dfrac{\pi t}{T_3}\right)$.

For the same value of V_0, the following parameters hold:

Case 1: $T_s = T_1$, $F = F_1$.
Case 2: $T_2 = 1.68T_1$, $F_2 = 1.18F_1$.
Case 3: $T_3 = 0.7T_1$, $F_3 = 2.8F_1$.

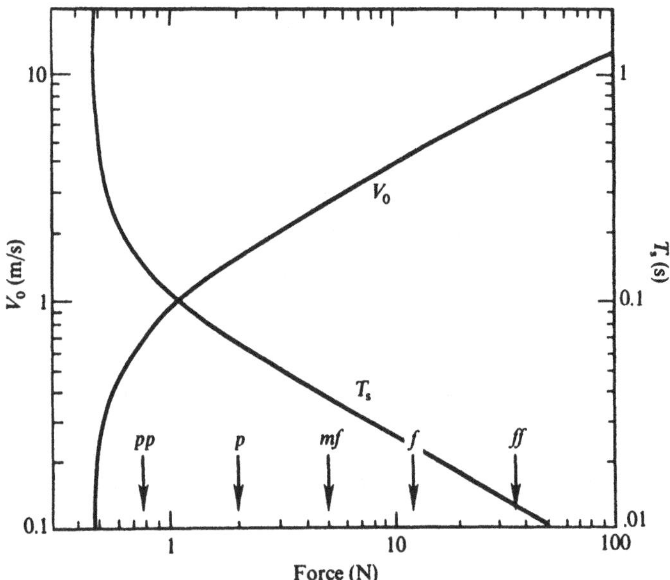

FIGURE 12.9. Hammer velocity (V_0) and key depressing time (T_s) for different values of force (assumed to be constant during the time T_s) (after Dijksterhuis, 1965).

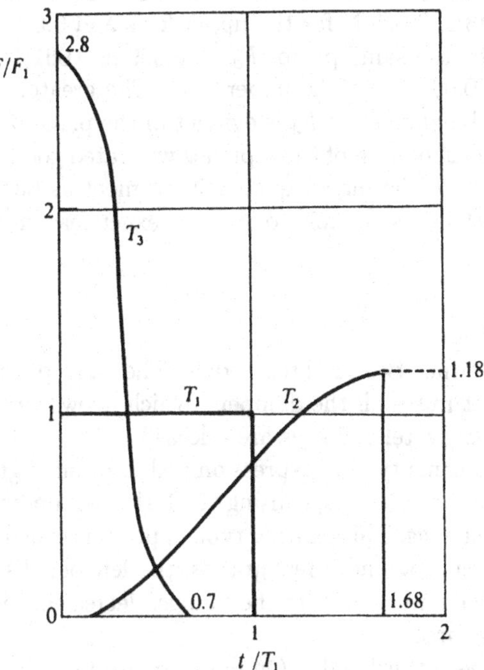

FIGURE 12.10. Computed key travel times for increasing force (T_2) and decreasing force (T_3) compared to constant force (T_1) during touch (after Dijksterhuis, 1965).

The results are shown in Fig. 12.10.

The three cases summarized in Fig. 12.10 show examples of how identical tones could be achieved with different touch. Although V_0 is the same, and therefore the tones will be equally loud, the tones will sound at different times. A skilled pianist could play a chord in such a way that a melody note sounds earlier or later (but at the same intensity) as the rest of the chord.

Henderson (1936) contends that, contrary to general opinion, accentuated notes are not played with greater intensity than unaccentuated notes; rather, the pianist is inclined to play the accentuated notes a little earlier. Vernon (1936) remarks that Padarewski played 56% of the chords asynchronously in Beethoven's *Moonlight Sonata* but only 20% in Chopin's *Polonaise Militaire.* The touch of a pianist appears to be of importance in piano performance.

Lieber (1985) compares the static force F_s necessary to press a key downward and the dynamic force F_d needed to just make a note sound (*ppp*). For an upright piano of high quality, $F_s = 0.480 \pm 0.013$ N for the lower keys and 0.488 ± 0.010 N for the upper keys. $F_d = 0.653 \pm 0.062$ N for the lower keys and 0.673 ± 0.059 N for the upper keys. For an upright of medium quality, $F_s = 0.443 \pm 0.025$ N for the upper keys and 0.532 ± 0.028 N for the lower keys. For the same piano, $F_d = 0.669 \pm 0.061$ N for the lower keys and 0.471 ± 0.030 N for the upper keys. The greater variation from key to key in the dynamic force F_d is evident in the piano of lower quality.

Although the touch of each of these pianos was rated good, some pianists described the touch of the higher quality instrument as being like that of a grand piano, without being able to give the exact reason why.

12.2.4 Pedals

Pianos may have either two or three pedals. The right pedal is called the sustaining pedal. It raises all the dampers, which allows the struck strings to continue vibrating after the keys are released.

The left pedal is some type of expression pedal. In most grand pianos, it shifts the entire action sideways, causing the treble hammers to strike only two of their three strings. This shifting type of pedal is called the *una corda* pedal. In vertical pianos, and a few grands, the left pedal is a soft pedal, which moves the hammers closer to the strings, decreasing their travel and thus their striking force.

Many pianos have a third pedal. On most grands and a few uprights, the center pedal is a *sostenuto* pedal, which sustains only those notes that are depressed prior to depressing the pedal, and does not sustain subsequent notes. On a few pianos, the center pedal is a bass sustaining pedal, which lifts only the bass dampers. On a few uprights, the center pedal is a practice pedal, which lowers a piece of felt between the hammers and strings, muffling the tone (Reblitz, 1976).

12.3 Piano Strings

The strings are the heart of the piano. They convert some of the kinetic energy of the moving hammers into vibrational energy, store this energy in normal modes of vibration, and pass it on to the bridges and soundboard in a manner that determines the sound quality of the instrument.

In Chapter 2, we discussed the motion of a string excited by striking it with a moving hammer. The impact causes a disturbance to propagate in both directions on the string. Reflected impulses return from both ends of the string and interact with the hammer in a complicated way. Eventually, the hammer is thrown clear of the string, and the string vibrates freely in its normal modes.

Piano strings make use of high-strength steel wire. Efficiency of sound production calls for the highest string tension possible, while at the same time minimizing inharmonicity calls for using the smallest string diameter (core diameter in a wrapped string) possible. This results in tensile stresses of around 1000 N/mm^2, which is 30–60% of the yield strength of high-strength steel wire. For steel with an elastic modulus of 2×10^{11} N/m^2, this results in an elongation of about $\frac{1}{2}$% when the string is under tension. Fortunately, strings usually break near the keyboard end so that the broken string recoils away from the pianist.

Although the treble strings of a piano are solid wire, the lower strings consist of a solid core wound with one or two layers of wire (usually copper). This minimizes the stiffness (and thus the inharmonicity) in the lower strings, where greater mass is required. The diameter of the copper winding may vary from twice (lowest string) to one-fourth (highest wrapped string), the diameter of the core.

The expressions for stiff strings with simply-supported (pinned) ends and clamped ends were given by Eqs (2.67a) and (2.67b), respectively. Note that they differ by a factor of $[1 + (2/\pi)B^{1/2} + (4/\pi^2)B]$. At a typical value for the inharmonicity coefficient in the middle register ($B = 0.0004$), this factor is about 1.014. The actual termination of a real piano string is somewhere between pinned and clamped, but this factor is small enough to make the difference unimportant, so the simpler expression in Eq. (2.67a) can generally be used. Both expressions lead to the same amount of stretching, which at $B = 0.0004$ is sufficient to shift the 17th partial one "partial position" to the frequency of the 18th partial of an ideal string without stiffness.

When relating the inharmonicity to the actual frequency of the stretched fundamental f_1 rather than to the frequency of the ideal string, f_n is given approximately by

$$f_n \approx nf_1[(1 + n^2B)/(1 + B)]^{1/2}. \tag{12.5}$$

It is sometimes convenient to express the deviation of the partial frequencies from a harmonic series by numbers of cents (100 cents = 1 semitone). The inharmonicity I_n, which expresses the ratio f_n/nf_1, is given by

$$I_n = 1200 \log_2[1 + n^2B]^{1/2} \approx 866n^2B = bn^2, \tag{12.6}$$

where $b = 866B$ is the inharmonicity coefficient in cents.

For the singly wound and doubly wound strings in Fig. 12.11, Sanderson (1983) gives the following formulas:

$$\text{Single-wound:} \quad b = b_{\text{core}} + b_{\text{end 1}} + b_{\text{end 2}}, \tag{12.7}$$

where

$$b_{\text{end}} = 0.287 \left(\frac{D_2^2 - d^2}{D_2^2 - 0.12d^2} \right) \left(4 \sin \frac{4\pi L_1}{L_s} - \sin \frac{16\pi L_1}{L_s} \right).$$

$$\text{Double-wound:} \quad b = b_{\text{core}} + b_{\text{end 1}} + b_{\text{end 2}} + b_{\text{step 1}} + b_{\text{step 2}}, \tag{12.8}$$

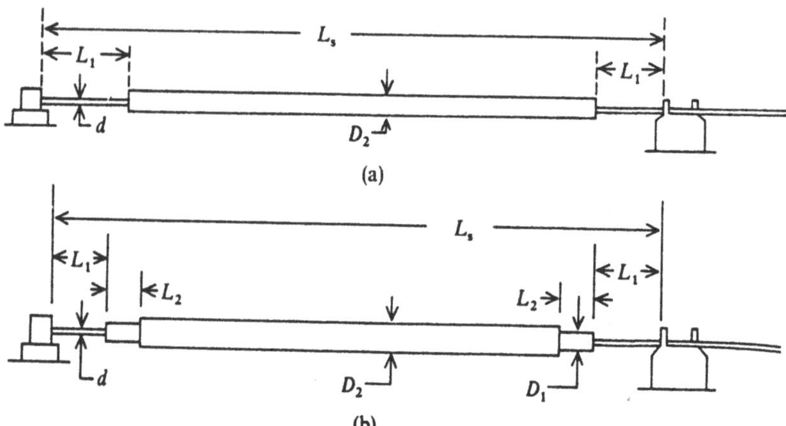

FIGURE 12.11. (a) Singly-wound string and (b) doubly-wound string.

where b_{end} is the same as for the single-wound string and

$$b_{\text{step}} = 0.287\left(\frac{D_2^2 - D_1^2}{D_2^2 + 0.12d^2}\right)$$

$$\cdot\left(4\sin\frac{4\pi(L_1+L_2)}{L_s} - \sin\frac{16\pi(L_1+L_02)}{L_s} - 4\sin\frac{4\pi L_1}{L_s} + \sin\frac{16\pi L_1}{L_s}\right).$$

Scaling is the term applied to selecting the string diameters for the various notes on the piano. Figure 12.12 shows how the tension and inharmonicity might vary with note number using high-tension and low-tension scales. Note that the tension decreases as we go from unicord (one string per note) to bicord (two strings per note) to tricord (three strings per note).

Kock (1937) modeled a piano string by an electrical transmission line, an approach that has been followed by several subsequent investigators as well. Mass and compliance per unit length are represented by inductance and capacitance per unit length, and transverse displacement is represented by current. The hammer is represented by an inductance with a voltage across its terminals.

More refined transmission line models, including an impedance analogy and a mobility analogy, are discussed by Nakamura (1982). The analog simulator shown in Fig. 12.13, which uses the impedance analogy, includes the hammer. Switches S_1 and S_2 operate in opposite directions. The condition when S_1 is closed and S_2 is open represents the hammer touching the string.

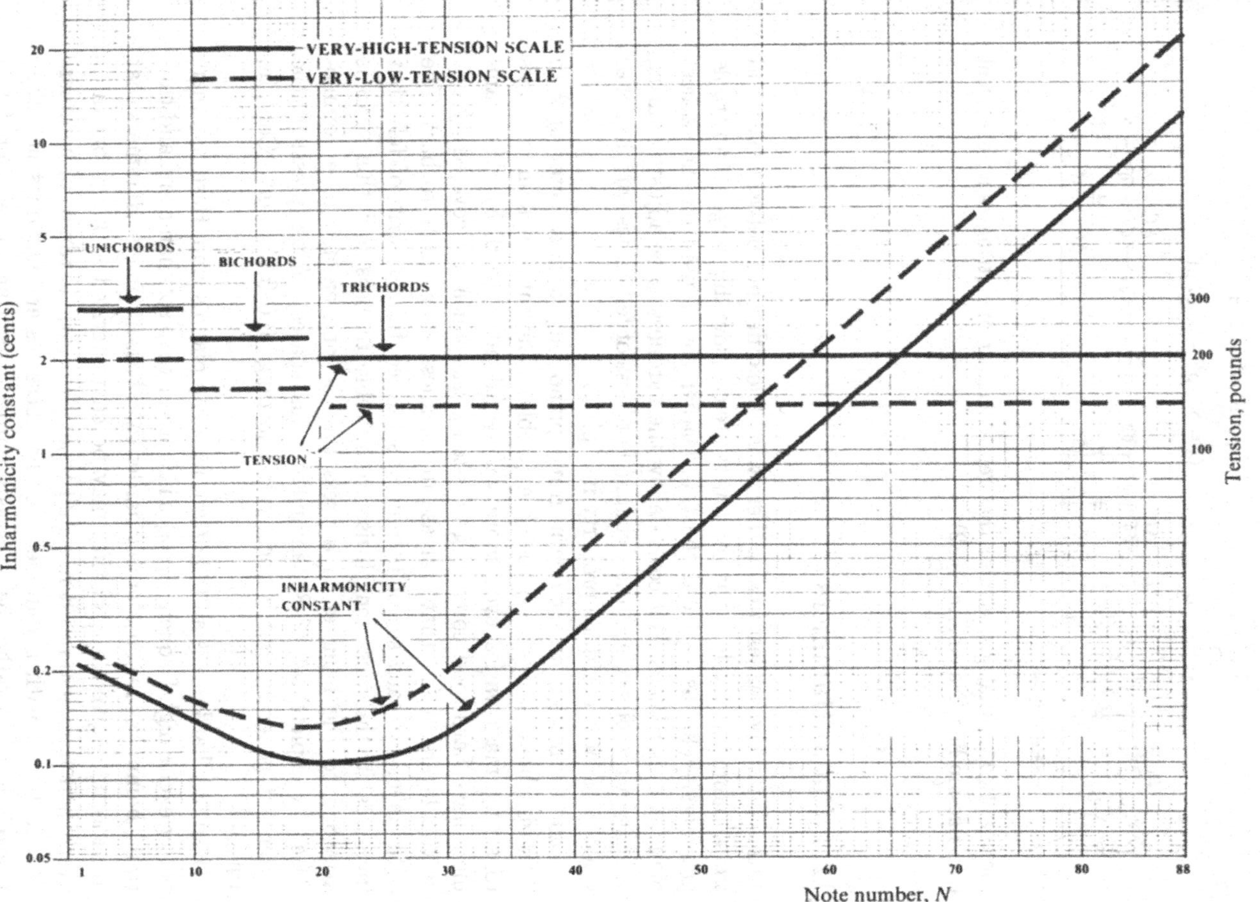

FIGURE 12.12. Piano string scales: variation of tension and inharmonicity with note number (Sanderson, 1983).

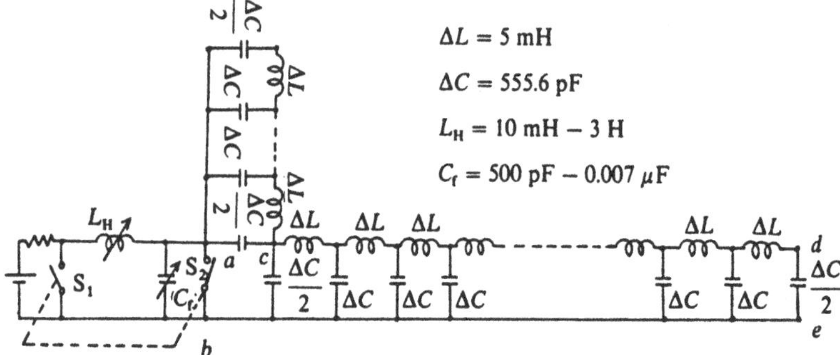

FIGURE 12.13. Analog simulator that represents a piano string as an electrical transmission line (Nakamura, 1982).

12.4 Piano Hammers

Although in the earliest pianos there was little difference between bass and treble hammers, modern bass hammers are considerably larger than treble hammers. Felt has replaced the thin leather covers used on the earliest piano hammer heads. Felt covers allow the hardness to be adjusted during voicing of the piano, and the dynamic hardness of felt varies with hammer velocity, a most desirable property in a piano.

Hammer masses in a modern grand piano vary over a factor of 2 to 3, from about 10 g (bass) to 3.8 g (treble). The ratio of hammer mass to string mass varies even more widely (about 8 to 0.08). The hammers have hardwood cores of graduated sizes, which are covered by one or two layers of felt that increase in thickness from treble to bass. The outer layer for each set of hammers is formed from a strip of felt that increases in thickness lengthwise from treble to bass and this strip is attached to the hammer in a press that is long enough to accommodate a complete set of 88 hammers (plus spares). After pressing, the outer felt generally has a density in the range 0.6–0.7 g/cm^3 (Conklin, 1996a).

Hammer hardness is an important factor in piano sound. Hard hammers better excite high-frequency modes in a string than soft hammers, and thus treble hammers are considerably harder than bass hammers. Hammers that are too hard give a sound that may be characterized as harsh or tinny; overly soft hammers result in a dull tone. The static hardness of hammers can be tested with a durometer or hardness tester. Felt does not obey Hooke's law but rather behaves as a hardening spring (see section 1.12). Force-compression measurements on piano hammers can be characterized by a power law:

$$F = K\xi^p, \tag{12.9}$$

where F is force, ξ is the compression, K is a generalized stiffness, and the exponent p, which describes how much the stiffness changes with force, ranges from 2.2 to 3.5 for hammers taken from pianos and 1.5 to 2.8 for unused hammers (Hall and Askenfelt, 1988).

Dynamic measurements of the force and compression show hysteresis: K and p take on different values for compression and relaxation. Hard hammers tend to have a larger value of the exponent p than soft hammers. Using a nonlinear hammer model along with a flexible string, Hall (1992) found that increasing exponent p leads to a smoother and more gradual rise of force as the hammer compression begins and a correspondingly steeper falloff in string mode amplitudes toward higher frequencies.

For the best tone, each hammer should have its hardness within a certain range, and the hardness should have a gradient such that the string-contacting surface is softer than the inner material. Figure 12.14 shows the hardness gradient for three different G_5 hammers measured with a durometer (Conklin, 1996a). The hardness was measured at five points across the outer felt layer from the inside (0) to the outer string-contacting surface (100% of the width). The center curve represents a hammer that was judged to be within the optimum quality range; the other two hammers produced tones that were judged a bit too dark (bottom curve) and bright but somewhat harsh (top curve).

It is well known that piano tone spectra change with dynamic level. The principal reason for this is found in the nonlinear mechanical behavior of piano hammers, in particular the change in dynamic hardness with force. At high velocity the hammer felt is effectively harder than at low velocity, and this results in considerably more excitation of the high-frequency modes in the string and the resulting radiation of more high-frequency partials. This spectral shift toward the higher partials is at least as important a component in a crescendo as an increase in total sound level.

One way to test the dynamic hardness of a piano hammer is to observe the force–time pulse shape of the hammer as it strikes a rigid surface. Hall and Askenfelt (1988) determined that the contact duration τ is related to the maximum force $F_{\max}$ by $\tau \propto (F_{\max})^{(1-p)/2p}$, where p is the same as in Eq. (12.9). The residual shock spectrum (Broch, 1984) provides related information about what string modal frequencies would be most effectively excited by a given hammer (Russell and Rossing, 1998).

12.4.1 String Excitation by the Hammer

The dynamics of the hammer–string interaction is one aspect of piano physics that has been the subject of considerable research, beginning with

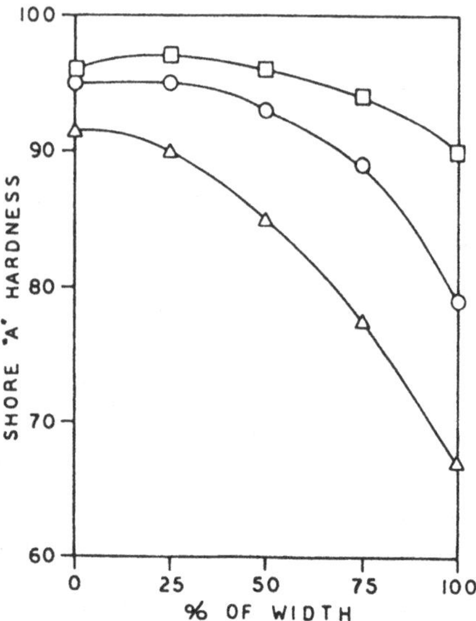

FIGURE 12.14. Hardness gradients for three G_5 piano hammers of similar size, weight, and shape. 0 is at the center of the outer felt layer; 100% is near the string-contacting surface. (from Conklin, 1996a).

von Helmholtz (1877). The problem drew the attention of a number of Indian researchers, including C.V. Raman, in the 1920s and 1930s. More recently, the subject has been reviewed by Askenfelt and Jansson (1985, 1993), Conklin (1996a), and Hall (1986, 1987a–c, 1988, 1992). These reviews are highly recommended, and this brief discussion will draw rather heavily from them.

In Section 2.9, we briefly discussed what happens when a string is struck by a hammer having considerably less mass than the string. In the most extreme case (which may apply at the bass end of the piano), the hammer is thrown clear of the string by the first reflected pulse. The theoretical spectrum envelope [shown in Fig. 2.10(a)] is missing harmonics numbered n/β (where β is the fraction of the string length at which the hammer strikes), but it does not fall off systematically at high frequency. When the hammer mass is a slightly greater portion of the string mass [as in Fig. 2.10(b)], the spectrum envelope falls off as $1/n$ (6 dB/octave) above a certain mode number.

A heavier hammer is less easily stopped and thrown back by the string. It may remain in contact with the string during the arrival of several reflected pulses, or it may make multiple contacts with the string. Analytical models of hammer behavior are virtually impossible to construct, but computer simulations can be of some value.

Hall (1987a) has considered the cases of a hard, narrow hammer and a soft, narrow hammer. In the case of the hard, narrow hammer, a mode energy spectrum envelope with a slope of −6 dB/octave at high frequency emerges, although the individual mode amplitudes are not easily predicted. Adding compliance to the hammer model reduces the likelihood of multiple contacts with the string and also adds the possibility of resonance in the hammer head. For the treble strings, where the hammer mass exceeds the string mass, the spectrum envelope may have a slope as great as −12 dB/octave at high frequency.

When struck by the hammer, the string is momentarily divided into two parts. By comparing the string velocities and displacements measured on both sides of the hammer, Askenfelt and Jansson (1993) learned much about the striking process on a grand piano. Repeated reflections take place between the hammer and the agraffe end of the string. Each time the pulse reflects from the hammer, it receives a downward impulse, which contributes to its release, and after release these pulses propagate down the string, soon to be joined by reflections from the bridge end of the string (even before release, some of the pulse passes the hammer, because the hammer is not a rigid support).

Since the initial part of the first wave traveling toward the bridge is not disturbed by the reflected wave, the hammer velocity can be estimated from the amplitude of the first velocity wave. In spite of the pronounced nonlinear compressional characteristics of hammers, the initial hammer velocity and the initial string velocity amplitude follow a linear relationship, within 10% deviation, at various dynamic levels (Askenfelt and Jansson, 1993).

String velocity waveforms and the corresponding spectra for treble, middle, and bass notes (C_7, C_4, and C_2), played mezzoforte, are shown in Fig. 12.15. The peak velocities are rather similar, ranging from 1.5 to 2 m/s, but the velocity waveforms and spectra are quite different, partly due to the differences in contact time, as noted in Fig. 12.15a. In the bass, the hammer–string contact duration is only $T/5$, where T is the period of the string fundamental. In the middle register, the duration is about half a fundamental period, while in the treble it increases to more than a period ($1.5T$ for the C_7 note). Hammer–string contact times are compared to fundamental period in Fig. 12.16.

Askenfelt and Jansson (1985) investigated the effect of hammer mass by exciting the C_4 string with three hammers: the original one, a heavy bass hammer, and a light treble hammer. They obtained a longer than normal contact time with the bass hammer and a shorter than normal contact time with the treble hammer, as expected. Although the shapes of the string vibrations did not look very different for the different hammers, the sounds they produced were markedly different. The lighter, harder treble hammer, for example, produced a sound somewhat like that of a harpsichord.

Figure 12.17 shows a rather clear relationship between hammer–string contact time and string velocity at different dynamic levels, that is, striking

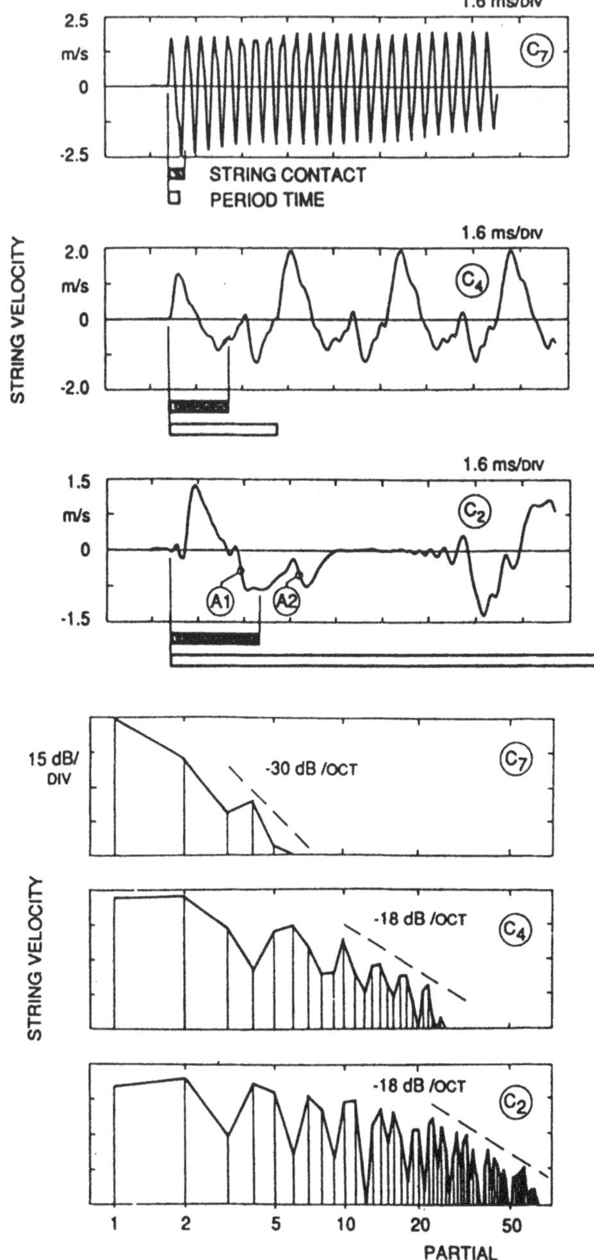

FIGURE 12.15. (a) String velocity waveforms for treble, middle, and bass notes (C_7, C_4, and C_2) played mezzoforte by means of a pendulum striker. Hammer–string contact times and fundamental periods are shown below each waveform. Reflections from the agraffe are indicated by A1 and A2. (b) String velocity spectra corresponding to the waveforms in (a) along with estimated spectral slopes (dB/octave) (Askenfelt and Jansson, 1993).

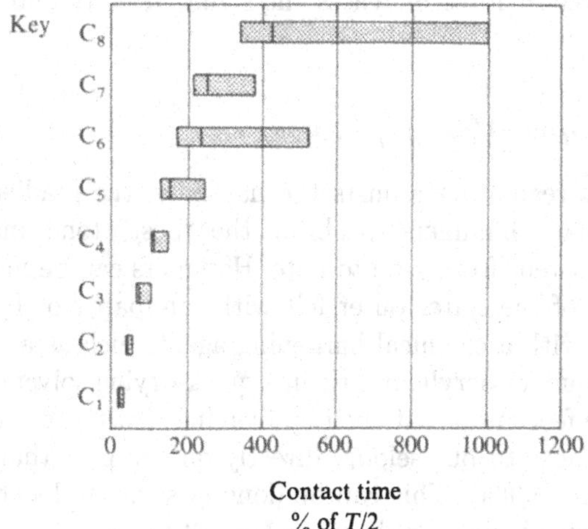

FIGURE 12.16. Hammer–string contact time as a percentage of half-period for various notes on a grand piano (Askenfelt and Jansson, 1987).

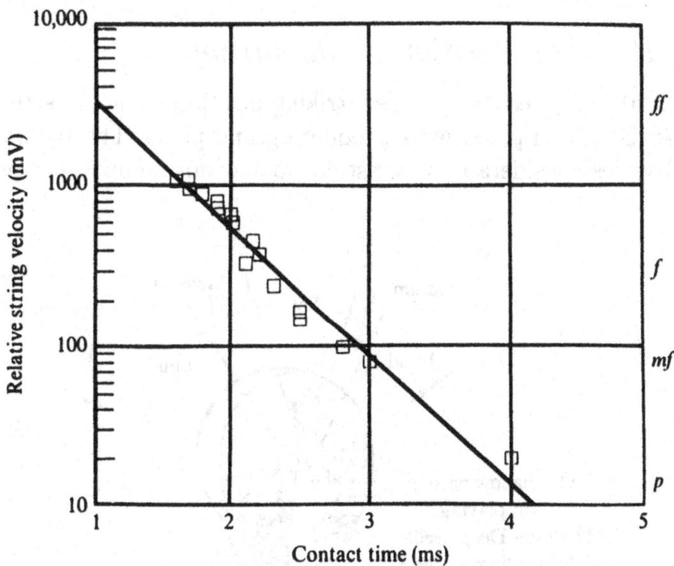

FIGURE 12.17. String velocity and hammer–string contact time at various dynamic levels for the C_4 note on a grand piano (Askenfelt and Jansson, 1987).

the key with greater force increases the string velocity and decreases the contact time.

12.4.2 Hammer Voicing

Voicing or tone regulating adjusts the hardness, the gradient, and other properties of piano hammers to obtain the "best" tone and one that is reasonably consistent from note to note. Hammers can be made harder by removing some of the softer outer felt with sandpaper or by treating the underlying felt with a chemical hardening agent, such as a solution of nitrocellulose lacquer or acrylic plastic in a quick-drying solvent. If a hammer is too hard, the felt can be softened by piercing it with needles in carefully selected areas, near to but seldom directly on the part that contacts the strings (Conklin, 1996a). This can be done in separate locations for loud, medium, and soft playing, as shown in Fig. 12.18.

Figure 12.19 shows the effect of hammer voicing on the piano sound spectrum. A normal hammer is compared to hammers that are too hard and too soft. Notice the difference in the strengths of the high harmonics.

12.4.3 Hammer Position on the String

Figure 12.20 compares the hammer striking position along the strings d/L for a 1720 Cristofori piano with a modern grand piano. The striking position d/L varies considerably in Cristofori pianos for no apparent purpose.

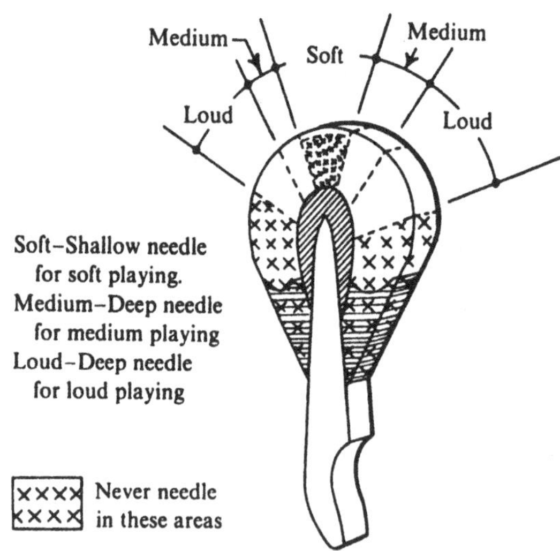

FIGURE 12.18. Hammer voicing: locations for loud, medium, and soft playing (Reblitz, 1976).

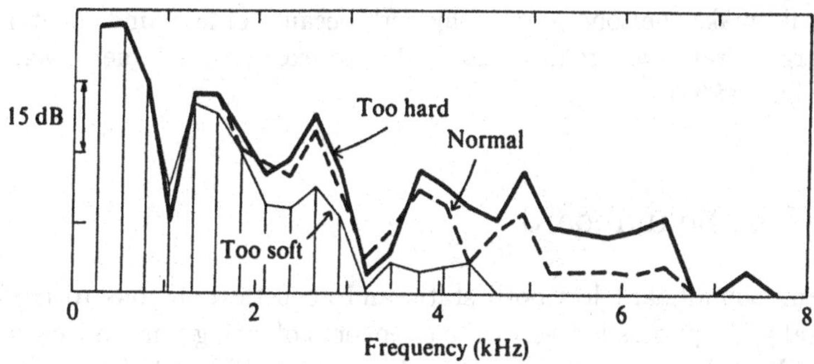

FIGURE 12.19. Piano sound spectra for C_4 note on a grand piano with hammers of three different hardnesses (Askenfelt and Jansson, 1987).

Later in the 18th century, a consensus developed that for the best tone the hammers should strike the strings between 1/7 and 1/9 of their speaking length. Figure 12.20 suggests that nowadays that criterion is followed in the lower half of the piano, but that the striking position d/L is smaller in the treble. The optimum d/L value at key 88 varies with hammer weight and shape and also with string size and contact conditions, but normally occurs for d/L between 1/12 and 1/17 (Conklin, 1996a).

In theory, an ideal stretched string struck at a certain point will not vibrate at frequencies for which vibrational nodes exist at the striking point. For actual piano strings, a large ($\sim$ 40 dB) but finite attenuation of the L/d th partial does occur, but this is not a major factor in selecting the striking position. In the mid-treble, the best d/L value is often a compromise between fundamental strength and best tone. Reducing d/L will

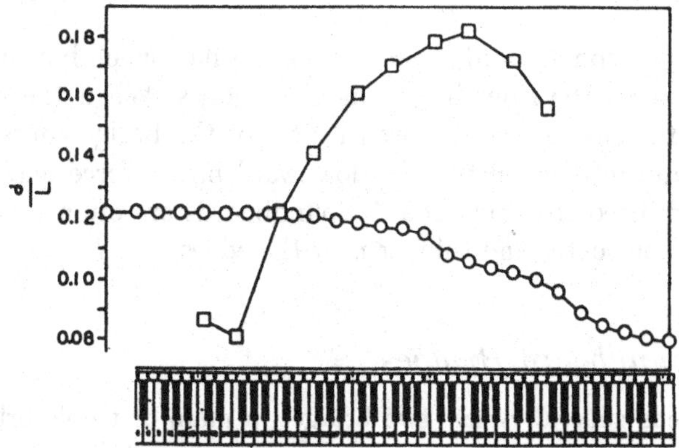

FIGURE 12.20. Hammer striking ratio d/L for a Cristofori piano □ and a modern grand piano ○ (Conklin, 1996a).

generally make the tone sound "thinner" because of less fundamental, but too large a ratio will reduce clarity due to excessive hammer dwell time (Conklin, 1996a).

12.5 The Soundboard

A piano soundboard has both static and acoustic functions to perform. Statically, it opposes the vertical components of string tension that act on the bridges. This vertical force is in the range of 10 to 20 N per string, or a total of about 900 to 1800 N. Acoustically, the soundboard is the main radiating member in the instrument, transforming some of the mechanical energy of the strings and bridges into acoustic energy. The static and acoustic functions are not totally independent of one another.

Nearly all piano soundboards are made by gluing strips of spruce together and then adding ribs that run at right angles to the grain of the spruce. The ribs are designed to add enough cross-grain stiffness to equal the natural stiffness of the wood along the grain, which is typically about 20 times greater than across the grain. In many pianos, the net cross-grain stiffness, including the contribution of the treble bridge, ends up being greater than the stiffness along the grain, a condition that is described as overcompensation (Lieber, 1979). However, Wogram (1981) cautions that overcompensation should be avoided in pianos of high quality.

Materials other than solid spruce have been used for piano soundboards. Laminated wood or plywood has been used, especially in low-cost pianos. Plywood soundboards tend to be less susceptible to splitting, but have lower acoustical efficiency and particularly less bass response than those of solid spruce. Various metals, such as steel, copper, and aluminum, have also been used with varying degrees of success. Sandwiches of various materials, including aluminum and composite materials, have been studied (Conklin, 1996b).

The unloaded soundboard is generally not a flat panel, but has a crown of 1 to 2 mm on the side that holds the bridges. When the strings are brought up to tension, the downward force of the bridge causes a dip in the crown, and in older pianos the downward bridge force may have permanently distorted the soundboard. Soundboards are often tapered to be thicker near the center and thinner near the edges.

12.5.1 Soundboard Bridges

Modern pianos generally have two bridges: a main or treble bridge and a shorter bass bridge. The bass bridge is made 2 or 3 cm taller than the treble bridge in order to raise the bass strings so they pass easily over the treble bridge and strings. In small pianos the bass bridge may be a cantilevered type that shifts the soundboard driving point away from the stiff outer edge of the soundboard but still allows the longest possible string lengths.

The bridges couple the strings to the soundboard; they function as impedance transformers, presenting a higher impedance to the strings than would exist if the strings were terminated directly on the soundboard. Piano bridges have an important effect on the tone production. By changing the bridge design, the piano designer can change the loudness, the duration, and the quality of the tone. Bridges allow the piano soundboard to respond to string vibrations in the plane of the soundboard (including longitudinal modes of the string) as well as vibrations normal to the soundboard.

12.5.2 Upright Piano Soundboard

The soundboard of a typical upright piano is shown in Fig. 12.21(a). Although rectangular in shape, the two frame members called trimming rims restrict the vibrations to a more or less trapezoidal section across which the bridges run diagonally. The vibrational modes of a soundboard without the trimming rim [Fig. 12.21(b)] were measured by Nakamura (1983).

At low frequencies, the individual resonances can be resolved, and the modal frequencies compared to those predicted by Eq. (3.17) for simply supported (hinged) boundaries. The modal frequencies and modal shapes (determined from Chladni figures) for 12 modes in a rectangular soundboard are shown in Fig. 12.22. The modal frequencies are compared to those calculated for simply supported and fixed boundaries in Fig. 12.23. Note that the measured values lie between those calculated for simply supported and fixed boundaries.

An important soundboard parameter is the driving-point impedance (see Chapter 3), since it plays a key role in determining the energy transfer from the strings to the soundboard. The driving-point impedance varies with frequency and from point to point on the soundboard. The most significant places to measure impedance, of course, are at various points on the main (treble) bridge and the bass bridge. Figure 12.24 indicates 14 measuring points selected by Wogram (1981) and the fundamental frequencies of the strings that drive the bridges at these points. Impedances at four of these points are shown in Fig. 12.25.

The impedance decreases with increasing frequency at a rate of about 5 dB per octave from $1 - 2 \times 10^3$ kg/s at 100 Hz and below to about $0.1 - 0.2$kg/s at 10,000 Hz. Only in the range from 100 to 1000 Hz does the impedance vary appreciably with the measuring point; this is the region in which the individual vibrational modes can be resolved.

On each impedance curve in Fig. 12.25, the fundamental frequency of the string that intersects the bridge at the measuring point [see Fig. 12.24(b)] is indicated. Since the string harmonics all lie above the fundamental, only this frequency range is of importance. While the driving point impedance

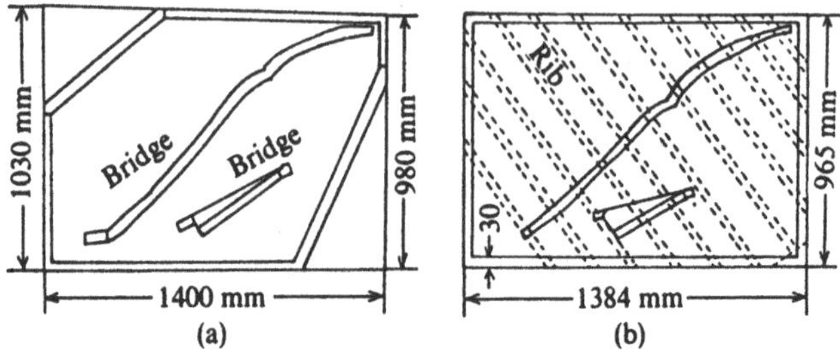

FIGURE 12.21. (a) Soundboard of an upright piano and (b) rectangular soundboard without trimming rims (Nakamura, 1983).

shows a rapid decrease above 1 kHz, Conklin (1996b) found no such decrease in grand piano soundboards, using separate force and acceleration sensors. To what extent this difference is due to the differences in measurement technique and how much it indicates a difference between upright and grand piano soundboards is not known.

12.5.3 Sound Radiation

An idealized plate, clamped at its edges, has a radiation resistance that increases with increasing frequency and equals the characteristic impedance of air at high frequency. This assumes an optimum impedance match between the plate and its surroundings and no internal losses. Losses do occur, however, and these result in a decrease in sound radiation in the upper tre-

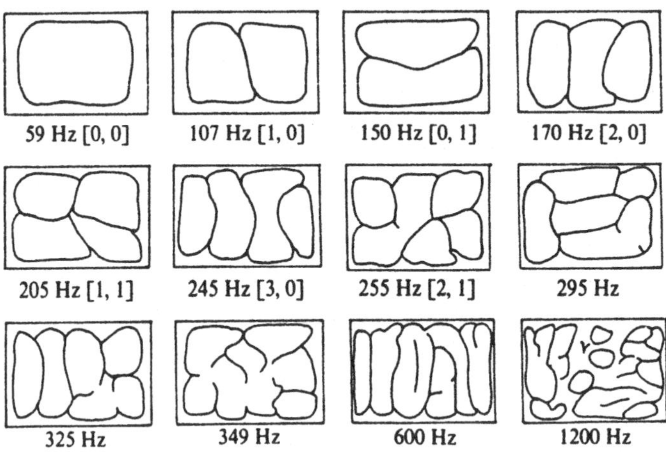

FIGURE 12.22. Chladni figures of a rectangular piano soundboard as shown in Fig. 12.21(b) (Nakamura, 1983).

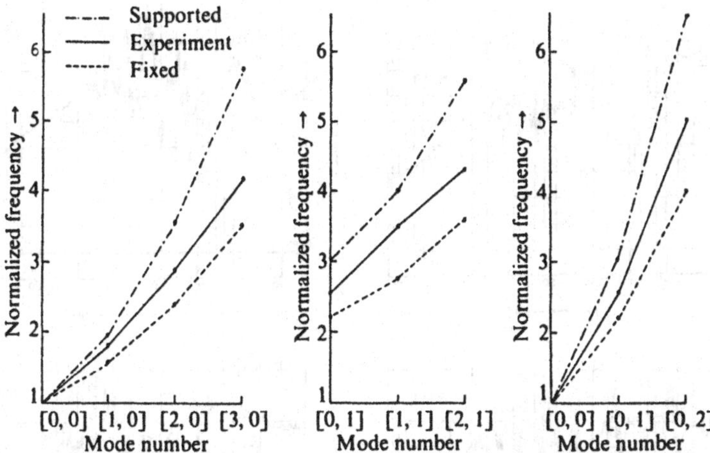

FIGURE 12.23. Modal frequencies of a rectangular piano soundboard, as measured and calculated for two different boundary conditions (Nakamura, 1983).

ble range. At the bass end, below the critical frequency (i.e., the frequency at which the speed of bending waves in the soundboard equals the speed of sound in air), the radiation efficiency drops dramatically. This results in a favored region for acoustic radiation between about 200 and 2000 Hz. This is easily seen in Fig. 12.26, which shows the driving-point impedance at measuring point 8, along with the sound radiation which results when the soundboard is driven at the point (Wogram, 1981).

The influence of the impedance peaks on the sound radiation depends upon several factors, one of which is the velocity resonance; that is, how

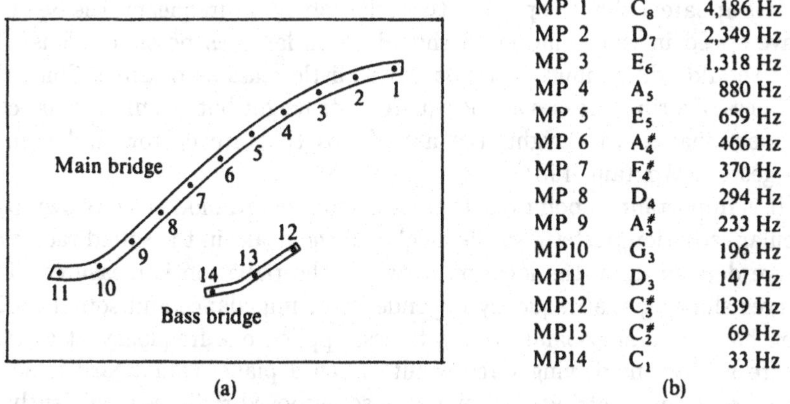

MP 1	C_8	4,186 Hz
MP 2	D_7	2,349 Hz
MP 3	E_6	1,318 Hz
MP 4	A_5	880 Hz
MP 5	E_5	659 Hz
MP 6	$A_4^\#$	466 Hz
MP 7	$F_4^\#$	370 Hz
MP 8	D_4	294 Hz
MP 9	$A_3^\#$	233 Hz
MP10	G_3	196 Hz
MP11	D_3	147 Hz
MP12	$C_3^\#$	139 Hz
MP13	$C_2^\#$	69 Hz
MP14	C_1	33 Hz

FIGURE 12.24. (a) Locations of 14 measuring points and (b) fundamental frequencies of strings that drive the bridge at these points (Wogram, 1981).

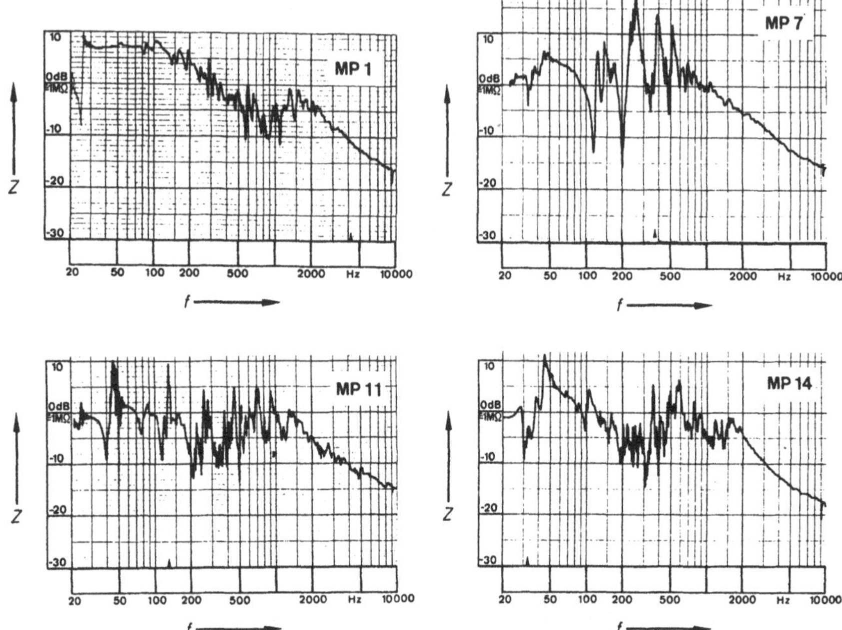

FIGURE 12.25. Driving-point impedances measured at measuring points 1, 7, 11, and 14 (0 dB = 1 MΩ = 10^3kg/s). Strings were muted with felt strips (Wogram, 1981).

nearly the speed of bending waves in the soundboard matches the speed of sound in air. Just above the critical frequency, the velocity resonance, or coincidence effect, results in enhanced radiation efficiency. The frequency band over which this enhancement takes place is broadened by internal damping in the soundboard.

It is apparent that to optimize the radiation at low frequency, the bending wave speed in the soundboard should be as large as possible. Thus, the ribs should add as much stiffness but as little mass as possible. Since the stiffness of a rib increases as the square of its height but the mass is directly proportional to the height, the use of ribs that are narrow and high is suggested (Wogram, 1981).

It is important to point out that each soundboard mode has its own particular radiation pattern, so the heights of the peaks in the sound radiation curve depend upon the location at which the radiation is measured. The curves showing the frequency dependence of impedance and sound radiation have used sinusoidal driving forces, applied one frequency at a time. Quite a different driving force results when a piano soundboard is acted on by a vibrating string. Although the soundboard radiates inefficiently at low frequency, the sound of a bass note is carried by its more efficiently radiated upper partials. Sound levels radiated by all 88 notes on a piano

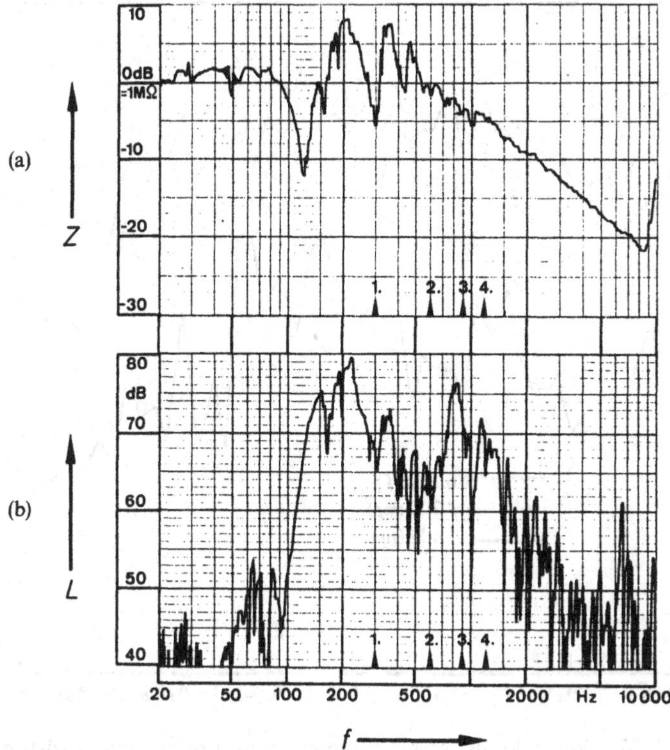

FIGURE 12.26. (a) Driving-point impedance at measuring point 8. (b) Sound radiation when the soundboard is driven at this point. The frequencies of the first four harmonics for the corresponding string are indicated (Wogram, 1981).

are shown in Fig. 12.27. Each curve represents a different playing force (dynamic level). The bass notes, carried by their upper partials, are in reasonably good balance with the treble notes.

12.5.4 Grand Piano Soundboard

Vibrational modes of a large grand piano soundboard are considerably different from those of a smaller upright piano soundboard. Seven modal peaks appear on the frequency response curve of a 6-foot grand piano soundboard in Fig. 12.28, and modal shapes of the first six modes are shown in Fig. 12.29 (Suzuki, 1986).

Kindel and Wang (1987) have compared the results of modal analysis and finite element analysis (see Appendix A.3, Chapter 4) to measurements on a 9-ft concert grand piano (Baldwin SD-10) with two different soundboards. The mode shapes and frequencies that they observe, some of which are illustrated in Fig. 12.30, are quite similar to those in Fig. 12.29,

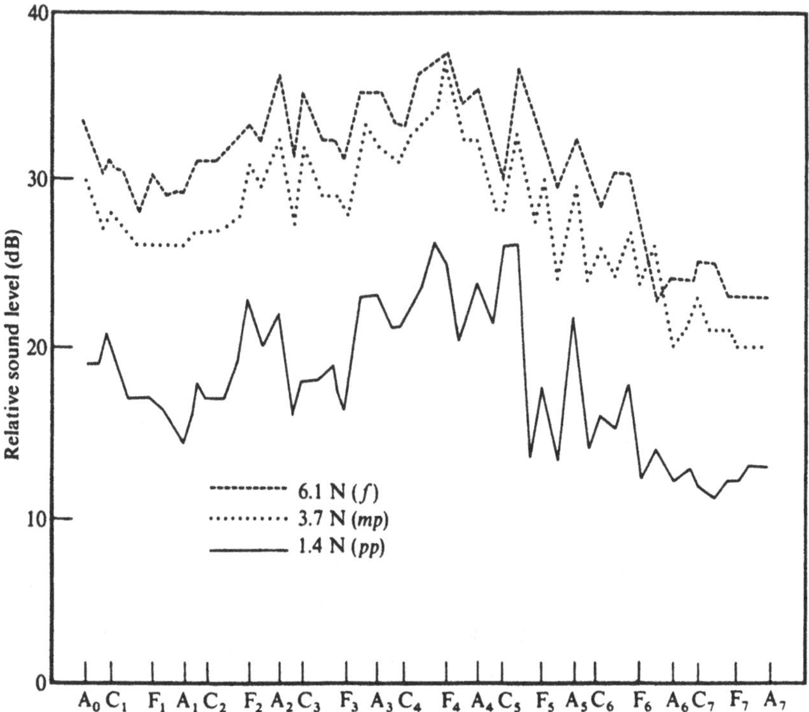

FIGURE 12.27. Sound levels for all 88 notes on an upright piano with different playing forces (Lieber, 1979).

even though the soundboards were considerably larger. The modal frequencies calculated by the finite element method were 7–12% greater than the measured frequencies, probably indicating that the ribs used in the finite element model were too stiff. Comparable results have been obtained by Wogram (1988) from modal analysis of a 2.9-m Bosendorfer grand piano.

Only a few distinct resonances are observed above 200 Hz. However, near-field sound intensity patterns can provide information about the vibrational motion in this frequency range. Figure 12.31 shows relative intensity patterns at several frequencies in the range of 200–500 Hz and 2–5 kHz. The radiation efficiency is quite large at several frequencies that are not at or near resonances of the soundboard for two reasons: (1) the cancellation of sound from different areas of the soundboard surface is a factor in determining radiation efficiency; (2) the volume velocity of the soundboard vibration becomes large when two modes interfere strongly. A few areas of negative intensity occur, indicating that sound energy radiated by one part of the soundboard is being absorbed by another part (Suzuki, 1986).

Figure 12.32 shows the mobility (velocity/force; mechanical admittance) normal to the soundboard at the terminating point for key 20 (E_2), with

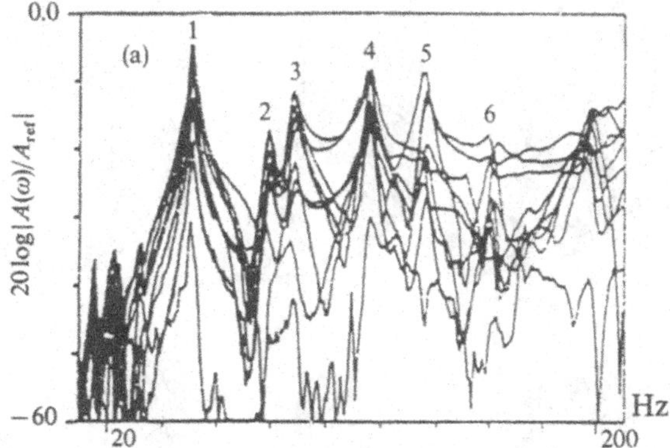

FIGURE 12.28. Frequency response functions for a 6-foot grand piano soundboard showing acceleration as a function of frequency. The soundboard was excited by impact at 10 different points to produce the 10 superimposed spectra. (Suzuki, 1986).

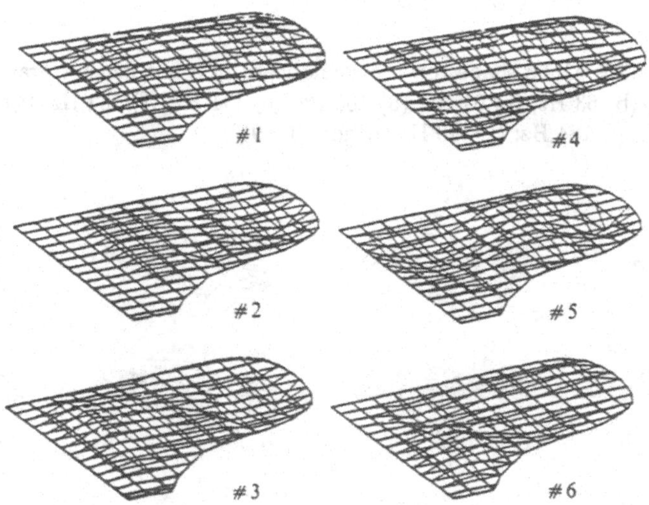

FIGURE 12.29. Mode shapes for the first six resonances in a 6-foot grand piano soundboard: (1) 49.7 Hz; (2) 76.5 Hz; (3) 85.3 Hz; (4) 116.1 Hz; (5) 135.6 Hz; (6) 161.1 Hz (Suzuki, 1986).

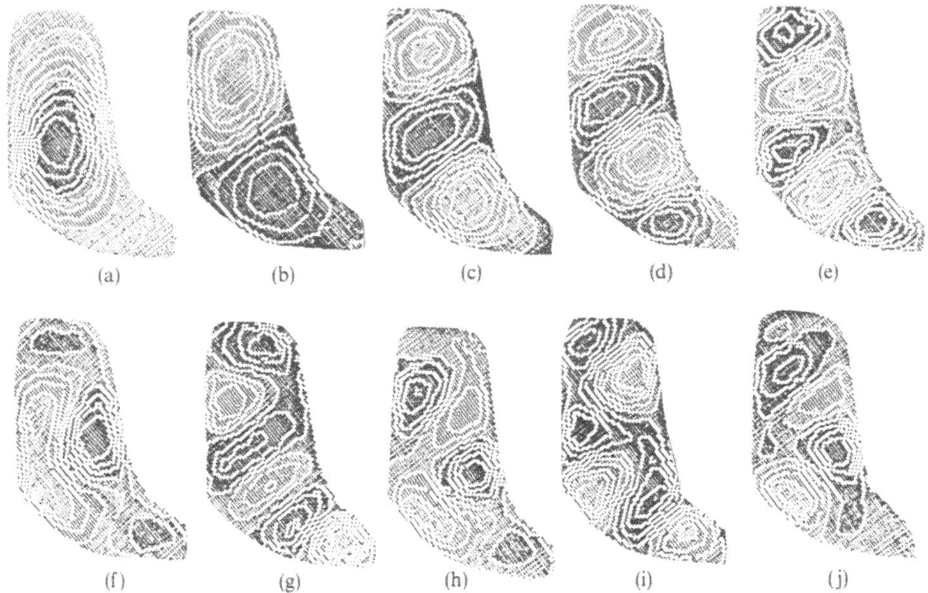

FIGURE 12.30. Mode shapes on the soundboard of a 9-ft concert grand piano: (a) 52 Hz; (b) 63 Hz; (c) 91 Hz; (d) 106 Hz; (e) 141 Hz; (f) 152 Hz; (g) 165 Hz; (h) 179 Hz; (i) 184 Hz; (j) 188 Hz (Kindel, 1989).

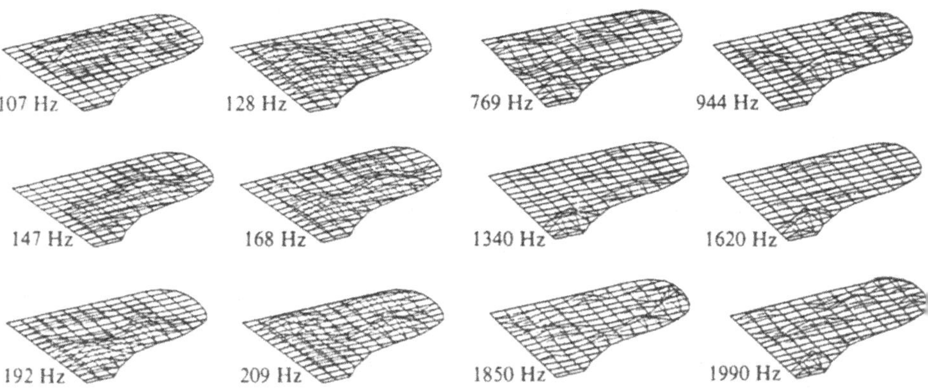

FIGURE 12.31. Relative intensity patterns at several frequencies in a grand piano soundboard (Suzuki, 1986).

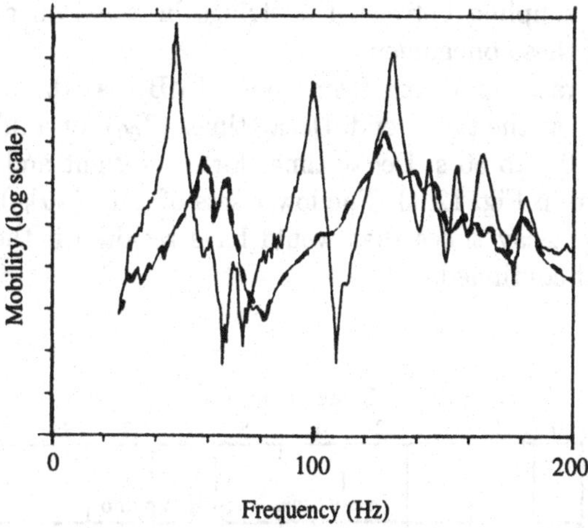

FIGURE 12.32. Mobility normal to the soundboard at the terminating point for strings of E_2 (key 20) on a 2.74-m grand piano. Solid curve is without strings; dashed curve is with strings (Conklin, 1996b).

and without the strings. Adding the strings moved the first modal frequency from 48 to 60 Hz, and the peak mobility decreased about 15 dB. The other mobility peaks are similarly smoothed out by adding the strings (Conklin, 1996b).

12.6 Sound Decay: Interaction of Strings, Bridge, and Soundboard

Since the strings are the principal reservoir for storing vibrational energy in a piano, the rate of sound decay is mainly determined by how rapidly energy is extracted from the strings. This depends upon the coupling between the strings, the bridge, and the soundboard.

The mechanical impedance of the soundboard, which varies considerably with location and with frequency (see Fig. 12.25), is always much greater than the impedance of the strings. Typically, the impedance ratio is on the order of 200:1. Thus, vibrational energy is transferred rather slowly from the strings to the soundboard. The mismatch is even greater for vibrations parallel to the soundboard, since the soundboard stiffness is very large in this plane.

Most piano notes show a compound decay curve: the initial decay rate is several times greater than the final decay rate. At least two different phenomena contribute to the compound decay: a change in the predominant direction of vibration of the string from perpendicular to parallel during decay and the coupling between the strings in a unison group. We will discuss both of these phenomena.

Initial decay rates may vary from about 4 dB/s at the bass end of the scale to 80 dB/s at the treble end. Decay times (T_{60}) for a full 60-dB decay may vary from 0.2 to 50 s. Decay times for an upright and a baby grand piano are shown in Fig. 12.33. The lower sets of data ($_AT_{60}$) in each curve are the shorter decay times that would have resulted if the faster initial decay rates had continued.

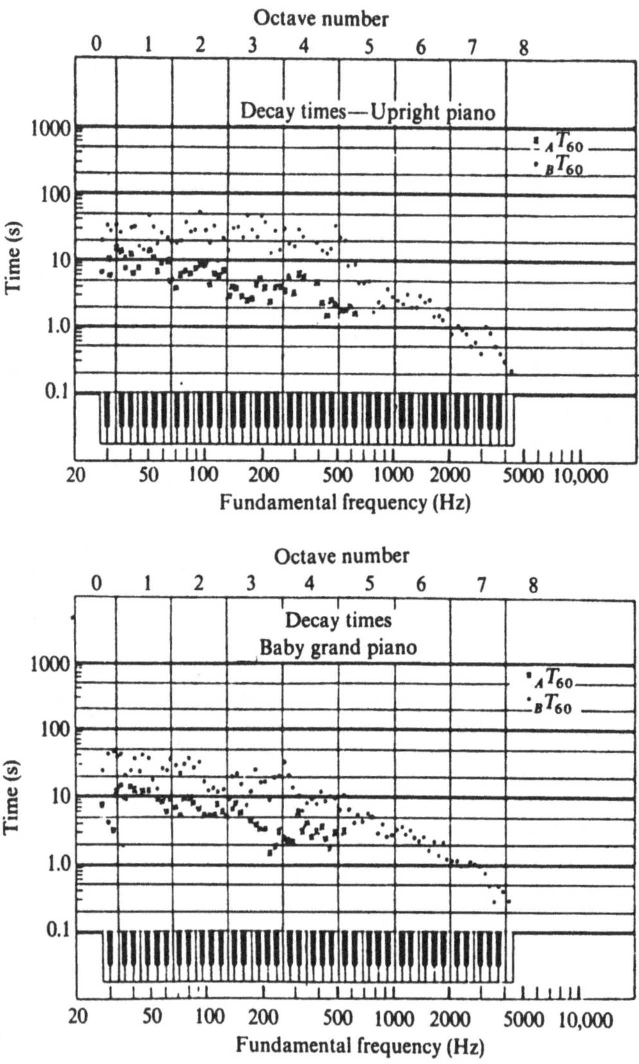

FIGURE 12.33. Decay time for an upright and a baby grand piano. $_AT_{60}$ denotes the decay times calculated from the faster initial decay rates (Martin, 1946).

Figure 12.34 compares the decay rates for vibrations perpendicular to and parallel to the soundboard in the same string. Note the slower decay rate for parallel vibrations due to the greater stiffness of the soundboard in its own plane. The hammer excites mainly perpendicular vibrations in the string, but because of the spiral wrapping and the mechanical termination of the string at the bridge, the polarization changes in the course of time. This leads to a compound decay curve with a rapid initial decay (when perpendicular vibrations are dominant) and a slower final decay (due to poor coupling of the parallel vibrations).

12.6.1 Coupling Between Unison Strings

One way to improve the coupling between the strings and the soundboard would be to increase the diameter (and hence the mass) of the strings, thereby decreasing the impedance mismatch; but this would increase the inharmonicity of the string (which depends on the fourth power of the diameter). So, a better way is to have multiple strings for each note, which is usually the case for all except the lowest bass notes.

When a hammer strikes a tricord (set of three unison strings), it sets all three strings into vibration with the same phase. Thus, they all exert vertical forces with the same phase on the bridge, and energy is transferred at a maximum rate. Because of very small differences in frequency, however, they soon get out of phase, and the resultant force on the bridge is diminished. Thus, the decay rate starts high and diminishes, a second reason for the compound decay curves. This process is discussed in Section 4.10 and also by Weinreich (1977).

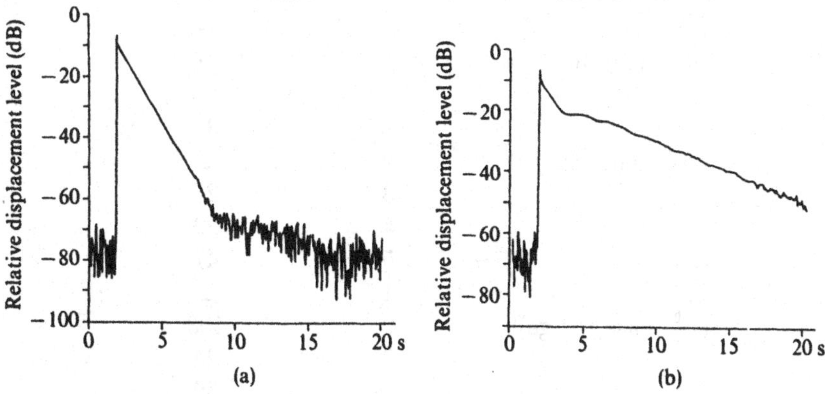

FIGURE 12.34. Decay curves for vibrations in a single $D_4^\sharp$ string (f = 311 Hz) of a grand piano: (a) perpendicular polarization and (b) parallel polarization (Weinreich, 1977).

In principle, a skilled piano tuner could, if desired, balance the initial sound and the aftersound in a piano by the tuning of the unisons. If the unison strings are closely tuned, the aftersound will develop more slowly since it will require a longer time for the unison strings to lose phase coherence.

By the same token, depressing the *una corda* pedal on a grand piano will favor the aftersound, since the string that is not struck by the hammer will be excited through coupling with the other unison strings. The unstruck string, which absorbs energy from the bridge, will vibrate out of phase with the struck strings.

12.6.2 Dependence on Soundboard Impedance

Wogram (1981) investigated the effect on decay time of peaks and valleys in the soundboard impedance curves. He attached a single string to points on the bridge at which the impedance was accurately known as a function of frequency (see Fig. 12.25). The tension of the string could be adjusted to tune it to peaks or valleys on the impedance curve. The decay times he measured are given in Table 12.1.

Measuring point 2 was located at the upper end of the treble bridge. Point 1 corresponds to the normal frequency, while points 2 and 3 are at lower frequencies (higher impedance). For measuring point 8 at the center of the main bridge, point 1 again represents the normal frequency. However,

TABLE 12.1. Decay times measured at various locations on an upright piano soundboard.

Measuring point on bridge	Point on impedance curve	Decay time T_{60}(s)
2	1	0.6
(D_7, main bridge)	2	1.0
	3	1.4
8	1	31
(D_4, main bridge)	2	2
	3	27
	4	22
	5	41
12	1	2
($C_3^\sharp$, bass bridge)	2	108
	3	24
	4	36
	5	34

[a] Wogram, 1981.

lowering the tension takes one through a series of peaks and valleys on the impedance curves, likewise for measuring point 12 on the bass bridge. From these data, Wogram concluded that substantially greater decay times occur at peaks in the impedance curves than at valleys. However, a good soundboard should have as uniform an impedance function as possible.

12.7 Scaling and Tuning

Although the scale of a piano refers to the relationship of all the major structures, including the strings, bridges, and soundboard, the main aspect that will concern us here is the scale of string lengths from treble to bass. The design of a string scale begins with the shortest treble strings, because these allow the least latitude in terms of length. They are usually the most highly stressed and therefore the ones most likely to break. In modern pianos, the strings for the highest key normally have a speaking length of 5 to 5.4 cm, and the string tension is about 65% of the tensile strength of the wire.

String scales for four different pianos are compared in Fig. 12.35. The speaking lengths are quite similar above middle C (C_4), where the lengths follow fairly well a scaling equation such as $L_n \propto f_n^{-s}$ with $s \approx 1$ (Conklin 1996c). Below C_4 the strings of larger pianos tend to follow the same equation, but their lengths must be shortened for smaller pianos. Strings for the bass keys, usually 20 to 28 in number, terminate on a separate bass bridge where different s values apply. At some point, the strings of two contiguous keys will terminate on different bridges. This discontinuity, known as the "bass break," may or may not also coincide with a discontinuity in speaking length, a change in the number of strings per key, or the change from plain to wrapped strings.

12.7.1 Longitudinal String Modes

Longitudinal string modes (see Section 2.14) are excited when a piano is played, and they contribute significantly to the character of the lower tones (Knoblaugh, 1944). A more pleasing tone and more uniform scales can be obtained by designing both scale and strings so as to place the longitudinal modes in certain more favorable relationships with the frequencies of the transverse modes (Conklin, 1970).

For plain steel strings, the first longitudinal mode occurs at a frequency of about $2500/L$, where L is the length in meters. For lengths in the range of 0.05–2 m, this leads to frequencies in the range of 1250 Hz to 50 kHz. Wrapping the string increases the mass of a string with almost no change in stiffness, so the lowest wrapped bass strings typically have longitudinal modes in the 400–500 Hz range. For real piano strings, plain or wrapped, the

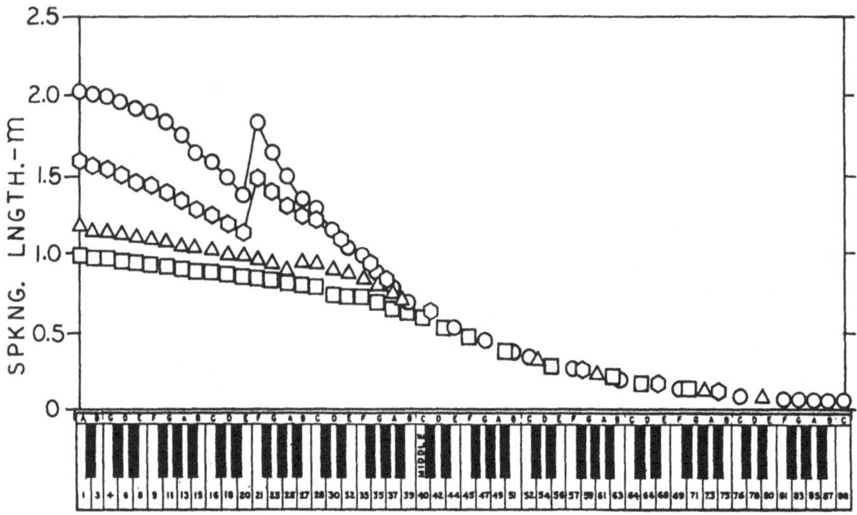

FIGURE 12.35. String scales for four different pianos: □ = 0.91 m upright piano; △ = 1.57 m grand piano; ⬡ = 2.13-m grand piano; ○ = 2.74-m grand piano (Conklin, 1996c).

frequency ratio of the first longitudinal mode to that of the first transverse mode is approximately the square root of the quotient of Young's modulus divided by the tensile stress (Conklin, 1996c)

In the lowest third of the piano scale, the first longitudinal mode produces a distinct tonal emphasis that is most noticeable at the onset. In small pianos, this tone can be the most prominent pitch identifier in the lower bass tones. In the bass register, the tuning of the longitudinal modes in relation to the transverse modes of a piano, whether intentional or not, gives the piano a characteristic timbre that cannot be changed readily by the tuner (Conklin, 1996c).

12.8 Tuning and Inharmonicity

It is a well-known fact that a piano sounds better if the highest and lowest octaves are stretched to more than a 2:1 frequency ratio. There is less than unanimity about the amount of stretching that is optimum, however. Without a doubt, it depends upon the size and other characteristics of the piano.

There appear to be both physical and psychological reasons for preferring stretched octaves. The physical reasons are related to the slight inharmonicity of the string partials ($f_n > nf_1^0$; see Section 12.3). Thus, in order to minimize beats between the fourth partial of A_4, for example, and A_6 two octaves higher, the octaves should be stretched. Since the inharmonicity constant is greater for short strings than for long strings, one would expect to find greater stretching in an upright piano than in a concert grand and the greatest stretching of all in a small spinet. This is found to be the case.

There appears to be another reason for stretch tuning, however, a psychoacoustical one. Several experiments have shown that listeners judge either sequential or simultaneous octaves as true octaves when their interval is about 10 cents (0.6%) greater than a 2:1 frequency ratio (Burns and Ward, 1982). When a melody is played in a high octave with an accompaniment several octaves lower, many listeners will judge them in tune when the intonation is stretched by a semitone (bass in C, melody in C♯); this demonstration, described by Terhardt and Zick (1975) is reproduced by Houtsma et al. (1987).

Figure 12.36 shows the deviations from equal temperament that resulted when a spinet piano was aurally tuned by a fine tuner at a piano factory. Also shown is the Railsback stretch, which is an average from 16 different pianos measured by O.L. Railsback. The aural tuning results follow the Railsback curve generally, but with a few deviations that are probably attributable to soundboard resonances. A jury of listeners showed little preference between tuning done electronically according to the Railsback curve and tuning done aurally by a skilled tuner (Martin and Ward, 1961).

The inharmonicities of five strings in a grand piano are shown in Fig. 12.37. Note the dependence of the inharmonicity on n^2 in each case [see Eq. (12.6)]. The curves have been given an arbitrary vertical displacement for clarity. The inharmonicities found in pianos of three different sizes

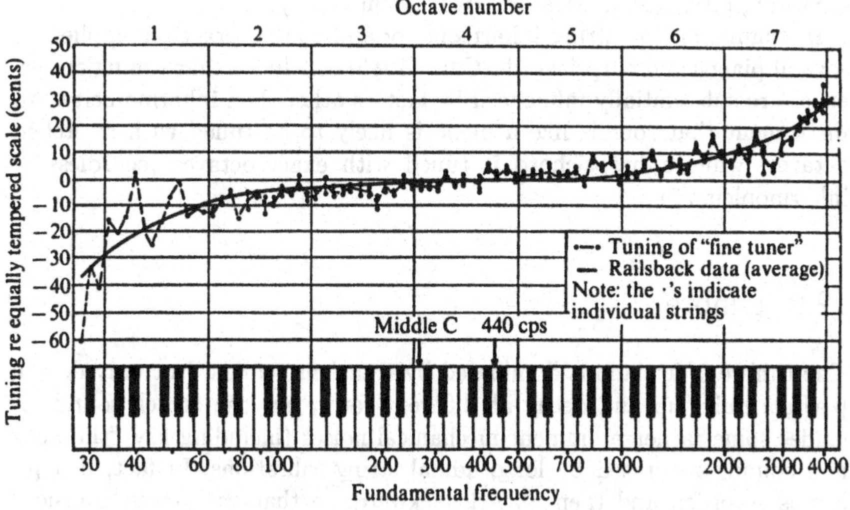

FIGURE 12.36. Deviations from equal temperament in a small piano (Martin and Ward, 1961).

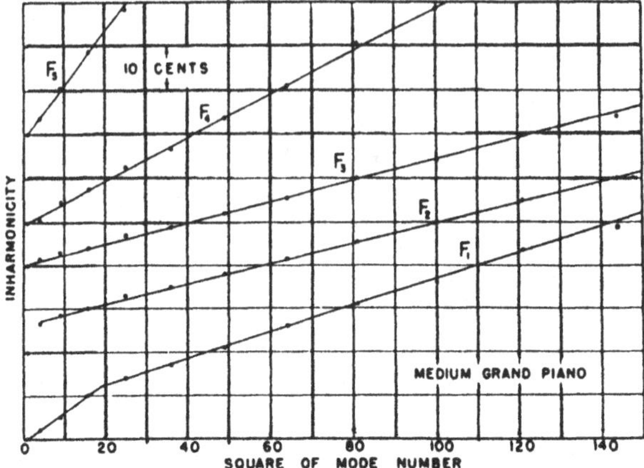

FIGURE 12.37. Dependence of inharmonicity on the square of mode number for five strings in a grand piano. Curves are displaced vertically for clarity (Schuck and Young, 1943).

are compared in Fig. 12.38. In general, the longer the strings, the less inharmonicity they will have.

Irregular patterns of inharmonicity, especially in the bass strings of small upright pianos, are a challenge to the piano tuner. Furthermore, the fundamentals of the low bass notes are usually weak, and do not contribute much to the pitch of the note. The upper treble notes, on the other hand, are of such short duration that beats between partials may no longer serve as a dependable guide for tuning. The tuner must rely on his or her judgment regarding pitch and musical intervals (Kent, 1982).

In examining the string inharmonicity and tuning practices applied to several pianos and harpsichords, Carp (1986) concluded that tuning preferences are substantially influenced by factors other than inharmonicity. An instrument that sounds like a piano is likely to be tuned with stretched octaves, while a harpsichord is tuned with exact octaves regardless of inharmonicity.

12.9 Timbre

The timbre of a piano is dominated by transient sounds. Not only do the partials build up and decay at different rates, but the attack sound includes some rather prominent mechanical noises (including key "thump") and sounds generated by longitudinal string vibrations. In fact, if a piano is recorded and then played backward, so that the attack transient occurs at the end of the note, it is hardly identifiable as piano sound (see Demonstration 29 in Houtsma et al., 1987).

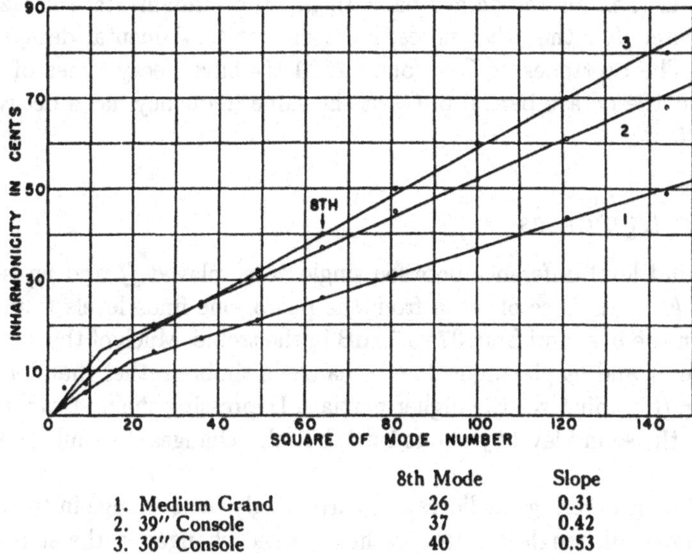

FIGURE 12.38. Comparison of the inharmonicity of F_1 for various pianos (Schuck and Young, 1943).

Although the fundamental dominates the piano sound spectrum over most of its range, it is very weak in the lower two octaves. For the lowest note A_0 (27.5 Hz), the fundamental may be 25 dB below the level of the strongest component. Above the strongest component (the fundamental throughout most of the range), the amplitude envelope falls off smoothly with frequency; sound coloring formants seldom appear. Bass tones have partials that extend out to about 3000 Hz, while treble tones may extend out to 10,000 Hz. For the highest note (C_8, 4186 Hz), this means that only one or two overtones are heard (Meyer, 1978).

When a string is struck (or plucked) at a fraction β of its length, harmonics that are multiples of $1/\beta$ will be weak or missing (see Section 2.8 and 2.9). In pianos, β typically varies from 1/7 at the bass end to 1/20 at the treble end. However, Conklin (1987) has found that the bass tone quality can be improved by increasing β to as much as 1/5, depending on the design parameters of each particular string.

Attack and decay curves for nine partials of C_1 (32.7 Hz) are shown in Fig. 12.39. Clearly, the sound spectrum of this tone changes with time.

Sound decay curves for several notes on a piano are shown in Fig. 12.40. In some notes, the transition from initial to aftersound is quite abrupt. Sometimes there are interference effects. The decay times for all notes on two grand pianos are shown in Fig. 12.41.

The curves in Fig. 12.42 show the decay times for the different partials in five notes on a grand piano. The top curve shows a maximum value of 43 s at 131 Hz (i.e., for the octave partial), but the fundamental (65 Hz) dies away faster. For the other notes, however, the fundamental decays most slowly. The overtones of C_2 around 2000 Hz have decay times of about 15 s, but the octave partial of C_6 at the same frequency has a decay time of only 5 s.

12.9.1 Dynamics

The sound level difference between single notes played *ff* and *pp* is 30 to 35 dB. At a distance of 10 m from the piano, one finds levels from 50 to 85 dB in the bass and from 37 to 70 dB in the treble. Much of the difference between *ff* and *pp* playing is due to change in timbre rather than loudness. Playing *ff* emphasizes the higher partials. Depressing the *una corda* pedal reduces the sound level by about 1 dB, but also changes the timbre (Meyer, 1978).

Raising or lowering the lid causes surprisingly little change in the overall sound level, although it causes rather marked changes in the strength of the high-frequency sound in certain directions.

12.9.2 Directional Characteristics

The sound radiation pattern of a piano is largely determined by the shape of the soundboard and the modal shapes of the various modes in which it vibrates. The pattern is further complicated by reflections from the lid and from other parts of the piano structure. The radiation patterns are more complicated at high than at low frequencies. Radiation patterns in the vertical plane are shown in Fig. 12.43.

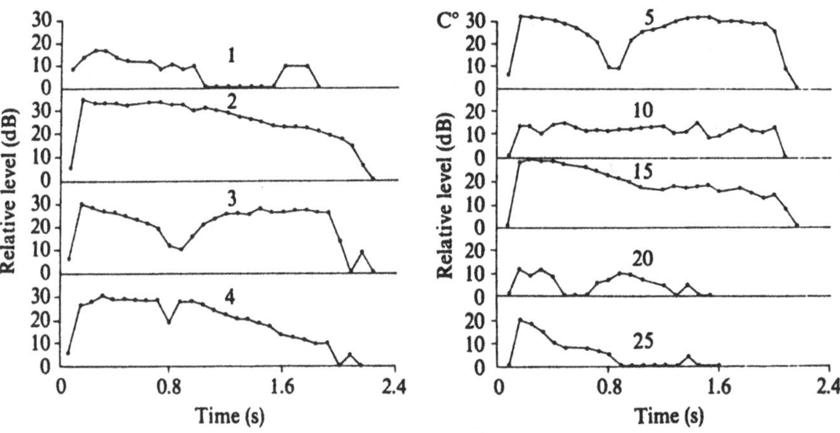

FIGURE 12.39. Attack and decay curves for partials 1, 2, 3, 4, 5, 10, 15, 20, and 25 of C_1 ($f = 32.7$ Hz) on a grand piano (Fletcher et al., 1962).

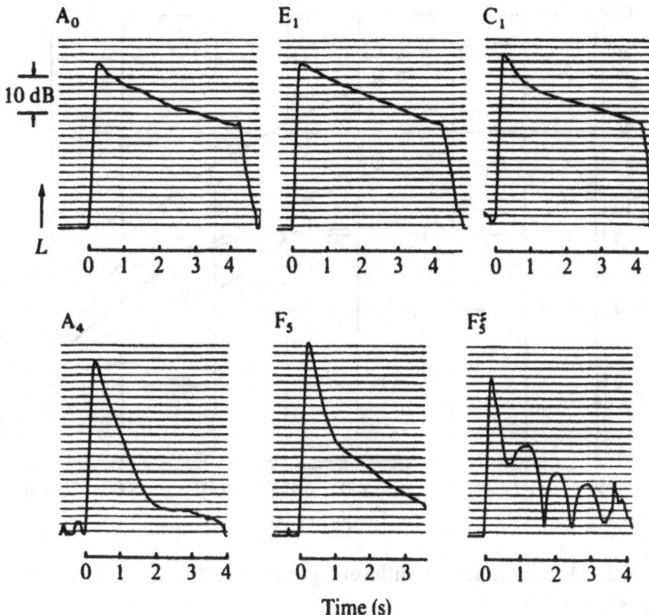

FIGURE 12.40. Sound decay curves for low and high notes on a piano (Meyer and Melka, 1983).

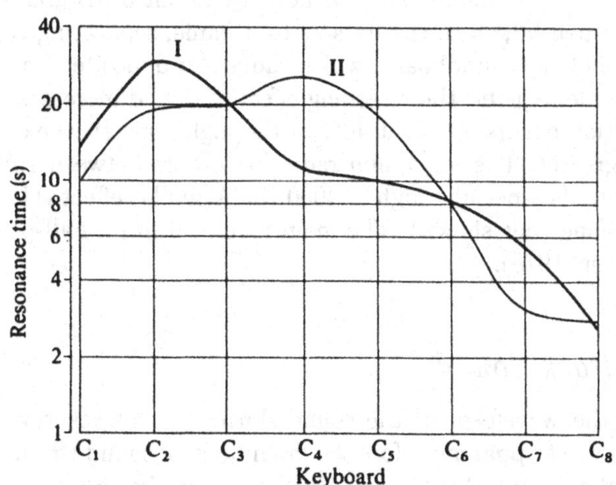

FIGURE 12.41. Decay times for notes on two grand pianos (Meyer, 1978).

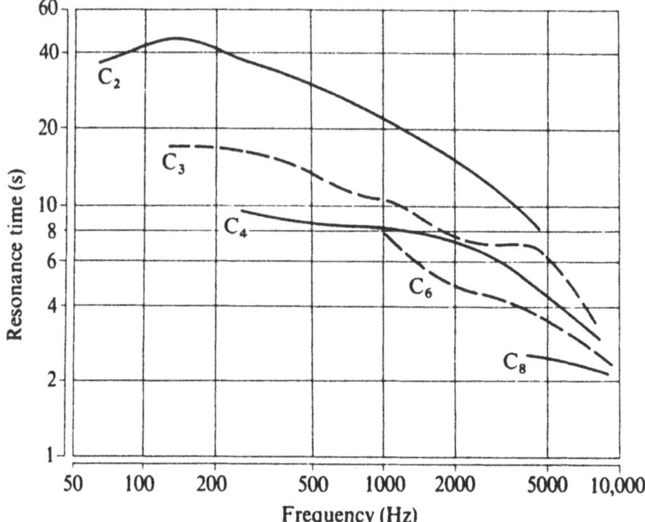

FIGURE 12.42. Decay times for different partials in five notes on a grand piano (Meyer, 1978).

In the low register (around C_2), the radiation is relatively symmetrical. The level behind the piano (180°) is actually about 5 dB greater than in front of the open lid, since the lid acts as a baffle, separating the top and bottom sides of the soundboard, which radiate in opposite phase.

In the middle register, the screening effect of the lid increases. The maximum radiation occurs at about 40°. In the high register, the effect of the lid is even greater. The maximum radiation occurs between 15° and 35°. The directivity is especially high at 4000 Hz. A similar effect is found in the horizontal plane (not shown). The main lobe at 0 has a half-power width of ±5° (Meyer, 1978).

12.9.3 Attack Sound

Examining the waveform of the sound during the attack reveals attack noise that is most apparent before the main body of sound (from transverse vibrations of the string) develops. Mostly, this is the result of noise from mechanical vibrations in the action (the hammer, for example, has modes of vibration in the range of 300 to 3000 Hz).

Podlesack and Lee (1987) have identified spectral components in the attack sound that match the longitudinal modes of bass strings. For the lowest bass notes, the sound level due to such modes may be only 10 to 20 dB below the main sound, but the decay rate is estimated to be 100 dB/s. Designing strings so that the longitudinal mode frequencies are in the range

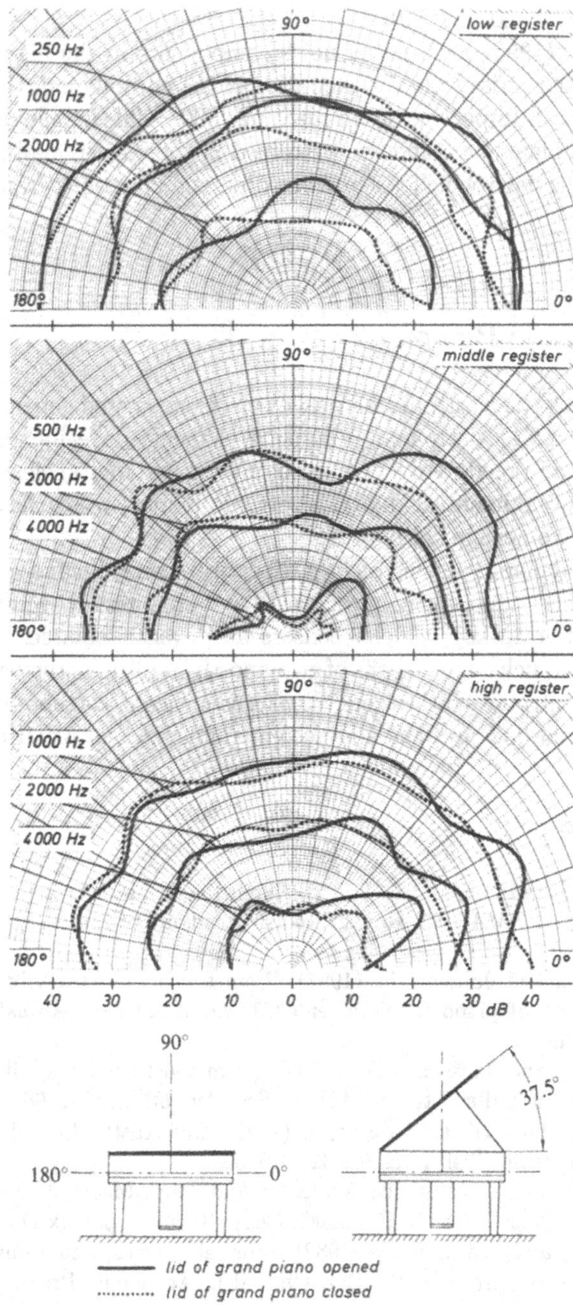

FIGURE 12.43. Radiation pattern (vertical plane) for a grand piano in the low, middle, and high registers. Note the increase in radiation between 15 and 35° in the high register when the lid is open (Meyer, 1978).

of 4300 to 5200 cents above the fundamental frequencies can improve the initial sound (Conklin, 1970).

Bork et al. (1995) have noted noise components in the high notes of a grand piano that arise from the excitation of the normal modes of the frame and soundboard. Impact noises in the range 300–1000 Hz influence the first 70 ms of the piano tone and are perceived in the concert hall mixed with side wall reflections which come from different directions from the main piano tone.

12.10 Electric Pianos

Electric pianos use electromechanical or electronic sound generators to emulate the sound of the piano. Vibrating elements have included strings, tuning forks, and reeds. Advances in computers and digital electronics have led to the development of digital electronic pianos, which have virtually replaced electromechanical pianos.

Most digital pianos depend on large libraries of stored piano sounds which can be recalled when the appropriate keys are pressed. Modern digital pianos incorporate key velocity sensors and even simulated hammers to imitate the "touch" and "feel" of a piano. In addition, they incorporate features of electronic synthesizers, such as sequencers, transposers, multiple voices, pre-recorded rhythms and accompaniments, and MIDI control.

References

Askenfelt, A., and Jansson, E. (1985). Piano touch, hammer action and string motion. *Proc. SMAC* 83. Royal Swedish Academy of Music, Stockholm, pp. 137–143.

Askenfelt, A., and Jansson, E. (1987). From touch to string vibrations—The initial course of piano tone. Paper CC3, *113th Meeting, Acoust. Soc. Am., Indianapolis.*

Askenfelt, A., and Jansson, E.V. (1993). From touch to string vibrations. III: String motion and spectra. *J. Acoust. Soc. Am.* **93**, 2181–2196.

Bork, I., Marshall, H., and Meyer, J. (1995). Zur Abstrahlung des Anschlaggeräusches beim Flüel. *Acustica* **81**, 300–308.

Broch, J.T. (1984). "Mechanical Vibration and Shock Measurements," 2nd ed., Brüel and Kjaer, Naerum, Denmark. Chap. 3 and Appendix D.

Burns, E.M., and Ward, W.D. (1982). Intervals, scales, and tuning. In "The Psychology of Music" (D. Deutsch, ed.). Academic Press, San Diego, California.

Carp, C. (1986). The inharmonicity of strung keyboard instruments. *Acustica* **60**, 295–299.

Conklin, H.A. Jr. (1970). Longitudinal mode tuning of stringed instruments. U.S. patent No. 3,523,480.

Conklin, H.A. Jr (1987). Augmented bass hammer striking distance for pianos. U.S. patent No. 4,674,386.

Conklin, H.A., Jr. (1990). Piano design factors—their influence on tone and acoustical performance. In "Five lectures on the acoustics of the piano" (A. Askenfelt, ed.), Swedish Acad. Music, Stockholm.

Conklin, H.A., Jr. (1996a). Design and tone in the mechanoacoustic piano. Part I. Piano hammers and tonal effects. *J. Acoust. Soc. Am.* **99**, 3286–3296.

Conklin, H.A., Jr. (1996b). Design and tone in the mechanoacoustic piano. Part II. Piano structure. *J. Acoust. Soc. Am.* **100**, 695–708.

Conklin, H.A., Jr. (1996c). Design and tone in the mechanoacoustic piano. Part III. Piano strings and scale design. *J. Acoust. Soc. Am.* **100**, 1286–1298.

Dijksterhuis, P.R. (1965). De piano. *Nederlandse Akoest. Genootschap* **7**, 50–65.

Fletcher, H., Blackham, E.D., and Stratton, R. (1962). Quality of piano tones. *J. Acoust. Soc. Am.* **34**, 749–761.

Hall, D.E. (1986). Piano string excitation in the case of small hammer mass. *J. Acoust. Soc. Am.* **79**, 141–147.

Hall, D.E. (1987a). Piano string excitation II: General solution for a hard narrow hammer. *J. Acoust. Soc. Am.* **81**, 535–546.

Hall, D.E. (1987b). Piano string excitation III: General solution for a soft narrow hammer. *J. Acoust. Soc. Am.* **81**, 547–555.

Hall, D.E. (1987c). Piano string excitation IV: The question of missing modes. *J. Acoust. Soc. Am.* **82**, 1913–1918.

Hall, D.E. (1992). Piano string excitation VI. Nonlinear modeling. *J. Acoust. Soc. Am.* **92**, 95–105.

Hall, D.E. and Askenfelt, A. (1988). Piano string excitation V. Spectra for real hammers and strings. textitJ. Acoust. Soc. Am. **83**, 1627–1638.

von Helmholtz, H.L.F. (1877). "On the Sensations of Tone," 4th ed. Translated by A.J. Ellis, Dover, New York, 1954.

Henderson, M.T. (1936). Rhythmic organization in artistic piano performance. In "Objective Analysis of Musical Performance" (C.E. Seashore, ed.), Iowa Studies in Piano Perf. 4, 281–305. Univ. Press, Iowa City, Iowa.

Houtsma, A.J.M., Rossing, T.D., and Wagenaars, W.M. (1987). "Auditory Demonstrations," Demonstration #16. Inst. Percept. Res., Eindhoven, The Netherlands.

Kent, E.L. (1982). Influence of irregular patterns in the inharmonicity of piano-tone partials upon tuning practice. *Das Musikinstrument* **31**, 1008–1013.

Kindel, J. (1989). Modal analysis and finite element analysis of a piano soundboard. M.S. thesis, University of Cincinnati.

Kindel, J., and Wang, I. (1987). Modal analysis and finite element analysis of a piano soundboard, *Proc. 5th International Modal Analysis Conf. (IMAC)*, pp. 1545–1549.

Knoblaugh, A.F. (1944). The clang tone of the pianoforte. *J. Acoust. Soc. Am.* **16**, 102.

Kock, W. (1937). The vibrating string considered as an electrical transmission line. *J. Acoust. Soc. Am.* **8**, 227–233.

Kubota, H., and Nagai, Y. (1986). The kinematical study on the initial behaviour of hammer striken piano string. *Proc. 12th ICA*, Toronto, K2-1.

Lieber, E. (1979). The influence of the soundboard on piano sound. *Das Musikinstrument* **28**, 304–316 [original German version in *Das Musikinstrument* **9**, 858 (1966)].

Lieber, E. (1985). On the possibilities of influencing piano touch. *Das Musikinstrument* **34**, 58–63.

Martin, D.W. (1946). Decay rates of piano tones. *J. Acoust. Soc. Am.* **19**, 535–547.

Martin, D.W., and Ward, W.D. (1961). Subjective evaluation of musical scale temperament in pianos. *J. Acoust. Soc. Am.* **33**, 582–585.

Meyer, J. (1978). "Acoustics and the Performance of Music." Verlag das Musikinstrument, Frankfurt am Main.

Meyer, J., and Melka, A. (1983). Messung and Darstellung des Ausklingverhaltens von Klavieren. *Das Musikinstrument* **32**, 1049–1064.

Nakamura, I. (1982). Simulation of string vibrations on the piano—a struck string. *Proceedings of Co-op Workshop on Stringed Muscial Instruments.* (A. Segel, ed.). Wollongong, Australia.

Nakamura, I. (1983). The vibrational character of the piano soundboard. *Proc. 11th ICA*, Paris, 385–388.

Podlesak, M., and Lee, A. R. (1987). Longitudinal vibrations in piano strings. Paper CC5, *113th Meeting, Acoust. Soc. Am., Indianapolis.* See also *J. Acoust. Soc. Am.* **83**, 305–317 (1988).

Reblitz, A. (1976). "Piano Servicing, Tuning, and Rebuilding." Vestal Press, Vestal, New York.

Rossing, T.D. (1982). "The Science of Sound," Addison-Wesley, Reading, Massachusetts. Chapter 14.

Russell, D. A., and Rossing, T. D. (1998). Testing the nonlinearity of piano hammers using residual shock spectra. *Acustica* (in press).

Sanderson, A. (1983). "Piano Technology Topics: # 5 Piano Scaling Formulas." Inventronics, Chelmsford, Massachusetts.

Schuck, O.H., and Young, R.W. (1943). Observations of the vibrations of piano strings. *J. Acoust. Soc. Am.* **15**, 1–11.

Suzuki, H. (1986). Vibration and sound radiation of a piano soundboard. *J. Acoust. Soc. Am.* **80**, 1573–1582.

Terhardt, E., and Zick, M. (1975). Evaluation of the tempered tone scale in normal, stretched, and contracted intonation. *Acustica* **32**, 268–274.

Vernon, L.N. (1936). Synchronization of chords in artistic piano music. In "Objective Analysis of Musical Performance" (C.E. Seashore, ed.) pp. 306–345. Univ. Press, Iowa City, Iowa.

Weinreich, G. (1977). Coupled piano strings. *J. Acoust. Soc. Am.* **62**, 1474–1484.

White, W.B. (1906). "Theory and Practice of Piano Construction." Edward Lyman Bill, New York. Reprinted by Dover, 1975.

Wogram, K. (1981). Acoustical research on pianos, part I: Vibrational characteristics of the soundboard. *Das Musikinstrument* **24**, 694–702, 776–782, 872–880 [original German version in *Das Musikinstrument* **23**, 380–404 (1980)].

Wogram, K. (1988), The strings and the soundboard. In "Piano Acoustics." Royal Academy of Music, Stockholm.

Part IV

Wind Instruments

13

Sound Generation by Reed and Lip Vibrations

Wind instruments are made to sound either by blowing an air jet across some sort of opening, as in whistles, flutes, and organ flue pipes, or by buzzing together the lips, or a thin reed and its support, as in trumpets, clarinets, or oboes, or indeed as in the human voice. In this chapter we discuss only the second class of sound generators—vibrating reed valves—and defer discussion of air-jet generators to Chapter 16.

In order to describe the flow of air through any wind instrument, it is necessary in principle to solve the Navier–Stokes equation, which describes such flow, including viscosity but neglecting compressibility. Unfortunately this equation is nonlinear, and the only way in which it can be solved for any but the most artificially simple cases is by numerical methods on a large computer. While such a solution would yield great detail about a particular flow, and could be extended to the time-varying case, it would not reveal a great deal about the general principles underlying the mechanism of sound generation and control. It is therefore appropriate to use much simpler models and to incorporate fluid-mechanical principles into general assumptions about fluid flow.

13.1 Pressure-Controlled Valves

The vibrating flow-control devices in typical wind instruments belong to one or other of the types shown in Fig. 13.1. We can either have a single reed, as in clarinets, saxophones, organ reed-pipes, and harmonicas, a double reed as in oboes, bassoons, crumhorns, and bagpipes, or the player's vibrating lips as in trumpets and other brass instruments. In each case, the motion of the valve is controlled by the pressure difference across it, and the frequency with which it vibrates is controlled partly by its own natural frequency and partly by the resonances of the instrument air column to which it is connected.

We can classify pressure-controlled valves into three simple types, depending upon the way they are deflected by steady pressure applied to their

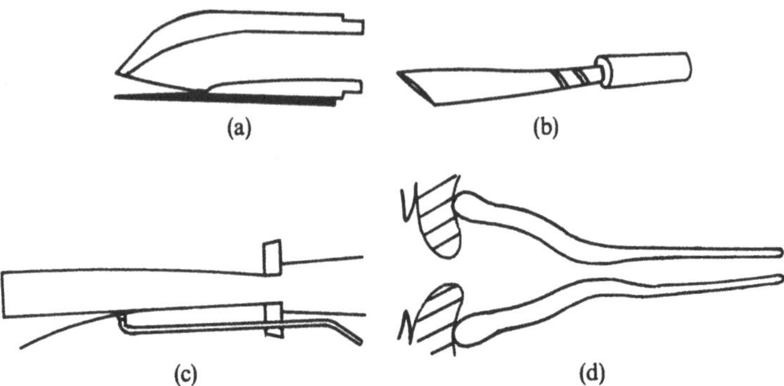

FIGURE 13.1. Typical reed-generator configurations: (a) clarinet or saxophone, (b) oboe or bassoon, (c) organ reed pipe, and (d) brass-instrument player's lips.

inlets and outlets, these two ports being defined by the direction of flow irrespective of whether it is caused by overpressure at the inlet or suction at the outlet. The three types of valve are shown in Fig. 13.2, the inlet pressure being p_0 and the outlet pressure p. In most musical instruments the valves are operated by blowing with a steady pressure, though reed organs often use suction and harmonicas have reeds operated both by overpressure and by suction. This makes no difference from the viewpoint of classification or valve action. Each valve is then characterized by a doublet symbol (σ_1, σ_2): σ_1 is $+1$ if a pressure excess p_0 applied at the inlet tends to open the valve, and -1 if it tends to close it. The symbol σ_2 refers similarly to the effect of an excess pressure p applied at the outlet. The common reed valves of woodwind instruments, cases (a), (b), and (c) of Fig. 13.1, have classification $(-,+)$, and were termed by Helmholtz "inward-striking reeds," though a more physical name is an "inward-swinging door." Lip valves and the human vocal folds are more complex but, on a simple model, have classification either $(+,-)$ or perhaps $(+,+)$. The former can be called "outward-striking" or "outward-swinging" and, by analogy, we might call the $(+,+)$ models "sideways-striking" or perhaps "sliding-door" valves. Valves with configuration $(-,-)$ do not appear to be useful.

We should extend this classification a little by noting that we could have a sliding-door valve of the type shown in Fig. 13.2(c) without any bevels on the slider. In such a case, the only pressure tending to open or close the valve comes from the Bernoulli effect in the narrow valve passage. The behavior of such a valve depends critically upon details of the flow. If we assume that the flow leaves the valve as a jet on the downstream side, and neglect viscosity, then the static pressure in the flow channel of the valve is equal to the downstream pressure, independent of variation of the upstream pressure. Because the Bernoulli pressure tends to close the valve, we should therefore adopt for it a classification $(0,+)$. If such a valve is

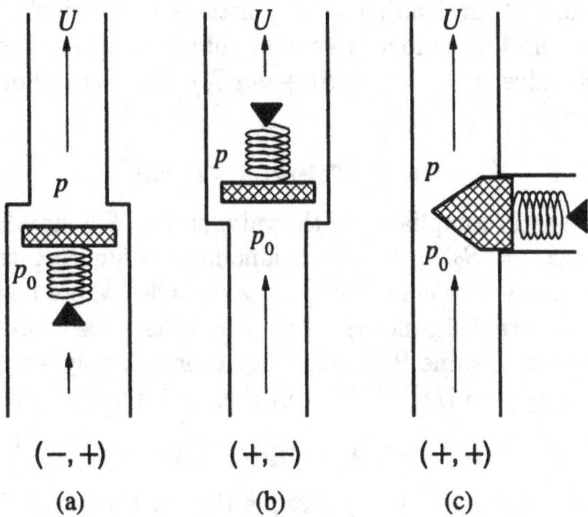

FIGURE 13.2. Simplified models of the three commonly occurring configurations of pressure-controlled valves, described by the couplet (σ_1, σ_2). (a) Configuration $(-,+)$, corresponding to a woodwind reed, (b) configuration $(+,-)$, corresponding to one model for a brass-instrument player's lips, and (c) configuration $(+,+)$ corresponding to an alternative model for the brass-player's lips. The human vocal folds and the syrinx of a bird may be either $(+,-)$ or $(+,+)$.

blown at constant pressure, its behavior is indistinguishable from that of $(-,+)$ or $(+,+)$ valves.

13.2 Quasi-Static Model

Our discussion begins with a consideration of steady flow through a valve supplied from an infinite reservoir at pressure p_0 when the outlet pressure is p. Detailed fluid-dynamic considerations (Hirschberg et al., 1990; Hirschberg, 1995) show that the situation is not ideally simple, but we neglect these complications here and assume that the flow separates cleanly from the valve at its exit. Let us suppose that x measures the opening of the valve, assumed to be of width W. The force tending to open the valve can be found, in principle, by integrating the static pressure $p_0 - \frac{1}{2}\rho v^2$ over the whole valve surface. Here v is the local flow velocity, and ρ is the density of air. If S is the area of valve flap over which the upstream and downstream pressures act, then the Bernoulli term $\frac{1}{2}\rho v^2$ can be taken into account by reducing these areas to effective values S_1 and S_2, respectively.

The result clearly depends upon the way in which the flow enters the valve and separates at its exit, and we make the assumption that the entry is rather abrupt and that the flow separates cleanly to form a jet at the exit from the valve itself, leading to very little of the downstream pressure recovery that we find in an ideally smooth tube transition. The force tending to open the valve is then $(\sigma_1 p_0 S_1 + \sigma_2 p S_2)$. If the static opening is x_0, then we can write

$$(x - x_0) = (\sigma_1 p_0 S_1 + \sigma_2 p S_2)C, \tag{13.1}$$

where C is the elastic compliance of the valve spring. For simplicity in what follows, we set $S_1 = S_2 = S$, which amounts to ignoring the Bernoulli term, which is in any case small for valve geometries where flow separation is clean. The pressure difference $(p_0 - p)$ across the valve is ordinarily large enough that we can assume Bernoulli's equation to apply, so that the flow velocity is $v = [2(p_0 - p)/\rho]^{1/2}$. We can then write the volume flow U as

$$U = Wxv = W[x_0 + (\sigma_1 p_0 + \sigma_2 p)SC][2(p_0 - p)/\rho]^{1/2}. \tag{13.2}$$

The interesting case, for steady flow, is that of the woodwind reed, for which $\sigma_1 = -1$ and $\sigma_2 = +1$. We can then write

$$U = \alpha(p_0 - p)^{1/2} - \beta(p_0 - p)^{3/2}, \tag{13.3}$$

where α and β are positive constants. This relation is plotted as a full curve OABC in Fig. 13.3, from which it can be seen that the flow rises from zero to a maximum at the point A as the blowing pressure is increased, then falls to zero again at the point C where the reed closes. We are interested in the acoustic admittance Y_r presented to the instrument by the reed. At the low frequencies considered in quasi-static flow, this admittance is purely a conductance G_r and is given by

$$G_r = -\frac{\partial U}{\partial p} = \left.\frac{\partial U}{\partial(p_0 - p)}\right|_{p=0}, \tag{13.4}$$

which is just the slope of the curve in the figure. At low blowing pressures, the conductance is positive, since the air is simply flowing through an aperture about equal to $x_0 \times W$. As the blowing pressure increases past the point A, however, the effect of reed closing becomes dominant and the conductance becomes negative, allowing the reed to act as an acoustic generator. In an instrument such as the clarinet, whose behavior is well described by this curve, the player sets his blowing pressure close to the point B, which then allows as much pressure excursion as possible around this operating point. A simple calculation based on Eq. (13.3) shows that, if $p_c = \alpha/\beta = x_0/SC$ is the closing pressure corresponding to point C, then the pressure for point A is $\frac{1}{3}p_c$ and that for point B is between this value and p_c.

The behavior detailed above is, of course, idealized. We examine in the next section the practically important case in which the frequency is not nearly zero so that the static flow equation is no longer applicable. We should also mention that, in instruments such as the clarinet, the geometry

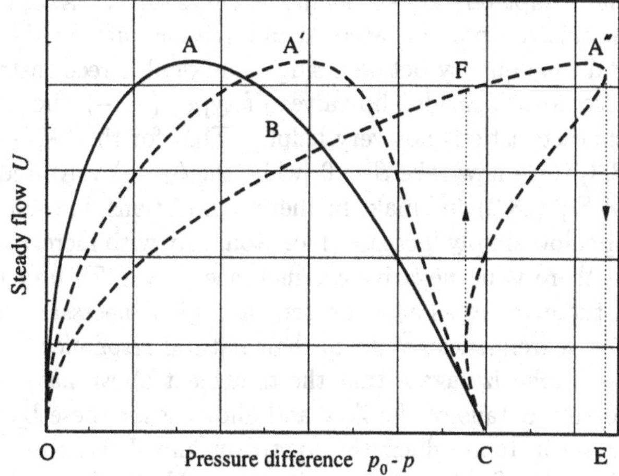

FIGURE 13.3. Static flow curves for a reed valve of configuration $(-,+)$, as in a clarinet or oboe. The full curve OABC is the normal single-reed characteristic and shows a negative conductance over the region AC. The operating point is generally set near B. In the case of a double reed, the characteristic follows one of the curves OAC, OA′C or OA″C as the channel resistance is increased from zero. (After Wijnands and Hirschberg 1995.)

is a little more complicated so that the aperture area is not simply xW but contains also a small contribution proportional to x^2 from the triangular side apertures. In the case of a very narrow reed, as is found in free-reed instruments such as the harmonium, this side-flow contribution will dominate and the aperture area will vary as x^2. Backus (1963) measured the flow behavior in the case of a clarinet and fitted it to a curve of the form $x^{4/3}W$, which is a qualitatively reasonable interpolation formula, but his curve-fitting also yielded a factor $(p_0 - p)^{2/3}$ rather than the expected $(p_0 - p)^{1/2}$, which is questionable.

An important modification of this behavior occurs in the case of double reeds such as those of the oboe and bassoon, in which there is a long narrow passage through the reed. Such a passage can introduce appreciable flow resistance, which can have a significant effect on reed behavior (Wijnands and Hirschberg 1995). Equation (13.3) can be simply modified to take this into account, for the pressure p acting on the downstream side of the reed is simply reduced by an amount RU^2, where R is the turbulent flow resistance of the reed channel. We can therefore simply replace p by $p - RU^2$ in Eq. (13.3) and evaluate the result numerically. The broken curves of Fig. 13.3 show the flow behavior in this case for several values of R. The result is a narrower and steeper region A′C of negative conductance, or even a hysteresis behavior in which the reed oscillates sharply between fully open and completely closed, as in the curve OA″C. As we see in the next chapter, this can perhaps account in large measure for the difference in playing and tone quality between single and double-reed instruments.

When we come to consider lip-valves of types $(+,-)$ and $(+,+)$, the static flow characteristic is not very helpful. That for the $(+,-)$ valve follows Eq. (13.3) if we now take $\beta < 0$, while the $(+,+)$ valve requires that we go back to Eq. (13.2) and make further modifications. In each case, however, the static flow simply increases monotonically with increasing blowing pressure and there is no negative-conductance region. To understand the operation of these valves as acoustic generators it is necessary to consider their behavior at frequencies close to their natural resonance.

Finally, we emphasize again that the treatment above has been simplified by neglecting details of the flow and allowing for these by modifying the effective areas upon which the upstream and downstream pressures are assumed to act. Such an artifice is acceptable in the analysis, but a detailed aerodynamic study is necessary to determine these effective area parameters. Hirschberg (1995) has given a detailed discussion of the aerodynamic theory involved, and has pointed out that quite minor changes in the geometry of the valve can have large aerodynamic and acoustic consequences.

13.3 Generator Behavior at Playing Frequency

The reason for spending time with the classification of pressure-controlled valves in the previous section is that each class of valve has a very distinct vibrational and acoustic behavior (Fletcher 1979b, 1993; Saneyoshi et al., 1987). To see how this comes about, we need to consider the vibrational behavior of the reed, and its interaction with the flow, at frequencies that are comparable with the natural resonance of the reed itself.

We assume that, because of their rather high internal flow resistance, the lungs provide a steady air flow, rather than a steady pressure, and that this feeds a vocal tract that presents an impedance Z_1 at the valve inlet. It is then simplest to analyze the behavior of the valve in terms of the pressure $p_1 = \bar{p}_1 - Z_1 U$ measured inside the player's mouth. Similarly the outlet pressure is $p_2 = \bar{p}_2 + Z_2 U$, where Z_2 is the input impedance of the instrument air column.

We can then rewrite Eq. (13.3) as

$$U = Wx[2(p_1 - p_2)/\rho]^{1/2}. \tag{13.5}$$

This equation is clearly a simplification of the true situation, since we have neglected the inertia of the air in the valve, which inserts a small phase lag into U and hence into the Bernoulli pressure, which has likewise been neglected. We have also taken the flow resistance R to be zero and neglected the contribution to the flow from the physical displacement of the reed surface. In most cases all these factors contribute only a small correction to the general behavior.

We expect the reed opening x to follow a simple equation like

$$\frac{d^2x}{dt^2} + 2\gamma\frac{dx}{dt} + \omega_r^2(x - x_0) = \frac{S}{m}(\sigma_1 p_1 + \sigma_2 p_2), \tag{13.6}$$

where m is the effective moving mass of the reed, ω_r its natural resonance frequency, and γ its damping, including both internal and aerodynamic effects. The quasi-equilibrium opening of the valve is

$$\bar{x} = x_0 + (\sigma_1\bar{p}_1 + \sigma_2\bar{p}_2)S/m\omega_r^2, \tag{13.7}$$

where $\bar{p}_1$ and $\bar{p}_2$ are the mean inlet and outlet pressures, and we can write a similar quasi-steady version of the flow equation (13.5) by putting bars over U, x, and the pressure quantities. To obtain a linearized solution we assume that x, p_1, p_2, and U all vary sinusoidally around their mean values with frequency ω, and combine the two equations (13.5) and (13.6). Without going into the details, which are given by Fletcher (1993), we find that these linearized equations specify the conditions under which oscillation can commence, because the total linear damping becomes negative. If we write $Z_1 = R_1 + jX_1$ and $Z_2 = R_2 + jX_2$, and assume that $R_1 = R_2 = 0$ for simplicity, these oscillation conditions can be summarized as follows:

$$\begin{aligned} &(-,+): \quad \omega < \omega_r \quad \text{and} \quad X_1 + X_2 > 0, \\ &(+,-): \quad \omega > \omega_r \quad \text{and} \quad X_1 + X_2 < 0, \\ &(+,+): \quad (\omega - \omega_r)/(X_1^2 - X_2^2) > 0 \quad \text{and} \quad X_1 - X_2 < 0. \end{aligned} \tag{13.8}$$

It should be noted that these are only necessary conditions—the impedance inequalities must be strongly satisfied by an amount that depends upon the damping of the reed and resistive losses in the ducts. The actual oscillation frequency ω has, in each case, a definite relationship to the valve resonance frequency ω_r as well as to the resonance frequencies of the air column. In each case there is, in addition, a threshold blowing pressure $\bar{p}_1$ that must be exceeded, but this is much less than the threshold pressure $\frac{1}{3}p_c$ derived from quasi-static flow considerations, provided the reed is free to vibrate at a frequency close to its natural resonance.

It is helpful to have these results expressed in a physically more meaningful way. For a clarinet-type reed valve $(-,+)$ as in woodwinds and organ reed pipes, the resonance frequency of the reed must be above the playing frequency. The impedance condition tells us that the total duct reactance $X_1 + X_2$ must be positive or inertive, which means that the compliance $X_1 < 0$ of the mouth cavity, which increases the generator damping, must be compensated by a large inertive reactance $X_2 > 0$ in the instrument tube, which must therefore play a little below some tube resonance. In organ reed pipes, the reed resonance frequency is only a little below the first resonance frequency of the pipe, but in woodwinds it is high and the instrument sounds just below one of the lower pipe resonances.

For a brass instrument, $(+,-)$ or $(+,+)$, the mouth cavity compliance $X_1 < 0$ contributes a negative damping and assists the oscillation. The frequency conditions, however, differ for the two configurations. In the $(+,-)$ case, the playing frequency will be above the lip resonance, and probably above the horn resonance, while for the $(+,+)$ configuration it will generally be below the lip resonance and below the horn resonance. These statements are conditional, in both cases, because they depend upon the relative magnitudes of the mouth and horn impedances, which means that a skilled player can generally push the pitch in either direction.

Some of these results have been investigated experimentally by Saneyoshi et al. (1987) for woodwind reeds and the lips of brass-instrument players. These experiments broadly confirm the analysis. There has been disagreement, however, on the model that is appropriate for the brass-player's lips. Fletcher (1979b) assumed a $(+,-)$ configuration but Saneyoshi et al. (1987) objected to this on experimental grounds and favored a clarinet-like $(-,+)$ model. Both of these papers, however, were based upon a simple "inward or outward striking" dichotomy, following the classification of Helmholtz, and neglected the "sideways striking" $(+,+)$ possibility. We can now see that the real choice of simple models is between outward striking $(+,-)$ and sideways striking $(+,+)$ configurations, since mouth pressure undoubtedly tends to force the lips open.

This particular question has been largely answered by recent experiments. Yoshikawa (1995) investigated the phase of upper-lip motion relative to mouthpiece pressure in brass instrument players, using a miniature strain gauge attached to the player's lips, and showed that the lip configuration varies between $(+,-)$ and $(+,+)$, depending on note pitch and playing technique. The outward-striking $(+,-)$ mode appears to dominate for low-pitched notes and the sideways striking $(+,+)$ mode at high pitches. Copley and Strong (1996) carried out a more detailed study for the trombone, and showed that there is in all cases a combination of the two motions, with longitudinal $(+,-)$ motion and transverse $(+,+)$ motion being approximately equal in magnitude at low frequencies while transverse motion predominates at high frequencies. The phase difference between the two motions was found to vary with frequency, but to be about $90° \pm 45°$. Another study by Chen and Weinreich (1996) used a single-mode Helmholtz resonator with active feedback instead of a real musical instrument and concluded that players can vary their playing technique to play either sharp or flat of the resonator frequency, though the most natural technique gave a sound frequency below the resonance and implied a $(+,-)$ motion at the frequency studied, which was in the range 200–350 Hz.

These studies are all in basic agreement. If we use a simple one-degree-of-freedom description of the lip valve, then its action is basically $(+,-)$ at low frequencies and $(+,+)$ at high frequencies. More realistically, the motion should be described by at least two parameters, corresponding to longitudinal and transverse displacements, respectively, and the vibration typically shifts from mostly longitudinal at low frequencies to mostly transverse at high frequencies. Factors influencing the transition between these two modes of vibration have been investigated theoretically by Adachi and Sato (1995a,b), who show that the distribution of the Bernoulli pressure component over the lip surface, neglected in our simplified analysis, may provide the key to the transition. A beginning to the description of this more complex two-parameter motion has been made by Adachi and Sato (1996). This approach shows promise, though we should recognize that an even more complex model, analogous to those that have been developed for the human vocal folds, is probably required. There is, however, a limit to the degree of sophistication that can reasonably be added to the model when we realize that performance technique may well vary significantly from one player to another.

An important feature of both lip-valve models $(+,-)$ and $(+,+)$ is that the associated pressure variations in a mouth cavity of limited volume ($X_1 < 0$) contribute negative damping to the valve and can lead to sustained oscillations even in the absence of an external resonator ($X_2 = 0$). This is important in brass-instrument playing because the player must sustain lip vibrations of the desired frequency for several cycles until the first reflection returns from the instrument bell. The theory (Fletcher 1993) allows the conditions under which this can happen to be investigated. If we assume that the lip valve exhausts to open air, then the behavior of the two valve configurations is identical because the outlet pressure p_2 is zero. Figure 13.4 then shows the threshold pressure required for oscillation as a function of reservoir volume, with valve damping as a parameter. The numerical quantities are applicable only in order of magnitude to the musical instrument case, but it is clear that the player's mouth volume plays a large part in ensuring sustained vibration of the lips. The optimal mouth volume is a function of the desired frequency of lip vibration, which is itself controlled by lip muscle tension.

The background to this figure can be understood by realizing that there are pressure oscillations in the mouth cavity, and that the impedance of this cavity influences their phase so that they contribute to the effective damping of the whole system. For lip valves of either configuration, the contribution to the damping is negative, so that the system can break into self-sustained oscillation. For a woodwind reed with configuration $(-,+)$, however, the situation is reversed and the pressure oscillations in the mouth add to the valve damping and inhibit oscillation. As we see below, however, this apparently negative effect actually contributes greatly to the operation of woodwind instruments. The fact that a reed can be made to "crow" when detached from the instrument comes about partly because of the small inertive loading contributed by the air flow (St. Hilaire et al., 1971) and partly because it has an exit tube that contributes a further inertive loading, so that $X_1 + X_2 > 0$ as required by the first of Eqs. (13.8).

In most playing situations, however, it is the large resonant impedance of the air column of the instrument that controls the oscillation. While we should properly include the impedance of the player's mouth, or more

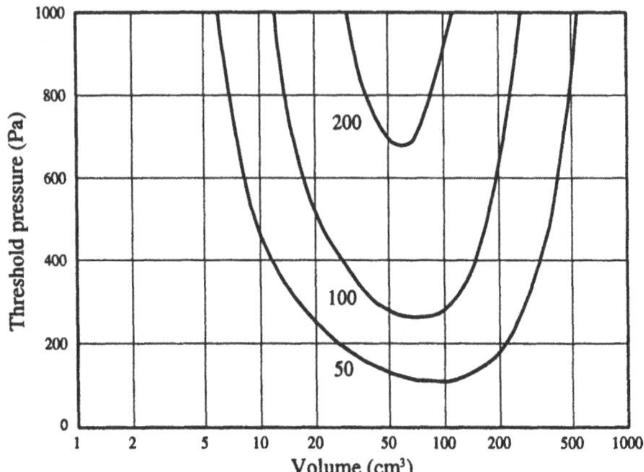

FIGURE 13.4. Threshold blowing pressure for a valve of configuration $(+,-)$ or $(+,+)$, as for a brass-player's lips, as a function of mouth cavity volume, assuming the lips to exhaust to open air. The parameter γ measures the valve damping. Numerical values (m = 100 mg, S = 1 cm^2, x_0 = 1 mm, ω_r = 1000 s^{-1}) are appropriate to the musical instrument situation only in an order-of-magnitude sense (Fletcher 1993).

accurately of the complete vocal tract (Johnston et al., 1986; Sommerfeld and Strong 1988), in our calculations, we can gain a good insight by neglecting this and assuming the blowing pressure to be constant so that $p_1 = p_0$. Under these conditions, there is no distinction between configurations $(+,+)$ and $(-,+)$, and what we calculate is the complex admittance Y_r of the reed generator as a function of frequency and blowing pressure. In the low-frequency limit this must agree with the quasi-static conductance calculated in Eq. (13.4) and displayed in Fig. 13.3.

Figure 13.5 shows the results of such calculations for representative reed-valve generators, where the generator admittance Y_r is plotted as a function of frequency for several values of the blowing pressure p_0 (Fletcher 1979b). These calculations have been confirmed by experiment (Fletcher et al., 1982). It is interesting to compare these plots with the Nyquist plots of Fig. 1.13(c) for the mechanical admittance of a simple damped mechanical oscillator, which is a circle passing through the origin for $\omega = 0$ and with its center on the positive real axis. To a first approximation this plot is simply rotated through +90° (multiplied by j) for the $(-,+)$ or $(+,+)$ cases, and through −90° (multiplied by $-j$) for the $(+,-)$ case. The valve can act as an acoustic generator only when its admittance lies in the left half-plane, where its real part $G_r(\omega)$ is negative. In both cases it is clear that p_0 must exceed some threshold value, about 300 Pa in the example calculated, for this to happen. In the case of the clarinet-type reed $(-,+)$ or the sideways striking lip-reed $(+,+)$, the frequency must also be

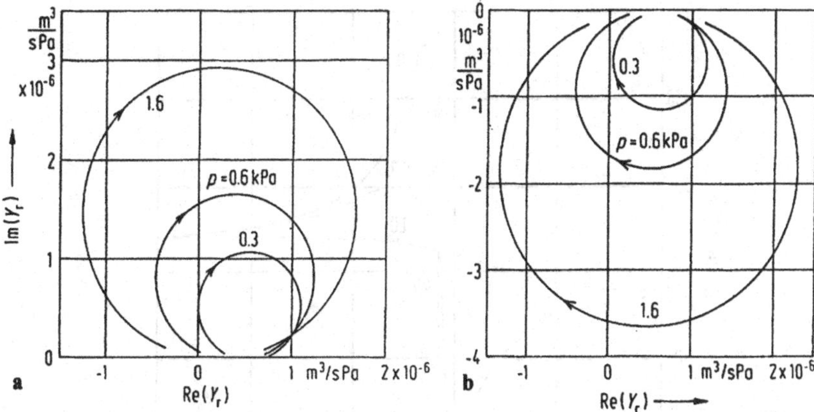

FIGURE 13.5. Complex admittance Y_r as a function of frequency (increasing in the direction of the arrow in each circle) for typical reed-generators in (a) inward-striking $(-,+)$ or sideways-striking $(+,+)$ configurations, and (b) outward-striking $(+,-)$ configuration, all fed from a constant-pressure source. Blowing pressure p_0 is shown as a parameter (Fletcher et al., 1982).

low, while for the outward-striking lip-reed $(+,-)$ the frequency must be high.

We gain additional insight into this behavior by choosing a blowing pressure somewhat greater than the threshold and plotting the generator conductance G_r as a function of frequency. This is done for the two cases of interest in Fig. 13.6. Clearly the woodwind reed $(-,+)$ and the sideways-striking lip reed $(+,+)$ have a negative conductance, and can therefore act as acoustic generators, at all frequencies from zero up to a little less than the reed resonance ω_r. The parameter on the curves gives a measure of the damping coefficient of the reed, which we expect to be large in woodwind reeds because of the extra damping contributed by the player's lips in contact with the reed and by the small reservoir volume of the mouth. For the lip reed $(+,+)$, however, the damping should be extremely low because of the negative damping contributed by the mouth cavity. The outward-striking lip reed $(+,-)$, in contrast, has a negative conductance only in a small frequency range just above its resonance frequency and we again expect this resonance to be very sharp because the mouth reservoir contributes negative damping.

From this figure we can see that a woodwind-type reed can act as a sound generator over a wide frequency range up to its own resonance frequency, which should therefore be set above the playing range. The high damping provided by the player's lips and mouth volume should be an advantage for ensuring smooth response and eliminating high-frequency squeaks, and the oscillation frequency is set primarily by the pipe resonance frequency. The organ reed pipe of Fig. 13.1(c) behaves rather differently. Being made

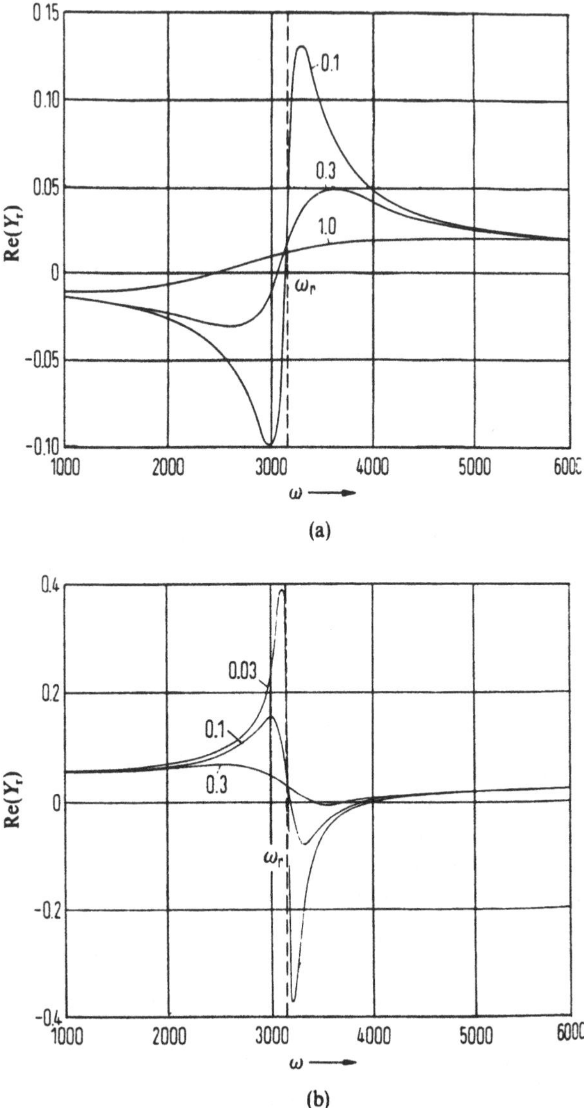

FIGURE 13.6. The real part G_r of the acoustic admittance Y_r for a reed-generator with (a) inward-striking $(-,+)$ or sideways-striking $(+,+)$ configurations, and (b) outward-striking $(+,-)$ configuration, when blown above the threshold pressure from a constant-pressure source. The parameter gives a measure of the valve damping, and the valve resonance frequency is in each case $\omega_r = 3142\ \text{s}^{-1}$ (Fletcher, 1979b).

of metal and constrained only by a metal tuning wire, the reed has little damping and thus a pronounced negative conductance maximum just below its resonance frequency. This tends to dominate the pipe impedance, with the result that the pipe can be made to sound over a considerable frequency range by adjusting the reed length, and hence its resonance frequency, with the tuning wire. The sound is weak, however, unless the pipe and reed resonances are brought into agreement.

For a lip-valve, by contrast, generation behavior is essentially limited to a narrow frequency range just above or just below the mechanical lip resonance, depending upon whether the configuration is $(+,-)$ or $(+,+)$. The player must therefore control this resonance frequency by adjustment of lip muscle tension so as to feed acoustic energy into the selected acoustic mode of the instrument horn. Players of the natural horn or the natural trumpet, which have no valves, can reliably produce semitone steps by exciting individual horn resonances 8 though 16, which implies that the mouth-enhanced Q value of the generator resonance must exceed about 20.

13.4 Free Reeds

A related but rather different type of reed is that found in the harmonium (reed organ), harmonica (mouth organ), and concertina or accordion. These are designated collectively as "free reeds" and their geometry is shown in Fig. 13.7(a) and (b). The reed itself is a thin brass tongue, in some cases with a complex curvature, which is attached to one side of a metal plate containing an aperture through which the reed tongue can actually pass. The static opening of the reed is generally toward the high-pressure side of the air supply, although the pressure difference may actually be created by suction on the outlet, as in the case of an "American organ," rather than by reservoir pressure. The reed therefore begins with a configuration $(-,+)$, but can be transformed in principle to configuration $(+,-)$ if the pressure differential is high enough so that the equilibrium position of the reed passes through the aperture of the plate. Since a typical free reed is long and narrow, the aperture area is proportional to x^2, where x is the tip opening, so that the flow curve is rather different from that in Fig. 13.3. The calculated quasi-static flow characteristic for the idealized reed of Fig. 13.7(a) is shown in Fig. 13.7(d); the curve for a realistic reed with a spoon-shaped curvature and moderately thick supporting plate will differ in detail. The change from $(-,+)$ to $(+,-)$ configuration near the point B is clearly evident, and it is also apparent that the threshold sounding pressure, represented by the point A, is at a much lower fraction of the closing pressure than in the case of a broad clarinet-like reed.

The only available studies of single free reeds appear to be those St. Hilaire et al., (1971) and a recent unpublished investigation by Koopman et al. (1996). St. Hilaire et al. showed that the inertia of the flow toward

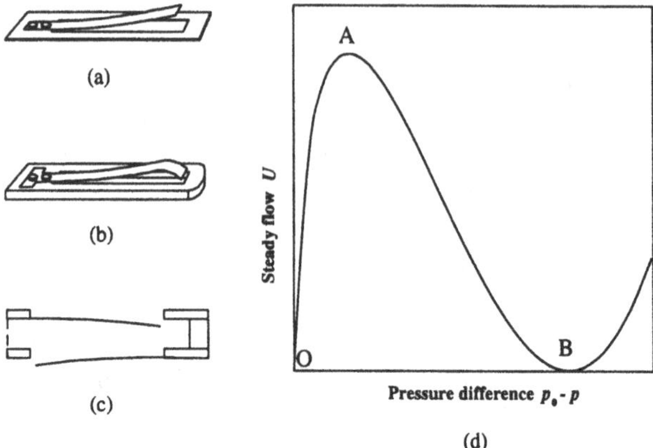

FIGURE 13.7. (a) An idealized free reed, consisting of a curved metal tongue attached to a thin metal plate in which there is an aperture through which it can just pass. (b) A typical harmonium reed. Note the curvature of the reed tongue and the appreciable thickness of the plate. (c) A pair of reeds of opposite geometry, closely coupled as in a harmonica (mouth organ). (d) Static flow characteristic of the idealized reed shown in (a).

the reed and through its channel adds a significant acoustic inertance that, according to our discussion above, results in a negative damping for a $(-, +)$ reed and a positive damping for a $(+, -)$ reed. In the usual arrangement, the reeds are also loaded by a rather narrow channel, which we may expect to enhance this effect. In practice, therefore, a blown-closed $(-, +)$ reed of harmonium geometry has a threshold oscillation pressure that is less than that of a blown-open $(+, -)$ reed, and is the one used in normal playing, the vibration frequency being a little less than the mechanical resonance frequency of the unblown reed. Because the reed operates close to its resonance frequency, its threshold pressure is a good deal less than the value predicted from static-flow considerations.

The studies of Koopman et al. (1996) on harmonium reeds showed that, once the threshold oscillation pressure had been reached, about 200 Pa in their case for a C_3 reed, the vibration amplitude of the reed achieved a large value (typically about 4 mm), which then increased only a little with further pressure increase before beginning to decrease at large pressures (above about 1000 Pa). This behavior can be readily understood in terms of the static flow curve of Fig. 13.7. Being close to resonance, the reed vibrates nearly sinusoidally about a center determined by the average applied pressure difference. The downward-sloping part AB of the flow characteristic contributes a negative acoustic conductance that supports the oscillation, but any excursion of the reed deflection past B into the rising part of the curve causes damping. Of course, this qualitative description requires refinement, since the reed vibrates only a little below its natural resonance and there are phase shifts to be considered. At very large pressures the reed may transform to $(+,-)$ configuration and vibrate at a frequency rather above its natural resonance, but this depends upon details of the impedances and damping.

In the harmonica or mouth organ, reeds are grouped in pairs, as shown in Fig. 13.7(c), one nominally operated by blowing pressure and one by suction. Because these two reeds are closely coupled by the small air volume in the reed channel, their operation is more complex than that of a single reed. This problem has been investigated in detail by Johnston (1987), using a small-displacement approximation. When the harmonica is blown, the sounding threshold pressure of the $(-,+)$ reed is much lower than that of the $(+,-)$ reed, despite their complementary geometry, so it is the $(-,+)$ reed that sounds, the other reed sounding when the harmonica is played by suction. As discussed above, this asymmetry is partly due to the acoustic impedance loading contributed by the flow through the reeds, and there is an additional asymmetry because the mean center of vibration of the $(+,-)$ reed leaves a much more open channel than in the case of the $(-,+)$ reed. Johnston investigated, in particular, the "pitch bending" that is characteristic of skilled harmonica playing, an effect produced by modifying the player's oral cavity. The only notes that can be bent are those for which the other note on the same channel has a lower pitch than the note being played, and the pitch can only be lowered. He showed that this can be accounted for by a detailed model including the coupling between the two reeds, which leads to the two curves shown in Fig. 13.8. The accessible frequency domain is that for which the acoustic conductance is negative, in each case.

13.5 Generators Coupled to Horns

Most wind instruments employ a reed or lip valve in conjunction with some sort of pipe resonator, and it is now appropriate to examine the coupling between these two systems, the one passive and the other active. As we will see later, nonlinear effects can become very important in determining this interaction, but for the present we will be content with a discussion of the first-order linear case. The first thing we must note is that since reed generators are pressure-controlled devices they must operate at the input to a resonator that presents to them a maximum pressure variation—an input impedance maximum for the resonator. It is simplest to discuss this for a cylindrical pipe and then to indicate how we can extend the argument to horns of other geometries.

As we saw in Eq. (8.35) of Chapter 8, the input impedance of an open cylindrical pipe of length L and cross section S that is damped

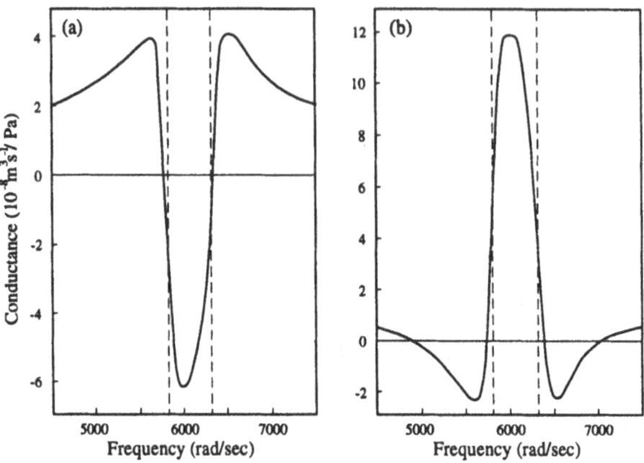

FIGURE 13.8. (a) The acoustic conductance of a coupled-reed harmonica model when the resonance frequency of the closing reed is higher than that of the opening reed, making pitch-bending possible for the closing reed; (b) the acoustic conductance for the case in which the resonance frequency of the closing reed is lower than that of the opening reed.

predominantly by wall losses can be written in the form

$$Z_p = Y_p^{-1} = \left(\frac{\rho c}{S}\right)\left(\frac{1 + jH \tan kL}{H + j \tan kL}\right), \tag{13.9}$$

where ρ is the density and c is the velocity of sound in air, $k = \omega/c$. The quantity $H = 1/\tanh \alpha L$, which decreases with increasing frequency, is the height of the impedance maxima, or depth of the admittance minima, relative to the characteristic pipe impedance $\rho c/S$. The impedance maxima near which oscillation can occur are the odd harmonics for which $kL = (2n + 1)\pi/2$ or

$$\omega_p^{(n)} = \frac{(2n + 1)\pi c}{2L}. \tag{13.10}$$

The real part of Y_p is always positive (dissipative), but, from Eq. (13.9), the imaginary part is negative just below an impedance maximum and positive just above. In order that the whole coupled system should be in resonance and that the oscillation should grow, we require that

$$\operatorname{Im} Y_r = -\operatorname{Im} Y_p, \tag{13.11}$$

and

$$-\operatorname{Re} Y_r > \operatorname{Re} Y_p. \tag{13.12}$$

Since, as we saw in Fig. 13.5(a), Im $Y_r > 0$ for a valve of configuration $(-, +)$, as in woodwinds, or $(+, +)$ for high notes on brass intruments, this requires that Im $Y_\mathrm{p} < 0$, which implies that the operating frequency ω should be slightly below a pipe resonance:

$$\omega = \omega_\mathrm{p}^{(n)} - \delta. \tag{13.13}$$

Conversely, for a $(+,-)$ valve as in low notes on brass instruments, Im $Y_\mathrm{r} < 0$ as shown in Fig. 13.5(b), so that we must have Im $Y_\mathrm{p} > 0$ and

$$\omega = \omega_\mathrm{p}^{(n)} + \delta. \tag{13.14}$$

Thus, for a properly cooperative oscillation, a woodwind reed instrument should sound slightly flat of the stopped-pipe resonance because of the dynamics of the reed, and a brass instrument slightly sharp. This is, however, a conclusion reached by considering only the linear approximation.

To determine whether the pipe will in fact sound and, if so, in which of its modes, we must use Eq. (13.12). Neglecting nonlinear effects, that mode will sound, preferentially, for which the inequality is most strongly satisfied. The behavior is very different for different types of reed generators.

The reed of a woodwind instrument is heavily damped by the player's lips, and the damping is enhanced by the mouth-volume effect discussed in Section 13.2. The real part of the reed admittance is therefore negative and nearly flat up to the reed resonance, as illustrated by the curve for $\gamma = 0.3$ in Fig. 13.6(a). The reed resonance frequency is also typically 6 to 10 times the frequency of the lowest mode of the pipe, so that the oscillation condition in Eq. (13.12) may be satisfied by several of the lower pipe modes. Typically, however, as given by Eq. (13.9) and illustrated in Fig. 8.11, the mode of lowest frequency has the highest relative impedance H (or lowest relative admittance $1/H$), so that the instrument sounds this lowest mode.

The metal reed of an organ pipe, though geometrically similar to a clarinet reed, behaves very differently. It has very little damping, so that its admittance is even more sharply peaked than is the curve for $\gamma = 0.1$ in Fig. 13.6(a). In these circumstances, the magnitude of the reed conductance $|G_\mathrm{r}|$ is sufficiently large at a frequency just below ω_r that the inequality of Eq. (13.12) can be satisfied over a large frequency range below the first impedance maximum of the pipe. The resonance frequency of the reed therefore dominates the behavior, and tuning is achieved by changing the vibrating length of the reed tongue rather than by adjusting the pipe resonator. The sound is strongest, however, when the resonances of the two parts of the system are aligned.

Finally, the behavior of a lip reed, as in a trumpet or horn, is as illustrated in the low-damping curve of Fig. 13.6(a), if the lips move transversely to give a $(+,+)$ valve configuration, or as in the low-damping curve of Fig. 13.6(b) if they move outwards to give a $(+,-)$ configuration. The frequency range over which G_r is negative is narrow in each case, since the effective damping is small because of the mouth-volume effect discussed in Section 13.3. The lip resonance once again dominates the behavior, and

Eq. (13.12) can be satisfied only in a narrow frequency range either just below ω_r for the $(+,+)$ case or just above for the $(+,-)$ case. There will generally be only one pipe resonance in this range, and the oscillation will drive it. It may be possible to satisfy Eq. (13.12) and produce an oscillation over a considerable range between the pipe resonances, as skilled trombone players are able to demonstrate, but the sound will be weak compared with that from a well-aligned cooperating pipe resonance.

We should note again from Eqs. (13.8) that if we include the effects of the player's vocal tract, its reactance X_2 simply adds to the pipe reactance X_1 in the case of a valve of configuration $(+,-)$ or $(-,+)$, but subtracts from it for the $(+,+)$ configuration. The player has many accessible variables with which to influence even the linear behavior of an instrument! We return to discuss this in a little more detail in Chapters 14 and 15.

All these remarks also apply qualitatively to reeds coupled to horn resonators of shapes other than cylindrical. Indeed, in the linear, small-amplitude approximation we have been using, there is nothing that need be added to cover this more general case. The difference arises when we consider vibrations of larger amplitude and with significant harmonic development, for then the nonlinearities of the reed generator assume an important or even a dominant role.

13.6 Large-Amplitude Behavior

While the linear approach used in the sections above reveals a great deal about the behavior of pressure-driven valve generators as far as oscillation thresholds are concerned, it tells us nothing about the large-signal behavior of the generator system, and this is, in reality, what we are interested in. Fortunately a relatively simple extension of the discussion gives us at least a first-order approximation to this large-amplitude situation.

13.6.1 Woodwind Single Reeds

Once again, the analysis is simplest for the case of the clarinet, so we begin here. In order for the oscillation to start at all, the blowing pressure p_0 must be less than the closing pressure p_c for the reed, and greater than the threshold value $\frac{1}{3}p_c$ corresponding to point A in Fig. 13.9(a). In essence, what we want to discover is the amplitude and shape of the pressure and flow waveforms when the valve oscillation has settled down to a steady state, and it is simplest to begin with an ideal system with precisely harmonic odd-mode resonances and zero tube impedance at the even-harmonic frequencies. We then know that the mouthpiece pressure waveform $p(t)$ will contain no even harmonics and will therefore be completely symmetrical, whatever the flow waveform. Suppose we fix the blowing pressure at a value p_0, then the pressure difference $p_0 - p$ must make equal excursions either side of this value, and the "load-line" for the resonator must be horizontal because it has no losses. Actually it is simpler to go in the opposite direction and draw a horizontal load-line such as BC and then deduce that the blowing pressure must be $p_0 = \frac{1}{2}(p_B + p_C)$. The locus of these blowing pressures, shown as a broken curve in the figure, can then be used in reverse to predict the pressure amplitude as a function of blowing pressure. We can guess that the reed will begin to beat when $p_0 > \frac{2}{3}p_C$, though possibly at a lower pressure.

In this lossless approximation, the mouthpiece pressure switches abruptly between its two extreme values, and the volume flow in each of the switched states is the same, though there is a switching transient, as shown at the right in Fig. 13.9(a). The surprising apparent continuity of the flow is appropriate, since no power is required to maintain the lossless tube resonator in oscillation.

As a step toward reality, Fig. 13.9(b) shows the system diagram when we allow for losses in the resonator. The load lines now slope downwards, and

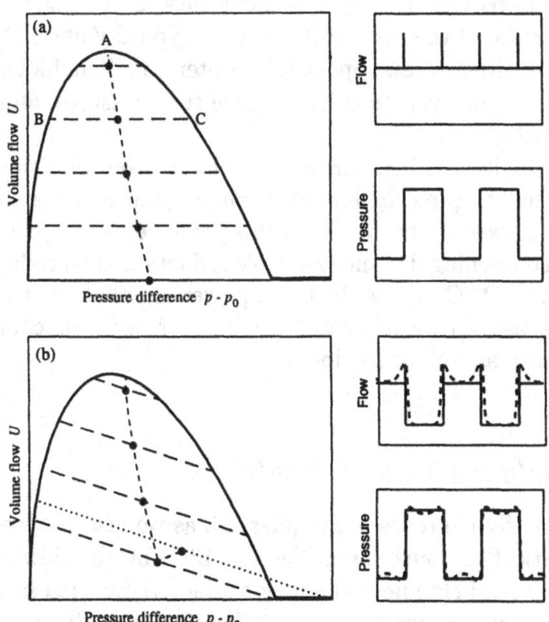

FIGURE 13.9. Large-amplitude behavior of a clarinet-like system. (a) State diagram for a lossless system; flow and pressure waveforms for a lossless system. (b) State diagram for a system with resistive losses; flow and pressure waveforms for a system with resistive losses; the dashed curves show the behavior if these losses are greater at high frequencies.

we note that oscillation can begin only if the load-line slope is less than the maximum negative slope of the reed flow characteristic, as we can deduce from the small-signal linear treatment. The argument goes just as before, but now the two parts of the flow cycle have different magnitudes, leading to a flow waveform that supplies energy to overcome the losses. Once the reed vibration reaches the beating point, the behavior changes somewhat, as indicated by the kink in the blowing-pressure curve. As a final refinement, we can allow qualitatively for the fact that losses are generally greater at high frequencies, so that the transitions between the two flow states will not be abrupt, leading to the sort of waveform shown with a broken line. Some of this analysis has been formalized by Kergomard (1995).

When we come to apply the same ideas to conical-bore instruments such as saxophones, we meet an immediate difficulty because the pressure waveform is no longer restricted to being symmetrical. Indeed Dalmont and Kergomard (1995) have argued that the input impedance of a truncated conical resonator has a form similar to that of a string excited at the same fraction along its length, so that, by analogy with the Helmholtz motion of a bowed string, we might expect the mouthpiece pressure waveform to consist of a short section of constant low pressure followed by a long section of constant high pressure, the ratio of the durations of these being equal to the truncation ratio of the resonator, which is typically about 1 : 10. There is experimental evidence to support this contention. The blowing pressure is then set at the time-average value of these two pressures, and thus closer to the lower pressure.

In all these single-reed instruments, while the reed vibration amplitude can be controlled to some extent by changing the blowing pressure, the major control is exerted by means of lip pressure, which can reduce the equilibrium reed opening distance x_0. This reduction in x_0 reduces the scale of the flow curve OAC and with it the power supplied by the generator. Skilled players use lip pressure and blowing pressure in combination to control tone color as well as loudness.

13.6.2 Woodwind Double Reeds

The behavior of double reeds is complicated, as we discussed in relation to their quasi-static flow behavior in Section 13.2, by the flow resistance in the long narrow reed channel. The flow characteristic then has the looped behavior shown by the curve through A″ in Fig. 13.3. This characteristic has not yet been examined experimentally but, if we regard it as a reliable guide to reed-valve behavior, then we should expect severely nonlinear hysteresis effects. Suppose that the pressure is raised slowly along the curve OFA″ in the figure, then the small-signal conductance is positive, and no oscillations can occur. When the pressure reaches the point A″, however, the reed closes abruptly and the operating point falls to E. The acoustic pulse produced by this flow discontinuity propagates along the resonator and returns to the mouthpiece, after a time delay depending on the resonator shape, in such a phase as to raise the mouthpiece pressure and lower $p_0 - p$ past the knee of the curve just below C. The reed then flips open to the point F and the cycle repeats.

Recent work by Gokhshtein (1995) has given new insight into this behavior. He shows that, in the case of a double-reed instrument with a conical horn, the reed is almost fully open for most of each cycle of the oscillation and then closes fully for a much shorter time. The ratio of the closed time to the open time is equal to the truncation ratio for the cone, which is typically about 0.1. This behavior can be understood by considering the reflection of a pressure pulse from the truncated end of the cone, terminated by a reed cavity, and the conclusion is the same as that reached on the basis of a different argument by Dalmont and Kergomard (1995) for the saxophone, as discussed in the previous subsection.

Clearly this behavior has no small-amplitude counterpart—once the reed reaches the critical pressure corresponding to A″ the vibration immediately attains large amplitude. The cycle A″ECFA″ represents a lower limit to this motion. If the resonator has sufficiently low losses, then the negative excursion of mouthpiece pressure p can take the operating point well to the right of E in the closed-reed situation, and the returning positive pulse can take it well to the left of F.

As in the case of single-reed instruments, the power of the generator, and hence the loudness of the sound, is controlled by lip pressure on the reed, which reduces the equilibrium opening x_0 and with it the whole scale of the flow characteristic.

13.6.3 Resonant Reeds

The reeds considered above all operate at frequencies well below their natural resonance, but this is not true of the reed tongues of organ reed pipes or of the free reeds of harmoniums and similar instruments. In these cases the reed has very little damping and vibrates, as we have discussed in Sections 13.3 and 13.4, at a frequency just below its natural resonance. The result is that the reed vibration is quite closely sinusoidal, except for the one-sided closing constraint of the reed against the shallot in the organ pipe.

Provided the blowing pressure is above the oscillation threshold, the oscillation amplitude is determined largely by the shallot constraint in the case of the organ-pipe reed and by the depth of the mounting plate in the case of the free reed. The sound power therefore does not vary much with blowing pressure within the normal operating range. We can get a first approximation to the flow waveform for a free reed by using the first

Eq. (13.2) in the form $U = Wxv$ and neglecting the effect of the back pressure p on the flow velocity v, but a more detailed calculation is required for an organ pipe.

13.6.4 Lip Valves

Lip valves share some of the properties of resonant reed valves, in that their oscillation is fairly closely sinusoidal. Because the lips are blown open by mouth pressure, however, whether the lip configuration is $(+,-)$ or $(+,+)$, the lip vibration amplitude increases approximately linearly with blowing pressure p_0 above some threshold p_T, the opening width rather less than linearly with p_0, and the flow velocity as $p_0^{1/2}$, if we neglect the back-pressure in the instrument mouthpiece. The volume flow U thus varies approximately as $p_0^{1/2}(p_0 - p_T)^2$. The power output P from the lip generator is RU^2, where R is the real part of the instrument input impedance at resonance, and P therefore behaves roughly as $p_0(p_0 - p_T)^4$. This quantity rises rapidly from zero at the threshold pressure p_T, and then behaves about as p_0^5 until the lip opening becomes limited by nonlinear effects, after which P rises only linearly with p_0. The radiated acoustic power follows a rather similar course, though there are large losses in the instrument. In the mid-pressure range, the output power thus rises by about 15 dB for a doubling of blowing pressure. While this analysis neglects the back-pressure in the mouthpiece, it actually describes the measured behavior of the lip generator remarkably well. We defer consideration of the flow waveform until Chapter 14.

13.7 Nonlinear Analysis

In musical instruments involving plucked strings or struck metal plates, nonlinearity came to our attention as an interesting addendum to the description of the behavior of a system that was well described by the linear approximation. The same is not true of instruments producing sustained notes, whether the fundamental oscillator be a bowed string, a vibrating reed, or a deflecting jet. In all these cases of self-regenerating oscillation, nonlinear phenomena determine the onset and decay of oscillation, the amplitude of the steady state, and the harmonic content of the sound. We shall attempt to cover only a few of these points here.

The relation between pressure p and flow U set out in Eqs. (13.5) and (13.6) is clearly a nonlinear one, as also is the simpler quasi-static expression of Eq. (13.2). This means that if we assume a simple variation of pressure at frequency ω,

$$p = p_1 \sin \omega t, \tag{13.15}$$

the resulting flow U will contain, as well as terms of frequency ω, terms with frequencies $n\omega$, where $n = 2, 3, 4 \ldots$. The amplitude U_n of the term of frequency $n\omega$ will vary as U_1^n, and thus as p_1^n, provided that $U_n < U_1$. These higher frequency terms will then interact with the horn, which is fortunately a very nearly linear system, so that the component $U_n \sin n\omega t$ in the flow will give rise to a pressure

$$p_n = Z_{\mathrm{p}}(n\omega)U_n \sin n\omega t \tag{13.16}$$

at the mouthpiece.

If the reed is of a woodwind type, then generally its resonance frequency ω_{r} will be much higher than the fundamental frequency ω, and the motion of the reed, as described by Eq. (13.6), will follow the pressure signal obtained by adding to Eq. (13.15) terms such as those from Eq. (13.16). When this reed deflection x is combined with this new expression for p and inserted in the flow equation [Eq. (13.2)], then a rich harmonic spectrum results with each component rigorously locked in frequency and phase to the fundamental, as discussed in Section 5.5 (Fletcher, 1978, 1990).

The situation is a little different for an organ reed pipe or the lip generator of a brass instrument, both of which operate with the fundamental ω almost equal to the reed resonance ω_{r}. From Eq. (13.6), the reed displacement response x is almost unaffected by these higher frequencies and remains sinusoidal. However, the flow equation [Eq. (13.5)] is sufficiently nonlinear that many phase-locked harmonics of large amplitude are generated, even when x itself is sinusoidal.

Work along these lines was first reported by Benade and Gans (1968) and was further developed for clarinet-like systems by Worman (1971), Wilson and Beavers (1974), and Schumacher (1978). The spectrum generated depends in detail on the geometry of the reed and the resonance properties of the instrument tube, and for that reason we defer its consideration to Chapter 15. It is worthwhile to note once again, however, the common property of moderately nonlinear systems whose behavior can be represented by a power series, that, when the amplitudes of the generated harmonics are small compared with that of the fundamental, the amplitude of the nth harmonic varies as the nth power of the amplitude of the fundamental. Increasing loudness is therefore always associated with increasing harmonic development. At large amplitudes, this rule no longer holds, and the relative harmonic levels in the spectrum approach a limiting distribution.

The case of brass-instrument-type lip generators has similarly been investigated by Backus and Hundley (1971), building on earlier work by Luce and Clark (1967) and on the experimental verification by Martin (1942) that the motion of the lips when playing a brass instrument is essentially sinusoidal. Again, we defer detailed discussion to Chapter 14.

As a brief example of the importance of the resonance properties of the instrument horn in determining reed behavior, we may contrast the case of the saxophone and the clarinet, both of which have flat cane reeds attached to rather similar mouthpieces. The horn of the saxophone is a nearly complete cone and therefore, as we saw in Chapter 8, it has a complete set of impedance maxima at frequencies near $n\omega_1$. The very general form of the nonlinearity of Eqs. (13.5) and (13.6) leads us to expect a complete set of harmonic components in the flow that can interact back to produce harmonic driving components through Eq. (13.16). This is what is observed.

For the clarinet, however, the instrument tube is essentially cylindrical and has impedance maxima only at frequencies near $(2n - 1)\omega_1$. Thus, though the reed motion produces flow components at all harmonics of the fundamental ω_1, only the odd harmonics can react back through Eq. (13.16) to produce appreciable driving pressures and so attain large amplitude. This effect is clear in the spectra of low notes on the clarinet which show a very weak second harmonic and, consequently, a nearly symmetrical waveform. High notes are less affected because the tube impedance behaves less regularly with complex fingerings.

Brass instruments, as we shall see in the next chapter, generally have a flaring bore so designed that impedance peaks are well aligned in frequency with harmonics of the played fundamental. This alignment is clearly important for a rich tone to be produced and for the pitch to remain stable as the loudness and harmonic content are varied (Benade and Gans, 1968; Benade, 1973).

Finally, we note that sometimes, in either brass or more particularly woodwind instruments, conditions may be modified by using special fingerings or lip positions so that the reed generator can drive, simultaneously, two horn resonances, ω_1 and ω_2, that are not harmonically related. The nonlinearity of the flow through the reed will then generate all sum and difference frequencies $n\omega_1 \pm m\omega_2$ to produce a broadband, multiphonic sound (Bartolozzi, 1981; Backus, 1978). Reliable production of such sounds requires a reed, such as in a clarinet, that can present a negative acoustic conductance over an appreciable frequency range, together with a set of horn resonances that is sufficiently inharmonic to defeat the natural tendency of nonlinear systems to achieve frequency locking (Fletcher, 1978).

13.8 Numerical Simulation

The nonlinearity that we discussed in the previous section is confined entirely to the reed or lip generator—the air column in the horn of the instrument operates at a sufficiently low level that its behavior is completely linear in most cases. This leaves open several alternative approaches to calculating the acoustic behavior of the instrument as a whole. Such a calculation is valuable from at least two points of view. In the first place, it allows us to calculate actual acoustic waveforms from our model instrument under different assumed playing conditions, and these can be compared with the output of a real instrument with a human player. If the agreement is good, then we can be reassured that our analytical model has captured most of the essential features governing the operation of the instrument. The second reason for pursuing such simulation calculations is that, as computer power increases, they provide a means of synthesizing realistic musical voices using just a few "physical" parameters to control the sound, rather than resorting to manipulation of artificial waveforms. A good numerical simulation would allow the composer not only to produce realistic instrument sounds, but also to extend these models in a convincing way to produce sounds characteristic of imaginary instruments that might be difficult to realize physically.

There are several approaches to this simulation problem that differ principally in the way in which they treat the air column (Fletcher, 1990). Such alternatives are possible because the air column behavior is linear, so that we can discuss it equally in the frequency domain, through its normal modes and resonances, or in the time domain in terms of its impulse response. These two approaches are related simply by a Fourier transform, as we discussed in Section 8.14. Because of its nonlinearity, flow through the reed valve is most easily treated in the time domain.

A completely frequency-domain approach has been developed by Fletcher (1976) and applied to organ flue pipes, though it can in principle be applied to reed-valve instruments as well. It relies upon a particular approach to solving the differential equations for the air-column modes, coupled together by the nonlinear flow characteristic of the valve, in which each mode is represented by an expression of the form $a_n(t) \sin[\omega_n t + \phi_n(t)]$, where a_n and ϕ_n are taken to be slowly varying functions of time. This method gives good results for both the steady state and for quite rapid transients, but is limited in the number of pipe modes that can be realistically included.

Another approach, often termed the harmonic balance method, combines a time-domain treatment of the valve with a frequency-domain treatment of the instrument horn (Gilbert et al., 1989). We can, for example, assume first a sinusoidal variation of the valve opening x and calculate the time-varying flow $U(t)$ through it using Eq. (13.5). The Fourier transform $U(\omega)$ of this flow takes us into the frequency domain and, since we know the input impedance $Z(\omega)$ of the horn, we can easily calculate the resulting mouthpiece pressure $p(\omega)$. The effect of this pressure on the motion of the valve can be calculated by transforming it back into the time domain as $p(t)$ and inserting this into the equation of motion (13.6) of the valve. The newly calculated motion can then be used to start the computation cycle again, and the process can be repeated until it converges. The computed steady oscillation appears to be a very satisfactory simulation of actual clarinet tone for the case considered. The need for iteration, however, makes the method unsuitable for computing rapid transients.

Finally, we can use an approach exclusively in the time domain, as pioneered by McIntyre et al. (1983), in which the impulse response of the horn is convolved with the nonlinear volume flow to give an integral equation, as discussed in Section 8.14. The impulse response itself can be either measured directly or else calculated by taking the Fourier transform of the measured input impedance spectrum. The problem of the long temporal extent of the impulse response can be overcome by a formal device due to Schumacher (1981), and the approach then gives good results. Some of the numerical problems inherent in the method have been discussed by Gazengel et al. (1995). Apart from the references listed here, which refer to the clarinet, perhaps the simplest of instruments in the time domain, the method has also been applied successfully to brass instruments by Adachi and Sato (1996). It is particularly well adapted for the calculation of transients, since it operates completely in the time domain in the same way that the instrument itself does.

References

Adachi, S., and Sato, M. (1995a). Time-domain simulation of sound production in the brass instrument. *J. Acoust. Soc. Am.* **97**, 3850–3861.

Adachi, S., and Sato, M. (1995b). On the transition of lip-vibration states in the brass instrument. *Proc. Internat. Symp. Musical Acoust.* (Dourdan, France), IRCAM, Paris pp. 17–22.

Adachi, S., and Sato, M. (1996). Trumpet sound simulation using a two-dimensional lip vibration model. *J. Acoust. Soc. Am.* **99**, 1200–1209.

Backus, J. (1961). Vibrations of the reed and air column in the clarinet. *J. Acoust. Soc. Am.* **33**, 806–809.

Backus, J. (1963). Small-vibration theory of the clarinet. *J. Acoust. Soc. Am.* **35**, 305–313; erratum (1977) **61**, 1381–1383.

Backus, J. (1978). Multiphonic tones in the woodwind instruments. *J. Acoust. Soc. Am.* **63**, 591–599.

Backus, J. (1985). The effect of the player's vocal tract on woodwind instrument tone. *J. Acoust. Soc. Am.* **78**, 17–20.

Backus, J., and Hundley, T.C. (1971). Harmonic generation in the trumpet. *J. Acoust. Soc. Am.* **49**, 509–519.

Bartolozzi, B. (1981). "New Sounds for Woodwinds." 2nd ed. Oxford Univ. Press, London and New York.

Benade, A.H. (1973). The physics of brasses. *Sci. Am.* **299**(1), 24–35.

Benade, A.H. (1986). Interactions between the player's windway and the air column of a musical instrument. *Cleveland Clinic Quart.* **53** (1), 27–32.

Benade, A.H., and Gans, D.J. (1968). Sound production in wind instruments. *Ann. N.Y. Acad Sci.* **155**, 247–263.

Chen, F-C., and Weinreich, G. (1996). Nature of the lip reed. *J. Acoust. Soc. Am.* **99**, 1227–1233.

Clinch, P.G., Troup, G.J., and Harris, L. (1982). The importance of vocal tract resonance in clarinet and saxophone performance—A preliminary account. *Acustica* **50**, 280–284.

Copley, D.C., and Strong, W.J. (1996). A stroboscopic study of lip vibrations in a trombone. *J. Acoust. Soc. Am.* **99**, 1219–1226.

Dalmont, J-P., and Kergomard, J. (1995). Elementary model and experiments for the Helmholtz motion of single reed wind instruments. *Proc. Int. Symp. Mus. Acoust.* (Dourdan). IRCAM, Paris. pp. 115–120.

Fletcher, N.H. (1976). Transients in the speech of organ flue pipes—A theoretical study. *Acustica* **34**, 224–233.

Fletcher, N.H. (1978). Mode locking in non-linearly excited inharmonic musical oscillators. *J. Acoust. Soc. Am.* **64**, 1566–1569.

Fletcher, N.H. (1979a). Air flow and sound generation in musical wind instruments. *Ann. Rev. Fluid Mech.* **11**, 123–146.

Fletcher, N.H. (1979b). Excitation mechanisms in woodwind and brass instruments. *Acustica* **43**, 63–72; erratum (1982) **50**, 155–159.

Fletcher, N.H. (1983). Acoustics of the Australian didjeridu. *Australian Aboriginal Studies* **1**, 28–37.

Fletcher, N.H. (1990). Nonlinear theory of musical wind instruments. *Applied Acoustics* **30**, 85–115.

Fletcher, N.H. (1993). Autonomous vibration of simple pressure-controlled valves in gas flows. *J. Acoust. Soc. Am.* **93**, 2172–2180.

Fletcher, N.H., Silk, R.K., and Douglas, L.M. (1982). Acoustic admittance of air-driven reed generators. *Acustica* **50**, 155–159.

Gazengel, B., Gilbert, J., and Amir, N. (1995). Time domain simulation of single reed wind instrument. From the measured input impedance to the synthesis signal. Where are the traps? *Acta Acustica* **3**, 445–472.

Gilbert, J., Kergomard, J., and Ngoya, E. (1989). Calculation of the steady-state oscillations of a clarinet using the harmonic balance technique. *J. Acoust. Soc. Am* **86**, 35–41.

Gokhshtein, A. (1995). New conception and improvement of sound generation in conical woodwinds. *Proc. Int. Symp. Mus. Acoust.* (Dourdan). IRCAM, Paris. pp. 121–128.

Hirschberg, A. (1995). Aeroacoustics of wind instruments. In "Mechanics of Musical Instruments," Ed. A. Hirschberg, J, Kergomard and G. Weinreich. Springer-Verlag, Vienna and New York. pp. 291–369.

Hirschberg, A., van de Laar, R.W.A., Marrou-Maurières, J.P., Wijnands, A.P.J., Dane, H.J., Kruijswijk, S.G., and Houtsma, A.J.M. (1990). A quasi-stationary model of air flow in the reed channel of single-reed woodwind instruments. *Acustica* **70**, 146–154.

Johnston, R., Clinch, P.G., and Troup, G.J. (1986). The role of vocal tract resonance in clarinet playing. *Acoustics Australia* **14**, 67–69.

Johnston, R.B. (1987). Pitch control in harmonica playing. *Acoustics Australia* **15**, 69–75.

Kergomard, J. (1995). Elementary considerations on reed-instrument oscillations. In "Mechanics of Musical Instruments," Ed. A. Hirschberg, J. Kergomard and G. Weinreich. Springer-Verlag, Vienna and New York, pp. 229–290.

Koopman, P.D., Hanzelka, C.D., and Cottingham, J.P. (1996) Frequency and amplitude of vibration of reeds from American reed organs as a function of pressure. *J. Acoust. Soc. Am.* **99**, 2506 (abstract only).

Luce, D., and Clark, M. (1967). Physical correlates to brass-instrument tones. *J. Acoust. Soc. Am.* **42**, 1232–1243.

McIntyre, M.E., Schumacher, R.T., and Woodhouse, J. (1983). On the oscillation of musical instruments. *J. Acoust. Soc. Am.* **74**, 1325–1345.

Martin, D.W. (1942). Lip vibrations in a cornet mouthpiece. *J. Acoust. Soc. Am.* **13**, 305–308.

Saneyoshi, J., Teramura, H., and Yoshikawa, S. (1987). Feedback oscillations in reed woodwind and brasswind instruments. *Acustica* **62**, 194–210.

Schumacher, R.T. (1978). Self-sustained oscillations of the clarinet: An integral equation approach. *Acustica* **40**, 298–309.

Schumacher, R.T. (1981). Ab initio calculations of the oscillations of a clarinet. *Acustica* **48**, 71–85.

Sommerfeld, S., and Strong, W. (1988). Simulation of a player-clarinet system. *J. Acoust. Soc. Am.* **83**, 1908–1918.

St. Hilaire, A.O., Wilson, T.A., and Beavers, G.S. (1971). Aerodynamic excitation of the harmonium reed. *J. Fluid Mech.* **49**, 803–816.

Wijnands, A.P.J., and Hirschberg, A. (1995). Effect of a pipe neck downstream of a double reed. *Proc. Internat. Symp. Musical Acoust.* (Dourdan, France), IRCAM, Paris pp. 148–151.

Wilson, T.A., and Beavers, G.S. (1974). Operating modes of the clarinet. *J. Acoust. Soc. Am.* **56**, 653–658.

Worman, W.E. (1971). Self-sustained nonlinear oscillations of medium amplitude in clarinet-like systems. PhD thesis, Case Western Reserve University, Cleveland, Ohio. University Microfilms, Ann Arbor, Michigan (ref. 71-22869).

Yoshikawa, S. (1995). Acoustical behavior of brass player's lips. *J. Acoust. Soc. Am.* **97**, 1929–1939.

14

Lip-Driven Brass Instruments

Lip-driven wind instruments have a very long history, dating back to those made from hollow plant stems, sea shells, and animal horns (Carse, 1939; Baines, 1966); even metal trumpets roughly similar to those of the present day existed as long ago as Roman times. It is not our purpose to go extensively into the history of the development of these instruments or even to describe the large variety of forms in contemporary use. Rather, we shall concentrate on general principles and simply mention examples as they arise.

14.1 Historical Development of Brass Instruments

Early instruments usually had a more or less conical bore, as dictated by their origin in sea shells or animal horns, and were generally not over about half a meter in length. This yielded a fundamental in the high male vocal range around 300 Hz, and limited the repertoire to single notes or bugle calls using the first two or three more-or-less harmonically related modes. Exceptions arise in the case of instruments derived from plant stems, which are much more nearly cylindrical. An example is the didjeridu of the Australian aboriginal people (Fletcher, 1983; 1996), which is simply a small tree trunk, typically 1 to 2 m long, hollowed out by the action of fire and termites to give a tube flaring gradually from 30 to 50 mm in internal diameter. The fundamental, played as a rhythmic drone, has a frequency around 60 Hz in typical cases, and only one upper mode is accessible, though the sound is enriched by a variety of performance techniques.

The development of these early instruments took two different paths in medieval times, both aimed at producing more flexible melodic possibilities. The small conical instruments were provided with side holes, similar to those found in the woodwinds, to allow production of notes in between the modes of the complete horn. Small cupped mouthpieces were developed to help with tone production and playing comfort, and the structure of the instrument was refined by making it from wood or metal, rather than relying on natural shapes. Sometimes two gouged wooden half-horns of semicircular section, bound together with leather, were used, though other construction methods were also common. Among instruments of this type that survived in refined form through the baroque period were the cornett, usually about half a meter long and slightly curved to reflect its historical origin as an animal horn, and the serpent, a bass instrument curved to the shape of an S to make playing and carrying possible. These instruments are illustrated in Fig. 14.1, and a brief acoustical discussion has been given by Campbell (1994). In the nineteenth century, many other variants were developed, lip-blown (Russian) bassoons, for example, and, particularly as mechanical keys began to make this practical for larger instruments, various keyed metal ophicleides and bugles for brass band use. The common construction material for these was a combination of drawn and rolled-and-soldered brass tubing. While the baroque instruments are now being recreated and used in increasing numbers, most of the later developments survive only as museum curiosities. The general principles of this chapter apply to all these instruments, though we shall defer discussion of finger holes until the chapter on woodwind instruments.

The other line of development was to increase greatly the length of the instrument and to narrow its bore so that the closely spaced upper modes became available for melodic use. Early military trumpets and bugles went some way in this direction using modes from the third to the sixth to produce familiar musical signals for camp life. Ceremonial trumpets used a greater mode range for their fanfares, and the coiled natural horns had

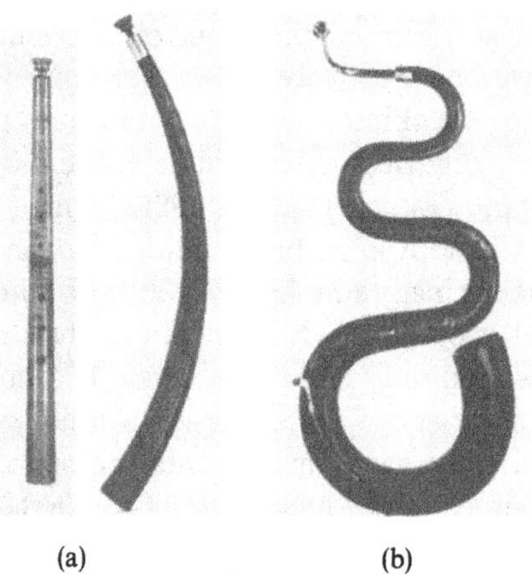

FIGURE 14.1. Some lip-blown instruments, such as the cornett (a) and the serpent (b) (to a different scale), have a conical bore and finger holes very much like those used in woodwind instruments. They are not in modern use, but modern replicas are often used in the performance of old music.

available to them modes up to about the sixteenth and were thus able to produce an octave of complete diatonic scale, which, because of their considerable length, lay at a musically useful pitch.

The lower modes of these instruments were still playable, of course, but produced notes quite widely separated in pitch. One of the earliest and most successful means developed to fill in these gaps was the slide mechanism of the medieval sackbut, the forerunner of the modern trombone, which increases the length of the narrow cylindrical part of the bore in a continuously variable manner. Most other instruments in current use, however, make use of mechanical valves, three to five in number, to switch extra lengths of tubing into the bore of the instrument. There are some problems and complexities associated with this mechanism, to which we return later, but at least the concept is simple.

It is to a discussion of the general principles underlying these long-horn, lip-driven instruments that most of the present chapter is devoted. The story falls naturally into three parts: the passive linear theory underlying the shaping of the horn flare and the associated mouthpiece to provide appropriately related resonance frequencies and impedances, the active nonlinear interaction between the lip generator and the air column, and the relation of these matters to the performance technique adopted by the player. We have, of course, already discussed the basics of the lip-valve acoustic generator in Chapter 13.

14.2 Horn Profiles

With the exception of a few instruments, such as the Swiss alphorn and the obsolete keyed bugles we mentioned previously, it is unusual to make a long horn entirely conical in shape. The bore deformations involved in curving such a long horn into a manageable form remove the basic simplicity associated with a simple cone, and the structural advantages of using nearly cylindrical tubing for much of the length of the instrument are very considerable. If the flaring part of the horn extends over a reasonable fraction of the total length, for example around one third, then there is still enough geometrical flexibility to allow the frequencies of all the modes to be adjusted to essentially any value desired. Quite apart from the structural simplicity of cylindrical tubing, its use is essential for the slide part of an instrument such as the trombone, and its use nearer to the mouthpiece enables the player of a horn or trumpet to insert extra lengths of tubing, if desired, to tune the instrument to different pitches.

A reasonable approximation to the shape of many brass instruments, particularly those from several centuries ago, is given by a cylindrical tube connected to a more-or-less conically expanding section of comparable length and terminated at the open end by a short section of more rapid flare. The proportions and flare angles of these sections vary quite widely from one instrument to another, with the more gently toned instruments typically having as much as two-thirds of their length conical (and thus having a small cone angle), while the more brightly toned trumpets and trombones have a shorter expanding section and a more pronounced flare at the bell. These proportions are, however, only one of the factors influencing tone quality (Bate, 1966; Morley-Pegge, 1960).

From our discussion in Section 8.9, we saw that such compound horns do not have a simple mode distribution, though the series of input impedance maxima is more-or-less harmonic, as shown in Fig. 8.19. This lack of simplicity leads us to proceed immediately to more complex but more realistic horn profiles.

A closer approximation to the shape of a real brass instrument is given by the so-called Bessel horn, which we discussed in Chapter 8. If we take the mouth of the horn to be located at a position x_0, then we can write Eq. (8.58), giving the relation between bore radius a and the distance x from the mouth of the horn, as

$$a = b(x + x_0)^{-\gamma}, \tag{14.1}$$

where b and x_0 are chosen to give the correct radii at the small and large ends and γ defines the rate of flare, as shown in Fig. 14.2. Benade (1976, p. 409) has shown that a good approximation to the frequencies f_n of the impedance maxima at the throat of such a horn is given by

$$f_n \approx \left[\frac{c}{4(l + x_0)}\right] \{(2n - 1) + \beta[\gamma(\gamma + 1)]^{1/2}\}, \tag{14.2}$$

where c is the speed of sound, l is the length of the horn, and β is a parameter equal to about 0.6 for $\gamma < 0.8$ and 0.7 for $\gamma > 0.8$. This shows

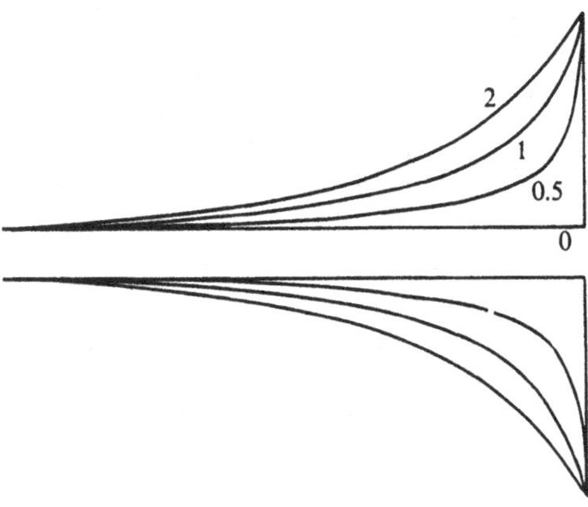

FIGURE 14.2. A family of Bessel horns given by the formula $a = b(x + x_0)^{-\gamma}$, with x_0 and b adjusted to give the same throat and mouth diameters in each case. The parameter is the flare constant γ.

that, to a good approximation, the resonances of a horn with $\gamma = 1$ are given by the harmonic relation

$$f_n \approx \frac{nc}{2(l + x_0)}. \tag{14.3}$$

If $\gamma = 0$ and $x_0 = 0$, then we have the odd-harmonic result for a cylinder [Eq. (14.1)], but for other values of γ there is no simple harmonic relation between the mode frequencies.

Such a Bessel horn can fit with very little geometric or acoustic mismatch onto a cylindrical tube of appropriate radius, provided the joint is at a distance x far enough from the mouth that the flare angle, which has the value

$$-\frac{da}{dx} = \frac{\gamma a}{x + x_0}, \tag{14.4}$$

is very small. For a typical brass instrument, x_0 is less than 10 cm and the required radius a is about 5 mm, so that the angle is less than half a degree at 50 cm from the mouth.

While some brass instruments have Bessel-like flare constants not far from unity, instruments of the trumpet and trombone families typically have γ closer to 0.7 (Young, 1960), so that they flare more abruptly at the mouth, as shown in Fig. 14.2. Such horns can fit even more smoothly onto cylindrical tubing, but they require shape adjustment to tune the resonances, as is indeed true for any such compound horn (Young, 1960; Kent, 1961; Cardwell, 1970; Pyle, 1975). Adjustment of the shape is usually carried out in the course of design to produce a mode series approximating $(0.7, 2, 3, 4, \ldots)f_0$. The first resonance is very much out of alignment, produces a very weak sound, and is not used in playing. Good players can, however, use the nonlinear effects we shall discuss later in this chapter to produce a pedal note at frequency f_0 by relying upon cooperation with the harmonically related higher resonances.

14.3 Mouthpieces

Before devoting further effort to this problem, however, we must introduce the effects of another important structural element in the complete instrument, the mouthpiece cup. The size of the cup follows roughly that of the instrument as a whole, a feature that we might expect on general grounds. Details of shape vary with the traditional style of the instrument, and to a much smaller extent with the preference of individual players, but all mouthpieces have the general design shown in Fig. 14.3. The player's lips press against the smooth surface of the cup, giving them comfortable support, and the cup itself communicates with the instrument proper through a constricted passage considerably narrower than the main bore of the instrument.

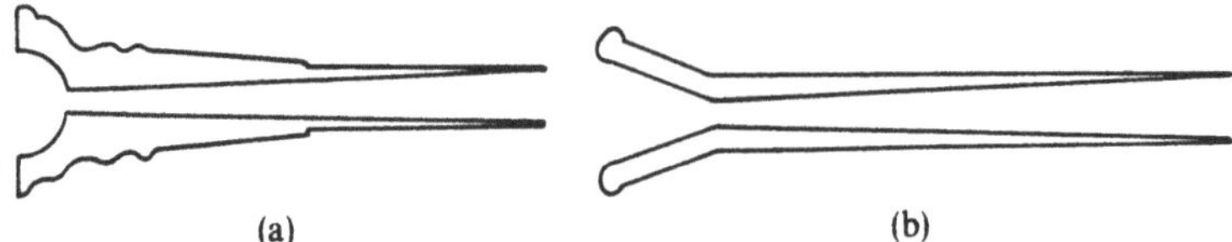

FIGURE 14.3. Mouthpieces for (a) a trumpet and (b) a horn.

Experience shows that both the cup volume and the diameter of the constricted passage have significant effects upon the performance of a given mouthpiece, with the shape being a much less important variable. As we shall see, it is not so much the cup volume and constriction diameter separately but rather a combination of the two that is important.

Figure 14.4(a) shows a system diagram of the whole instrument, while Fig. 14.4(b) gives a little more detail. The volume V of the cup presents an acoustic compliance

$$C = \frac{V}{\rho c^2}, \tag{14.5}$$

where ρ is the density and c is the velocity of sound in air. The constriction behaves as a series inertance:

$$L = \frac{\rho l_c}{S_c}, \tag{14.6}$$

where l_c is the length and S_c is the cross-sectional area of the constriction. There is an additional dissipative element R in series with this inertance representing viscous and thermal losses. The instrument horn is represented by the two-port impedance Z_{ij}, and finally the radiation impedance at the open end of the horn can be reasonably set equal to zero, since most of the losses are those to the walls of the horn.

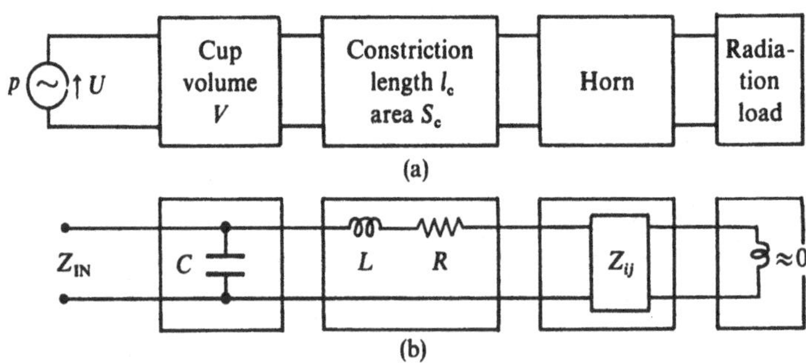

FIGURE 14.4. The linear system diagram for a mouthpiece connected to a brass instrument. Except for a small reactive component (the end correction), the radiation impedance can be taken to be nearly zero.

To gain an appreciation of the behavior of this system, we calculate the input impedance Z_{IN} for a mouthpiece coupled to a simple cylindrical tube of length l and cross section S. From Eq. (8.25), the input impedance presented by the pipe to the mouthpiece is

$$Z_{\mathrm{p}} = jZ_0 \tan kl, \tag{14.7}$$

where the characteristic impedance of the pipe is

$$Z_0 = \frac{\rho c}{S}, \tag{14.8}$$

and we can include wall losses by writing

$$k = \frac{\omega}{c} - j\alpha. \tag{14.9}$$

To a reasonable approximation, the attenuation coefficient α for waves in the tube is given from Eq. (8.15) as

$$\alpha \approx 2 \times 10^{-5} \left(\frac{\omega}{S}\right)^{1/2}. \tag{14.10}$$

It is now quite easy to evaluate the input impedance Z_{IN} presented to the lips of the player, and we find

$$Z_{\mathrm{IN}} = \frac{R + Z_{\mathrm{p}} + j\omega L}{1 - \omega^2 LC + j\omega C(R + Z_{\mathrm{p}})}. \tag{14.11}$$

We can dissect this expression graphically as shown in Fig. 14.5. Figure 14.5(a) gives the calculated input impedance for the tube alone (by setting $V = 0$ and $l_{\mathrm{c}} = 0$); Fig. 14.5(b) gives the input impedance for the mouthpiece alone (by setting $l = 0$) when loaded by the characteristic (resistive) tube impedance Z_0; and Fig. 14.5(c) gives the input impedance for the assembled instrument. The tube shows the expected series of impedance maxima, declining in amplitude with increasing frequency because of increasing wall losses. The mouthpiece in isolation behaves like an internally excited Helmholtz resonator with a peak driving-point impedance at the normal resonance frequency

$$\omega_0 = (LC)^{-1/2} \tag{14.12}$$

and a height and half-width determined by the magnitude of the loss coefficient R (here Z_0) in series with its radiation resistance, which is not present in the combined system. The combined response shows influences from both these components, with pipe resonances that lie near the mouthpiece resonance being emphasized. The mouthpiece response is widened now, not by its own radiation resistance, but by the losses in the tube.

Because the lip generator is a pressure-controlled system that functions best when working into a very high impedance, we can see that it is important to match the mouthpiece to the instrument so that the mouthpiece resonance emphasizes impedance peaks in the major playing

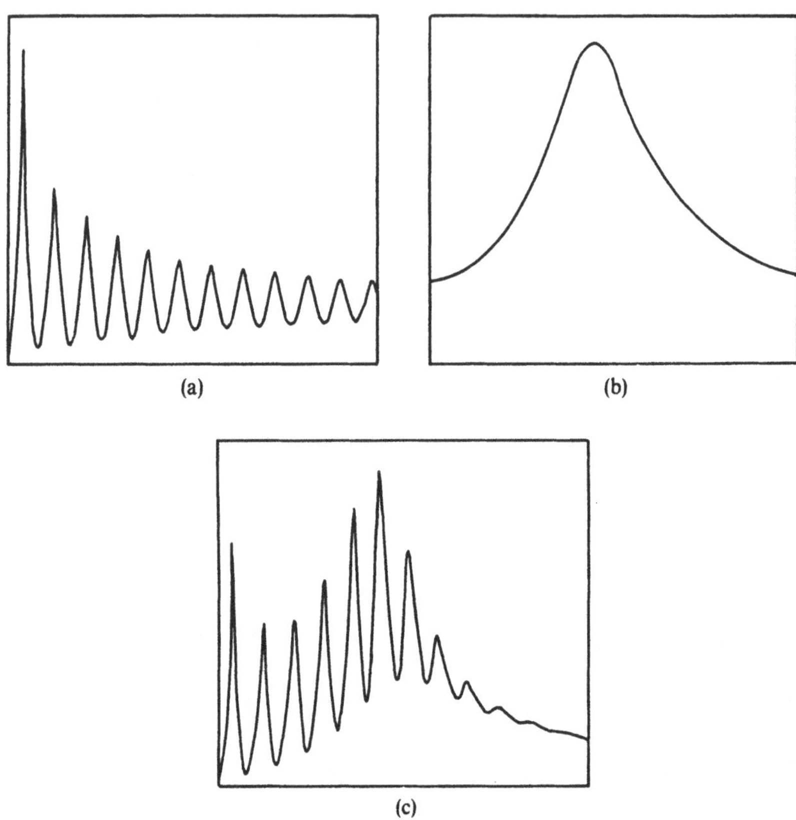

FIGURE 14.5. Calculated input impedance of (a) a cylindrical tube of radius 10 mm and length 2 m; (b) a simple mouthpiece of volume 5 cm^3, choke diameter 3 mm, and choke length 20 mm, loaded by the characteristic impedance of the tube; and (c) the mouthpiece fitted to the tube. The frequency scale is 0–1000 Hz.

range. It is in fact the mouthpiece resonance frequency ω_0, determined by Eq. (14.12), that is important, so that we can trade off increased mouthpiece volume V (and hence compliance C) by increasing the diameter of the constriction and hence decreasing L. This is, however, not an exactly balancing trade, since the characteristic impedance of the mouthpiece varies as $(L/C)^{1/2}$, and the ratio of this quantity to the characteristic tube impedance $Z_0 = \rho c/S$ determines the relative importance of the mouthpiece resonance.

The impedance of the mouthpiece also influences, to a small extent, the tuning of the horn resonances, and so must be taken into account when the complete instrument is being designed (Cardwell, 1970; Backus, 1976). Equally important is the balance of resonance peak heights over the playing range, since easy transition from one resonance to another is essential if the instrument is to be musically flexible. For a careful consideration of this point, account must be taken of the radiation impedance at the flaring mouth of the horn, and this is sensitively dependent upon details of the flare characteristic, as we saw in Chapter 8.

Even in the linear domain, it is not possible to neglect the effects of the lip-valve generator on the tuning of the modes of the instrument as played, since the source impedance of the generator has a reactive component, as shown in Chapter 13. From that discussion, and in particular from Fig. 13.6(b), we see that the imaginary part of a $(+,-)$ lip-valve admittance Y_r is always negative, corresponding to an inductive rather than a compliant or capacitive behavior. For exact resonance and a maximum acoustic pressure in the mouthpiece, the instrument must present a balancing positive acoustic admittance at the open face of the mouthpiece cup. Now, the admittance of a horn, or indeed of any multimode oscillator, has an imaginary part that rises through zero as the frequency passes through each admittance minimum (or impedance maximum). To fulfill the resonance condition, therefore, the operating frequency must lie a little above the passive resonance frequency of the mode in question. From Fig. 13.6(a), we see that exactly the opposite conclusion would be drawn for the case of a valve of configuration $(+,+)$.

Since the operating point for the mode being excited must be to the left of the imaginary axis, we see from Fig. 13.6(b) that all higher modes are also presented with a negative imaginary admittance component, though of smaller magnitude. Their frequencies will therefore be raised also, but by a smaller amount.

The actual shift in mode frequency between an instrument rigidly stopped at the mouthpiece face and one that is blown will depend on the relative magnitudes of instrument and lip impedances. Because of nonlinear effects, the pitch of a well-designed instrument is relatively stable against variations in lip admittance, but the attainable frequency variation provides both a problem for the novice and a subtle performance variable for the experienced player.

14.4 Radiation

It is all very well to calculate, as we have done, the internal linear characteristics of the instrument, but these must be extended to encompass the nonlinear domain and to include the radiation efficiency before we can pretend to a reasonable understanding of the instrument as a whole. Before we enter the nonlinear domain, it is worthwhile to point out a few features of the radiation behavior since this can be done within the linear approximation. Suppose that nonlinearities supplement the internal spectrum of each mode, as measured in the mouthpiece or near the throat of the bore, with a wide spectrum of harmonics. Can we say anything about the efficiency with which these harmonics are radiated and, therefore, anything about the brightness of the sound of the instrument?

If R is the resistive part of the radiation impedance at the mouth of the horn, which we take to have radius a, then a reasonable approximation to the detailed behavior shown in Fig. 8.8 is

$$R \approx \frac{Z_0(ka)^2}{4} \quad \text{for} \quad ka \leq 2,$$

and (14.13)

$$R \approx Z_0 \quad \text{for} \quad ka \geq 2,$$

where $k = \omega/c$ and $Z_0 = \rho c/\pi a^2$. Now, for an acoustic particle velocity of rms amplitude u in the horn near its mouth, the rms acoustic flow is $U = \pi a^2 u$, and the total radiated power is $P = RU^2$, which gives

$$P \approx \frac{\rho c(\pi a^2 u^2)(ka)^2}{4} \quad \text{for} \quad ka \leq 2,$$

and (14.14)

$$P \approx \rho c(\pi a^2 u^2) \quad \text{for} \quad ka \geq 2.$$

In Eq. (14.14), the factor $\pi a^2 u^2$ simply indicates that the radiated power is proportional to the size of the horn mouth and to the square of the acoustic velocity amplitude, which is what we might expect, but the additional factor $k^2a^2/4$ or $\omega^2 a^2/4c^2$ for $ka < 2$ shows that low frequencies are radiated increasingly inefficiently, the low-frequency cut varying as ω^2, or 6 dB per octave. This general behavior is shown in Fig. 14.6. It is fairly clear from this discussion that, if the horn mouth is narrow so that all important frequencies lie below the transition $\omega a/c = 2$, the output will show a constant treble boost of 6 dB/octave giving a very bright though not very loud sound. On the other hand, if the horn mouth is wide, it will radiate more energy and sound louder, but the higher partials may lie above the transition and so not share in the treble boost, giving a consequently less brilliant sound. The frequency for which $ka = 2$, in this simple model, is called the cutoff frequency.

The shape of the horn, as well as its diameter, can also influence the brightness of the sound. As we saw in Chapter 8, a sharply flaring horn imposes a transmission barrier that is greater for low than for high frequencies. Thus, while a simple conical horn may let waves of all frequencies reach the mouth before being reflected, a more sharply flaring horn will begin to block the low frequencies some distance before they reach its mouth. Thus, as another general rule, sharply flaring horns will produce a brighter and thinner sound than more nearly conical horns with the same mouth diameter.

Finally, the resonance frequency of the mouthpiece cup and its choke bore, as determined with the face of the cup rigidly covered, has a considerable influence on the distribution of impedance peak heights for the

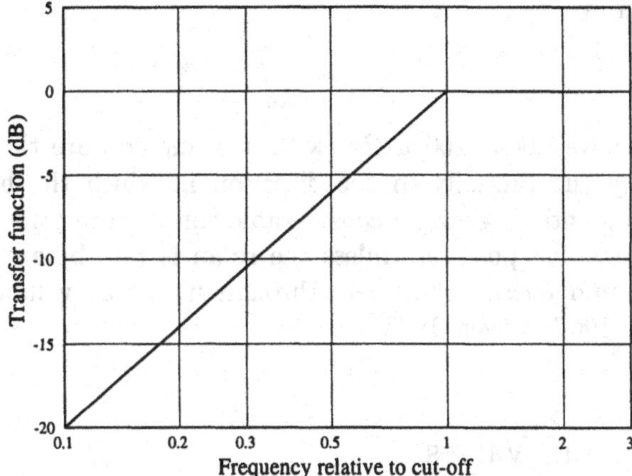

FIGURE 14.6. Schematic representation of the transfer function between the radiated pressure amplitude and the internal pressure amplitude for a wind instrument with cut-off frequency $\omega^* = 2c/a$, a being the radius of the horn mouth.

instrument, as we saw before. The allowable variation is limited if the instrument is to remain playable, but if the cup resonance frequency, which can be determined by slapping the mouth of the cup against the palm of the hand, is high because of a small cup size or a large choke bore, then high components will be favored. The opposite is true for a mouthpiece with a low cup frequency.

Again, however, we must remember that so far our discussion is entirely in the linear domain, while the mechanism of harmonic generation is essentially nonlinear. It is not until we consider this nonlinearity in detail that we can gain any real understanding of tonal quality (Benade, 1966).

We can also say something useful about the directional properties of the radiated sound using the results outlined in Chapter 8, and particularly Eq. (8.34), which gives a good approximation to the sound intensity distribution in the region in front of the horn mouth, though failing in the backward half-space. Recapitulating, we see that the radiation intensity at angle θ away from the axis of a horn with mouth radius a (perhaps not including the outer part of the bell flare) has the form

$$\left[\frac{2J_1(ka\sin\theta)}{ka\sin\theta}\right]^2, \tag{14.15}$$

where $k = \omega/c$ and J_1 is a Bessel function of order 1. At low frequencies ($ka \ll 1$), this distribution is nearly independent of the angle θ, but, as the frequency and hence ka increases, the energy is largely concentrated into a primary lobe centered on the horn axis and with an intensity null at an angle θ_0 given by

$$\sin\theta_0 \approx \frac{3.8}{ka}. \tag{14.16}$$

Frequencies above about $200/a$ Hz (with a in meters) are therefore concentrated very substantially in the direction in which the horn of the instrument is pointing, giving a considerable variation in tone quality depending on listening position, unless the room is reverberant enough to diffuse the sound energy effectively through its whole volume (Martin, 1942b; Olson, 1967; Meyer, 1978).

14.5 Slides and Valves

The final linear part of the design of the instrument that we need to consider is the means to be used to fill in the spaces between the lower resonances so as to produce a complete range of notes in the lower compass. The natural horns and trumpets of baroque times (Smithers et al. 1986) did not possess this refinement, except for an extension shank or crook that could be inserted between the mouthpiece and the main instrument to lower the whole pitch and allow it to be played in a different key. Provided the pitch change was not too great, this did not upset the harmonic relationships between the modes, but composers had to recognize that a complete scale was available only in the upper range of the instrument. Such a limitation was not acceptable to later composers, and means had to be devised to overcome it.

If we neglect consideration of the pedal note, then the largest gap to be filled is that between the second and third modes—a musical fifth with six extra notes lying within it. We have already noted that the slide of the

FIGURE 14.7. A tenor trombone. Note the considerable length of the cylindrical tube, of which the slide forms a part, and the approximate Bessel-horn shape of the flaring bell.

trombone, or its ancestor the sackbut, provides an elegant solution to this problem, made possible since at least half of the length of this instrument is cylindrical, as shown in Fig. 14.7. Since the frequency ratio of the second and third modes is 2:3, it is necessary for the slide to be able to increase the effective length of the horn by about 50% between its closed and extended positions. Allowing an arm throw of about a meter, this gives a maximum acoustic length of around 3 m for mode 2 and a mode frequency of around 70 Hz, corresponding to about C_2, as the lowest usable normal note on a trombone-type instrument. Below this lie the pedal notes that have more limited use. The normal tenor trombone is built with a range down to about E_2, while the range of the bass trombone, without any supplementary valves, extends down to about C_2. In the upper parts of the range, where the modes are more closely spaced, there will generally be two or more combinations of mode number and slide position for any given note, giving flexibility in playing. The infinitely variable slide position allows precise control of intonation and the easy production of vibrato effects. Its only drawback is the large movement required between notes, which restricts speed of execution and makes a true legato without a gliding tone nearly impossible. Many details of construction and playing for the trombone are given by Bate (1966).

The alternative approach to the problem, used on the majority of brass instruments, involves a system of valves, operated by the fingers, to insert extra lengths of tubing into the cylindrical part of the horn. If we accept the limitation that each valve should have only two positions, then we need at least three valves to be used in various combinations to produce the six notes needed to fill in the large gap between modes 2 and 3. Most instruments are in fact provided with just three valves that nominally lower the pitch by one, two, or three semitones and thus, in combination, by four, five, or six semitones as well. The valves of a trumpet are shown clearly in Fig. 14.8, while those of a French horn with its longer tube length are shown in Fig. 14.9.

In fact, there is an arithmetic snag to this scheme. An equal-tempered semitone involves a frequency change of very nearly 6%, so that to lower

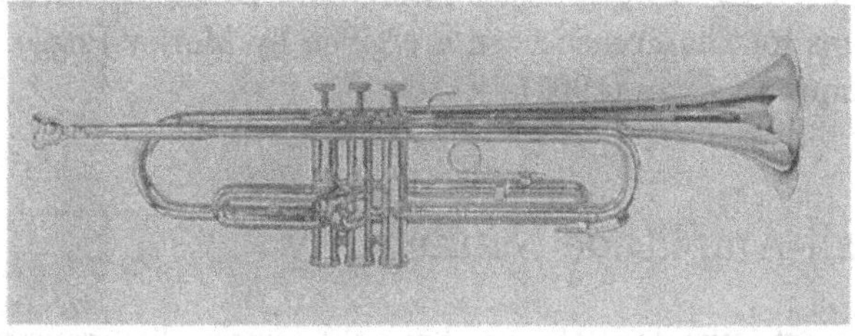

FIGURE 14.8. A trumpet. Note the three valves, of piston design, which insert extra lengths of tubing into the cylindrical part of the bore.

FIGURE 14.9. A French horn. Note the length of tubing involved and the extra sections inserted by the four valves, which in this case are of the French rotary design.

the frequency of a note by one semitone we must increase the tube length to 1.06 times its initial length. Writing this as $1+x$, we see that a frequency shift of n semitones requires an increase in tube length by $(1+x)^n$, which is significantly greater than $1+nx$. Thus, if the 1, 2, and 3 semitone valves are tuned exactly, semitones 4, 5, and 6 will be too sharp. A compromise must be reached, for example by tuning valve 3 flat, using 1 + 2 for the 3 semitone shift, and reserving valve 3 for the 4, 5, and 6 semitone shifts. Such compromises are satisfactory for the higher pitched instruments in which the lip resonance can influence the sounding frequency appreciably, but in many of the lower pitched instruments it is common to find a fourth or even a fifth valve to give truer intonation. Details of these and many other features for the French horn are given by Morley-Pegge (1960) and for the trumpet by Bate (1966).

14.6 Small-Amplitude Nonlinearity

One of the most important things about the buzzing lip acoustic generator is that it is nonlinear. We discussed this in some detail in Chapter 13, and we now go on to apply this theory to elucidate the behavior of a real lip-excited instrument. We do this first in the nearly linear approximation, which applies for rather gently blown notes, and then in the large-amplitude limit when the nonlinearity is extreme.

The relation between the mouthpiece pressure and the flow through a lip valve is given by Eq. (13.5), an equation that incorporates Bernoulli flow through the lip orifice and also the effect of changes in the size of the lip orifice itself, this latter quantity being described by Eq. (13.6). For any detailed quantitative discussion, it is necessary to go back to these equations. For our present purpose, however, it will be adequate to combine them symbolically and then invert the form so that the flow U into the instrument is expressed in terms of the pressure p inside the mouthpiece by an equation of the form

$$U = a_1 p + a_2 p^2 + a_3 p^3 + \cdots, \tag{14.17}$$

where the quantities a_n are not just numbers but rather operators involving phase shifts that are functions of frequency. To put this properly in the frequency domain, we assume that

$$p = \sum_n p_n \cos(n\omega t + \phi_n), \tag{14.18}$$

with a similar expansion for U. If we consider for the moment just a single flow component U of frequency ω and extract from the right-hand side of Eq. (14.17) only the resultant that is in phase with U, then we recall from the discussion in Chapter 13 that for a lip reed $a_1 > 0$ in a small region of ω just above the lip resonance frequency ω_r, but otherwise $a_1 < 0$. This means, with regard to the defined direction of the flow U, that the resistive part of the impedance of the lip generator (a valve that is blown open) is negative for $\omega = \omega_r + \delta$ and otherwise positive.

Looking now at Eq. (14.17) and the expansion, Eq. (14.18), we can see that the mth order term gives rise to expanded terms of the form

$$p_i p_j \cdots p_k \cos[(i \pm j \pm \cdots \pm k)\omega + (\phi_i \pm \phi_j \pm \cdots \pm \phi_k)], \tag{14.19}$$

where i, j, ... define the harmonics and there are m indices involved, not all necessarily different. If we collect all the terms of frequency $n\omega$, then we have an expression of the form

$$U_n = a_{n1} p_n + \sum_k a_{n2} p_{n\pm k} p_k + \sum_{jk} a_{n3} p_{n\pm j\pm k} p_j p_k + \cdots, \tag{14.20}$$

and if the blowing frequency ω is just a little greater than ω_r, then $a_{11} > 0$ and $a_{n1} < 0$ for all $n > 1$. Further, the third-order coefficients a_{n3} are all negative, as again we see from the discussion in Chapter 13.

Now this lip generator is to be coupled to an instrument of input impedance $Z(\omega)$ so that we can write, in an obvious notation,

$$p_n = Z_n U_n, \tag{14.21}$$

and this can be substituted back in Eq. (14.20), the Z_n being the values of the input impedance of the instrument at the set of harmonically related frequencies $n\omega$.

If we suppose that only the fundamental $n = 1$ is present, then Eq. (14.20) becomes

$$Z_1^{-1} p_1 = a_{11} p_1 + a_{13} p_1^3 + \cdots , \tag{14.22}$$

and we can neglect the higher terms in the series. This leads to the relation

$$p_1 \approx \left[\frac{a_{11} - Z_1^{-1}}{-a_{13}} \right]^{1/2} , \tag{14.23}$$

from which, since $a_{11} > 0$, $a_{13} < 0$, and the real part of Z_1 is necessarily positive, we can see that p_1 attains its maximum value when the input impedance Z of the instrument is a maximum. Indeed, the behavior of Z so dominates the situation that oscillation can be maintained only very close to the maximum in Z. We emphasize, however, that this argument is grossly oversimplified by the omission of phase shifts.

Now let us introduce the equation for the second harmonic $n = 2$ and suppose only p_1 and p_2 are nonzero. Then,

$$Z_2^{-1} p_2 = a_{21} p_2 + a_{22} p_1^2 + \cdots , \tag{14.24}$$

so that

$$p_2 \approx \frac{a_{22} p_1^2}{Z_2^{-1} - a_{21}} . \tag{14.25}$$

The potential problem that $Z_2^{-1} = a_{21}$ is averted because these quantities are of opposite sign, but it is clear once again that p_2 has its maximum value if Z_2 is a maximum so that the frequency of the second harmonic also coincides closely with an impedance maximum of the instrument.

Indeed, if we now consider p_2 to be large enough to be worth including in Eq. (14.22), then this equation becomes

$$Z_1^{-1} p_1 = a_{11} p_1 + a_{12} p_1 p_2 + a_{13} p_1^3 + \cdots , \tag{14.26}$$

so that

$$p_1 \approx \left[\frac{a_{11} - Z_1^{-1}}{-a_{13} + a_{12} a_{22} / (a_{21} - Z_2^{-1})} \right]^{1/2} , \tag{14.27}$$

which is clearly a generalization of Eq. (14.23). If Z_1 is large, then this represents only a small modification of Eq. (14.23). If Z_1 is small, however, as in the case of pedal notes, the expression in brackets on the right of Eq. (14.23) may be negative so that there is no real solution for p_1. Provided that Z_2 is large, the extra terms in the denominator of Eq. (14.27) may be able to reverse this sign change and give a real solution based on p_1 and p_2 together.

Details of this argument obviously require careful derivation, but it is clear that, for a note to be playable, either its fundamental frequency must coincide with an impedance maximum of the horn or, less satisfactorily, at least one of its low harmonics must coincide with such a resonance.

In the normal situation, the fundamental and several of its harmonics lie close in frequency to prominent horn resonances, so that equations such as Eq. (14.23) and Eq. (14.25) can be derived for the amplitudes of each harmonic. If a particular harmonic does not lie close to an impedance maximum, for example the second harmonic in the low range of an instrument with a cylindrical tube, then it will encounter a small value of the relevant impedance Z_2 in Eq. (14.25) and this will give it a very small amplitude. After allowing for this effect, however, we find quite generally within the small-amplitude nonlinear approximation that the amplitude of the nth harmonic varies as the nth power of the amplitude of the fundamental.

The exact frequency at which the note sounds depends upon the lip resonance frequency ω_r, upon the frequencies and quality factors (Q values) of all the horn resonances lying near harmonics of ω_r, and upon the general amplitude level as determined by the blowing pressure p_0. The part played by the upper harmonics becomes more important as their amplitudes relative to the fundamental increase at loud playing levels. Calculation of the resulting sounding frequency is, understandably, not a simple matter!

14.7 Large-Amplitude Nonlinearity

As the oscillation amplitude is increased by greater blowing pressure, it becomes more and more difficult to solve the set of equations, Eqs. (14.20) and (14.21), or even to determine the coefficients $a_{nj} (j = 1, 2, \ldots)$ in Eq. (14.20). All we can say is that the interactions between harmonics, or harmonically driven pipe modes, become stronger and stronger, as the relative amplitudes of high harmonics increase. In the limit of very large amplitudes, however, we can approach the description in a rather different and more physical way and gain some insight into the behavior. The nonlinear link in the chain is the vibrating lip generator, so let us examine it first.

Our assumption has been that the player's lips move like a simple oscillator with a single resonance frequency, or perhaps it is more accurate to say that they behave like two taut strings moving with opposite phase in their lowest vibration mode with half a wavelength across the mouthpiece width or the lip opening. This has been confirmed by Martin (1942a) for normal playing by viewing the player's lips under stroboscopic illumination through a transparent mouthpiece.

Now, since the quality factor Q [see Eq. (1.42)] of the lip vibrator is effectively very high because of impedance effects from the player's mouth, as discussed in Chapter 13, and because the lips are being driven closed by the mouth-cup pressure at a frequency slightly above their natural resonance, the lip vibration closing velocity lags nearly 90° in phase behind the mouth-cup pressure. The lip opening therefore goes through its maximum value as the mouth-cup pressure reaches its maximum. The two are nearly in phase. The lip vibration cannot be exactly sinusoidal in the large-amplitude limit because of the confining effects of the cup, so that the lip opening will reach a limiting maximum value at one end and will decrease to zero at the other, no matter how large the mouth-cup pressure amplitude may become. The lips do not spend an appreciable fraction of the cycle in the closed position, however, since large oscillation amplitudes are produced by large blowing pressures, which also increase the average opening between the lips.

The general form expected for the mouth-cup pressure, and measured with real players, can now be appreciated. When the mouth-cup pressure is high, the lips are wide open, but the flow is limited by the reduced pressure drop across them. The positive-going mouth-cup pressure thus tends to saturate. The negative-going mouth-cup pressure, however, encounters nearly closed lips and can fall to a sharp minimum. Such a pressure curve measured in a trombone mouthpiece by Elliott and Bowsher (1982) is shown in Fig. 14.10. The peak amplitude of the pressure variation is nearly 3 kPa, which corresponds to a sound pressure level of nearly 160 dB in the mouthpiece. The positive-going peak is only about 1 kPa above the average level, but even this corresponds to a water-gauge pressure of 10 cm, which is an appreciable fraction of the blowing pressure, as we shall see presently. Very similar waveforms were found for the trumpet by Backus and Hundley (1971).

This discussion can be approximately quantified by using Eq. (13.2) to write the relation between volume flow U and mouthpiece pressure p as

$$U \approx \gamma x (p_0 - p)^{1/2}, \tag{14.28}$$

where x is the lip opening and γ is a constant. If we assume that the instrument is played exactly at resonance, then it presents a simple large acoustic resistance R at the mouthpiece and

$$p \approx UR. \tag{14.29}$$

Substituting this into Eq. (14.28), we get an equation quadratic in U with the solution

$$U \approx \left(\frac{R\gamma^2 x^2}{2}\right)\left[\left(1 + \frac{4p_0}{R^2\gamma^2 x^2}\right)^{1/2} - 1\right]. \tag{14.30}$$

The behavior can now be understood by examining two extreme cases. If the lip opening x is small and the blowing pressure p_0 is large so that $4p_0 \gg R^2\gamma^2 x^2$, then the solution is approximated by

$$U \approx \gamma x p_0^{1/2}, \tag{14.31}$$

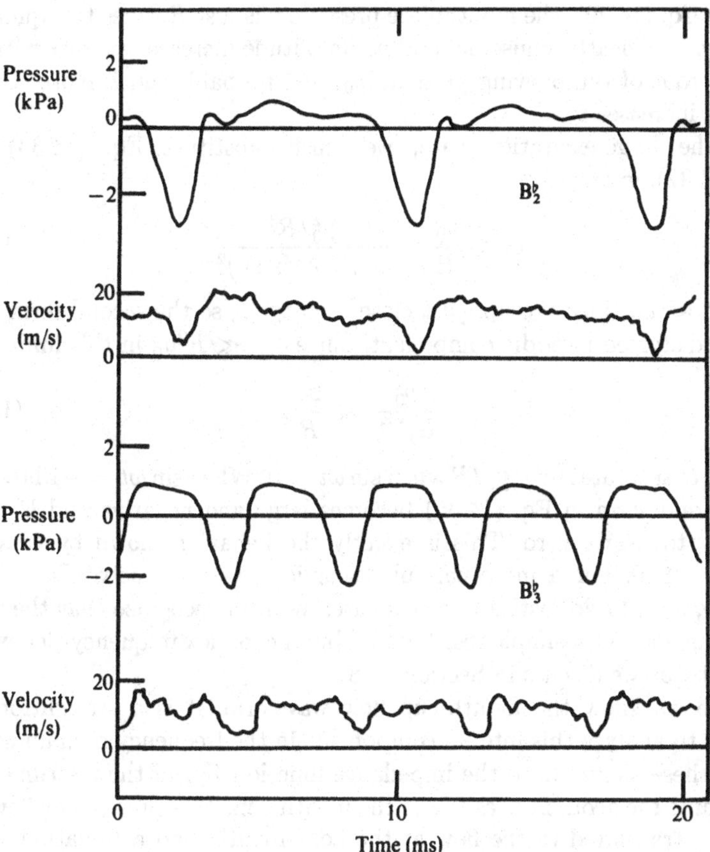

FIGURE 14.10. Mouth-cup pressure and flow velocity curves measured by Elliott and Bowsher (1982) for two notes played on a trombone.

while if the lip opening is large and the blowing pressure moderate so that $4p_0 \ll R^2\gamma^2x^2$, then

$$U \approx \frac{p_0}{R} - \frac{p_0^2}{R^3\gamma^2x^2}. \tag{14.32}$$

Now, the area of the lip opening, represented here by γx, is known by observation to vary nearly sinusoidally with time (Martin, 1942a), so we can write

$$\gamma x \approx a_0 + a \sin \omega t. \tag{14.33}$$

Thus, in the small-excitation limit represented by Eq. (14.31), the flow U into the instrument is

$$U \approx a_0 p_0^{1/2} + a p_0^{1/2} \sin \omega t, \tag{14.34}$$

and by Eq. (14.29), the mouthpiece pressure p is just R times this quantity. Clearly, U is nearly sinusoidal and its amplitude increases at least with the square root of the blowing pressure p_0, and probably much more steeply, since a increases too.

In the large-excitation case, we must substitute Eq. (14.33) into Eq. (14.32), giving

$$U \approx \frac{p_0}{R} - \frac{p_0^2/R^3}{(a_0 + a\sin\omega t)^2}. \tag{14.35}$$

Now, $a \approx a_0$, since the lips just close each cycle, so the second term gives a very distorted periodic component. Since $4p_0 \ll R^2a_0^2$ in this limit,

$$\frac{p_0^2}{a_0^2R^3} \ll \frac{p_0}{R}, \tag{14.36}$$

so that U saturates near p_0/R when $\sin\omega t > 0$. When $\sin\omega t = -1$ however, the second term in Eq. (14.35) becomes large and negative and U drops sharply to nearly zero. This is exactly the behavior shown by U and p in Fig. 14.10. For a more careful discussion, we must solve Eq. (14.30) exactly, and, to do justice to a real case, we must recognize that the input impedance is not a simple resistance R but a complex frequency-dependent quantity, as we discuss in Section 14.8.

Once we know the mouth-cup flow waveform, it is a straightforward matter to analyze this into its components in the frequency domain and to apply these as inputs to the impedance function Z_{IN} of the instrument to determine the acoustic pressure in the mouth cup. This pressure or flow can then be translated to the flow at the horn mouth, and a radiation transformation similar to that in Fig. 14.6 can be applied to find the radiated sound intensity and spectrum. Because of phase shifts and high-frequency emphasis, the radiated waveform bears little resemblance to the pressure waveform in the mouthpiece.

There is one further gradation in nonlinearity in brass instruments that must also be considered. At the upper extreme of playing level, the tone quality of instruments such as trumpets and trombones undergoes a change associated with a great increase in upper harmonic development. It was reported by Long (1947) that the sound pressure level in the mouthpiece constriction of a trumpet can be as high 175 dB, which corresponds to an amplitude of nearly 20 kPa—a pressure level a good deal higher than the 160 dB measured by Backus and Hundley (1971) for ordinary loud playing. At such a high sound pressure, the wave equation is no longer linear, and shock waves can be generated. More recently, Hirschberg et al. (1996) have re-examined this possibility for the case of the trombone, and conclude that shock waves generated in the long cylindrical part of the bore are responsible for the change of tone color at extreme fortissimo level. They show that this phenomenon is associated with progressive sharpening of an already large pressure gradient in the propagating wave, and that it requires a significant length cylindrical tube in order to become a shock wave. For this reason, such behavior is confined to instruments such as the trombone and trumpet, and is absent from predominantly conical instruments such as the cornet or Saxhorn.

14.8 Input Impedance Curves

It is clear, from the discussion in Section 14.6 and 14.7, that the alignment of input impedance peaks for a brass instrument is very important for its acoustic response and harmonic generation, quite apart from any question of playing different notes accurately in tune. For this reason, a good deal of effort has been expended to develop accurate means for measuring the acoustic input impedance curves of brass instruments at the position defined by the entry plane of the mouthpiece (Benade, 1973; Backus, 1976; Pratt et al., 1977; Elliott et al., 1982; Caussé et al., 1984).

Many of the reported studies are concerned largely with the general shape and alignment of resonances, and these do indeed follow the pattern discussed in Sections 14.2 and 14.3, with the horn profile determining the resonance frequencies and the mouthpiece cup the envelope of the impedance maxima. Instruments of good quality have several well-aligned resonances for each note so that the nonlinearity of the lip generator can bring these into cooperative response to produce a rich and stable tone. There is, however, not a great deal of difference between curves for one instrument and another (Pratt et al., 1977), so that the distinctions are very subtle.

Only a few studies report quantitative results, one of these being the work of Pratt et al. (1977), who used a hot-wire anemometer to measure flow velocity directly and a horn-loaded probe microphone to measure pressure. For various reasons, they measured the input impedance at the throat of the mouthpiece rather than at its entry face, so that the effects of the mouth-cup resonance were removed. For two trombones with a bore diameter (defined conventionally as the inside diameter of the inner tube of the slide) of 12.7 mm, the first eight impedance maxima had the characteristics shown in the following table. Note that the frequency of the first impedance maximum is much less than half that of the second, as we have already remarked, while the spacing between the other resonances is close to 63 Hz. The impedance values are given in acoustic megohms (10^6 Pa m^{-3} s), and the characteristic impedance of the bore is about 3.5 MΩ.

Resonance number	1	2	3	4	5	6	7	8
Frequency (Hz)	40	115	178	241	306	367	425	488
Impedance (MΩ)	42	26	22	20	17	24	23	19
Q value	51	42	49	49	58	64	65	68

14.9 Transients

As with all musical instruments, the starting transient of brass instrument sound is an important identifying feature. Lip-reed instruments differ from all others by virtue of the fact that the lip generator can operate autonomously, without the assistance of the tube resonances, and indeed it is largely this feature that makes it possible to play reliably the notes in the upper registers.

Luce and Clark (1967) measured the general properties of the initial transient for brass instruments and concluded that there was a distinction in behavior between partials with frequencies, respectively, below or above the cutoff frequency of the instrument. They found that the relative amplitudes of partials below cutoff all built up together and reached their steady-state values nearly simultaneously. Partials with frequencies above cutoff, on the other hand, built up in amplitude more slowly and reached their steady states at times that were longer the higher their frequencies lay above cutoff.

Risset and Mathews (1969) reached a similar conclusion both by analysis of instrument sounds and by subsequent synthesis of subjectively matched sounds using a computer. The delayed buildup of higher harmonics that they found did not separate so neatly into different behavior for partials below and above cutoff, but the first three harmonics for a short trumpet note reached their maximum values after only about 20 ms, while harmonics above the fifth took from 40 to 60 ms, with the higher harmonics being longer delayed. They found that this form of transient is a more important subjective clue to the identity of an instrument than is the steady-state spectrum.

A well-conceived graphical method of displaying the evolution of the sound of an instrument during the attack transient by plotting the path of a point on a tristimulus diagram describing the relative intensities of the fundamental, the midfrequency partials, and the high-frequency partials, has been devised by Pollard and Jansson (1982). We discuss this in more detail in Chapter 17 since these authors have applied their method primarily to the sounds of pipe organ ranks.

Regarded in the time domain, the vibrating lip generator sees essentially just the characteristic impedance $Z_0 = \rho c/S$ of the instrument tube when playing of a note begins, and it is not until the return of the first reflection from the open bell that the horn influences the note in any way. The time taken for this return trip is about equal to the period of the first mode, irrespective of the note actually being played, so that for a high-register note the lips may go through many oscillation periods before they receive any help from the horn. No such problem exists in woodwind instruments, or even in the finger-holed lip-reed instruments, because the tube length to the first open hole is short and the first reflection returns to the reed after only at most one cycle time for the note being played.

It would seem easiest, perhaps, to treat this transient problem in the time domain rather than the frequency domain, because of the long delay associated with building up a standing wave in the horn. Conceptually, this is simple, as our previous qualitative discussion indicates, but actual implementation of the description involves a computer simulation based on the equations for pulse propagation, dispersion, and reflection in the horn. Once a program correctly embodying this physics has been written, it is very convenient to be able to explore the effects of different physical parameters without having to make actual operating physical systems.

We shall not go into the details of this here because of the complications of pulse dispersion during travel along the horn. We note, however, that for a well-tuned horn having many harmonically related resonances, the equivalent physical outcome in the time domain is that the reflection phase shifts at the open bell vary with frequency in such a way as to cancel the transit-time dispersion and generate a coherent reflected pulse at the instrument mouthpiece. This may be one of the major features distinguishing a really good instrument.

Any detailed discussion in the time domain clearly requires knowledge of the impulse response of the instrument. While this impulse response can be obtained by inverse Fourier transformation of the input impedance curve, provided both phase and amplitude are known, it is more satisfying in some ways to measure it directly in the time domain. This has been done for the trombone by Elliott et al. (1982), who showed that various features of the impulse response can be correlated qualitatively with reflections expected from discontinuities in the bore. The pattern is not simple to either measure or interpret, however, and no one appears to have carried a time-domain analysis through to a thorough treatment of the behavior of an instrument.

We have outlined the general method for treating transient behavior in the frequency domain in Chapter 5. Basically, what is involved is to write down the differential equations (13.6), and (13.7), which describe, respectively, the flow through the lips and the lip motion, and to combine these with an equation for the pressure variation in the mouth to give a differential equation equivalent to the flow relation of Eq. (14.20) set out previously. Instead of the harmonic expansion of Eq. (14.18) for the pressure, however, we use the form

$$p(t) = \sum_n p_n \cos(\omega_n t + \phi_n), \tag{14.37}$$

where the ω_n are the frequencies of the horn modes giving maximum input impedance at the lip position. As we have already seen, the frequencies ω_n usually approximate well to a harmonic series except for ω_1, which lies at too low a frequency. The form of Eq. (14.37) can be regarded as a quasi-Fourier series describing p, extended so that it does not necessarily represent a repeating waveform. The amplitudes p_n and phases ϕ_n are regarded as quantities that vary slowly with time, so that the actual frequency at which mode n is forced to oscillate is not ω_n but rather $\omega_n + \dot{\phi}_n$.

To these equations, we finally add the impedance relationship for the horn and mouthpiece, expressed in frequency space:

$$p(\omega) = Z(\omega)U(\omega). \tag{14.38}$$

This equation contains all the complexity of the instrument when measured with steady sinusoidal signals and is simply another representation of the time-domain equation expressing the delay and distortion suffered by a sharp pressure pulse moving from the lips to the open bell and back again.

Calculation in the frequency domain, as in the time domain, involves computer-generated solutions of the equations, but it is straightforward to appreciate in a general way the effects involved. If the blowing pressure p_0 is suddenly raised from zero to a steady value, or indeed if some other form of pressure onset is assumed, then the lips begin to vibrate at very nearly their resonant frequency ω_r. The flow associated with this vibration has components at all combination frequencies $(i \pm j \pm \cdots \pm k)\omega_r$, because of the nonlinearity, together with a continuous spectrum associated with the sudden onset of the vibration.

The horn equation [Eq. (8.38)] can be separated into a large (strictly infinite) number of normal mode equations representing damped linear oscillations at frequencies ω_n, and these are driven both by the spectrum of the flow transient and by the harmonics of the lip frequency ω_r. When the interaction of these horn modes back on the reed is taken into account, we have a system of equations such as Eq. (5.41), one for each horn mode.

The outcome of this excitation is that the modes for which $\omega_n \approx m\omega_r$ grow to a steady amplitude and lock in phase so that

$$\omega_n + \dot{\phi}_n = m_n(\omega_r + \delta), \tag{14.39}$$

where m_n is an integer and δ is a frequency shift in ω_r. This is nearly always possible if the ω_n are in a nearly harmonic relationship (Fletcher, 1978), but, if this is not the case or if δ is too large relative to ω_r then the modes may split into two or more categories for each of which a mode-locking equation such as Eq. (14.39) applies. The result is a nonharmonic multiphonic sound.

At the same time as the modes supported by the lip valve are building up, the other modes, excited by the initial lip transient at their natural frequencies ω_n, decay with a time constant characteristic of the losses and stored energy in the horn. The characteristic time for this process is typically around 10 cycles for each of the modes, so that the initial transient extends over something like 10 cycles of the horn fundamental, or 50–100 ms, depending on instrument size. This is just the same conclusion we would have arrived at had we examined the initial transient in the time domain.

14.10 Acoustic Spectra

The steady-state and transient characteristics of the acoustic spectra of a variety of brass instruments have been measured by Luce and Clark (1967). It is useful to cite their results directly in summary form.:

- The duration of the attack transient is typically 50 ± 20 ms, independent of the note or the instrument.
- The steady-state spectral envelope is characterized by a cutoff frequency below which all radiated spectral components are approximately equal in amplitude or increase gradually with frequency, and above which the amplitude decreases sharply with increasing frequency. The rate of rise below cutoff is typically 2 to 4 dB/octave, and the rate of fall above cutoff 15 to 25 dB/octave.
- As the instrument is played more loudly, a greater fraction of the radiated power is contained in the partials near and above cutoff. The slope below cutoff increases and the slope above cutoff decreases as the intensity level is increased.
- The cutoff frequency is located in about the same part of the range of each instrument, so that their spectral envelopes can be appropriately scaled and superposed as shown in Fig. 14.11.
- The spectral behavior of an open French horn is broadly similar to that of the other brass instruments. When the horn is hand stopped, the roll-off rate above cutoff increases to about 30 dB/octave.

14.11 Mutes

Mutes are especially effective in brass instruments, since all the sound is radiated from the horn mouth and so can be modified in a simple, consistent way. French horn players normally play with one hand inside the horn bell in such a position that its shape can be modified to produce particular tuning or tone quality adjustments in the instrument, which is itself designed with this playing technique in mind. Other instruments are normally played with an open bell, but it is possible to insert a conical obstruction in such a way as to partially block it.

As the name implies, one purpose of a mute is to reduce the sound level of the instrument, but this reduction is almost inevitably frequency dependent so that the sound quality is also changed. Mutes of various designs therefore exploit this effect to achieve characteristic sounds that are often quite different from that of the instrument when played normally.

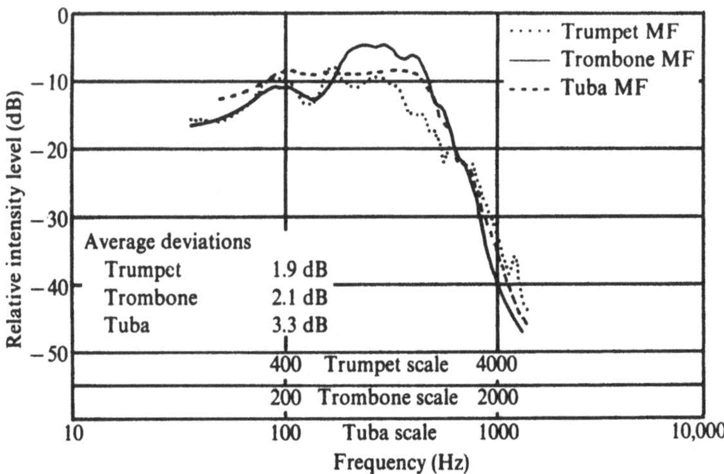

FIGURE 14.11. The spectral envelopes of the trumpet, trombone, and tuba, scaled in frequency as indicated, for mezzoforte playing (Luce and Clark, 1967).

The variety of construction of modern mutes is illustrated by the selection shown in Fig. 14.12. Some fit tightly into the bell of the horn and must therefore have some sort of open tube leading through them, while others have cork spacers that leave an annular gap between the mute and the bell and generally have no open passage through the mute, though they are usually hollow. It is clear that both types reduce the effective area of the bell opening, which we would expect to reduce the radiated intensity, and also possess resonances associated with their tubes and cavities, which we would expect to modify the shape of the radiated spectrum. A badly designed mute might also affect the tuning of valved instruments.

Ancell (1960) has measured the acoustic behavior of several mutes applied to a cornet, and these results are typical of similar mutes on other brass instruments. Most of the mutes show an admittance maximum, or Helmholtz resonance, associated with their internal cavity at a frequency

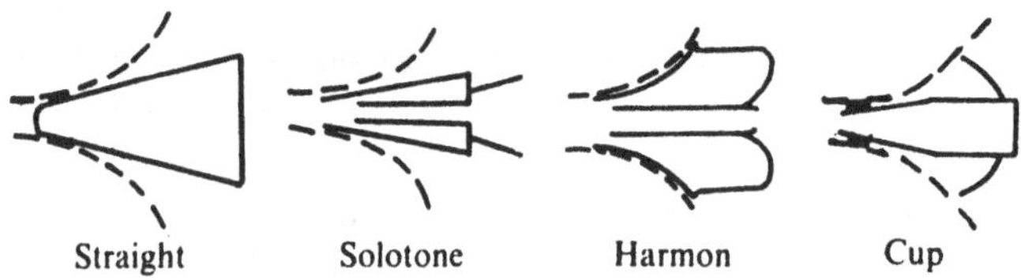

FIGURE 14.12. Structural details of several commonly used brass instrument mutes. The open mutes fit tightly in the instrument bell, while the closed mutes are supported on cork pads, leaving an annular opening.

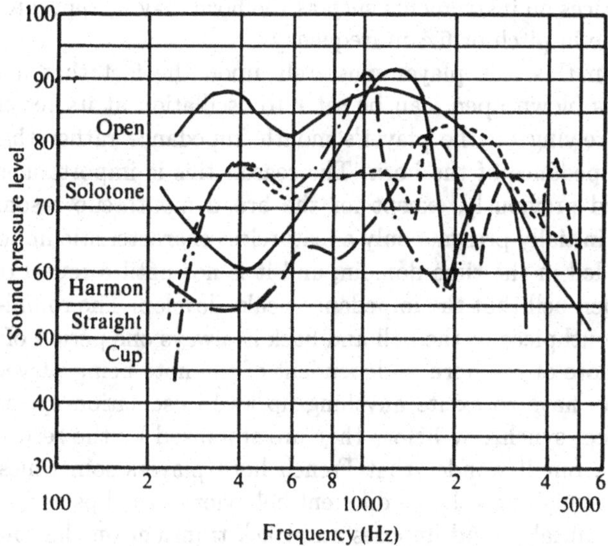

FIGURE 14.13. Spectral envelopes of the sound of a normal B♭ cornet and of the same instrument played with various mutes (Ancell, 1960).

in the range 200–300 Hz. This resonance causes a broad dip in the radiated spectrum of the instrument in the region of the first and second harmonics over most of its compass and makes the sound thin and reedy, as well as relatively soft.

All the mutes have admittance maxima and minima at higher frequencies, associated with standing waves inside their tubes and cavities. These resonances and antiresonances all lie above about 1000 Hz and impart particular tonal qualities to the sound. Some typical spectral envelopes for cornet mutes are shown in Fig. 14.13. The sound radiated by some open-tube mutes can be further modified by the player by moving a hand or cup over the open end while playing, producing a wah-wah effect. A study of the muted trumpet by Backus (1976) led to similar conclusions.

14.12 Performance Technique

We shall not attempt to go into detail about performance technique, but it is useful to consider some of the essentials in order to illustrate the physical principles set out in this chapter.

The first and most obvious thing is that the player must adjust his or her lips to vibrate at a frequency very close to that of the note to be played. This requires considerable muscular tension for high notes on instruments such as the trumpet and a great deal of skilled muscular control, since the high resonances on instruments such as the horn may be separated by only one semitone in pitch or 6% in frequency.

More than this, the player must rely upon the fact that a lip valve, because it is blown open, can be set into oscillation at its resonance frequency by relying on the player's mouth impedance rather than on the resonant impedance of the horn. The reason this is important is that, as we discussed previously, except for the broad mouth-cup resonance, the instrument initially presents only a featureless characteristic impedance Z_0 when regarded in the time domain, and it is not until a reflection returns from the open bell that the impedance peaks develop. The round-trip time from the mouthpiece to the bell and back is always one period of the nominal pedal note of the horn, independent of the note being played, so that the lips may have to excite anything up to 16 oscillation cycles unaided and in perfect synchrony before they are stabilized by the returning wave in the horn. Small wonder that French horn players sometimes fluff the beginnings of high notes! The different behavior of the lips and air column during this initial period imposes an attack transient on the sound of the instrument that is one of its principal identifying auditory characteristics (Luce and Clark, 1967).

To control loudness, and indeed to make a sound at all, the player must also control blowing pressure. Figure 13.5(b) shows that the negative real part of the admittance of the lip generator increases with increasing blowing pressure, and this is equivalent to increasing the negative magnitude of the coefficient a_{11} in Eq. (14.20). Because the mean lip opening also increases, the coefficient a_{13} does not increase to the same extent as does a_{11}, so that by Eq. (14.23) the amplitude of the fundamental mode p_1 also increases. This increase in p_1 increases the amplitudes of the harmonics p_n by the nonlinear mechanism we have discussed, so that as well as increasing in loudness, the sound also has increased harmonic development.

The actual blowing pressures used for brass instruments vary widely with the instrument and the music being played (Bouhuys, 1965). A soft, low-pitched note on the French horn can be produced with an air pressure as low as 3 cm water-gauge (0.3 kPa), while a high note played loudly on the same instrument may require more than 60 cm water-gauge (6 kPa). Trumpet players use even higher blowing pressures, up to 15 kPa, but an upper limit is set by normal systolic blood pressure in the arteries of the neck, which is not much higher than this. Attempts to play even louder lead to dizziness or even collapse!

As we discussed in Chapter 13, the lips of a brass player are, however, not really simple valves of one of the classes $(+,-)$ or $(+,+)$. In the very simplest realistic approximation, they can transform their motion between these two configurations, with $(+,-)$ behavior dominating at low frequencies and $(+,+)$ at high (Adachi and Sato, 1995; Yoshikawa, 1995; Copley and Strong, 1996). The change in lip vibration mode brings with it a change in the relation between the frequencies of lip resonance and horn resonance, and hence allows adjustment of the sounding pitch. More realistically, the player's lips must be recognized to have more than one degree of motional freedom, so that their behavior is even more complex (Adachi and Sato, 1996).

The maximum acoustic power output from a brass instrument is about a watt (Meyer, 1978). The dynamic range is typically about 30 dB and this is reinforced psychophysically by a considerable increase in harmonic development as the level increases. The pneumatic power involved in blowing an instrument is simply the product of blowing pressure and volume flow. The pressure can be easily measured with a simple, water-filled U tube connected to a fine plastic tube inserted in the corner of the lips, while the volume flow can be estimated from the length of time a note can be sustained, reckoning that the total vital capacity of the human lungs is 3–5 liters. A more sophisticated measurement technique can, of course, be used. Since loud notes can be sustained for 5 to 10 s, even on the lower instruments, and soft notes for 30 s or more, this gives flow rates in the range 10^{-4} to 10^{-3} m^3 s^{-1}, while blowing pressures range from 0.3 to perhaps 15 kPa, with low pressures being associated with low flow rates. The pneumatic input power range is thus typically from 30 mW to 10 W. Acoustic power output, measured either in a calibrated reverberant room or using multiple microphone positions in an anechoic room, typically ranges from less than 1 mW to about a watt, so that the maximum acoustic efficiency is no more than a few percent, which is reached for loud playing (Bouhuys, 1965). There is some disagreement about efficiencies at lower loudness levels, and of course there is considerable variation depending on the skill of the player, but the average efficiency may be as low as 0.1%. Detailed exampination of the flow expression Eq. (14.30) shows that nearly all the power loss occurs in flow through the vibrating lips, though there are additional small losses to the instrument walls. We have already discussed some aspects of the dependence of acoustic power on blowing pressure in Section 13.6.4.

One of the more subtle points of performance technique is the extent to which the sound of the instrument can be modified by changing the configuration of the player's mouth and vocal tract. From a system viewpoint, of course, the vibrating lips separate two ducts each of which has its own resonances, so that some interaction is possible and indeed the two systems are acoustically in series. Direct measurements by Elliott and Bowsher (1982) show that the oscillating component of pressure in the player's mouth may have an amplitude almost one quarter of that in the instrument mouthpiece. The extent to which the amplitude and spectral formant of these internal oscillations can be altered by the player depends on the relative magnitudes of the instrument impedance and the vocal tract impedance, as we discussed in Chapter 13.

The characteristic impedance $\rho c/S$ of a typical brass instrument is of order 4×10^6 Pa m^{-3} s, since S is about 1 cm^2, and the resonance peaks are at least a factor 10 higher than this, as we have seen. The characteristic impedance of the vocal tract is only about 10^6 Pa m^{-3} s because of its larger cross section, and its lowest resonances are at about 500, 1500, and 2500 Hz, though these frequencies can be varied considerably by changing tongue configuration, as in speech, giving typical vowel formants. The resonance quality factors are not high because of the nature of biological tissue, but they do occur in the frequency range of the first few harmonics for most instruments. While vocal tract resonances can have an effect on instrument tone quality, this effect is perhaps less pronounced than believed by the player, because the vibrations of air in the vocal tract are communicated internally to the ears and thus have a prominence not accorded to them by an independent listener.

An extreme example of the effect of vocal tract resonances occurs in the case of primitive instruments such as the didjeridu of the Australian aboriginal people, which is a simple slightly flaring wooden tube, about 1.5 m long and with a diameter increasing from about 30 mm at the blowing end to about 50 mm at its open mouth (Fletcher, 1983, 1996). It has no mouthpiece and its characteristic impedance is thus significantly less than that of the player's vocal tract. Although the lowest pipe resonance determines the drone pitch, there is a prominent vocal tract resonance that can be varied between about 1.5 and 2.5 kHz to provide a strong and clearly audible formant in the radiated sound.

A further interesting involvement of the vocal tract in playing technique has been identified by Mukai (1989, 1992). His studies of the configuration of the larynx during playing show a distinct difference between beginning players, who hold their vocal folds wide open, and experienced players, for whom the vocal folds are almost closed. This appears to be true of players on virtually all wind instruments.

Finally, we should note the traditional technique of French horn players, who insert the right hand into the horn bell and then modify its geometry for particular notes by varying the hand shape. The position of the hand is well within the bell at the place where it begins its rapid expansion toward the mouth, and therefore just at the place where the horn function, discussed in Section 8.8 and illustrated in Fig. 8.18, rises to create a reflection barrier for waves propagating in the horn. There is some argument about the precise effect of different hand positions and the way they are used in performance—there may well be some variance between what players actually do and what they think they do! A brief discussion from a physical point of view is given by Roberts (1976).

References

Adachi, S., and Sato, M. (1995). On the transition of lip-vibration states in the brass instrument. *Proc. Internat. Symp. Musical Acoust.* (Dourdan, France), IRCAM, Paris pp. 17–22.

Adachi, S., and Sato, M. (1996). Trumpet sound simulation using a two-dimensional lip vibration model. *J. Acoust. Soc. Am.* **99**, 1200–1209.

Ancell, J.E. (1960). Sound pressure spectra of a muted cornet. *J. Acoust. Soc. Am.* **32**, 1101–1104.

Backus, J. (1977). "The Acoustical Foundations of Music," 2nd ed., Chap. 12. Norton, New York.

Backus, J. (1976). Input impedance curves for the brass instruments. *J. Acoust. Soc. Am.* **60**, 470–480.

Backus, J., and Hundley, T.C. (1971). Harmonic generation in the trumpet. *J. Acoust. Soc. Am.* **49**, 509–519.

Baines, A. (1966). "European and American Musical Instruments." Viking, New York.

Bate, P. (1966). "The Trumpet and Trombone." Norton, New York.

Benade, A.H. (1959). On woodwind instrument bores. *J. Acoust. Soc. Am.* **31**, 137–146.

Benade, A.H. (1966). Relation of air-column resonances to sound spectra produced by wind instruments. *J. Acoust. Soc. Am.* **40**, 247–249.

Benade, A.H. (1973). The physics of brasses. *Scientific American* **229**(1), 24–35.

Benade, A.H. (1976). "Fundamentals of Musical Acoustics," pp. 391–429. Oxford University Press, New York.

Benade, A.H., and Gans, D.J. (1968). Sound production in wind instruments. *Ann. N.Y. Acad. Sci.* **155**, 247–263.

Bouhuys, A. (1965). Sound-power production in wind instruments. *J. Acoust. Soc. Am.* **37**, 453–456.

Campbell, M. (1994). The sackbut, the cornett and the serpent. *Acoustics Bulletin* May/June, pp. 10–14.

Cardwell, W.T. (1970). Cup-mouthpiece wind instruments. U.S. Patent No. 3507181; reprinted in Kent (1977), pp. 205–215.

Carse, A. (1939). "Musical Wind Instruments." Macmillan, London. Reprinted by Da Capo Press, New York, 1965.

Caussé, R., Kergomard, J., and Lurton, X. (1984). Input impedance of brass musical instruments—Comparison between experiment and numerical models. *J. Acoust. Soc. Am.* **75**, 241–254.

Copley, D.C., and Strong, W.J. (1996). A stroboscopic study of lip vibrations in a trombone. *J. Acoust. Soc. Am.* **99**, 1219–1226.

Elliott, S.J., and Bowsher, J.M. (1982). Regeneration in brass wind instruments. *J. Sound Vib.* **83**, 181–217.

Elliott, S.J., Bowsher, J., and Watkinson, P. (1982). Input and transfer response of brass wind instruments. *J. Acoust. Soc. Am.* **72**, 1747–1760.

Fletcher, N.H. (1978). Mode locking in nonlinearly excited inharmonic musical oscillators. *J. Acoust. Soc. Am.* **64**, 1566–1569.

Fletcher, N.H. (1983). Acoustics of the Australian didjeridu. *Australian Aboriginal Studies* **1**, 28–37.

Fletcher, N.H. (1996). The didjeridu (didgeridoo). *Acoustics Australia* **24**, 11–15.

Hirschberg, A., Gilbert, J., Msallam, R., and Wijnands, A.P.J. (1996). Shock waves in trombones. *J. Acoust. Soc. Am.* **99**, 1754–1758.

Kent, E.L. (1961). Wind instrument of the cup mouthpiece type. U.S. Patent No. 2987950; reprinted in Kent (1977), pp. 194–204.

Kent, E.L. (1977). "Musical Acoustics: Piano and Wind Instruments." Dowden, Hutchinson & Ross, Stroudsburg, Pennsylvania.

Long, T.H. (1947). The performance of cup-mouthpiece instruments. *J. Acoust. Soc. Am.* **19**, 892–901.

Luce, D., and Clark, M. (1967). Physical correlates of brass-instrument tones. *J. Acoust. Soc. Am.* **42**, 1232–1243.

Martin, D.W. (1942a). Lip vibrations in a cornet mouthpiece. *J. Acoust. Soc. Am.* **13**, 305–308.

Martin, D.W. (1942b). Directivity and the acoustic spectra of brass wind instruments. *J. Acoust. Soc. Am.* **13**, 309–313.

Meyer, J. (1978). "Acoustics and the Performance of Music," pp. 37–49, 75–80, 160–173. Verlag Das Musikinstrument, Frankfurt am Main.

Morley-Pegge, E. (1960). "The French Horn." Ernest Benn, London.

Mukai, S. (1989). Laryngeal movement during wind instrument play. *J. Otolaryngol. Jpn* **92**, 260–270.

Mukai, S. (1992). Laryngeal movement while playing wind instruments. *Proc. Int. Symp. Mus. Acoust. 1992* Acoust. Soc. Japan, Tokyo. pp. 239–242.

Olson, H.F. (1967). "Music, Physics and Engineering," pp. 161–171, 235–236. Dover, New York.

Pollard, H.F., and Jansson, E.V. (1982). A tristimulus method for the specification of musical timbre. *Acustica* **51**, 162–171.

Pratt, R.L., Elliott, S.J., and Bowsher, J.M. (1977). The measurement of the acoustic impedance of brass instruments. *Acustica* **38**, 236–246.

Pyle, R.W. (1975). Effective length of horns. *J. Acoust. Soc. Am.* **57**, 1309–1317.

Risset, J.-C., and Mathews, M.V. (1969). Analysis of musical-instrument tones. *Physics Today* **22**(2), 23–30.

Roberts, B.L. (1976). Some comments on the physics of the horn and right-hand technique. *The Horn Call* **6**(2), 41–45.

Smithers, D., Wogram, K., and Bowsher, J. (1986). Playing the baroque trumpet. *Scientific American* **254**(4), 108–115.

Yoshikawa, S. (1995). Acoustical behavior of brass player's lips. *J. Acoust. Soc. Am.* **97**, 1929–1939.

Young, F.J. (1960). The natural frequencies of musical horns. *Acustica* **10**, 91–97.

15

Woodwind Reed Instruments

Woodwind instruments, whether excited by vibrating reeds or by air jets, have the common characteristic of using finger holes to change the pitch of the note being played. They share this feature, as we remarked in Chapter 14, with several more or less obsolete lip driven instruments such as cornetts, ophicleides, and serpents. We begin this chapter, therefore, with a rather general discussion of possible bore shapes and finger hole dispositions before looking at particular instruments in more detail.

Like nearly all musical instruments, woodwinds have evolved slowly over the centuries rather than being designed to a finished state, and it is the peculiar features left by this evolution that give them their individuality and musical character (Carse, 1939; Baines, 1967; Benade, 1986). The two real exceptions to this statement are the modern flute, designed by Theobald Boehm in the middle of last century, and the family of saxophones developed by Adolph Sax at about the same time. We should not, therefore, expect real instruments to show a complete logic of structure, though clearly they must conform to certain general rules to be musically effective. In this chapter, we discuss only reed-excited woodwinds, leaving flutes to Chapter 16.

15.1 Woodwind Bore Shapes

The brass-type instruments discussed in Chapter 14 have a variety of horn profiles, and each instrument maintains its musical integrity because nearly the whole length of the horn, and in particular the flaring bell mouth, is actively involved as a resonator for every note. In the case of an instrument with finger holes, the situation is quite different because the lower part of the horn resonator is partly decoupled by open holes between it and the mouthpiece. Musical coherence in the tone quality therefore requires that the acoustic properties of the horn resonator should scale simply as its length is reduced, and this requirement becomes even stronger if the resonator is to be reused in its second mode to produce an upper register.

The only bore profile to fulfill this requirement approximately is the simple power law

$$S(x) = ax^{\varepsilon}, \quad 0 \leq x \leq L, \tag{15.1}$$

where $S(x)$ is the bore area at position x, measured from the mouthpiece at $x = 0$, and the open end of the instrument is at $x = L$. If the point $x = 0$ is always included, then a change in the length L is equivalent to rescaling of the width and length with no fundamental change in geometrical shape. Clearly, the mouthpiece must be at the fixed end $x = 0$, and this requires $\varepsilon \geq 0$ so that the mouthpiece is small rather than extremely large. We met this profile [Eq. (15.1)] in a slightly different guise in Chapter 14 where, with a negative rather than a positive value of ε and x measured from the opposite direction, it served as a model for the flaring bell of a brass instrument. We discussed the behavior of these Bessel horns in detail in Section 8.8 and showed that for $\varepsilon \geq 0$ the acoustic pressure in a horn extending to $x = 0$ has the form

$$p(x) = Ax^{(1-\varepsilon)/2} J_{-(1-\varepsilon)/2}(kx), \tag{15.2}$$

where $J_\nu(z)$ is a Bessel function of order ν and $k = \omega/c$ as usual. Since the bore is essentially open at $x = L$, neglecting radiation impedance, the normal modes occur at frequencies ω_n given by

$$J_{-(1-\varepsilon)/2}\left(\frac{\omega L}{c}\right) = 0. \tag{15.3}$$

This verifies formally that, for a given positive flare exponent ε, the mode frequencies ω_n simply scale inversely with length L. No other horn shape has this property.

The mode frequencies derived from Eq. (15.3) correspond to impedance maxima at the mouthpiece, as is readily appreciated from the fact that the flow must vanish at $x = 0$, while the pressure does not vanish. The horn can therefore be driven by a pressure-controlled reed of the type discussed in Chapter 13.

The solutions of Eq. (15.3) follow from the properties of Bessel functions and, as Benade (1959) shows, there are three potential candidates for musically useful horns. If $\varepsilon = 0$, the horn becomes cylindrical and the mode series is 1, 3, 5.... Strictly speaking, the cylinder must be infinitely long to satisfy our assumption that the point $x = 0$ is included, but the result is still correct for the input impedance maxima of a finite cylinder. If $\varepsilon = 2$, the horn is conical and the mode series is 1, 2, 3.... Finally, if ε is a little greater than 7, the mode series is very nearly 1, $(1\frac{1}{2})$, 2, $(2\frac{1}{2})$, 3.... The half-integral values are not exact but the integral values could form the basis of a musical horn, though the large flare rate would lead to practical difficulties.

We are left with the conclusion that useful bore profiles for reed- or lip-driven instruments using finger holes should approximate fairly closely to cylinders or cones, and this is indeed what is found in practice. For lip-driven instruments with finger holes, such as cornetts, serpents, and ophicleides, the bore is in all cases conical, leading to an instrument with a complete harmonic mode series and a true fundamental rather than the pedal note of normal brass instruments. Both cylindrical and conical bores are found in the reed-driven woodwinds, however. Members of the clarinet family, with their ancestor the chalumeau, are all essentially cylindrical and possess an odd-harmonic mode series, despite the small flaring bell at the foot of the instrument. Oboes and bassoons and their ancestors the shawms, on the other hand, are close to conical in bore and possess a complete-harmonic mode series. Saxophones also have conical bores and a complete mode series. No substantial variations from these two basic horn shapes exist among the reed-driven instruments.

To be more quantitative we must now refer to our discussion of the input impedance of cylindrical and conical horns in Chapter 8. From Eqs. (8.25) and (8.33), we saw that the input impedance of a cylinder of length L and area $S = \pi a^2$ and open at its far end is

$$Z_{\rm IN} \approx jZ_0 \tan kL', \tag{15.4}$$

where $Z_0 = \rho c/S$ and $L' \approx L + 0.6a$ is the acoustic length, including the end correction $0.6a$ at the open mouth. From Eq. (15.4), the maxima of $Z_{\rm IN}$ occur at frequencies ω_n given by

$$\omega_n = \left(n - \frac{1}{2}\right)\frac{\pi c}{L'}. \tag{15.5}$$

Those resonances are at 1, 3, 5 ... times the fundamental $\omega_1 = \pi c/2L'$.

For a cone of geometrical length L with a throat opening of area S_1 at a distance x_1 from the conical vertex and open at the mouth, Eq. (8.53) similarly gives

$$Z_{\rm IN} \approx jZ_1 \frac{\sin kL' \sin k\theta_1}{\sin k(L' + \theta_1)}, \tag{15.6}$$

where $k\theta_1 = \tan^{-1} kx_1$, $Z_1 = \rho c/S_1$, and $L' \approx L + 0.6a$ as before. Provided $x_1 \ll L$ and we are dealing with just the first few modes, $\theta_1 \approx x_1$ and $Z_{\rm IN}$ has its maximum value when the denominator vanishes in Eq. (15.6), that is, for

$$\omega_n = \frac{n\pi c}{L' + x_1}. \tag{15.7}$$

To a first approximation, therefore, removal of the tip of the cone to insert the reed or mouthpiece has no effect on the resonant frequencies, which are close to those of the complete cone of length $L' + x_1$. The mode frequencies form a complete harmonic series at 1, 2, 3 ... times the fundamental $\omega_1 = \pi c/(L' + x_1)$.

A more careful calculation shows that the end correction varies slightly with frequency, and of course, our treatment of Eq. (15.6) was not exact.

TABLE 15.1. Bore dimensions of common woodwinds (in mm).

	musical range	tube length	top diam.	bell diam.	semi-angle	trunc. ratio
clarinet	D_3–G_6	664	14	16 → 60	0°	—
oboe	$B^\flat_3$–G_6	644	3	13 → 37	0.7°	0.13
bassoon	$B^\flat_1$–C_5	2560	4	40	0.4°	0.09
alto sax.	D_3–A_6	1062	12	70 → 123	1.6°	0.12

Neither the modes of a cylinder nor a cone are thus in a perfect harmonic relationship. This is a fact of life for all real musical instruments. What we do see from Eqs. (15.5) and (15.7), however, is that the fundamental frequency of a cylindrical instrument of given length is only about half that of a conical instrument of the same physical length. Instruments of the clarinet family thus play an octave lower than the same sized instruments of the oboe or bassoon families. Bass clarinets are therefore particularly convenient for playing the lower parts of wind-ensemble music. The conical-bore equivalent, the bassoon, must be folded double to bring its 2.6-meter length within reach of the player's fingers.

Real instruments do not in fact adhere rigidly to cylindrical or conical bores even though this is their basic shape. Most have some sort of flaring bell at the foot, though the oboe d'amore and cor anglais have egg-shaped resonators instead. In addition, there are small variations in bore throughout all instruments to improve intonation or alignment of resonances. Finally, the reeds of conical instruments are clearly not of zero size and some form of mouthpiece is necessary for the cylindrical instruments, involving further departures from idealized geometry. Details of the bores and mouthpiece shapes of many representative woodwind instruments are given by Nederveen (1969). Table 15.1 summarizes some of this information for typical examples of common instruments. The lengths are measured to the reed tip, the bell dimensions give the flare from the end of the cylindrical or conical main bore to the bell mouth, the angle is the semi-angle of the cone, and the truncation ratio is the fraction of the bore truncated at the reed position.

15.2 Finger Holes

The note played by a woodwind instrument is changed by opening one or more finger holes, thus changing the acoustic length of the air column. To illustrate the simplest possible situation, consider a cylindrical tube with a single hole as shown in Fig. 15.1. If the hole is closed, then, neglecting the small perturbation caused by the air under the finger, the input impedance at the mouthpiece is

$$Z = jZ_0 \tan kL', \tag{15.8}$$

where $Z_0 = \rho c/S$ and L' includes the end correction Δ, which is due to non-zero radiation impedance at the open end. Now, we examine the effect of a hole of cross section S_1 and acoustic length l placed at a small distance D from the open end of the pipe. At the position of the hole, the admittance presented to the pipe leading to the mouthpiece is

$$Y = Y_{\text{hole}} + Y_{\text{pipe}} = -j\frac{S_1}{\rho c}\cot kl - j\frac{S}{\rho c}\cot k(D+\Delta), \tag{15.9}$$

and, if we suppose that both l and D are small compared with the wavelength, this can be written

$$Y \approx -\frac{j}{\rho\omega}\left(\frac{S_1}{l} + \frac{S}{D+\Delta}\right), \tag{15.10}$$

or, expressed as an impedance,

$$Z \approx j\rho\omega\left[\frac{l(D+\Delta)}{S_1(D+\Delta)+Sl}\right]. \tag{15.11}$$

Now, a simple continuation of the main tube past the position of the hole by an amount Δ' would present an impedance

$$Z \approx \frac{j\rho\omega\Delta'}{S}, \tag{15.12}$$

so that the effective end correction at the position of the hole is

$$\Delta' \approx \frac{Sl(D+\Delta)}{S_1(D+\Delta)+Sl}. \tag{15.13}$$

The acoustic length of the tube with its open hole is therefore reduced below that of the original tube by an amount

$$\delta = D + \Delta - \Delta' = \frac{S_1(D+\Delta)^2}{S_1(D+\Delta)+Sl}. \tag{15.14}$$

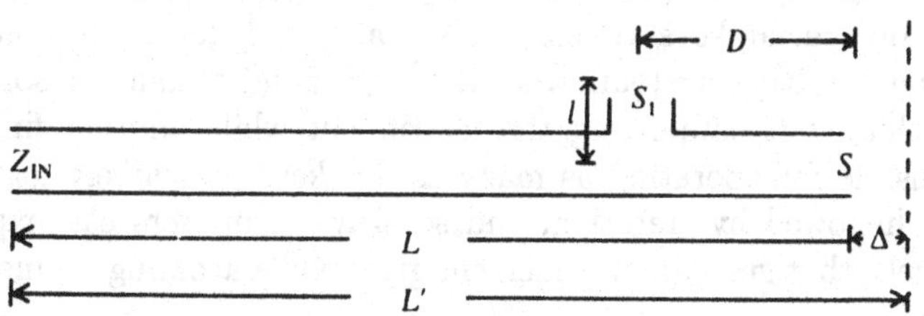

FIGURE 15.1. Length relations in a tube with a side hole.

If we make the area of the hole equal to that of the pipe, in which case $S_1 = S$ and $l \approx \Delta$, then from Eq. (15.14) the reduction in acoustic length is about

$$\delta \approx D + \frac{\Delta^2}{D + 2\Delta}. \tag{15.15}$$

Now, Δ is about 0.3 times the pipe diameter, and D is necessarily greater than this diameter if we are going to make a row of holes, so that the length reduction given by Eq. (15.15) is very close to D.

For smaller finger holes, however, the acoustic length reduction given by Eq. (15.14) is considerably less than D. Clearly, from this equation, a reduction in the area S_1 of the hole can be compensated by an increase in the distance D of the hole from the open end (Benade, 1960b), and this allows a greater separation between the finger holes. At higher frequencies, however, when D is no longer small compared with the wavelength, we must return to Eq. (15.9) and these simple approximations are no longer valid.

The objective of a system of finger holes in a modern wind instrument is to allow the production of a full chromatic scale throughout the compass of the instrument. We have already seen that there are basically two practical instrument bore shapes: the cone with its mode frequencies forming a complete harmonic series and the cylinder whose modes are in an odd-harmonic relationship. Instruments with conical bores have the first and second modes separated by an octave containing 12 semitones, while those with cylindrical bores have their lowest two modes separated by a twelfth or 20 semitones. The simplest chromatic instrument therefore requires 11 finger holes in the first case and 19 in the second if the chromatic scale is to be simply realized.

Woodwind instruments, however, are the product of cultural evolution and their early forms were designed to play simple 7-tone or even 5-tone scales, thus requiring a number of finger holes comfortably less than the 10 available fingers. Various compromise fingerings allowed the missing semitones to be produced with moderate accuracy and instruments retained basically this simple form until the eighteenth century.

After that time the increasing use of chromatic harmonies and the adoption of equal temperament demanded that instruments be able to play with equal facility and good intonation in any key, and this led to the insertion of special keys to control extra tone holes to produce these notes. The resulting system, with the right-hand thumb in some cases doing nothing but supporting the instrument while another finger may be responsible for operating as many as five keys, is scarcely logical, but its use is hallowed by tradition, and sensitive composers can exploit the tonal variety that such an arrangement gives while avoiding its mechanical problems.

15.2.1 Fingering Systems

In this section, we discuss briefly some of the basic principles of fingering systems while leaving details for presentation in relation to particular instruments.

The basic disposition of finger holes in all wind instruments involves three fingers of each hand, omitting the little fingers, the left hand being that closest to the mouthpiece. These six holes provide between them enough flexibility to produce a diatonic major scale of good tuning, as shown in Fig. 15.2. The actual disposition of the semitones varies from one instrument type to another so that the fingerings shown here as F♯ and C♯ could perhaps be F♮ and C♮. At the end of the chart are shown two forked fingerings that could give F♮ and C♮ if the normal fingerings gave sharps for these notes. Other semitones can be interpolated in a similar way.

More sophisticated schemes then added a hole, or a pair of closely spaced holes, near the foot of the instrument for the right-hand little finger, thus extending the compass downward by two semitones. The left-hand thumb, not generally used to hold the instrument, was also given a hole, and some of the primary holes were sometimes made double to aid with the semitones. Such a scheme could easily fill out the 12 semitones of a chromatic octave scale.

In the case of instruments with conical bores overblowing an octave to the second mode, essentially the same fingerings can be used to produce the notes of the upper octave if a means can be found to produce the second rather than the first mode. For capped-reed instruments, such as bagpipe chanters, this is not attempted and a single octave must suffice.

Instruments with a cylindrical bore overblow, if at all, to a second mode that is one and a half octaves, or a musical twelfth, above the fundamental. Early instruments, such as the capped-reed krumhorns or even the chalumeau, which is the ancestor of the clarinet, did not attempt this and

	D	E	F(♯)	G	A	B	C(♯)	F(♮)	C(♮)
LH	●	●	●	●	●	●	○	●	○
	●	●	●	●	●	○	○	●	●
	●	●	●	●	○	○	○	●	○
RH	●	●	●	○	○	○	○	●	○
	●	●	○	○	○	○	○	○	○
	●	○	○	○	○	○	○	●	○

FIGURE 15.2. Typical fingering chart for a wind instrument with six finger holes. Many variants exist, particularly for the semitone steps on F and C.

kept a compass extended to only a note or two past the octave. The classical clarinet, however, used a small vent hole for the left thumb to reach the upper register and then required two small keys above the normal top hole and several extra keyed holes below the lowest of the normal finger holes to fill in the missing notes and give a complete scale. We defer discussion of these complications until later.

It is easy to appreciate in a general sense the way in which a register hole can aid overblowing to the second mode in a reed instrument, though, as we see later, the real situation is more complex than initially appears. For a cylindrical instrument, all modes have a pressure maximum at the reed. The first mode has a quarter wavelength along the tube, and the second has three quarters of a wavelength, as shown in Fig. 15.3. Thus, if a small hole is opened in the bore at the point R, one-third of the way down from the reed toward the first open hole, it should have no effect on the second mode but should displace and, because of its viscous losses, reduce the Q of the first mode resonance. These effects should encourage preferential sounding of the upper mode. In principle, this mechanism requires a different register hole for each note, but in practice a single register hole can be made to suffice (Benade, 1976).

The principle is the same for a conical instrument except that there is more compromise involved in placing the register hole. The oboe actually has two register holes, one for the lower and one for the higher notes in the second register, while bassoons often have three.

Returning to the simplest six-hole arrangement shown in Fig. 15.2, let us look briefly at the way in which hole sizes and positions can be calculated once the fingering for the scale has been defined. This is simple in concept, though the details, as set out for example by Benade (1960a) and by Nederveen (1969), are quite complicated. In fact, all we have to do for the bottom register is to begin with the lowest finger hole of area S_1 at a distance D from the foot of the instrument and adjust S_1 and D so that the reduction in acoustic length given by Eq. (15.14) is either 12% of the original acoustic length for a whole-tone step or 6% for a semitone. Starting from the position of this hole, with its new end correction, we then use Eq. (15.14) to find an appropriate size and position for the next hole, and so on. The freedom allowed by the existence of two parameters, S_1 and D, for each hole allows them to be made of convenient size and spacing for the fingers, but this still leaves some freedom in choice of precise values for S_1 and D. This freedom is used in the first place to adjust the scale in the second register, where D is no longer small compared with the wavelength, and to optimize forked fingerings for interpolated semitones. For instruments with extra keys, this last step can be omitted.

Clearly, the solution of this hole-positioning problem is unlikely to give an exact scale over the whole compass of the instrument and there may be considerable variation between different traditions, particularly for nonkeyed instruments. Tuning deficiencies are generally adjusted by the maker

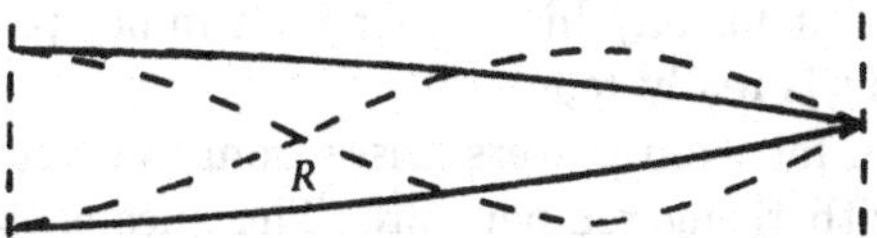

FIGURE 15.3. Positioning of a register hole in a cylindrical instrument, such as the clarinet. The curves show pressure standing waves for the first and second modes of the pipe.

by undercutting certain holes, not necessarily symmetrically, and by making slight enlargements to various sections of the bore. Modern instruments with their independent key work for all semitones have now evolved to standard patterns with little variation from one maker to another, but critical and complex instruments, such as bassoons, still benefit from careful handwork after manufacture.

15.2.2 Tuning, Temperament, and Temperature

The standard of pitch to which instruments are built has changed considerably over the years. The pitch is generally referenced to that of A_4 and, by international agreement, this is now set at $A_4 = 440$ Hz. Pitch in earlier times was generally lower, around $A_4 = 435$ Hz, but military wind bands of last century used a pitch that was nearly a semitone higher. Even today there is some disagreement on standards, and a few major orchestras tune to $A_4 = 442$ Hz, ostensibly to give greater brilliance to the tone. Orchestras generally tune to a note given by the oboe—not because it is stable in pitch, which it is not, but rather because of its clearly heard tone quality.

String instruments can adjust to new pitches without any difficulty, except to the ear of the player, and there is certainly a change in tone quality since the instrument resonances remain fixed. The pitch of a woodwind instrument can be lowered by extending it at one or more of its joints, but this has disastrous consequences for the scale temperament if the shift exceeds about a quarter of a semitone. It is easy to see why this is so. Suppose we tune to the lowest note in low register and flatten it by half a semitone (about 3%) by lengthening the bore at the reed end by 3%. Then this lengthening will actually be 6% for the top note of the register in an oboe and 9% for a clarinet, making the semitone step from the top note of the lower register to the first note of the second register too large by half a semitone for the oboe and a whole semitone for the clarinet! While a skilled player can adjust the pitch of each note to some extent, this is much more than can be tolerated. In the eighteenth century, some woodwind instruments were provided two different middle joints to help with this problem, and even today bassoons are supplied with at least two bocals or crooks of different taper, but for anything other than minor pitch adjustments a different instrument is really required.

Another problem for wind players arises from the fact that the speed of sound increases with rising temperature. This means that a cold instrument will play flat, and its pitch will gradually rise as it warms up. This is, unfortunately, a drift in the opposite direction from that of string instruments, so that orchestras must take particular care with tuning. Another problem is that the warming of an instrument is uneven—the upper bore near the reed will settle down to a higher temperature than the lower bore, and this gradient will vary with the temperature of the room. If the reed end is at about 35°C and the lower end at 20°C, the difference in absolute temperature is about 5% and the difference in sound speed about 2.5%. Averaging over the bore length, this can make notes with nearly all finger holes closed flat with respect to those with nearly all holes open by as much as 1%, or one sixth of a semitone. Clearly woodwind players cannot take the pitch and temperament of their instruments as fixed and reliable, but must rely upon their ears and their abilities to make subtle adjustments as they play.

15.3 Impedance Curves

The approach outlined above can give, at best, only a first-order approximation to the design and acoustic behavior of a woodwind instrument, since all that is calculated is the acoustic length, and thus the resonance frequency, for the first one or two modes. What we need to know, however, is a complete impedance curve, as a function of frequency, at some convenient reference point close to the reed—it is not simply the location of the major resonances that is important but also their actual impedance values and their relation to other significant resonances. Armed with this information and an appropriate description of the nonlinear behavior of the reed generator, we can then discover something about the sound-producing properties of the instrument as a whole.

There are two complementary approaches to this problem, experimental and theoretical. In both cases, we must decide on an appropriate point at which to define the impedance and thereby conceptually separate the reed generator from the instrument tube. This point, which is selected more from convenience than from any fundamental consideration, is usually taken to be the place at which the removable part of the reed and mouthpiece is attached, since the instrument below this point remains invariant as reeds and mouthpieces are changed. The closer the point is to the actual reed opening, however, the easier the curves are to interpret.

Experimental measurements of this type on many woodwind instruments have been made by Benade (1960b) and Backus (1974), among others. What is required is to measure both the acoustic pressure and acoustic volume flow at the selected point, or rather across a plane through that point. Acoustic pressure presents no problem, since all that is required is a small microphone with an acoustic impedance high enough not to upset the measurements. Acoustic volume flow is more difficult to measure, however, and several approaches have been used. The hot-wire anemometer method (Pratt et al., 1977) described for brass instruments in Chapter 14 could perhaps be used, but is complicated and not very sensitive. A second method uses a solid piston located in the measurement plane and tracked by an appropriate accelerometer or velocity transducer. The third, which is the simplest of all but loses information about the phase of the flow, involves producing and measuring a relatively high acoustic pressure in a closed cavity and using this to drive a proportional flow through a high acoustic resistance. This resistance can be produced by packing a narrow tube with fine wool or, better still, by constructing an annular capillary by centering a solid rod in a narrow metal tube (Backus, 1974). While for complete analysis we require to know both the magnitude and phase of the impedance, a knowledge of its magnitude fortunately suffices for qualitative discussion.

Another elegant method uses two spaced microphones in a tube connecting a noise source to the input of the instrument to give the acoustic impedance directly (Chung and Blaser, 1980). This has been used by Ishibashi and Idogawa (1987), though in a rather different context, which we take up in Section 15.4.

Calculations can give similarly detailed information about the variation of input impedance with frequency. The essence of the method is to divide the bore of the instrument into a large number of small lengths, in each of which the geometry is adequately represented by a short length of cone or cylinder. The acoustic impedance coefficients Z_{ij} representing the relationship between the pressures and acoustic volume flows p_1, U_1 and p_2, U_2 at the upper and lower ends of such a section are, as discussed in Section 8.15,

$$p_1 = Z_{11}U_1 + Z_{12}U_2, \tag{15.16}$$

and

$$p_2 = Z_{21}U_1 + Z_{22}U_2, \tag{15.17}$$

from which we easily derive that if

$$Z_2 = \frac{p_2}{U_2} \tag{15.18}$$

is the acoustic load attached to the end 2, then the impedance

$$Z_1 = \frac{p_1}{U_1} \tag{15.19}$$

seen at the entrance to end 1 is

$$Z_1 = Z_{11} + \frac{Z_{12}^2}{Z_2 - Z_{22}}, \tag{15.20}$$

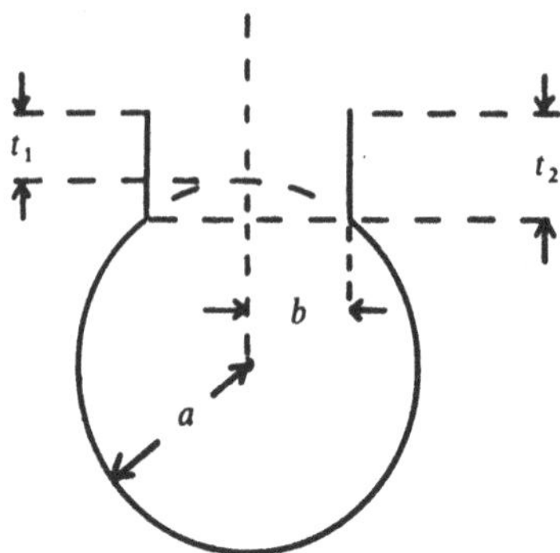

FIGURE 15.4. Dimension definitions for a side hole in a cylindrical tube.

where we have used the fact that $Z_{21} = Z_{12}$. The coefficients Z_{ij} all have lossy resistive parts contributed by viscous and thermal effects at the tube walls.

The influence of an open or closed finger hole presents a complication since the topology is then quite different from that of a simple cone. Keefe (1982a, 1982b), however, has shown that an open or a closed hole can be adequately represented by a section of zero geometric length, described by two equations, such as Eqs. (15.16) and (15.17), and inserted in the bore at the position of the hole. Clearly, a different section is required depending upon whether the hole is open or closed. For a circular finger hole of radius b and chimney height t in a tube of radius a, as shown in Fig. 15.4, the equivalent network has the form shown in Fig. 15.5, and the imaginary parts of the series and shunt elements have the forms

$$Z_s = \left(\frac{\rho c}{\pi a^2}\right)\left(\frac{a}{b}\right)^2 \times \begin{cases} (-j \cot kt) & \text{(closed)} \\ jkt_e & \text{(open)} \end{cases}, \tag{15.21}$$

and

$$Z_a = \left(\frac{\rho c}{\pi a^2}\right)\left(\frac{a}{b}\right)^2 \times (-jkt_a) \quad \text{(closed or open)}. \tag{15.22}$$

The exact values of t, t_e, and t_a relative to the geometrical heights t_1 and t_2 are rather complicated and depend on whether the hole is open or closed and, if it is open, whether or not there is a key pad or finger poised over it. These impedances also have real parts, contributed by radiation and viscous losses, but these are not easily calculated and must generally be estimated from measurements.

The form of Eq. (15.21) reflects the fact that an open hole behaves like a shunt inertance, as we have already discussed, while a closed hole behaves like a shunt compliance because of its enclosed volume. The series element Z_a, on the other hand, reflects the effective increase in tube cross section in

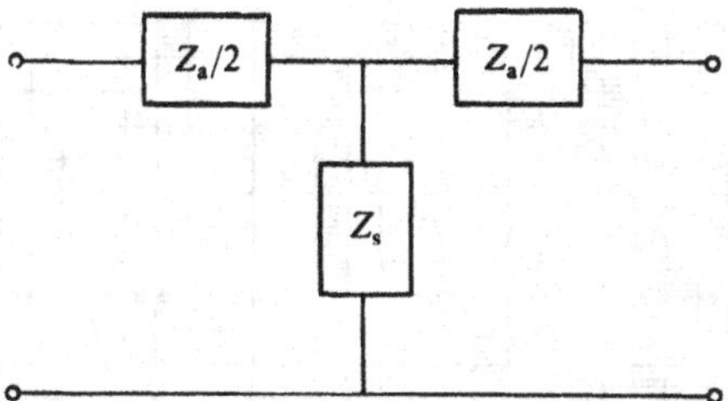

FIGURE 15.5. T-section network representing the acoustic effect of a side hole in a tube.

the vicinity of the hole, whether it is open or closed. Detailed expressions for t, t_e, and t_a are given by Keefe (1982a).

Armed with this sort of information it is possible, as Plitnik and Strong (1979) showed for the oboe, to start with the radiation impedance at the open end of the instrument and work steadily up the air column toward the reference plane near the reed, using measured bore diameters at each step and inserting appropriate networks at each finger hole, depending on whether it is open or closed. Input impedance curves calculated in this way agree closely with curves measured on the same instrument.

Figure 15.6 shows impedance curves, measured at the face of the reed, for a low note and a high note fingered on the clarinet, a typical, nearly cylindrical instrument, by Backus (1974). For the low note (written E_3), the resonance peaks are quite close to an odd-harmonic series 1, 3, 5 ... based on the first maximum, which itself corresponds to the fundamental of the blown note. This is very much as expected, though we note that only the first few harmonics of the fundamental are really well aligned with the impedance maxima. For the high register note (written C_6), it is the second impedance peak upon which the note is based. At first sight, there is little to indicate why it is this resonance that is the important one. This is a problem to which we return later, since it is partly a matter of performance technique.

Similarly, Fig. 15.7 shows impedance curves for a low and a high note on the oboe. For the low note $B^{\flat}_3$, the curve shows a complete and nearly harmonic series of resonance peaks, 1, 2, 3, ..., as we expect for a nearly conical horn. The lower peaks are asymmetric, which is characteristic of a cone, as we discussed in Section 8.7. The alignment of resonances with the harmonics of a note corresponding to the first peak is reasonable for the first few harmonics, but then it deteriorates. For the high-register fingering D_6, the note is based on the third resonance, which is certainly the most prominent, but there are two significant lower resonances and no peak is aligned with the second harmonic. Again, this presents a problem for understanding.

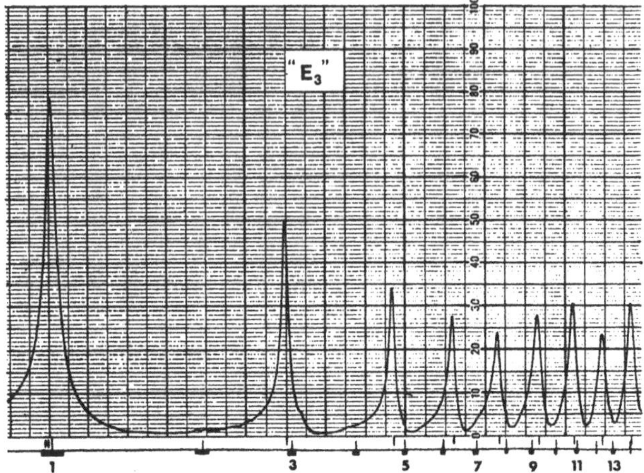

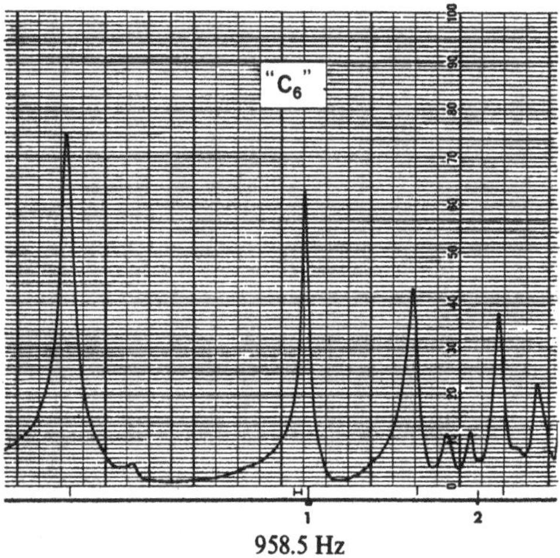

FIGURE 15.6. Measured impedance curves for a low note and a high note fingered on a clarinet. The frequencies of the fundamental and its harmonics for the note sounded with these fingerings are shown at the foot of each graph. Note the shifted first-mode peak in the lower graph (Backus, 1974).

There is one other aspect of finger-hole acoustic behavior that we should mention here, though its effects become apparent only when we examine the total acoustic spectrum produced by an instrument. This is the acoustic cutoff frequency of the air column, considered as that frequency above which sound is radiated freely to the environment, rather than being reflected back to produce resonances.

In the case of a brass instrument, as discussed in Sections 8.6 and 14.4, this cutoff behavior is relatively simple and is determined by the mouth

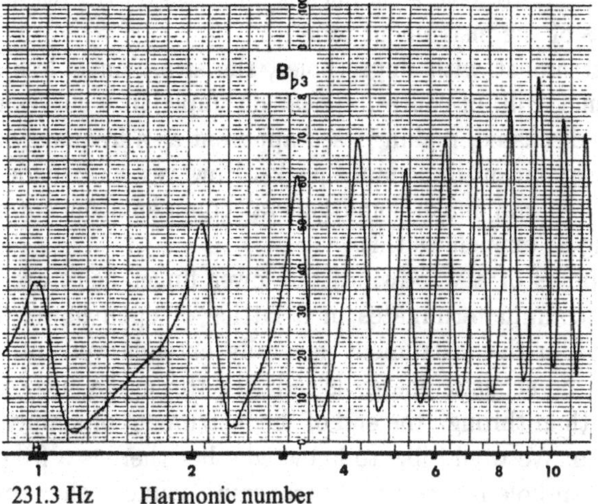

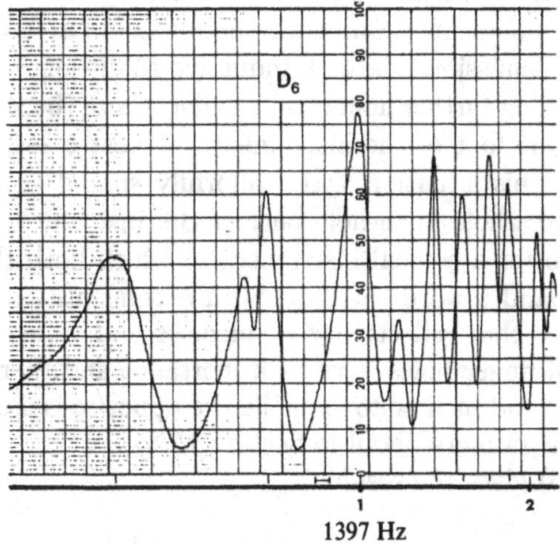

FIGURE 15.7. Measured impedance curves for a low note and a high note fingered on an oboe. Note the asymmetric shape of the lower peaks, which is a consequence of the conical bore. The frequencies of the fundamental and its harmonics for the note sounded with these fingerings is shown at the foot of each graph. Note the lower-mode peaks in the lower graph (Backus, 1974).

diameter and flare rate at the open end of the horn. Below cutoff, the radiation resistance increases as the square of the frequency, giving a treble boost of 6 dB/octave in the radiated spectrum relative to that inside the instrument, as shown in Fig. 14.6. Above cutoff however, the radiation resistance is constant at $\rho c/S$, where S is the mouth area, and the upper parts of internal and external spectra are the same. Clearly, the cutoff frequency is likely to have a significant influence on the overall brightness or mellowness of the instrument tone.

For an instrument with finger holes, the situation is less simple. The cutoff frequency will, of course, appear on impedance curves measured or calculated as discussed, making itself apparent as a frequency above which there are no significant impedance peaks. However, as Benade (1960a) has pointed out, we can make some general comments on cutoff frequency without being as explicit as this.

For most notes of a woodwind, the lower finger holes are all open, so that the air column terminates in a lattice of open holes. The internal standing wave in the instrument generally extends well into this region, and much of the acoustic energy is radiated through a succession of these holes, rather than through the end of the instrument or simply through the first open hole. We would like to know whether there is a frequency above which this open-hole lattice simply transmits sound rather than reflecting it.

In fact, the open finger holes, as we have already seen, act as inertances, or inductive elements, in shunt with the air column. We therefore have essentially a transmission line with inductive shunt loads along it, the inductances becoming further apart and smaller in size (because the holes are larger) toward the foot of the instrument. In most cases, it is reasonable to approximate this situation with uniformly spaced equal shunt inductances along the whole open-hole part of the instrument. Indeed, this is appropriate also for a conical instrument in which both the bore and the finger holes expand toward the foot. Such a transmission line constitutes a high-pass filter that reflects low frequencies and transmits those above its cutoff. Small finger holes have high inertance and therefore give a low cutoff frequency and mellow tone, while the opposite is true of large holes. We examine this in more detail in Section 15.8.

Benade (1976) gives the result

$$f_c \approx 0.11c \left(\frac{b}{a}\right)\left(\frac{1}{sl}\right)^{1/2} \tag{15.23}$$

for the cutoff frequency in hertz of an open tone-hole lattice in a pipe of diameter $2a$, each hole having diameter $2b$ and acoustic length l, and the separation between holes being $2s$. For small tone holes, as found on most woodwind instruments, the acoustic length is approximately equal to the physical length plus 1.5 times the hole radius, to account for the internal and external end corrections. His measurements show that musically satisfactory instruments have cutoff frequencies within the range 1000–2000 Hz for oboes, 1200–2000 Hz for clarinets, and 300–600 Hz for bassoons. Clearly the value of f_c limits the highest note that is readily playable on the instrument.

15.4 Reed and Air Column Interaction

In Section 13.1, we discussed the behavior of pressure-controlled reed valves as acoustic generators and divided them into two classes, those blown open by the static pressure and those blown closed. The first class is represented by the lip-blowing mechanism in brass instruments and is characterized by the fact that it can act as an acoustic generator over only a very narrow frequency range just above the natural frequency of the lips. The second class contains the single-reed generators of clarinets and saxophones and the double reeds of oboes and bassoons and is characterized by a broad frequency range of operation extending from low frequencies up to nearly the resonance frequency of the reed.

The fingered, lip-driven instruments, such as the cornett and serpent, clearly belong to the first family, and, apart from the fact that finger holes rather than valves are used to change the tube resonances, their sound production mechanism and performance techniques are essentially the same as those for normal valve or slide-tuned brass instruments. Their behavior has therefore been covered in the discussion of Chapter 14.

The acoustical behavior of a woodwind reed generator is entirely different from the lip generator, since it can present a negative resistance to a significant number of harmonics of the note being played. Before we consider this complication, however, we should understand the linear behavior.

Figure 15.8 shows the linear system diagram for a reed generator coupled to a pipe. The pipe, with whatever finger-hole openings are specified, is completely determined by its input impedance $Z_p(\omega)$ or, equivalently, by the inverse of this quantity, the input admittance $Y_p(\omega)$. Similarly, for a given blowing pressure p_0, the flow through the reed generator is related to the pressure acting on it from inside the mouthpiece as specified by a linear approximation $Y_r(\omega)$ to the admittance at the operating point. Since the reed acts as a generator, the real part of Y_r is negative at the operating frequency, while the real parts of all the other admittances are positive. Finally, we must allow for the fact that the vibration of the reed itself displaces a small volume within the mouthpiece, this displacement being in phase with the pressure inside the mouthpiece at frequencies below the reed resonance, so that the associated admittance Y_s has the character of a slightly lossy compliance.

Now, for a steady oscillation of the system, we must have

$$Y_r(\omega) + Y_s(\omega) + Y_p(\omega) = 0. \tag{15.24}$$

This condition implies a relation between the imaginary parts jB of these admittances:

$$B_r(\omega) + B_s(\omega) + B_p(\omega) = 0, \tag{15.25}$$

which serves to determine the allowable oscillation frequencies. For a growing oscillation, the real parts G must satisfy:

$$-G_r(\omega) - G_s(\omega) \geq G_p(\omega), \tag{15.26}$$

and the oscillation will grow until, because of nonlinear effects, the equality is satisfied. If Eq. (15.26) is not satisfied, then the oscillation will decay to zero. Equation (15.25) can be satisfied most easily at a frequency close to an impedance maximum or minimum of the resonator, since $B_p(\omega)$ changes sign there and goes through a large range of values. To satisfy Eq. (15.26) we require that $G_p(\omega)$ be very small, corresponding to an impedance maximum with a high Q value. Stable oscillation of large amplitude thus occurs very close to a prominent narrow impedance peak of the resonator.

In the first correction for nonlinearity, we note that, from the discussion in Section 13.6, the large-amplitude average value of G_r is essentially the slope of the line joining two extremes of the pressure excursion about the operating point in Fig. 13.9. The value of $-G_r$ therefore decreases as the oscillation amplitude increases, and this amplitude builds up until the inequality in Eq. (15.26) becomes an equality.

Before we consider other aspects of nonlinearity, we can push the linear analysis a little further. From Fig. 13.6(a), the imaginary part B_r of the reed admittance is always positive for a woodwind-type reed below its resonance frequency, and the same is true for the parallel admittance component B_s contributed by reed motion. Equation (15.25) therefore requires that $B_p(\omega)$ should be negative at the operating frequency. Now, B_p behaves essentially like

$$B_p(\omega) \approx -Y_0 \cot kl, \tag{15.27}$$

so that $B_p(\omega)$ goes from negative to positive as ω increases through an admittance minimum (impedance maximum) for the pipe. This means that the sounding frequency for a woodwind instrument will always be, in this linear approximation, somewhat below the frequency of the impedance peak on which it is operating.

Indeed, we can go a little further and note that, again in this linear approximation, from Fig. 13.5(a) the value of B_r increases when the blowing pressure p_0 is increased and also, as it happens, when the lips are relaxed so

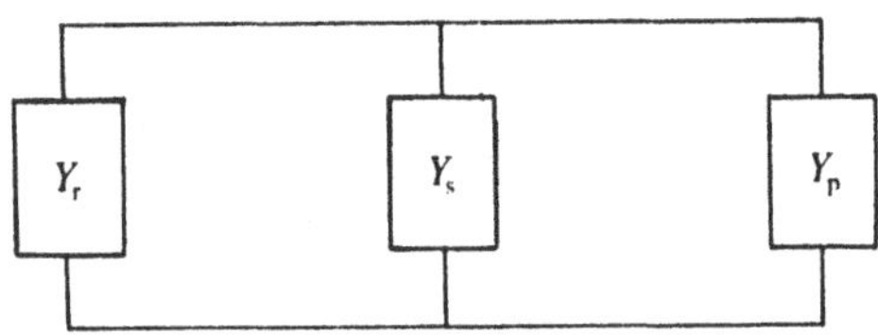

FIGURE 15.8. Admittance network for an instrument pipe Y_p, the flow through the reed generator Y_r, and the motion of the reed Y_s.

that the unblown opening of the reed is increased. Both these adjustments, though mainly the second, are made in loud playing, so that there is a tendency for the instrument to play flat unless other effects intervene or other adjustments are made.

When we come to consider nonlinear effects and harmonic generation in more detail, we see that the situation is complicated by the fact that the reed can act as a generator over a wide frequency range, which may include several impedance maxima of the instrument tube for which Eqs. (15.25) and (15.26) can be satisfied. In the linear approximation, this would simply lead to independent oscillation at an appropriate amplitude at each of the frequencies involved and a consequent nonharmonic or multiphonic sound (Backus, 1978). Such sounds can indeed be produced on most woodwinds by employing an unconventional fingering so that the instrument impedance maxima are very far from being harmonically related (Bartolozzi, 1981). Even in this nearly linear multiphonic domain, the nonlinearity of the reed flow produces very significant audible effects. Thus, if two modes with frequencies ω_1 and ω_2, respectively, are independently excited, the reed nonlinearity will generate components with frequencies $n\omega_1 \pm m\omega_2$, where n and m are small integers. Low-frequency beats and combination tones, which may be quite strong, are produced and the sound has a chordal character. These playing techniques are used to advantage by some contemporary composers.

In normal playing, however, such effects are unwanted and the desired sound has only harmonic components. Fortunately, it is a feature of the sort of nonlinearity exhibited by musical generators that the oscillations of different pipe modes become rigidly locked together into a precisely harmonic frequency relationship within a few cycles of the fundamental, provided the resonances are not too far from integral frequency relationships (Fletcher, 1978). This mode-locking phenomenon was explored briefly in Chapter 5. Mode locking is enhanced by increased nonlinearity, and the final mode-locked frequency is a compromise between those of the implied fundamentals of the various resonances, the weightings depending on the prominence of each component in the final sound.

These phenomena provide some idea of the complexity of the behavior of a real instrument. As the level of loudness is increased, so are the relative levels of the harmonics of the fundamental, so that upper instrument resonances contribute to a greater extent to determination of the frequency. In all registers of the instrument, nonlinearity is vital in allowing stable production of the tone.

The resonance frequency of the reed may also be an important factor in the production of some notes and in the general tone quality of the instrument, as pointed out by Thompson (1979). The real part of the reed admittance G_r has its maximum negative value just below the reed resonance, as shown in Fig. 13.4(a), and the player is able to vary the frequency of this resonance over a significant range, typically between about 1 and 3 kHz, depending on the instrument. This range is such that the reed resonance can be aligned with a relatively minor peak when playing in the topmost register to produce a stable preferred oscillation. In the lower parts of the upper register, it can be aligned with one of the harmonics of the fundamental to enhance its stability and thus to select the appropriate peak from the choices shown in Figs. 15.6 or 15.7. Once this has been done, the nonlinear coupling between the two modes will lock them into coherence, particularly if there is a maximum in pipe impedance near each of the first few harmonics. The third factor assisting the player in stabilizing the oscillation upon the desired resonance peak is the impedance of the player's mouth and vocal tract. This too can be adjusted to favor the particular resonance involved.

As we discussed in Chapter 8, it is possible to treat the behavior of the air column of a wind instrument by considering, in the time domain, the reflection of a sharp flow pulse injected at the mouthpiece. The behavior of the air column is then described by the integral equation (8.80). If the reflection coefficient $r(t)$ is known from measurement or calculation, and the flow characteristic of the reed generator is known in the form of Eqs. (13.5) and (13.6), then these equations can be combined and solved numerically on a computer to give the flow and pressure waveforms in the instrument. Schumacher (1981) has done this for the initial transient of the clarinet obtaining good agreement with observation, but the approach has not yet been applied in detail to any other instrument.

Direct measurement of impulse response is possible but difficult. The time and frequency domains are, however, related by a Fourier transform in the case of a linear system, and it is possible to use a frequency-domain measurement of input impedance to deduce the impulse response of the air column of an instrument, provided that phase as well as amplitude is measured. This has been done for the bassoon by Ishibashi and Idogawa (1987), who have shown that small acoustic differences between nominally identical instruments are easily detected and localized to particular parts of the bore. The impulse response is, however, very complex, with superposed reflections from all of the key holes and finger holes, whether open or closed, so that simple interpretation is nearly impossible.

15.5 Directionality

The directional characteristics of radiation from woodwind instruments are less simple than those of brass instruments because radiation from the open mouth of the bell is supplemented by radiation from open finger holes. For harmonics with frequencies below the cutoff frequency of the tone-hole lattice, which is typically about 1500 Hz for oboes and clarinets and 400 Hz for bassoons (Benade, 1976), the radiation is mostly from the first one or two open tone holes, and is therefore nearly isotropic. For harmonics above this cutoff frequency, however, the sound propagates along the open-hole part of the bore and is radiated both from the open bell and from the open tone-holes. Radiation from the tone-holes combines to give a conical beam around the axis of the instrument, as discussed in Section 7.3, and radiation from the bell is also concentrated along this axis. The overall result is therefore that the tone is "brighter" along the axis of the instrument than in other directions, at least in anechoic conditions.

Because it is small, each open finger hole acts as an isotropic source. The phase of each source is determined by the internal standing wave, and they must therefore all be exactly in phase or 180° out of phase, with the sign-change positions separated by about half a wavelength for the harmonic concerned. The source amplitude may vary irregularly from one hole to the next, depending on the shape of the internal standing wave mode.

For the fundamentals of notes in the bottom register, the sources at all the finger holes are in phase. Radiation intensity is then greatest in directions at right angles to the axis of the instrument and is significantly reduced in the direction of the axis, as discussed in Section 7.3 and illustrated in Fig. 7.5(a).

For notes in the second register, or for the upper harmonics of lower notes, it is necessary to consider the fact that different groups of finger-hole sources have opposite phase. Because the holes are small in diameter relative to the bore, they are shifted from their ideal positions and the result is generally to concentrate the radiated intensity into a conical shell about the axis of the instrument as illustrated in Fig. 7.5(b) and (c). In principle this cone has two lobes directed, respectively, toward the reed end and bell end of the instrument, though the player's body may interfere with the upward-directed cone (Meyer, 1978).

These peculiarities of radiation mean that the sound quality of an instrument can differ appreciably, depending on the position of the listener or microphone, if the environment is nearly anechoic. Fortunately, most listening environments have reflecting surfaces close to the instrument and a moderately strong diffuse reverberant sound field. In these circumstances, the sound quality that is heard is quite close to the spectral balance of the total radiated power, provided the listener is at a reasonable distance from the instrument.

These general principles apply to all reed-driven woodwind instruments. In the sections that follow, we shall examine some of these instruments individually; but, first, we examine a few further principles of performance technique that are common to all reed woodwinds.

15.6 Performance Technique

From Fig. 13.3, which shows the static flow characteristic of a reed and, by implication, also the essence of this characteristic at any frequency below the reed resonance, it is clear that the static blowing pressure must be high enough to bring the operating point B into the negative resistance region AC of the curve. If we neglect the acoustic pressure p for the moment, then we can rewrite Eq. (13.5) for the static curve in the form

$$U = \gamma p_0^{1/2}(p_c - p_0)^\alpha, \tag{15.28}$$

where p_c is the static pressure required to close the reed, γ is a constant, and the parameter α may be somewhat greater than 1 because of the shape of the reed opening. We then find that the lowest operating blowing pressure, corresponding to the point A on the curve, is

$$p_0^{(A)} = \frac{p_c}{2\alpha + 1} \approx \frac{p_c}{3}. \tag{15.29}$$

As we discussed in Section 13.6, the operating point in Fig. 13.9 must be somewhat to the right of the flow maximum A in order to give a negative generator conductance. If the blowing pressure is significantly greater than $p_0^{(A)}$, then the reed will beat, giving a considerable change in tone quality. For a double-reed instrument, this beating of the reed is perhaps always present because of the shape of the flow curve, as shown in Fig. 13.3. In all cases, the blowing pressure must be less than the closing pressure p_c.

Blowing pressure thus gives some measure of control over loudness, but this is coupled to a change in timbre. A more direct control can be exerted by squeezing the reed more nearly closed with the lips. This generally reduces the closing pressure p_c as well, so that the blowing pressure to be used is also reduced. There is thus a very great change in the acoustic flow through the vibrating reed opening.

Measurements on several clarinet, alto saxophone, oboe, and bassoon players have been reported by Fuks and Sundberg (1996). In all cases there is an increase in blowing pressure with dynamic level, and in the case of double-reed instruments there is also an increase of pressure with the pitch of the note being played. For single-reed instruments, the change of pressure with pitch is less systematic. The measurements, in each case on just two professional players, are summarized in Figure 15.9. Note that 1 kPa corresponds closely to 10 cm water gauge pressure.

We have already mentioned that an adjustment with the lips of the reed resonance frequency is desirable if upper register notes are to be played stably and with good tone. It seems likely that a similar adjustment to the mouth and throat to change vocal tract resonances is even more important (Clinch et al., 1982; Backus, 1985; Johnston et al., 1986; Sommerfeld and Strong, 1988). The vocal tract presents a series of impedance minima at the lips, the lowest of which can be adjusted over a range from about 300 Hz to 1 kHz and the second from about 1 to 3 kHz. Stable sounding of a high note is promoted if one of these resonances is aligned with its fundamental, while in the low register the tone quality can be considerably enhanced

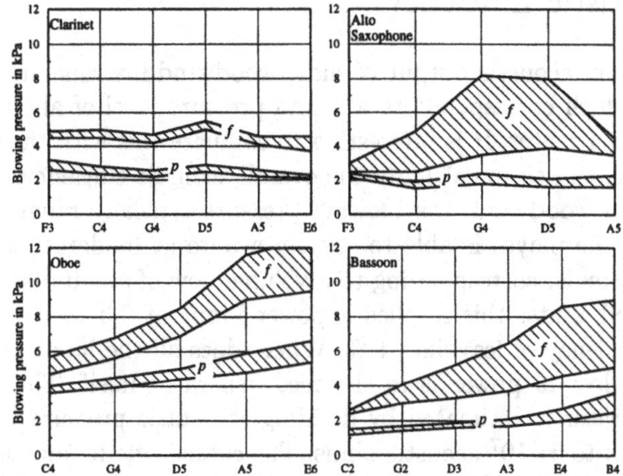

FIGURE 15.9. Typical blowing pressures across the main compass of clarinet, alto saxophone, oboe, and bassoon, for *piano* and *forte* playing. (Data from Fuks and Sundberg, 1996.)

by aligning a resonance with one of the harmonics of the note. We should perhaps remark, however, that the change in tone quality perceived by the player is probably greater than that heard by the listener, because the vocal tract resonance has a larger effect within the mouth than in the instrument, and the acoustic pressure in the mouth is conducted rather directly to the ear through the bones of the skull.

The subtleties of tuning and tone quality permitted by these adjustments are clearly great, and an expert player will exploit all the available adjustments of blowing pressure, reed resonance frequency, and vocal tract resonances to achieve the desired effect.

Finally, we should make some mention of transients, particularly initial transients, in woodwind instrument performance. To sound an isolated note, the player normally places the tongue against the tip of the reed to prevent vibration, increases air pressure in the mouth to normal sounding level, and then removes the tongue. Such a tongued note begins abruptly at nearly its steady level. At the other extreme, the player may compress the lips and apply blowing pressure so that the note begins very softly and then reaches its normal level as lip pressure is relaxed. Other possibilities exist, such as gradual or sudden application of pressure without tonguing or lip contraction, and, of course, one note may simply follow on from its predecessor in a legato fashion.

Our discussion allows us to appreciate the physical phenomena underlying these and other techniques. It is their subtle use by a skilled player that distinguishes music from mere sound.

15.7 Acoustic Efficiency

The maximum acoustic output of most woodwind instruments is about 1 milliwatt, which corresponds to a sound pressure level of about 80 dB at a distance of 1 meter. It is interesting to compare this with the pneumatic power input, and so to calculate the overall acoustic efficiency.

For a reed woodwind, the blowing pressure typically ranges from 3 to 5 kPa, and the player is able to sustain a note at moderate loudness for about 30 seconds, corresponding to a volume flow of about 100 ml/s. Converting to SI units, this implies a power input of 0.3 to 0.5 W and an acoustic efficiency of less than 1 %. Where does all the input power go?

The first loss of power is in the reed generator itself. The impedance of the generator is estimated by dividing the static pressure by the flow, and is thus 3 to 5×10^7 acoustic ohms. The characteristic impedance of the instrument tube is $\rho c/S$, which is typically about 10^6 acoustic ohms, and the Q value at resonance is typically about 10, so that the load impedance is only about 10^7 acoustic ohms. This means that most of the pneumatic power is dissipated in the flow resistance of the valve, rather than being transferred to the air column. This could be analyzed in detail by considering the aerodynamics of the flow, but the main loss is by viscosity and turbulent diffusion. The energy in the pipe oscillations is itself dissipated in two ways—by viscous and thermal losses to the tube walls and by acoustic radiation. For the rather long narrow tube of a typical woodwind instrument, wall losses exceed radiation by about a factor 10, as discussed in Section 8.5, so that the maximum radiated power is only about 1 percent of the input power.

15.8 The Limiting Spectrum

Following a discussion by Benade and Kouzoupis (1988), we can describe the limiting spectrum for reed instruments in quite a general way. In discussing individual instruments, it is then the manner in which they deviate from this idealized behavior that is interesting and that perhaps decides musical quality.

The sound radiated by the instrument is determined, as we might expect, by interactions between the air column and the vibrating reed source. As we discussed in Section 15.3, the behavior of a horn resonator with open finger holes along its length can be divided into two frequency ranges (Benade, 1960a). At low frequencies, the tone-hole lattice is acoustically reflecting and the air column, in acoustic length a little greater than the distance to the first open finger hole, produces distinct resonances. At higher frequencies, however, the impedance of the open finger holes is high, and the lattice of open holes transmits the acoustic wave rather than reflecting it. The transition frequency $\omega_c = 2\pi f_c$ between these two ranges is called the tone-hole cutoff frequency. It is typically about 1500 Hz for treble instruments and proportionally lower for bass instruments, in each case corresponding roughly to the frequency of the highest easily playable note. For $\omega < \omega_c$, the input impedance Z_{in} of the air column shows resonance peaks, the heights of which decline with increasing frequency roughly as $\omega^{-1/2}$ because the damping is due mainly to viscous and thermal losses in the boundary layer at the walls. The impedance in the troughs similarly rises as $\omega^{1/2}$. Above the cutoff, there are no well-organized resonance peaks and the impedance is nearly constant at Z_0, the characteristic impedance of the tube.

Nonlinearity in flow through the reed generates all harmonics of the basic frequency ω_1, which is itself determined by a low resonance of the instrument horn. As discussed in Section 13.4, the amplitude U_n of the nth harmonic of the flow is proportional to $(U_1/U_0)^n$, where U_0 is some reference amplitude, and not necessarily the steady flow. As the vibration amplitude of the reed increases, so do the amplitudes of all the harmonics, that of the nth harmonic increasing by n dB for every 1 dB increase in the level of the fundamental—a rule that Benade referred to as Worman's Theorem, although it was well known in other contexts. Clearly this increase cannot go on indefinitely, and ultimately saturates when $U_1 = U_0$ and the levels of all harmonics are equal, as in a pulse-like waveform. There is a high-frequency limit to this behavior, too. For an ideally sharp delta-function pulse, the spectrum is uniform with an envelope like ω^0. A flow with an ideal rectangular-wave character and thus a flow discontinuity, which could be obtained by integrating a series of pulses of alternating signs, has a limiting spectrum varying as ω^{-1} or -6 dB/octave. If we integrate again to obtain a discontinuity only in the derivative of the flow, as for example in a beating reed, then the limiting spectrum behaves as ω^{-2} or -12 dB/octave. If beating is just avoided, the limiting spectrum is ω^{-3} or -18 dB/octave, and so on.

We can now put these two components together to derive the form of the spectral envelope. In the case of loud playing, where $U_1 = U_0$ we could say that the flow spectrum is fully developed, and the mouthpiece pressure for the nth harmonic with frequency $\omega_n = n\omega_1$ is $Z_{in}(\omega_n)U_n$ at all frequencies up to the tonehole cutoff ω_c, Z_{in} being the input impedance of the air column. For $\omega > \omega_c$, there is no reinforcement from tube resonances, and the spectrum falls at -12 dB/octave for a beating reed and at -18 dB/octave if the reed just fails to beat. Rather surprisingly, the spectrum of the sound radiated from the toneholes of the instrument largely follows the internal spectrum, because the pressure signal extends partly down the tube past the first open hole, and the inertive impedance of each hole contributes an ω^{-1} term to the flow through it that just balances the ω^2 behavior of the radiation resistance (Benade, 1960a).

We can write an approximate generalized expression for the envelope of the radiated power spectrum as

$$P(\omega = n\omega_1) \sim \frac{(U_1/U_0)^n Z_{\text{in}}(n\omega_1)}{1 + (\omega/\omega_c)^m}, \tag{15.30}$$

where $m = 4$ for a beating reed and $m = 6$ for a reed that just fails to beat. In the case of an instrument with a conical bore, or for the odd harmonics of an instrument with a cylindrical bore, the harmonics coincide approximately with resonance peaks, and Z_{in} falls at about -3 dB/octave for $\omega < \omega_c$, after which it remains constant at the characteristic tube value Z_0. The frequencies of the even harmonics of a cylindrical-bore instrument, on the other hand, lie near the minima of the impedance curve, so that for them Z_{in} rises at about 3 dB/octave while $\omega < \omega_c$ [although Benade (1960a) suggests that the rise is more like 6 dB/octave because the even harmonic frequencies are significantly misaligned from the impedance minima]. In the idealized case there is a discontinuity in behavior at $\omega = \omega_c$, while in reality we would expect a smooth transition. The spectral envelopes for both odd and even harmonics are illustrated in Fig. 15.10. It must be emphasized again, however, that this discussion is only a sort of prototype—it works well for the simplest case of the clarinet, but requires modification for other instruments.

For quiet playing, if the reed can be induced not to beat, the factor $(U_1/U_0)^n$ becomes dominant over all others, and the radiated spectral envelope simply falls smoothly above the fundamental. As we see later, this playing mode is straightforward on the clarinet, more difficult on the saxophone, and virtually impossible on double-reed instruments such as the oboe and bassoon.

15.9 The Clarinet

For various reasons of convenience and relative simplicity, the clarinet has been more extensively studied by acousticians than any of the other woodwinds (Backus, 1961, 1963; Worman, 1971; Wilson and Beavers, 1974; Schumacher, 1978; Stewart and Strong, 1980; Benade and Larson, 1985; Benade and Kouzoupis, 1988; Gilbert et al., 1989). It is essentially an instrument with a cylindrical bore and a single reed and in this form cannot be traced back further than the late seventeenth century, when it was known as the chalumeau. Earlier instruments, whatever their names, seem to have had double reeds.

As we have already discussed in some detail, a cylindrical pipe overblows to the musical twelfth so that if a complete scale over two registers is to be produced, even without chromatic semitones, a large number of finger holes are required. The modern clarinet, as shown in Fig. 15.11,

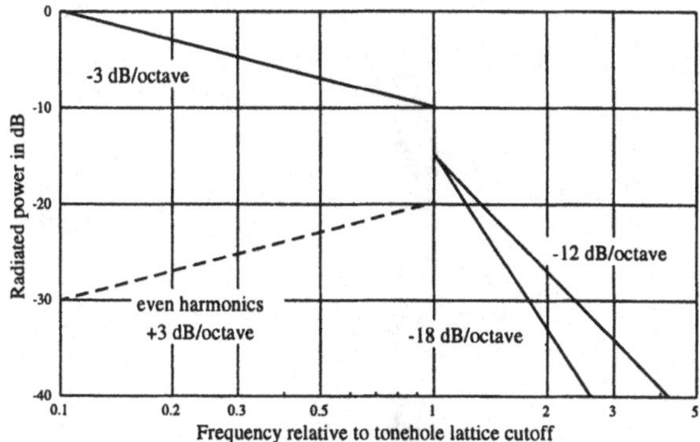

FIGURE 15.10. The spectral envelope for an idealized reed-woodwind instrument played loudly (solid line). Below the tone-hole lattice cutoff frequency ω_c, the radiated power in the harmonics falls at about -3 dB/octave, while above ω_c the radiated power falls at -12 dB/octave if the reed beats, and at -18 dB/octave if it just fails to beat. In the case of an instrument with a cylindrical bore, the power in the even harmonics of the sound rises at about 3 dB/octave for $\omega < \omega_c$ (broken line), while behaving in the same way as the odd harmonics for $\omega > \omega_c$. There is, of course, a smooth transition near ω_c in each case.

uses a considerable number of keys, both above and below the six normal finger holes, to accomplish this. The clarinet of classical times, as shown in Fig. 15.12, necessarily possessed these extra keys, though not the associated mechanism and chromatic semitone keys (Carse, 1939, pp. 148–175), the chromatic semitones being produced using "forked" fingerings with one or more holes closed below an open hole (Rice, 1984).

The key system of the modern clarinet, as illustrated in Fig. 15.11, shows most of the features used on other woodwinds. There are three basic finger holes closed by the left-hand fingers on the upper part of the instrument, two surrounded by rings and one left bare. The three finger holes for the right hand on the lower part of the instrument are all surrounded by rings. Each ring is normally in a raised position. Pressing a finger to close the hole depresses the ring so that, through a system of axle rods and levers, it can be made to open or close a hole somewhere else on the instrument by use of a softly padded cap. Sometimes the fingering plan may require that two ring keys both open or close a particular complementary hole, but we do not wish the primary holes to interact. Rings allow this, since the motion of a ring by itself has no effect on the hole that it surrounds.

This part of the clarinet mechanism, designed following principles developed for the flute by Theobald Boehm in the mid-nineteenth century,

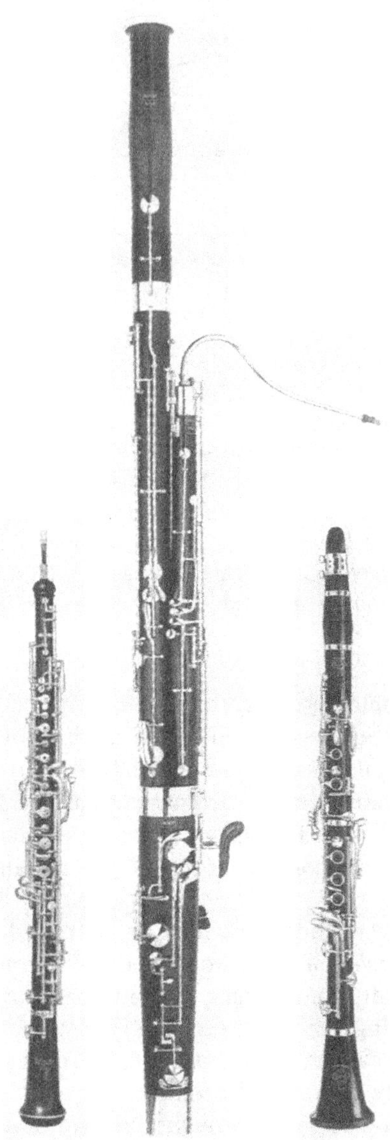

FIGURE 15.11. A modern oboe, bassoon, and clarinet.

is of relatively recent origin. The other keys on the instrument, clustered for operation by the little fingers of the two hands or by the side of the index finger, simply open and close individual extra holes and are of much greater antiquity in concept. These simple keys provide several holes of normal size, normally closed, above the top finger hole, and several larger holes, normally open, below the lowest finger hole. These tone holes fill the

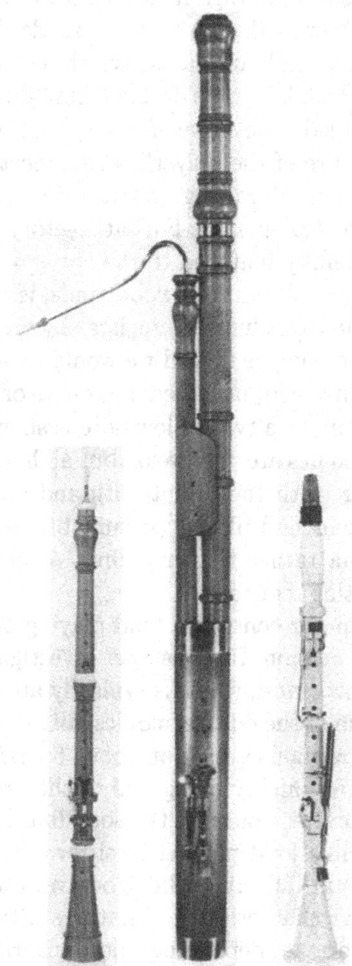

FIGURE 15.12. Eighteenth century oboe and bassoon, and early nineteenth century clarinet.

gap between first and second registers and add useful low notes below the basic compass.

The clarinet family has a number of members, extending to low-pitched bass clarinets, and is popular because of its tone, flexibility, and large musical range—two registers cover three octaves and there is a half-octave third register above these—and because the cylindrical tube means that it need be only half as long as a conical instrument of the same pitch. For traditional reasons, common clarinets are pitched in the key of B♭, which means that a written C sounds a whole tone lower as B♭. To overcome playing difficulties in keys with several sharps, orchestral musicians also have clarinets pitched in A.

The clarinet reed, as shown in Figs 13.1(a) and 15.13, is a flat rectangular piece of cane, thinned toward the tip on the outside. In use, it is clamped against a mouthpiece, generally of plastic, which has a gently curving flattened side, the lay, in which is a more or less rectangular window extending nearly to the tip of the reed. The curve of the lay on the mouthpiece is vital since it allows the pressure of the player's lip to shorten the freely vibrating part of the reed, thus both raising its natural resonance frequency and decreasing the size of the tip opening. For satisfactory playing, the stiffness of the reed must be carefully matched to the curve of the lay.

The tone of the clarinet, like all reed woodwinds, is rich in harmonics and is characterized in the low (chalumeau) register by an almost complete absence of the second harmonic, as indeed we would expect from the absence of this resonance peak in the input impedance curve of Fig. 15.6(a). The radiated pressure waveform for a typical low note is shown in Fig. 15.14. The second and fourth harmonics are very weak, but all harmonics from the fifth to the sixteenth appear, with the seventeenth and nineteenth missing, for this particular instrument and player, presumably because of coincidence with impedance minima rather than maxima. A similar general pattern occurs for all notes in this register.

The limiting spectrum for controlled loud playing on the clarinet is very much as described in Section 15.8, as was investigated by Benade and Kouzoupis (1988). Below cutoff, which is typically about 1500 Hz for B♭ instruments, the spectrum of the odd harmonics falls at about −3 dB/octave. The level of the even harmonics rises at about 6 dB/octave below cutoff, the extra rate of rise presumably being due to the progressive shift of the minima away from exact harmonics of the sounding frequency. Above cutoff there is a reproducible loud playing level for which the reed just fails to beat and the spectrum falls at −18 dB/octave. Individual instruments have, of course, their own characteristic signatures within this envelope, and these vary from note to note, depending upon fingerings. This behavior is illustrated in Fig. 15.14.

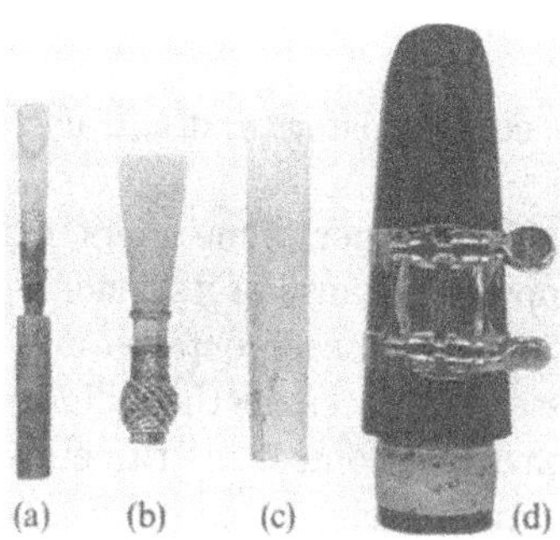

FIGURE 15.13. Reeds for (a) an oboe, (b) a bassoon, and (c) a clarinet, with (d) a clarinet mouthpiece.

The clarinet has a typical maximum power output of a few milliwatts, and actually has the largest dynamic range of any woodwind instrument—about 50 dB (Meyer 1978)—and the psychophysical impression of loudness change is greatly enhanced by the change in harmonic development. As illustrated in Fig. 15.9, blowing pressure is typically about 3 kPa (30 cm water gauge) for soft playing and 4–5 kPa for loud playing throughout the compass of the instrument.

Because of the relative simplicity of the behavior of its cylindrical bore in the time domain, the clarinet has been extensively studied by computer simulation (Schumacher 1978, 1981; Sommerfeld and Strong 1988; Gilbert et al. 1989; Gazengel et al. 1995) with generally good results. Much of the work of Nederveen (1969) has also been updated in a recent paper by Dalmont et al. (1995). There have also been extensive studies of multiphonics and other nonlinear effects by Kobata and Idogawa (1993) while Keefe and Laden (1991) have examined these sounds using the techniques of nonlinear dynamics.

For a high note in the upper (clarion) register, as shown in Fig. 15.14, the second harmonic is quite strong, perhaps because it has been reinforced by reed or vocal-tract resonances, as we discussed previously, but at least partly because of the irregularity of the impedance curve at this frequency. In this upper register, the clarinet has largely lost its characteristic hollow tone.

The bore of the clarinet, while closely cylindrical over most of its length, differs significantly from this simple shape both within the mouthpiece and in the flaring bell at its foot. The bell has a significant effect on the radiation of the lower partials of the sound, particularly for low notes, and, of course, shape variations near the reed affect the tuning of all the notes. Either the perturbation methods discussed in Chapter 8 or direct calculation, as discussed in Section 15.3, can be used to determine the effects of these features (Nederveen, 1969, Chapter 4).

15.10 The Oboe

The modern oboe, also shown in Fig. 15.11, is of almost exactly the same length as the clarinet but has a bottom note nearly an octave higher in pitch because of its conical bore. It has a narrow double reed, as shown in Figs 13.1(b) and 15.13, bound to a cork-jacketed narrow brass tube that is inserted tightly into the upper end of the instrument, continuing the conical bore area moderately accurately to nearly the reed tip.

Instruments of this structure date from remote antiquity and were called by many names. One term covering many instruments from the late middle ages is shawm. In the early seventeenth century, no less than seven sizes of shawms were in popular use, each with six basic finger holes, sometimes

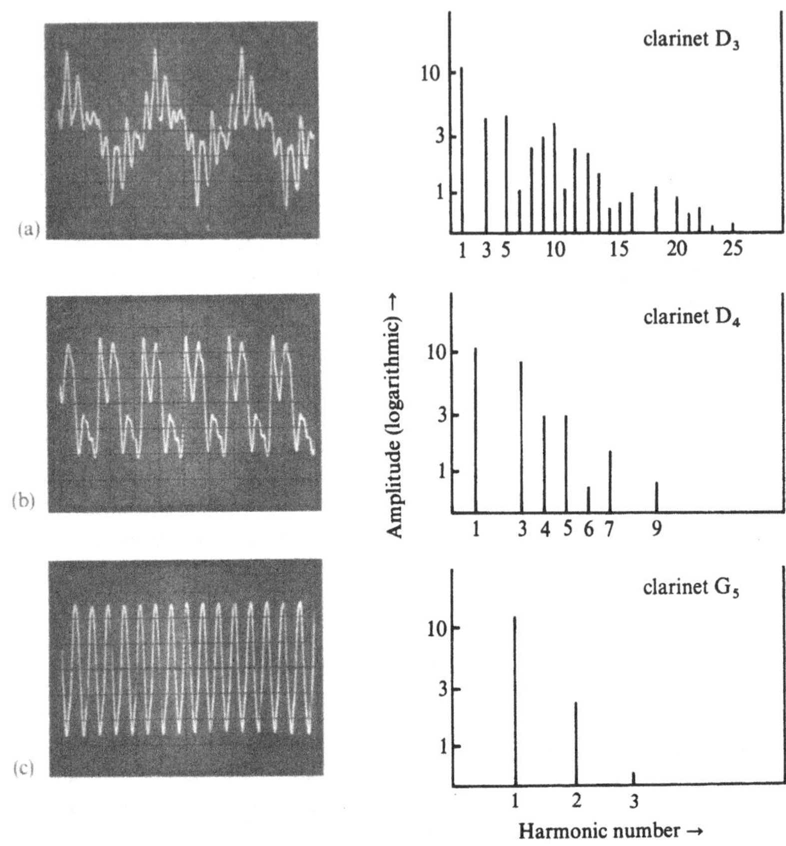

FIGURE 15.14. Radiated sound pressure waveform and corresponding spectrum for three notes played on a clarinet (Fletcher, 1976).

a thumb hole, and sometimes one or more extension keys below the bottom finger hole. All overblew to the octave and had essentially identical fingerings (Carse, 1939, pp. 120–147).

The classical oboe of the late eighteenth century, as shown in Fig. 15.12, was essentially a refinement of the earlier shawm, with a narrower reed, better intonation, and a less strident tone. The modern oboe differs from this largely by the addition of complicated keywork to extend the range and provide conveniently for well-tuned chromatic semitones.

In the modern oboe of Fig. 15.11, the six basic finger holes are covered by perforated plates rather than being surrounded by rings as in the clarinet, though oboes are also made with ring-based key systems. The keys between the uppermost finger hole and the reed are not for extra notes, as in the clarinet, but rather comprise two register holes, with semiautomatic coupling, and two extra holes for the convenient playing of trills across the register break. The basic instrument uses only about two-thirds of the total length, but extra open keys are provided, to be closed by the little fingers of either hand, which extend the compass some four semitones below the normal six-finger range to $B^{\flat}_3$.

The conical semiangle for an oboe is small, typically only about 0.7°, and the effect of the reed staple must be taken into account when discussing impedance curves in detail. Nevertheless, it is clear from Fig. 15.7 that impedance maxima are fairly well aligned with the first few harmonics of the fundamental for notes in the low register. We thus expect the sound output to exhibit a full harmonic spectrum and this is, indeed, what is shown in Fig. 15.15. This feature together with the relative weakness of the fundamental and the higher level of upper harmonics gives the oboe a lighter, brighter sound than that of a clarinet playing the same note. Even in its upper register, the oboe shows a high level of overtone components in its sound.

Relatively less research has been done on the oboe than on the clarinet, and the geometry of its reed makes analysis more difficult (Benade and Richards, 1983). In particular, Bernoulli forces in the narrow reed channel cannot be neglected, the flow characteristic is highly nonlinear because of flow resistance in the narrow channel, as shown in Fig. 13.3, and the asymmetry of the resonances of the conical horn adds further complexity. The tuning of an oboe is not particularly stable, and its use as a tuning standard by orchestras is rather for reason of its clear and penetrating tone.

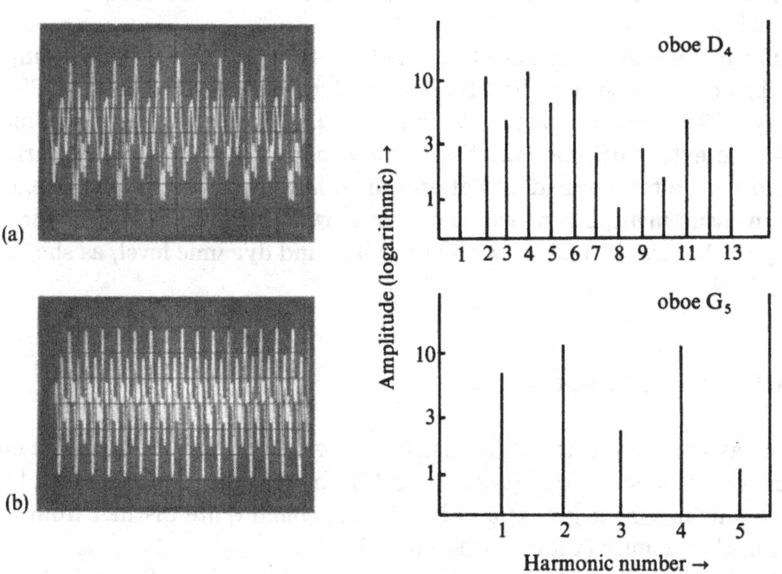

FIGURE 15.15. Radiated sound pressure waveform and corresponding spectrum for two notes played on an oboe (Fletcher, 1976).

The modern oboe family is not as extensive as the clarinet family, but two other members exist, the oboe d'amore and the cor anglais. Both are lower pitched than the oboe, the oboe d'amore by three semitones and the cor anglais (or English horn) by a fifth (seven semitones), and both have globular rather than simply flaring bells. This latter feature undoubtedly contributes to the mellow tone of these instruments. Both have a relatively long history but have now been brought into close accord with the oboe by use of closely similar keywork.

The spectral envelope of oboe sound is well described by the discussion in Section 15.8 and particularly by Eq. (15.30), once we recognize that the input impedance typically has a form like that shown in Fig. 15.7, with impedance peaks initially rising in height with increasing frequency, rather than falling. The likely reason for this is discussed when we come to the case of the saxophone, where the situation is quite clear. The radiated spectrum, as shown in Fig. 15.15, therefore initially rises with increasing frequency until, above cutoff, it typically falls at about −12 dB/octave, almost irrespective of playing level, giving a bright "reedy" tone and suggesting that the oboe reed may always beat. It is easy to see how this comes about. The reed channel in the oboe reed is long and narrow and, as well as inhibiting flow separation and inducing a Bernoulli effect, this long channel introduces appreciable flow resistance. As discussed in Section 13.6.2 and illustrated in Fig. 13.3, this has the effect of steepening the negative-resistance part of the flow curve, while the fact that all harmonics of the flow are reinforced by resonances means that the pressure waveform can be quite asymmetrical. In combination these factors cause the reed to close completely once in each cycle.

The maximum power output of an oboe is about 1 mW, which is similar to that of other woodwinds. Its limited dynamic range of only 15–20 dB (Meyer 1978), and the fact that the reed is always in a beating mode, mean that the extent of tone variation is very much less than for the clarinet. Performance technique does not appear to have been the subject of extensive measurement, but we can note that blowing pressure is typically in the range 4–12 kPa, increasing with both pitch and dynamic level, as shown in Fig. 15.9.

15.11 The Bassoon

The bassoon, too, is a conical instrument with a double reed, but its even narrower cone semiangle (typically 0.4°), folded tube, and traditional fingering arrangement give it an identity and sound quite distinct from that of the oboe family (Carse, 1939, pp. 181–209).

As shown in Fig. 15.11, the bassoon has a tube about 2.6 m in length, folded back at about the midpoint and provided with a metal crook or bocal, to which the reed, shown in Fig. 15.13, is attached. The usual six finger holes are all on the descending part of the tube, and early instruments had holes for both thumbs as well. Because of the great length of the tube, the finger holes are bored at a considerable angle through a purposely thickened wall, and this feature probably contributes significantly to the typical bassoon tone quality.

Early bassoons, as shown in Fig. 15.12, and their higher pitched relatives, the kortholt and dulcian, had only a few keys in addition to finger and thumb holes, and even the modern bassoon has gained extra keys of only a simple type and mostly for the two thumbs, to provide some of the semitones (though some still use forked fingerings), to open vent keys, or to close keys for low notes that use the long upward extension of the tube.

The spectrum of the bassoon contains a complete range of harmonics, as expected, but the radiated fundamental is weak in the low notes because of the small tube diameter, and the tone is mellow rather than bright because of the relatively low cutoff frequency, typically 300–600 Hz. Lehman (1964) has tentatively identified two formants, a strong one at 440–500 Hz and a weaker one at 1220–1280 Hz, which may perhaps describe the characteristic timbre of the bassoon.

From our discussion above, we can identify the lower-frequency formant with the transition at the tone-hole cutoff frequency. The higher formant is probably due to the reed, which truncates the conical bassoon bore in much the same way as does the saxophone mouthpiece to be discussed in the next section. The length of the truncation is about 250 mm, which implies a reed-cavity resonance at about 1400 Hz, in general agreement with the position of the higher formant peak. This interpretation differs somewhat from that of Smith and Mercer (1974), who attributed the formants rather more generally to the influence of cone angle rather than truncation ratio, though the two are clearly related.

Measurements and simulation studies on the bassoon are not common. Ishibashi and Idogawa (1987) measured the impulse response, and later Shimizu et al. (1989) examined the reed vibration and output spectrum for a simplified bassoon without tone holes. This later study showed that a large variety of oscillation regimes can be established, depending upon the attack transient and precise reed adjustment, a finding that accords with the difficulties experienced by beginning bassoonists! The radiated spectrum, normalized to the much lower tone-hole cutoff frequency which is around 500 Hz, has a shape broadly similar to that of the oboe, and the double reed normally beats, as we might expect. The maximum power output is not large—about 1 mW—and the dynamic range is only about 20 dB. Because the reed always beats, there is not a great change in tone color between soft and loud playing. As shown in Fig. 15.9, the required blowing pressure rises with both pitch and dynamic level, and typically ranges from about 1.5 to 8 kPa.

15.12 The Saxophone

In many ways the family of saxophones, developed by Adolph Sax in the mid-nineteenth century, is at the opposite extreme to the bassoon. These instruments, as shown in Fig. 15.16, have a wide conical bore with a flare semiangle of nearly 2°, a single clarinet-like reed, and a unified logical system of padded keys covering very large holes. The saxophone was designed, rather than having evolved, and the family has soprano, alto, tenor, and baritone members.

As expected, the saxophones all overblow to the octave and produce a spectrum with all harmonics present. The lattice cutoff frequency is relatively high, but the broad bell mouth radiates well, so that there is a strong fundamental component and strong radiation damping for higher partials. The tone is thus loud and full, though some playing techniques favor a raucous rather than a mellow sound (Benade and Lutgen, 1988). Typical blowing pressures for an alto saxophone are shown in Fig. 15.9.

While the saxophone has a single reed like the clarinet, the mouthpiece effectively truncates the conical taper of the main bore and introduces significant changes in tone color. In order that the horn modes be as nearly harmonic as possible, it is desirable that the mouthpiece mimic the acoustic behavior of the missing apex of the cone. This can be done at two frequencies, and then fits reasonably well over the whole range. At low frequencies,

FIGURE 15.16. A modern saxophone.

the matching is achieved if the internal volume of the mouthpiece is equal to that of the missing conical apex, which requires that the mouthpiece have a slightly bulbous internal shape so that it actually constitutes a sort of Helmholtz resonator. The high-frequency match can then be achieved by arranging the shape of the constriction where it joins the main part of the instrument so that the Helmholtz resonance frequency of the mouthpiece is the same as the first resonance of the missing conical apex, at which it is half a wavelength long.

This mouthpiece cavity has an important effect on the spectrum of the saxophone (Benade and Lutgen, 1988), which differs in important aspects from the form illustrated in Fig. 15.10. The cavity acts rather like the mouthcup of a brass instrument, as shown in Fig. 14.5, and imparts an extra rise of 6 dB/octave below its resonance and a fall of −6 dB/octave above. The mouthpiece resonance frequency is typically comparable with the tonehole cutoff frequency ω_c, so these extra slopes contribute simply to the behavior above and below ω_c, which is typically about 850 Hz for an alto saxophone. A normalized spectral envelope for the radiated power $P(\omega)$, derived from Eq. (15.30) as modified by inclusion of the mouthpiece cavity effect, and applicable to all instruments of the saxophone family, has the form

$$P(\omega) \sim \left(\frac{U_1}{U_0}\right)^n \frac{\omega/\omega_c}{1+(\omega/\omega_c)^m}. \tag{15.31}$$

At low and moderate playing levels, the envelope is dominated by the term $(U_1/U_0)^n\omega$, and falls rapidly after the first few harmonics. For loud playing with a non-beating reed, $U_1 \approx U_0$ and $m = 9$, while for very loud playing with a beating reed $m = 7$.

15.13 Capped Reed Instruments

In all the modern woodwinds, the reed is held firmly between the player's lips, and its opening and resonance frequency can be adjusted readily by the player, giving a good deal of control over the behavior. Indeed such control of the reed is almost essential for producing notes other than those of the lowest register of the instrument. Some early reed instruments from Medieval and Renaissance times, however, removed the reed from contact with the player's lips and did not take advantage of this extra control. The same is true of bagpipes. In this section we review these instruments briefly.

15.13.1 Medieval Instruments

One of the earliest reed instruments was the shawm, which is essentially an early oboe, with the usual conical bore and double reed. The reed, however, which was short and wide, was taken completely inside the player's mouth and the lips served simply to seal the mouth pressure against a flat lip plate. Shawms were outdoor instruments, with a load strident tone, and were scarcely adapted for more refined musical use. Their compass was essentially limited to the fundamental octave.

It is interesting to note the difference in behavior between a shawm, in which the pipe resonances dominate that of the reed, and the reed organ pipe, in which the opposite is true. The obvious difference between the two cases is that the organ pipe has a clarinet-like single metal reed, which has very low damping, while the shawm has a bassoon-like double reed of cane. One might surmise that the internal damping of the cane plays a part in damping the reed resonance and broadening its operating frequency range, but perhaps more important is the extreme nonlinearity induced by the flow resistance of the double-reed channel, as discussed in Sections 13.2 and 13.6.2 and illustrated in Fig. 13.3. This does not appear to have been investigated.

Another important renaissance instrument, or rather family of instruments, were the krumhorns. Because these blend well with voices and other renaissance consort instruments, they have been revived and are used often by early music groups. The instrument tube is essentially cylindrical, with normal finger holes, but is bent around at it lower end into a shape like the handle of an umbrella. Several sizes are made, and a krumhorn ensemble is often used. The krumhorn is a true capped-reed instrument and is blown through a small slit in a wooden cap that is fixed over the top of the reed. The reed is again short and double and made of cane, but the design of both the reed and the instrument are such that its sound is soft and "buzzy," making a delightful contrast to the sound of recorders.

15.13.2 Bagpipes

Another early instrument that has maintained its following is the bagpipes. Many different types have been developed, ranging from the musette of France to the Uillean pipes of Ireland, but the most popular today are the Highland bagpipes of Scotland, shown in Fig. 15.17. The essential feature of all bagpipes is a wind reservoir that can be filled in a semicontinuous manner by blowing or pumping through a valved inlet, thus maintaining a constant playing pressure and allowing continuous sounding for long periods. In the case of the Scottish bagpipes, air is provided by blowing into a bladder held under the left arm, and the pressure is maintained by the pressing that arm against the player's body. This arrangement has the advantage that air pressure can be suddenly applied or released, allowing clean starts and finishes to musical items.

Many bagpipes, including the Highland pipes of Scotland, the Uillean pipes of Ireland, and the French musette, have several drone pipes of fixed pitch to provide a background against which the melody is played on a

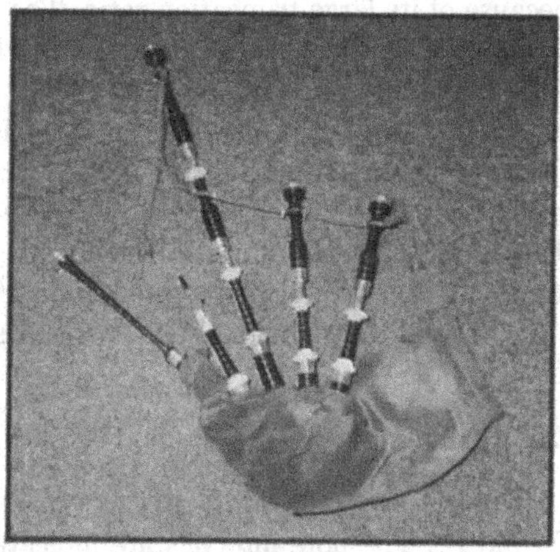

FIGURE 15.17. The Highland bagpipes of Scotland; a set by Kilberry of Edinburgh. From the left we see the conical chanter, the mouthpipe, the bass drone, and two tenor drones.

chanter pipe with normal finger holes. In the case of the Scottish bagpipes, which are those heard most often, there are three drone pipes, two tenors tuned an octave below the lowest note of the chanter, and a bass tuned two octaves below the chanter. An excellent discussion of the construction and acoustics of this instrument has been given by Harris et al. (1963), while the chanter in particular has been investigated by Firth and Sillitto (1978). The three drone pipes are basically cylindrical, though the bore has several large steps in diameter which are brought about by the need for slides to adjust the pipe length for tuning purposes. The fundamental of the bass drone is about 115 Hz and that of the two tenors about 230 Hz. The drones are driven by rather primitive single reeds, made by cutting a broad tongue in the side of a cane tube. The reed resonance frequency is initially adjusted by sliding a wire loop along the cane until it matches that of the drone pipe, but subsequent tuning is done by adjusting the pipe length. The chanter, in contrast, is approximately conical, though with rather a large truncation, and is driven by a broad double reed, rather like a small bassoon reed. As with the shawm, the extreme reed nonlinearity and large damping allow the pipe resonances to determine the pitch. Finger holes allow for about an octave of scale to be played, and there is no overblowing.

The Highland pipes have an acoustic output of 10–20 mW, which is large compared with the 1 mW typical of other woodwinds. Most of the sound comes from the chanter. The spectrum of the drones contains both odd and even harmonics at comparable intensities, presumably because of the irregularity of the bore, and their relative levels vary greatly as the tuning is adjusted. The resonances of the chanter tube are similarly very far from being harmonic, because of its large truncation ratio. The sound is rich in harmonics, with an apparent formant peak around 2 kHz which is probably due to the finger-hole cutoff, as discussed in Section 15.7.

There have been several studies of the scale of the Highland bagpipe (Lenihan and McNeil, 1954; Firth and Sillitto, 1978). Certainly the scale differs from the ordinary familiar keyboard scale but, while there is reasonable consistency, the pitch variations between instruments are such that a precise specification is not meaningful. A representation of the scale measured by Lenihan and McNeil (though not their conclusions) approximates G major, in the form

$$\text{G } 200 \text{ A } 200 \text{ B } 150 \text{ C}^{+} \ 150 \text{ D } 200 \text{ E } 200 \text{ F}^{+} \ 150 \text{ G } 200 \text{ A}$$

where the note C^{+} lies mid-way between C and C♯, and the note F^{+} lies midway between F and F♯. The note intervals are indicated in cents (100 cents equals 1 equal-tempered semitone). Firth and Sillitto found a rather different and more irregular scale for the instrument they studied, but the reason for this is not clear.

References

Backus, J. (1961). Vibrations of the reed and air column in the clarinet. *J. Acoust. Soc. Am.* **33**, 806–809.

Backus, J. (1963). Small-vibration theory of the clarinet. *J. Acoust. Soc. Am.* **35**, 305–313.

Backus, J. (1974). Input impedance curves for the reed woodwind instruments. *J. Acoust. Soc. Am.* **56**, 1266–1279.

Backus, J. (1978). Multiphonic tones in the woodwind instruments. *J. Acoust. Soc. Am.* **63**, 591–599.

Backus, J. (1985). The effect of the player's vocal tract on woodwind instrument tone. *J. Acoust. Soc. Am.* **78**, 17–20.

Baines, A. (1967). "Woodwind Instruments and their History." Faber and Faber, London.

Bartolozzi, B. (1981). "New Sounds for Woodwinds," 2nd ed. Oxford Univ. Press, London.

Benade, A.H. (1959). On woodwind instrument bores. *J. Acoust. Soc. Am.* **31**, 137–146.

Benade, A.H. (1960a). On the mathematical theory of woodwind finger holes. *J. Acoust. Soc. Am.* **32**, 1591–1608.

Benade, A.H. (1960b). The physics of woodwinds. *Scientific American* **203**(4), 145–154.

Benade, A.H. (1976). "Fundamentals of Musical Acoustics," pp. 430–504. Oxford Univ. Press, London and New York.

Benade, A.H. (1986). Woodwinds: the evolutionary path since 1700, *Proc. 12th Intl. Congress Acoustics*, Toronto.

Benade, A.H., and Kouzoupis, S.N. (1988). The clarinet spectrum: Theory and experiment. *J. Acoust. Soc. Am.* **83**, 292–304.

Benade, A.H., and Larson, C.O. (1985). Requirements and techniques for measuring the musical spectrum of the clarinet. *J. Acoust. Soc. Am.* **78**, 1475–1498.

Benade, A.H., and Lutgen, S.J. (1988). The saxophone spectrum. *J. Acoust. Soc. Am.* **83**, 1900–1907.

Benade, A.H., and Richards, W.B. (1983). Oboe normal mode adjustment via reed-staple proportioning. *J. Acoust. Soc. Am.* **73**, 1794–1803.

Carse, A. (1939). "Musical Wind Instruments." Macmillan, London. Reprinted by DaCapo Press, New York, 1965.

Chung, J.V., and Blaser, D.A. (1980). Transfer function method of measuring in-duct acoustic properties: I. Theory; II. Experiment. *J. Acoust. Soc. Am.* **68**, 907–913 and 914–921.

Clinch, P.G., Troup, G.J., and Harris, L. (1982). The importance of vocal tract resonance in clarinet and saxophone performance—A preliminary account. *Acustica* **50**, 280–284.

Dalmont, J.P., Gazengel, B., Gilbert, J., and Kergomard, J. (1995). Some aspects of tuning and clean intonation in reed instruments. *Appl. Acoust.* **46**, 19–60.

Firth, I.M., and Sillitto, H.G. (1978). Acoustics of the Highland bagpipe chanter and reed. *Acustica* **40**, 310–315.

Fletcher, N.H. (1976). "Physics and Music." Heinemann, Melbourne.

Fletcher, N.H. (1978). Mode locking in nonlinearly excited inharmonic musical oscillators. *J. Acoust. Soc. Am.* **64**, 1566–1569.

Fuks, L., and Sundberg, J. (1996). Blowing pressures in reed woodwind instruments. *Quart. Prog. Status Rep., Speech Music and Hearing* Royal Inst. Tech., Stockholm. No.3/1996, pp. 41–56.

Gazengel, B., Gilbert, J., and Amir, N. (1995). Time domain simulation of single reed wind instrument. From the measured input impedance to the synthesis signal. Where are the traps? *Acta Acustica* **3**, 445–472.

Gilbert, J., Kergomard, J., and Ngoya, E. (1989). Calculation of the steady-state oscillations of a clarinet using the harmonic balance technique. *J. Acoust. Soc. Am.* **86**, 35–41.

Harris, C.M., Eisenstadt, M., and Weiss, M.R. (1963). Sounds of the Highland bagpipe. *J. Acoust. Soc. Am.* **35**, 1321–1327.

Ishibashi, M., and Idogawa, T. (1987). Input impulse response of the bassoon. *J. Acoust. Soc. Japan.* **(E) 8**, 139–144.

Johnston, R., Clinch, P.G., and Troup, G.J. (1986). The role of vocal tract resonance in clarinet playing. *Acoustics Australia* **14**, 67–69.

Keefe, D.H. (1982a). Theory of the single woodwind tone hole. *J. Acoust. Soc. Am.* **72**, 676–687.

Keefe, D.H. (1982b). Experiments on the single woodwind tone hole. *J. Acoust. Soc. Am.* **72**, 688–699.

Keefe, D.H., and Laden, B. (1991). Correlation dimension of woodwind multiphonic tones. *J. Acoust. Soc. Am.* **90**, 1754–1765.

Kobata, T., and Idogawa, T. (1993). Pressure in the mouthpiece, reed opening, and air-flow speed at the reed opening of a clarinet artificially blown. *J. Acoust. Soc. Japan* (E) **14**, 417–428.

Lehman, P.R. (1964). Harmonic structure of the tone of the bassoon. *J. Acoust. Soc. Am.* **36**, 1649–1653.

Lenihan, J.M.A. and McNeil, S. (1954). An acoustical study of the Highland bagpipe. *Acustica* **4**, 231–232.

Meyer, J. (1978). "Acoustics and the Performance of Music," pp. 153–160. Verlag Das Musikinstrument, Frankfurt am Main.

Nederveen, C.J. (1969). "Acoustical Aspects of Woodwind Instruments." Frits Knuf, Amsterdam. (Reprinted by Nothern Illinois University Press, Dekalb, 1998.)

Plitnik, G.R. and Strong, W.J. (1979). Numerical method for calculating input impedances of the oboe. *J. Acoust. Soc. Am.* **65**, 816–825.

Pratt, R.L., Elliott, S.J., and Bowsher, J.M. (1977). The measurement of the acoustic impedance of brass instruments. *Acustica* **38**, 236–245.

Rice, A.R. (1984). Clarinet fingering charts, 1732–1816. *Galpin Soc. J.* **37**, 16–41.

Schumacher, R.T. (1978). Self-sustained oscillations of the clarinet: An integral equation approach. *Acustica* **40**, 298–309.

Schumacher, R.T. (1981). Ab initio calculations of the oscillations of a clarinet. *Acustica* **48**, 71–85.

Shimizu, M., Naoi, T., and Idogawa, T. (1989). Vibrations of the reed and air column in the bassoon. *J. Acoust. Soc. Japan* **10**, 269–278.

Smith, R.A., and Mercer, D.M.A., (1974). Possible causes of woodwind tone colour, *J. Sound Vibr.* **32**, 347–358.

Sommerfeld, S. and Strong, W. (1988). Simulation of a player-clarinet system. *J. Acoust. Soc. Am.* **83**, 1908–1918.

Stewart, S.E., and Strong, W.J. (1980). Functional model of a simplified clarinet. *J. Acoust. Soc. Am.* **68**, 109–120.

Thompson, S.C. (1979). The effect of the reed resonance on woodwind tone production. *J. Acoust. Soc. Am.* **66**, 1299–1307.

Wilson, T.A., and Beavers, G.S. (1974). Operating modes of the clarinet. *J. Acoust. Soc. Am.* **56**, 653–658.

Worman, W.E. (1971). Self-sustained nonlinear oscillations of medium amplitude in clarinet-like systems. Ph.D. Thesis, Case Western Reserve University, Cleveland, Ohio.

16

Flutes and Flue Organ Pipes

Instruments of the flute type, which are made to sound by blowing a stream of air across a hole in some hollow body, are of extremely ancient origin. Their principal modern descendants are the orchestral transverse flute and the flue organ pipe, but the baroque recorder has been revived during the past 50 years, and other instruments, such as the end-blown Japanese shakuhachi and the Balkan nai or panpipes, are widely heard.

The physical structure varies considerably from one instrument to another. Some have finger holes and some do not; some use the player's lips to form an air jet and some use a mechanically defined flue or windway. However, they all rely upon the effect of an air jet striking a sharp edge for their acoustical excitation. We therefore begin by considering such a system, coupled to an acoustic resonator, in some detail. There are at least two ways in which our discussion might proceed. The first, and the one that we in fact adopt, is based upon an approximate model that is readily visualized and that comes quite close to accounting for the behavior of the instruments of the family. The second, which we mention only briefly, is a rigorous aerodynamic approach to the problem. Ultimately this second approach should prevail, but at the present stage of its development it is not only more difficult to follow but also much less satisfactory in accounting for the sounding behavior of the instruments.

16.1 Dynamics of an Air Jet

The mechanics of a fluid jet entering an infinite body of the same or a different fluid is a fundamental physical and mathematical problem to which a great deal of attention has been given during the past 100 years. The literature is very extensive, and the range of the relevant parameters, such as velocity, stretches from infinitesimal through supersonic to relativistic. Only a small fraction of this range concerns us and only a small selection of the possible phenomena.

In the absence of viscosity, compressibility, or other complications, the steady flow of one semi-infinite layer of fluid across another is unstable, as was shown by Rayleigh (1894) nearly a century ago. The behavior is most simply appreciated by supposing that the two layers are separated by the plane $z = 0$ and move, respectively, in the positive and negative x-directions with equal speeds $\pm V/2$. If a sinusoidal displacement in the z-direction of the form

$$h = H \cos\left(\frac{2\pi x}{\lambda}\right) \tag{16.1}$$

is imposed on the dividing surface, then the laws of fluid flow dictate that h decreases exponentially with distance away from the dividing surface, and the pressure gradients are in such directions as to cause the initial displacement to grow exponentially with time. The length scale of the physical situation is set by the wavelength of the disturbance, and the time scale is set by the time $t_0 = \lambda/V$ for two points moving with the two layers to separate by a wavelength λ.

Detailed analysis shows that the amplitude h increases by a factor e^{π}, or about 23, in the characteristic time t_0, so that growth is rapid, with short-wavelength disturbances growing more rapidly than those of long wavelength. By symmetry, the disturbance remains stationary in our coordinate system, which means that if we move along with one of the streams so that it appears stationary and the other stream has a velocity V, then the growing wave will move relative to our observing point with velocity $V/2$, half that of the moving stream.

Rayleigh's analysis can be extended to the case in which one of the streams has a finite depth b. By imposing appropriate conditions at this boundary, the solution represents the behavior of a planar jet of thickness $2b$, constrained to have displacements that are either symmetric or antisymmetric about its midplane—disturbances that may be referred to as varicose or sinuous, respectively. Most of our concern will be with sinuous jets, as shown in Fig. 16.1, rather than with varicose jets, but the latter are acoustically important in some types of whistles in which the jet passes through an aperture rather than meeting an edge (Chanaud, 1970; Wilson et al., 1971).

If we suppose the jet to move with velocity V through a body of stationary fluid, then the length scale is fixed by the undisturbed jet width $2b$, and the behavior is more complicated than in our original problem. In particular, the propagation speed and growth rate of a small disturbance depend significantly on the ratio of its wavelength λ to the jet thickness $2b$. If $k = 2\pi/\lambda$, then Rayleigh (1894) showed that the propagation velocities u are

$$u_{\text{sinuous}} = \frac{V}{1 + \coth kb}, \tag{16.2}$$

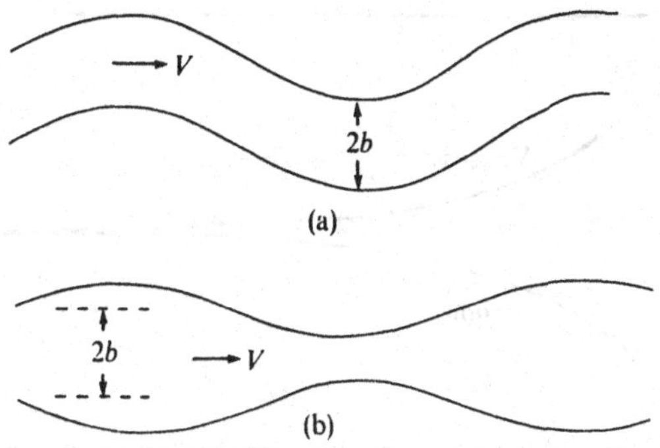

FIGURE 16.1. (a) Sinuous and (b) varicose disturbances of a plane jet.

and

$$u_{\text{varicose}} = \frac{V}{1 + \tanh kb}. \tag{16.3}$$

For very short wavelengths, the disturbances are effectively confined to the immediate environment of the two dividing surfaces and the jet thickness is irrelevant. In this case, $kb \gg 1$ and both u_{sinuous} and u_{varicose} approach $V/2$. For long wavelengths, however, the spread in jet influence in the z-direction, which is of order λ, becomes much greater than the jet width. Equations (16.2) and (16.3) show that in this limit, as $kb \to 0$, $u_{\text{sinuous}} \to 0$ and $u_{\text{varicose}} \to V$. Details of the behavior are shown in Fig. 16.2, where u/V is plotted as a function of the Strouhal number kb.

The temporal growth coefficient derived by Rayleigh is more usefully converted to a spatial one so that we follow the growth of the wave as it moves along. The relevant growth relation is then $\exp(\pm\mu x)$, which includes $\sinh(\mu x)$ or $\cosh(\mu x)$, depending on the initial conditions. Rayleigh's results in this form become

$$\mu_{\text{sinuous}} = k(\coth kb)^{1/2}, \tag{16.4}$$

and

$$\mu_{\text{varicose}} = k(\tanh kb)^{1/2}, \tag{16.5}$$

implying a zero growth rate in the limit of long wavelengths, and with μb tending to kb, and thus to very large values, for very short wavelengths.

To go a further step toward physical reality, we must include the effects of viscosity. The steady flow of a laminar jet in a fluid of nonzero viscosity has been studied theoretically by Bickley (1937). The result for a jet issuing from a slit of zero width at $x = 0$ has the form

$$V(x, z) = 0.4543\cdots \left(\frac{J^2}{\nu x}\right)^{1/3} \operatorname{sech}^2\left(\frac{z}{b}\right), \tag{16.6}$$

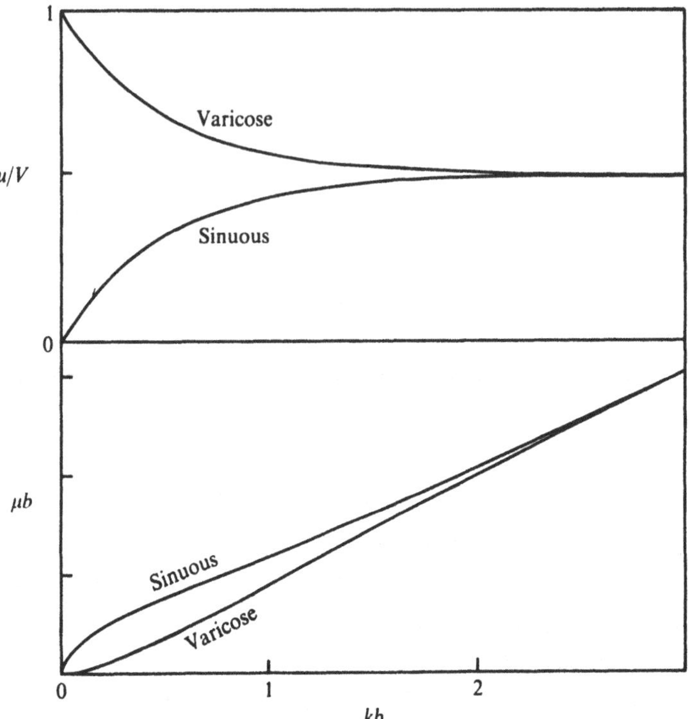

FIGURE 16.2. Propagation velocity u and growth parameter μ for disturbances on a plane jet of thickness $2b$ and having a top-hat velocity profile with center velocity V.

where the local semithickness b is given by

$$b = 3.635\cdots\left(\frac{\nu^2}{J}\right)^{1/3} x^{2/3}. \tag{16.7}$$

Here, ν is the kinematic viscosity ($\nu \approx 1.5 \times 10^{-5}$ m^2 s^{-1} for air) and J is the flow integral

$$J = \int_{-\infty}^{\infty} V^2 \, dz, \tag{16.8}$$

which is just the total momentum flux in the jet, divided by the air density. Clearly, from Eq. (16.6), the jet has a bell-shaped velocity profile that spreads with distance according to Eq. (16.7).

We can use this velocity profile in an analysis similar to that of Rayleigh to refine the predictions, without introducing any of the other effects of viscosity, such as the spreading of the jet. This was done by Savic (1941) who deduced a disturbance speed of the form

$$u = 1.016\cdots(J\omega)^{1/3} \tag{16.9}$$

for a sinuous jet, while Drazin and Howard (1966) produced the numerical results shown in Fig. 16.3 for the same situation. Over most of the range, the wave velocity u behaves rather like Eq. (16.9) and is indeed qualitatively similar to the Rayleigh result. The growth parameter μ, however, behaves very differently, going through a maximum for $kb \approx 0.6$ and becoming negative for $kb > 2.0$. A very similar behavior was calculated by Mattingly and Criminale (1971), except that their curve for μb has a lower maximum value of about 0.27. These more realistic analyses thus remove the catastrophic instability at high frequencies exhibited by a top-hat velocity profile and introduce a preferred wavelength $\lambda \approx 10b$, near which instability growth of the sinuous type is most rapid.

The expansion of the jet as it travels in the x-direction can be included in the treatment by simply regarding b and V, and hence u and μ, as local variables. Growth will not now be simply exponential along the jet but rather it will be concentrated along that part of the jet for which $\lambda \sim 10b$, or equivalently $\omega b/u \sim 0.6$. Further along the jet, we will reach a condition for which $kb > 2$, and disturbance will die away with further propagation.

While this simple analysis contains many elements of the real behavior of disturbed jets, it overlooks the fact that the model applies only where the disturbance amplitude is small compared both to its wavelength and to the width of the jet. For a jet of width 2 mm, however, the peak value of μ is about 800 m^{-1}, so that the amplitude increases by a factor $e^{0.8}$ or about 2.2 for every millimeter of travel, or about a factor 3000 for a jet 10 mm long. Even a small disturbance may therefore easily violate this condition. When this happens, the deflection peaks break up to give an alternating street

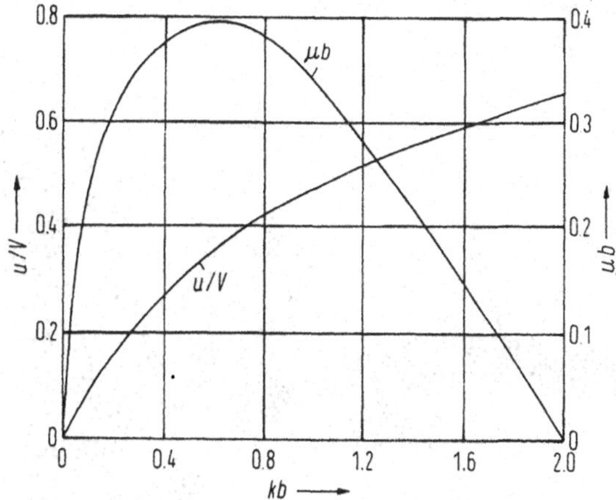

FIGURE 16.3. Propagation speed u and growth parameter μ for sinuous disturbances on a jet with a bell-shaped velocity profile of half-width b and center velocity V. (After Drazin and Howard, 1966.) The calculations of Mattingly and Criminale (1971) give about half this magnitude for μb.

of counter-rotating vortices, as shown in Fig. 16.4. A complete discussion should clearly take this behavior into account, but the necessary theory is complicated, as we discuss briefly in Section 16.5.

Finally, we must note that not all air jets used to excite musical instruments are laminar, and indeed many of them may be fully turbulent. The velocity profile of a turbulent jet is still bell-shaped, but not exactly the same as the $V(x,0)\text{sech}^2(z/b)$ form shown by a laminar jet. However, this is still a good approximation. The half-width b of the jet turns out to be proportional to the distance x and the central velocity $V = V(x,0)$ varies as $x^{-1/2}$, as would be expected from conservation of the flow integral J of Eq. (16.8).

The theory of turbulent jets is not well enough developed to allow a prediction of wave velocity and growth rate, though we might reasonably expect the behavior to be similar to that for a laminar jet with the ordinary viscosity replaced by a scale-dependent eddy viscosity (Fletcher and Thwaites, 1979). Measurements by Thwaites and Fletcher (1980, 1982) show that this expectation is, in the main, fulfilled. Their results show that u/V has a nearly constant value of about 0.5 over a frequency range of a factor 100, with some suggestion of a rise to about 0.6 at the highest frequencies measured. The growth parameter μ, in the combination μb, shows a peak value $\mu b \approx 0.5$ for $kb \approx 0.5$ and drops to zero for $kb > 2$ in very much the same way as illustrated in Fig. 16.3, provided the wave amplitude is less than about $0.1b$. For wave amplitudes about equal to b, the peak value of μb is about 0.25 near $kb \approx 0.2$, and μb falls to zero for $> kb \neq 1$.

A real jet emerging from a flue slit shares some of the properties of all the jets we have discussed. Its initial velocity profile is top-hat shaped but

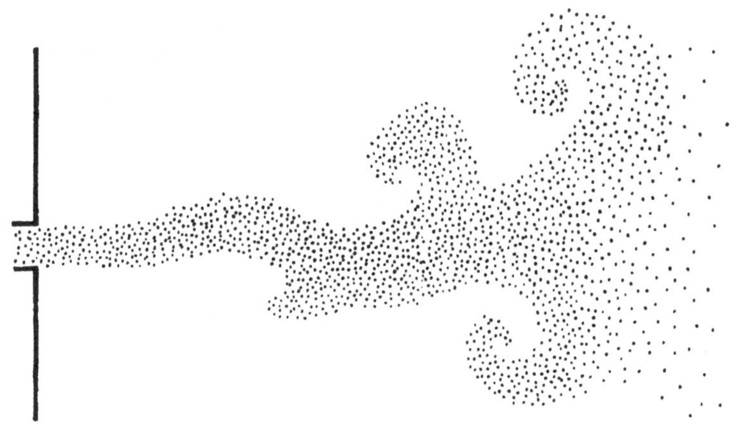

FIGURE 16.4. Vortex street on a laminar jet emerging from a slit into an acoustic flow field normal to the center plane of the jet.

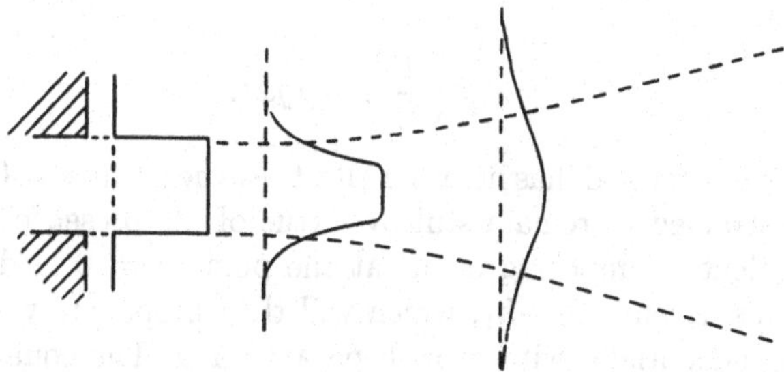

FIGURE 16.5. Evolution of the velocity profile of a plane jet emerging from a slit.

tends toward a bell shape as vorticity diffuses in from its edges. It also gradually develops turbulence, so that its evolution in space is rather as shown in Fig. 16.5. The total behavior is therefore complex and depends on jet length.

16.2 Disturbance of an Air Jet

All we have written so far concerns the growth of a sinusoidal disturbance once it has been impressed upon a jet. We now look at the way in which the disturbance can be initiated by an acoustic field. The first point to note is that we will be concerned with a jet in a standing wave field near an acoustic resonator, rather than in a free field. It is therefore useful to separate out the effects of acoustic pressure and acoustic flow. The former is a scalar quantity and can affect the jet only in the sense of inducing varicose modes. The acoustic flow, however, is a vector and so will induce sinuous modes, unless the flow is parallel to the jet, in which case varicose modes can be produced. In nearly all musical instruments, though not in some aperture whistles, the resonator lies on one side of the jet and it is the acoustic flow and its related sinuous modes with which we are concerned (Brown, 1935, 1937; Chanaud, 1970).

There is not yet complete agreement about the way in which the flow of a jet emerging from a slit into a transversely directed acoustic flow field should be treated. The discussion that we give here is therefore not definitive, but does lead to a fairly realistic model.

An acoustic flow field normal to an unconstrained jet has very little effect upon it but simply moves the whole jet back and forth. What is required to initiate a disturbance on the jet is some sort of shearing motion, and this is best produced by having the jet enter the flow field through a fixed slit located, let us say, at $x = 0$. If $v\exp(j\omega t)$ is the acoustic flow velocity in the z-direction at frequency ω, which we suppose to be uniform over the whole plane of the jet, then the acoustic displacement of the jet by the flow can be written

$$h_1 = -j\left(\frac{v}{\omega}\right)\exp j\omega t. \tag{16.10}$$

This, as we have remarked, has no other effect. At the point $x = 0$, however, the jet is constrained to remain still by virtue of the presence of the slit. This is equivalent to imposing on it, at the point $x = 0$, a disturbance or "negative displacement" $-h_1$, which will then propagate with velocity u and grow exponentially with growth parameter μ. The equations allow growth as both $\exp \mu x$ and $\exp -\mu x$, and the appropriate combination is $\cosh \mu x$, since the flow angle of the jet cannot change discontinuously as it emerges from the slit. The total displacement of the jet at a point x will therefore be

$$h(x) = -j\left(\frac{v}{\omega}\right)\left\{\exp j\omega t - \cosh \mu x \exp\left[j\omega\left(t - \frac{x}{u}\right)\right]\right\}. \tag{16.11}$$

The form of the jet displacement for the simple case in which u and μ are constant along the jet is shown in Fig. 16.6. The form of the curve is fairly flat for $x < 1/\mu$, but to the right of this point the second term in Eq. (16.11) dominates, and we have an exponentially growing wave moving to the right with a velocity u that is about half the jet velocity V. In a real jet, we allow that u and μ in Eq. (16.11) are generally functions of x once the jet begins to spread, and the important effect of this is to reduce both the wavelength of the disturbance and its growth rate as it propagates along the jet.

We can immediately see a weakness in this model, for it treats the action of the acoustic flow field on the jet, but not the action of the jet in disturbing the acoustic flow field, which it must do because of its constrained entry point and non-zero inertia. There are, however, more important uncertainties, so we shall not pursue this here. It is interesting, however, to look briefly at the behavior in the low-frequency limit. To do this, we use Eq. (16.4) for μ and then expand the right-hand side of (16.11), retaining only terms up to second order in x. The result is

$$h(x) \approx \frac{vx}{u} - j\frac{vx^2}{2bu}. \tag{16.12}$$

The first term on the right in this equation is a simple deflection of the jet by the acoustic flow field, while the second can be interpreted as the deflection of the jet under the action of a pressure force associated with the acoustic field and resisted by the inertia of the jet. Each of these terms has been used separately by various authors, albeit with a phase change of 90° in the case of the second term, as a basis for predicting jet behavior.

The behavior of jets under conditions applicable to wind instruments has been studied experimentally by Coltman (1968a, 1968b, 1976a, and 1981) by Fletcher and Thwaites (1979, 1983), and also by Thwaites and Fletcher, (1980, 1982). The results confirm the general correctness of the theory

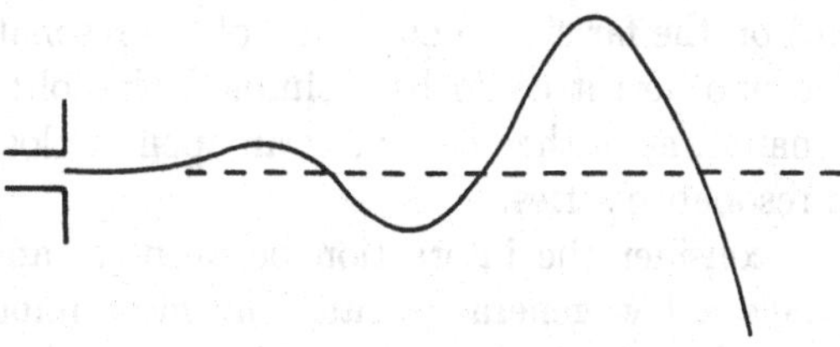

FIGURE 16.6. A growing sinuous wave on a plane jet emerging from a slit.

leading to Eq. (16.11), though there are still some uncertainties. In the downstream region, more than about one wavelength from the flue slit, the second term in Eq. (16.11) is dominant and the growing wave propagates with a phase velocity u, which is about half the jet velocity V. Near the flue, however, both terms must be included and their partial cancellation leads to an apparent increase in the phase velocity of the wave. Coltman used a simple velocity-drive mechanism, rather like the first term in Eq. (16.12), in much of his analysis, and a later experiment appeared to support this, though suggesting that the real situation is rather complex. The argument leading to this conclusion, however, omitted the variation of the amplification factor μ with frequency, as shown in Fig. 16.3, and is therefore not convincing. Later Adachi and Sato (1997) compared simulations based on the full equation (16.11), which they termed "negative-displacement drive," with those based on "velocity drive" and "force drive", as discussed in relation to Eq. (16.12). They concluded that Eq. (16.11) gives the best approximation to experiment.

All this analysis is based initially upon a linear approximation for the jet behavior. Once the disturbance amplitude becomes significantly greater than the local jet width, however, nonlinear effects intervene and, as we remarked in relation to Fig. 16.4, the wave breaks up into a vortex street. In real musical instruments, the edge with which the jet interacts is generally placed at a sufficiently small distance from the slit that this breakup has not yet occurred, so that the simple wave picture is usually adequate. It is not reasonable to assume, however, that exponential growth continues once the jet wave amplitude becomes comparable with the jet thickness. Indeed we expect that, for large amplitudes, growth will become linear with distance. We return to the role of vortices in the next section.

16.3 Jet-Resonator Interaction

The essence of the interaction between an air jet and a resonator, as in musical instruments of the flute family, is as shown in Fig. 16.7. The jet emerges as a more-or-less plane sheet from a flue slit, or from the player's lips, crosses an opening in the resonator, and strikes against its more-or-less sharp edge (or lip) on the far side. The nature of the resonator varies from one instrument to another; it could be a simple Helmholtz cavity with a single major resonance, as in the ocarina, or an open or closed pipe with a nearly harmonic resonance series.

Even before we consider the interaction between jet and resonator in detail, we can make a few general points. The most important is that, since the deflection of the jet is the acoustic driving mechanism and since this is driven by the acoustic flow out of the mouth of the resonator, the system will work best when this flow is a maximum and therefore at an acoustic admittance maximum. This is in contrast to the pressure-driven reed generators discussed in the previous two chapters, which operate at an impedance maximum of the resonator.

The other point we should note is that there is a phase shift along the jet caused by the finite wave propagation speed. For a typical blowing pressure p_0 of 1 kPa (10 cm water gauge), the initial jet speed V is given by

$$V = \sqrt{\frac{2p_0}{\rho}}, \tag{16.13}$$

where ρ is the density of air, and turns out to be about 40 m s^{-1}. The distance from the flue slit to the far edge of the hole in the resonator is typically 1 cm, so that, since the wave speed u is about $V/2$, the delay time is about half a millisecond. Allowing for jet spreading, the delay may be even longer, so that it is comparable with the period of a resonator oscillation. Clearly, this will be important to the mechanism of excitation.

If we consider just a single mode of frequency ω in the resonator, this will induce a wave of the same frequency on the jet. The displacement associated with the wave is amplified as it propagates along the jet, with the result that the jet tip flips in and out of the resonator at its sharp edge and injects a volume flow at the resonator frequency. It is clear that this process can lead to a sustained oscillation if the phase relations are right, so we now investigate the matter in two stages. First, we look at the behavior

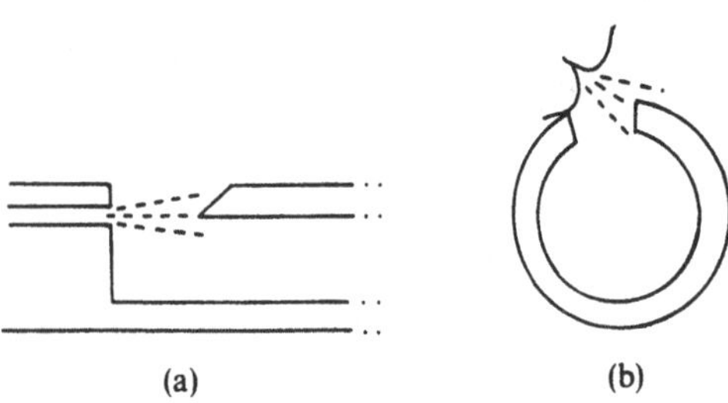

FIGURE 16.7. Typical jet geometries when the jet is defined by (a) a mechanical windway and (b) the player's lips.

of a resonator driven by an oscillating jet, and then we connect the phase and amplitude relations for the whole system.

Several people have studied in detail the way in which a pulsating jet can drive a pipe, chief among the pioneers being Cremer and Ising (1967/1968) and Coltman (1968a,b). Basically, there are two ways in which the drive can be communicated from the jet to the pipe. Either the jet is brought to rest by mixing and thereby generates an acoustic pressure that acts on the flow inside the resonator lip, or else the jet essentially contributes a volume flow that acts on the acoustic pressure inside the resonator lip, which is not zero because of the end correction. Actually, both these driving mechanisms contribute, as shown later by Elder (1973) and by Fletcher (1976a). A further discussion has been given by Yoshikawa and Saneyoshi (1980).

To see what happens, let us adopt the simplifications shown in Fig. 16.8. A jet of air of steady velocity V meets the edge of the resonator so that a fraction of cross section S_j enters the resonator. This blends with an acoustic flow of average velocity v_m into the mouth area S_m of the resonator and, after a small mixing length Δx, becomes an acoustic flow velocity v_p into the main pipe of the resonator, the area of which is S_p. It is best for us to define S_m to be the part of the mouth not inside the jet, so that in our simple figure

$$S_p = S_m + S_j; \quad S_p v_p = (S_m + S_j) v_m + S_j V. \tag{16.14}$$

If we consider the momentum balance of the small parcel of air between the two planes M and P, then, after a little algebra and using Eqs. (16.13) and (16.14), we find

$$\rho \Delta x \left(\frac{dv_p}{dt} \right) + (p_p - p_m) = \rho A_{jp}(1 - A_{jp}) V^2, \tag{16.15}$$

where ρ is the air density, p_p and p_m are the pressures, assumed uniform, across planes P and M, respectively, and

$$A_{jp} = \frac{S_j}{S_p}. \tag{16.16}$$

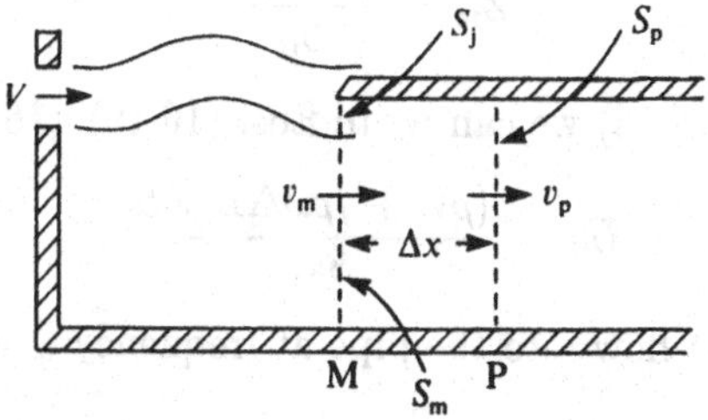

FIGURE 16.8. Definition of geometrical parameters for a jet-driven pipe.

The acoustic flows $U_m = S_m v_m$ and $U_p = S_p v_p$ can be related to the pressures p_m and p_p through equations of the form

$$p_m = p_{m0} - Z_m v_m S_m, \tag{16.17}$$

and

$$p_p = p_{p0} + Z'_p v_p S_p, \tag{16.18}$$

where Z_m and Z'_p are impedances at frequency ω looking out from the volume between planes M and P. If we substitute these equations into Eq. (16.15) and assume that the jet oscillates with frequency ω so that A_{jp} in Eq. (16.16) has this same frequency variation, then we find that the pipe flow $U_p = S_p v_p$ can be divided into three parts:

$$U_p = U_I + U_{II} + U_{III}, \tag{16.19}$$

where

$$U_I = \left(\frac{Z_m}{Z_p + Z_m}\right) S_p V A_{jp}, \tag{16.20}$$

and

$$U_{II} = \left(\frac{\rho V^2}{Z_p + Z_m}\right) A_{jp}, \tag{16.21}$$

both at frequency ω, and U_{III} is a small nonlinear term at frequency 2ω that we can neglect. The term in Δx in Eq. (16.15) has been used to convert the pipe impedance Z'_p on the plane P to a new value Z_p evaluated at plane M.

It is clear that U_I is the pipe flow produced by the jet flow $S_j V$ acting inside the mouth impedance Z_m, while U_{II} is the effect produced by the dynamic jet pressure ρV^2. Both of these terms, and hence U_p itself, are at a maximum when the series impedance $Z_p + Z_m$ is at a minimum, and this is just the resonance condition for the pipe loaded by its mouth impedance Z_m.

If the frequency is not too high, then Z_m can be related to the end correction ΔL at the open mouth by

$$Z_m = \frac{j\rho\omega\Delta L}{S_p}, \tag{16.22}$$

so that using Eq. (16.16) we can write Eqs. (16.19)–(16.21) in the form

$$U_p = \frac{(\rho V + j\rho\omega\Delta L) V S_j}{S_p Z_s}, \tag{16.23}$$

where VS_j is the jet flow into the pipe at frequency ω and Z_s is the series impedance,

$$Z_s = Z_p + Z_m, \tag{16.24}$$

also at frequency ω. From Eq. (16.23), it is clear that there are two drive mechanisms, associated with the two terms in brackets, the first constituting a momentum drive and the second a volume flow drive. These two driving forces differ by 90° in phase as well as behaving differently with jet velocity and frequency. In many practical situations, $\omega \Delta L > V$, so that the volume-flow mechanism dominates, but we cannot really neglect the momentum-drive term.

We can extend the model a little by allowing that the jet may not enter the pipe in a direction parallel to its acoustic axis—a simple model of the transverse flute might imply an angle of 90° to the axis, though we see in the next chapter that this is not actually correct. Suppose the angle between the jet flow and the pipe axis is θ, then this modifies the momentum balance equation (16.15) by inserting a factor $\cos\theta$ on the right-hand side. This factor follows through, so that the generalization of Eq. (16.23) is

$$U_\mathrm{p} = \frac{(\rho V \cos\theta + j\rho\omega\Delta L) V S_\mathrm{j}}{S_\mathrm{p} Z_\mathrm{s}}. \tag{16.25}$$

The momentum part of the jet drive thus depends significantly upon jet direction, and we might expect it to vanish in the case of a transverse flute, were it not for the influence of the chimney in the embouchure hole. This angular dependence has been checked, at least in the limits $\theta = 0$ and 180° by Coltman (1981).

Before going on, it is only right to point out that the concept of a mixing volume, as sketched in Fig. 16.8, really conceals a great deal of ignorance about the aerodynamic details of the system. It is quite clear from flow-visualization studies that vortices are generated where the jet strikes the pipe lip, even if they have not been present on the jet itself (Brown, 1935; Cremer and Ising, 1967/68; Coltman, 1968a). These vortices exchange energy with the acoustic field as they move across it and thus either maintain or damp the oscillation. We return to discuss this briefly in a later section, but for the moment we simply note that, though mass and momentum are both conserved in the mixing region, kinetic energy is not. In fact, since a typical jet velocity is around 30 m s^{-1} and the acoustic flow velocity in the instrument pipe is only a few meters per second, something like 90 % of the initial jet energy is dissipated in the mixing process. Of the small fraction of energy that is transferred to air-column oscillations, by far the greater part is then dissipated by viscous and thermal losses to the walls, so that the overall efficiency with which pneumatic energy is converted to acoustic energy is only of order 1%.

Finally, mention should be made of a further excitation mechanism that is not included in this model (Yoshikawa and Saneyoshi 1980; Elder 1992). The mechanism that we have considered derives its energy from the flow of the jet into the resonator at the lip, and this is analogous to the air flow through a reed aperture. Just as with a reed there is a subsidiary drive caused by the physical displacement of the reed surface during its vibration, so there is an equivalent "air-reed" type of drive generated by the deflection of the jet along its length that is independent of any actual jet flow into the pipe. A consideration of magnitudes involved shows that this component is very small for normal flutes and organ pipes, but it can become important in the case of small enclosed resonators, in which the condition of zero mean jet flow into the resonator precludes significant actual jet flow. The jet is then effectively fixed at the flue and the lip and simply oscillates transversely as its instability drives the resonator flow.

16.4 The Regenerative Excitation Mechanism

We can now put this whole analysis together into a model describing the behavior of a resonator driven by an air jet. To do this, it is most convenient to separate the generator part of the system from the resonator by a plane, such as P or M in Fig. 16.8, and work out the effective acoustic admittance of the generator as we did in the case of a vibrating reed. Our analysis will necessarily lack something in rigor, in the interests of removing nonessential complication, but it still gives a good insight into the actual behavior.

Assume, then, that the jet is short enough that it does not break up into vortices before it reaches the lip, and that the mixing of the jet with the resonator flow is so rapid that the pressure p_{p} and flow U_{p} are both uniform over the plane P in Fig. 16.8. Then, assume further that the mixing length Δx is very small, so that, from the momentum balance equation, Eq. (16.15), and the definition Eq. (16.16), we have in the limit $\Delta x \to 0$

$$p_{\mathrm{p}} - p_{\mathrm{m}} \approx \frac{\rho V^2 S_{\mathrm{j}}}{S_{\mathrm{p}}} = \frac{\rho V U_{\mathrm{j}}}{S_{\mathrm{p}}}, \tag{16.26}$$

where p_{m} is the pressure on the plane M, assumed uniform outside the jet, S_{j} is the sinusoidally varying area of the jet flowing into the resonator at its lip, and U_{j} is the magnitude of this flow.

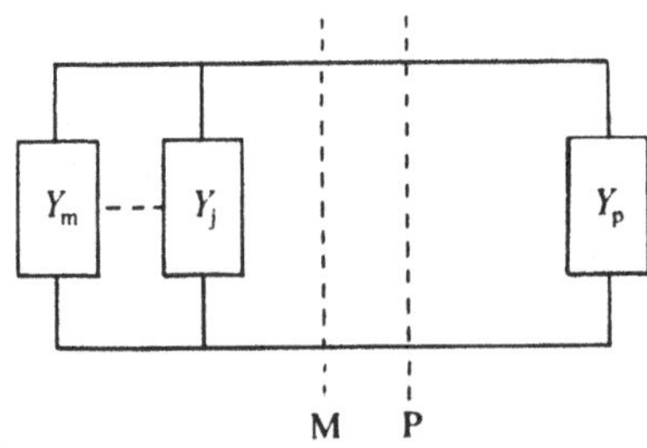

FIGURE 16.9. Equivalent network defining the mouth admittance Y_{m} and jet admittance Y_{j} connected to the pipe admittance Y_{p}.

Now, as shown in Fig. 16.9, we conceptually divide the generator admittance Y_g into two parts Y_m and Y_j, referring to the mouth and jet, respectively, but note, as shown by the coupling line, that the mouth flow controls the jet flow as we have already discussed.

The acoustic flow out through the mouth of the resonator is

$$U_m = v_m S_m = p_m Y_m \approx \frac{p_m S_p}{j\omega\rho\Delta L}, \tag{16.27}$$

where ΔL is the end correction at the mouth and we have neglected the dissipative part of Y_m.

Now, since v_m is the same as v in Eq. (16.11), we can use that expression to give the outward displacement $h(l)$ of the jet where it meets the lip at a distance l from the flue slit. If we assume that $\mu l \gg 1$, then we can simplify Eq. (16.11) and, omitting the implicit time variation $\exp j\omega t$, we have

$$h(l) \approx j\left(\frac{v_m}{\omega}\right)\cosh \mu l \exp\left(-\frac{j\omega l}{u}\right), \tag{16.28}$$

where $u \approx V/2$. If we assume a top-hat velocity profile for the jet and further assume that $h(l)$ is small compared with the jet width, then the oscillating part of the jet flow into the resonator is

$$U_j = -VWh(l), \tag{16.29}$$

where W is the width of the jet in its planar transverse direction.

If we use Eqs. (16.27) and (16.28), then we can write Eq. (16.29) as

$$U_j = -\frac{VWG}{\rho\omega^2\Delta L}\left(\frac{S_p}{S_m}\right)p_m, \tag{16.30}$$

where we have simplified matters by writing

$$G = \cosh \mu l \exp\left(-\frac{j\omega l}{u}\right). \tag{16.31}$$

Now, the total flow out of the resonator across the plane P is

$$U = U_m - U_j = p_m\left[\frac{S_p}{j\rho\omega\Delta L} + \frac{VWG}{\rho\omega^2\Delta L}\left(\frac{S_p}{S_m}\right)\right]; \tag{16.32}$$

but, from Eqs. (16.26) and (16.30), p_p can be written

$$p_p = p_m\left(1 - \frac{V^2WG}{\omega^2\Delta L S_m}\right), \tag{16.33}$$

so that the total generator admittance as seen from plane P is

$$Y_g = U/p_p \approx \frac{S_p}{j\rho\omega\Delta L} + \frac{VWG}{\rho\omega^2\Delta L}\left(\frac{S_p}{S_m}\right)\left[1 + \frac{V}{j\omega\Delta L}\right], \tag{16.34}$$

where we have assumed that the second term within the parentheses in Eq. (16.33) is small and have neglected still smaller terms. If we write

$$Y_{\mathrm{g}} = Y_{\mathrm{m}} + Y_{\mathrm{j}}, \tag{16.35}$$

thus defining the parallel combination in Fig. 16.9, then Y_{m} is just the first term in Eq. (16.34), and the jet admittance Y_{j} has the value

$$Y_{\mathrm{j}} \approx \frac{VW}{\rho\omega^2\Delta L}\left(\frac{S_{\mathrm{p}}}{S_{\mathrm{m}}}\right)\left[1 + \frac{V}{j\omega\Delta L}\right]\cosh\mu l \exp\left(-\frac{j\omega l}{u}\right). \tag{16.36}$$

The term in square brackets, representing the flows I and II of Eq. (16.19), is never much different from unity in magnitude, but it has a phase $-\phi$, where

$$\phi = \tan^{-1}\left(\frac{V}{\omega\Delta L}\right), \tag{16.37}$$

so we can write

$$Y_{\mathrm{j}} \approx \frac{VW}{\rho\omega^2\Delta L}\left(\frac{S_{\mathrm{p}}}{S_{\mathrm{m}}}\right)\cosh\mu l \exp\left[-j\left(\frac{\omega l}{u} + \phi\right)\right], \tag{16.38}$$

where ϕ is generally small.

There is obviously a problem with this expression in the low-frequency limit as $\omega \to 0$. This arises because we have not defined a thickness for the jet, so that its flow into the pipe becomes infinitely large. We have also neglected the first term in Eq. (16.11), which is not justified in this limit since, by Eq. (16.4), $\mu \to 0$ also. Rectification of these omissions makes Y_{j} finite at all frequencies, but the resulting expression is complicated and amplitude dependent. We therefore restrict ourselves, in what follows, to frequencies comparable with, or greater than, the frequency of the pipe fundamental.

We can immediately see that Y_{j} is a negative real quantity when $\omega l/u = \pi - \phi$, and there is very nearly a half-wavelength of the disturbance on the jet, an experimental fact that seems to have been first pointed out by Coltman (1968a). Under these conditions, the resonator will sound at a frequency determined by the condition $\mathrm{Im}(Y_{\mathrm{m}} + Y_{\mathrm{p}}) = 0$, which is the same as the condition the magnitude of Z_{s} should be a minimum, as implied in Eqs. (16.23) and (16.24). If the resonator is a pipe of length L with open-end correction $\Delta L'$, it will thus sound at a frequency appropriate to a doubly open pipe of acoustical length $L + \Delta L' + \Delta L$.

The form of the jet admittance Y_{j} as a function of frequency for a given blowing pressure, and hence fixed V and u, is a spiral as shown in Fig. 16.10(a). For zero frequency, the operating point on the spiral extrapolates to a phase $-\phi$ and rotates steadily clockwise around the spiral as the frequency is raised. This has interesting consequences, for the jet can act as a negative resistance generator whenever the operating point is in the left half of the plane. If ω^* is the frequency of the first crossing of the negative real axis and ϕ is small, then these generating ranges are approximately $\frac{1}{2}\omega^*$ to $\frac{3}{2}\omega^*$, $\frac{5}{2}\omega^*$ to $\frac{7}{2}\omega^*$, etc., with nongenerating or dissipative frequency ranges in between.

Another way of looking at the behavior is to keep the frequency fixed and vary the blowing pressure p_0, which is related to V and u by

$$p_0 = \frac{1}{2}\rho V^2 \approx 2\rho u^2. \tag{16.39}$$

Again, this gives a spiral, as shown in Fig. 16.10(b), and there are several pressure ranges over which the jet acts as a generator at a given frequency. The operating point rotates anticlockwise increasingly slowly as p_0 is raised and extrapolates to the real axis at infinite blowing pressure, though the actual high-pressure behavior may be different since the approximations used in deriving Eq. (16.36) are then no longer valid.

The form of the jet admittance curve has been examined experimentally by Cremer and Ising (1967/1968) and, more extensively, by Coltman (1968a, 1968b, 1976a, 1981). Coltman's results are not directly interpretable in terms of our treatment because he used a model for his analysis that treated the mouth and jet as two impedances in series rather than two admittances in parallel, but he did find the expected spiral behavior. Measurements confirming our theory more directly were carried out by Fletcher and Thwaites (1979, 1983) and by Thwaites and Fletcher (1980, 1982) using an active impedance tube technique. The agreement with theory is moderately good at low blowing pressures and reasonably high frequencies, but

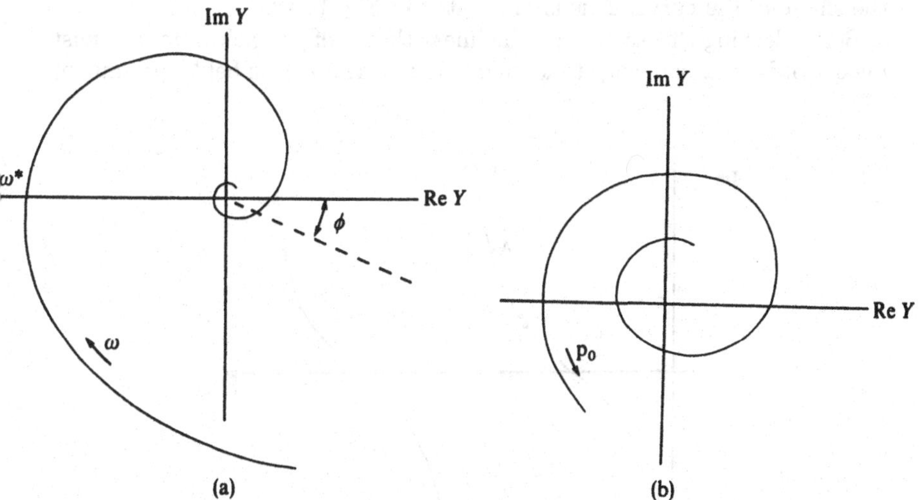

FIGURE 16.10. General form of (a) jet admittance $Y_j(\omega)$ for a fixed blowing pressure as a function of frequency ω and (b) jet admittance $Y_j(p_0)$ as a function of blowing pressure p_0 for a fixed frequency ω.

deteriorates somewhat at low frequencies and high blowing pressures, a result that is not altogether unexpected in view of the approximations in the theory.

It is interesting and important to note the effect of blowing pressure on sounding frequency in the linear approximation. Suppose we have a cylindrical pipe of length L, then its admittance has the form

$$Y_{\mathrm{p}} = -j\left(\frac{S_{\mathrm{p}}}{\rho L}\right)\cot kL, \tag{16.40}$$

as shown in Fig. 16.11, if we neglect the real part. Similarly, the mouth admittance is

$$Y_{\mathrm{m}} = -\frac{jS_{\mathrm{p}}}{\rho\omega\Delta L}, \tag{16.41}$$

and the stability condition for the system is

$$Y_{\mathrm{j}} + Y_{\mathrm{m}} + Y_{\mathrm{p}} = 0. \tag{16.42}$$

Now, if the blowing pressure p_0 is adjusted so that Y_{j} is real and negative—the ideal situation—then, since Im $Y_{\mathrm{m}} < 0$, we must have Im $Y_{\mathrm{p}} > 0$, giving a situation such as that at point A in Fig. 16.11. If the blowing pressure is increased, then, from Fig. 16.10(b), Im Y_{j} becomes negative and the operating point must move to B in Fig. 16.11, giving a small increase in sounding frequency. The extent of the frequency increase necessary to achieve balance is limited because of the increasing steepness of the $Y_{\mathrm{p}}(\omega)$ curve. For a lower blowing pressure, conversely, the frequency will fall to C, and the extent of the frequency shift may be considerable because of the shape of the curve. This is illustrated in Fig. 16.12.

Before leaving this section on the linear theory of jet excitation, we must make brief mention of edge tones and vortices. These both held a prominent

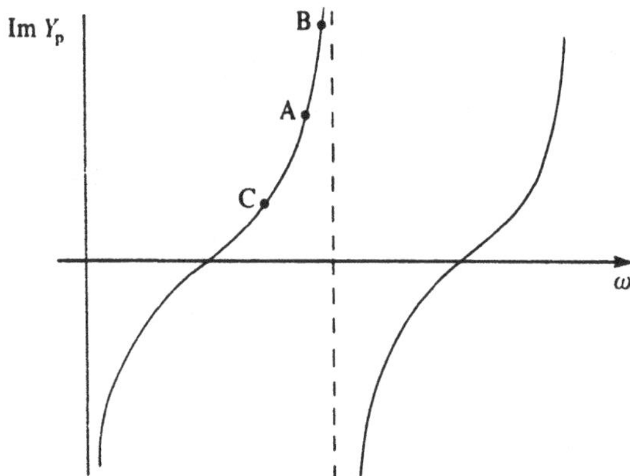

FIGURE 16.11. Admittance of an open pipe as a function of frequency.

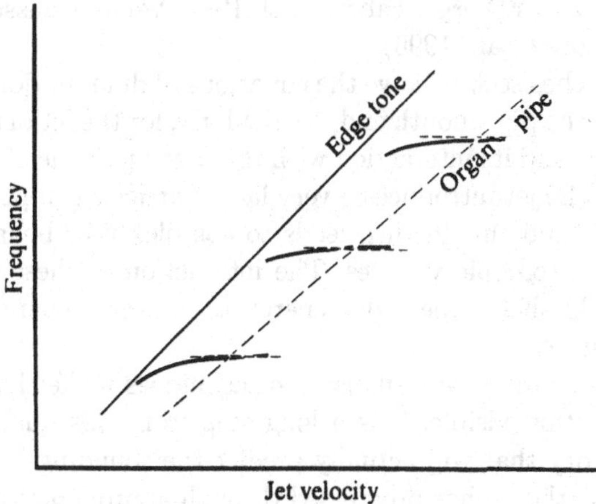

FIGURE 16.12. Sounding frequencies of the modes of a low-Q organ pipe as a function of jet velocity (Coltman, 1976a).

place in the jet excitation literature prior to about 1970, and vortices are now reappearing in the detailed aerodynamic theory of sound generation.

Edge tones are produced when a jet impinges more or less directly on a sharp edge that is physically remote from any acoustic resonant structure (Brown, 1937; Powell, 1961; Coltman, 1976a). The edge divides the jet, and the resultant flow is unstable because any fluctuation of the jet to one side of the edge or the other changes the flow pattern in the environment and this feeds back to excite the jet as it leaves the flue, as we have already seen. Phase relations for reinforcement of this instability appear to involve a phase delay on the jet that differs by $\pi/2$ between the edge-tone and organ-pipe cases because of the acoustic effect of the resonator, but the matter is not yet completely clear (Coltman, 1976a). It is thus appropriate to state that the mechanism of edge-tone production is very similar to that for the jet drive of a resonator, but there is not generally an identifiable edge-tone component in the normal mechanism for the sounding of such a resonator.

16.5 Rigorous Fluid-Dynamics Approaches

As mentioned at the beginning of this chapter, the theory developed above is far from rigorous and includes many conceptual approximations. A rigorous theory should be based upon the Navier–Stokes equations of fluid dynamics and must consider the vorticity field of a real fluid in the complex geometry of the instrument. A beginning was made along these lines by Howe (1975, 1981) and has since been followed in some detail by Hirschberg and his collaborators (Verge, Fabre, et al. 1994; Verge, Caussé, et al. 1994; Verge, 1995; Fabre et al., 1996).

These approaches seek to solve the equations of fluid motion both for the acoustic flow in the pipe mouth and, particularly, for the jet in its emergence from the flue slit and its interaction with the pipe lip. None of this behavior is simple, since the jet introduces a very large vorticity component through its shear layers, and this in turn leads to complex flow behavior and the formation of macroscopic vortices. The interaction of these vortices with the acoustic field then either adds energy to, or takes energy away from, the pipe oscillation.

While sophisticated flow visualization techniques make the vortices and their time evolution visible, it is a long step from this qualitative observation to a theory that will actually predict the sounding behavior of the instrument. The theory has progressed through approximations such as replacing the oscillating jet by a pair of acoustic point sources spanning the pipe lip, but there is still uncertainty about the conditions under which the vortex motion is generating or dissipative. This is, indeed, an understandable difficulty, since we have already seen that vortices are responsible for the viscous mixing of the jet with the pipe fluid in our simplified mixing region model, and we have noted that this process dissipates around 90% of the total jet energy. The visualization approach is, however, particularly useful in studying the initial transient behavior of the jet in the pipe.

16.6 Nonlinearity and Harmonic Generation

In the analysis of jet excitation given in Section 16.3, we have already neglected certain nonlinear terms, such as U_{III} in Eq. (16.19), on the basis that their contribution to total system nonlinearity is small. We now consider the major cause of nonlinearity, which arises from the shape of the velocity profile of the jet.

We have already seen, for example in Eq. (16.6), that the velocity profile of a jet tends toward a smooth bell-shaped form that is well described by

$$V(z) = V_0 \operatorname{sech}^2\left(\frac{z}{b}\right), \tag{16.43}$$

where V_0 and b are local parameters that vary along the jet. If we take V_0 and b to be the values appropriate to the x-coordinate of the resonator lip and suppose that this lip is offset by a distance h_0 from the undisturbed center plane of the jet, then the flow into the resonator when the jet tip is displaced by an amount h is

$$U_{\mathrm{j}} = W \int_{-\infty}^{h_0} V_0 \operatorname{sech}^2\left(\frac{z-h}{b}\right) dz = WbV_0 \left\{\tanh\left(\frac{h_0-h}{b}\right) + 1\right\}. \tag{16.44}$$

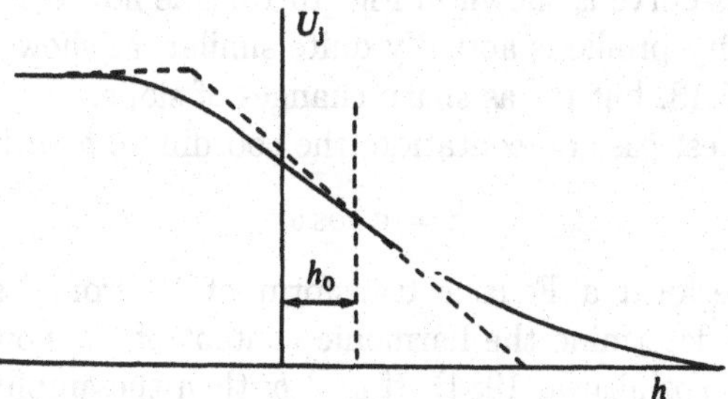

FIGURE 16.13. Flow U_j of a jet with deflection h into the mouth of a pipe. The steady-state offset is h_0. The broken curve shows the flow for a jet with a top-hat velocity profile.

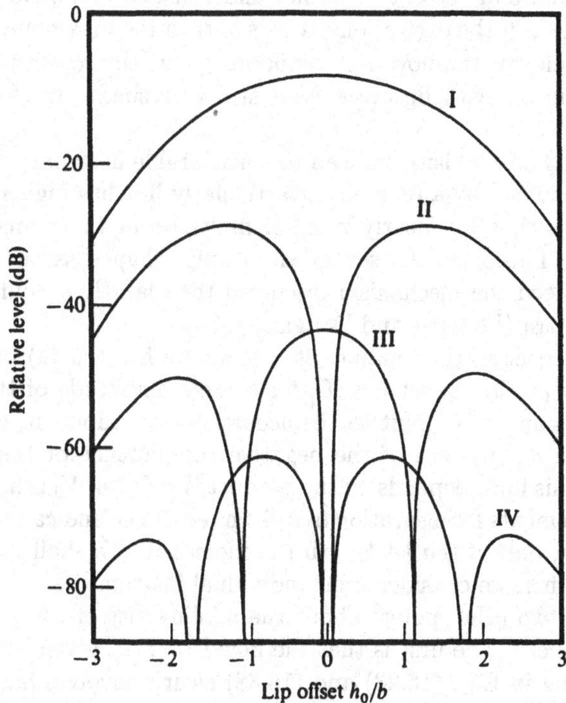

FIGURE 16.14. Relative levels of the first four harmonics of the jet drive of an organ pipe as a function of the ratio of the offset h_0 of the jet relative to the jet half-width b. For clarity, the curve for each harmonic has been displaced downwards by 10 dB relative to the harmonic before it (Fletcher and Douglas, 1980).

The form of this curve is shown in Fig. 16.13. The flow curve for a jet with a top-hat velocity profile is actually quite similar, as shown by the broken curve in Fig. 16.13, but it has sharp changes of slope.

For the simplest case of excitation, the coordinate h varies sinusoidally:

$$h = a \cos \omega t, \tag{16.45}$$

and we can perform a Fourier transform of U_{j} from its explicit form Eq. (16.44), to determine the harmonic content of the flow (Fletcher and Douglas, 1980; Yoshikawa, 1984). If $a < b$, then the amplitude of the nth harmonic varies as $(a/b)^n$ and there is also a marked dependence on the lip offset h_0/b, as shown in Fig. 16.14. If $h_0 = 0$, so that the jet strikes the lip symmetrically, then no even harmonic components are generated. Conversely, for $h_0 \approx 1.1b$, there is almost no third harmonic but the first and second harmonics are strong (Fletcher and Douglas, 1980; Nolle, 1983). If the amplitude a of the excitation exceeds b, then the maximum amplitudes of all harmonics in the flow are comparable, but the relative behavior of various harmonics with lip offset h_0 is still very similar to that shown in Fig. 16.14.

Clearly, this mechanism can lead to considerable harmonic development in the tone emitted by a resonator, particularly if it has higher resonances at frequencies that are nearly integral multiples of the frequency of the fundamental. This occurs for several simple pipe shapes, as we have already seen, and indeed the mechanism discussed then largely accounts for their radiated spectra (Fletcher and Douglas, 1980).

Another aspect of the nonlinearity shown by Eq. (16.44) or Fig. 16.13 is the saturation in the jet flow U_{j} for a large amplitude of the jet wave. There is thus an upper limit to the acoustic power input or, equivalently, a progressive fall to zero of the negative conductance of the jet-excited generator. This limit depends on the product WbV_0, but V_0 is limited by the phase requirements for excitation of a given resonance and can be increased, within limits, only if the jet length l is increased. We shall return to this point later when we consider some individual instruments.

There are two other points about the nonlinearity of the jet drive that require comment. The first is that the flow U_{II} of U_{j}, given by Eq. (16.21) and appearing in Eqs. (16.23) and (16.38) clearly involves harmonic components very similar to those associated with U_{I} since the profiles $V(z)$ and $V(z)^2$ are qualitatively very similar. This term, however, is generally small compared with U_{I} and alters the behavior very little.

Second, and rather more important, the presence of any upper harmonics in the pipe flow will be reflected in propagating waves of appropriate frequency on the jet and consequently in further drive terms. The effect of this interaction is relatively small unless the jet length and velocity are such that it acts as a generator at the harmonic frequency. When this happens, there will be competition between the two modes and the possibility of a transition to the higher mode (Fletcher, 1974).

The relative level of upper harmonics in the sound is greatly influenced by the frequency alignment and Q-values of the upper pipe resonances. Good alignment is favored, in simple cylindrical pipes, by a small ratio of pipe radius to length. Exact harmonicity of the resonances is, however, not always the correct solution for strong harmonic development, since a strongly blown pipe will sound at a frequency slightly above that of its first resonance. In such cases, exemplified by the flute, it is desirable that the first resonance be set slightly flat relative to the fundamental implied by the higher resonances. This is achieved, as we shall see, by a slight tapering of the head of the tube relative to the body. In some pipes or resonators of more complex geometry, particular harmonics or groups of harmonics may be particularly reinforced by resonances of the air column. We will discuss this again in relation to certain types of organ pipes.

16.7 Transients and Mode Transitions

The formal theory of transient behavior is easy to write down in terms of the formalism of Chapter 5, each resonator mode obeying an equation of the form

$$\ddot{x}_n + 2\alpha_n \dot{x}_n + \omega_n^2 x_n = g(\dot{x}_1, \dot{x}_2, \ldots, \dot{x}_n, \ldots), \tag{16.46}$$

where the function g gives essentially the jet forcing term acting on the pipe for an acoustic flow $\dot{x}_n$ in the pipe mouth. The form of g clearly must depend on the blowing pressure p_0 and must involve a different phase delay for each mode x_n. There will, dominantly, be a tendency for harmonic generation by the mechanism discussed in the previous section, but in any case the jet nonlinearity will tend to lock the modes together as discussed in Chapter 5.

The nature of the attack transient in the acoustic output of a jet-driven pipe depends on the form of the buildup of pressure in the mouth cavity below the flue. This may be either gradual, abrupt, or even plosive (with an overshoot on the pressure), depending on the way in which the blowing is accomplished. For mechanical systems, such as organ pipes, this depends on the key-valve mechanism, as we shall discuss in Chapter 17, while for mouth-blown instruments, it depends on the tonguing technique used by the player.

Figure 16.15 shows the calculated behavior of the first few modes of a simple organ pipe under these three types of pressure onset (Fletcher, 1976b). In each case, it takes something like 30 periods of the fundamental before the pipe settles into steady speech, though the abrupt application of pressure with no overshoot clearly gives the briefest transient. For the pipe calculated, with a fundamental of about 150 Hz, the transient duration is 0.2–0.3 s, in good agreement with experience.

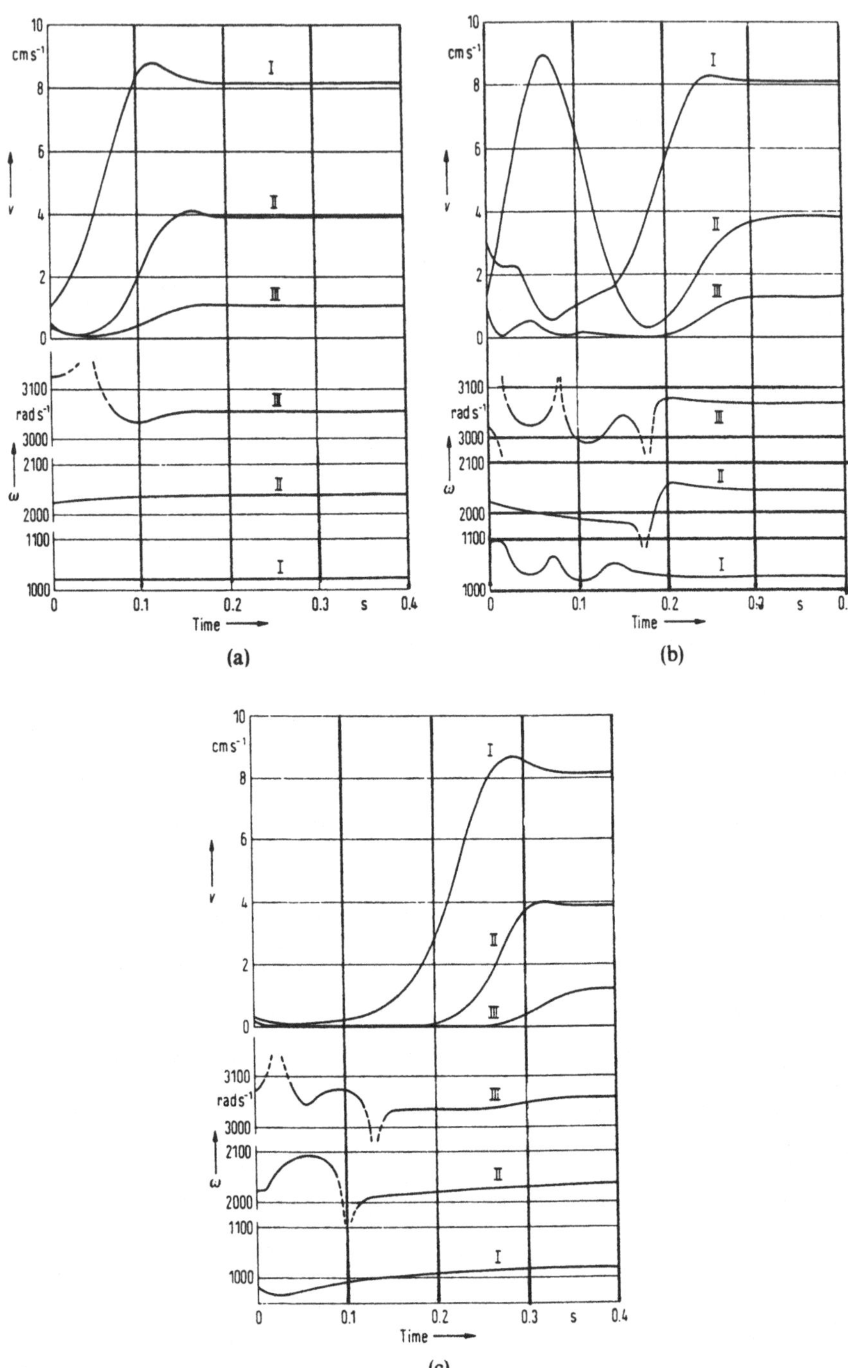

FIGURE 16.15. Development in time of the first three components of a flue organ pipe sound for (a) abrupt, (b) plosive, and (c) slow applications of air pressure (Fletcher 1976b).

If we look at the possibility of exciting the various modes of a simple open pipe as the blowing pressure is raised, we find pressure regimes for each nearly harmonic mode, as discussed in Section 16.4 and shown in Fig. 16.16. The allowable range is in each case a little less than shown in the figure because of losses in the pipe, but it is clear that there is a good deal of overlap, particularly in the higher modes. Each mode sounds most efficiently near the point of maximum negative conductance of the jet—the negative real axis crossing in Fig. 16.10.

The measured behavior of a typical flue pipe as the pressure is gradually raised is shown in Fig. 16.17. The behavior calculated from the theory is very similar (Fletcher, 1976c). It should be noted that there is a certain amount of hysteresis in the behavior—the pipe tends to remain in the mode in which it is sounding as the pressure is changed, rather than making an early transition to another mode. The low-pressure regions shown with broken lines correspond to excitation on the second loop of the spiral of Fig. 16.10 and are not significant in normal playing.

Our earlier discussion shows that it is the phase shift along the jet that determines the mode in which a pipe will sound—a longer jet can be balanced by a greater blowing pressure. It is also clear, however, that changing jet length has other effects. A long jet requires a high blowing pressure, and we can therefore have a large jet flow into the resonator and a consequently large acoustic output. However, a long jet is necessarily broad when it reaches the lip, and it is therefore relatively inefficient in generating high harmonics, so that the tone is broad rather than keen. We will come to consider these adjustments in more detail in connection with organ pipes in Chapter 17, but they apply also to other jet-excited instruments.

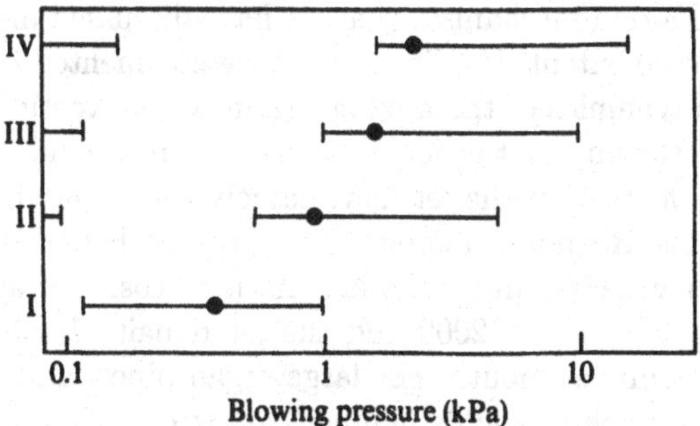

FIGURE 16.16. Pressure ranges over which each of the first four modes of a typical organ pipe or recorder can be made to sound. The points suggest preferred blowing pressures.

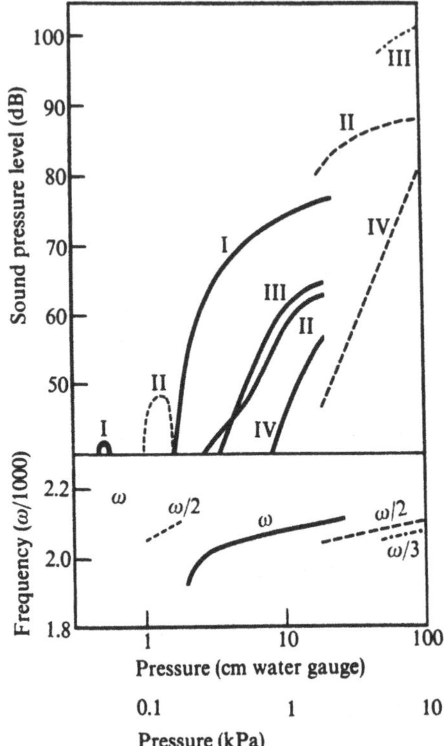

FIGURE 16.17. Measured overblowing behavior of a simple organ pipe (Fletcher, 1976c).

16.8 Aerodynamic Noise

The sound-production mechanism that we have discussed in the preceding sections is, to some extent, idealized. We have commented on this in relation to the aerodynamics of the mixing region, where vorticity is certainly involved in the slowing of the jet, but there is also a more direct effect associated with vortices in the jet flow, namely aerodynamic noise.

In recorders, the Reynolds number ($Re = dv/\nu$, where d is the jet thickness, v is the jet velocity, and ν the kinematic viscosity of air) is typically in the range $1000 < Re < 2000$ and the jet remains laminar across the width of the instrument mouth. For large organ pipes, and for high notes on flutes, however, the Reynolds number can exceed 3000, and turbulence noise makes a significant contribution to the sound. This has been discussed in some detail by Verge and Hirschberg (1995), though the whole subject is so complex that we cannot yet claim to have a full understanding.

An unconfined turbulent jet is a wide-band noise source, as we know from experience with jet aircraft. The noise intensity increases as a high power of the jet velocity, and is influenced greatly by the presence of any obstacles in the flow path. Flutes and organ pipes have such an obstacle, in the form of the pipe lip, and the exact geometry of both jet and lip can have a strong influence on the radiated power. The important thing, from a musical point of view, is that the radiated noise power is filtered by the resonances of the pipe, so that it has a pronounced tonal quality. Players of the panpipes in the South American tradition make particular use of this turbulence noise, which they can do by using a jet length and speed well away from that which matches the phase requirements for regenerative sounding and indeed often favor it over the normal sounding of the pipe. More normally, turbulence noise simply accompanies the normal tonal sound as a sort of undertone, which can be pleasant in moderation but is generally regarded as a defect in performance technique if it is too pronounced.

Figure 16.18 illustrates the spectrum of turbulence noise generated in an organ pipe that is overblown to the second (octave) mode by using a rather high blowing pressure. The sharp and precisely harmonic partials of the normal tone are clearly apparent, as also is the noise background contributed by the turbulence. It is noticeable that the peaks in the turbulence spectrum are only approximately harmonically related, because of the frequency dependence of the mouth correction of the pipe. It is not usual, of course, to overblow organ pipes in this way, but the technique is used for all the upper notes on flutes and recorders. The top-octave fingerings for these instruments produce strongly inharmonic pipe resonances, some of which lie well below the note being sounded, and in no particular relationship to it. Turbulence noise exciting these resonances is generally undesirable.

16.9 Simple Flute-Type Instruments

Acoustically, the simplest type of flute-like instrument is the panpipes, one example of which is shown in Fig. 16.19. There is a single pipe resonator for each note, made from bamboo cut so that its lower end is closed by a septum in the plant stem. The adjustment of the pipe length is then accomplished by pouring wax into the pipes.

The instrument, which generally produces a simple diatonic major scale over a compass of about $1\frac{1}{2}$ to 2 octaves, is sounded by blowing across the top of each pipe with the player's lower lip in contact with one edge of the pipe opening. The pipes are scaled so that their cross sectional area is roughly proportional to length—long pipes are relatively narrower than short pipes—a feature that we discuss again in Chapter 17 in relation to organ pipes.

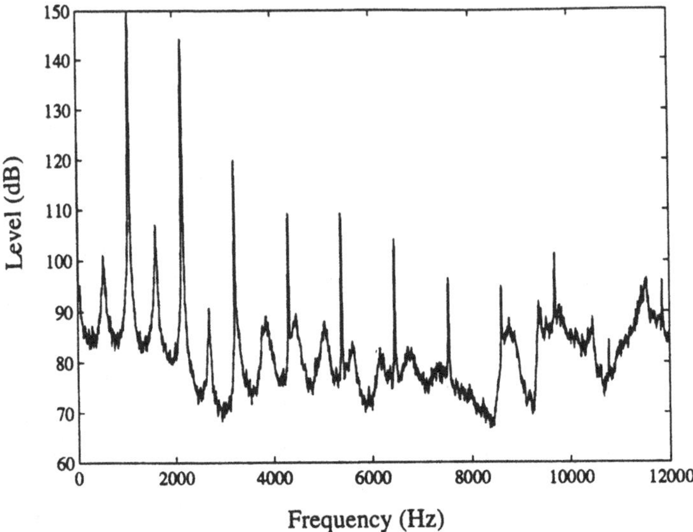

FIGURE 16.18. Power spectrum of tonal and turbulence noise measured inside a small recorder-like organ pipe, overblown to its second mode. Note how the turbulence noise spectrum is shaped by the slightly inharmonic resonances of the pipe (Verge and Hirschberg, 1995).

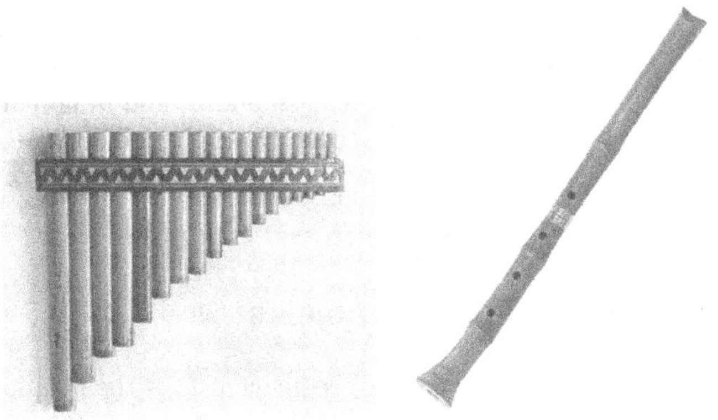

FIGURE 16.19. Panpipes and shakuhachi.

An incidental effect of the scaling of pipe diameter is a consequent scaling of jet length, which aids in the sounding of high notes. Since the jet length is decreased by a factor of $1/\sqrt{2}$ for an octave change in pitch because of this geometric constraint, it is necessary to increase the jet speed by a factor of $\sqrt{2}$ or the blowing pressure by a factor of 2 to maintain the appropriate phase relations. We shall see later that this is a common feature in playing many flute-like instruments.

Since the pipe resonators are stopped at their lower ends, their resonance frequencies follow approximately an odd-harmonic series so the second harmonic is very weak in the radiated sound, giving a characteristic hollow quality. The blowing pressure can be varied and the lip used to partly block the pipe opening, thus increasing the end correction and lowering the pitch of the sound dramatically, giving a very flexible performance repertoire in the hands of an expert player. Percussive breathy sounds can also be produced by deliberately failing to satisfy the normal regeneration conditions and blowing with a wide, turbulent jet. The acoustic spectrum then resembles the noise background in Fig. 16.18, though modified to allow for the stopped nature of the pipe.

The second traditional instrument we should mention is the Japanese shakuhachi (literally a reference to its length: 1 shaku is about a foot or 30 cm, hachi means 8, with the unit sun, about an inch, implied, so a "one-foot-eight," though both longer and shorter instruments are made) also shown in Fig. 16.19. This is made from the root end of a bamboo plant and constitutes a nearly cylindrical, slightly curved, open tube. The mouth end has a bevel cut on one side that is fashioned to a sharp edge, or may even have an ivory or agate wedge inserted. The player blows across the open end but rather more into the instrument than for the panpipes. Since the tube is open, odd and even harmonics are comparably strong in the radiated sound.

The shakuhachi has just five finger holes, two for the fingers of each hand and one for the left thumb, and is naturally adapted for the playing of traditional Japanese pentatonic music. Lip position variations, however, along with the use of forked fingerings and partly closed holes, allow most of the notes of the diatonic scale to be produced. These forced notes add characteristic nuances to the sound of shakuhachi music (Terada, 1906, 1907; Ando and Ohyagi, 1985; Ando, 1986; Yoshikawa and Saneyoshi, 1980).

16.10 The Recorder

The next family of instruments to which we direct our attention is that in which the air jet is defined by an extended slit or windway, the shape of the mouth and lip are defined by the construction of the instrument, and the notes of the scale are produced by opening finger holes. With the exception of the ocarina, an instrument in which the resonator is a globular vessel that acts as a single-mode Helmholtz resonator the frequency of which is raised as holes are opened, the instruments of this family use a cylindrical or a slightly tapering rather than flaring air column. Many whistle flutes of this type have flourished at various times, including the flageolet and the still-common penny whistle, but the only instruments to which we need give serious attention are those of the recorder family, also referred to as the English flute, Blockflöte, flûte-à-bec, or flauto dolce.

Recorders have a very long history, but they are conveniently divided into Renaissance recorders, reflecting the construction details up to the late 17th century, and Baroque recorders, incorporating the new design features introduced by the Hotteterre family and others at about that time. Renaissance recorders have a nearly cylindrical bore and a simple outside form, while instruments of the baroque style have a tapering conical bore, two or more separately jointed sections, and decorative turning near the head and foot. Many instruments, particularly of the baroque type, have survived in playable condition and, as a result of the work of Dolmetsch and others, accurate modern reproductions are now available (Hunt, 1972; Morgan, 1981). Relatively little has been published on the acoustics of the recorder (von Lüpke, 1940; Herman, 1959; Bak, 1969/70; Lyons, 1981; Wogram and Meyer, 1984; Martin, 1994).

Baroque recorders are made in a wide range of sizes, as shown in Fig. 16.20. The normal consort consists of soprano (descant), alto (treble), tenor, and bass, though above the soprano there is also a sopranino, only about 24 cm in total length, and below the bass a great bass about 130 cm long. The alto recorder is the usual solo instrument, with a compass of about two and a half octaves upward from F_4. The sopranino lies an octave above this and the bass an octave below. The alternate instruments lie a fourth below (or a fifth above) these pitches, the range of the descant extending down to C_5. The recorders are not transposing instruments in the sense, for example, that clarinets are. Rather, the alto, sopranino, and bass have a fingering system in which the middle note (all left-hand finger holes closed) is a C, as on the bassoon, while for the descant, tenor, and great bass, this fingering is a G as on the flute or oboe. (Alternatively, the fingerings can be thought of as the same as those of the lower and upper registers of the clarinet.) The finger-hole disposition is the same on all instruments, consisting of three main finger holes for each hand, a thumb hole for the left hand, and a little finger hole for the right hand. On some instruments, the lowest two holes are doubled, while the largest instruments use simple keys to bring the finger holes within reach. A generalized fingering chart is shown in Fig. 16.21 (Hotteterre, 1707; Hunt, 1972). Note that the thumb hole serves both as a normal hole and, when half open, as an octave vent.

Even though a tapered conical tube belongs to the Bessel horn family, which we identified in Section 15.1 as being suitable for finger-hole instruments, a tapered incomplete cone excited at its wide end certainly

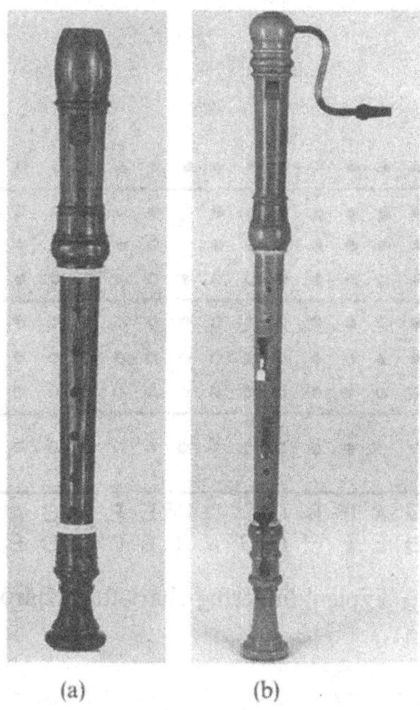

(a) (b)

FIGURE 16.20. Modern recorders based upon some of the best surviving Baroque instruments: (a) alto, and, to a different scale, (b) bass (Moeck).

does not fulfill the conditions expected for a suitable bore profile, so its behavior requires further discussion.

Equation (8.51) gives a general expression for the input impedance of an incomplete cone that we can apply to the present case most simply by neglecting the radiation load impedance Z_L at the open end. With this simplification, the input impedance is, as in Eq. (8.53),

$$Z_{in} = \frac{j\rho c}{S_1} \times \frac{\sin kL \sin k\theta_1}{\sin k(L + \theta_1)}, \tag{16.47}$$

where S_1 is the cross section at the input, L is the bore length (including the end correction at the open end), and θ_1 is related to the distance x_1 from the open end to the truncated apex of the cone by

$$k\theta_1 = \tan^{-1} kx_1. \tag{16.48}$$

An input impedance curve for a typical case in which x_1 is equal to L is shown in Fig. 16.22. Since $k\theta_1 < \pi/2$ by Eq. (16.48), it is clear from Eq. (16.47) that the input impedance zeros, which are the operating resonances for a jet-driven instrument, occur for frequencies

$$\omega_n = \frac{n\pi c}{L}, \tag{16.49}$$

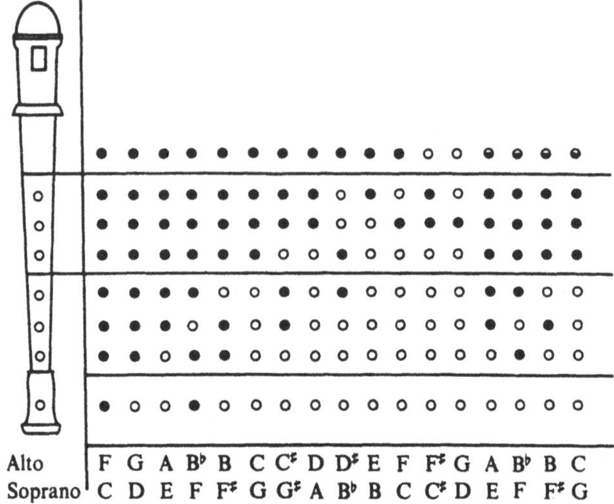

FIGURE 16.21. Typical fingering chart for a Baroque recorder.

which are independent of the taper of the pipe and are the same as those of a cylindrical pipe of the same length. The impedance maxima, however, occur when $\sin k(L + \theta_1)$ vanishes, and this raises the frequencies of the lower maxima relative to those of a cylindrical pipe so that they lie closer to one of the neighboring minima rather than midway between them.

We might speculate that one of the advantages this form of impedance curve confers on a tapered instrument relative to a cylindrical one is an increased pitch stability with blowing pressure, for the asymmetrical shape of the lower resonances tends to reduce the drop of pitch as the blowing pressure is lowered, while a rising pitch at high blowing pressure is self-limiting, as we saw earlier in this chapter. However, this effect is very small.

A more important practical advantage may, however, have been a constructional one. The careful tuning of the best instruments requires adjustments in the diameter of the bore at various places along its length, and this can be achieved for an initially cylindrical bore only by careful and difficult handwork. When the bore is tapered, however, the necessary variations can be built into the tapered reamer used to give the final finish to the bore, with consequent ease of adjustment and reproducibility from one instrument to the next.

The actual taper of the main bore in a typical baroque recorder gives a conical semiangle between 0.5 and 1° and the diameter of the bore at the foot end is typically 0.5 to 0.7 times that at the head (Morgan, 1981). The head joint itself, which makes up about one third of the total instrument length, is usually very nearly cylindrical. This change of taper, which we

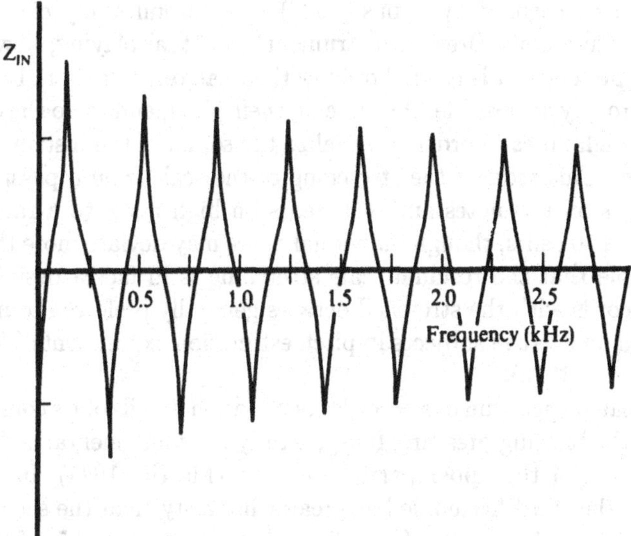

FIGURE 16.22. Input impedance at the large end of an open, tapering, truncated conical pipe, the truncated part being of the same length as the pipe itself.

shall discuss in more detail in relation to transverse flutes, has the effect of flattening slightly the lower resonances of the tube.

As we can see from Eqs. (16.23) and (16.24), the resonance condition for the whole instrument is not actually that its impedance be a minimum at the point of entry of the jet, but rather that the impedance of the air column in series with the impedance of the mouth window be a minimum. This mouth impedance Z_{m} is essentially an inertance $j\omega M$, which can be evaluated from the mouth dimensions. If ΔL is the end correction at the mouth, then

$$\tan k\Delta L = \frac{kMS}{\rho}, \tag{16.50}$$

where S is the bore cross section at the mouth. From Eq. (16.50), since $k\Delta L \ll \pi/2$ in general,

$$\Delta L = \frac{MS}{\rho}, \tag{16.51}$$

but ΔL decreases slightly with increasing frequency. For an alto recorder, ΔL is typically about 40 mm in the low-frequency limit and decreases to about 30 mm at the top of the range of the instrument.

This variation of ΔL, together with the effect of the cylindrical form of the head joint, which we discuss in Section 16.11, tends to stretch the intonation of the whole instrument so that high note resonances are sharp and low note resonances flat. Such a stretching of the passive resonance behavior was measured by Lyons (1981) in a careful study of a modern copy of a high-quality Bressan instrument. In actual playing, however, as we discuss presently, it is normal to blow the relatively weakly radiating low notes as strongly as possible, thus raising their pitch, and to do the opposite for strong high notes in order to equalize the sound of the instrument over its compass. This reduces the stretching of the scale from a passive range of −30 cents on low notes and +30 cents on high notes to a range more like −10 to +10 cents, though individual notes may deviate more than this. It is possible that the residual scale stretching is in fact a desired result since it accords with the stretched octaves generally preferred in music, as determined on average subjects in pitch estimation experiments (Sundberg and Lindqvist, 1973).

The radiated spectrum of a recorder depends on details of its construction and upon the blowing pressure, but generally the fundamental is dominant and the levels of the upper partials are low (Martin, 1994). In some instruments, the third harmonic has greater intensity than the second. This can be understood in terms of our discussion, in Section 16.5, of the offset of the dividing edge or upper lip from the center plane of the jet, though the maker generally uses this variable to ensure prompt attack and an adequately flexible blowing pressure range, rather than for tonal control. There does not seem to be any reason to associate the prominence of the third harmonic with the tapering conical bore, as some writers have done (Herman, 1959).

Performance technique on the recorder involves many subtleties, but the basic variable accessible to the player is blowing pressure. We have already seen that for a fixed windway and lip configuration there is just one blowing pressure that will make the jet present a pure negative conductance, with no reactive component, to the pipe resonator. Since the jet length is fixed and the velocity of the jet varies as the square root of the blowing pressure, this "optimal" pressure varies as the square of the frequency of the note being played and so increases by a factor of 4 when an octave leap is made. However, this is not what a competent player does. Each note has a considerable range of blowing pressure over which it will sound satisfactorily, as we showed in Fig. 16.16, and, in order to equalize sound output, a skilled player uses a high blowing pressure for low notes and a rather low blowing pressure for high notes, as shown in detail in Fig. 16.23. The general practice is to use a blowing pressure roughly proportional to frequency, and thus increasing by a factor 2 for an octave jump, but deviations from this rule are necessary to keep the notes in tune and to allow for the sounding characteristics of the individual instrument (Martin, 1994).

The onset of each note can also be controlled by tonguing to interrupt the air supply so that the attack is plosive, abrupt, or gradual, giving characteristic audible effects. Legato playing, in which the pipe resonance is varied by altering the fingering while making an appropriate small change in blowing pressure, is also possible. It is difficult to play large leaps in a

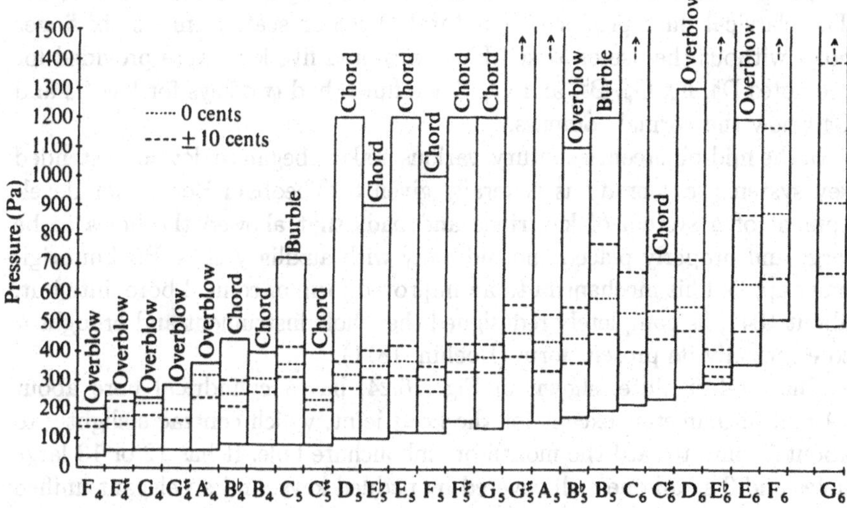

FIGURE 16.23. The range of possible blowing pressures p_0 for each note on a typical alto recorder and the normal pressure used by a competent player (Martin, 1994).

legato fashion because of the hysteresis associated with the overlapping pressure ranges for sounding of different notes.

16.11 The Flute

These days, when we speak of the flute we generally mean the transverse flute, German flute, or Querflöte. It, too, is an instrument of great antiquity and many forms, but consists of a more or less cylindrical pipe, stopped at one end, with a blowing hole near the stopped end and finger holes along its length. Its history is discussed in detail by Rockstro (1890), Baines (1967) and Bate (1975). Before Baroque times, flutes were generally cylindrical, but this form was revised in the late seventeenth century, at the same time that the Renaissance recorder developed to its Baroque counterpart. The Baroque flute, like the recorder, had a slightly tapered conical main bore and was divided into two or more sections, the head being nearly cylindrical. It had six main finger holes, an extra hole for the right-hand little finger, and no thumb hole. A detailed discussion of the scale intonation of such early flutes has been given by Coltman (1993).

Unlike the recorder, the flute continued to develop into the eighteenth and early nineteenth centuries along lines similar to those followed by the oboe, and it progressively acquired keys between and below the six normal finger holes to allow the accurate and even playing of chromatic semitones. The classical flute produced a natural D major scale from its six finger holes without the use of forked fingerings, and five keys were provided for the notes D♯, F♮, G♯, B♭, and C♮. Some flutes had two keys for low C♯ and C♮ below the normal compass.

In the mid-nineteenth century, various makers began to develop extended key systems, but credit is generally given to Theobald Boehm for development of a system of key rings and pads that allowed the holes to be large and properly placed and did away with auxiliary keys. Boehm originally applied his mechanism to an improved form of conical bore flute but, about 1847, he completely redesigned the whole instrument and brought it to essentially its present form (Boehm, 1871).

The modern flute, shown in Fig. 16.24, has a cylindrical bore about 19 mm in diameter, except for the head joint, which contracts slightly to about 17 mm toward the mouth or embouchure hole. It has 12 or 13 large holes and 3 small ones, all covered by padded keys, and an elegant unified system of axles and clutches that allows virtually any combination of holes to be closed without moving the fingers from their standard positions. Details of the mechanism are shown by Boehm (1871) and by Bate (1975). Most modern flutes are made of metal—silver or gold in the more expensive models—but many fine instruments were made to the Boehm design until about 1930 using a wooden body and silver keywork.

The large toneholes of a Boehm flute give it a much brighter tone than that of a Baroque wooden flute with its necessarily rather small finger holes. This change raises the tonehole lattice cutoff frequency of the Boehm flute significantly, so that the upper partials are stronger and the tone brighter. A similar contrast can be heard with the sound of the keyless wooden flute still used in Irish folk music, which has a mellow recorder-like sound.

Boehm also developed an alto flute, a transposing instrument sounding a fourth below the normal flute and now often used in jazz as well as classical music. Its mechanism is identical with that of the standard flute except that the position of the finger keys has been shifted to allow convenient handling. Boehm likened the tone of an alto flute to that of a female contralto voice, and the analogy is apt. The tonal difference between a standard flute and an alto flute comes mainly from the difference in size, which results in a corresponding lowering of the tonehole cutoff frequency. Modern bass flutes, sounding an octave below the normal flute, are also made but are used only rarely. They have a rather soft dull tone. We are not aware of any published acoustical studies on either of these isntruments.

The piccolo is a small flute sounding an octave above the ordinary flute and traditionally lacking the extension keys to C♯ and C♮ found on the foot joint of the standard flute. Piccolos are made with either cylindrical and conical bores, and the body may be of metal, wood, or plastic material. Orchestral players generally prefer conical-bored wooden piccolos, while band musicians use cylindrical metal instruments because of their brighter tone.

The tone of a piccolo is lighter and brighter than that of a standard flute playing the same note, partly because its smaller size raises the tonehole cutoff frequency by nearly a factor 2, and partly because its small size limits the radiation efficiency for low partials of the sound. Coltman (1991) has given a brief and informal discussion of piccolo acoustics.

The modern flute is the most logically designed of all the woodwinds, and, because it is usually made of metal, good manufacturing control and reproducibility can be achieved. The tone hole spacings set out by Boehm on the basis of acoustic theory supplemented by practical trials are still recognized as being close to correct, though minor refinements were made by the nineteenth French maker Louis Lot, who introduced the perforated keys now used on many of the best professional flutes, and recently by Albert Cooper, who slightly revised the tone hole positions to achieve an improved scale. A discussion of general principles has been given by Coltman (1994).

Most important to the tuning and tonal quality of a flute are the bore profile of the head joint, the size of the cavity between the embouchure hole and the stopped end, and the detailed shape of the embouchure hole (Benade and French, 1965; Benade, 1976, Coltman 1985). Of these, the most significant is the cavity above the embouchure hole, since gross mispositioning of the stopper cork defining this cavity can easily mistune the two lower registers by as much as half a semitone, relative to each other.

The physical situation is shown in Fig. 16.25 along with an analog network representation. The embouchure impedance Z_e is that of an inertance M, the magnitude of which depends on the area S_e and acoustic length l_e of the embouchure hole:

$$Z_e = jM\omega = \frac{j\rho l_e \omega}{S_e}, \tag{16.52}$$

where the shielding effect of the player's lips must be taken into account in evaluating l_e and S_e. The impedance Z_c is simply that of an acoustic compliance C, so that

$$Z_c = \frac{1}{j\omega C} = -\frac{j\rho c^2}{\omega V}, \tag{16.53}$$

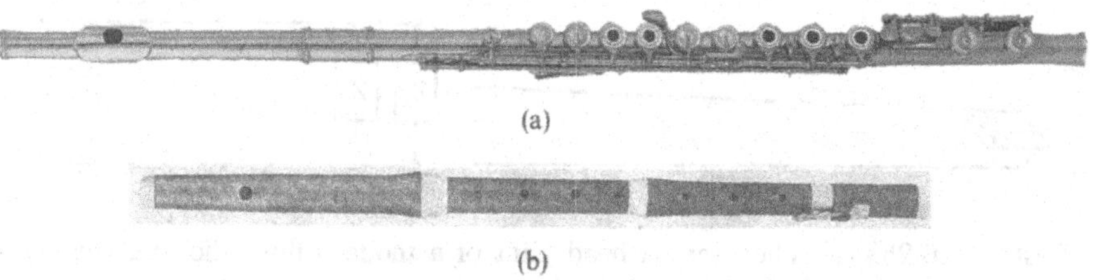

FIGURE 16.24. (a) A modern flute. The design and key mechanism is very close to that originally established by Boehm, and (b) a Baroque flute.

where V is the cavity volume. Now, if Z_0 is the characteristic impedance of the flute tube,

$$Z_0 = \frac{\rho c}{S}, \tag{16.54}$$

where S is the tube cross-sectional area, then the impedance presented by the embouchure hole and cavity in combination is equivalent to an end correction ΔL, where

$$jZ_0 \tan k\Delta L = (Z_e^{-1} + Z_c^{-1})^{-1}, \tag{16.55}$$

or

$$\Delta L = \frac{c}{\omega} \tan^{-1}\left[\frac{M\omega}{Z_0(1 - MC\omega^2)}\right]. \tag{16.56}$$

The behavior of ΔL depends critically upon the cavity volume V, as shown in Fig. 16.26. It can be made to be nearly constant over the range 0–2 kHz if the stopper is set about 17 mm from the center of the embouchure hole in a normal head joint. The end correction ΔL is then about 42 mm with the player's lips in normal position. From the form of Eq. (16.56), however, it is clear that the optimum stopper position depends on the geometry of the embouchure hole and also upon the lip position used by the player.

The effect of the contraction of the bore of the head joint toward the embouchure end can be calculated from the discussion of bore perturbations set out in Section 8.10. If we measure x from the embouchure hole, then the unperturbed pressure modes have the form

$$p_0 = \sin k(x + \Delta L), \tag{16.57}$$

and the bore has the form

$$S(x) = S_0 + \Delta S(x). \tag{16.58}$$

The normalization integral N of Eq. (8.70) has the value $LS_0/2$ if we integrate over the whole tube including ΔL, and the frequency shift $\delta\omega$ is

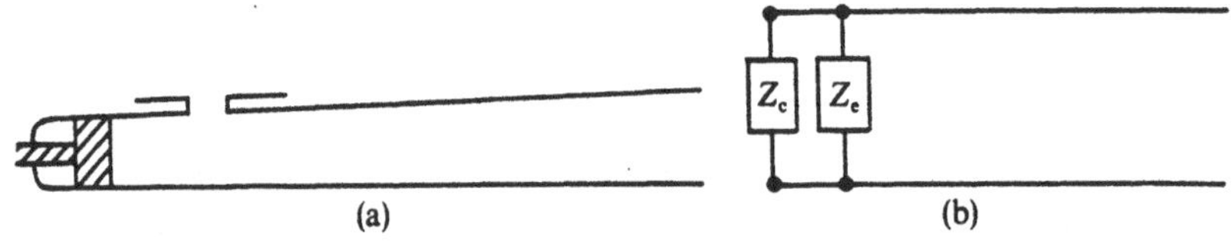

FIGURE 16.25. (a) The tapered head joint of a modern flute showing the embouchure hole and the cavity between the hole and the stopper. (b) An equivalent network showing the cavity impedance Z_c and the embouchure hole impedance Z_c.

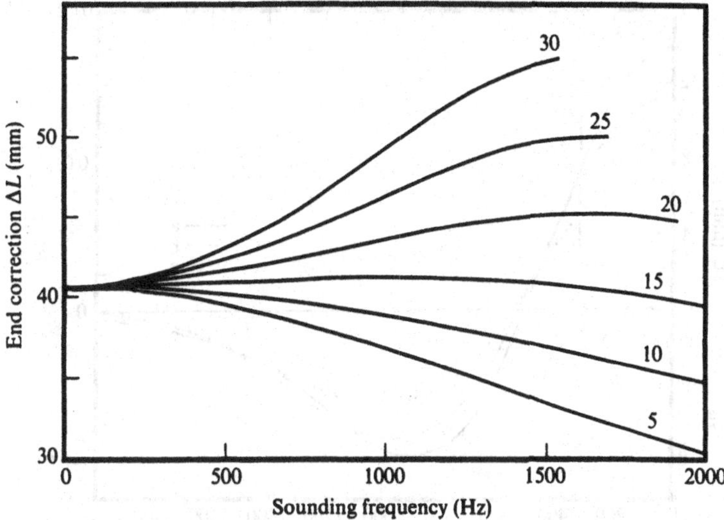

FIGURE 16.26. The dependence upon frequency of the end correction ΔL, measured from the center of the embouchure hole, as a function of the distance in millimeters between the embouchure hole and the stopper, shown as a parameter.

given, after a little algebra, as

$$\frac{\delta\omega}{\omega_0} = -\frac{\Delta L'}{L} = \frac{1}{L}\int_0^L \frac{\Delta S(x)}{S_0}\cos[2k(x+\Delta L)]\,dx, \tag{16.59}$$

where $\Delta L'$ is the additional end correction due to the head-joint taper. Since $\Delta S(x)$ is appreciably negative near $x = 0$ and increases to zero in the length of the head, $\delta\omega/\omega_0$ is negative at all frequencies but has its greatest negative value at low frequencies, for which the cosine term is nearly unity over the whole range of x where ΔS is nonzero.

It is clear that by adjusting the form of $\Delta S(x)$, the bore profile in the head joint, the flute maker can vary the relative positions of the mode frequencies in a systematic way. Boehm stated that the head shape should be parabolic, but measurements of flutes made by him show that his profiles were only approximately of this form. Figure 16.27 shows tuning curves calculated by Benade and French (1965) for several head shapes, and it is clear that the differences, while not large, are certainly significant. Related curves can readily be measured (Fletcher et al., 1982). Contemporary flute makers place great store by the tonal qualities of the particular profiles they use, the alignment of resonances perhaps being more important to tone quality than to tuning, but the reasons for preferring one profile and the corresponding $\Delta L'/L$ curve to another are still obscure.

The final crucial component of the flute head joint is the embouchure hole across which the air jet is directed. Early flutes of wood had a wall thickness

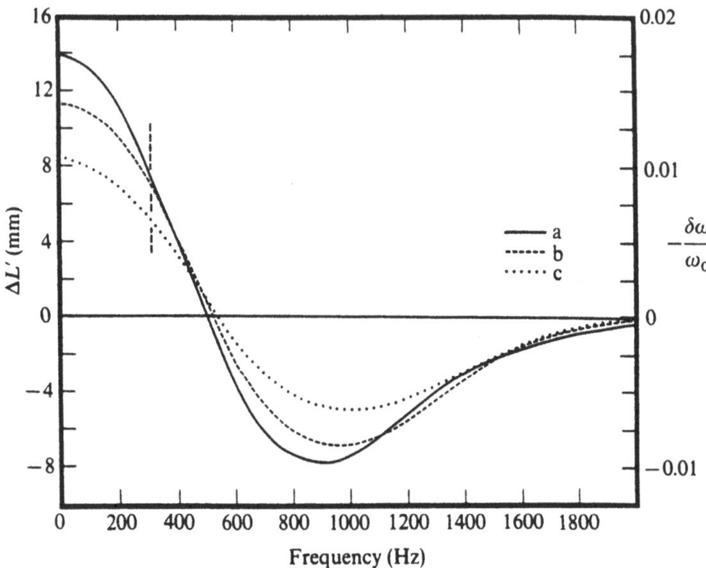

FIGURE 16.27. Frequency corrections caused by head-joint tapering for three different designs of a flute head joint: (a) Quantz; (b) Rudall Carte; (c) Boehm (Benade and French, 1965).

at the head joint of about 5 mm and almost any hole of reasonable size, for example, 7 to 12 mm diameter, could function as an embouchure hole. For an average player, a hole about 10 mm in diameter is best, a smaller one giving a soft, constricted sound and a larger one being loud but less controlled. The reasons for this difference can be understood in part from the change in jet length and, consequently, in the blowing pressure required.

Later flutes often had an elliptical embouchure hole about 10 mm by 12 mm, but Boehm introduced a hole of nearly rectangular shape, as shown in Fig. 16.28, that has since become standard on essentially all flutes, the dimensions being quite closely 10 mm by 12 mm. In a metal flute, the wall thickness is only about 0.3 mm and an embouchure hole simply cut into the wall works very inefficiently, particularly for low notes. The hole is therefore built up using a metal chimney about 5 mm in height, on

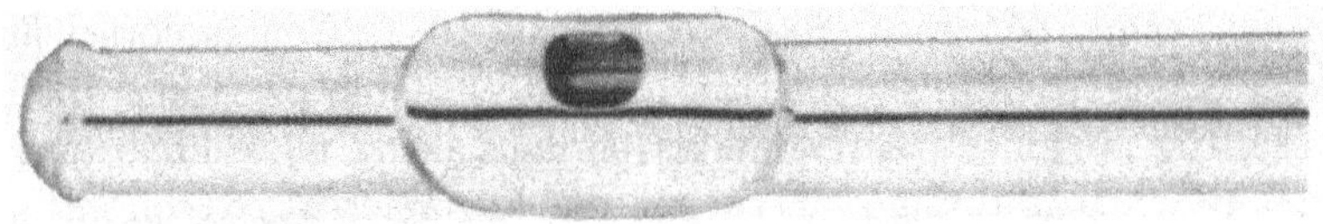

FIGURE 16.28. The embouchure hole of a modern flute. The larger dimension of the hole is 12 mm, and its sides are undercut at an angle of about 7°.

top of which is the embouchure plate against which the player's lower lip rests. This arrangement recreates the effective wall thickness of a wooden flute, while the use of a curved metal embouchure plate allows subtle shape adjustments to give optimal lip support. Boehm recommended that the walls of the embouchure hole should not be straight, but rather undercut at an angle of about 7°, and this is done also by most modern makers. There is, however, scope for much individual hand finishing in this most critical and subtle part of the flute, and slight variations in matters such as embouchure plate curvature, edge sharpness, and undercutting angle undoubtedly affect the finer aspects of the response of an instrument.

Although the modern flute has been designed logically, and its key mechanism allows complete freedom in the size and placement of the tone holes, it is not possible to produce a set of primary resonances that is exactly in tune for all notes of the scale. Many flutes have less than perfect scales, being somewhat flat at the bottom of each register and sharp at the top so that the register-break semitone C♯-D is too small (Coltman, 1966, 1968b, 1976b). Modern rescalings of key positions help to correct this problem, though indeed it is well within the range of adjustment available to a competent player. Detailed calculation of the acoustic effect of open and closed tone holes is in fact a very complex business (Coltman, 1979; Strong et al., 1985).

16.11.1 Materials

Wooden flutes have generally been made from cocus wood or grenadilla, but novelty flutes in the eighteenth century were often made of materials as diverse as ivory, glass, or porcelain. In 1847, Boehm began making flutes from metal and chose as ideal a hard silver alloy, 900/1000 fine in silver, similar to US coin silver. (Sterling silver is 925/1000 fine.) Pure silver would, of course, be far too soft to use. The flute tubing is hard drawn and seamless and about 0.3 mm in wall thickness. The tone hole chimneys are either spun from the tube wall material or else separately cut and soldered on. The head joint is generally made from the same tubing and drawn onto a steel mandrel to give the precise shape desired. The other components of the flute are generally of comparable quality. The keywork is forged from the same silver alloy, and the springs returning the keys are, in the best flutes, made from 12 carat gold.

Of course, not all flutes use materials of this quality, and students' flutes are generally made from nickel silver, a hard copper–nickel–zinc alloy, which is then silver plated. At the other end of the price scale, some solo players have flutes made entirely of gold (usually 14 carat for strength and hardness) or even from platinum tubing with gold keys.

Experiments by Coltman (1971) confirm the view that there is no audible distinction between the sounds produced by simple flutes with tubes made from materials as diverse as copper and cardboard, so that the most likely source of any tonal differences between keyed flutes made from different metals lies with simple differences in the geometry of the head joint and embouchure hole. We return to discuss this whole matter in more detail in Chapter 22.

16.11.2 Directional Characteristics

The radiation characteristics of a flute-like instrument are more complicated than those of reed-driven woodwinds. The reasons for this are that not only do we have a row of phase-related sources at the finger holes and pipe foot as we discussed in Section 15.5, but there is also an intense source associated with the motion of the jet in the pipe mouth.

In the case of a simple organ pipe, or a flute-like instrument with all finger holes closed, Eq. (16.14), with Fig. 16.8, predicts, and experiments (Coltman, 1969; Franz et al., 1969/1970) confirm, that the pipe mouth and open end act as acoustic sources of approximately equal strength. These sources are in phase for odd harmonics and 180° out of phase for even harmonics. We have already discussed the radiation from such a doublet source in Section 7.2, and it is clear that the radiation pattern is complex, with a minimum for even harmonics in a plane normal to the pipe length and through its midpoint. We shall return to this in Section 17.7.

When some of the tone holes are open, then, since the standing wave pattern has appreciable amplitude along the whole length of the instrument tube, particularly for high notes and the upper partials of low notes (Coltman, 1979; Strong et al., 1985), radiation from these holes will contribute significantly.

As we discussed in Section 7.3, a row of open tone hole sources, all with the same phase, as we find in the low register of the instrument, concentrates the radiation pattern to some extent away from the axis of the instrument. For the second harmonic, there is some measure of the same effect with, superimposed upon it, a minimum in the normal plane, as we discussed previously. For higher notes or higher harmonics of low notes, minima develop in other directions as well (Meyer, 1978). Fortunately, in a performance situation, reflections from surrounding objects and the diffuse reverberant field in the listening room tend to average out these directional effects and a reasonably distant listener hears a sound closely related to the total radiated power spectrum.

16.11.3 Performance Technique

Details of the performance technique used by competent flute players have been examined by Coltman (1966, 1968b) and by Fletcher (1975) and are found to accord very well with what we might expect from our knowledge of the physics of sound generation by air jets. The parameters at the disposal of the player in a sustained note are the blowing pressure, the jet length, and the jet cross section, or equivalently the size and shape of the lip opening. The extent to which the player's lower lip covers the embouchure hole is also a parameter, but it is closely related to the jet length. The jet direction in relation to the edge of the embouchure hole can also be adjusted over a small range, but this is a rather subtle variation compared with the others.

Blowing pressure is easy to measure, using a fine tube inserted in the mouth and connected to an appropriate pressure gauge, such as, for example, a simple U-tube containing water. It is found that players universally adjust the blowing pressure p_0 when producing a note of nominal frequency f, so that

$$p_0 \approx 0.8f, \tag{16.60}$$

where p_0 is in pascals (100 Pa = 1 cm water gauge) and f is in hertz. This implies a blowing pressure ranging from 2 cm water gauge for C_4 up to 15 cm for C_7. While some individuals consistently adopt somewhat higher or somewhat lower pressures, these differ by less than a factor 2 from Eq. (16.60), as is shown in Fig. 16.29(a). Players use slightly higher pressures for loud than for soft playing, but the total pressure range for a given note is usually within ±30% of the value given by Eq. (16.60).

In the same way, players are very consistent about the jet length they use, this parameter being easily evaluated from photographs (Fletcher, 1975).

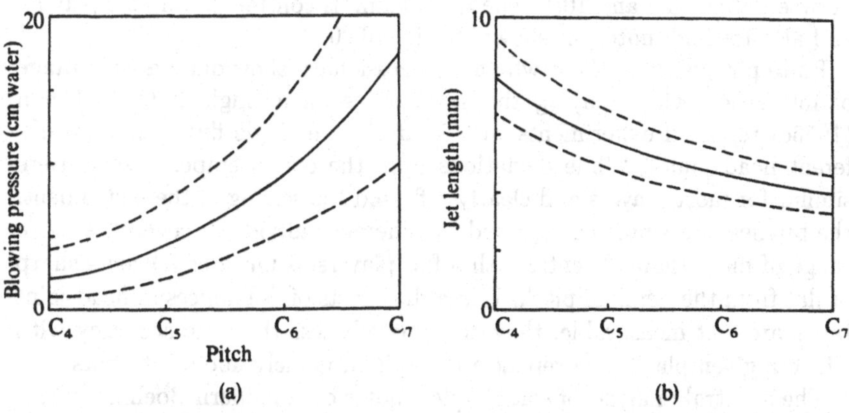

FIGURE 16.29. (a) Blowing pressure p_0 used by experienced flute players for notes of different pitches. All measured pressures lay between the broken lines, and most were within about 30% from a curve characteristic of the player and parallel to that drawn. (b) Jet length l used by experienced flute players for notes of different pitches. All measured values lie between the broken lines (Fletcher, 1975).

As shown in Fig. 16.29(b), the average relation is something like

$$l \approx 0.14 f^{-1/2}, \tag{16.61}$$

where l is the jet length in meters for a nominal note frequency f in hertz. Equations (16.60) and (16.61), together with the fact that the wave velocity on the jet is about half the actual jet velocity, give a wave propagation delay of about a quarter of a period between the player's lip opening and the sharp edge of the embouchure plate. This is surprising at first glance since our discussion would lead us to expect nearly a half-period of delay. Two factors probably contribute. The first is that as the jet emerges from the lips it spreads and slows down, so that the real time delay is longer than that given by this simple calculation. The second is that flute players normally use a blowing pressure that is as high as possible without jumping to the next mode, apparently in the interests of tone quality and pitch stability.

It is interesting to note that a competent flute player can play at will any of the first six modes on a flute with all finger holes closed, simply by adjusting blowing pressure and jet length. It is thus possible to play simple bugle calls on a flute pipe with no finger holes, though only the first few modes are used in normal playing (Sawada and Sakaba, 1980). The frequency selectivity of the air jet generator is quite effective, as shown in Fig. 16.16, although not nearly as sharp as that of a lip-valve brass instrument.

Since blowing pressure is not available as a primary parameter to control loudness, the player must do this by changing the volume flow in the jet through a change in its cross section. Photographs show that a good player uses a lip opening roughly in the form of an ellipse with an axial ratio between about 5:1 and 10:1. The lip opening is contracted for soft playing and also for high notes, as shown in Fig. 16.30.

Flute players, even those who are professionals, show quite a wide range of intonation when playing the same notes on a single flute. Coltman (1976c) reported experiments with 50 flutists, a single flute, and two different head joints. While deviations from the equal-tempered scale were similar for most players and clearly reflected the scaling of the instrument, the pitches of a single note played by different individuals covered a mean range of more than 22 cents, with a few players as much as 30 cents sharp or flat from the nominal pitch. While the effects of differences in head-joint taper are just measurable, they are generally less than the accuracy with which a given player can repeat a note pitch, namely about ± 6 cents.

The spectral analysis of some typical notes on a modern Boehm-pattern flute is shown in Fig. 16.31. It is clear that the harmonic development is quite considerable for low notes, though the higher harmonics are much weaker than for the reed woodwinds. Spectral analysis shows that in soft playing of low notes the level of the fundamental is nearly as high as in loud playing but the harmonic development is much reduced. For higher notes, the spectral envelope is reduced more uniformly. Benade's analysis

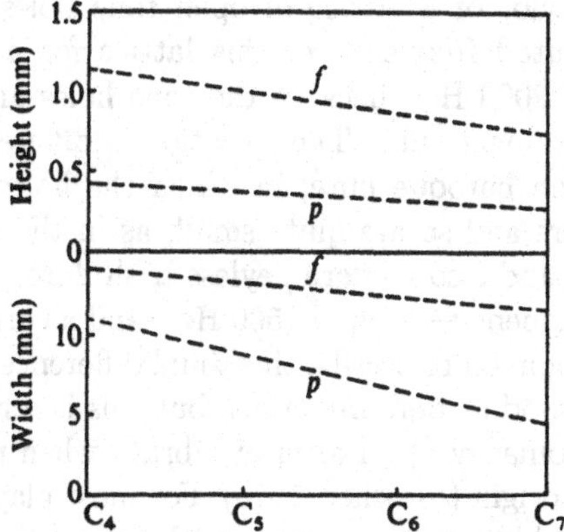

FIGURE 16.30. Lip opening width W and height h used by experienced flute players to produce notes at different pitches and volume levels. All measured values lay between the broken lines (Fletcher, 1975).

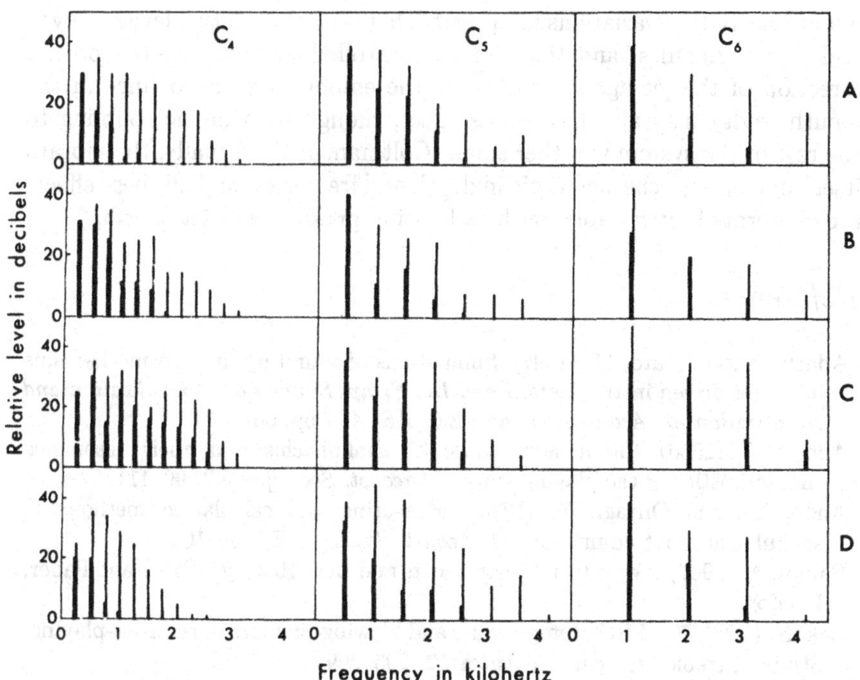

FIGURE 16.31. Typical spectra for loud and soft notes of various pitches played by four different flutists. The same reference level is used in each case (Fletcher, 1975).

of the acoustic behavior of a lattice of open tone holes (Benade, 1960) indicates that the cutoff frequency of this lattice for a modern flute is rather high—around 2000 Hz—because the tone holes are comparable in diameter to the instrument tube. The situation is different for the more-or-less keyless wooden baroque flute, in which the holes must be closed by the player's fingers and so are quite small, as in the oboe or clarinet. Such baroque flutes, and also modern keyless Irish flutes, therefore have a much lower cutoff frequency—around 1500 Hz—and a correspondingly less brilliant tone. It is common to ascribe this tonal difference to the difference between silver and wood as bore materials, but this is erroneous.

Flute players customarily use a form of vibrato when playing, and it is interesting to see its origin (Fletcher, 1975). For most players, this lies in a rhythmic variation of blowing pressure p_0, with an amplitude around $0.1p_0$ and a frequency of 5–6 Hz. Analysis of the sound output shows that the level of the fundamental varies but little, while there are large variations in the levels of most of the upper harmonics. The frequency changes very little, and we may characterize the effect as being "timbre vibrato."

We recognize that some flute players do have very characteristic tonal quantities. Subtle variations are possible between the relative levels of even and odd harmonics, and this can be controlled by changing the precise direction of the jet against the lip of the embouchure. Resonance in the mouth cavity may also have some effect, though its acoustic coupling to the rest of the system is rather small (Coltman, 1973). Finally, the vibrato itself may be very characteristic in depth and frequency, and all these effects are moderated by variables such as blowing pressure and jet length.

References

Adachi, S., and Sato, M. (1997). Simulations of sound production mechanisms in air-jet driven instruments. *Proc. Int. Symp. Simulation, Visualization and Auralization for Acoust. Res. and Ed., ASVA97* pp. 667–674.

Ando, Y. (1986). Input admittance of shakuhachis and their resonance characteristics in the playing state. *J. Acoust. Soc. Jpn.* **E7**, 99–111.

Ando, Y., and Ohyagi, Y. (1985). Measuring and calculating methods of shakuhachi input admittance. *J. Acoust. Soc. Jpn.* **E6**, 89–101.

Baines, A. (1967). "Woodwind Instruments and their History." Faber and Faber, London.

Bak, N. (1969/70). Pitch, temperature and blowing-pressure in recorder-playing. Study of treble recorders. *Acustica* **22**, 295–299.

Bate, P. (1975). "The Flute." Ernest Benn, London.

Benade, A.H. (1960). On the mathematical theory of woodwind finger holes. *J. Acoust. Soc. Am.* **32**, 1591–1608.

Benade, A.H. (1976). "Fundamentals of Musical Acoustics," pp. 489–504. Oxford Univ. Press, London and New York.

Benade, A.H., and French, J.W. (1965). Analysis of the flute head joint. *J. Acoust. Soc. Am.* **37**, 679–691.

Bickley, W.G. (1937). The plane jet. *Phil. Mag.* **28**, 727–731.

Boehm, T. (1871). "The Flute and Flute-playing." Translated by D.C. Miller, Dover, New York, 1964.

Brown, G.B. (1935). On vortex motion in gaseous jets and the origin of their sensitivity to sound. *Proc. Phys. Soc. (London)* **47**, 703–732.

Brown, G.B. (1937). The vortex motion causing edge tones. *Proc. Phys. Soc. (London)* **49**, 493–507.

Chanaud, R.C. (1970). Aerodynamic whistles. *Scientific American* **222** (1), 40–46.

Coltman, J.W. (1966). Resonance and sounding frequencies of the flute. *J. Acoust. Soc. Am.* **40**, 99–107.

Coltman, J.W. (1968a). Sounding mechanism of the flute and organ pipe. *J. Acoust. Soc. Am.* **44**, 983–992.

Coltman, J.W. (1968b). Acoustics of the flute. *Physics Today* **21** (11), 25–32.

Coltman, J.W. (1969). Sound radiation from the mouth of an organ pipe. *J. Acoust. Soc. Am.* **46**, 477.

Coltman, J.W. (1971). Effects of material on flute tone quality. *J. Acoust. Soc. Am.* **49**, 520–523.

Coltman, J.W. (1973). Mouth resonance effects in the flute. *J. Acoust. Soc. Am.* **54**, 417–420.

Coltman, J.W. (1976a). Jet drive mechanisms in edge tones and organ pipes. *J. Acoust. Soc. Am.* **60**, 725–733.

Coltman, J.W. (1976b). Flute scales—Pitch and intonation. *The Instrumentalist* (May 1976)

Coltman, J.W. (1976c). Fifty flutists play one flute—An acoustic experiment. *Woodwind World* **15**(March), 31–33

Coltman, J.W. (1979). Acoustical analysis of the Boehm flute. *J. Acoust. Soc. Am.* **65**, 499–506.

Coltman, J.W. (1981). Momentum transfer in jet excitation of flute-like instruments. *J. Acoust. Soc. Am.* **69**, 1164–1168.

Coltman, J.W. (1985). The role of the head-joint in flute intonation. *Flute Notes* (London) 5–7.

Coltman, J.W. (1991). Some observations on the piccolo. *The Flutist Quarterly* Winter 1991, pp. 17–18.

Coltman, J.W. (1993). The intonation of antique flutes. *The Woodwind Quarterly* August 1993 pp. 76–91. [Reprinted from *The Instrumentalist* Dec 1974, Jan–Mar 1975.] (Refer www.winworld.com.)

Coltman, J.W. (1994). Designing the scale of the Boehm flute. *The Woodwind Quarterly* February 1994 pp. 24–41. (Refer www.winworld.com.)

Cremer, L., and Ising, H. (1967/1968). Die selbsterregten Schwingungen von Orgelpfeifen. *Acustica* **19**, 143–153.

Drazin, P., and Howard, L.N. (1966). Hydrodynamic stability of parallel flow of inviscid fluid. *Adv. Appl. Mech.* **9**, 1–89.

Elder, S.A. (1973). On the mechanism of sound production in organ pipes. *J. Acoust. Soc. Am.* **54**, 1554–1564.

Elder, S.A. (1992). The mechanism of sound production in organ pipes and cavity resonators. *J. Acoust. Soc. Jpn (E)* **13**, 11–23.

Fabre, B, Hirschberg, A., and Wijnands, A.P.J. (1996). Vortex shedding in steady oscillation of a flue organ pipe. *Acustica* **82**, 863–877.

Fletcher, N.H. (1974). Nonlinear interactions in organ flue pipes. *J. Acoust. Soc. Am.* **56**, 645–652.

Fletcher, N.H. (1975). Acoustical correlates of flute performance technique. *J. Acoust. Soc. Am.* **57**, 233–237.

Fletcher, N.H. (1976a). Jet-drive mechanism in organ pipes. *J. Acoust. Soc. Am.* **60**, 481–483.

Fletcher, N.H. (1976b). Transients in the speech of organ flue pipes—A theoretical study. *Acustica* **34**, 224–233.

Fletcher, N.H. (1976c). Sound production by organ flue pipes. *J. Acoust. Soc. Am.* **60**, 926–936.

Fletcher, N.H., and Douglas, L.M. (1980). Harmonic generation in organ pipes, recorders and flutes. *J. Acoust. Soc. Am.* **68**, 767–771.

Fletcher, N.H., Strong, W.J., and Silk, R.K. (1982). Acoustical characterization of flute head joints. *J. Acoust. Soc. Am.* **71**, 1255–1260.

Fletcher, N.H., and Thwaites, S. (1979). Wave propagation on an acoustically perturbed jet. *Acustica* **42**, 323–334.

Fletcher, N.H., and Thwaites, S. (1983). The physics of organ pipes. *Scientific American* **248** (1), 94–103.

Franz, G., Ising, H., and Meinusch, P. (1969/1970). "Schallabstrahlung von Orgelpfeifen." *Acustica* **22**, 226–231.

Herman, R. (1959). Observations on the acoustical characteristics of the English flute. *Am. J. Phys.* **27**, 22–29.

Hotteterre, J.-M. (1707). "Rudiments of the Flute, Recorder and Oboe." Translated by P.M. Douglas, Dover, New York, 1968.

Howe, M.S. (1975). Contributions to the theory of aerodynamic sound, with applications to excess jet noise and the theory of the flute. *J. Fluid Mech.* **71**, 625–673.

Howe, M.S. (1981). The role of displacement thickness fluctuations in hydroacoustics, and the jet-drive mechanism of the flue organ pipe. *Proc. Roy. Soc. London* **A374**, 543–568.

Hunt, E. (1972). "The Recorder and its Music." Barrie and Jenkins, London.

Lyons, D.H. (1981). Resonance frequencies of the recorder (English flute). *J. Acoust. Soc. Am.* **70**, 1239–1247.

Martin, J. (1994). "The Acoustics of the Recorder." Moeck-Verlag, Celle.

Mattingly, G.E., and Criminale, W.O. (1971). Disturbance characteristics in a plane jet. *Phys. Fluids* **14**, 2258–2264.

Meyer, J. (1978). "Acoustics and the Performance of Music," pp. 83–84, 151–153. Verlag das Musikinstrument, Frankfurt am Main.

Morgan, F. (1981). "The Recorder Collection of Frans Bruggen." Zen-On Music Co., Tokyo.

Nolle, A.W. (1983). Flue organ pipes: Adjustments affecting steady waveform. *J. Acoust. Soc. Am.* **73**, 1821–1832.

Powell, A. (1961). On the edgetone. *J. Acoust. Soc. Am.* **33**, 395–409.

Rayleigh, Lord (1894). "The Theory of Sound," Vol. 2, pp. 376–414. Macmillan, New York. Reprinted by Dover, New York, 1945.

Rockstro, R.S. (1890). "A Treatise on the Flute." Reprinted by Musica Rara, London, 1967.

Savic, P. (1941). On acoustically effective vortex motion in gaseous jets. *Phil. Mag.* **32**, 245–252.

Sawada, Y., and Sakaba, S. (1980). On the transition between the sounding modes of a flute. *J. Acoust. Soc. Am.* **67**, 1790–1794.

Strong, W.J., Fletcher, N.H., and Silk, R.K. (1985). Numerical calculation of flute impedances and standing waves. *J. Acoust. Soc. Am.* **77**, 2166–2172.

Sundberg, J. and Lindqvist, J. (1973). Musical octaves and pitch. *J. Acoust. Soc. Am.* **54**, 922–929.

Terada, T. (1906). On syakuhati. *Proc. Tokyo Math-Phys. Soc.* **3**, 83–87.

Terada, T. (1907). Acoustical investigation of the Japanese bamboo pipe, syakuhati. *J. College of Sci., Tokyo* **21** (10), 1–34. Reprinted in "Scientific Papers," Vol. 1 (1904–1909), Nos. 18 (pp. 81–85) and 31 (pp. 211–232). Iwanami Syoten, Tokyo, 1939.

Thwaites, S., and Fletcher, N.H. (1980). Wave propagation on turbulent jets. *Acustica* **45**, 175–179.

Thwaites, S., and Fletcher, N.H. (1982). Wave propagation on turbulent jets: II. Growth. *Acustica* **51**, 44–49.

Verge, M-P. (1995), *Aeroacoustics of Confined Jets, with Applications to the Physical Modeling of Recorder-Like Instruments.* Thesis, Technical University of Eindhoven.

Verge, M-P., Caussé, R., Fabre, B., Hirschberg, A., Wijnands, A.P.J., and van Steenbergen, A. (1994). Jet oscillations and jet drive in recorder-like instruments. *Acta Acustica* **2**, 403–419.

Verge, M-P., Fabre, B., Mahu, W.E.A., Hirschberg, A., van Hassel, R., Wijnands, A.P.J., De Vries, J.J., and Hogendoorn, C.J. (1994). Jet formation and jet velocity fluctuations in a flue organ pipe. *J. Acoust. Soc. Am.* **95**, 1119–1132.

Verge, M-P., and Hirschberg, A. (1995), Turbulence noise in flue instruments. *Proc. Int. Symp. Mus. Acoust.* Dourdan. IRCAM, Paris. pp. 94–99.

von Lüpke, A. (1940). Untersuchungen an Blockflöten. *Akust. Z.* **5**, 39–46.

Wilson, T.A., Beavers, G.S., DeCoster, M.A., Holger, D.K., and Regenfuss, M.D. (1971). Experiments on the fluid mechanics of whistling. *J. Acoust. Soc. Am.* **50** 366–372.

Wogram, K., and Meyer, J. (1984). Zur Intonation bei Blockflöten. *Acustica* **55**, 137–146.

Yoshikawa, S. (1984), Harmonic generation mechanism in organ pipes. *J. Acoust. Soc. Jpn (E)* **5**, 17–29.

Yoshikawa, S., and Saneyoshi, J. (1980). Feedback excitation mechanism in organ pipes. *J. Acoust. Soc. Jpn.* **E1**, 175–191.

17

Pipe Organs

The pipe organ was known to the Romans and Greeks nearly 2000 years ago and began to develop toward its modern form in the late Middle Ages, the organ in Winchester Cathedral in the tenth century already having 400 pipes and a compass of 40 notes. By the seventeenth and early eighteenth centuries the art of organ building in Europe had reached an artistic and technical peak, particularly in Germany, and some of the greatest music of all time, particularly that of Bach, was strongly influenced by the style and structure of organs of this Baroque period.

With the rise in popularity of the orchestra in the Classical and Romantic periods, the organ suffered an artistic decline because of a desire on the part of its builders to imitate orchestral sounds, though many fine organs continued to be built for cathedrals and halls throughout the world. Fortunately, the workmanship of many early organs was excellent, and some large eighteenth century instruments have survived nearly unaltered to the present day.

In the past 50 years, there has been a return to the organ as an instrument in its own right, and organs being built today for churches and concert halls have largely returned to the principles that guided the master builders of the eighteenth century. Even the mechanical technology of the organ mechanism has largely returned to its eighteenth century form—for good reason, as we shall see later—though the use of modern materials and electrical aids is rejected only by purists aiming at the reproduction of historical instruments.

In this chapter, we will be concerned almost exclusively with the acoustical aspects of organ building, and the reader is referred to one of the many fine books on the organ (e.g., Bonavia-Hunt, 1950; Andersen, 1969; Sumner, 1973; Lottermoser, 1983) for historical and technical details. Sumner (1973), in particular, gives the specifications of some 144 representative organs from around the world, built between 1497 and the present day, while Andersen (1969) presents 90 specifications illustrated by 80 plates.

17.1 General Design Principles

The pipe organ is essentially a mechanized wind instrument of the pan-pipe type, in the same way that a harpsichord is a mechanized zither and the piano a mechanized dulcimer. Each pipe is a simple sound generator optimized to produce just one note with a particular loudness and timbre, and the organ mechanism directs air to particular combinations of pipes to produce the desired sound. A set of pipes of uniform tone quality with one pipe for each note over the compass of the organ keyboard is called a rank. A small, portable ("portative") organ may have only a single rank of about 25 pipes, a somewhat larger transportable "positive" organ designed for chamber music may have three to 10 ranks, each of about 50 pipes, while a large modern concert organ may have as many as 10,000 pipes in more than 200 ranks, each of 61 pipes.

The essence of the mechanical arrangement is shown in Fig. 17.1. The pipes are set out logically, and generally to a large extent physically, in a matrix. The rows of the matrix are the individual ranks, while the columns are the notes of the keyboard. Figure 17.2 shows how this sometimes looks in practice. To give a symmetrical appearance, columns 1, 3, 5,... of the matrix (C, D, E...) are set out in order on the left, as viewed from the front of the organ, and columns 2, 4, 6 ... (C♯, D♯, F...) in a mirror-symmetric arrangement on the right. Some modern builders, however, prefer the asymmetric soaring appearance of a rank set out in the order 1, 2, 3, 4.... There is often a further physical disturbance to the matrix layout through the displacement of selected pipes from different ranks to the front of the organ case to provide an architecturally appropriate facade, as shown in Fig. 17.3.

While these physical arrangements may add some complication to the mechanism, they do not affect the logic of its design. In medieval organs, the only controls were keys, one for each note, which could be pressed or slid—they were quite large and medieval organs required several organists—to admit air to all the pipes in that matrix column. These pipes were flue pipes, normally tuned in octaves and fifths above the main rank, and constituted what we would now call a mixture, as we see later. This loud rich sound was the only timbre available to the organist. It was wondrously impressive in a large cathedral with a long reverberation time but was of little use for more delicate music.

The next development was the introduction of stops to turn off the air to particular ranks. This was usually accomplished by means of a wooden lath, known as a slider, with holes in it at positions corresponding to each pipe of the rank. When in register with the pipes, it allowed free access of air to that rank from any key that was pressed. When slightly withdrawn so that solid sections of slider were opposed to each pipe, the air access to that rank was blocked.

At about the time that the slider stop system was introduced, the note action was also stabilized in design. All the pipes sat in holes on top of

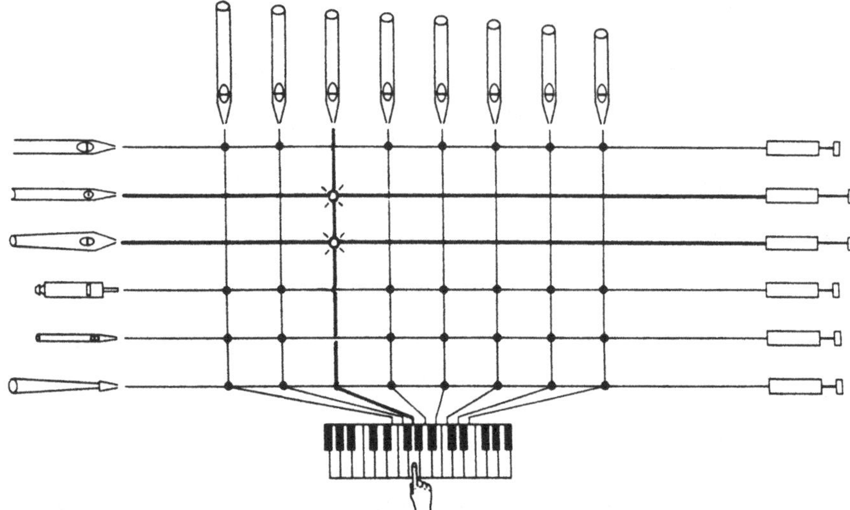

FIGURE 17.1. The pipes of an organ are laid out as a matrix. Each row of the matrix contains the pipes of a single rank; each column of the matrix contains all the pipes for a single note. Drawstops control the admission of air to the pipes of a rank and keys of a keyboard control the admission of air to the pipes of a note. Pipes on all active intersections produce sound (Fletcher and Thwaites, 1983). (Copyright ©Scientific American.)

a windchest, as shown in Fig. 17.4, which represents one column of the matrix. The windchest itself was divided into separate key channels, one for each matrix column or note, and these operated quite independently except for the influence of each stop slider on all pipes of its rank. The whole windchest can be made, indeed must be made, wider than the keyboard to accommodate the larger pipes. The action can easily be spread laterally by means of rollers, essentially axles fitted with levers at each end, and vertically by extended pull-downs or trackers attached to the lever arms of these rollers. The whole mechanism, known as a slider chest with tracker action, was the basis of organ building up to the mid-nineteenth century and has recently come back into favor for organs of all sizes because of the intimate control of the admission of air to the pipes that it affords to the organist.

We have so far described a single organ controlled by a single keyboard. To allow variety and contrast on larger instruments, it is invariable practice to distribute the ranks on different windchests, each controlled by a different keyboard or pedalboard. Except for small chamber organs, the lower limit is 2 manual keyboards of 5 octaves compass and a pedal keyboard of $2\frac{1}{2}$ octaves. The upper limit for a large concert or cathedral organ is 5 manuals and a pedalboard, the compass of each being 5 octaves for the manuals and

FIGURE 17.2. Pipe ranks in the Swell division of the organ of the Sydney Opera House. Each rank is split and arranged symmetrically (Fletcher and Thwaites, 1983). (Copyright ©Scientific American.)

$2\frac{1}{2}$ octaves for the pedals. An example of such a console is shown in Fig. 17.5.

In the German organ-building tradition, all these separate organs, or divisions of the organ, stood open within the organ case and differed from each other in power and timbre. In the British tradition, it became usual in the nineteenth century to enclose one or more of the divisions (and often all but the main great organ) in solid boxes of masonry or timber, fitted with louvres that could be controlled by the organist to produce a swell effect. Modern organs combine the best features of both traditions with several unenclosed and several enclosed divisions.

With the rise of ingenious mechanisms in the nineteenth century, it was natural that these should be applied to the organ. The first were pneumatic levers that enabled the manuals to be coupled to each other and to the pedals without requiring extra playing force on the keys. Pneumatic action was then extended to the primary mechanism as well, so that the only force required on the key was that necessary to open a small valve and allow a pneumatic bellows motor to collapse. With the advent of reliable electrical supplies, key actions using electromagnets were also introduced, either in conjunction with a pneumatic action or as a direct valve actuator for each pipe. Along with these innovations came electropneumatic (or now even microprocessor-controlled) stop actions, with combination buttons to

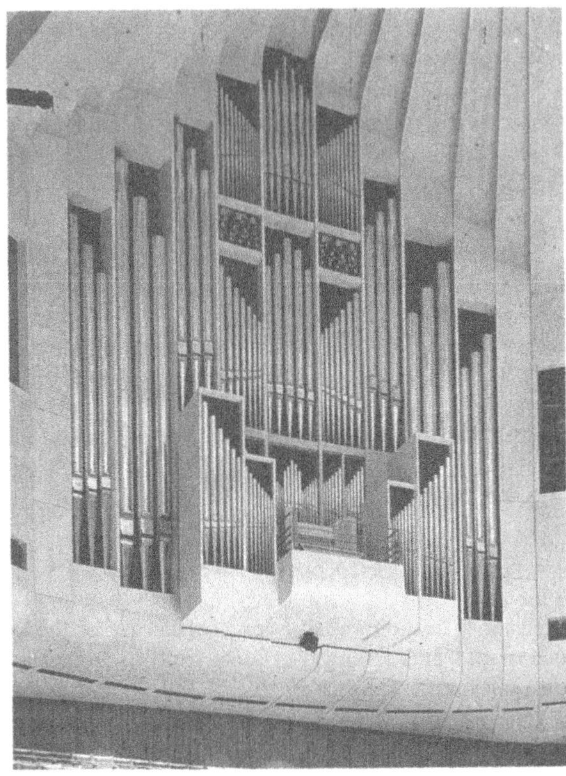

FIGURE 17.3. Display pipes of the organ at the Sydney Opera House. Pipes of appropriate size are borrowed from various ranks of the organ and moved to positions at the front of the case. Display pipes are linked to their normal positions by wind trunks and produce sound in the normal way (Fletcher and Thwaites, 1983). (Copyright ©Scientific American.)

select particular sets of stops, and many other devices. Some of these have become standard on modern organs, even those with mechanical tracker action, and some of them have faded into oblivion.

Other aspects of the organ have also benefited from modern technology. The tenth-century Winchester Cathedral organ was reputed to have required the services of 70 men working a total of 26 bellows. This may have been an exaggeration, but certainly several bellows pumpers were required until comparatively recent times. Hydraulically driven bellows were in vogue in the 19th century and served well, but it is now universal practice to use electric blowers and pressure regulators to provide a steady air supply at the different pressures required for the various windchests of the organ.

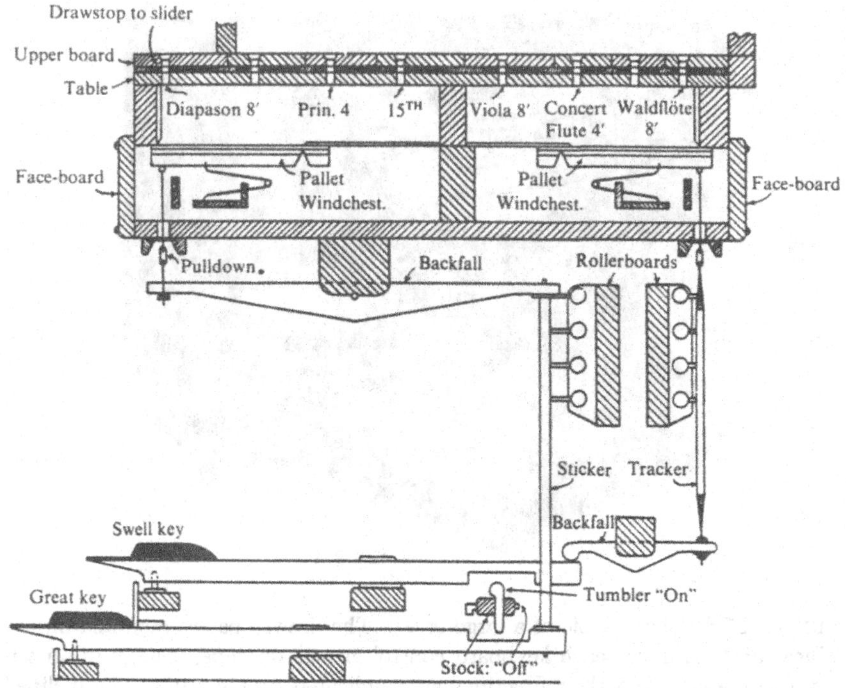

FIGURE 17.4. A typical organ tracker action and slider chest. Two independent manuals with slightly different action are shown (Bonavia-Hunt, 1950).

17.2 Organ Pipe Ranks

The normal compass of a manual keyboard on a modern organ is from C_2 (65 Hz) to C_7 (2093 Hz). Organs of Bach's time lacked only the seven notes above F_6. A convenient terminology has developed for specifying organ ranks because, as we remarked above, they are not all tuned to produce the note nominally being played. For such a nominal or unison rank, the lowest pipe, C_2, has a sounding length of about 2.6 m, which is a little more than 8 feet. Allowance for end corrections makes the physical length very close to 8 ft. Such a unison rank is thus referred to as an 8-ft rank. A rank sounding an octave higher than nominal is a 4-ft rank, a twelfth higher a $2\frac{2}{3}$-ft rank, two octaves a 2-ft rank, and so on, these being ranks that reinforce the second, third, and fourth harmonics of the nominal fundamental, respectively.

The pedal keyboard is written in the score as though it had a compass from C_2 to G_4 but the standard pedal stop is a 16-ft rank sounding an

FIGURE 17.5. The console of a large organ. The drawstops are mounted on the wings, buttons under each keyboard control groups of stops, usually in an adjustable manner, and their functions are duplicated by toe studs. The inclined flat pedals control the opening of the various swell shutters (Aeolian-Skinner).

octave lower than written. Large organs may also have a 32-ft rank, sounding two octaves lower than written and with a lowest note of about 16 Hz. There has even been one full-length 64-ft rank constructed, with mammoth conical wooden pipes of square cross section and a top opening more than a meter across for the lowest pipe, which has a frequency of 8 Hz. This rank is a reed stop, labeled "contra trombone" on the fine organ built in 1886 for the Town Hall in Sydney, Australia, by William Hill of London. At the time it was built, this organ was the largest in the world—it has recently been lovingly restored to its original condition, pneumatic action and all.

We will consider particular stops later, but for the present we should recognize two basically different types of ranks—flues and reeds. Flue pipes, also called labials because it is the upper lip of the mouth that is important in sound production, belong to the flute-instrument family discussed in Chapter 16. Construction is shown in Fig. 17.6. Open flue pipes are historically the basis of the pipe organ and still provide its foundation sound. We can also have stopped flue pipes, which have the economic advantage that a stopped pipe of 16 ft pitch has a physical length of only 8 ft. In addition, there are various partly stopped pipes in which the stopper has a vent or chimney to produce special effects.

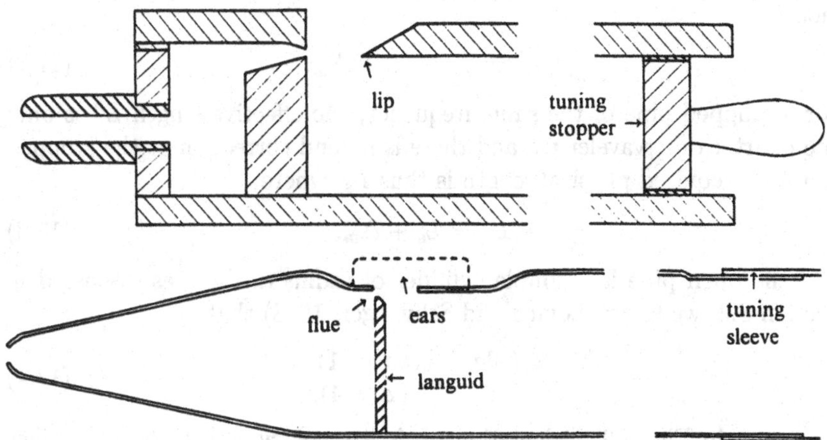

FIGURE 17.6. Sectional views of a stopped wooden flue pipe and an open metal flue pipe. Details of construction vary from one builder to another.

Reed pipes, or linguals, have a metal tongue vibrating against a rather clarinetlike structure called a shallot. There are two major classes of reeds, those with full-length conical resonators supporting all harmonics and those with half-length cylindrical resonators supporting primarily the odd harmonics. In addition, we find short reed pipes with cavity resonators rather like trumpet mutes, but they are quite unusual in modern organs.

17.3 Flue Pipe Ranks

Flue pipes may be open, stopped, or partly stopped and may be made of wood or metal. The wooden pipes are generally nearly square but some are appreciably rectangular, the main effect of this being to alter the ratio of the mouth width to the circumference of the pipe. Metal pipes are generally cylindrical but may taper to narrower open ends. The mouth opening is cut into a flattened part of the pipe wall and can have any desired ratio to the circumference of the pipe.

One of the first things to be determined by the builder is the length of pipe necessary for a note of a particular frequency. The effective length L' for an open pipe of frequency f is half a wavelength, so that

$$L' = \frac{c}{2f}, \tag{17.1}$$

where c is the velocity of sound in air at the room temperature at which the organ is to be played. This effective length consists of the physical length L_0 together with the end corrections Δ_e at the open end and Δ_m at the mouth:

$$L' = L_0 + \Delta_e + \Delta_m. \tag{17.2}$$

For a stopped pipe of the same frequency, the effective length L'' is only one quarter of a wavelength and there is no end correction at the stopped end. The required physical length is thus L_s, where

$$L'' = L_s + \Delta_m. \tag{17.3}$$

If the open pipe is a simple cylinder of radius a, then, as discussed in Section 8.3, we know (Levine and Schwinger, 1948) that

$$\begin{aligned} \Delta_e &\approx 0.6a \qquad (ka \ll 1), \\ &\approx 0 \qquad (ka > 4), \end{aligned} \tag{17.4}$$

as shown in Fig. 8.9. Calculation of Δ_m is not rigorously possible since the mouth is a rectangle cut into the side of the pipe and the acoustic flow geometry is very complex (Dänzer and Kollmann, 1956). Various approximate formulas have been given however (Jones, 1941). Ingerslev and Frobenius (1947) have derived a result based on approximating the rectangular mouth by an ellipse of the same area and related eccentricity that reduces approximately to

$$\Delta_m \approx \frac{2.3a^2}{(lb)^{1/2}} \tag{17.5}$$

for a mouth of width b and height l, with $l \approx b/4$, cut in a cylindrical tube of radius a. Generally, $\Delta_m \gg \Delta_e$.

In practice, neither Δ_e nor Δ_m may be as simple as given by Eqs. (17.4) and (17.5), for the open end may have a tuning slot, tongue, or sleeve and the mouth may have ears or other obstructions. Organ builders have therefore traditionally relied upon rules of thumb or cut-and-try methods, supplemented by centuries of practical experience.

The tone quality of a flue pipe depends on many things, some of them fixed by the dimensions of the pipe body and some influenced by the subtle voicing adjustments made to the pipe mouth when the whole rank is being brought into balance. As we saw in Chapter 16, the flue pipe behaves as an active element in the sound generation process as far as its fundamental is concerned, but largely as a passive resonant filter for the upper harmonics of the sound (Fletcher and Douglas, 1980). It is therefore important to understand the general influence of pipe dimensions on this passive resonance behavior.

Two aspects of this behavior require attention—the frequencies of the resonances relative to the harmonics of the fundamental and the Q-factors or damping coefficients associated with each resonance.

The open-end correction Δ_e decreases smoothly with increasing frequency (Levine and Schwinger, 1948) and is essentially zero for $ka > 3.8$ as was shown in Fig. 8.9. The mouth correction Δ_m also decreases with increasing frequency and at a rather more extreme rate (Meyer, 1961; Wolf, 1965) as characterized by the parameter $k\Delta_{\rm m}^0$, where $\Delta_{\rm m}^0$ is the low-frequency value of $\Delta_{\rm m}$, rather than by ka. Since from Eq. (17.5), using typical mouth dimensions, $\Delta_{\rm m}^0 \approx 3a$, the sharpening of upper resonances can be quite large and is accentuated by a small mouth area. A useful parameter to quantify this mouth-detuning effect is

$$\frac{\Delta_{\rm m}^0}{L} \propto \frac{a^2}{(lb)^{1/2}L} \approx \frac{2a^2}{bL} = g, \tag{17.6}$$

where the last form follows from the fact that typically $l \approx b/4$. For representative types of flue pipes, the parameter g ranges from about 0.05 for narrow-scaled soft pipes producing a soft violinlike sound to as much as 3 or even 6 for broad-scaled, narrow-mouthed pipes giving a very dull flutelike sound with little harmonic content.

Since this discussion suggests that the parameter g should be held constant for all the different sized pipes of a single rank and since the mouth width b is usually made a constant fraction of the pipe circumference, around one quarter for a pipe of normal diapason or principal tone, it seems that the ratio a/L should be constant, making all pipes geometrically similar. Such a scaling rule, however, gives bass pipes that are loud and broad in tone and treble pipes that are thin and weak. The problem of finding a scaling rule that gives tonal coherence and balance across a rank of pipes is of central importance in organ building and one to which the great builders have found satisfactory empirical solutions (Andersen, 1969; Mahrenholz, 1975).

The similarity scaling suggested above would result in a doubling of the pipe radius every octave, "doubling on the 12th pipe" as builders often call it. For a satisfactory scale, however, the bass pipes must be made narrower than this and the treble pipes wider. A scaling with doubling at the fifteenth to eighteenth pipe is generally satisfactory for diapason ranks, but in fact the scalings used historically generally depart from such a rule over at least part of their compass.

Some theorists of the past have advocated scaling rules of the form

$$a(f_1) = b_0 + \left(\frac{f_0}{f_1}\right) a_0 \tag{17.7}$$

for the radius $a(f_1)$ at frequency f_1, relative to the lowest pipe at f_0. More generally applicable, however, are rules of the form

$$a(f_1) = \left(\frac{f_0}{f_1}\right)^x a_0, \tag{17.8}$$

which correspond to the simple doubling on the nth pipe rules (where $n = 12/x$). The similarity scaling has $x = 1$, while a typical modern scaling has $x \approx 0.75$. Numerologically inclined theorists have argued for octave ratios 2^x with particular values such as $5 : 3 = 1.667$, $\sqrt[4]{8} : 1 = 1.682$, or the "Golden Ratio" 1.618 of Renaissance art theory. The scaling with $x = 0.75$ is in fact $^4\sqrt{8} : 1$, and the others will be scarcely distinguishable from it, but there is no basis for the numerological arguments. Instead, we must look for physical and psychophysical reasons behind the scaling (Fletcher, 1977).

Physically, we must consider the quality factors Q of the pipe resonances as functions of frequency, or preferably as functions of mode number n for the pipe concerned. There are two sorts of loss mechanisms contributing to the damping: radiation loss from the mouth and open end and losses to the pipe walls through viscosity and thermal conductivity. There may also be losses caused by turbulence at the sharp edges of the mouth, but we shall neglect these as they are mostly associated with the air jet from the flue.

The energy loss rate from the open end of the pipe is proportional to the square of the volume flow $\pi a^2 v$ multiplied by the frequency squared; the loss to the walls is proportional to the total wall area $2\pi aL$ and to the product of the acoustic flow velocity v, with the gradient of this velocity across the boundary layer, the thickness of which is proportional to $f^{-1/2}$. The total loss rate D can therefore be written as

$$D = (A'a^4 f^2 + B'aLf^{1/2})v^2, \tag{17.9}$$

where A' and B' are constants. The total stored energy E can be similarly written as

$$E = Ca^2 Lv^2, \tag{17.10}$$

so that the quality factor Q at the resonance in question is

$$Q = \frac{2\pi f E}{D} = (A''L^{-1}a^2 f + B''a^{-1}f^{-1/2})^{-1}, \tag{17.11}$$

where A'' and B'' are new constants. However, for the nth resonance of a pipe with fundamental frequency f_1, $f \approx 2ncL^{-1} \approx nf_1$, so that its Q value can be written

$$Q_n = (Aa^2 nf_1^2 + Ba^{-1}n^{-1/2}f_1^{-1/2})^{-1}. \tag{17.12}$$

Now, if we substitute the scaling law, Eq. (17.8), we find that

$$Q_n = (Af_0^{2x} f_1^{2-2x} a_0^2 + Bf_0^{-x} n^{-1/2} f_1^{x-(1/2)} a_0^{-1})^{-1}. \tag{17.13}$$

If x has the particular value $\frac{5}{6} \approx 0.83$, then this expression is homogeneous in f_1, and a factor $f_1^{-1/3}$ can be taken outside the brackets. For such a scaling, which corresponds to an octave scaling ratio 2^x of 1.78, the relative Q values of all resonances remain the same from one pipe to the next, though the absolute values decline as we ascend the scale. This should give tonal similarity across the whole rank, though the basses may be rather loud compared to the trebles.

A reduction of x to the usual value near 0.75 should remedy this unbalance in loudness across the rank at the expense of weakening the fundamentals of the lowest pipes. This tendency is further enhanced for any $x < 1$ by the behavior of the parameter g in Eq. (17.6). The total effect, therefore, is that a tonally balanced rank has greater harmonic development in the bass than in the treble, a feature that we remarked on in Chapter 11 in relation to the harpsichord and one that accords with the region of greatest human auditory discrimination being between about 500 and 3000 Hz.

17.4 Characteristic Flue Pipes

The diapason or principal ranks of a pipe organ produce its major characteristic tone color. They are open pipes, generally made from some form of pipe metal, a tin-rich lead–tin alloy, and have a moderate scaling, around a 4-cm radius at tenor C (C_3, where the pipe length is 4 ft). Most organs have diapason ranks at 8 ft, 4 ft, and 2 ft pitch on the principal manual (the Great Organ in English terminology or Hauptwerk in German) and may supplement these by ranks at $2\frac{2}{3}$ ft and 1 ft, in addition to the mixture ranks we will mention later. The mouth width of a diapason pipe is usually about one quarter of the circumference, and the cut-up distance from the flue to the upper lip is about one quarter of the mouth width.

Principal ranks in older organs and in many modern instruments are blown with quite low wind pressure, 0.5 to 0.8 kPa (5 to 8 cm or 2 to 3 in. water gauge). In some cases, the edge of the languid is lightly nicked to produce homogeneous turbulence in the jet and ensure stable speech. During the low point of classical organ building in the earlier years of this century, blowing pressures were raised to 2 kPa (20 cm water gauge) or more, nicking was very heavy, and the cut-up was necessarily greater, giving a loud but dull sound, which has now gone out of favor.

Principal ranks also occur in the pedal divisions of larger organs where they may be made either of wood or metal. A 16-ft rank represents the foundation tone, but 8-ft and 4-ft ranks can be provided to reinforce this or for use as solo stops, and pedal mixtures are found on large organs.

Flute stops of many tone colors are found in most of the organ divisions where they provide a softer and less incisive sound, either as chorus or solo voices. They are typically provided at 8-, 4-, and 2-ft pitches and can be supplemented by other ranks, as we shall see later, in the form known as mutation stops. Flute pipes may be of wood or metal, and may be open, tapered, or stopped, but are built with generally wider scales, narrower mouths, and softer voicing than diapason ranks. A common stopped flute found on the pedal division of most organs—it is sometimes the only rank on small organs—is the 16-ft bourdon. It is generally an undistinguished rank but, being stopped, it is inexpensive and takes relatively little space. A 16-ft bourdon or other flute is sometimes found as a subharmonic rank on manuals as well. A quiet 32-ft flute is sometimes found in the pedal division of large instruments.

Chimney flutes, in which the stopper is pierced by a narrow pipe, produce interesting and characteristic solo sounds. The chimney presents a rather low inertive impedance at low frequencies, rising to infinity at its quarter-wave resonance. Above this the chimney impedance is capacitive, and falls to zero at its half-wave resonance. The chimney impedance thus upsets the initially harmonic tuning of the pipe resonances, and its length can be adjusted to bring any one of the upper resonances into tune with some harmonic of the pipe fundamental, thus bringing it into prominence in the resulting tone. Generally a pipe mode is adjusted to match the frequency of the fifth harmonic of the fundamental, giving a pleasant and characteristic sound. One such rank is usually called a gedackt on German or English organs. In other solo flutes, various types of resonant caps or other devices can be used to produce interesting new voices.

Another common rank is the harmonic flute, a double-length pipe with a small hole near its center so that it always overblows to its second mode, though with a suggestion of the suboctave mode and its odd harmonics to enrich the tone.

Finally, we should mention the so-called string ranks with names such as violone, gamba, or salicional. These are all soft-toned ranks of very narrow scaling, voiced to give a weak fundamental and a large range of well-developed harmonics. These ranks are not found on baroque organs but have their origin in the earlier years of the orchestral imitative period. However, they do represent a useful and distinctive quiet chorus effect and are present on most modern organs of medium to large size.

The wealth of variety among these ranks is so large that it is impossible to document their structure or sound in detail. Some examples are shown in Fig. 17.7. An extensive glossary is given by Sumner (1973), while some measurements have been presented by Boner (1938), Tanner (1958), Fletcher et al. (1963), and Angster and Miklós (1995).

17.5 Mixtures and Mutations

In the earliest organs, as we have already remarked, ranks of pipes were tuned in octaves and fifths above the fundamental for each note, and all sounded together. This synthesis principle is the basis of the mixture and mutation stops of baroque and modern organs.

In virtually all natural sounds, increasing loudness is associated not simply with a uniform increase in sound pressure level at all frequencies, but rather with a change in the slope of the frequency spectrum to give more weight to components of higher frequencies. This is a natural consequence of the nonlinearities associated with the production of such sounds. Because of masking in the human auditory system and other well-established

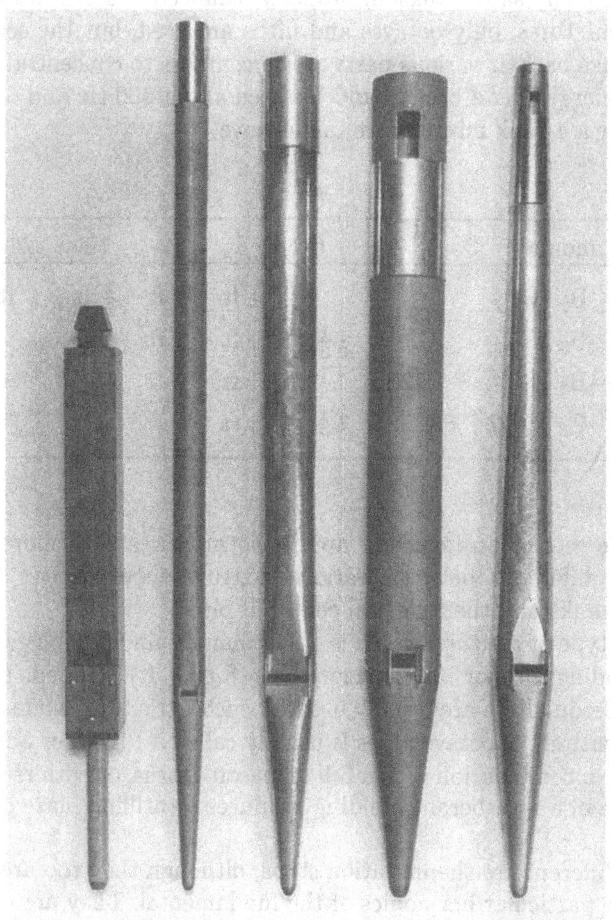

FIGURE 17.7. Varieties of flue pipes. From left to right: a stopped wooden flute; a string-toned dulciana; a principal or diapason (note the tin-rich spotted metal pipe alloy); a dull-toned flute; and a soft waldflöte. All pipes sound the same note C_4; the length differences are due to differing end corrections (Fletcher and Thwaites, 1983). (Copyright ©Scientific American.)

psychophysical phenomena (Stevens and Davis, 1938; Plomp, 1976), the ear also perceives sounds of wide bandwidth as being louder than sounds with only a few strong components. An organ mixture stop achieves this end not by simply adding more and louder unison ranks with reasonable harmonic development, but by adding higher pitched ranks to selectively reinforce the upper harmonics (Pollard, 1978a, 1978b).

Mixtures are generally ranks of diapason-scaled pipes designed to augment the normal 8-ft + 4-ft + 2-ft diapason chorus. They generally have from 3 to as many as 10 ranks all properly balanced and coupled together. In normal mixtures, only octaves and fifths are used, but the selection of pitches breaks back at various parts of the compass to concentrate most of the sound energy over a broad band between about 500 Hz and 6 kHz. For example, for a 4-rank mixture, we might have

Harmonic	3	4	6	8	12	16	24
C_2–B_2				1 ft	$\frac{2}{3}$ ft	$\frac{1}{2}$ ft	$\frac{1}{3}$ ft
C_3–B_3			$1\frac{1}{3}$ ft	1 ft	$\frac{2}{3}$ ft	$\frac{1}{2}$ ft	
C_4–B_4		2 ft	$1\frac{1}{3}$ ft	1 ft	$\frac{2}{3}$ ft		
C_5–C_6	$2\frac{2}{3}$ ft	2 ft	$1\frac{1}{3}$ ft	1 ft			

The same sort of arrangement applies to mixtures with more or fewer ranks; the difference between various mixture types derives from the number of ranks and their general center of pitch.

Another type of mixture, which is of German origin, also exists, in which a rank sounding a 5th or 10th harmonic ($1\frac{3}{5}$ ft or $\frac{4}{5}$ ft) is added. The names Zimbel or Sesquialtera are usually used for such a mixture, while a mixture composed entirely of octave pipes is usually called a Piffero or Schreipfeife. When used in combination with a full diapason chorus, or with reed stops, a full mixture in a reverberant building produces a thrilling blaze of acoustic color.

Rather different are the mutation stops, although they too are designed to reinforce particular harmonics of the fundamental. They are used, however, as solo voices with component ranks of flute character and do not have the breaks characteristic of mixtures. Generally, the third, fourth, and fifth harmonics of the fundamental ($2\frac{2}{3}$ ft, 2 ft, $1\frac{3}{5}$ ft) will be available as separate ranks to be added in any combination to the 8-ft flute rank to produce a reed-like solo voice called a Kornett, much used for playing decorated melodies in baroque music. Larger Kornetts with more ranks also exist. It is interesting that such synthetic stops, anticipating the principles of modern electronic music, go back in history to the Middle Ages!

17.6 Tuning and Temperament

Mixture ranks bring into sharp focus the insoluble problems of tuning and temperament for keyboard instruments, and it is worthwhile to examine these briefly. Unlike a piano string, the overtone components of an organ pipe sound are strictly harmonic and phase locked to the fundamental.

Beats and other interference effects are therefore very obvious. There is no problem in tuning octaves on the keyboard or between octave ranks, because the overtones of all organ pipes are exact harmonics, and the organ has exact octaves throughout its entire compass, unlike the stretched octaves characteristic of the piano.

When the tuner lays out the other notes of the scale, however, he or she meets the classic problem of tuning and temperament (see Chap. 9 in Rossing, 1990). For perfect concord and no beats, we require simple integer frequency ratios for all important musical intervals: 2:1 for the octave, 3:2 for the perfect fifth, 5:4 for the major third, etc. A cycle of fifths, F $\rightarrow$ C $\rightarrow$ G $\rightarrow$ D $\rightarrow$ A $\rightarrow$ E $\rightarrow$ B $\rightarrow$ F♯ $\rightarrow$ C♯ $\rightarrow$ G♯ $\rightarrow$ D♯ $\rightarrow$ A♯ $\rightarrow$ E♯, should ideally bring us back to an E♯, that is the enharmonic equivalent of the F from which we started. However, the prime number theorem shows that the frequency ratio $(\frac{3}{2})^{12}$ reached by this progression cannot be an exact number of octaves 2^N, with $N = 7$ (since it is not possible for 3^{12} to precisely equal 2^{19}). The error, which is called the Pythagorean comma, is about 1.3% or nearly a quarter of a semitone. The system of equal temperament used universally today distributes this error equally over all 12 of the fifths by tuning each one flat (tempering it) by about 0.1%. This gives a slow beat that is detectable but not really objectionable since the audible beat rate (between the third harmonic of the lower note and the second harmonic of the upper) is about one per second in the midrange of the keyboard.

The situation with major thirds is, however, very much worse. Three equal-tempered major thirds (C $\rightarrow$ E $\rightarrow$ G♯ $\rightarrow$ C) make up an octave, so the frequency ratio of each is $2^{1/3} = 1.260\ldots$, which is nearly 1% greater than the just major third, $\frac{5}{4} = 1.25$. There is thus a rather fast beat between the upper components of two notes a major third apart, giving a sense of roughness to the interval. The situation is much worse when a mixture such as the Sesquialtera, which contains a fifth-harmonic rank tuned a just major third above one of the octaves of the fundamental, is added to the sound, for there is then a direct beat between this rank and one of the octave ranks from the higher note. The components of one note of the mixture cannot be tuned to tempered intonation for they would then beat together, producing just as bad an effect.

Organists in the Baroque and Classical periods generally minimized this problem by tuning their organs to some form of meantone tuning (Barbour, 1953; Backus, 1977; Rossing, 1982). In this system, the important fifths are tuned slightly flatter than in equal temperament so that the important major and minor thirds are exact, with frequency ratios of $\frac{5}{4}$ and $\frac{6}{5}$. This system concentrates most of the mistunings into rarely used intervals and, of course, cannot close the cycle of fifths. Some organs even had a split A♭/G♯ key to accommodate the worst of this problem. Though meantone tuning gave sweet-sounding mixtures and major triads, its limitations eventually made it obsolete, and modern ears have come to tolerate the clash of sounds inevitable in equally tempered full chords. A few new organs are, however, now being tuned to the best of these older systems.

17.7 Sound Radiation from Flue Pipes

The radiation of sound from stopped pipes is a relatively simple matter, since there is only a single radiating source at the pipe mouth. Because the mouth is very small compared with the pipe length, and therefore with the wavelength of the fundamental, the radiation pattern of the lower harmonics of the sound is nearly isotropic. For higher harmonics, when the wavelength becomes less than a few times the width of the mouth, the radiation pattern becomes concentrated to the front of the pipe, with a large angular spread in the vertical plane and a smaller spread in the horizontal plane.

Much more significant is the radiation pattern for an open pipe, for then we have two coherent sources, at the mouth and open end, respectively, which are acoustically in phase for odd harmonics and out of phase for even harmonics. It is not immediately clear that the strength of these two sources is the same, for the area of the mouth is typically less than one fifth of the area of the open end, and the flow of the jet must also be considered. The discussion of Section 16.3, however, showed us that the acoustic flow in the pipe, and thus at the open end, is exactly equal to the sum of the jet flow and the mouth flow, and, since the power radiated depends only on the total acoustic flow and not on the aperture size, we expect the strengths of the two sources to be exactly equal. This point has been confirmed experimentally by Coltman (1969).

The consequences of this balance are quite significant. If we consider the nth harmonic of the pipe, then the phase relation between the two sources is $(-1)^{n+1}$. If L is the physical length of the pipe, slightly less than an integral number of half-wavelengths, then the radiation pattern is symmetrical about a plane through the midpoint of the pipe and perpendicular to its length, and the intensity radiated in a direction making an angle θ to this plane is proportional to

$$\begin{aligned} I(\theta) &= [\cos(kL\sin\theta) + (-1)^{n+1}]^2 + \sin^2(kL\sin\theta) \\ &= 2[1 - (-1)^n \cos(kL\sin\theta)], \end{aligned} \tag{17.14}$$

where $k = 2\pi n f_1/c$ and f_1 is the sounding frequency of the pipe. Calculated polar diagrams for several harmonics of a typical pipe are shown in Fig. 17.8, and these are confirmed by measurement, the near cancellation at certain angles being very pronounced for the first few harmonics in an anechoic environment.

Fortunately, most organs are heard in relatively live environments and at a considerable distance from the instrument, so that it is the total radiated

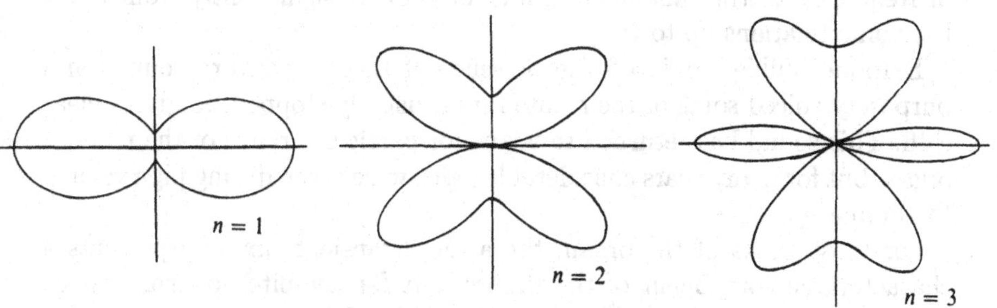

FIGURE 17.8. Typical radiation patterns for three harmonics of an open flue pipe, standing vertically. The total end correction is assumed to be 20% of the pipe length.

power that is important rather than the intensity radiated in a given direction. Indeed, standing waves in the hall are likely to be of more audible importance than these directional effects. The phenomenon is, however, of importance if we are recording an organ using microphones close to the pipes or, even more, if we are examining the tonal spectra of organ pipes on the basis of such measurements.

17.8 Transients in Flue Pipes

In Chapter 16, we discussed briefly the starting transients of jet-blown instruments, organ pipes being those best understood because of the reliably reproducible nature of the air supply mechanism. Experimental studies have been reported by Trendelenburg et al. (1936, 1938), Nolle and Boner (1941), Caddy and Pollard (1957), Franz et al. (1969-1970), Keeler (1972), Pollard and Jansson (1982a, 1982b), and Nolle (1983). Fletcher (1976) has provided a theoretical basis.

All these studies show that the initial transient stage lasts for 20 to 40 periods of the fundamental of the pipe, during which time the components of the sound build to their steady levels, but the behavior of these components during the transient depends on the voicing of the pipe and the nature of the blowing pressure transient in the pipe foot. For a pipe voiced and blown near the middle of its stable regime of normal speech and for a relatively slow rise in blowing pressure, the fundamental and overtones rise in level together and rapidly achieve locking into harmonic relationship. If the pipe is voiced so that it is close to overblowing to its next mode, however, and particularly if the blowing pressure is applied rather abruptly, there may be an initial burst of sound at the second mode frequency—either the octave or twelfth depending on whether the pipe is open or stopped—before the steady sound develops. This premonitory sound is not locked in frequency to the fundamental and may differ significantly from exact harmonic relationship to it.

Baroque builders enjoyed the presence of these transitory sounds and purposely voiced some of their flute ranks, usually stopped, to give a clear chiff. This sound has returned to vogue for particular ranks of the modern organ, but for many years considerable pains were taken during pipe voicing to eliminate it.

For most ranks of the organ, the attack transient simply represents a characteristic component of the timbre, but for low-pitched pedal ranks, the delay time, approaching a second, before the pipe settles to steady speech can present musical difficulties. There is little that can be done about this for a single pipe, for the jet must feed acoustic energy to the air column and this takes time. However, when the fundamental pipe has its sound augmented by octaves and mixtures at much higher pitches, these pipes speak relatively quickly with their own typical initial transients, and the attack is prompt. The ability of the human auditory system to imply a fundamental when an adequate range of its upper harmonics is present gives tonal coherence and a consistent low pitch to the perceived sound.

17.9 Flue Pipe Voicing

One of the great advantages of lead-tin alloys for pipe making is their relative softness, which means that pipes can be readily finished and adjusted by hand using simply a small, sharp knife. Parameters available to the voicer for adjustment, though some of them only on a one-time basis, are

- the opening in the foot of the pipe, effectively controlling the blowing pressure at the flue slit;
- the width of the flue slit, and whether or not there is nicking at the edge of the languid;
- the cut-up and shape of the upper lip, effectively controlling the jet length;
- the height of the languid, effectively controlling the direction of the jet relative to the upper lip; and
- the upper cap, tongue, or sleeve, effectively controlling the pitch for tuning.

In all these adjustments, the voicer must have regard to the steady sound of the pipe, to its promptness of speech, and to the presence or absence of any desired starting transient or chiff. The mouth of a diapason pipe after voicing, in this case with particularly heavy nicking, is shown in Fig. 17.9.

Obviously, these are all matters at the very heart of the organ builder's art, but the physical effect of many of the adjustments can be easily understood in terms of our discussion in Chapter 16 (Bonavia-Hunt, 1950; Mercer, 1951, 1954; Fletcher, 1974; Nolle, 1979, 1983).

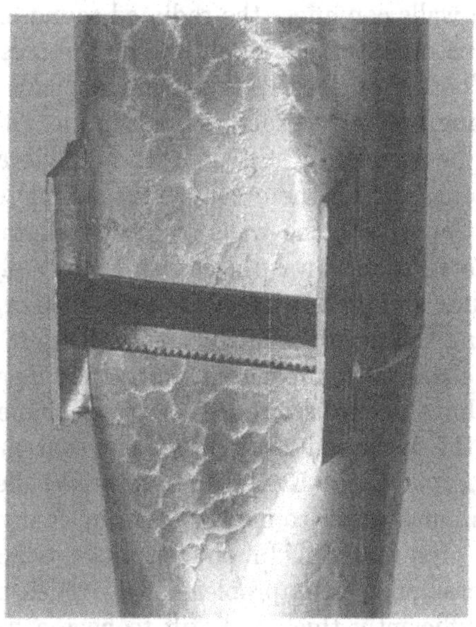

FIGURE 17.9. The mouth of a metal open diapason pipe after voicing with unusually heavy nicking of the languid. Note the ears to either side of the mouth and the characteristic surface pattern of the tin-rich spotted metal pipe alloy (Fletcher and Thwaites, 1983). (Copyright ©Scientific American.)

17.10 Effect of Pipe Material

There is a long organ-building tradition that holds that the best material for bright-toned organ pipes is a tin-rich lead–tin alloy and that too much lead gives a dull-toned pipe, as may be required for a flute rank. Such an alloy is, however, quite expensive and not very strong, so that zinc has often been used for the larger pipes. Some pipe ranks are, of course, made from wood, but there does not appear to have been much discussion about the merits of various different timbers, provided only that they are reasonably thick and smooth.

Scientific discussion has persisted on the subject for more than a century with many ingenious explanations being advanced to explain the superiority of one material over another. Miller (1909), in particular, devised a demonstration with a double-walled metal pipe and showed that the tone quality varied greatly as the space between the walls was progressively filled with water.

Most of the uncertainty has now effectively been laid to rest by the work of Boner and Newman (1940), which showed little effect even for pipes made of paper, and by the more thorough analysis of Backus and Hundley (1965), which confirmed this result and provided at the same time an explanation of the effects observed by Miller and others.

Briefly, the pipe walls can affect the radiated sound only if they are set into vibration of reasonably large amplitude by the pressure variations in the air column. Whether or not this is possible depends essentially on the geometry of the pipe walls. An exactly circular pipe with walls of reasonable thickness will be sufficiently stiff against radial vibrations that almost no motion can occur, and so there can be no audible effect of wall material properties. Conversely, the walls of a pipe of square cross section made from thin metal, such as was the case in Miller's pipe, can vibrate to quite large amplitude and greatly affect the sound. No organ builder would contemplate making square pipes of thin metal for just this reason, since the effects are unreliable and unpleasant. In the middle ground, nominally round pipes that have been distorted to elliptical shape can vibrate to a small extent, but the measured levels are so low that there is a negligible audible result for normal pipes. Even if these low-level effects were regarded as significant, it is much easier to modify them by changing wall thickness than by changing pipe material. In wooden pipes, which have the possibly susceptible rectangular form, problems are avoided by using timber or plywood that is sufficiently thick and stiff to have a negligible vibration amplitude. No organ builder has been persuaded to use thin plywood for bass pipes!

Perhaps more important than any of the virtually negligible acoustic differences are the properties of the pipe material in relation to pipe construction, voicing ease, and appearance. The tin–lead alloys, common as pewter in drinking vessels back to Roman times, are ideal from this point of view, being hard enough to stand unsupported as pipes of reasonable size and soft enough to be cut with ease by the voicer's knife. Of these alloys, those with a high tin content can be burnished to a bright finish, which is durable in time, while those rich in lead are dull and tend to grow oxide films. Typical compositions range from 30 to 90% tin, with the higher tin fractions generally being used for diapasons and the lower fractions for flutes.

These virtues still commend tin-rich tin–lead alloys to modern organ builders, but improved manufacturing methods allow the use of zinc for large pipes, copper for particular display pipes, and even spun brass for show pipes of the trumpet family mounted in heraldic fashion on the front of the organ case. The art of the pipe voicer is completely dominant in determining the sound quality.

Having said this, however, there is generally a clear audible distinction between the tone of metal pipes and of wooden pipes, even when both are unstopped. The reason is, however, geometrical. Figure 17.6 shows typical cross-sections for a wooden and a metal pipe, and we can forget about the tuning stopper in the former case if we are dealing with an open pipe. The geometrical differences are forced upon the maker by the nature of the pipe materials—easily rolled thin metal sheet and rigid flat planks of wood. The difference in cross-sectional shape makes only a minor difference to acoustic behavior, but the thick walls of a wooden pipe impose restrictions upon the geometry of the mouth, particularly the flue, languid, and lip, that can have a major influence on air jet behavior. Of particular significance is the difference between the wedge-shaped upper lip of a wooden pipe and the knife-blade form of the lip of a metal pipe.

17.11 Reed Pipe Ranks

Although it is possible to build a very satisfactory small organ using flue pipes only, reed-driven pipes, which were introduced into organ building about the fourteenth or fifteenth century, provide an exciting variant in timbre and are an essential part of any large modern organ.

The reed-driven acoustic generator of the pipe has the structure shown in Fig. 17.10. The reed is a curved tongue of brass closing against a matching cavity called a shallot. The vibrating length of the reed is determined by a stiff wire pressing it against the shallot. The whole arrangement is superficially similar to a clarinet mouthpiece, except that the face of the shallot is quite flat and the reed curved, while in a clarinet the reed is flat and the mouthpiece has a curved lay. From an acoustic point of view, however, this resemblance is only superficial. In a clarinet, the reed resonance is at a frequency much higher than the note to be played, the reed is highly damped, and the playing frequency is controlled mostly by the acoustic impedance maxima of the pipe. In an organ reed, on the other hand, the reed is very lightly damped and is tuned to the frequency of the note to be sounded. The pipe is also tuned to this note, but its function is largely that of a passive acoustic resonator determining the loudness and tone quality of the sound produced by the reed. Both these possibilities are treated quite explicitly in our discussion of reed generators in Sections 13.2 and 13.3; the difference between the two situations is only a matter of differences in the values of various parameters describing the reed.

We do not need to repeat our discussion of reed generators here, but only to emphasize some particular points. The reed itself, since it operates at very nearly its resonance frequency, moves in a nearly sinusoidal manner, though there is some distortion because of the way in which the curve of the reed unrolls against the face of the shallot. This curve, the shape of which is crucial to the operation of the pipe, is imparted to the tongue by burnishing it against a flat plate during the voicing operation. A well-voiced reed does not usually quite close the opening in the shallot at the extreme of its motion.

In very large bass pipes, the reed tongue may be weighted to lower its frequency in a convenient way. Some weighted tongues, such as the 64-ft reed on the Sydney Town Hall organ with its fundamental of 8 Hz, have a small pneumatic actuator to set the reed into initial motion and ensure prompt speech, but this is not normally necessary.

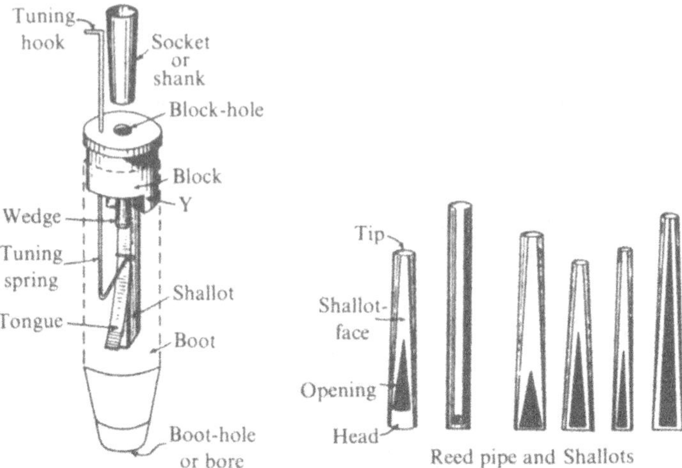

FIGURE 17.10. The reed assembly in a typical oboe or trumpet stop, with shallots for ranks producing other different tone qualities (Bonavia-Hunt, 1950).

The flow of air through the reed into the pipe is controlled by the motion of the reed in conjunction with the shape of the opening in the shallot. This opening is basically triangular but, as shown in Fig. 17.10, it may be short or long. Different shapes clearly impart different waveform characteristics to the acoustic flow.

The pipe part of the reed pipe is, as remarked previously, largely a passive resonator since the Q-value of the reed is quite high and its frequency is little affected by the tuning of the pipe. Since the reed works near an acoustic impedance maximum for the pipe in order to achieve maximum energy transfer, a normal conical resonator is half a wavelength long, as in an oboe, and the physical pipe length is the same as an open flue pipe of the same pitch. A cylindrical resonator, on the other hand, need only be one quarter of a wavelength long and will couple essentially to only the odd harmonics of the reed generator, as in a clarinet. Such half-length reed pipes, like the clarinet, were a rather recent development and are generally used for characteristic solo voices with names such as schalmei and clarinet. The sound in the lower part of the compass is often quite like that of a clarinet because of the characteristic weak second harmonic. These ranks generally occur at 8-ft pitch only.

More central to organ reed color are those ranks with conical resonators of full length that support all harmonics generated by the reed. Generically, they are called reeds, but the stop names are more often those of brass instruments: trumpets, tubas, etc. These ranks occur at 16-ft, 8-ft, and 4-ft pitches on large organs and together make a chorus of great power and impact that can be used either alone, with mixtures, or with a full diapason chorus. Some very large organs have 32-ft or even 64-ft reed ranks.

The loudness and harmonic development of a reed rank are governed by the same sort of scaling rules as apply to flue ranks, with variation in the reed and shallot replacing variations in mouth configuration. Wide-scaled pipes are broad and loud in tone, while narrow-scaled ranks are keen and softer. These softer ranks are often given woodwind names, such as oboe or bassoon.

As well as the chorus reeds, we also find solo reeds of various types, from the great tuba stops of English cathedral organs to the horizontally mounted fanfare trumpets of French and Spanish organs. As with flue pipes, the variety is too large for us to survey here, and organ builders strive to achieve their own individual balances of tone colors.

The blowing pressure requirements for reed pipes are similar to those for flues, though there is a general tendency to use rather higher pressures—the organ built in 1929 for the Atlantic City Convention Hall had a battery of reed stops operating at immense power on 25 kPa or 2.5 meters water gauge!

17.12 Analysis of Timbre

Several proposals have been investigated for the more or less objective evaluation of the timbre of organ pipes, when played either one at a time or as complete ranks, that rely upon fewer parameters than those needed to specify the complete spectrum and its timbre evolution. We mention just three of these here.

Sundberg and Jansson (1976) evaluated the long-time average spectra (LTAS) of complete ranks for different stops on a particular organ and found characteristic spectral shapes for each stop. The LTAS was evaluated by playing a full-compass scale on the rank, recording it at 10 different microphone positions, and then averaging the spectra through a bank of filters set at one-third-octave spacings. Such an analysis shows up characteristic timbre differences, as is illustrated in Fig. 17.11.

Pollard and Jansson (1982a,1982b) used a different method to assign a particular point in a two-dimensional tristimulus diagram to the sound of an individual pipe, as shown in Fig. 17.12(a). The three vertices of the diagram represent a strong fundamental, strong midfrequency partials (actually harmonics 2, 3, and 4) and strong upper partials (harmonics 5 and above), respectively, and the position of the representative point on this diagram is constrained by the requirement that total loudness should be normalized. This diagram has the advantage that the path of the representative point during the initial transient can also be plotted, as shown for several representative pipes in Fig. 17.12(b). The loudness of the sound is, of course, not represented on the diagram.

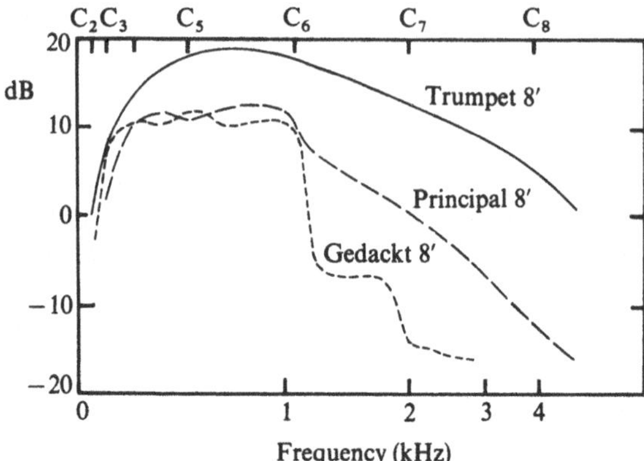

FIGURE 17.11. LTAS analysis of the tone of three typical organ pipe ranks (after Sundberg and Jansson, 1976).

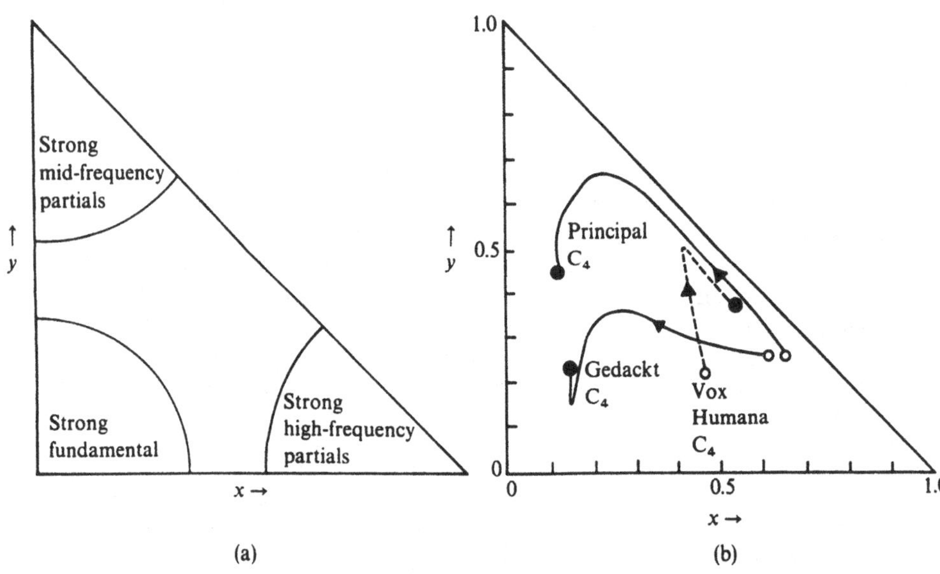

FIGURE 17.12. (a) Tristimulus diagram for the representation of musical timbre. (b) The initial transient and steady-state representation of principal, flute (gedackt), and imitative reed (vox humana) organ tones on the diagram together with a trumpet, clarinet, and viola (after Pollard and Jansson, 1982a).

Finally, Padgham (1986) has adopted a two-dimensional subjective scale with tone Θ ranging from 0 to 24 through flute → diapason → string → trumpet and complexity C given a value between 0 and 100. (In his diagrams, Θ is plotted as an angle on a 24-segment circle and C as a radius, but this is rather misleading since the tone scale is not closed—24 is at the opposite extreme to 0 rather than being identical to it.) The significant thing for our present discussion is that Padgham finds the subjective tone coordinate is well correlated linearly with the sound pressure level of the second harmonic relative to the fundamental for $0 < \Theta < 12$ (flute → string) and with the relative level of the third harmonic for $12 < \Theta < 24$ (string → trumpet). The subjective complexity C is similarly correlated with the arithmetic sum of the sound pressure levels in third-octave bands from 9 to ∞.

17.13 Tonal Architecture

An organ is so much a part of the building in which it is played that it is impossible to separate tonal design from architecture. An organ for a large cathedral with a reverberation time approaching 12 s will need to be entirely different from an organ in a concert hall with a reverberation time of 3 s or from a small organ for chamber music. Much has been written about this subject, about different national traditions in organ building, and about individual organ specifications (Bonavia-Hunt, 1950; Andersen, 1969; Sumner, 1973). We have space here for only a few remarks.

The organ is an instrument in its own right; it is not a substitute for an orchestra. A satisfying organ must therefore have, as a first requirement, a satisfying main chorus built from properly scaled diapason ranks, if possible including a mixture. Attempts to take short cuts, for example by using electric action to borrow higher pipes from an extended rank to produce nominal 4-ft, 2-ft, and even mixture pitches produces a characteristic theater organ sound.

For most music, a balancing smaller chorus of softer diapasons or even flutes is required, and this is best located on a second manual to give flexibility. Contrapuntal music requires an equally independent pedal division with its own chorus of voices, a requirement met for centuries on German organs but typically given scant attention in small English-style organs until comparatively recently.

With an increase in organ size, chorus reed ranks are added, then solo ranks of various types and contrasting choruses on different keyboards are added. It is here that national traditions enter, with German organs concentrating on bright flue pipes, mixtures, and mutations, Spanish organs providing ceremonial trumpet stops, and English organs concentrating on full, rich choruses with an emphasis on enclosed divisions. There is a continuing interest in documenting the sound of organs by distinguised builders

(Angster and Miklós, 1995). Some organs of moderate size have survived almost unaltered for nearly 300 years, quite a number of large organs date from the time of Bach, and there are many great instruments built more recently. Because of the complexity of the instrument, and its intimate relation to the building in which it is housed, such documentation presents difficulties that are not met in smaller instruemnts.

Modern large organs attempt to combine the best of many of these traditions so that they are effectively several instruments in one, the player being able to select just that range and distribution of ranks required for the performance of each particular piece of music. A good example is the organ in the Sydney Opera House, completed in 1979, the design of which has been described by its builder (Sharp, 1973). It has some 10,000 pipes in 205 ranks grouped into 127 stops that are distributed over five manual keyboards and a pedalboard. The action is entirely mechanical, but there is a microprocessor-controlled stop action and a duplicate electrical action used to couple the manuals or even for remote playing. Organs of this type attempt to blend tradition with the best of modern scholarship, insight, and technology to perpetuate a noble instrument.

References

Andersen, P-G. (1969). "Organ Building and Design." George Allen and Unwin, London.

Angster, J., and Miklós, A. (1995). Documentation of the sound of a historical pipe organ. *Applied Acoustics* **46**, 61–82.

Backus, J. (1977). "The Acoustical Foundations of Music," 2nd ed. Norton, New York.

Backus, J., and Hundley, T.C. (1965). Wall vibrations in flue organ pipes and their effect on tone. *J. Acoust. Soc. Am.* **39**, 936–945.

Barbour, J.M. (1953). "Tuning and Temperament." Michigan State College Press, East Lansing, Michigan.

Bonavia-Hunt, N.A. (1950). "The Modern British Organ." A. Weekes & Co., London.

Boner, C.P. (1938). Acoustic spectra of organ pipes. *J. Acoust. Soc. Am.* **10**, 32–40.

Boner, C.P., and Newman, R.B. (1940). The effect of wall materials on the steady-state acoustic spectrum of flue pipes. *J. Acoust. Soc. Am.* **12**, 83–89.

Caddy, R.S., and Pollard, H.F. (1957). Transient sounds in organ pipes. *Acustica* **7**, 277–280.

Coltman, J.W. (1969). Sound radiation from the mouth of an organ pipe. *J. Acoust. Soc. Am.* **46**, 477.

Dänzer, H., and Kollmann, W. (1956). Über die Strömungsverhaltnisse an der Lippenöffnung von Orgelpfeifen. *Z. Physik* **144**, 237–243.

Fletcher, H., Blackham, E.D., and Christensen, D.A. (1963). Quality of organ tones. *J. Acoust. Soc. Am.* **35**, 314–325.

Fletcher, N.H. (1974). Nonlinear interactions in organ flue pipes. *J. Acoust. Soc. Am.* **56**, 645–652.

Fletcher, N.H. (1976). Transients in the speech of organ flue pipes—A theoretical study. *Acustica* **34**, 224–233.

Fletcher, N.H. (1977). Scaling rules for organ flue pipe ranks. *Acustica* **37**, 131–138.

Fletcher, N.H., and Douglas, L.M. (1980). Harmonic generation in organ pipes, recorders, and flutes. *J. Acoust. Soc. Am.* **68**, 767–771.

Fletcher, N.H., and Thwaites, S. (1983). The physics of organ pipes. *Sci. Am.* **248** (1), 94–103.

Franz, G., Ising, H., and Meinusch, P. (1969/1970). Schallabstrahlung von Orgelpfeifen. *Acustica* **22**, 226–231.

Ingerslev, F., and Frobenius, W. (1974). Some measurements of the end-corrections and acoustic spectra of cylindrical open flue organ pipes. *Trans. Dan Acad. Tech. Sci.* **1**, 1–44.

Jones, A.T. (1941). End corrections of organ pipes. *J. Acoust. Soc. Am.* **12**, 387–394.

Keeler, J.S. (1972). The attack transients of some organ pipes. *IEEE Trans. Audio Electroacoust.* **AU-20**, 378–391.

Levine, H., and Schwinger, J. (1948). On the radiation of sound from an unflanged circular pipe. *Phys. Rev.* **73**, 383–406.

Lottermoser, W. (1983). "Orgeln, Kirchen und Akustik." Verlag Erwin Bochinsky/Das Musikinstrument, Frankfurt am Main.

Mahrenholz, C. (1975). "The Calculation of Organ Pipe Scales from the Middle Ages to the Mid-nineteenth Century." Positif Press, Oxford.

Mercer, D.M.A. (1951). The voicing of organ flue pipes. *J. Acoust. Soc. Am.* **23**, 45–54.

Mercer, D.M.A. (1954). The effect of voicing adjustments on the tone quality of organ flue pipes. *Acustica* **4**, 237–239.

Meyer, J. (1961). Uber die Resonanzeigenschaften offener Labialpfeifen. *Acustica.* **11**, 385–396.

Miller, D.C. (1909). The influence of the material of wind instruments on the tone quality. *Science* **29**, 161–171.

Nolle, A.W. (1979). Some voicing adjustments of flue organ pipes. *J. Acoust. Soc. Am.* **66**, 1612–1626.

Nolle, A.W. (1983). Flue organ pipes: Adjustments affecting steady waveform. *J. Acoust. Soc. Am.* **73**, 1821–1832.

Nolle, A.W., and Boner, C.P. (1941). The initial transients of organ pipes. *J. Acoust. Soc. Am.* **13**, 149–155.

Padgham, C. (1986). The scaling of the timbre of the pipe organ. *Acustica* **60**, 189–204.

Plomp, R. (1976). "Aspects of Tone Sensation." Academic Press, London.

Pollard, H.F. (1978a). Loudness of pipe organ sounds I. Plenum combinations. *Acustica* **41**, 65–74.

Pollard, H.F. (1978b). Loudness of pipe organ sounds II. Single notes. *Acustica* **41**, 75–85.

Pollard, H.F., and Jansson, E.V. (1982a). A tristimulus method for the specification of musical timbre. *Acustica* **51**, 162–171.

Pollard, H.F., and Jansson, E.V. (1982b). Analysis and assessment of musical starting transients. *Acustica* **51**, 249–262.

Rossing, T.D. (1990). "The Science of Sound." (2nd ed.). Addison-Wesley, Reading, Massachusetts.

Sharp, R. (1973). The grand organ in the Sydney Opera House. *J. Proc. Roy. Soc. NSW* **106**, 70–80.

Stevens, S.S., and Davis, H. (1938). "Hearing: Its Psychology and Physiology." Reprinted by the Acoustical Society of America, Woodbury, New York, 1983.

Sumner, W.L. (1973). "The Organ: Its Evolution, Principles of Construction and Use." MacDonald and Jane's, London.

Sundberg, J., and Jansson, E.V. (1976). Long-time-average-spectra applied to analysis of music. Part II: An analysis of organ stops. *Acustica* **34**, 269–274.

Tanner, R. (1958). Etude expérimentale de divers type de tuyaux sonores bouches, du point de vue de la pureté du timbre. *Acustica* **8**, 226–236.

Trendelenburg, F., Thienhaus, E., and Franz, E. (1936). Klangeinsätze an der Orgel. *Akust. Z.* **1**, 59–76.

Trendelenburg, F., Thienhaus, E., and Franz, E. (1938). Klangübergänge bei der Orgel. *Akust. Z.* **3**, 7–20.

Wolf, D. (1965). Über die Eigenfrequenzen labialer Orgelpfeifen. *Z. Physik* **183**, 241–248.

Part V

Percussion Instruments

18

Drums

Drums are practically as old as the human race. With the exception of the human voice, they are our oldest musical instruments. The earliest drums were probably chunks of wood or stone placed over holes in the earth. Then it was discovered that more sound could be obtained from hollow tree trunks, the ancestors of our contemporary log drums.

No one knows where the first membrane drum was made, but some ancient drums are at least 5000 years old. Skins of animals or fish were stretched across hollow tree trunks, and the drums were probably struck with the bare hands. Later progress in skin drums led to bowls or shells with special shapes and various means for applying tension to the drum heads. Drums acquired religious as well as musical importance, and drum making became an important ritual in several cultures. A detailed history of drums is given by Blades (1970).

Modern drums can be divided into two groups: those that convey a strong sense of pitch and those that do not. In the former group are the kettledrums, tabla, boobams; in the latter group are the bass drum, snare drum, tenor drum, tom-toms, bongos, congas, and countless other drums, mainly of African and Oriental origin.

As vibrating systems, drums can be divided into three categories: those consisting of a single membrane coupled to an enclosed air cavity (e.g., kettledrums); those consisting of a single membrane open to the air on both sides (e.g., tom-toms, congas); and those consisting of two membranes coupled by an enclosed air cavity (e.g., bass drums, snare drums).

Other ways of categorizing drums are according to their ethnic origin (e.g., Oriental, African, Latin American) or according to the types of musical performance with which they are associated (e.g., symphonic, military, jazz, ethnic dance, and marching bands).

Steel drums and wooden slot drums, whose vibrating members are more platelike than membranelike, will be discussed in Chapter 19.

18.1 Kettledrums

Kettledrums, or timpani, are usually considered to be the most important drums in modern orchestras. While ancient in origin, their preeminence in orchestras dates from the invention of screw tensioning devices in the seventeenth century. The timpanist in a symphony orchestra plays on three to five timpani of various sizes, each tunable over a range of about a musical fifth. Berlioz's "Grande Messe des Morts" calls for 16 timpani to be played by 10 players. During the last century, various mechanisms were developed for changing the tension to tune the drumheads rapidly. Most modern timpani have a pedal-operated tensioning mechanism in addition to six or eight tensioning screws around the rim of the kettle. The pedal typically allows the player to vary the tension over a range of at least 3:1, which corresponds to a tuning range in excess of a musical sixth.

At one time most timpani heads were calfskin, but this material has gradually given way to Mylar (polyethylene terephthalate). Calfskin heads require a great deal of hand labor to prepare and great skill to tune properly. Some orchestral timpanists prefer them for concert work under the conditions of controlled humidity, but use Mylar when touring. Mylar is insensitive to humidity and is easier to tune because of its homogeneity. A thickness of 0.19 mm (0.0075 in.) is considered standard for Mylar timpani heads. Timpani kettles are roughly hemispherical; copper is the preferred material, although fiberglass and other materials are frequently used.

Although the modes of vibration of an ideal membrane are not harmonic, a carefully tuned kettledrum is known to sound a strong principal note plus two or more harmonic overtones. Rayleigh (1894) recognized the principal note as coming from the (1,1) mode (see Section 3.3) and identified overtones about a perfect fifth ($f/f_1 = 1.50$), a major seventh (1.88), and an octave (2.00) above the principal tone. He identified these overtones as originating from the (2,1), (3,1), and (1,2) modes, respectively, which in an ideal membrane should have frequencies of 1.34, 1.66, and 1.83 times the (1,1) mode. Rayleigh's results are quite remarkable, considering the equipment available to him. More recent measurements (Benade, 1976; Rossing and Kvistad, 1976) have indicated that the (1,1), (2,1), and (3,1) modes in a timpani have frequencies nearly in the ratios 1:1.5:2. The (4,1) and (5,1) modes typically have ratios of 2.44 and 2.90 times the fundamental (1,1) mode, and these are within about half a semitone of the ratios 2.5 and 3. Thus, the family of modes having one to five nodal diameters radiate prominent partial tones having frequency ratios nearly 2:3:4:5:6, which give the timpani a strong sense of pitch.

How are the inharmonic modes of an ideal circular membrane (see Section 3.3) shifted in frequency so that a series of prominent harmonic partials appears in the sound of a carefully tuned kettledrum? Four effects appear to contribute (see Section 3.4):

1. the membrane vibrates in a sea of air, and the mass of this air sloshing back and forth lowers the frequencies of the principal modes of vibration;
2. the air enclosed by the kettle has resonances of its own that will interact with the modes of the membrane that have similar shapes;
3. the bending stiffness of the membrane, like the stiffness of piano strings, raises the frequencies of the higher overtones; and
4. drumheads have a rather large stiffness to shear, so they resist the type of distortion needed to deflect a membrane without wrinkling it (or to wrap it around a bowling ball).

Our studies have shown that air loading, which lowers the modes of low frequency, is mainly responsible for establishing the harmonic relationship of kettledrum modes; the other effects only fine tune the frequencies but may have considerable effect on the rate of decay of the sound (Rossing, 1982a). The stiffness of the air enclosed in the kettle raises the frequencies of the axial-symmetric modes, especially the (0,1) mode (Morse, 1948).

It is appropriate here to mention a theoretical analysis by Aebischer and Gottlieb (1990) of the acoustics of an annular kettledrum. They were able to show that there are many combinations of dimensions for an annular membrane coupled to an air cavity that yield mode frequencies in very good approximation to a harmonic series. Although these designs have yet to be developed into real musical instruments, they appear to have significant potential.

18.1.1 Bending Stiffness

In Section 3.12, the effect of bending stiffness was estimated by adding a plate-like term to the membrane equation. For a typical timpani drum head, modal frequencies were found to be raised by about 0.5%.

A more-detailed analysis considers the radial part of the modal function:

$$U_m(r) = AJ_m(k_1 r) + BJ_m(k_2 r), \tag{18.1}$$

where J_m is the cylindrical Bessel function of order m. The vanishing of the membrane displacement at $r = a$ gives

$$U_m(a) = AJ_m(k_1 a) + BJ_m(k_2 a) = 0. \tag{18.2}$$

The assumption of a "hinged" boundary condition at $r = a$ is reasonable and implies that

$$U''_m(a) = Ak_1^2 J''_m(k_1 a) + Bk_2^2 J''_m(k_2 a) = 0, \tag{18.3}$$

with

$$J''_m(\xi) = \frac{d^2}{d\xi^2} J_m(\xi). \tag{18.4}$$

For nontrivial solutions of Eqs. (18.2) and (18.3) for A and B, we must have

$$\frac{J_m(k_2a)}{k_2^2 J_m''(k_2a)} = \frac{J_m(k_1a)}{k_1^2 J_m''(k_1a)} = \frac{-a^2 J_m(k_1a)}{k_1aJ'm(k_1a) + (k_1^2a^2 - m^2)J_m(k_1a)}, \tag{18.5}$$

where the last equality follows from the Bessel differential equation.

Since $k_2a \gg 1$, we may use asymptotic behavior,

$$J_m(k_2a) \approx \sqrt{\frac{2}{\pi k_2 a}} \cos\left(k_2a - \frac{m\pi}{2} - \frac{\pi}{4}\right), \tag{18.6}$$

so that in Eq. (18.5), we have (using terms from Section 3.12)

$$\frac{J_m(k_2a)}{k_2^2 J_m''(k_2a)} \simeq \frac{1}{k_2^2} \simeq \frac{S^4}{c^2}. \tag{18.7}$$

In the absence of stiffness effects, the membrane modal wave numbers k_{mn}^0 are determined by the equation

$$J_m(k_{mn}^0 a) = 0. \tag{18.8}$$

The shifts in modal wave numbers may be estimated by writing

$$k_1 = k_{mn}^0 + \Delta k_{mn} \tag{18.9}$$

and using the approximation

$$J_m(k_1a) \simeq J'm(k_{mn}^0 a)(a\Delta k_{mn}) \tag{18.10}$$

in Eq. (18.5). Using the values for S, a, and c in Section 3.12, we find

$$\frac{\Delta k_{mn}}{k_{mn}^0} \simeq -\frac{S^4}{c^2a^2} = -8.1 \times 10^{-6}. \tag{18.11}$$

Finally, we evaluate

$$\frac{\Delta\omega_{mn}}{\omega_{mn}^0} = \frac{\Delta\omega_{mn}}{k_{mn}^0 c} \tag{18.12}$$

using

$$\omega_{mn} = \sqrt{c^2k_1^2 + k_1^4S^4} = \sqrt{c^2(k_{mn}^0)^2 + (k_{mn}^0)^4S^4} + \left(\frac{d\omega}{dk_1}\right)_{k_1} \Delta k_{mn} + \ldots$$

to find

$$\frac{\Delta\omega_{mn}}{\omega_{mn}^0} \simeq \left(1 + 2\frac{S^4}{c^4}\left(\omega_{mn}^0\right)^2\right)\left(-\frac{S^4}{c^2a^2}\right) \simeq 6.2 \times 10^{-4} - 8.1 \times 10^{-6} \tag{18.14}$$

for $\omega^0_{mn}/2\pi = 600$ Hz.

This more-detailed analysis of bending stiffness effects thus gives essentially the same result as the analysis in Section 3.12. These effects are indeed negligible for modes of musical interest.

18.1.2 Piston Approximation for Air Loading

In order to estimate the effects of air loading, we consider the effective air mass loading for a piston in an infinite baffle. It is given by (see Section 7.6)

$$m = \frac{\rho_0 c_a \pi a^2 X_m}{\omega}, \tag{18.15}$$

where X_m is the piston reactance function, ρ_0 is the density of air, and c_a is the velocity of sound. For frequencies $f < c_a/4\pi a$, $X_m \simeq (16/3)(fa/c_a)$, so that $m = (8/3)\rho_0 a^3$. This is equivalent to the air in a disc with the radius of the piston and a thickness $8a/3\pi$. At higher frequency, the air load decreases as $1/f^2$, becoming negligible compared with the membrane mass above 500 Hz or so. The surface density of the air load for a piston in a baffle is given in Fig. 18.1. The effective air loading on a rigid baffled piston the size and mass of a typical kettledrum would shift the frequencies of the lowest modes by as much as 40% at low tension, somewhat less at high tension.

The velocities of transverse waves in a 0.19-mm-(0.0075-in.-)thick Mylar timpani membrane are shown in Fig. 18.2. The observed wave velocities are determined from the frequencies of the principal vibrational modes, and the wave velocities have been calculated using the mass of the membrane alone and by adding the air mass predicted by the piston model (as shown in Fig. 18.1).

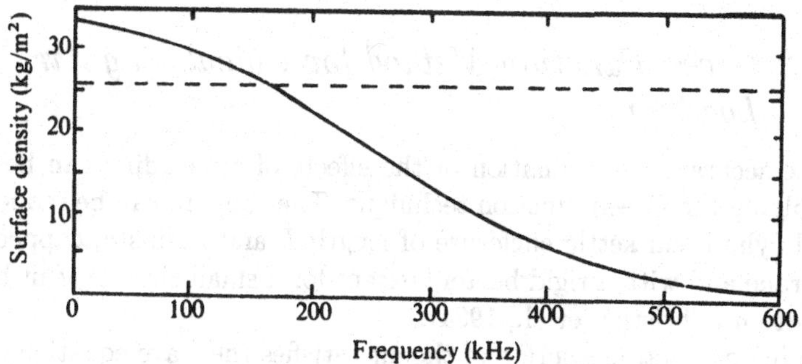

FIGURE 18.1. The effective surface density of the air load for a circular piston in a very large baffle. The dashed curve gives the surface density of a timpani head for a comparison.

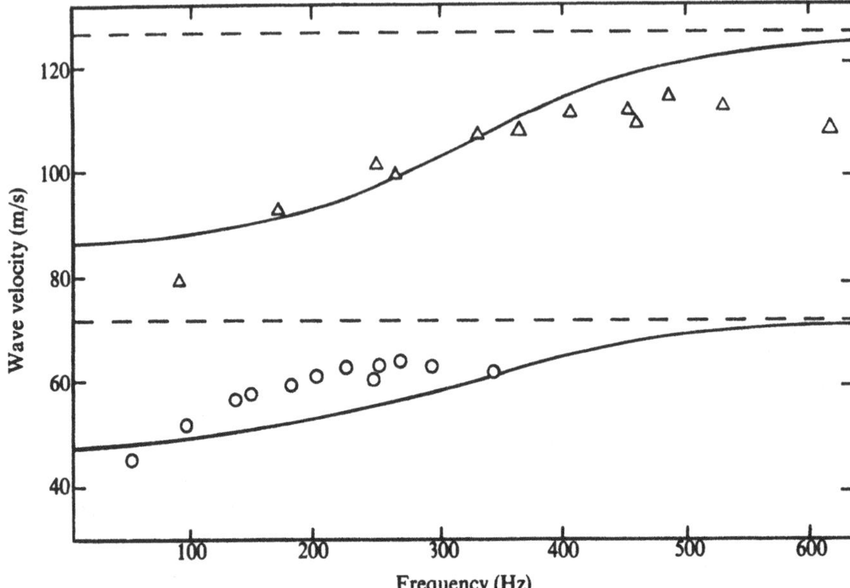

FIGURE 18.2. Wave velocity as a function of frequency for a timpani membrane at two different tensions without a kettle. The dashed curves are the wave velocities in unloaded membranes [$c = (T/\sigma)^{1/2}$], while the solid curves include air loading estimated from the equivalent piston model (Mills, 1980).

The equivalent piston model gives us a ball-park estimate for the effect of air loading, but it appears to overestimate the effect below about 350 Hz [except for the fundamental (0,1) mode] and to underestimate it above 350 Hz, as seen from Fig. 18.2. Thus we look for a more exact method of calculating air loading.

18.1.3 Green Function Method for Calculating Air Loading

A more-accurate determination of the effects of air loading can be made by applying the Green function technique. The timpani can be modeled as a rigid cylindrical kettle enclosure of length L and radius a, capped by a membrane and with a rigid bottom except for a small circular vent hole at the bottom (Christian et al., 1984).

The incremental pressure p of the air satisfies the wave equation

$$\left(\nabla^2 \frac{1}{c_a^2} - \frac{\partial^2}{\partial t^2}\right) p = 0, \tag{18.16}$$

where c_a is the speed of sound in air. For a given normal mode of the timpani, the pressure may be written as $p = p(\mathbf{r})e^{-i\omega t}$ with

$$\left(\nabla^2 + \frac{\omega^2}{c_a^2}\right) p(\mathbf{r}) = 0. \tag{18.17}$$

In order to take into account the effect of inside-air loading on the membrane, it is convenient to use the Neumann-Green function, $G_{\text{in}}(\mathbf{r}|\mathbf{r}')$. For points inside the kettle enclosure, $G_{\text{in}}(\mathbf{r}|\mathbf{r}')[= G_{\text{in}}(\rho, \phi, z|\rho', \phi', z')]$ satisfies the equations

$$\begin{aligned}
&\left(\nabla^2 + \frac{\omega^2}{c_a^2}\right) G_{\text{in}}(\mathbf{r}|\mathbf{r}') \\
&= \left[\frac{1}{\rho}\frac{\partial}{\partial\rho}\left(\rho\frac{\partial}{\partial\rho}\right) + \frac{1}{\rho^2}\frac{\partial^2}{\partial\phi^2} + \frac{\partial^2}{\partial z^2} + \frac{\omega^2}{c_a^2}\right] G_{\text{in}}(\rho, \phi, z|\rho', \phi', z') \\
&= -4\pi\delta(\mathbf{r} - \mathbf{r}') = -4\pi\frac{\delta(\rho - \rho')}{\rho}\delta(\phi - \phi')\delta(z - z');
\end{aligned}$$

$$G_{\text{in}}(\mathbf{r}|\mathbf{r}') = G_{\text{in}}(\mathbf{r}'|\mathbf{r}); \tag{18.19}$$

and

$$\left.\frac{\partial G_{\text{in}}}{\partial \rho'}(\mathbf{r}|\mathbf{r}')\right|_{\rho'=a} = \left.\frac{\partial G_{\text{in}}}{\partial z'}(\mathbf{r}|\mathbf{r}')\right|_{z'=0,L} = 0. \tag{18.20}$$

Neglecting sound reflections from the walls of the room and assuming an infinite and rigid plane baffle, expressions can be developed for the pressure inside and outside the membrane.

The pressure at points $\mathbf{r}$ inside the kettle enclosure can be written

$$\begin{aligned}
p(\mathbf{r}) = {}& \frac{1}{4\pi}\int_0^a \rho'\,d\rho' \int_0^{2\pi} d\phi' G_{\text{in}}(\rho, \phi, z|\rho', \phi', L)\frac{\partial p}{\partial z'}(\rho', \phi', L_-) \\
&- \frac{1}{4\pi}\int_0^{a_v} \rho'\,d\rho' \int_0^{2\pi} d\phi' G_{\text{in}}(\rho, \phi, z|\rho', \phi', 0)\frac{\partial p}{\partial z'}(\rho', \phi', 0),
\end{aligned} \tag{18.21}$$

where $p(\rho', \phi', L_-)$ is the pressure on the kettle side of the membrane, and the Green's function inside the kettle is given by

$$\begin{aligned}
G_{\text{in}}(\rho, \phi, z|\rho', \phi', z') = {}& -4\pi \sum_{m=-\infty}^{\infty} \frac{e^{im(\phi-\phi')}}{2\pi} \\
&\times \sum_{n=1}^{\infty} \frac{J_m[y_{mn}(\rho/a)]J_m[y_{mn}(\rho'/a)]}{a^2(1 - m^2/y_{mn}^2)J_m^2(y_{mn})} \\
&\times \frac{\cos(\gamma_{mn}z_<)\cos[\gamma_{mn}(L - z_>)]}{\gamma_{mn}\sin(\gamma_{mn}L)}.
\end{aligned}$$

Similarly, the pressure outside can be written

$$p(\rho, \phi, L_+) = -\frac{1}{4\pi} \int_0^a \rho' \, d\rho' \int_0^{2\pi} d\phi' G_{\text{out}}(\rho, \phi, L_+|\rho', \phi', L) \frac{\partial p}{\partial z'}(\rho', \phi', 0), \tag{18.23}$$

where the Green's function above the membrane is given by

$$G_{\text{out}}(\mathbf{r}|\mathbf{r}') = \frac{e^{i(\omega/c_a)[(x-x')^2+(y-y')^2+(z-z')^2]^{1/2}}}{[(x-x')^2+(y-y')^2+(z-z')^2]^{1/2}}$$

$$+ \frac{e^{i(\omega/c_a)[(x-x')^2+(y-y')^2+(z+z'-2L)^2]^{1/2}}}{[(x-x')^2+(y-y')^2+(z+z'-2L)^2]^{1/2}}.$$

These results can be used with the membrane equation of motion to calculate the modes of vibration of the air-loaded membrane. A similar calculation can be made to determine the modes of a timpani membrane without a kettle and without a baffle (Christian et al., 1984).

The modal frequencies calculated for a membrane with and without a kettle are given in Table 18.1. Also given are the frequencies of an ideal membrane at the same tension. Note the downward shift in frequency of all the modes, especially those of low frequency.

TABLE 18.1. Calculated timpani modal frequencies (with and without kettle) for tension $T = 3990\,\text{N/m}$. For comparison, ideal membrane frequencies for the same tension are listed.

Mode	Ideal	Ideal	With kettle		Without kettle	
mn	f_{mn} (Hz)	f_{mn}/f_{11}	f_{mn} (Hz)	f_{mn}/f_{11}	f_{mn} (Hz)	f_{mn}/f_{11}
0,1	143	0.63	131	0.87	89	0.54
1,1	228	1.00	150	1.00	165	1.00
2,1	306	1.34	227	1.51	237	1.44
0,2	328	1.44	253	1.68	257	1.55
3,1	380	1.66	299	1.99	308	1.92
1,2	417	1.83	352	2.34	343	2.08
4,1	452	1.98	370	2.46	377	2.28
2,2	501	2.20	411	2.74	424	2.57
5,1	522	2.29	434	2.93	445	2.69
3,2	581	2.55	492	3.28	501	3.04
6,1	591	2.61	507	3.38	512	3.10
1,3	605	2.66	507	3.38	525	3.18
4,2	658	2.89	570	3.80	578	3.50

18.1.4 Comparison of Calculated and Measured Modes for a Timpani Membrane

The modal frequencies calculated with the Green function technique agree very well with those measured in a timpani membrane both with and without the kettle, as shown in Table 18.2. Note that the tension is different in the two cases, although the frequency of the fundamental (1,1) mode is essentially the same.

Note that the (1,1), (2,1), (3,1), and (4,1) modes, which had frequencies in the inharmonic ratios of 1.00:1.34:1.66:1.98 in the ideal membrane, are shifted into the nearly harmonic ratios 1.00:1.50:1.97:2.44 in the case of the membrane with the kettle and to a tolerable 1.00:1.47:1.91:2.36 ratio without a kettle. Even without the kettle, an air-loaded timpani membrane conveys a fairly definite sense of pitch.

The interaction of the air enclosed by the kettle with the various modes of the membrane is not unlike that encountered in the guitar in Chapter 9. Each membrane mode may be thought of as driving kettle modes having the same symmetry in the plane of the membrane. If the membrane and kettle mode frequencies are close together, the interaction will be strong.

The normal modes of a 26-in.-diameter Ludwig kettle are shown in Fig. 18.3. These modes were observed by covering the kettle with a rigid cover having small holes through which a driving tube and probe microphone could be inserted. In every case, the kettle modes are higher in frequency than the membrane modes (see Table 18.1) to which they couple (Rossing et al., 1982).

Also shown in Fig. 18.3 are the air modes in the kettle when the volume is reduced to one-half and one-quarter of the original volume by partly filling the kettle with water. The effect on the membrane modal frequencies of reducing the kettle volume was similarly determined (Rossing et al., 1982).

The effect on the membrane modal frequencies of reducing the volume of the kettle is shown in Table 18.3. Note that reducing the kettle volume raises the frequencies of the axial-symmetric (0,1), (0,2), and (0,3) modes but lowers the frequencies of the other modes. This can be understood by noting that the average air flow velocity (and hence the effective momentum) will be slightly greater when the available volume is decreased. The harmonic tuning of the $(m, 1)$ modes is pretty well retained when the volume is reduced by 25%, but it is lost when the volume is reduced to half its original volume. Note that again the modal frequencies, calculated by the Green function technique, show good agreement with the experimentally measured frequencies.

From the results in Table 18.3 it might be concluded that the size of a timpani kettle is quite near its optimum value insofar as fine tuning the most prominent partials into a harmonic series.

TABLE 18.2. Calculated and experimental timpani modal frequencies (with and without the kettle).[a]

	With kettle ($T = 5360\,N/m$)				Without kettle ($T = 4415\,N/m$)			
Mode m, n	Calculated f_{mn} (Hz)	Experimental f_{mn} (Hz)	Calculated f_{mn}/f_{11}	Experimental f_{mn}/f_{11}	Calculated f_{mn} (Hz)	Experimental f_{mn} (Hz)	Calculated f_{mn}/f_{11}	Experimental f_{mn}/f_{11}
0,1	138	140	0.80	0.81	93	92	0.54	0.54
1,1	172	172	1.00	1.00	173	173	1.00	1.00
2,1	261	258	1.52	1.50	249	253	1.44	1.47
0,2	291	284	1.69	1.65	270	266	1.56	1.54
3,1	344	340	2.00	1.97	322	330	1.86	1.91
1,2	390	344	2.27	2.00	367	365	2.08	2.12
4,1	427	420	2.48	2.44	394	408	2.28	2.36
2,2	471	493	2.74	2.86	445	443	2.57	2.63
0,3	511	467	2.97	2.71	458	459	2.65	2.66
5,1	506	501	2.94	2.91	467	485	2.70	2.81

[a] From Christian et al., 1984.

TABLE 18.3. Experimental and calculated timpani modal frequencies for four different kettle volumes.[a,b]

	$V_0 = 0.14\,\mathrm{m}^3$[c]			$0.75V_0$[c]		
Mode m, n	Experimental f_{mn} (Hz)	Experimental f_{mn}/f_{11}	Calculated F_{mn}/f_{11}	Experimental f_{mn} (Hz)	Experimental f_{mn}/f_{11}	Calculated F_{mn}/f_{11}
0,1	141	0.83	0.80	160	0.94	0.92
1,1	170	1.00	1.00	171	1.00	1.00
2,1	255	1.50	1.52	260	1.52	1.54
0,2	281	1.65	1.68	295	1.73	1.75
3,1	337	1.98	2.00	345	2.02	2.03
4,1	416	2.45	2.48	425	2.49	2.51
2,2						
0,3	488	2.87	2.97	486	2.84	2.90
5,1	495	2.91	2.94	507	2.96	2.99
3,2	556	3.27	3.29			
6,1	573	3.37	3.40	585	3.42	3.45

[a] From Christian et al., 1984. [b] The tensions in the four cases are slightly different. The theoretical tensions were adjusted in each case so as to give f_{11}(calc.) = f_{11}(expt.). [c] $V_0 = 0.14\,\mathrm{m}^3$.

TABLE 18.3. *continued.*

	$V_0 = 0.5V_0^c$			$0.25V_0^c$		
Mode m, n	Experimental f_{mn}(Hz)	Experimental f_{mn}/f_{11}	Calculated F_{mn}/f_{11}	Experimental f_{mn}(Hz)	Experimental f_{mn}/f_{11}	Calculated F_{mn}/f_{11}
0,1	182	1.12	1.07			
1,1	165	1.00	1.00	148	1.00	1.00
2,1	259	1.57	1.56	248	1.67	1.65
0,2	301	1.82	1.82			
3,1	346	2.10	2.08	338	2.28	2.23
4,1	427	2.59	2.57	423	2.86	2.79
2,2				465	3.14	3.05
0,3	491	2.98	3.01	496	3.35	3.31
5,1	509	3.08	3.05	507	3.43	3.32
3,2				587	3.97	3.71
6,1	588	3.56	3.53	604	4.08	3.85

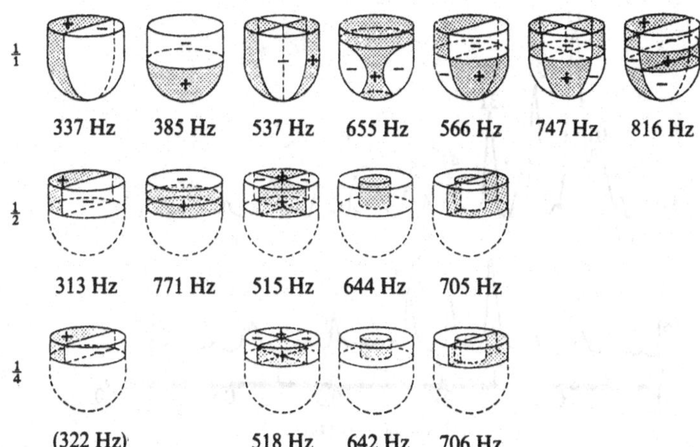

FIGURE 18.3. Normal modes of the air in a rigidly capped kettle for three different air volumes.

18.1.5 Timpani Sound

The sound spectra obtained by striking a 26-in.-diameter kettledrum in its normal place (about one-fourth of the way from the edge to the center) and at the center are shown in Fig. 18.4 (also see Table 18.4). Note that the fundamental mode (0,1) appears much stronger when the drum is struck at the center, as do the other symmetrical modes [(0,2), (0,3)]. These modes damp out rather quickly, however, so they do not produce much of a drum sound. In fact, striking the drum at the center produces quite a dull, thumping sound.

A blow at the normal strike point, however, excites the harmonic (1,1), (2,1), (3,1), and (4,1) modes, and these modes decay more slowly than the (0,1), (0,2), and (0,3) modes. Even in the case of the center blow, the (1,1) and (2,1) partials are more prominent after 1s [Fig. 18.4(d)].

Most observers identify the pitch of the timpani as corresponding to that of the (1,1) partial. It is a little surprising that the pitch of timpani corresponds to the pitch of the principal tone rather than the missing fundamental of the harmonic series, which would be an octave lower. Apparently, the strengths and durations of the overtones are insufficient, compared to the principal tone, to establish the harmonic series of the missing fundamental. Some timpanists report that a gentle stroke at the proper spot with a soft beater can produce a rather indistinct sound an octave below the nominal pitch (Brindle, 1970). It is possible to make the following observations from Table 18.4:

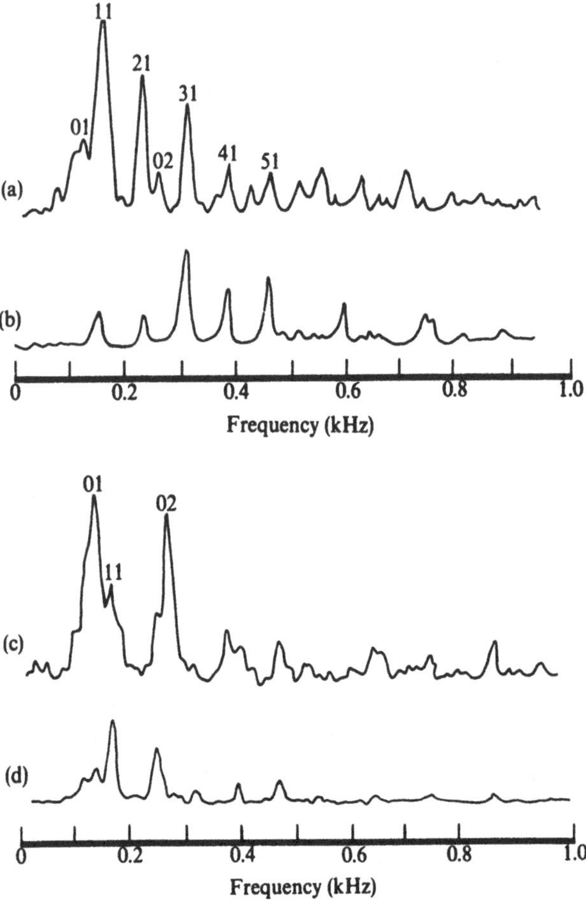

FIGURE 18.4. Sound spectra from a 65-cm timpani tuned to E_3: (a) approximately 0.03 s after striking at the normal point, (b) approximately 1 s later, (c) approximately 0.03 s after striking at the center, and (d) approximately 1 s later (from Rossing and Kvistad, 1976).

1. The harmonically tuned (1,1), (2,1), (3,1), and (4,1) modes decay much more slowly than the other modes, especially the (0,1), (0,2), and (0,3) modes.
2. The (0,1) mode, which acts as a monopole source, decays rapidly; the baffled (1,1) mode and the unbaffled (0,1) mode, which act as dipole sources, decay less rapidly; the baffled (2,1) mode and the unbaffled (1,1) mode decay still less rapidly.

TABLE 18.4. Decay times for a 26-in.-diameter timpani membrane with and without a kettle.[a]

	With kettle				Without kettle			
	$T = 5360$ N/m		$T = 3710$ N/m		$T = 4415$ N/m		$T = 2820$ N/m	
Mode m, n	f_{mn} (Hz)	τ_{60} (s)	f_{mn} (Hz)	τ_{60} (s)	f_{mn} (Hz)	τ_{60} (s)	f_{mn} (Hz)	τ_{60} (s)
0,1 monopole	140	< 0.3	128	0.4	93	0.8	73	1.5
1,1 dipole	172	0.8	145	2.3	173	2.5	139	3.4
2,1 quadrupole	258	1.7	218	3.7	249	3.3	204	3.4
0,2	284	0.4	235	0.3	270	0.4	214	< 0.3
3,1	340	2.7	287	4.6	322	2.6	267	4.6
1,2	344	0.5	303	2.5	367	< 0.3	295	1.3
4,1	420	1.7	354	4.3	394	2.8	330	4.2
2,2	493	0.5	394	0.9	445	0.7	364	1.2
0,3	467	< 0.3	383	0.5	458	< 0.3	353	< 0.3
5,1	501	2.6	421	4.1	467	2.1	392	4.2

[a] After Christian et al. (1984).

3. Increasing the tension increases the frequency of a given mode and thus the radiation efficiency, resulting in a shorter decay time. (Monopole radiation efficiency goes as f^2; dipole as f^4; and quadrupole as f^6.)

The Green function technique can be used to predict decay times of various modes with and without a kettle over a wide range of tension (Christian et al., 1984). The calculated values are in reasonably good agreement with the measured decay times.

18.1.6 Radiation and Sound Decay

There are four possible mechanisms for energy loss and the resulting damping of a membrane: (1) radiation of sound, (2) mechanical loss in the membrane, (3) viscothermal loss in the confined air, and (4) mechanical loss in the kettle walls. Our investigations indicate that radiation accounts for a major part of the damping of a kettledrum membrane. Nevertheless, kettle loss cannot be ruled out completely in light of the general feeling that timpani with copper kettles sound different from those with kettles of fiberglass or other synthetic material. By the same token, the preference for the sound of calfskin drumheads suggests that the energy loss in the membrane itself may not be ignorable.

A baffled membrane vibrating in its (0,1) mode (a monopole source) radiates its energy very rapidly, and hence this mode damps out rapidly, as compared with modes that act as dipole or quadrupole sources (see Chapter 7). If the kettle (which acts as a baffle) is removed, the (0,1) mode damps out less rapidly since it acts as a dipole source, as shown in Fig. 18.5. The (1,1) mode normally acts as a dipole source, but when the kettle is removed, it acts as a quadrupole source and radiates away its energy less rapidly.

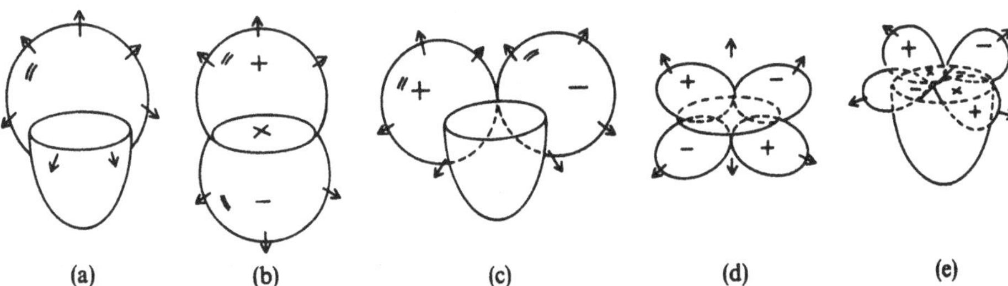

FIGURE 18.5. Radiation from a vibrating membrane: (a) monopole radiation from a baffled membrane vibrating in its (0,1) mode; (b) dipole radiation from an unbaffled membrane vibrating in its (0,1) mode; (c) dipole radiation from a baffled membrane vibrating in its (1,1) mode; (d) quadrupole radiation from an unbaffled membrane vibrating in its (1,1) mode; and (e) quadrupole radiation from a baffled membrane vibrating in its (2,1) mode.

Polar plots showing the sound radiation pattern from a small (40-cm-diameter) kettledrum in an anechoic room are shown in Fig. 18.6 (Fleischer, 1988). Note that the number of maxima for each mode equals twice the number of nodal diameters. Compare the first three plots to Fig. 18.5(a), (c), and (e).

18.1.7 The Kettle

We have already shown that the kettle fine-tunes the modes of the musically significant partials (Section 18.5) and increases their decay times by acting as a baffle (Section 18.7). Although the fine-tuning of the modes depends upon the kettle volume, the effect of kettle shape does not appear to be very significant (Tubis and Davis, 1986; Davis, 1989).

The bottom of the kettle nearly always has a small vent hole to equalize the average air pressure inside and outside. It has been suggested (Benade, 1976) that viscous friction in the air at the vent hole is the principal cause of the large damping of the (0,1) mode, and that the open vent hole prevents the rise in frequency that would ordinarily occur due to the added stiffness of the air enclosed by the kettle. Experimental observations by Christian et al. (1984) were in disagreement with both of these theories, however.

Closing the vent hole with a rubber stopper has little or no effect on the decay time of the (0,1) mode and lowers the modal frequency by a very small amount, typically 0.4%. At one tension, for example, the observed frequency of 135.4 $\pm$ 0.1 Hz and the decay time of 0.29 $\pm$ 0.05 s with the vent hole open only changed to 134.9 $\pm$ 0.1 Hz and 0.29 $\pm$ 0.05 s with the vent hole plugged. This is consistent with the observations of several experienced timpanists who could hear no consistent difference in the sound of the timpani with the vent hole closed or open.

18.2 Bass Drums

The bass drum is capable of radiating the most power of all the instruments in the orchestra. [A peak acoustical power of 20 W was observed by Sivian et al., (1931).] A concert bass drum usually has a diameter of 80–100 cm (32–40 in.), although smaller drums (50–75 cm or 20–30 in.) are popular in marching bands. Most bass drums have two heads, set at different tensions, but single-headed "gong" drums are widely used when a more defined pitch is appropriate. Mylar heads with a thickness of 0.010 in. (0.25 mm) are widely used, although calfskin heads are preferred by some percussionists for large concert bass drums.

Most drummers tune the *batter* or beating head to a greater tension than the *carry* or resonating head; some percussionists suggest that the difference be as much as 75% [giving an interval of about a fourth; see

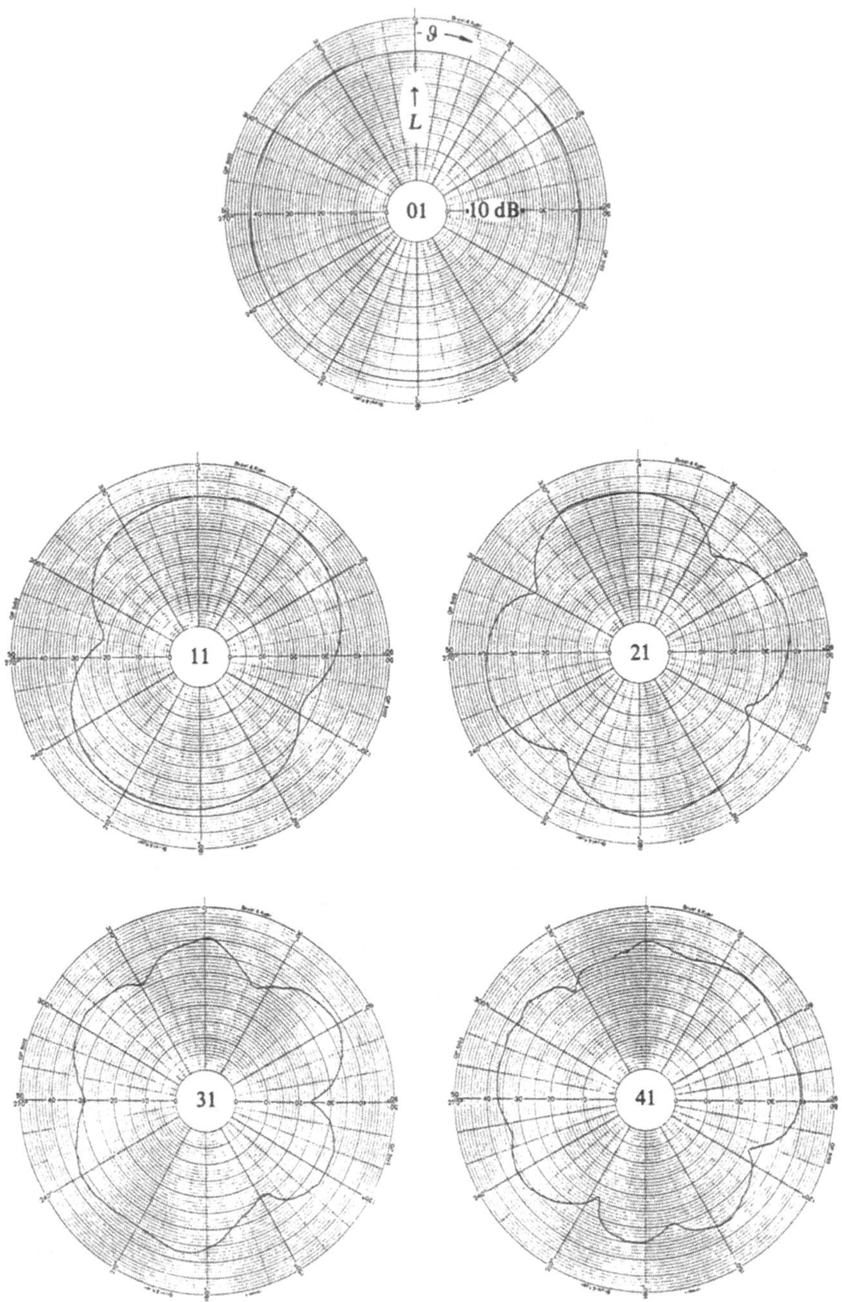

FIGURE 18.6. Sound radiation from a 40-cm-diameter kettledrum. The number of maxima is twice the number of nodal diameters (Fleischer, 1988).

Noonan, (1951)]. Levine (1978) recommends tuning the carry head higher than the batter head for orchestral music but lower than the batter head in solo or small ensemble playing. A distinctive timbre results from setting both heads at the same tension, but the prominent partials in the 70–300-Hz range appear to be stronger (initially) and to decay faster when the carry head is tuned below the batter head (Rossing, 1987).

Modal frequencies of an 82-cm-diameter bass drum are given in Table 18.5. Frequencies of the (0,1), (1,1), (2,1), (3,1), and (4,1) modes fall surprisingly near a harmonic series, and if their partials were the only ones heard, the bass drum sound would be expected to have a rather definite pitch. In the frequency range above 200 Hz, however, there are many inharmonic partials that sound louder since the ear discriminates against sounds of low frequency. Fletcher and Bassett (1978) found 160 partials in the frequency range 200–1100 Hz, although their scheme for associating these partials with vibrational modes of the membrane appears to be incorrect (Rossing, 1987).

Coupling between the two heads gives rise to an interesting doublet effect in the lowest modes. When the two heads were at the same tension, for example, we observed a pair of (0,1) modes at 44 and 104 Hz and a pair of (1,1) modes at 76 and 82 Hz. The (0,1) pair are the modes of a simple two-mass oscillator whose frequencies are given in Section 4.4 as $f_1 = f_0$ and $f_2 = \sqrt{f_o^2 + 2f_c^2}$, where f_o is the frequency of either head, and the coupling frequency f_c depends upon the stiffness of the air volume and the membrane masses. Setting the heads to the same tension tends to maximize this coupling.

From the two frequencies for the (0,1) modal pair given in Table 18.5, we calculate $f_c = 67$ Hz. However, the equivalent piston model (assuming a piston with area equal to one-half the total membrane area and with a mass equal to one-half the mass of the head) gives a somewhat higher value

TABLE 18.5. Modal frequencies in the bass drum head.[a]

Mode	Carry head at lower tension	Heads at same tension
0,1	39	44, 104
1,1	80	76, 82
2,1	121	120
3,1	162	160
4,1	204	198
5,1	248	240

[a] Rossing (1987).

of f_c = 89 Hz. Obviously, a more refined calculation of air loading using Green's functions is called for (compare Section 18.4).

Removing the carry head changes the modal frequencies but little from the values in the first column in Table 18.5 (for the carry head tuned below the batter head). Decay rates vary from 3 to 9 dB/s when the carry head is at a lower tension than the batter head; this increases to 6–11 dB/s when the heads are at the same tension. Removing the carry head gives decay rates from 3 to 8 dB/s, about the same as the preferred playing arrangement of unequal tensions in the two heads. Decay times are quite dependent upon the acoustical environment, however, because of the long wavelengths at these low frequencies.

The average surface tension of a membrane increases when it vibrates at finite amplitude. The increase in the average surface tension ΔT is proportional to the square of the displacement amplitude d, and the frequency is proportional to the square root of $T_0 + \Delta T$:

$$f(z) = \frac{1}{2\pi}\sqrt{\frac{T_0 + Kd^2}{\sigma}}. \tag{18.25}$$

Thus, each mode will have a greater frequency just after the drum is struck, decreasing as the amplitude dies down. When a bass drum is struck a full blow, a typical value for the initial amplitude is 6 mm, which results in an upward frequency shift of about 10%, or nearly a whole tone on the musical scale (Cahoon, 1970). Of course, the pitch shift is made less noticeable by the downward pitch shift with increasing sound intensity, a well-known psychoacoustical effect that is especially strong at low frequency [see, for example, Chapter 7 in Rossing (1982b)].

18.3 Snare Drums

The orchestral snare drum is a two-headed instrument about 35 cm in diameter and 13–20 cm deep. Strands of wire or gut stretch across the lower (snare) head. When the upper (batter) head is struck, the snare head vibrates against the snares. Alternatively, the snares can be moved away from the head to give a totally different sound

In the snare drum, like the bass drum, there is appreciable coupling between the two heads, especially at low frequencies. This coupling may take place acoustically through the enclosed air or mechanically through the shell, and it leads to the formation of mode pairs. In the first two modes of vibration of the drum, the batter and snare heads move in the manner of the (0,1) mode of an ideal membrane (see Fig. 3.6), as shown in Fig. 18.7. In the lower-frequency member of the pair, both heads move in the same direction, and in the higher-frequency mode they move in opposite directions.

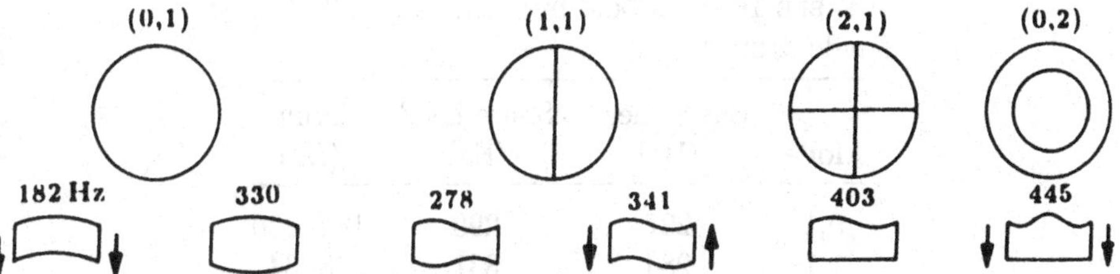

FIGURE 18.7. The lowest six vibrational frequencies of a snare drum. The modal designations (m, n) refer to the membrane modes shown in Fig. 3.6. Mode frequencies are for a typical snare drum.

A simple two-mass model describes the first two modes reasonably well (Rossing et al., 1992). The batter head is represented by a mass m_b and a spring with stiffness K_b, the snare head by a mass m_s and a spring constant K_s; the enclosed air constitutes a third spring with constant K_c connecting the masses. This system has two modes of vibration, whose angular frequencies are given by

$$\omega^2 = \tfrac{1}{2}(\omega_b^2 + \omega_s^2 + \omega_{cb}^2 + \omega_{cs}^2) \pm \tfrac{1}{2}\sqrt{[(\omega_b^2 + \omega_{cb}^2) - (\omega_s^2 + \omega_{cs}^2)]^2 + 4\omega_{cb}^2\omega_{cs}^2} \tag{18.26}$$

where

$$\omega_b^2 = \frac{K_b}{m_b}, \quad \omega_s^2 = \frac{K_s}{m_s}, \quad \omega_{cb}^2 = \frac{K_c}{m_c}, \quad \omega_{cs}^2 = \frac{K_c}{m_s}. \tag{18.27}$$

If the snare head (or batter head) is damped so that it cannot vibrate, one obtains a single frequency ω_b' for the batter head, or ω_s' for the snare head:

$$\omega_b' = \sqrt{\frac{K_b + K_c}{m_b}}, \qquad \omega_s' = \sqrt{\frac{K_s + K_c}{m_s}}. \tag{18.28}$$

The third and fourth modes in Fig. 18.7, in which the heads move in the manner of the (1,1) membrane mode, are more difficult to model. In the lower-frequency (1,1)-like mode, in which the heads move in opposite directions, air "sloshes" from side to side, and the mass of the air acts to lower the frequency. In the higher-frequency antisymmetric (1,1)-like mode, the air moves a smaller distance, essentially normal to the plane of the heads. Because the mass loading is thus diminished, the frequency of this mode is higher.

Frequencies of a few modes of vibration of a 36-cm-diameter snare drum are given in Table 18.6. In order to measure the modes of each head, the opposite head was damped with sandbags; thus, the frequencies are for the drumhead backed by the enclosed air.

The lowest modes of a free drum shell are the cylindrical shell modes having m nodal lines parallel to the axis and n circular nodes, as shown in Fig. 18.8(a). Holographic interferograms of two $(m, 0)$ modes and two $(m, 1)$

TABLE 18.6. Modal frequencies in a 36-cm side drum.[a]

Mode	Batter head (Hz)	Snare head (Hz)	Drum (Hz)
0,1	227	299	182, 330
1,1	284	331	278, 341
2,1	403	507	403
0,2	445	616	442
3,1	513	674	512
1,2	555	582	556
4,1	619	859	619

[a] Zhao, 1990.

modes are shown in Fig. 18.8(b). (The outside and inside surfaces were recorded simultaneously using two object beams.) Figure 18.8(c) shows two modes of a complete drum that are mainly shell modes.

18.3.1 Snare Action

The coupling between the snares and the snare head depends upon the mass and the tension of the snares. At a sufficiently large amplitude of the snare head, properly adjusted snares will leave the head at some point during the vibration cycle and then return to strike it, thus giving the snare drum its characteristic sound. The larger the tension on the snares, the larger the amplitude needed for this to take place.

Velocities of the snare head and snares in one snare drum are shown in Fig. 18.9. The snare velocity initially follows a sine curve whose period is greater than that of the snare head. Therefore, the snare head reverses its direction first, and the snares lose contact with the head. The smooth snare curve is disturbed when the snares, vibrating back ($v_s < 0$), meet the head which is already moving in the opposite direction ($v_h > 0$). Through the impact, higher modes of vibration are excited in the snares as well as the snare head.

For the snares to sound at all requires a certain amplitude of the snare head. This critical amplitude increases with the snare tension. The snare tension is optimum when both the head and the snares are moving at maximum speed in opposite directions at the moment of contact. In this case, the impact is the greatest.

Figure 18.10 shows sound spectra for a drum with three blow strengths and two snare tensions. At low snare tension, the medium blow causes the head to exceed its critical amplitude [Fig. 18.10(b)], whereas at the

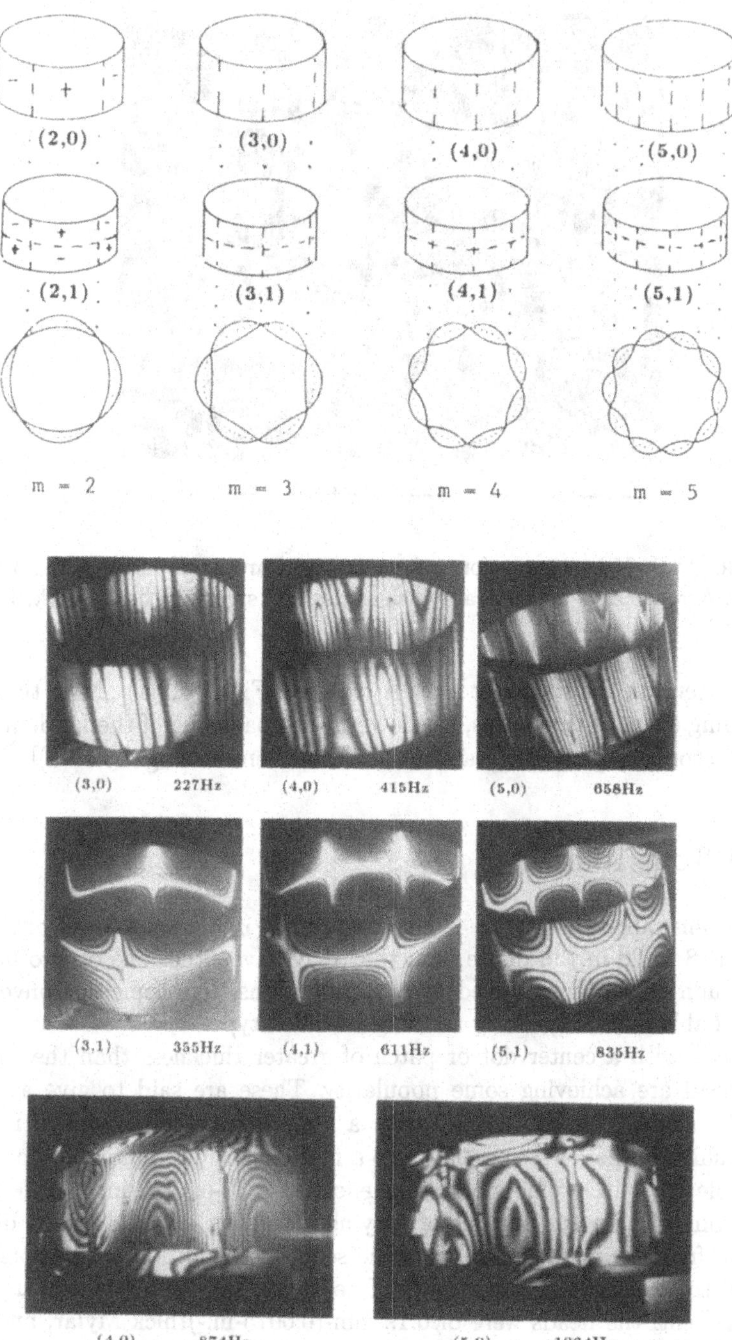

FIGURE 18.8. (a) Vibrational modes of a cylindrical drum shell. (b) Holographic interferograms of a snare drum shell without the heads. (c) Two modes of a complete drum that are mainly shell modes (after Rossing et al., 1992).

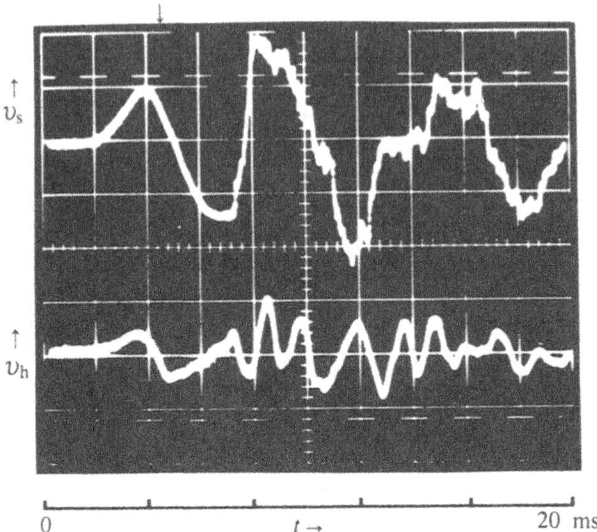

FIGURE 18.9. Velocity waveforms for snares (v_s) and snare head (v_h) in a snare drum. Arrow marks the time at which snares first strike the head (Bork, 1983).

higher tension the strongest blow is required [Fig. 18.10(f)]. Note that the damping effect of the snares, as shown by a broadening of the fundamental head resonance peak, is greater at low snare tension [Fig. 18.10(a)].

18.4 Tom-Toms

Tom-toms range from 20 to 45 cm (8 to 18 in.) in diameter and from 20 to 50 cm (8 to 20 in.) in depth, and they may have either one or two heads. Although generally classified as untuned drums, tom-toms do convey an identifiable pitch, especially the single-headed type.

Heads with a center dot or patch of greater thickness than the rest of the head are achieving some popularity. These are said to give a "centered," slightly "tubby" sound with a more distinct pitch, which results from shifting the lowest partials into a more nearly harmonic relationship.

Table 18.7 gives some modal frequencies of a 30-cm (12-in.), single-head tom-tom with center patches of varying diameters. Note the shift of the modal frequencies toward a harmonic sequence. Addition of the dots was found to increase the decay times of nearly all the modes as well. In this experiment, the heads were of 0.19-mm-(0.0075-in.-)thick Mylar, and the dots were of the same material 0.25-mm-(0.010-in.-)thick (Rossing, 1977).

Whereas a tom-tom with one head has a single fundamental mode of the (0,1) type, adding a second head adds a second (0,1) mode, as in the side drum. Since the volume is greater than the side drum, however, the

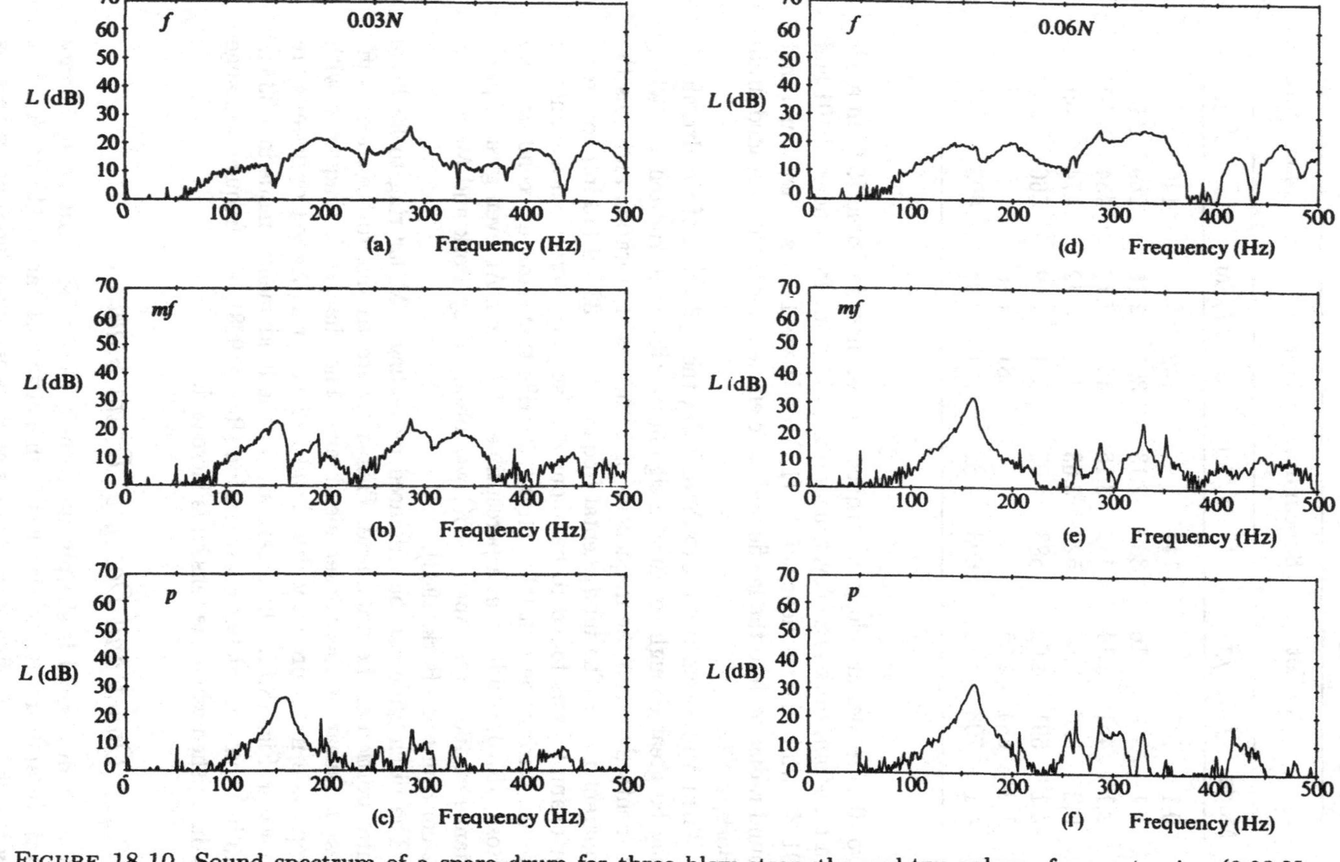

FIGURE 18.10. Sound spectrum of a snare drum for three blow strengths and two values of snare tension (0.03 N and 0.06 N) (Bork, 1983).

TABLE 18.7. Modal frequencies of a 30-cm tom-tom with and without center dots.

	No dot		8.9-cm dot		11.4-cm dot		14-cm dot	
Mode	f	f/f_{01}	f	f/f_{01}	f	f/f_{01}	f	f/f_{01}
0,1	147	1	144	1	142	1	140	1
1,1	318	2.16	307	2.13	305	2.15	289	2.06
2,1	461	3.14	455	3.16	450	3.17	434	3.10
0,2	526	3.58	522	3.63	485	3.42	474	3.39
3,1	591	4.02	583	4.05	581	4.09	566	4.04
1,2	684	4.65			681	4.80		
4,1	703	4.78	693	4.81	701	4.94	693	4.95

two (0,1) modes are closer in frequency than those in Table 18.6. In a 32-cm tom-tom, Bork and Meyer measured (0,1) mode frequencies of 101 and 191 Hz, while the frequency of the (1,1) mode was 179 Hz. They found that sound radiation into the far field is more efficient for the two-headed drum (Bork, 1983).

When a tom-tom is struck a hard blow, the deflection of the drumhead may be great enough to cause a significant change in the tension, which momentarily raises the frequencies of all modes of vibration and thus the apparent pitch. The fundamental frequency in a 33-cm (13-in.) tom-tom, for example, was found to be about 8% (slightly more than a semitone) greater during the first 0.2 s after the strike than after a second or more (Rose, 1978), resulting in a perceptible pitch glide. An even greater pitch change of 160 cents (about 10%) was observed by Bork and Meyer in a 32-cm tom-tom (Bork, 1983).

The pitch glide can be enhanced by adding a Mylar ring to load the outer portion of the drumhead. This is apparently due to the added stiffness to shear of the thicker membrane. The changes in frequency with amplitude in an ordinary drumhead and in a ring-loaded drumhead are shown in Fig. 18.11. These data are for the fundamental mode in a 33-cm (13-in.) tom-tom at fairly low tension (Rose, 1978). The frequency change is diminished when the tension is increased.

18.4.1 Onset and Decay of Drum Sound

When a drumhead is sharply struck near the center, most of the energy initially appears in the circularly symmetric (0,1) and (0,2) modes, but by the end of the first second, the sound spectrum includes many partials radiated by modes that received energy from the (0,1) and (0,2) modes to which they were coupled by nonlinear processes (see Chapter 5). At the

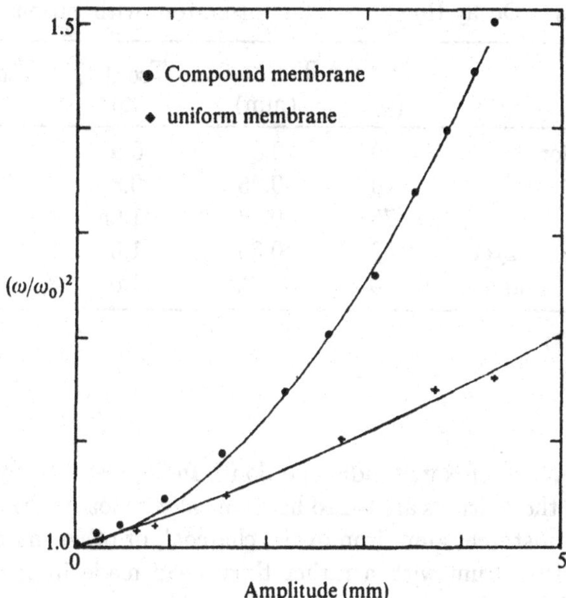

FIGURE 18.11. Dependence of frequency on amplitude for a 33-cm tom-tom head vibrating in its fundamental mode. The vertical axis is the square of the ratio of the frequency to the low-amplitude frequency. The tension was 351 N/m, which is at the low end of the normal playing range for this drum (Rose, 1978).

same time the spectral peaks associated with the (0,1) and (0,2) modes narrow substantially.

The sound decay time depends on a number of factors, such as the type of drumhead, the tension, the kettle weight and material, and especially the way in which the drum is supported. Changing the length of the arm on which a drum is supported, for example, can change the decay time of the fundamental from 5.5 s to 0.6 s, and adding mass to the kettle tends to lengthen the decay time (Bork, 1983). Decay times measured in the same drum with several different types of heads are given in Table 18.8. The 60-dB decay time of the higher (0,1) mode and that observed in the one-third-octave band at 500 Hz are given. The double-layer Pin Stripe head and the center-loaded Controlled Sound (CS) head are seen to have shorter decay times, especially at the higher frequency.

18.5 Indian Drums

Nowhere in the world is the drum considered to be a more important musical instrument than in India. Foremost among the drums of India are the

TABLE 18.8. Decay times of a 32-cm tom-tom with various heads.[a]

Head	Mass (g)	Thickness (mm)	T_{60}:(01) (s)	T_{60}:500 Hz (s)
Ambassador	50	0.3	0.9	2.5
Diplomat	40	0.25	0.8	2.7
Emperor	75	0.46	1.05	2.5
Pin stripe (2 layers)	62	0.35	1.0	1.4
Controlled sound	50	0.3-0.4	1.0	1.1

[a] Bork, 1983.

tabla and *mrdanga* of North India and South India, respectively. The overtones of both these drums are tuned harmonically by loading the drumhead with a paste of starch, gum, iron oxide, charcoal, or other materials.

The *tabla* is a drum with a rather thick head made from three layers of animal skin (calf, sheep, goat, or buffalo skins are apparently used in different regions). The innermost and outermost layers are annular, and the layers are braided together at their outer edge and fastened to a leather hoop. Small straws or strings are placed around the edge between the outer and middle head.

Tension is applied to the head by means of a long leather thong that weaves back and forth (normally 16 times) between the top and bottom of the drum. The tension in the thong can be changed by moving small wooden cylinders up or down, and fine-tuning of the head is accomplished by upward or downward taps on the hoop with a small hammer.

To the center of the head is applied a circular patch of black paste that is built up in many thin layers. The paste consists of boiled rice and water with heavy particles, such as iron oxide or manganese dust, added to increase the density. Each layer is allowed to dry and is then rubbed with a smooth stone until tiny cracks appear in the surface. The patch ends up with a slightly convex surface.

The tabla we have described is usually played along with a larger drum, called by various names: banya, bayan, bhaya, dugga, or left-handed tabla. The head of this larger drum is also loaded, but slightly off center, and the shell may be of clay, wood, or metal. In the usual mode of playing, the player has the edge of the palm resting on the widest portion of the unloaded membrane, and this constraint causes the nodal patterns to be quite symmetrical. Releasing palm pressure produces a sound of different quality. A tabla pair is shown in Fig. 18.12(a).

The *mrdanga* or mrdangam is an ancient, two-headed drum that functions, in many respects, as a tabla and banya combined into one. The smaller head, like that of the tabla, is loaded with a patch of dried paste,

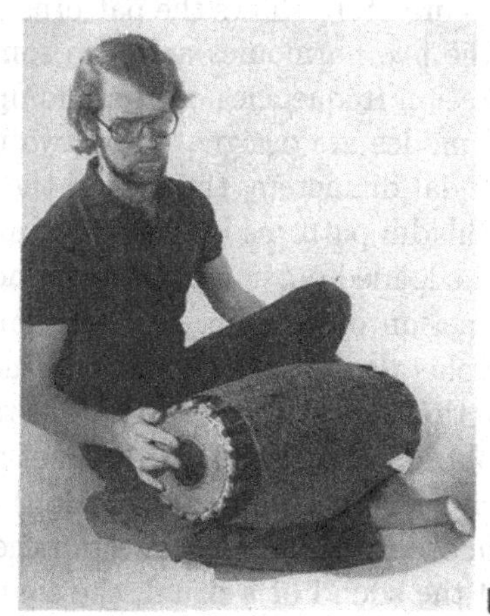

a b

FIGURE 18.12. (a) Tabla and (b) mrdanga.

while the larger head is normally loaded with a paste of wheat and water shortly before playing. A mrdanga is shown in Fig 18.12(b).

The acoustical properties of these drums have been studied by a succession of Indian scientists beginning with C.V. Raman. Raman and his colleagues recognized that the first four overtones of the tabla are harmonics of the fundamental. Later they identified the five harmonics as coming from nine normal modes of vibration, several of which have the same frequencies. The fundamental is from the (0,1) mode; the second harmonic is from the (1,1) mode; the (2,1) and (0,2) modes supply the third harmonic; the (3,1) and (1,2) modes similarly supply the fourth harmonic; and three modes, the (4,1), (0,3), and (2,2), contribute to the fifth harmonic.

Figure 18.13 shows Chladni patterns of six different modes, all of which have frequencies near the third harmonic. These patterns, published by Raman in 1934, were obtained by sprinkling fine sand on the membrane before or immediately after the stroke. The sand gathers along the nodes (lines of least vibration) and thus forms a map of the vibration pattern excited by that stroke. Figure 18.13(a) shows the (0,2) mode and Fig. 18.13(f) the (2,1) mode; the other four modes are combinations of the (0,2) and (2,1) normal modes. Note that exciting any of the modes in Fig. 18.13 will result in a third harmonic partial.

Figure 18.14, also from Raman's paper, shows how the mode in Fig. 18.13(c) can be excited by touching the membrane with two fingers and striking it with a third. It is easy for us to excite these modes by driving the membrane at the proper frequency with an audio amplifier and a suitably placed loudspeaker, but these were not available to Raman in 1934.

Figure 18.15 shows the patterns of the nine normal modes corresponding to the five harmonics and also some of the combination modes that have vibration frequencies corresponding to the five tuned harmonics. The normal modes are designated by two numbers: the first indicates the number of nodal diameters, the second the number of nodal circles.

Chladni patterns indicate that most of the vibrational energy is confined to the loaded portion of the drumhead. This is accentuated by the restraining action of a rather stiff annular leather flap in loose contact with the peripheral portion of the drumhead. Thus, the head is essentially divided into three concentric regions. The tabla player uses these concentric regions to obtain three distinctly different sounds, which can be described as "tun" (center), "tin" (unloaded portion), and "na" (outermost portion).

In order to study the effect of the center patch on the modes of vibration and the sound of a drum, the sound spectrum was measured at 32 stages (roughly each three layers) during application of a patch to the head of a mrdanga. The paste was prepared by kneading together roughly equal volumes of overcooked rice and a black powder composed of manganese and iron oxide. The area at the center of the head was cleaned, dried, and scraped with a knife to raise the nap and provide good adhesion. A thin layer of overcooked rice was first smeared onto this surface as glue, and then a small lump of the paste was applied. It was smeared out evenly with a swirling motion by the thumb. The excess was scraped away with a knife, and a smooth rock was used to pack and polish the mixture by rubbing the surface. The resulting spectral frequencies are shown in Fig. 18.16.

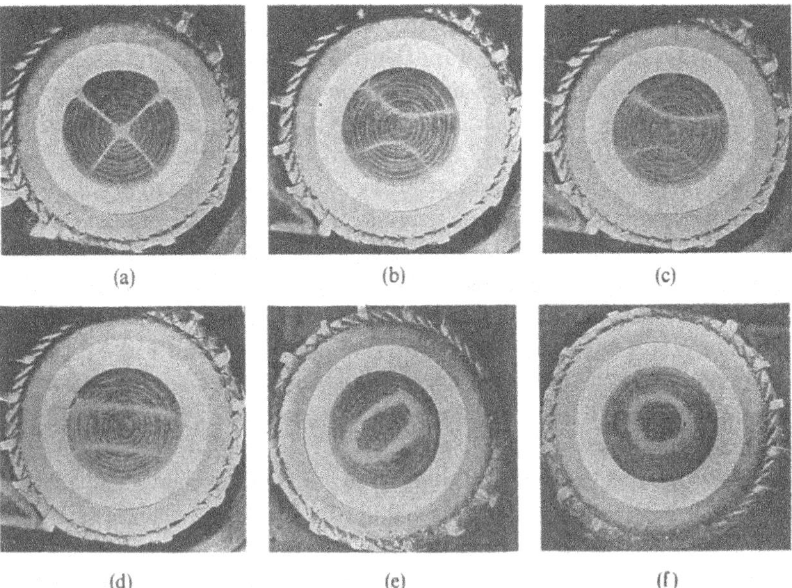

FIGURE 18.13. Chladni patterns of six different modes of the tabla, all of which have frequencies near the third harmonic. The (0,2) and (2,1) normal modes are shown in (a) and (f), respectively; the other modes are combinations of these (from Raman, 1934).

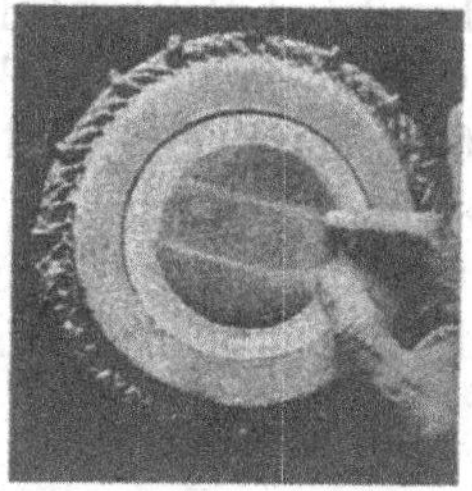

FIGURE 18.14. Method used to excite the mode in Fig. 18.13(c). Two fingers damp the membrane at the two nodes, and a third finger strikes it at a point 90° away (from Raman, 1934).

Note the five harmonic partials in the sound of the fully loaded membrane. Note also that several of these partials originate from two or three modes of vibration tuned to have the same frequency by appropriately loading the membrane. This is similar to the behavior Raman noted in the tabla.

Studies by Ramakrishna and Sondhi (1954) at the Indian Institute of Science in Bangalore and by De (1978) in Santiniketan, West Bengal, have

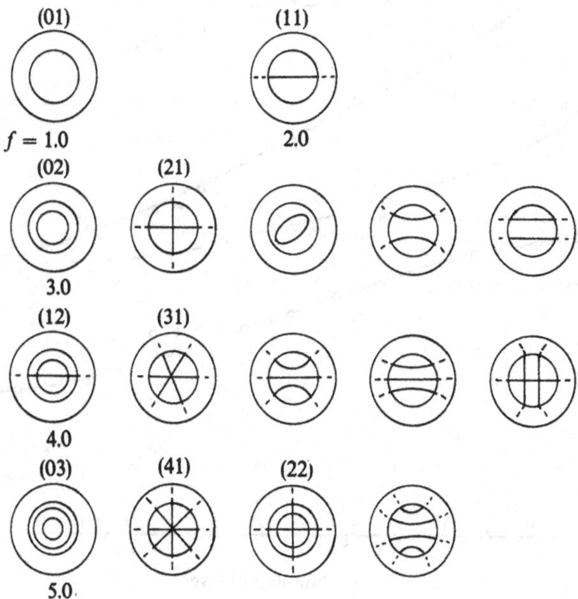

FIGURE 18.15. Nodal patterns of the nine normal modes plus seven of the combination modes that correspond to the five harmonics of a tabla or mrdanga head.

indicated that the areal density of the loaded portion of the membrane should be approximately 10 times as great as the unloaded portion (which is typically around 0.02–0.03 g/cm^2). The total mass of the loaded portion is in the range of 9–15 g for different tabla. We estimate that the mrdanga patch in Fig. 18.16, which was about 3 mm thick at the center, had a total mass of 29 g and an areal density of 0.8 g/cm^2. The density of the dry paste was about 2.8 g/cm^3.

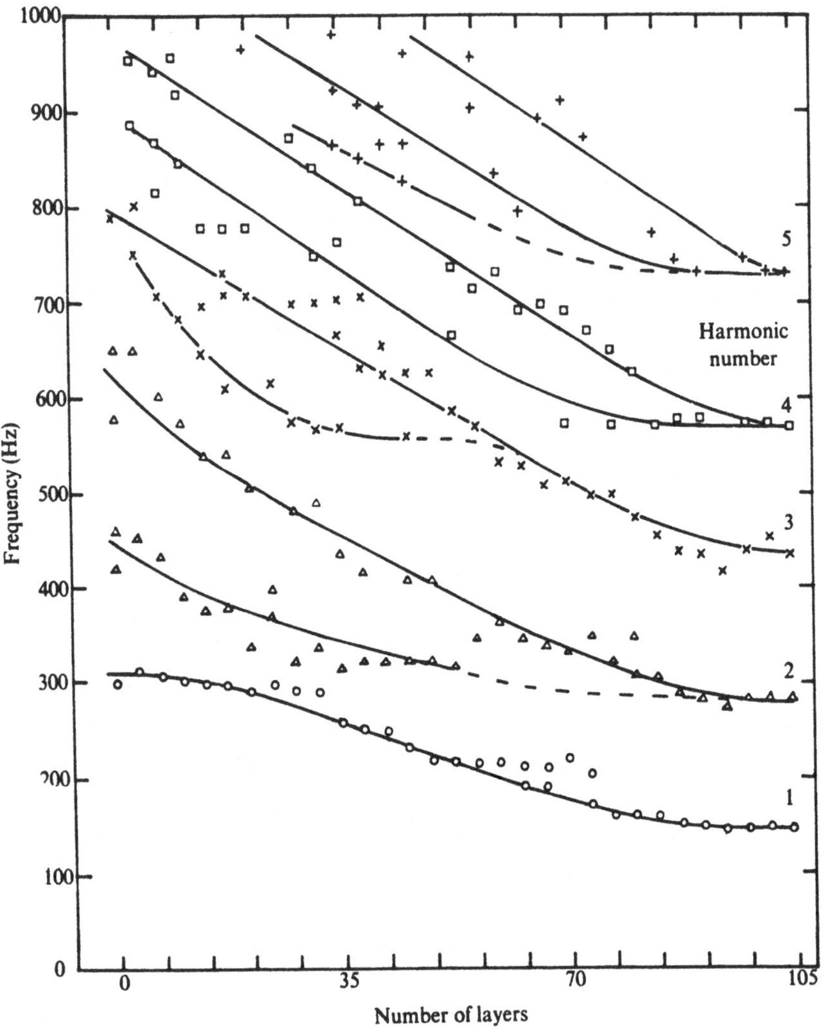

FIGURE 18.16. Frequencies of prominent partials in the sound spectra of the mrdanga with no patch and at various stages during application of the patch (from Rossing and Sykes, 1982).

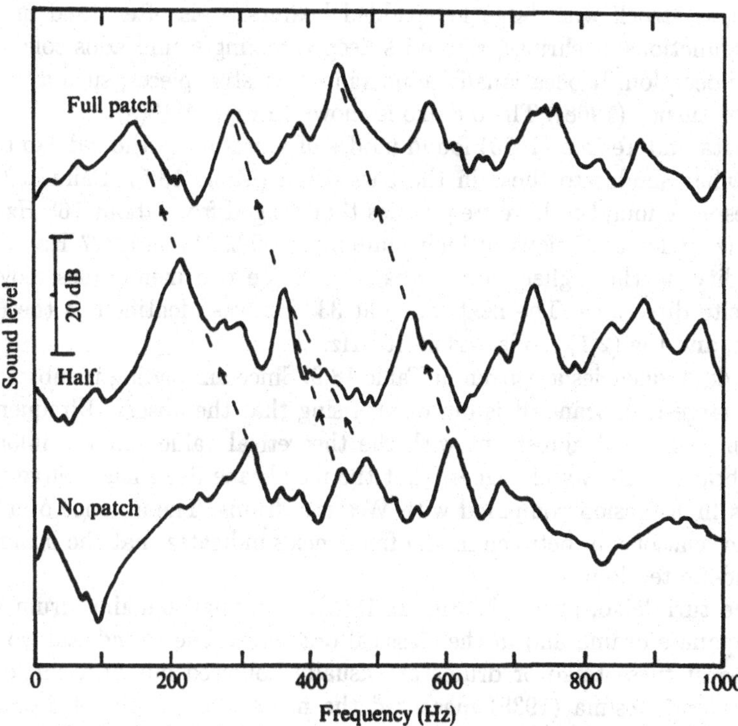

FIGURE 18.17. Sound spectra from the *din* stroke with no patch, with about half the layers in place, and with the finished patch.

Figure 18.17 compares the sound spectra from the "din" stroke with no patch, with about half the layers in place, and with the finished patch. The differences are rather striking. The fundamental (0,1) mode has moved down in frequency from 295 Hz to 144 Hz, slightly more than one octave. The other modes have moved by varying amounts into a nearly harmonic frequency relationship.

18.6 Japanese Drums

More than 50 years ago, Juicha Obata and his colleagues at the Tokyo Imperial University made acoustical studies of several Japanese drums. It appears that little has been published, at least in Western journals, since that time.

The o-daiko is a large drum consisting of two cowhide membranes stretched tightly across the ends of a wooden cylinder 50–100 cm in diameter and about 1 m in length. The drum, hanging freely in a wooden frame, is struck with large felt-padded beaters. It is often used in religious functions at shrines, where its deep rumbling sound adds solemnity to the occasion. It occasionally appears in orchestral pieces, such as Orff's "Prometheus" (1968). The o-daiko is shown in Fig. 18.18(a).

Obata and Tesima (1935) found modes of vibration in the o-daiko to be somewhat similar to those in the bass drum (Table 18.5). Pairs of (0,1) modes were found to have frequencies that ranged from about 168 Hz and 193 Hz under conditions of high humidity to 202 Hz and 227 Hz at low humidity; in the higher mode of each pair, the two membranes move in opposite directions. The next mode at 333 Hz was identified as the (1,1) mode, and the (2,1) mode was at 468 Hz.

Mode frequencies are given in Table 18.9. Since air loading is substantial for a large membrane, it is quite surprising that the observed frequencies are in such good agreement with the theoretical values for an unloaded membrane. This would suggest that the membrane itself has a large mass and a high tension compared with Western drums. The absence of a harmonic relationship between modal frequencies indicates that the drum has an indefinite pitch.

The turi-daiko, shown in Fig. 18.18(b), is a small hanging drum used in Japanese drama and in the classical orchestra. The cylindrical wooden bodies of these taiko or drums are usually hollowed out of a single log. Obata and Tesima (1935) measured the modal frequencies of a shallow turi-daiko 29 cm in diameter with a length of 7 cm. A single (0,1) mode was observed to have a frequency of 195 Hz. Their experiments showed that

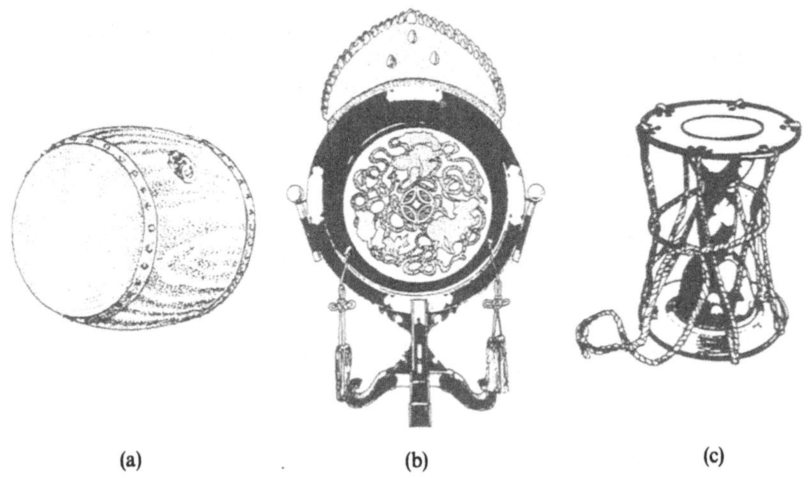

FIGURE 18.18. Japanese drums: (a) O-daiko, (b) turi-daiko, and (c) kotodumi.

TABLE 18.9. Modal frequencies of the Japanese o-daiko.[a]

Mode	Frequency	f/f_{01}	f/f_{01} (ideal)	f/f_{11}	f/f_{11} (ideal)
0,1	202	1.00	1.00	0.61	0.63
0,1	227	1.12		0.68	
1,1	333	1.65	1.59	1.00	1.00
	383	1.90		1.15	
2,1	468	2.32	2.13	1.41	1.34
0,2	492	2.44	2.29	1.48	1.44
3,1	544	2.69	2.65	1.63	1.66
1,2	621	3.07	2.92	1.86	1.83
4,1	695	3.44	3.15	2.09	1.98
	739	3.66		2.22	
	865	4.28		2.60	
	905	4.48		2.72	
	1023	5.06		2.07	

[a] Adapted from Obata and Tesima, 1935.

the fundamental frequency is lowered by increasing the length (volume) of the drum.

The wooden body of the tudumi or tsuzumi has a dumbbell shape, and the faces, which are larger in diameter than the body, consist of leather heads stretched over iron rings and secured by stitching [see Fig. 18.18(c)]. The faces are connected by hemp cords, which allow the drummer to vary the tension. There are two different sizes of tudumi, called o-kawa and kotodumi. The faces of the larger o-kawa are usually cowhide, while the smaller kotudumi has deerhide faces.

The kotudumi produces four different tones, designated as "ta," "ti," "pu," and "po," according to the manner of grasping the cords and striking the face. For the ta and ti tones, the cords are tightly grasped, giving the membrane its maximum tension. The pu and po tones begin at high membrane tension, but immediately after the stroke they are loosened, causing a gradual lowering in pitch of the tone. Differences between ta and ti or between pu and po are created by differences in the manner of striking, namely, the number of fingers used and the length of the stroke.

Obata and Ozawa (1931) found the fundamental frequencies of the ta and ti (around 300 Hz) to be about one tone higher than the pu and po (around 270 Hz). They found a prominent partial about three times the fundamental frequency, but their assumption that this is from the (1,2) mode [which has a frequency 2.9 times that of the (0,1) mode in an unloaded membrane] is probably not justified.

The larger o-kawa is usually struck with a gloved hand. In spite of its larger size, it has a fundamental frequency about an octave higher than the kotudumi, because it has greater tension. It, too, has a prominent third-harmonic partial in its sound spectrum (Obata and Ozawa, 1931).

18.7 Indonesian Drums

Drums are very important instruments in both Balinese and Javanese gamelans. The drummer in a gamelan sets the tempo and thus, in a sense, conducts the gamelan. Nearly all kendang or drums in the gamelan are two headed, and they come in various sizes. Shown in Fig. 18.19 are a large kendang gending and a small kendang ciblon. Note the difference in the cross sections of the carved shells. The sound spectrum of a Balinese kendang is shown in Fig. 18.20. The nearly harmonic overtones give the drum a fairly definite pitch.

18.8 Latin American Drums

Musicologists point out that much of what is considered Latin American music had its roots in Africa. Perhaps that is one reason why various types of drums play such an important role. Although many interesting drums are found throughout Central and South America, we will consider only three familiar ones that have found their way into dance orchestras and jazz ensembles: bongo drums, conga drums, and timbales.

The conga drum evolved from an African drum that was constructed from a hollowed-out tree. Nowadays, they are usually constructed of long

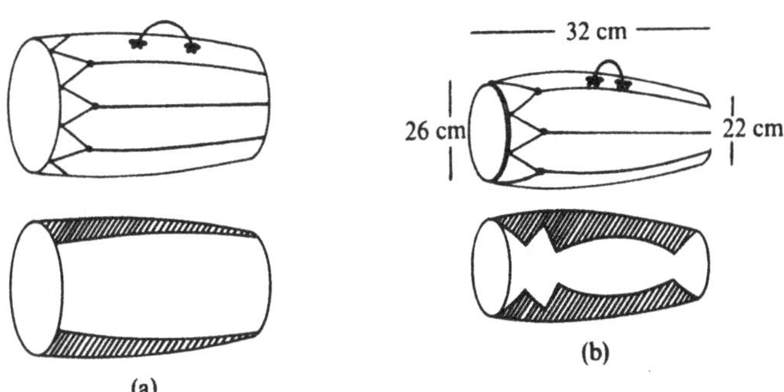

FIGURE 18.19. Javanese drums: (a) kendang gending and (b) kendang ciblon (from Lindsay, 1979).

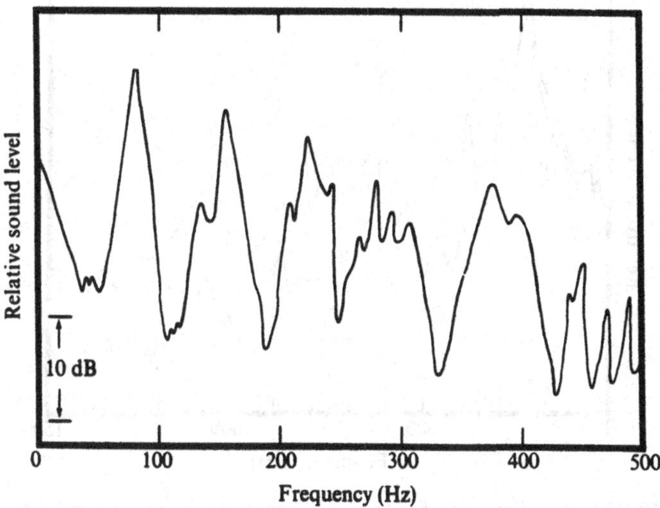

FIGURE 18.20. Sound spectrum of a Balinese drum (from Rossing and Shepherd, 1982).

strips of wood, grooved and glued together, and a tensioning mechanism is added to tune the heads. Head diameters measure from 23 cm to 30 cm (9 in. to 12 in.), and their pitches are in the range of C_3–C_4 (131–262 Hz). The individual drums are sometimes designated "quinto," "conga," and "tumba" (from small to large, respectively).

There are three basic sounds on the conga:

1. an open tone produced by striking the rim with the palm of the hand, allowing the fingers to slap on the head;
2. a slap produced by striking the drum with the palm of the hand near the rim, allowing the fingers to be slapped onto the head; and
3. a bass tone, in which the fleshy part of the hand strikes the center of the head and rebounds, allowing the head to vibrate freely in its fundamental mode.

Other important sounds are called the "wave" (or heel-and-toe), in which the striking hand makes a rocking motion, and the closed tone, produced by striking the drum at or near the center with the palm of the hand, which remains in contact long enough to damp the head.

Sound spectra of the open tone, slap, and bass tone are shown in Fig. 18.21. The first two peaks, which occur around 80 Hz and 240 Hz, appear to be caused by piston-like (0,1) motion of the head coupled to the first two pipe-like modes of the tubular body. The next peak, around 420 Hz (not present in the bass tone where the drum is struck at the center), corresponds to the (1,1) mode in the head. The peak at 620 Hz probably results from the (2,1) mode, but we have not verified this experimentally.

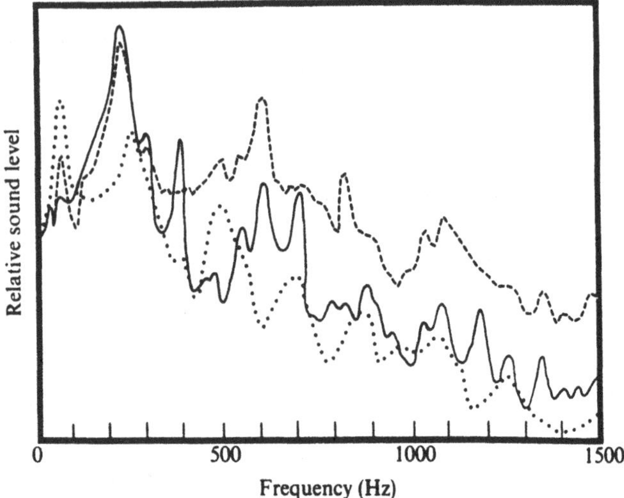

FIGURE 18.21. Sound spectra for three different strokes on a conga drum: open tone —, slap - - - -, and bass tone · · · ·.

Bongo drums have the highest pitch among Latin American drums. They are played in pairs, and range from about 15 cm to 25 cm (6 in. to 10 in.) in diameter. Their wooden shells are conical in shape, and the heads are usually goat skin. In the original version, the skins were nailed to the shell, but most modern bongos have tensioning screws.

Timbales were originally small kettle drums, consisting of animal hide stretched over wooden bowls. Modern timbales usually have metal shells with about 30–35-cm (12–14-in.) diameters. The shells are either totally open at the bottom or they have a sound hole about the size of a hand. They are played with light wooden sticks, and they may be struck on the shell as well as on the head.

18.9 Tambourines

The tambourine is a very old instrument. Mentioned in the Old Testament, it was well known in ancient Asian cultures. It became a favorite instrument of the Gypsies in Spain, and has remained so in modern times. Shells of modern tambourines may be either wood or metal, and the heads, which range from 15 cm to 30 cm (6 in. to 12 in.) in diameter, may be either calfskin or plastic. Tensioning screws may or may not be provided for tuning the heads.

What distinguishes tambourines from other drums are the metal discs or jingles, about 5 cm in diameter, attached to wires set in a series of slots around the drum. The instrument is played by shaking, striking, or even rubbing the head with a moistened thumb. Needless to say, the tambourine produces a wide variety of different sounds.

References

Aebischer, H.A., and Gottlieb, H.P.W. (1990). Theoretical investigations of the annular kettledrum as a new harmonic musical instrument. *Acustica* **72**, 107–117; Errata: *Acustica* **73**, 171–174 (1991)

Benade, A.H. (1976). "Fundamentals of Musical Acoustics." Oxford Univ. Press, London and New York.

Blades, J. (1970). "Percussion Instruments and Their History." Faber and Faber, London.

Bork, I. (1983). "Entwicklung von akustischen Optimierungsverfahren für Stabspiele und Membraninstrumente." PTB report, Project 5267, Braunschweig, Germany (unpublished).

Brindle, R.S. (1970). "Contemporary Percussion." Oxford Univ. Press, London and New York.

Cahoon, D.E. (1970). "Frequency-Time Analysis of the Bass Drum Sound." M.S. thesis, Brigham Young University, Provo, Utah.

Christian, R.S., Davis, R.E., Tubis, A., Anderson, C.A., Mills, R.I., and Rossing, T.D. (1984). Effects of air loading on timpani membrane vibrations. *J. Acoust. Soc. Am.* **76**, 1336–1345.

Davis, R.E. (1989). "Mathematical Modeling of the Orchestral Timpani." Ph.D. thesis, Purdue University, West Lafayette, Indiana.

De, S. (1978). Experimental study of the vibration characteristics of a loaded kettledrum. *Acustica* **40**, 206–210.

Fleischer, H. (1988). "Die Pauke: Mechanischer Schwinger und akustiche Strahler." München, Univ. der Bundeswehr.

Fletcher, H., and Bassett, I.G. (1978). Some experiments with the bass drum. *J. Acoust. Soc. Am.* **64**, 1570–1576.

Levine, D. (1978). Drum tension and muffling techniques. *The Instrumentalist*, June 1978. Reprinted in "Percussion Anthology," pp. 545–547. The Instrumentalist, Evanston, Illinois, 1980.

Lindsay, J. (1979). "Javanese Gamelan." Oxford, Singapore.

Mills, R.I. (1980). "The Timpani as a Vibrating Circular Membrane: An Analysis of Several Secondary Effects." M.S. thesis, Northern Illinois University, DeKalb, Illinois.

Morse, P.M. (1948). "Vibration and Sound," 2nd ed. McGraw-Hill, New York.

Noonan, J.P. (1951). The concert bass drum. *The Instrumentalist*, October 1951. Reprinted in "Percussion Anthology," pp. 53–54. The Instrumentalist, Evanston, Illinois, 1980.

Obata, J., and Ozawa, Y. (1931). Acoustical investigation of some Japanese musical instruments. Part III. The tudumi, drums with dumb-bell-shaped bodies. *Proc. Phys.-Math. Soc. Jpn.* **13** (3), 93–105.

Obata, J., and Tesima, T. (1935). Experimental studies on the sound and vibration of drum. *J. Acoust. Soc. Am.* **6**, 267–274.

Ramakrishna, B.S., and Sondhi, M.M. (1954). Vibrations of Indian musical drums regarded as composite membranes. *J. Acoust. Soc. Am.* **26**, 523–529.

Raman, C.V. (1934). The Indian musical drum. *Proc. Indian Acad. Sci.* **A1**, 179–188. Reprinted in "Musical Acoustics: Selected Reprints," ed. T.D. Rossing, Am. Assn. Phys. Teach., College Park, MD, 1988.

Rayleigh, Lord (1894). "The Theory of Sound," 2nd ed., Vol. I, p. 348. Macmillan, New York. Reprinted by Dover, New York, 1945.

Rose, C.D. (1978). "A New Drumhead Design: An Analysis of the Nonlinear Behavior of a Compound Membrane." M.S. thesis, Northern Illinois University, DeKalb, Illinois.

Rossing, T.D. (1977). Acoustics of percussion instruments—Part II. *The Physics Teacher* **15**, 278–288.

Rossing, T.D. (1982a). The physics of kettledrums. *Scientific American* **247** (5), 172–178.

Rossing, T.D. (1982b). "The Science of Sound." Addison-Wesley, Reading, Massachusetts.

Rossing, T.D. (1987). Acoustical behavior of a bass drum. *114th Meeting Acoust. Soc. Am., Miami* (abstract in *J. Acoust. Soc. Am.* **82**, S69).

Rossing, T.D., Anderson, C.A., and Mills, R.I. (1982). Acoustics of timpani. *Percussive Notes* **19** (3), 18–31.

Rossing, T.D., Bork, I., Zhao, H., and Fystrom, D. (1992). Acoustics of snare drums. *J. Acoust. Soc. Am.* **92**, 84–94.

Rossing, T.D., and Kvistad, G. (1976), Acoustics of timpani: Preliminary studies. *The Percussionist* **13**, 90–96.

Rossing, T.D., and Shepherd, R.B. (1982). Acoustics of gamelan instruments. *Percussive Notes* **19** (3), 73–83. Reprinted in "Musical Acoustics: Selected Reprints," ed. T.D. Rossing, Am. Assn. Phys. Teach., College Park, MD, 1988.

Rossing, T.D., and Sykes, W.A. (1982). Acoustics of Indian drums. *Percussive Notes* **19** (3), 58–67.

Sivian, L.J., Dunn, H.K., and White, S.D. (1931). Absolute amplitudes of spectra of certain musical instruments and orchestras. *J. Acoust. Soc. Am.* **2**, 330.

Tubis, A., and Davis, R.E. (1986). Kettle-shape dependence of timpani normal modes, Paper K2-7. *12th Int'l. Congress on Acoustics, Toronto.*

Zhao, H. (1990). "Acoustics of Snare Drums: An Experimental Study of the Modes of Vibration, Mode Coupling and Sound Radiation Pattern." M.S. thesis, Northern Illinois University.

19

Mallet Percussion Instruments

An idiophone (self-sounder) is a musical instrument, usually of wood, metal, or plastic, that makes a musical sound when struck, plucked, or rubbed. From a musical standpoint, tuned idiophones (e.g., the xylophone, marimba, vibraphone, bells, glockenspiel, chimes, celesta, gong) produce tones of a definite pitch, whereas untuned idiophones (e.g., the cymbal, tam-tam, triangle, and wood block) do not. From a mechanical standpoint, we could classify them into one-dimensional and two-dimensional vibrating systems, as discussed in Chapters 2 and 3, respectively. This chapter deals with one-dimensional idiophones, mostly of the tuned type and played with various types of mallets.

In Sections 2.14–2.17, we discussed the vibrations of bars or rods, which are the physical basis of the musical instruments discussed in this chapter. Mostly, we dealt with thin bars having both ends free. A bar of uniform thickness with free ends vibrates in a series of normal modes whose frequencies, given by Eq. (2.63), are approximately in the ratios: $(3.01)^2 : 5^2 : 7^2 : 9^2 : 11^2 : 13^2 \ldots = 9.1 : 25 : 49 : 81 : 121 : 169 \ldots$. These are quite inharmonic, although the fourth, fifth, and sixth modes are not too far removed from the ratio 2:3:4.

Elastic properties of a number of common materials are given in Table 19.1. Bars of identical size and shape will have vibration frequencies that scale with the velocities of sound in the materials. Note that the sound velocities in aluminum, steel, glass, and Sitka spruce are within 3% of one another, whereas those of brass, copper, and maple are about 30% lower, and thus bars of these materials would vibrate at frequencies about 14% lower.

19.1 Glockenspiel

The glockenspiel, or orchestra bells, uses rectangular steel bars 2.5–3.2 cm (1–$1\frac{1}{4}$ in.) wide and 6–9 mm ($\frac{1}{4}$–$\frac{3}{8}$ in.) thick. Its range is customarily from G_5(f = 784 Hz) to C_8(f = 4186 Hz), although it is scored two octaves lower than it sounds. The glockenspiel is usually played with brass or hard plastic mallets. The bell lyra is a portable version that uses aluminum bars, usually covering the range $A_5(f = 880$ Hz) to $A_7(f = 3520$ Hz), and is used in some marching bands.

When struck with a hard mallet, a glockenspiel bar produces a crisp, metallic sound, which quickly gives way to a clear ring at the designated pitch. Because the overtones have very high frequencies and die out rather quickly, they are of relatively less importance in determining the timbre of the glockenspiel than are the overtones of the marimba or xylophone, for example. For this reason, little effort is made to bring the inharmonic overtones of a glockenspiel into a harmonic relationship through overtone tuning.

The frequencies for transverse vibrations in a bar with free ends were shown to be inharmonic; thus, the glockenspiel has no harmonic overtones. However, even its lowest note has a frequency of 2160 Hz, so its overtones occur in a range in which the pitch discrimination of human listeners is diminished.

The vibrational modes of a glockenspiel bar are shown in Fig. 19.1. Besides the transverse modes, which are labeled 1, 2, 3, 4, and 5, there are torsional or twisting modes labeled a, b, c, and d; a longitudinal mode l; and transverse modes in the plane of the bar, $1x$ and $2x$.

The transverse modes of vibration of the glockenspiel bar are spaced a little closer together than predicted by the simple theory of thin bars (see Section 2.17). Note that the four torsional modes in Fig. 19.1 have frequencies in the ratios 1.01:2.00:3.00:3.94. Similarly, the frequencies of the longitudinal modes (of which only the lowest mode is shown in Fig. 19.1) form a harmonic series [Eq. (2.53)]. The frequency ratios between the torsional and longitudinal modes are given by the factor $\sqrt{K_T/2I(1+\nu)}$, where I is the polar moment of inertia, K_T is the torsional stiffness, and ν is Poisson's ratio (see Section 2.19). For a metal bar whose width W is more than six times its thickness h, the torsional and longitudinal modes will have frequencies approximately in the ratios of $1.2h/W$. For the bar in Fig. 19.1, the ratios are 0.32.

19.2 The Marimba

The marimba typically includes 3–4$\frac{1}{2}$ octaves of tuned bars of rosewood or synthetic fiberglass, graduated in width from about 4.5 to 6.4 cm ($1\frac{3}{4}$ to $2\frac{1}{2}$ in.). Beneath each bar is a tubular resonator tuned to the fundamental frequency of that bar. When the marimba is played with soft mallets, it produces a rich mellow tone. The playing range of a large concert marimba is commonly A_2 to $C_7(f = 110$–2093 Hz), although bass marimbas extend down to $C_2(f = 65$ Hz). A few instruments cover a full five-octave range.

TABLE 19.1. Elastic properties of materials.

Material	Density $\rho(\mathrm{kg/m^3})$	Young's modulus $E(\mathrm{N/m^2})$		Sound velocity $v(\mathrm{m/s})$		Reference
Aluminum	2700	7.1×10^{10}		5150		Kinsler et al., 1982
Brass	8500	10.4×10^{10}		3500		
Copper	8900	10.4×10^{10}		3700		
Steel	7700	19.5×10^{10}		5050		
Glass	2300	6.2×10^{10}		5200		
Wood		$\parallel$ grain	$\perp$ grain	$\parallel$ grain	$\perp$ grain	
Brazilian rosewood	830	1.6×10^{10}	2.8×10^{9}	4400	1800	Haines, 1979
Indian rosewood	740	1.2×10^{10}	1.7×10^{9}	4000	1500	
African mahogany	550	1.2×10^{10}	1.2×10^{9}	5000	1500	
European maple	640	1.0×10^{10}	2.2×10^{9}	4000	1800	
Redwood	380	0.95×10^{10}	0.96×10^{9}	5000	1600	
Sitka spruce	470	1.3×10^{10}	1.3×10^{9}	5200	1700	

[a] Adapted from Obata and Tesima, 1935.

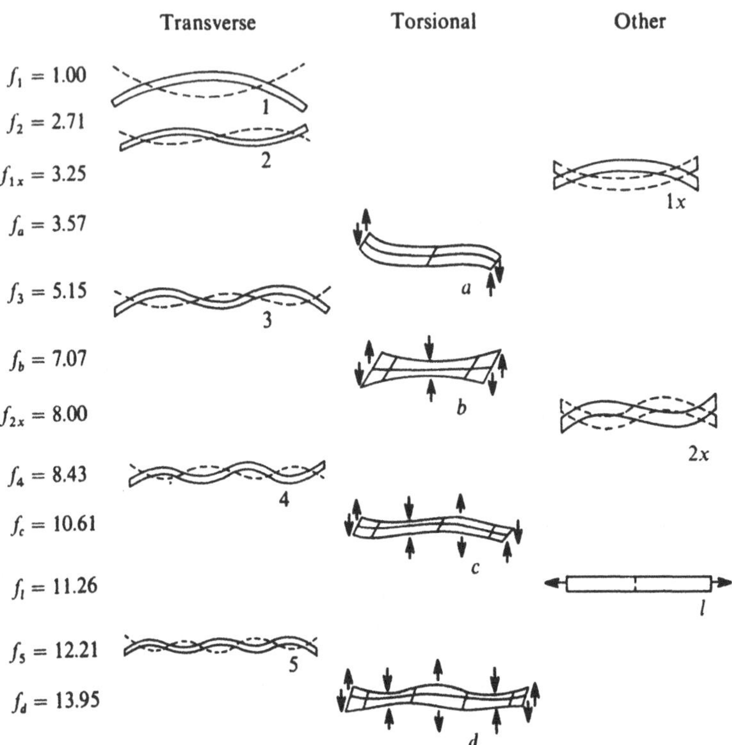

FIGURE 19.1. Vibrational modes of a glockenspiel bar. Relative frequencies in a C_6 bar are given.

A deep arch is cut in the underside of marimba bars, particularly in the low register. This arch serves two useful purposes: it reduces the length of bar required to reach the low pitches, and it allows tuning of the overtones (the first overtone is nominally tuned two octaves above the fundamental). Figure 19.2 shows a scale drawing of a marimba bar, and also indicates the positions of the nodes for each of the first seven modes of vibration. [These can be compared with the uniform bar shown in Fig. 19.1(a).] The ratios of the frequencies of these modes are also indicated. Note that the second partial (first overtone) of this bar has a frequency 3.9 times that of the fundamental, which is close to a two-octave interval (a ratio of 4.0).

Figure 19.3 compares the vibrational behavior of a marimba bar and a uniform rectangular bar for the first five bending modes. The upper row shows the local bending force, and the second row the bending moment (see Section 2.15). The slope of the bar is shown in the third row and the displacement in the fourth row. Note that the maximum displacement for all modes of the uniform bar occurs at the ends, but in the marimba bar it

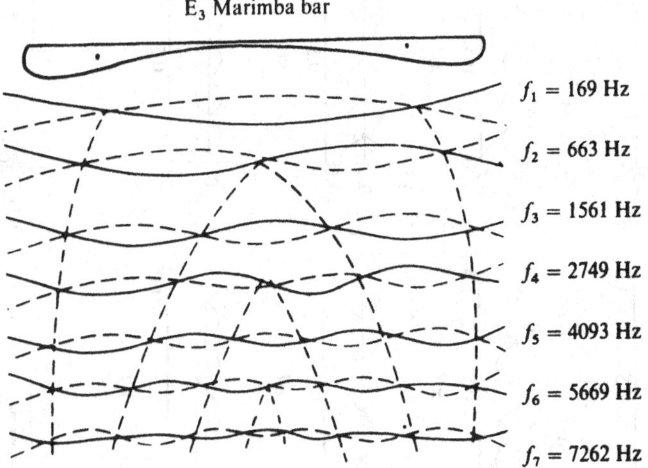

FIGURE 19.2. Scale drawing of a marimba bar tuned to $E_3(f = 169$ Hz). The dashed lines locate the nodes of the first seven modes (from Rossing, 1976).

occurs near the center. The maximum bending moment, however, occurs in the thickest part of the bar near the ends.

The sound spectrum of the E_3 marimba bar, shown in Fig. 19.4 along with the partials on musical staves, indicates the presence of a strong third partial, which has a frequency of 9.2 times the fundamental (about three octaves plus a minor third above it). The relative strengths of the partials, of course, depend on where the bar is struck and what type of mallet is used. To emphasize a particular partial, the bar should be struck at a point of maximum amplitude for that mode, as indicated in Fig. 19.2.

Oscillograms of the first three partials from an A_3 (220 Hz) marimba bar are shown in Fig. 19.5. Note that the second and third partials appear and also disappear much more quickly than the fundamental. In fact, the third partial has completely decayed before the fundamental has reached its maximum amplitude. It has been found that the tuning of the inharmonic third partial has some slight influence on the perceived pitch of the bar (Bork and Meyer, 1982).

19.3 Tuning the Bars

Removing material from any point on a bar affects all the modal frequencies to some extent. However, from Fig. 19.3, it is clear that removal from certain places affects certain modes more than others, and thus it is possible to tune the individual partials. In general, removing material from a place where the bending moment M(x) for a given mode is large will lower the frequency

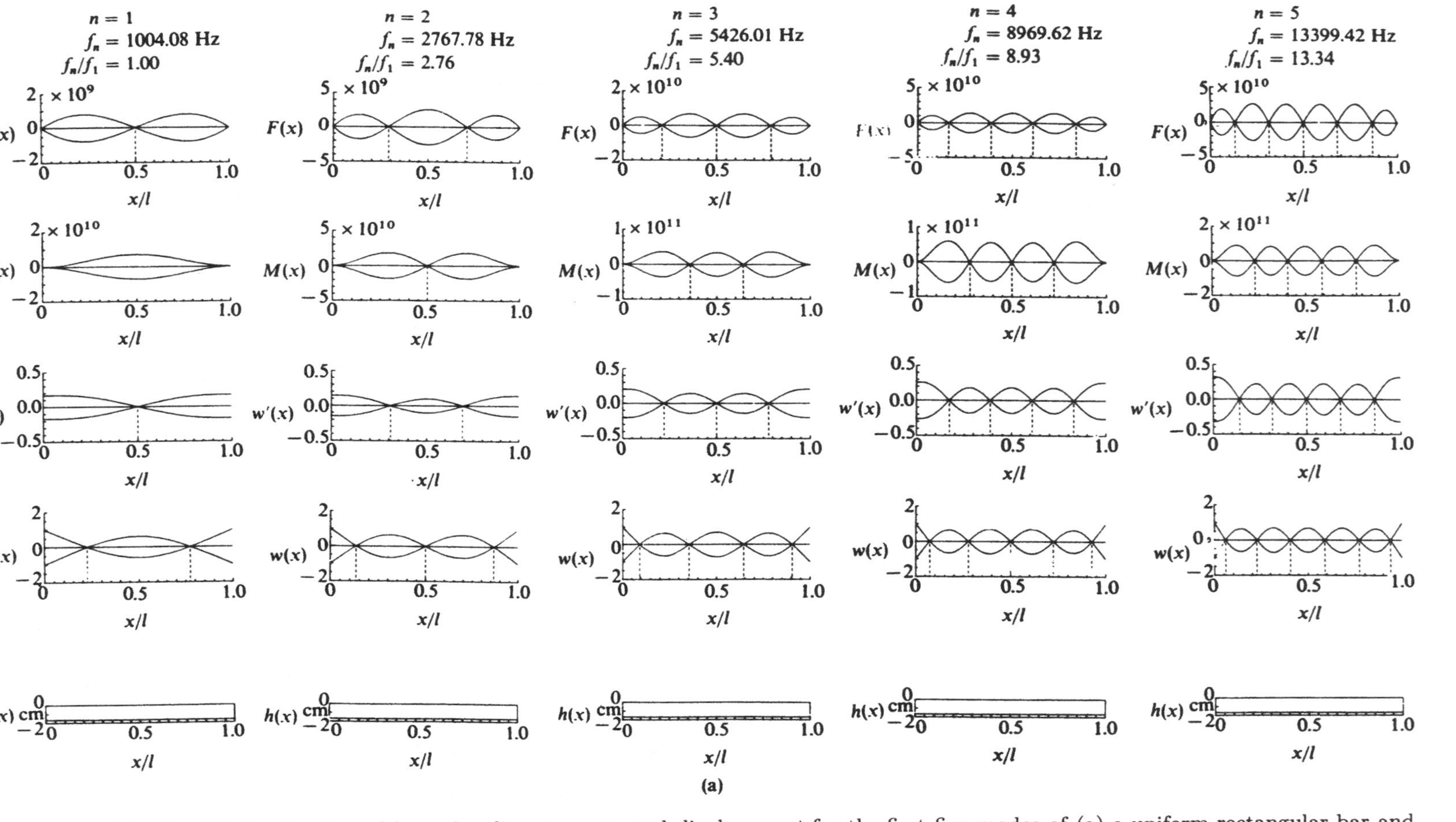

(a)

FIGURE 19.3. Spatial distribution of force, bending moment, and displacement for the first five modes of (a) a uniform rectangular bar and

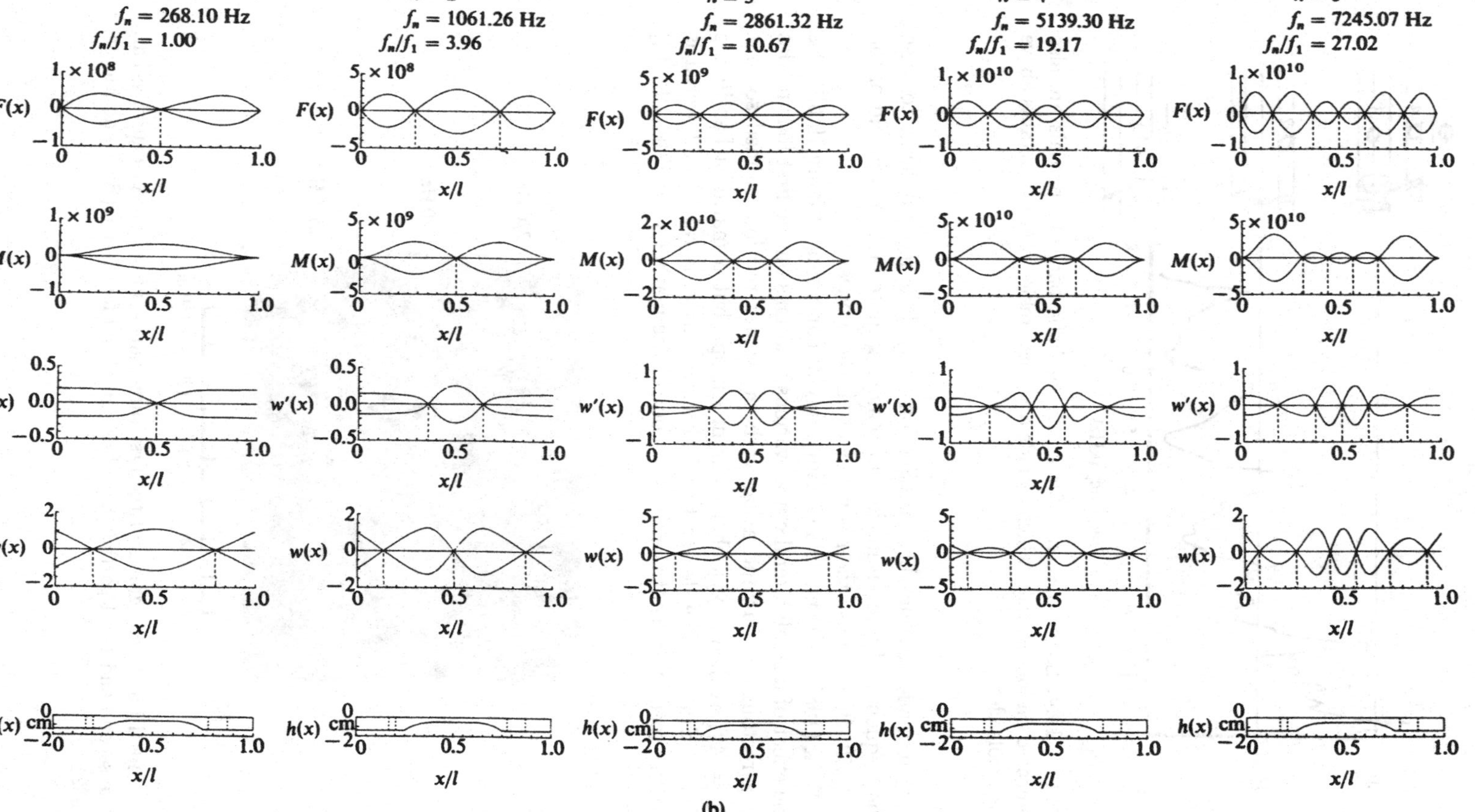

(b)

Fig. 19.3 ***(continued)***

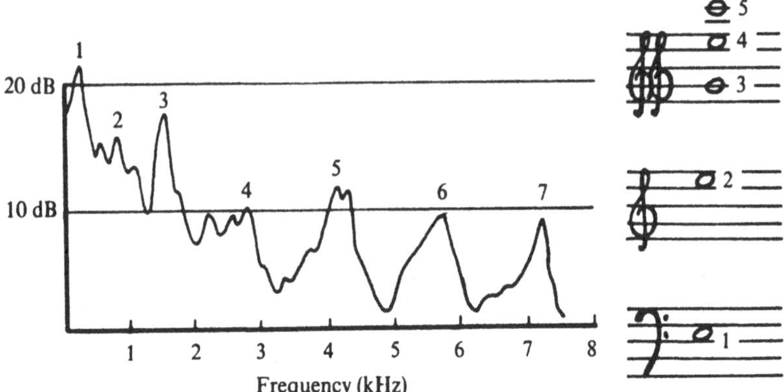

FIGURE 19.4. Sound spectrum for an E_3 marimba bar. The partials are also indicated on music staffs, which include a "super-treble" clef, two octaves above the treble clef.

of that mode considerably (see Section 2.15). It is more difficult to raise the frequency of a given mode, but removing material near the end of the bar will slightly raise the frequency of all the modes.

Bork (1983a) has made a careful study of bar tuning. Figure 19.6 shows the effect of a small lateral cut at various positions on the first four modes of a rectangular bar and a bar with the center portion thinned so that $f_2 = 4f_1$ (called a xylophone bar in Europe but a marimba bar in the United States). The degree to which each partial is raised or lowered is readily seen. Note that the second mode of the marimba bar [Fig. 19.6(b)]

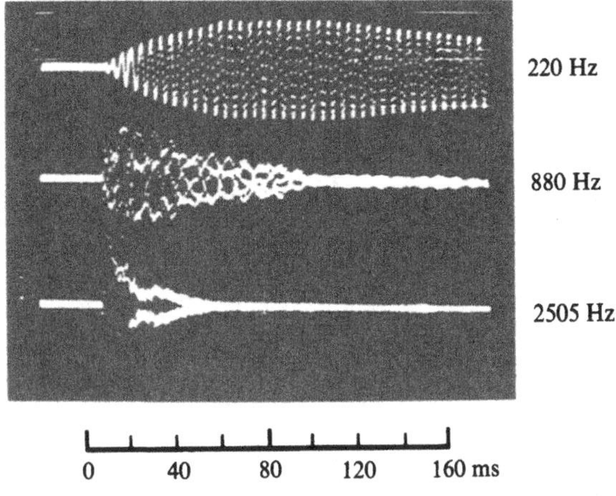

FIGURE 19.5. Oscillograms of the first three partials from an A_3 marimba bar. The second and third partials have been amplified for clarity (Bork and Meyer, 1982).

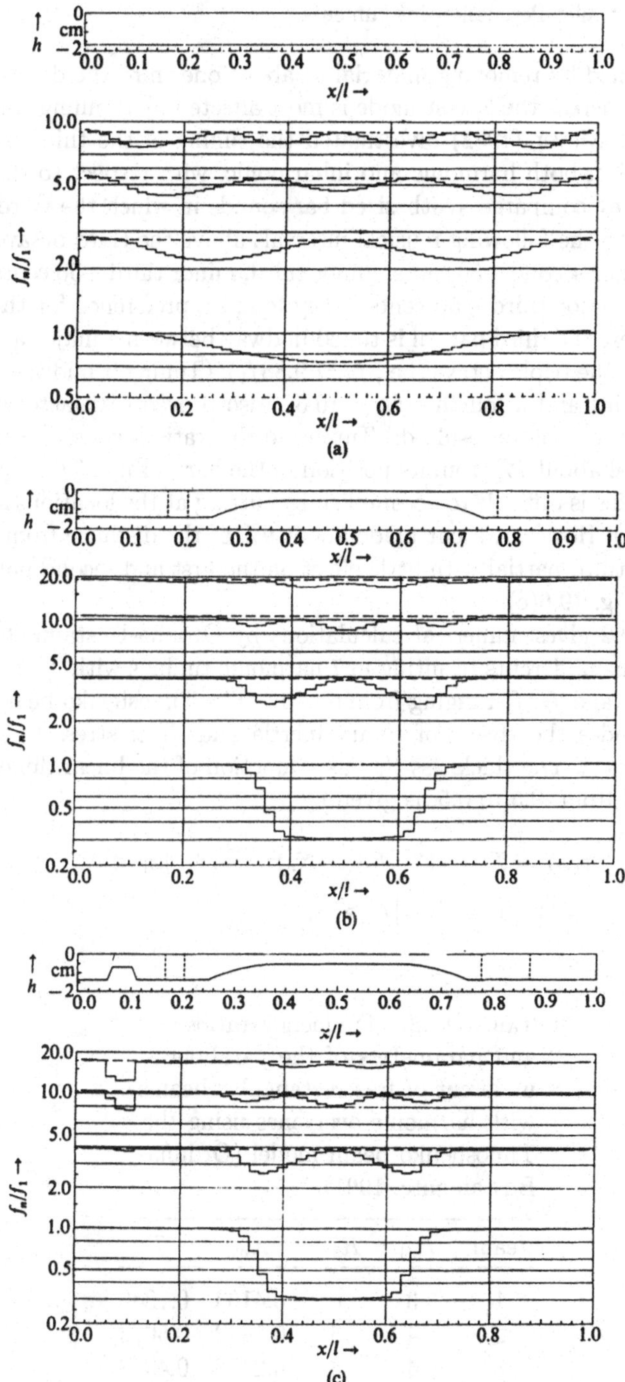

FIGURE 19.6. Effect of removing material from various locations along (a) a uniform rectangular bar; (b), (c) a marimba bar. The effect on the frequencies of the first four modes are shown. Tuning the third mode can best be done by cutting at a point about 9% of the distance from one end, as shown in (c) (Bork, 1983a).

is easily tuned by removing material at about one-third the distance from each end, whereas the lowest mode is most affected by thinning the center.

Bork and Meyer (1982) investigated the tuning of the third partial to various ratios, both harmonic and inharmonic, with respect to the fundamental. They compared synthesized bar sounds in which the third partial was tuned to the following musical intervals above the third octave: major second, major second +50 cents, minor third, minor third +50 cents, major third, and major third +50 cents. They found a preference for the fourth choice, where the third partial is tuned midway between a major and minor third above the triple octave (i.e., $f_3 = 9.88 f_1$). (Tuning it midway between a major third and a fourth—$f_3 = 10.3$—also proved satisfactory; in this case a brighter timbre resulted.) Tuning to this ratio involves lowering the third partial about 7% from its position in the bar in Fig. 19.6(b) [and Fig. 19.3(b)]. This is difficult to accomplish by cutting at the locations shown in Fig. 19.6(b). However, a cut made about 9% of the distance from the end lowers the third partial with little effect on the first and second partials, as shown in Fig. 19.6(c).

Table 19.2 gives numerical calculations by Orduña-Bustamante (1991) for parabolic undercuts resulting in 7 harmonic tunings with f_2/f_1 ranging from 3 to 5 and f_3/f_1 ranging from 6 to 13. The Timoshenko beam model, which includes the effects of rotary inertia and shear stress (see section 2.17) was used. The thickness $t(x)$ as a function of the linear dimension x, normalized from -0.5 to 0.5, is given by

$$\begin{aligned} t(x) &= T_c + (1 - T_c)(x/X_c)^2 \qquad \text{if} \quad |x| < X_c, \\ &= 1 \qquad \text{if} \quad |x| > X_c. \end{aligned}$$

TABLE 19.2. Frequency ratios and parameters of the parabolic undercut of free rectangular beams with harmonic overtones using the Timoshenko beam model (Orduña-Bustamante, 1991).

Beam	R_{21}	R_{31}	X_c	T_c
1	3	6	0.4163	0.7340
2	4	8	0.1642	0.5073
3	4	9	0.2478	0.4518
4	5	10	0.1282	0.3722
5	5	11	0.1595	0.3590
6	5	12	0.1898	0.3361
7	5	13	0.2236	0.3020

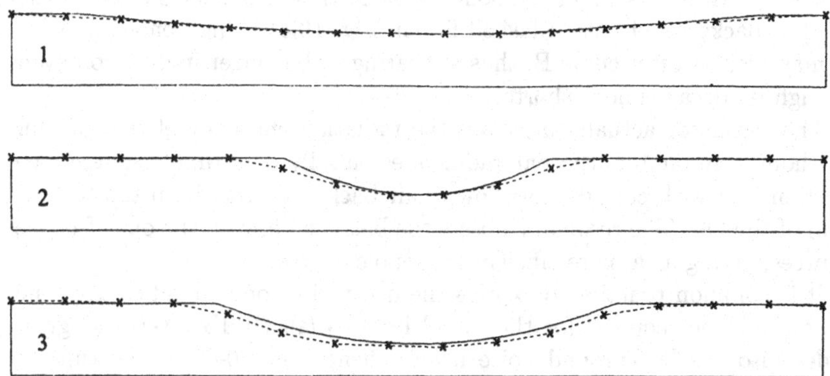

FIGURE 19.7. Contours of beams calculated by the Timoshenko beam model (solid lines) and experimentally tuned for exact harmonic ratios (dashed lines). Beam 1) 1:3:6; beam 2) 1:4:8; beam 3) 1:4:9 (Orduña-Bustamante, 1991).

While the parabolic contours in Table 19.2 can easily be cut with a numerically controlled milling machine, in practice fine tuning is necessary, especially with bars of natural wood. Orduña-Bustamante gives curves showing contours of equal frequency that can be used as a guide to fine tuning the bars. Figure 19.7 shows calculated parabolic contours and final contours for bars having harmonic ratios 1:3:6, 1:4:8, and 1:4:9, respectively.

Summers, et al. (1993) have used Rayleigh theory to show that f_2/f_1 ratios of 3 or 4 can be obtained with a simple rectangular cut. Discussions of practical marimba tuning can be found in MacCallum (1968), Moore (1970), and Bork (1995).

19.4 Resonators

Marimba resonators are cylindrical pipes tuned to the fundamental mode of the corresponding bars. A pipe with one closed end and one open end resonates when its acoustical length is one-fourth of a wavelength of the sound. The purpose of the tubular resonators is to emphasize the fundamental and also to increase the loudness, which is done at the expense of shortening the decay time of the sound. The statement is sometimes made that the resonator prolongs the sound, but this is incorrect. In a noisy environment or when played with other instruments in an ensemble, that impression may be conveyed, as can be understood by referring to Fig. 19.8.

Curve A, which represents the more rapid decay of a bar with a resonator, begins at a higher sound level than curve B, which represents the decay of a bar with no resonator. At some point in time, the curves cross. If the level of background noise (dashed line in Fig. 19.8) is high enough, curve A may cross it after curve B, thus appearing to be longer in duration even though its decay time is shorter.

The resonator actually increases the radiation efficiency of the system. Without a resonator, the bar radiates essentially as a dipole source (see Section 7.1) with considerable flow of air back and forth from top to bottom of the bar. The resonator upsets the balance between the out-of-phase sources, acting as a more efficient monopole source.

It is common practice to express the decay time of a sound as the time that would be required for the sound level to fall 60 dB (even though it fades into the background noise before then). The 60-dB decay time of a typical rosewood marimba bar in the low register (E_3) is about 1.5 s with the resonator and 3.2 s without it. Decay times in the upper register are generally shorter; we measured 0.4 s and 0.5 s for an E_6 bar with and without the resonator, respectively. The corresponding decay times for synthetic bars are somewhat longer.

Tubes that are closed at one end and open at the other have resonances that are nearly the odd-numbered harmonics of the fundamental mode (i.e., the frequencies would be f, $3f$, $5f$, ...). Thus, a resonator tuned to the fundamental frequency of a marimba bar would not affect the first overtone, because the frequency of that mode is four times the fundamental frequency. As an experiment, a marimba having a second set of resonators tuned to the first overtone of the bars was constructed. Each resonator was equipped with a vane that can partially or completely close the mouth of the tube; thus, the timbre can be varied by adjusting the amount of closure (Kvistad and Rossing, 1977).

The resonance frequency of a tube depends, to a small extent, on the environment near the open end. This is because the reflection of standing

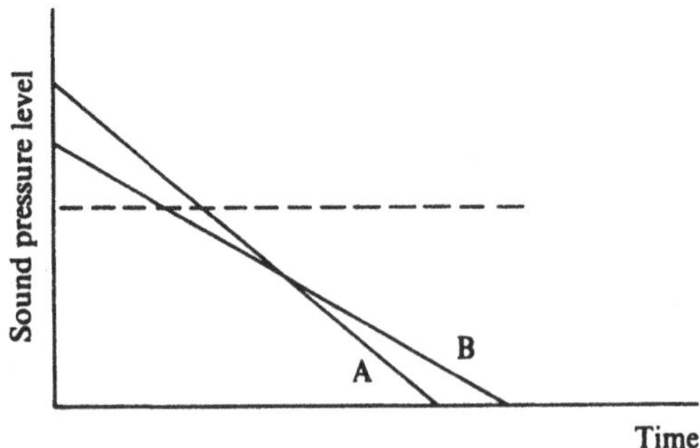

FIGURE 19.8. Sound decay for a marimba bar with a resonator (curve A) and a bar without a resonator (curve B). If the background level (dashed line) is high enough, the sound represented by curve A may appear to be longer in duration.

sound waves within the tube does not take place precisely at the end of the tube walls. The tube acts as if it were a little longer than it really is; the difference between the effective length and the actual length is called the end correction. The end correction for a cylindrical tube without a flange is 0.61 times its radius; if it has a large flange, the end correction is 0.85 times the radius (see Section 8.3).

If the bar is close (within a distance of one radius, for example) to the resonator tube, the resonance frequency will be affected by its presence. In particular, the resonance frequency will be lowered as the bar is moved closer to the tube. This effect can be used to compensate for changes in resonator frequency caused by changes in ambient temperature. It works as follows: if the temperature rises, the speed of sound increases, and so does the resonance frequency of the tube. Moving the resonator closer to the bar, however, lowers the resonance frequency of the tube, so that it once again matches that of the bar. Some marimbas incorporate an adjustment of this nature. As yet, the theory describing the resonances of the coupled bar–resonator system has not been worked out in detail.

Resonators with two open ends, having twice the length of conventional resonators, offer an interesting alternative at the higher frequencies. Because both ends of the bar radiate sound, both constructive and destructive interference occur, and the sound field becomes more directional. Figure 19.9 compares the sound pressure levels at three measuring points with a conventional $\lambda/4$-resonator and a $\lambda/2$-resonator with both ends open. Owing to constructive interference, the $\lambda/2$-resonator increases the sound field

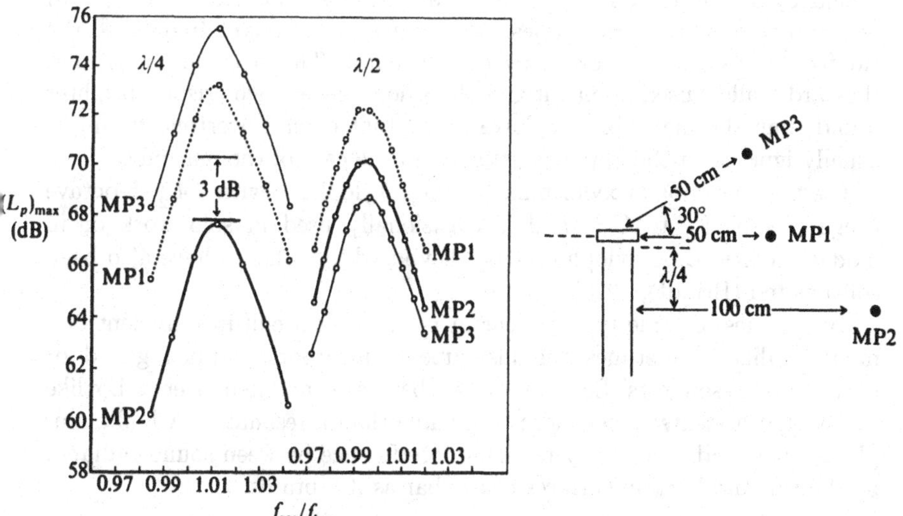

FIGURE 19.9. Sound pressure levels as functions of frequency at three different points for $\lambda/4$ (one closed end) and $\lambda/2$ (both ends open) resonators (Bork, 1983a).

by approximately 3 dB at MP2, which is in the far field in the mid-plane of the resonator. However, the sound level at the position of the performer (MP3) is lowered considerably. Relatively little difference between the $\lambda/4$- and $\lambda/2$-resonators occurs at MP1 in the plane of the bar (Bork, 1983a).

Bork also studied the effect of neighboring resonators on the sound radiation. He found that a nearby resonator tuned a semitone below the bar can decrease the sound output by as much as 2 dB; a nearby resonator tuned a semitone higher raised the sound output by about 1 to 1.5 dB. This interaction would be most noticeable, therefore at the semitone intervals E to F and B to C. A similar interaction becomes quite important in bass marimbas with trough resonators.

19.5 The Xylophone

The word xylophone derives from Greek words meaning "wood sound." In modern musical terminology, it usually refers to a particular mallet percussion instrument with bars of either wood or synthetic material. Xylophones typically cover a range of 3–3$\frac{1}{2}$ octaves extending from F_4 or C_5 to C_8(f = 349–4186 Hz). Like marimbas, modern xylophones are nearly always equipped with tubular resonators to increase the loudness of the tone.

Xylophone bars are also cut with an arch on the underside, but the arch is not as deep as that of the marimba, since the first overtone is usually tuned to a musical twelfth above the fundamental (that is, three times the frequency of the fundamental).[1] Since a pipe closed at one end (Fig. 5.6) can also resonate at three times its fundamental resonant frequency, the twelfth will also be reinforced by the resonator. This overtone boost and the hard mallets used to play it give the xylophone a much crisper, brighter sound than the marimba. We have found that careful overtone tuning is usually ignored in the upper register, as in the case of the marimba.

The xylomarimba or xylorimba is a large xylophone with a 4$\frac{1}{2}$–5-octave range (C_3 or F_3 to C_8), and is occasionally used in solo work or in modern scores. Bass xylophones and keyboard xylophones have also been constructed (Brindle, 1970).

Xylophones of some type appear in most musical cultures, ancient and modern alike. Sometimes the instruments incorporate tuned gourd or bottle-type resonators. Sometimes the bars are mounted over a boxlike cavity that appears to act more as a baffle than a resonator. A baffle amplifies the sound of a bar by reducing interference between sound radiated by the top and bottom surfaces of the bar as it vibrates.

[1]In Europe, it is customary to tune the second partial of the xylophone to the double octave, as in marimbas (Bork and Meyer, 1982).

The vibrational frequencies of a xylophone bar tuned to $F_4^\sharp$ ($f = 370$ Hz) are shown in Fig. 19.10. The frequencies of the bending and torsional modes have been lowered by cutting the arch on the underside, and this lowering is greatest in the lowest modes in each family. Thus, the slopes of curves connecting the lowest torsional and bending modes in Fig. 19.10 have greater slopes than those for a uniform rectangular bar, although they approach the latter slopes (1 and 2, respectively) with increasing mode number.

Studies of decay times in wooden xylophone bars without resonators have shown that the decay process is mainly determined by internal losses; radiation losses and friction at the cord support appear to be insignificant. The frequency dependence of the damping constant α or the decay time t_d can be expressed by $\alpha(f) = 1/t_d = a_0 + a_2 f^2$, where a_0 and a_2 depend on the wood species (Chaigne and Doutaut, 1997).

Holz (1996) compares the acoustically important properties of xylophone bar materials and concludes that the "ideal" wood is characterized by a density of 0.80 to 0.95 g/cm^3 and a Young's modulus of 15 to 20 GPa.

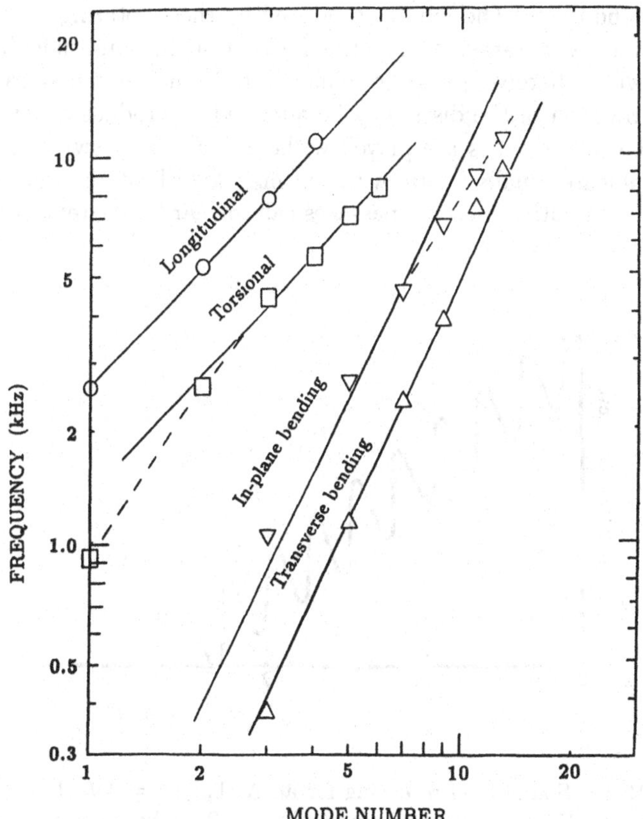

FIGURE 19.10. Modal frequencies of an $F_4^\sharp$ xylophone bar. Solid lines have slopes of 1 or 2. The mode number is $2n + 1$ for bending modes [cf. Eq. (2.63)].

These conditions are met by several Palissandre species, some other tropical woods, and also by cherry wood. In order to be suitable substitutes for wood, glass-fiber reinforced (GPR) plastics should be pressed and molded, rather than hand lay-up laminates, Holtz cautions.

19.6 Vibes

A very popular mallet percussion instrument is the vibraphone or vibraharp, as they are designated by different manufacturers. Vibes, as they are popularly called, usually consist of aluminum bars tuned over a three-octave range from F_3 to F_6 (f = 175–1397 Hz). The bars are deeply arched so that the first overtone has four times the frequency of the fundamental, as in the marimba. The aluminum bars tend to have a much longer decay time than the wood or synthetic bars of the marimba or xylophone, and so vibes are equipped with pedal-operated dampers.

The most distinctive feature of vibes, however, is the vibrato introduced by motor-driven discs at the top of the resonators, which alternately open and close the tubes. The vibrato produced by these rotating discs or pulsators combines a rather substantial fluctuation in amplitude (intensity vibrato) with a barely detectable change in frequency (pitch vibrato). The speed of rotation of the discs may be adjusted to produce a slow vibe or a fast vibe. Often vibes are played without vibrato by switching off the motor. They are usually played with soft mallets or beaters, which produce a mellow tone, although some passages call for harder beaters.

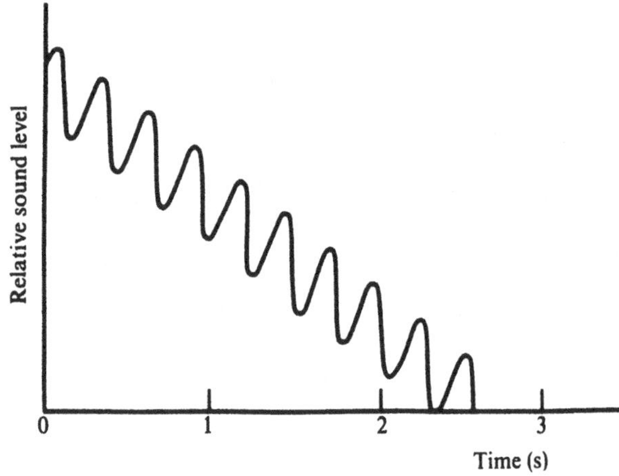

FIGURE 19.11. Sound level recording for an A_3 bar (f = 440 Hz) with a vibe rate of about 4 Hz. The level fluctuates about 6 dB, and the decay time (60 dB) is 7 s.

TABLE 19.3. Mode frequencies of three vibe bars; the ratios to the fundamental are also given.

n	f_n	f_n/f_1	f_n	f_n/f_1	f_n	f_n/f_1
1	175	1	394	1	784	1
2	700	4.0	1578	4.0	2994	3.8
3	1708	9.7	3480	8.9	5995	7.6
4	3192	18.3	5972	15.2	9400	12.0
5	4105	23.5	8007	20.2	14,014	17.9
6	6173	35.4	11,119	35	18,796	24.0
7	8080	46.3			21,302	27.2

Because vibraphone bars have a much longer decay time than do marimba and xylophone bars, the effect of the tubular resonators on decay time is more dramatic. At 220 Hz (A_3), for example, we measure a decay time (60 dB) of 40 s without the resonator and 9 s with the tube fully open. For A_5, we measure 24 s with the resonator closed and 8 s with it open. In the sound level recording shown in Fig. 19.11, the intensity modulation, as well as the slow decay of the sound, can be clearly seen.

The mode frequencies for three vibe bars tuned to F_3, G_4, and G_5 are shown in Table 19.3. Note that the frequencies of the higher modes change quite markedly with respect to the fundamental. In fact, for G_5 and higher notes, the second partial is only approximately tuned to the double octave. The same is true in most marimbas and xylophones (Moore, 1970).

19.7 Mallets

The modern percussion player can select from a wide variety of mallets that differ in mass, shape, and hardness. Through intelligent selection of a mallet, a player can greatly influence the timbre of the instrument being played. Striking a marimba or xylophone with a hard mallet, for example, produces a sound rich in overtones that emphasizes the woody character of the instrument. A soft mallet, on the other hand, which excites only the harmonically tuned lower partials, gives a dark sound to the instrument.

The interaction of the mallet with the bar, tube, or drumhead of the instrument is quite similar to the interaction between the piano hammer and string discussed in Sections 2.9, and 12.4. A mallet whose mass nearly equals the dynamic mass of the struck vibrator (typically about 30% of the total mass for a marimba bar in its fundamental mode) transfers the maximum amount of energy to the vibrator. A lighter mallet rebounds after a short contact time. A heavier mallet remains in contact for a longer time, which results in considerable damping of the higher partials. This is a desirable effect in drums, where a large amplitude and short decay time are desired, but is not so desirable in bar percussion instruments.

According to Hertz's law, the impact force is proportional to the 3/2 power of the mallet deformation δ: $F = R^{1/2}D\delta^{3/2}$, where R is the radius of the mallet head and D depends on the elastic constants of the mallet and bar. From measurements of impact force and impact time, Chaigne and Doutaut (1997) conclude that Hertz's law describes the impact over a fairly wide range of mallet hardness and impact time.

An effective way to test mallets is to strike a piezoelectric force transducer and to use the force waveform $F(t)$ to calculate the shock spectrum of the blow (Kittelson, 1966; Morrow, 1957). For a given impulse, the maximum in the shock spectrum indicates the eigenfrequency f_{max} of the vibrator that is best excited (for an impulse having a shape given by $\sin^2 2\pi t/\tau$, for example, $f_{\text{max}} = 0.85\tau$, where τ is the impulse duration) (Bork, 1983b).

For most mallets, $F(t)$ depends upon the force of the blow, and so also does the frequency f_s at which the maximum excitation occurs. Figure 19.12 shows the frequency of maximum excitation as a function of mallet

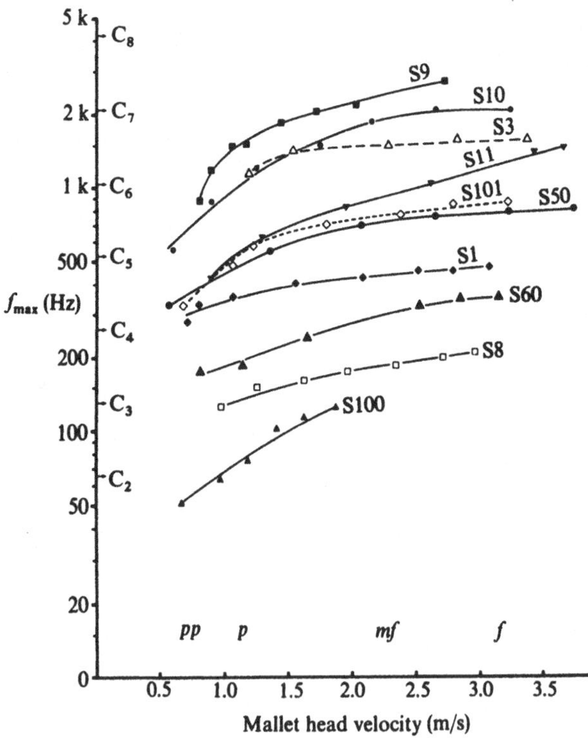

FIGURE 19.12. Dependence of the frequency of maximum excitation on the mallet head velocity of different mallets (Bork, 1983b).

head velocity for 10 mallets of widely different hardness. The slope of each curve indicates the variation of $f_{\max}$ with the blow strength. A flat curve indicates that the impulse duration (and thus the optimum playing range) changes very little with mallet velocity. A large slope, on the other hand, means that the timbre will change noticeably over the dynamic range of the instrument.

For the mallets in Fig. 19.12, the spectral maximum changes more at low velocity than at high velocity. This is especially clear for mallet S9, which is a wooden mallet with a rubber ring primarily intended for the glockenspiel. With a weak blow, the deformation of the rubber ring is dependent on the force, but with increasing blow strength a point is reached where further deformation is hindered by the solid core. Mallet S1, on the other hand, is an all-rubber mallet with little dependence of spectral maximum on blow strength. Mallet S10, a hard core wrapped with yarn, has a playing range of over two octaves; because of its hard surface it is especially suited for instruments such as the xylophone or vibraphone, which play in the high range (Bork, 1983b).

19.8 Chimes

Chimes or tubular bells are generally fabricated from lengths of brass tubing 32–38 mm ($1\frac{1}{4}$–$1\frac{1}{2}$ in.) in diameter. The upper end of each tube is partially or completely closed by a brass plug with a protruding rim. The rim forms a convenient and durable striking point.

The modes of transverse vibration in a pipe are essentially those of a thin bar, as given by Eq. (2.63). In a pipe, $K = \sqrt{a^2 + b^2}$, where a and b are the inner and outer radii.

Figure 19.13 is a graph of the frequencies of a G_4 chime as a function of m, along with those predicted by a thin-bar theory (denoted Rayleigh) and the more detailed thick-bar theory (denoted Flügge). Also shown are the frequencies of vibration with the end plug removed. Note that the end plug lowers the frequencies of the first few modes but has little effect on the higher modes. The strike tone, which lies one octave below the fourth mode, is also indicated.

One of the interesting characteristics of chimes is that there is no mode of vibration with a frequency at, or even near, the pitch of the strike tone one hears. This is an example of a subjective tone created in the human auditory system. Modes 4, 5, and 6 appear to determine the strike tone. This can be understood by noting that these modes for a free bar have frequencies in the ratios 9^2:11^2:13^2, or 81:121:169, which are close enough to the ratios 2:3:4 for the ear to consider them nearly harmonic and to use them as a basis for establishing a pitch. The largest near-common factor in the numbers 81,121, and 169 is 41.

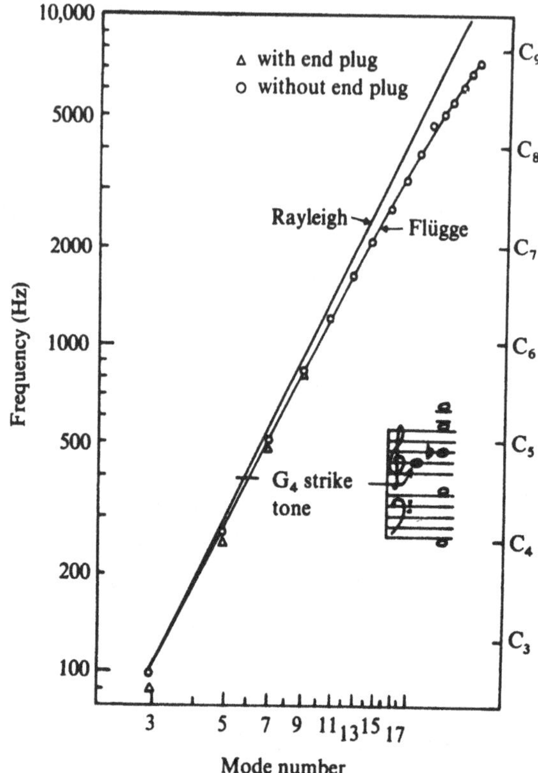

FIGURE 19.13. Frequencies of a G_4 chime as a function of mode number m, along with those predicted by a thin-bar theory (Rayleigh) and thick-bar theory (Flügge). The first five modes are shown on musical staves, along with the subjective strike tone (from Rossing, 1976).

The ratios of the modal frequencies of a chime tube with and without a load at one end are given in Table 19.4. Also given are the ratios considered desirable for a tuned carillon bell. Note the similarity between the tuning of partials 3–8 of a chime and those of a carillon bell. Adding a load to one end of a chime lowers the frequencies of the lower modes more than the higher ones and thus stretches the modes into a more favorable ratio. The end plug also adds to the durability of the chime and helps to damp out the very high modes.

19.9 Triangles and Pentangles

Because of their many modes of vibration, triangles are characterized as having an indefinite pitch. They are normally steel rods bent into a triangle (usually, but not always, equilateral) with one open corner. Triangles are

TABLE 19.4. Ratios of mode frequencies for loaded and unloaded chime tube.

			Loaded with			
n	Thin rod	Tube	193 g	435 g	666 g	Tuned bell
1	0.22	0.24	0.24	0.23		0.5
2	0.61	0.64	0.63	0.62	0.61	1
3	1.21	1.23	1.22	1.22	1.22	1.2
4	2	2	2	2	2	2
						2.5
5	2.99	2.91	2.93	2.95	2.94	3
6	4.17	3.96	4.01	4.04	4.03	4
7	5.56	5.12	5.21	5.21	5.18	5.33
8	7.14	6.37	6.50	6.43	6.37	6.67

suspended by a cord from one of the closed corners, and are struck with a steel rod or hard beater.

Triangles are typically available in 15-cm, 20-cm, and 25-cm (6-, 8-, and 10-in.) sizes, although other sizes are also used. Sometimes one end of the rod is bent into a hook, or the ends may be turned down to smaller diameters than the rest of the triangle to alter the modes of vibration. The sound of the triangle depends on the strike point as well as the hardness of the beater. Single strokes are often played on the base of the triangle and perpendicular to the plane (Brindle, 1970) to emphasize vibrations in the plane of the triangle. A grazing stroke in the upper third of the open leg of the triangle is recommended for the most even onset and decay of the many partials (Peinkofer and Tannigel, 1969). Figure 19.14 shows sound spectra for a 25-cm (10-in.) triangle at two different strike points with strokes parallel and perpendicular to the plane of the triangle.

Some of the modal frequencies observed in a 10-in. triangle are shown in Fig. 19.15, along with those calculated for a steel rod of the same length and diameter. The triangle modes show a surprisingly close correspondence to those of a straight rod.

In normal orchestral use, only the higher modes of the triangle are important, since the radiation efficiency of the lower modes is very small because of the small bar diameter. Recently, however, Australian composer Moya Henderson has developed a new family of instruments based upon triangles or similar bent-rod vibrators, coupled to resonators or radiators so arranged as to enhance the lower vibrational modes. These instruments are known as alembas.

The first alemba was based on metal triangles coupled by means of light cords to thin diaphragms closing the ends of appropriately tuned pipes

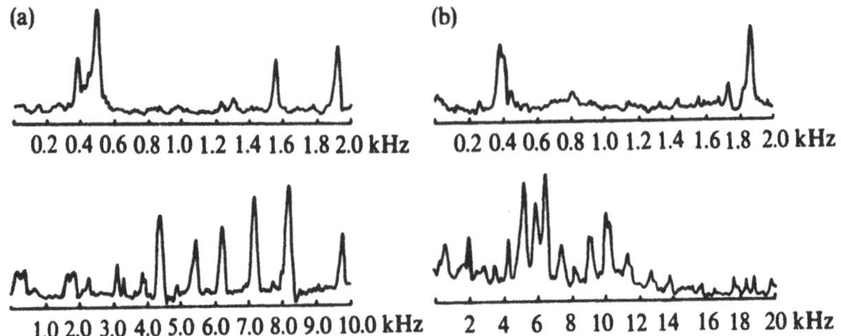

FIGURE 19.14. Sound spectra for a 25-cm steel triangle (a) struck in the plane and (b) struck perpendicular to the plane. Two frequency and amplitude ranges are shown in each case (Rossing, 1976).

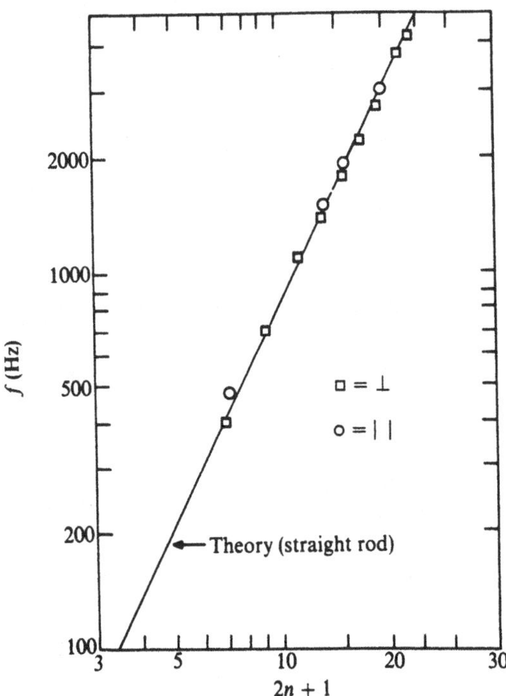

FIGURE 19.15. Mode frequencies for a 25-cm steel triangle driven in the plane and perpendicular to the plane. The line gives the predicted frequencies for a steel rod of the same diameter and length.

(Henderson, 1984; Dunlop, 1984a). As a part of this development, Dunlop (1984b) used finite element methods to calculate vibrational frequencies of five in-plane modes (corresponding to $n = 4$ to 8 in Fig. 19.15) for triangles with base angles of 30°, 45°, and 60° and various radii of curvature. His calculations indicate that the modes of a triangle would have frequencies somewhat lower than those of a straight rod, although the differences are least, in most cases, when the parameters of an orchestral triangle are approached (60°, small radius of curvature). In a 6-in. orchestral triangle, he measured in-plane frequencies for modes with four, five, and seven nodes that have ratios within 5% of those of a straight rod of the same dimensions.

More recently, Henderson and Fletcher (1994) have devised another alemba using five-segment pentangles as vibrating elements and coupling these to a broad-band soundboard used as a radiating element. Fletcher (1993) had previously shown that the five degrees of geometrical freedom of a symmetrical pentangle (three side lengths and two angles) could be used to tune the second mode to a desired pitch and to arrange the frequencies of a further four in-plane modes in concordant ratios to it, either in a sequence 1:2:3:4.8:6 or 1:1.5:2:3:4.8 according to the basic pentangle shape chosen. The untuned first-mode frequency was either 0.35 or 0.33 times the second-mode reference frequency. Note the ratio 4.8 appearing in each of these sequences and giving a minor-third character to the sound. These instruments, along with the earlier triangle-based alemba, produce an interesting new bell-like sound for chamber-music use, their sound output being rather limited.

19.10 Gamelan Instruments

The center of musical art in Indonesia is the gamelan, a generic term for ensemble. Important melodic instruments in every gamelan are various metallophones consisting of bronze bars suspended over bamboo resonators. Although generally trapezoidal in cross section, these bars have modes of vibration not much different from the bars found in a glockenspiel.

The frequencies of the modes of vibration in a four-bar jegogan are given in Table 19.5, and the sound spectrum of bar III is shown in Fig. 19.16. This instrument is tuned in "saih angklung," a type of slendro scale (Rossing and Shepherd, 1982).

19.11 Tubaphones and Gamelan Chimes

Metal tubes have been used in various types of metallophones. The tubaphone, developed in England, consists of brass or steel tubes in a keyboard

TABLE 19.5. Vibration frequencies of jegogan bars.

Bar	f_1(Hz)	f_2(Hz)	f_3(Hz)	f_4(Hz)	f_5(Hz)
I	341	903	1796	2953	3998
II	288	771	1497	2444	3519
III	261	694	1348	2199	3186
IV	228	610	1187	1964	2797
Ratio	1	2.7	5.2	8.4	12.2

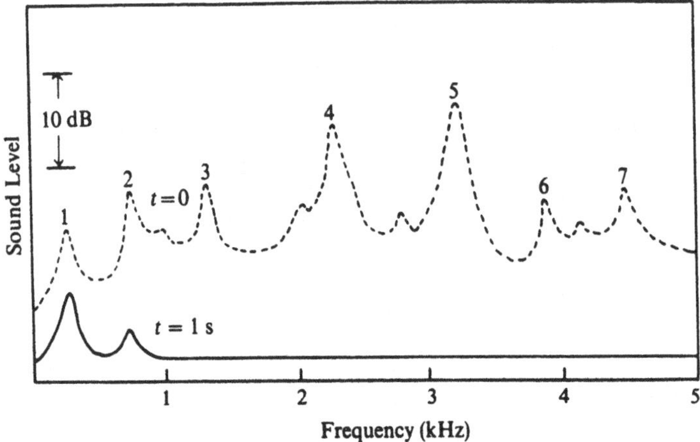

FIGURE 19.16. Sound spectrum from one of the jegogan bars (bar 3 in Table 19.4). The spectrum is shown at the time of striking (t = 0) and 1 s later. Note the rapid decay of the higher modes of vibration (Rossing and Shepherd, 1982).

arrangement similar to the bars of a glockenspiel or xylophone. Although scored in Khachaturian's "Gayané" ballet, the instrument is quite rare in modern orchestras.

Tubes of aluminum are used in a rather simple instrument called gamelan chimes, which was developed by percussionist Garry Kvistad especially for home music makers. A set of tubes, which will be used in a particular mode or selection to be played, are placed in plastic cradles and struck with soft mallets. The modes of vibration are similar to those of glockenspiel bars, but the cradles in which the bars rest tend to damp out all but the fundamental mode rather rapidly.

References

Bork, I. (1983a), "Zur Abstimmung und Kopplung von Schwingen den Stäben und Hohlraumresonatoren." Dissertation, Tech. Univ. Carolo-Wilhelmina, Braunschweig.

Bork, I. (1983b). "Entwicklung von akustischen Optimierungsverfahran für Stabspiele und Membraninstrumente." PTB report, Project 5267, Braunschweig, Germany (unpublished).

Bork, I. (1995). Practical tuning of xylophone bars and resonators. *Applied Acoust.* **46**, 103–127.

Bork, I., and Meyer, J. (1982). Zur klanglichen bewertung von Xylophonen. *Das Musikinstrument* **31** (8), 1076–1081. English translation in *Percussive Notes* **23** (6), 48–57 (1985).

Brindle, R.S. (1970). "Contemporary Percussion." Oxford Univ. Press, London and New York.

Chaigne, A., and Doutaut, V. (1997). Numerical simulations of xylophones. I. Time-domain modeling of the vibrating bars. *J. Acoust. Soc. Am.* **101**, 539–557.

Dunlop, J.I. (1984a). The acoustics of the alemba. *Acoust. Australia* **12**(1), 12–14.

Dunlop, J.I. (1984b). Flexural vibrations of the triangle. *Acustica* **55**, 250–253.

Fletcher, N.H. (1993). Tuning a pentangle—A new musical vibrating element. *Appl. Acoust.* **39**, 145–163.

Haines, D.W. (1979). On musical instrument wood. *Catgut Acoust. Soc. Newsletter*, No. 31, 23–32.

Henderson, M. (1984). The discovery of a new musical sound. *Acoust. Australia* **12**(1), 12–14.

Henderson, M., and Fletcher, N.H. (1994). The Tosca alemba: Ringing the changes. *Acoust. Australia* **22**(1), 11–14.

Holz, D. (1996). Acoustically important properties of xylophone-bar materials: Can tropical woods be replaced by European species? *Acustica* **82**, 878–884.

Hueber, K.A. (1972). Nachbilding des Glockenklanges mit Hilfe von Rohrenglocken und Klavierklungen. *Acustica* **26**, 334–343. English translation in "*Acoustics of Bells*" (T.D. Rossing, ed.), pp. 340–357. Van Nostrand-Reinhold, Princeton, New Jersey, 1984.

Kinsler, L.E., Frey, A.R., Coppens, A.B., and Sanders, J.V. (1982). "Fundamentals of Acoustics," 3rd ed. Wiley, New York.

Kittelson, K.E. (1966). Measurement and description of shock. *Brüel and Kjaer Tech. Rev.* No. 3-1966.

Kvistad, G., and Rossing, T.D. (1977). Variable timbre in mallet percussion instruments. *J. Acoust. Soc. Am.* **61**, S21 (abstract).

MacCallum, F.K. (1968). "The Book of the Marimba." Carlton Press, New York.

Moore, J. (1970). "Acoustics of Bar Percussion Instruments." Ph.D. thesis, Ohio State University, Colombus, Ohio.

Morrow, C.T. (1957). The shock spectrum as a criterion of severity of shock impulses. *J. Acoust. Soc. Am.* **29**, 596–602.

Orduña-Bustamante, F. (1991). Nonuniform beams with harmonically related overtones for use in percussion instruments. *J. Acoust. Soc. Am.* **90**, 2935–2941.

Peinkofer, K., and Tannigel, F. (1969). "Handbook of Percussion Instruments." English translation by K. and E. Stone, Schott, London, 1976.

Rossing, T.D. (1976). Acoustics of percussion instruments—Part I. *The Physics Teacher* **14**, 546–556.

Rossing, T.D., and Shepherd, R.B. (1982). Acoustics of gamelan instruments. *Percussive Notes* **19** (3), 73–83.

Summers, I.R., Elsworth, S., and Knight, R. (1993). Transverse vibrational modes of a simple undercut beam: an investigation of overtone tuning for keyed percussion instruments. *Acoustics Letters* **17**, 66–70.

20

Cymbals, Gongs, Plates, and Steel Drums

In Chapter 3, we discussed the vibrations of circular and rectangular plates and also shallow spherical shells. In this chapter, we consider various percussion instruments that are essentially flat or nearly flat metal plates. Most of the instruments are circular in shape. Some give a strong sensation of pitch; some do not. These instruments belong to a family called percussion idiophones (Marcuse, 1975), with the implication that their sound is determined "of themselves."

Although these instruments are usually set into vibration by striking with a mallet or beater, this is not always the case. Sustained tones may be obtained by bowing the edges of cymbals, tam-tams, and especially musical saws, with bass or cello bows; but even when thus excited, they generally retain their classification as idiophones.

20.1 Cymbals

Cymbals are among the oldest of musical instruments and have had both religious and military use in a number of cultures. Many different types of cymbals are used in orchestras, marching bands, concert bands, and jazz bands. Orchestral cymbals are often designated as French, Viennese, and Germanic in order of increasing thickness. Jazz drummers use cymbals designated by such onomatopoeic terms as crash, ride, swish, splash, ping, and pang.

Cymbals are normally made of bronze, and they range from 20 cm to 74 cm in diameter. The Turkish cymbals generally used in orchestras and bands are saucer shaped with a small spherical dome in the center, in contrast to Chinese cymbals, which have a flattened dome and a turned-up edge.

20.1.1 Modes of Vibration of Cymbals

The low-frequency modes in a cymbal are quite similar to those of a flat circular plate. When the cymbal is supported at its center, the first five or six modes have radial nodes extending from the cup to the edge. These may be thought of as being due to bending waves propagating around the cymbal in both directions. What we call the (m, n) mode has m nodal diameters and n nodal circles.

At higher frequencies, the modes of vibration often mix with one another, and mode identification becomes somewhat difficult. Figure 20.1 shows hologram interferograms of four modes of a 38-cm cymbal. The first two represent single modes with three and five radial nodes, respectively. The third one is largely due to the (6, 0) mode, but with a slight amount of mixture with another mode. In the fourth interferogram, however, the (13, 0) and (2, 2) modes have mixed together in nearly equal proportions, creating a combination mode. This mixing occurs because these two modes have nearly the same frequency. The first 15 resonances of this cymbal are shown in Fig. 20.2. Modal frequencies of cymbals can be described by a modified form of Chladni's law: $f = c(m + 2n)^p$ (see Section 3.6), as illustrated by the modes of a 60-cm cymbal in Fig. 20.3.

It is clear in Fig. 20.3 that two different values of p are required to properly fit the equation to the $n = 0$ family of modes. In most cymbals, a distinct change in slope is noted at some particular value of m, which is

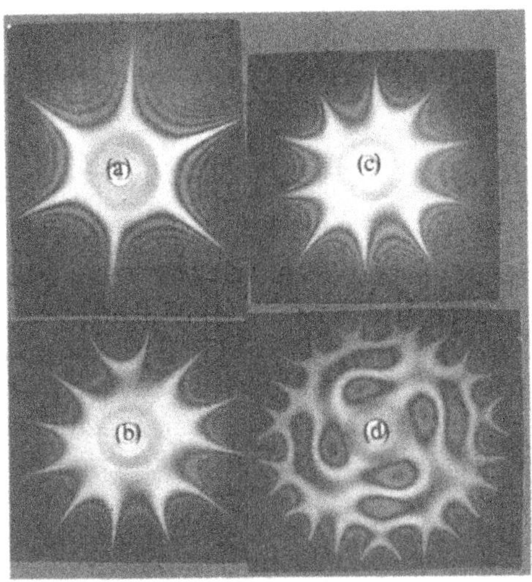

FIGURE 20.1. Hologram interferograms of four of the modes of vibration in a 15-in. cymbal: (a) (3, 0) mode; (b) (5, 0) mode; (c) (6, 0) mode; and (d) (13, 0 + 2, 2), a combination of two modes (Rossing and Peterson, 1982).

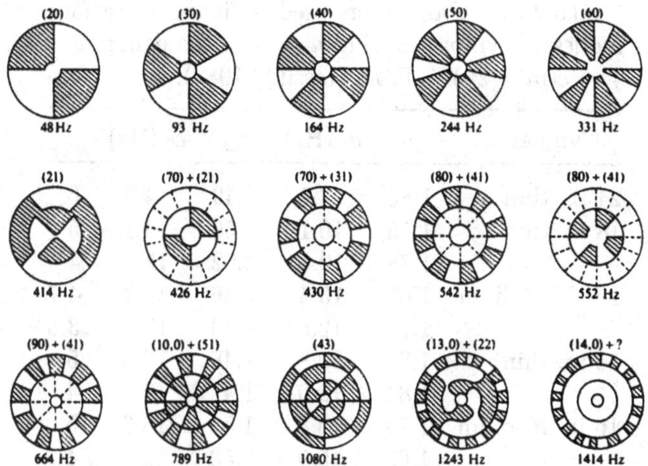

FIGURE 20.2. Modes of vibration of a 38-cm cymbal. The first six modes resemble those of a flat plate, but after that the resonances tend to be combinations of two or more modes (Rossing and Peterson, 1982).

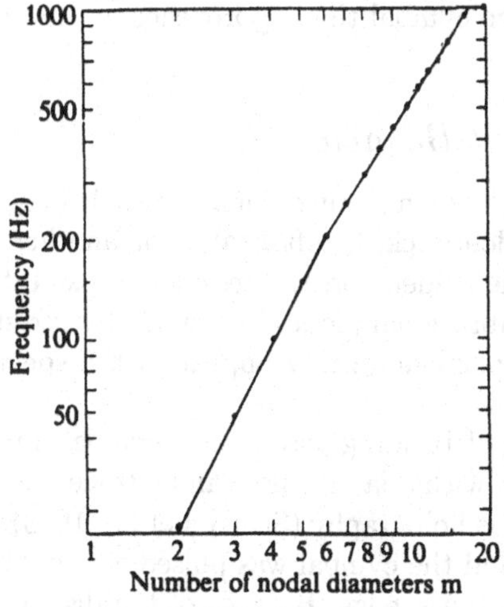

FIGURE 20.3. Modal frequencies of a 60-cm cymbal (Rossing, 1982).

TABLE 20.1. Parameters used to fit the vibration modes of cymbals having $n = 0$ to the equation $f = c(m + 2n)^p$. (From Rossing, 1982).

Cymbal	p_1	c_1 (Hz)	p_2	c_2 (Hz)	m_c
24 in. thin	1.86	7.7	1.49	14.2	6
18 in. thin	1.75	10.1	1.56	14.1	5
	1.78	10.6	1.52	15.7	4.5
18 in. medium	1.65	13.4	1.46	18.2	4.7
	1.70	12.6	1.43	17.8	3.6
16 in. thin	1.81	10.8	1.48	18.8	5
	1.84	12.0	1.47	19.5	4
16 in. medium	1.70	13.8	1.53	18.3	5
	1.65	15.9	1.53	19.4	4.3
15 in. thick	1.47	20.6			

denoted as m_c in Table 20.1. The largest p values occur when the cymbal is large and thin; the cymbal then approaches flat-plate behavior.

20.1.2 Transient Behavior

A cymbal may be excited in many different ways. It may be struck at various points with a wooden stick, a soft beater, or another cymbal. The onset of sound is quite dependent on the manner of excitation. The nonlinear coupling between vibrational modes in a cymbal is strong, however, so that a large number of partials quickly appear in the spectrum, however it is excited.

The propagation of bending waves on a cymbal immediately after momentary excitement with a laser pulse can be traced, as shown in Fig. 20.4, by using pulsed video holography (Schedin et al., 1998). The field of view is about 20 × 15 cm and the cymbal was pulsed at a point 10 cm (about half its radius) from the outer edge. Because of the dispersive propagation behavior of bending waves, the first to be seen have short wavelength, about 5 mm in this case, a frequency of about 340 kHz and a propagation speed of about 1700 m/s. These are followed by waves of longer wavelength and greater amplitude, which are subsequently reflected from the outer edge of the cymbal (at the far right-hand side) and from the central dome (near the left-hand side) to interfere with the circular wave-fronts. This wave-propagation behavior can also be modeled exactly as impulsive excitation of all the normal modes of the cymbal, with phases and amplitudes set by the initial conditions of the laser impulse so as to localize the initial transverse velocity. The normal modes do not, however, become distinguishable until a time Δt after the initial excitation that is fixed by the separation Δf between the modes near the frequency f of interest by the relation $\Delta f \Delta t \approx 1$. This relation is analogous to the Heisenberg uncertainty principle in the energy form $\Delta E \Delta t \approx h$. Physically this allows time for waves to be reflected from the boundary and to interfere with one another as in the later frames of Fig. 20.4.

Decay times have been measured both of single modes of vibration and also of octave bands of sound when the cymbal is struck. In order to measure the decay times of single modes, the mode of interest is excited with a magnetic driver, which is then switched off and the sound level is recorded as the vibration decays. Decay times (60 dB) for nine radial modes in a 40-cm medium crash cymbal are shown in Fig. 20.5.

Also shown in Fig. 20.5 are decay times of the octave bands of sound when the cymbal is struck in three different ways. The first is a crash stroke in which the shoulder of a wooden drumstick contacts the cymbal near the edge with a hard, glancing blow. The second is a ride stroke in which the plastic tip of the stick impacts the cymbal about halfway between the edge

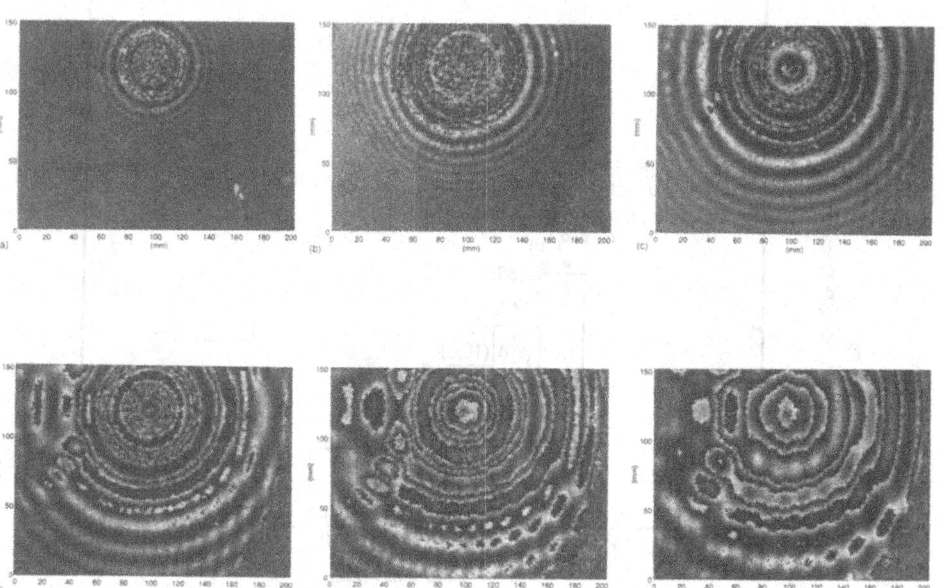

FIGURE 20.4. Phase maps showing wave propagation outward from a point 10 cm from the edge of a 41-cm diameter Zildjian cymbal. Time after impulse: (a) 30 μs; (b) 60 μs; (c) 120 μs; (d) 180μs; (e) 240μs; (f) 300 μs (Schedin et al. 1998).

and the dome of the cymbal. In the third, the cymbal is struck softly near the edge with a timpani mallet. The sound from the ride hit appears to be more sustained in the lower octave bands than from the crash, but the reverse is true at 16 kHz.

Sound spectra at various times after striking were recorded by means of a real-time spectrum analyzer. Figure 20.6 shows the spectra from a 40-cm cymbal at the time of striking and after intervals of 0.05, 1.0, and 2.0 s. Note that the sound level in the range of 2–10 kHz builds up by 10 dB or more during the first 50 ms, whereas the spectrum above 10 kHz shows little change.

From Fig. 20.6 and many other similar spectra, we can make the following observations:

1. The sound level below about 700 Hz shows a rather rapid decrease during the first 200 ms, after which it decays slowly. This is apparently caused by the conversion of energy into modes of higher frequency.

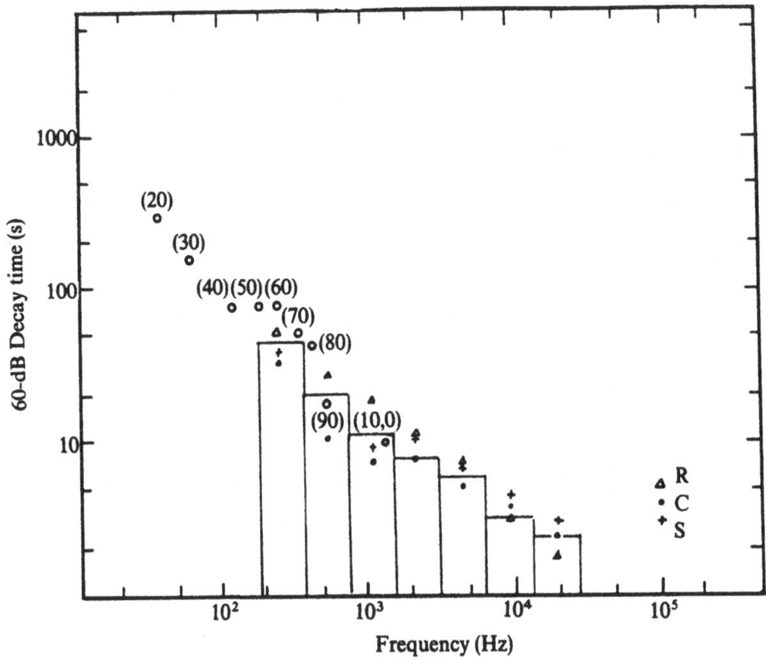

FIGURE 20.5. Decay times of single modes of vibration and octave bands of sound in a 40-cm-diameter cymbal. Octave-band decays were measured for three different strokes (see text), designated as crash (C), ride (R), and soft (S) (Rossing and Shepherd, 1983).

2. Several strong peaks in the 700–1000 Hz range build up between 10–20 ms, then decay.
3. Sound energy in the important 3–5 kHz range peaks about 50–100 ms after striking.
4. Sound in the range of 3–5 kHz, which gives the cymbal its shimmer, is often the most prominent feature from about 1–4 s after striking.
5. The low frequencies again dominate the lingering sound, but at a much lower level, so that they are rather inconspicuous.

20.1.3 Chaotic Vibration

There is now increasing evidence that the fully developed sound of cymbals and a tam-tams, which we discuss later, is actually the result of a chaotic vibration (Legge and Fletcher, 1989; Fletcher et al., 1989; Fletcher, 1993; Touzé et al., 1998). If we excite a thin flat circular plate sinusoidally at its center, it behaves linearly for small amplitudes and exhibits the expected mode resonances. Large-amplitude excitation at a frequency near one of these resonances, however, generally leads to bifurcation of the vibration with associated doubling or tripling of the vibration period. The same behavior is observed for an orchestral cymbal excited sinusoidally at its center, the particular case studied giving a five-fold increase in period and a consequent major-chord-like sound based upon the fifth subharmonic of the excitation frequency (Fletcher, 1993). For a slightly different excitation frequency, the transition was to a fully chaotic vibration with a typical

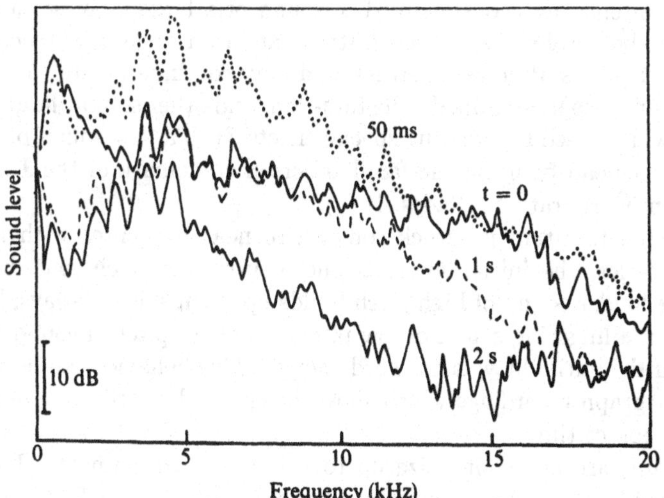

FIGURE 20.6. Sound spectra of a 40-cm cymbal immediately after striking and after intervals of 0.05, 1.0, and 2.0 s (Rossing and Shepherd, 1983).

cymbal-like sound. Studies by Legge (Legge and Fletcher, 1989) on a large gong show that the bifurcation map is often very complicated, with minor frequency shifts giving rise to very different behavior.

These results can be modeled to first order by assuming that the cymbal behaves like a simple nonlinear Duffing oscillator with a single degree of freedom, of the type we discussed in Section 5.2, appropriate choice of degree of nonlinearity, damping coefficient, and excitation frequency leading to various types of subharmonic (bifurcation) behavior or to chaotic oscillation when it is excited externally.

While such a system with only a single geometrical degree of freedom (generally referred to as having two degrees of freedom in phase space, since position and velocity are counted separately) cannot exhibit chaotic behavior in the absence of external excitation, because of the limitations imposed by conservation of energy, this restriction no longer applies if we increase the number of geometrical parameters or, equivalently, increase the number of possible vibration modes. A continuous system has infinitely many degrees of freedom, so there is no restriction on chaotic behavior. Essentially what can happen is that a low-order mode can be excited to large amplitude by an external impulse, and this vibration in turn drives other modes to bifurcation or chaotic behavior.

20.2 Tam-Tams

Among the many percussion instruments of Oriental origin that have been adopted into Western music, the tam-tam, shown in Fig. 20.7, is one of the most interesting acoustically. The sound of a large Chinese tam-tam, which can be the loudest of any instrument in an orchestra, reaches full brilliance 1 or 2 s after being struck and may continue for up to 1 min if the instrument is not damped. The tam-tam, and other similar large gongs, are also widely used for creating special effects, such as the sound produced by the Herculean figure in the familiar screen trademark of the J. Arthur Rank Film Corporation (Blades, 1970).

When the tam-tam is struck somewhere near its center with a large padded mallet, the initial sound is one of very low pitch, but in a few seconds a louder sound of high pitch builds up, then slowly decays, leaving once again a lingering sound of low pitch. The high-pitched sound fails to develop if the initial blow is not hard enough. This behavior can be verified from Sonagraph recordings which show the spectral distribution of energy as a function of time.

Tam-tams are of varying size up to about 1 m in diameter. They are usually made of bronze (approximately 80% copper and 20% tin, with traces of lead or iron). Although the center is usually raised slightly, they do not have a prominent central dome as do gongs and cymbals. They do,

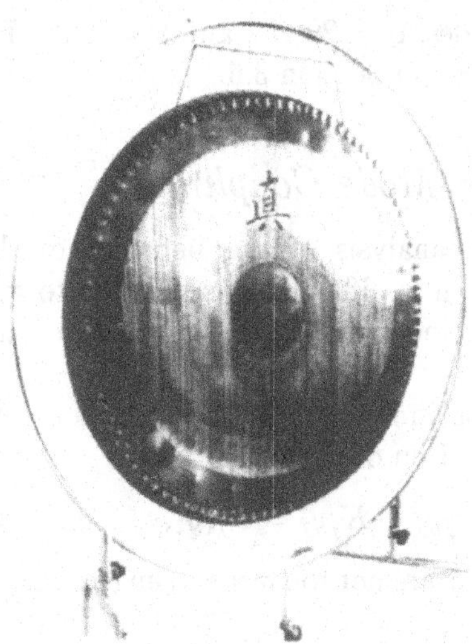

FIGURE 20.7. A large Chinese tam-tam, as used in Western symphony orchestras. The diameter is typically about 80 cm.

however, have one or more circles of hammered bumps and a fairly deep rim. They are considerably thinner than most large gongs.

20.2.1 Modes of Vibration

Because of their large size, their rim, small metal thickness, and numerous bumps and hammer marks, the modes of vibration of tam-tams show rather marked variations from the modes of a flat circular plate. The low-frequency domain has several prominent axisymmetric modes, which absorb much of the energy of the initial blow. In one large tam-tam of Japanese manufacture, for example, we observed prominent modes with frequencies of 39, 162, 195, 318, 854, 879, and 1000 Hz when the tam-tam was driven at its center by an electromechanical vibrator (Rossing and Fletcher, 1982). In a large (36-in-diameter) Paiste tam-tam, however, the lowest axisymmetric modes were found to lie somewhat higher in frequency (68, 184, 219, 257, and 362 Hz).

Other families of vibrational modes of considerable interest are those that have a number of radial modes equal to the number (or an even multiple) of hammered bumps in one of the circles. These modes would be favored in the delayed sound if the bumps played a prominent role in the conversion of energy from axisymmetric to nonsymmetric modes, as will be described in Section 20.2.2. The frequencies of these radial modes in one large tam-tam could be fitted to an empirical relationship $f = cm^p$, where m is the number of radial modes, $c = 29$ Hz, and $p = 1.17$. This is a variation on Chladni's law discussed in Section 3.6.

20.2.2 Nonlinear Mode Coupling

Although a detailed analysis of the behavior of the higher modes of vibration in a tam-tam is difficult, it is possible to write down some general results that are independent of the precise nature of the physical nonlinearities involved.

From the discussion in Section 5.4, the jth mode of a tam-tam or gong has an equation of motion of the form

$$m_j \ddot{y}_j + R_j \dot{y}_j + K_j y_j = F_j(y_1, \ldots, y_i, \ldots, y_j, \ldots). \tag{20.1}$$

If the mode amplitudes are not too large, then each mode is nearly harmonic and it behaves like

$$y_j(t) = a_j(t) \sin(\omega_j t + \phi_j), \tag{20.2}$$

where $a_j(t)$ is a slowly varying amplitude. For the lowest modes of the gong, for example, mode i, which are initially excited to large amplitude,

$$a_i(t) \approx a_i \exp(-t/\tau_i). \tag{20.3}$$

The upper modes, for example, mode j, may initially have nearly zero amplitude and receive energy through being driven by nonlinear terms in F_j. The most important such term is that for which

$$\omega_j \approx n\omega_i, \tag{20.4}$$

where mode i is one of those strongly excited. Use of the general solution method detailed in Section 5.1 then shows that the amplitude of mode j behaves like

$$a_j(t) = \frac{A_{ij} a_i^n \tau_j (e^{-nt/\tau_i} - e^{-t/\tau_j})}{1 - n\tau_j/\tau_i}, \tag{20.5}$$

where A_{ij} measures the strength of the coupling between modes i and j. The amplitude of mode j begins at zero, rises to a maximum after a time

$$t^* = \frac{\tau_i \tau_j \ln(\tau_i/n\tau_j)}{(\tau_i - n\tau_j)}, \tag{20.6}$$

and then decays toward zero. The maximum amplitude is proportional to a_i^n. This is essentially what is observed (Rossing and Fletcher, 1982).

When we seek a physical origin for the extreme nonlinearity of the tam-tam, we find this first in the fact that its metal is thin and not under significant tension. This means that the stiffness and tension restoring forces are very small, so that the quadratic tension generated by mode displacements has a large effect (Fletcher, 1993). Another source of mode coupling and nonlinearity arises from the hammered bumps in the surface, since abrupt changes in slope are known to generate mode coupling and nonlinear frequency multiplication (Legge and Fletcher, 1987, 1989).

The buildup and decay of vibrations at various frequencies during the first 0.4 s after vigorous excitation of a tam-tam is shown in Fig. 20.8. These waveforms were recorded with an accelerometer placed approximately halfway between the center and edge. The delay in excitation of higher modes is clearly apparent. The other significant thing is the irregular amplitude variation in each frequency band. The origin of this is unclear, but the most likely explanation is that the vibration is chaotic, as we have already discussed in Section 20.1.3 for the cymbal. Since the frequency band accepted by the analyzer in this measurement was only 100 Hz, such a chaotic vibration would be expected to show wide amplitude fluctuations.

The overall behavior of the sound of a tam-tam is shown in Fig. 20.9, which shows the radiated acoustic spectrum immediately after excitation and again after about 3 s. The initial large excitation of low-frequency modes is apparent, as is the subsequent transfer of vibrational energy to modes in the range 1–5 kHz, which contribute the late-developing "sheen" to the sound.

In summary, the distinctive timbre of a tam-tam arises from the relatively slow buildup of modes of vibration having high frequencies. This slow buildup occurs because energy is fed to these modes from the modes of low frequency excited initially. The nature of nonlinear coupling between

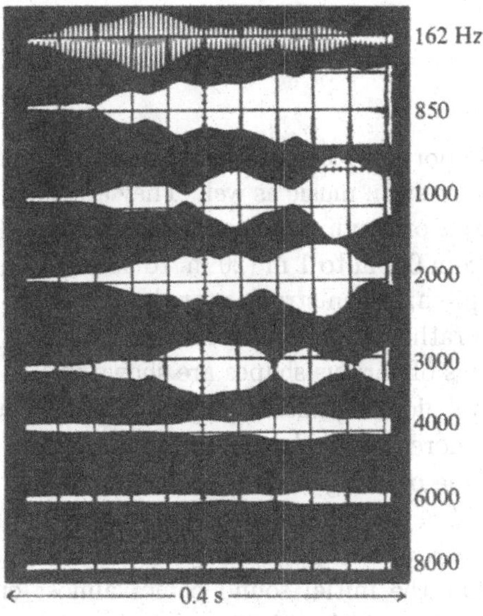

FIGURE 20.8. Buildup and decay of vibrations in different frequency bands during the first 0.4 s (Rossing and Fletcher, 1982).

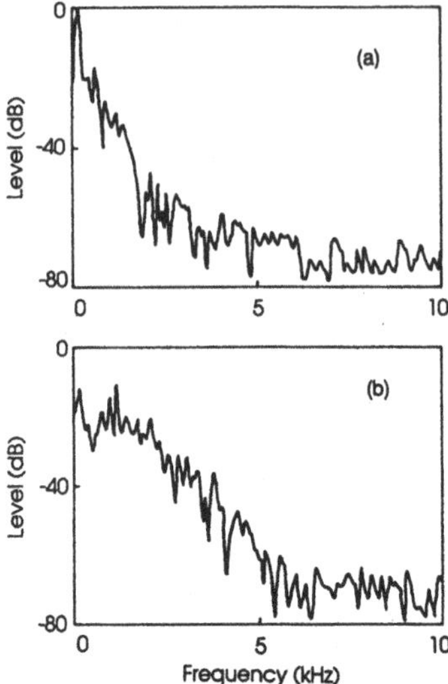

FIGURE 20.9. Intensity spectrum of the radiated sound from a tam-tam (a) immediately after excitation by a stroke with a padded striker, and (b) after a further delay of about 3 s. (Legge and Fletcher, 1989)

modes is not well understood at present, but the large number of hammered bumps spaced around the tam-tam appear to play a significant role in transferring energy from axisymmetric to modes of lower symmetry. The harder the blow, the more significant the nonlinear coupling becomes.

20.3 Gongs

Gongs play a very important role in Oriental music, but they enjoy considerable popularity in Western music as well. They are usually cast of bronze with a deep rim and a protruding dome. Gongs used in symphony orchestras usually range from 0.5 m to 1 m (20 in. to 38 in.) in diameter, and are tuned to a definite pitch. When struck near the center with a large mallet, the sound builds up rather slowly and continues for a long time if the gong is not damped. Gongs of various shapes are shown in Fig. 20.10.

Gongs have deep religious significance in many Eastern cultures. Nowhere is the gong revered more deeply than in Indonesia, where gongs of various sizes are the backbone of the gamelans of Java and Bali. A large gong is used to mark the end of each melodic section in gamelan music.

The sound spectrum of a small Balinese tawa tawa gong is shown in Fig. 20.11. Note that the initial sound comes almost entirely from two axially symmetric modes. Although they appear to be the same in the sketches, the one lower in frequency has a node where the face joins the rim, whereas the higher one has a node about one-third of the way

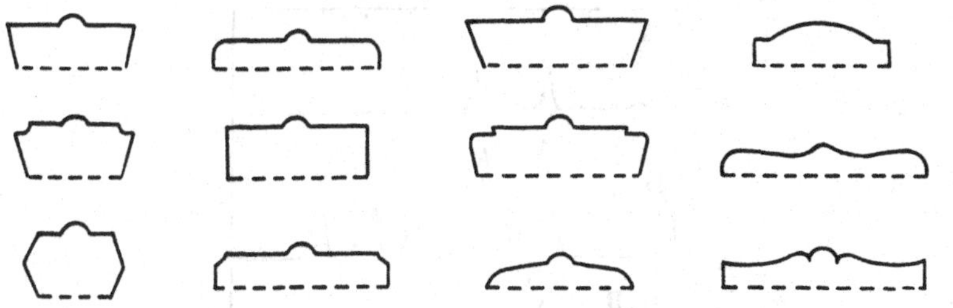

FIGURE 20.10. Gongs of various shapes. Note the protruding dome at the center.

down the rim. The second spectrum, recorded a half second after the gong is struck, shows that many other modes of vibration quickly develop and persist for a second or two. The top spectrum shows that 5 s after the strike the fundamental and a pair of modes with one nodal diameter dominate. Note that the two axisymmetric modes that produce the two prominent peaks in Fig. 20.11 have frequencies in a 2:1 (octave) ratio. We have found this to be the case in large gamelan gongs as well.

Figure 20.12 shows the spectrum of a large (59-cm-diameter) gamelan gong struck at its center. The modal frequencies are 67 Hz and 135 Hz, and the resulting pitch is identified as C_2. If the gong is struck off center, a peak appears at 115 Hz, which is due to a mode with one nodal diameter. In this large gong, too, many modes develop a short time after the initial strike.

A common feature in both the large gamelan gong and the smaller tawa tawa is the octave interval between the initial partials. The mass of the central dome compared to the rest of the gong is apparently an important factor in determining the frequency ratio of these two axisymmetric modes. Our best estimates indicate that the dome in the large gamelan gong has about 10% of the total mass of the gong, whereas in the smaller tawa tawa gong the dome mass is 7% of the total. Further experiments with a 66-cm Turkish gong showed that loading the center with 600 g (about 10% of the total mass) also brought the first two partials into an octave relationship (Rossing and Shepherd, 1982).

20.3.1 Pitch Glide

Of considerable interest, from an acoustical as well as a musical point of view, are the gongs used in Chinese opera orchestras, shown in Fig. 20.13. The pitch of the larger gong glides downward as much as three semitones after striking, whereas that of the smaller gong glides upward by about two semitones.

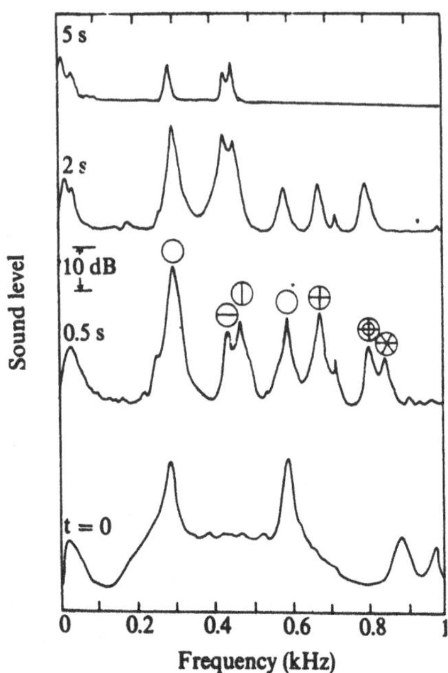

FIGURE 20.11. Sound spectrum of a tawa tawa gong. The initial sound ($t = 0$) comes mainly from two prominent axisymmetric modes, but after 0.5 s many modes of vibration have been excited, which decay at varying rates. Some of the modes are identified at the peaks (Rossing and Shepherd, 1982).

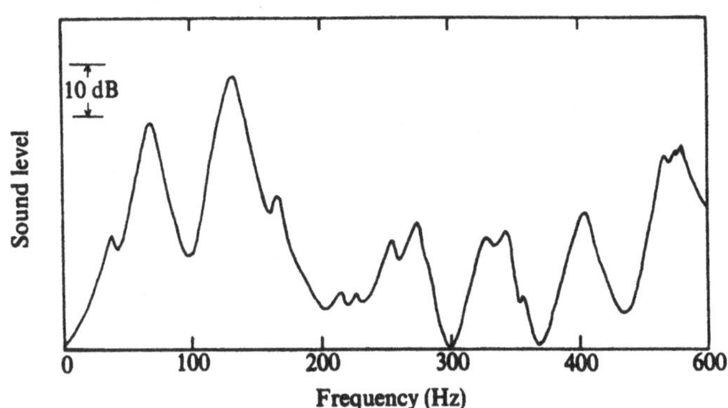

FIGURE 20.12. Sound spectrum of a large gamelan gong. The principal modes of vibration have frequencies of 67 Hz and 135 Hz, and their corresponding partials are about an octave apart (Rossing and Shepherd, 1982).

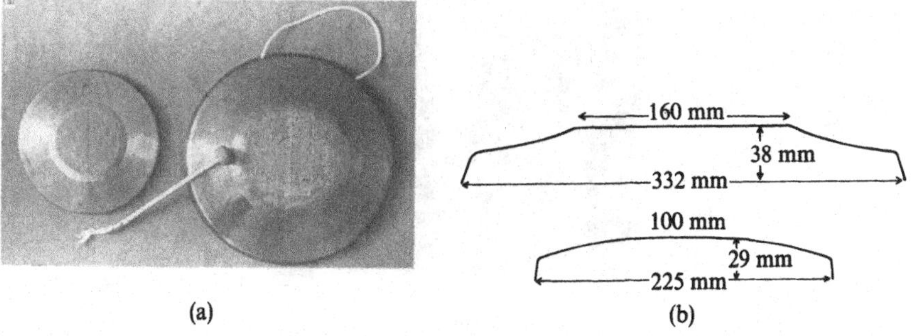

FIGURE 20.13. Two Chinese opera gongs that show a marked pitch glide after being struck a substantial blow. The smaller gong at the left glides upward, while the larger gong at the right glides downward in pitch.

The nonlinear change in modal frequency with amplitude in these Chinese opera gongs is a good example of the nonlinear behavior of plates and shallow shells discussed in Section 5.7. The central section of the larger gong is nearly flat, and the hardening spring behavior, characteristic of flat plates, dominates. The central part of the smaller gong, on the other hand, is sufficiently convex to behave as a spherical cap shell that has softening spring behavior at large amplitude (Rossing and Fletcher, 1983; Fletcher, 1985). The sloping sides and the rim of the gongs normally vibrate relatively little, but they provide an edge condition that appears to be somewhere between clamped and simply-supported (hinged).

20.4 Crotales

Although the term crotales originally applied to small antique finger cymbals or castanets, modern composers use the term for different types of small cymbals varying in shape from a small Turkish cymbal to a small bell plate played with a mallet.

The set of thirteen crotales shown in Fig. 20.14, covering the range C_6 through C_7 (1047–2093 Hz), consists of circular bronze plates 4.5 mm thick and with diameters from 101 mm to 129 mm. At the center is a stem about 32 mm in diameter and 18 mm long.

Some of the principal modal frequencies of the crotales in Fig. 20.14 are given in Table 20.2. The heavy stem (which comprises roughly one-fifth of the total mass) shifts the modal frequencies in such a way that the (3,0) mode has nearly twice the frequency of the (2,0) mode. This is similar to the effect of loading the center of a large gong (see Section 20.8), curiously enough, although there we were dealing with axisymmetric modes and here we are dealing with modes having mainly nodal diameters.

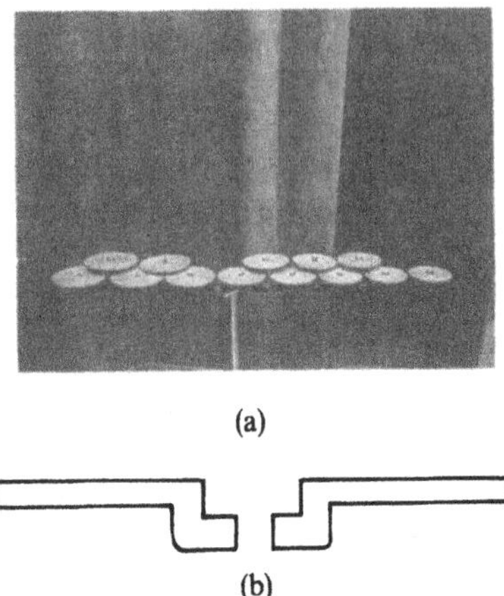

(a)

(b)

FIGURE 20.14. (a) A set of crotales tuned to notes on the diatonic scale from C_6 through C_7. (b) Cross section of one of the crotales.

TABLE 20.2. Frequency ratios for principal modes of vibration in a set of tuned crotales.

Crotale	f_{01}/f_{20}	f_{30}/f_{20}	f_{11}/f_{20}	f_{40}/f_{20}	f_{50}/f_{20}
C_6	1.45	2.08	3.05	3.05	
$C_6^\sharp$		2.09	2.89	3.63	
D_6		2.01	2.99	3.44	
$D_6^\sharp$		2.13	3.38	3.66	
E_6	1.43	1.93	2.86	3.31	
F_6	1.39	2.01	2.81	3.51	
$F_6^\sharp$	1.42	1.98	2.85	3.41	5.19
G_6	1.42	1.97	2.81	3.38	5.17
$G_6^\sharp$	1.49	1.96	2.97	3.34	5.07
A_6		1.98	2.76	3.42	5.14
$A_6^\sharp$		2.00	2.88	3.41	5.15
B_6		1.94	2.71	3.29	4.97
C_7	1.42	1.92	2.67	3.28	4.92
Flat plate	1.73	2.33	3.91	4.11	6.30

20.5 Bell Plates

Large metal plates are sometimes used in orchestras as substitutes for cast bells. Rectangular steel plates as large as 3 × 2 ft and weighing over 60 lb are commonly used to simulate the sounds of church bells. The spectrum of partials emitted by rectangular plates depends upon the length to width ratio, as can be seen in Fig. 3.14. For a free plate having length and width in the ratio 3:2, for example, Fig. 3.14 predicts modal frequencies approximately in the ratios 1 : 1.15 : 2.4 : 2.7 : 3.2 : 4.3, which are not very much like those of a carefully tuned church bell (1 : 2 : 2.4 : 3 : 4 : 5).

It is reported that large bell plates sound about a quarter tone sharp on impact, gliding to their final pitch in about three seconds (Holland, 1978). This would be due, of course, to the hardening spring behavior of flat plates (see Section 5.7).

20.6 Flexed Plates: The Musical Saw

Although the musical saw is mainly a regional folk instrument, the physics underlying it is so interesting that it merits inclusion. The instrument may consist of an ordinary carpenter's saw—perhaps with the teeth ground off for safety—the handle of which is held between the knees and the blade bent with the left had into a S-shape, as shown in Fig. 20.15. The right hand is then used to bow the saw, with a cello or bass bow, along one edge. This excites a nearly pure sinusoidal tone of useful musical pitch, which is radiated efficiently by the relatively large surface of the vibrating blade. The player controls the pitch of the sound by changing the curvature of the blade, increasing curvature giving a higher pitch. Because of the vibrato and pitch glide, the sound is reminiscent of an early electronic instrument, the theremin.

Our brief treatment is based on the sophisticated mathematical discussion given by Scott and Woodhouse (1992). As far as this discussion is concerned, the slight taper of the saw blade is irrelevant, though it is important in the playability of the instrument. An earlier treatment by Cook (1991) gives interesting information, but its mathematical analysis is restricted to the case of uniform blade curvature, which is not the geometry used when making the saw sound.

The transverse vibrational modes of a long flat metal strip can be divided into groups, as discussed in Section 3.8 and illustrated in Fig. 3.9. The $(0, n)$ modes are simple bar-like vibrations, while modes of the type (m, n) have m nodal lines parallel to the long sides of the strip and are of higher frequency if the strip is narrow. The modes with which we are concerned in the musical saw are those of the second symmetrical group of classification $(2, n)$, which can be excited by bowing the edge of the strip. The important thing is the behavior of these modes when the strip is curved.

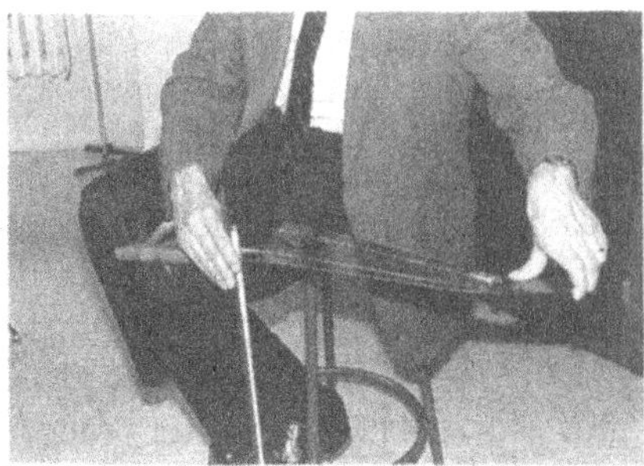
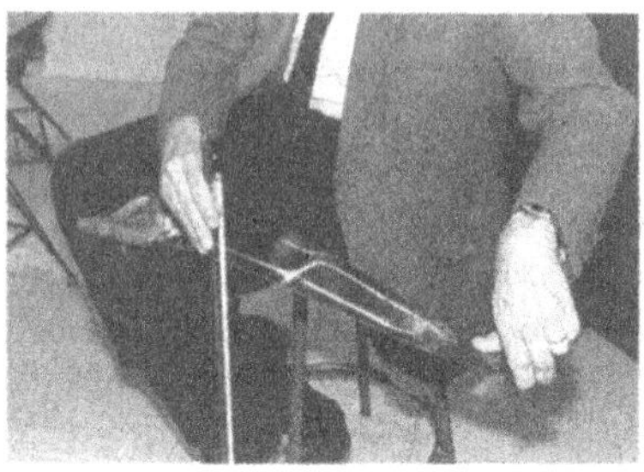

FIGURE 20.15. Chladni patterns showing two modes of vibration of a musical saw. Note that the handle is firmly held by the player's knees while the left hand applies the desired amount of stress by bending the blade (photographs courtesy of Arnold Tubis).

If the ends of the strip are free to move, then curvature has negligible effect on the frequencies of the bar-like modes $(0, n)$, but this is not true of the modes $(2, n)$, which have structure in the direction of the width of the strip. The presence of curvature introduces additional tensile stresses that raise the frequencies of these modes, as shown in Fig. 20.16. In practice it is necessary to curve the saw to a S-shape, in which the curvature increases linearly with distance away from the middle of the blade. If a vibration of frequency ω is excited, then a wave propagating away from the middle of the blade begins from a point such as A on the zero-curvature line in the figure. It propagates toward regions of increasing curvature, corresponding to point B, until finally it reaches a place where the intersection of ω with

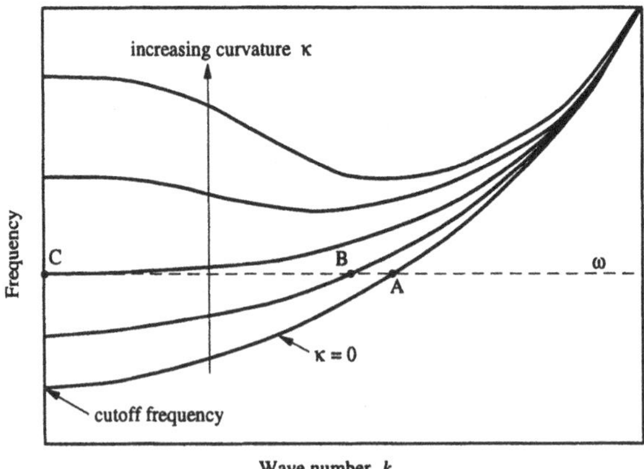

FIGURE 20.16. Dispersion curves giving the relation between wave number k and frequency ω for the second symmetrical mode $m = 2$ on a long strip, with curvature κ as a parameter.

the frequency curve corresponds to point C. Any propagation to regions of higher curvature would cause the wave number k to become imaginary, so the wave is reflected. The same thing then happens at the other end of the blade, so that the wave is effectively trapped near the middle of the blade by the curvature. The allowed frequencies are then determined by the condition that there should be an integral number of wavelengths on the round trip. This mode trapping is essential to the sounding of the instrument, since the vibrations never reach either end of the saw and so are preserved from damping. Two such trapped modes with $n = 1$ and 2, respectively, are shown in Fig. 20.15, but other modes are also used. The frequency of each trapped mode increases with increasing saw curvature and, as is clear from Fig. 20.16, there is a lower bound to the frequency that can be produced.

20.7 Steel Pans

An instrument of rather recent origin is the Caribbean steel pan or steel drum. In addition to being the foremost musical instrument in its home country, Trinidad and Tobago, the steel pan is becoming increasingly popular in North America and Europe. The modern family of steel pans covers a five-octave range, and steel bands of today use them to perform calypso, jazz, popular and classical music.

Steel bands originated in Trinidad in the 1940s, partly because of the ban on tamboo bamboo bands, whose strong rhythms had become a feature of the annual Carnival festivities. (Unfortunately, the large bamboo sticks were being used as weapons as well as musical instruments!) Not to be denied, the enterprising musicians turned to garbage cans, buckets, brake drums, or whatever was available; the first steel drums were rhythmic rather than melodic.

The development of tuned steel drums took place in the years following the end of World War II, when the annual celebration of Carnival was resumed with great enthusiasm. Thousands of 55-gallon oil drums left on the beach by the British navy provided ample raw material for experimentation.

Many claims have been made about the invention of the tuned steel pan. Undoubtedly, it resulted from a lot of trial and error on the part of the musicians and metalsmiths, and it is still evolving. Pioneer tuners included Bertie Marshall, Anthony Williams, and Ellie Mannette.

Modern steel pans are known by various names, such as tenor or lead, double second, double tenor, guitar, cello, quadrophonics, and bass (see Fig. 20.17). The tenor pan has from 26 to 32 different notes, but each bass pan has only 3 or 4; hence the bass drummer plays on six pans in the manner of a timpanist. Different makers use different patterns for their pans, although there is a movement in Trinidad and Tobago to standardize

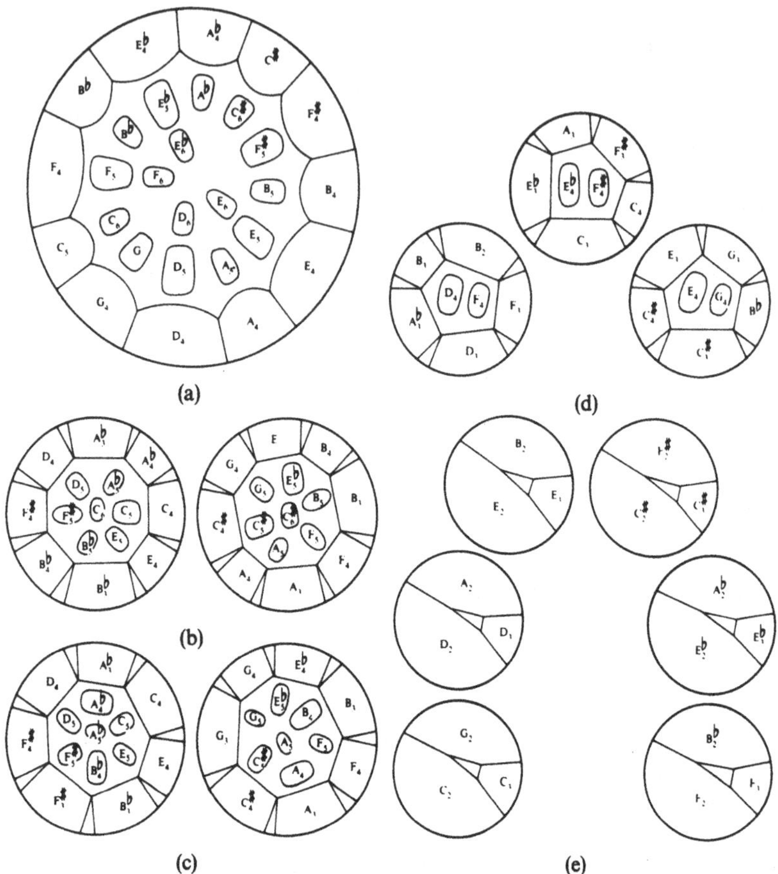

FIGURE 20.17. Design of a typical steel drum set, showing locations of various notes: (a) single tenor (lead), (b) double tenor, (c) double second (alto), (d) cello, and (e) bass (courtesy of Clifford Alexis).

designs (Roach, 1992). Note layouts in several steel pans are shown in Fig. 20.17.

20.7.1 Sound Spectra and Modes of Vibration

The sound spectra of steel pans are surprisingly rich in harmonic overtones. These harmonic overtones appear to have three different physical origins:

1. radiation from higher modes of vibration of a given note area, tuned harmonically by the tuner;

2. radiation from nearby notes whose frequencies are harmonically related to the struck note; and
3. nonlinear motion of the note area, vibrating at its fundamental frequency.

Sound spectra of two notes on a double second steel pan are shown in Fig. 20.18.

Modal shapes of various modes in three different B$^\flat$ note areas in a pan are shown in Fig. 20.19. Note that in all three of these notes, the two modes having a single nodal diameter [similar to (1,1) modes in a circular plate] have been tuned to the second and third harmonics of the fundamental note. In the B$_3^\flat$ and B$_4^\flat$ notes in the outer ring, the mode with the nodal line parallel to the circumference supplies the second harmonic, and the one with its node along a diameter of the drum is tuned to the third harmonic. In the circular B$_5^\flat$ note in the inner ring, however, the opposite is true.

20.7.2 Mechanical Coupling Between Note Areas

Striking the B$_3^\flat$ note in a steel pan excites both the B$_4^\flat$ and B$_5^\flat$ note areas nearby. The B$_4^\flat$ area radiates a very strong second harmonic; damping this note area produces a substantial change in the timbre of the B$_3^\flat$ played note.

The coupling between notes, which is nonlinear in character, can be studied by means of holographic interferometry by driving them sinusoidally at their fundamental frequencies as well as the frequencies of their first and

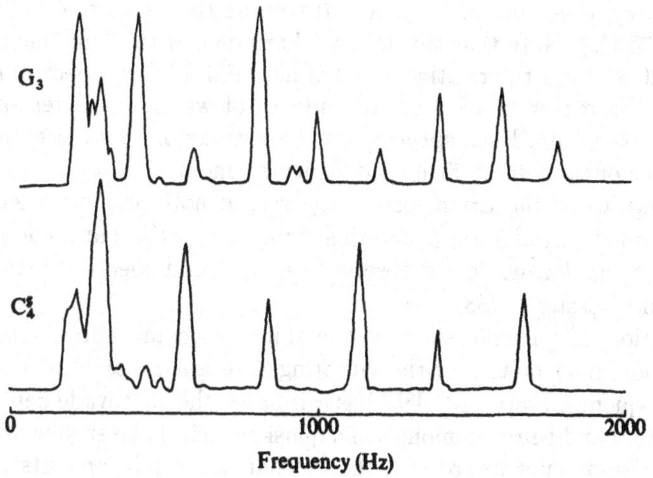

FIGURE 20.18. Sound spectra from two notes on a double-second steel pan. Note the large number of harmonics (Rossing et al., 1986).

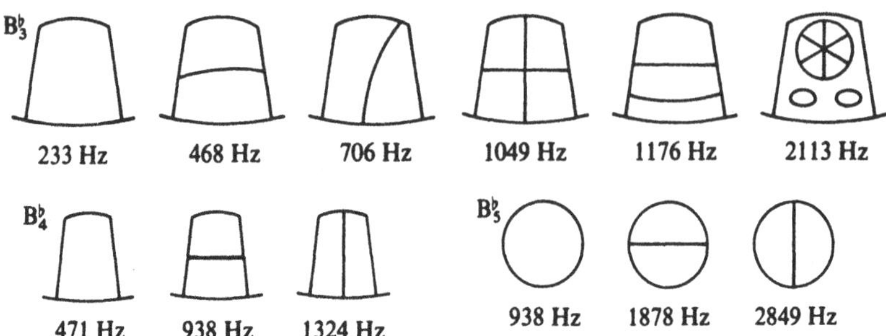

FIGURE 20.19. Modal shapes observed in three different B^b notes on a double-second steel pan. (Rossing et al., 1986).

second harmonics. The drive amplitude is varied in order to observe the effect on other note areas in the drum (Hampton et al., 1987).

Figure 20.20(a) shows the $G_5^\sharp$ note area driven at its fundamental frequency (815 Hz). As the drive amplitude increases, the second (octave) mode of the $G_4^\sharp$ note area (directly above) and the third mode of the C_4 note area (above and to the right) are excited. A further increase in amplitude excites additional notes, until at the largest amplitude (comparable to the amplitude given to the note by the player during performance), almost the entire drum vibrates and radiates sound.

Figure 20.20(b) shows the same $G_5^\sharp$ note area driven at the frequency of its second mode (1605 Hz). At this frequency, a node forms parallel to the circumference of the drum, as in the $B_5^\flat$ note described earlier. In Fig. 20.20(c), this same note area is driven at the frequency of its third mode (1878 Hz). Note that this is not a harmonic of the fundamental.

Table 20.3 shows the relative levels of harmonics in the spectrum of the $F_3^\sharp$ note when it is struck with hard and soft blows at the center and near the side of the note. Each spectrum is the average of 16 strikes. Note the prominence of the fourth, fifth, and eighth harmonics.

Damping all of the drum except the struck note area with sandbags all but eliminates radiation from other note areas. Still, harmonic partials up to the tenth harmonic are observed in the sound spectra of steel pans (Leung and Rossing, 1988).

Generation of harmonics above the third harmonic appears to occur through nonlinear motion of the vibrating note area, mainly at its fundamental frequency. Fletcher (1985) has estimated the amplitude dependence of the second and third harmonics of a quasi-spherical cap shell with a fixed edge, but we are not aware of a theory that accurately predicts the amplitude of higher harmonics. The irregular shapes of steel pan note areas and the variable boundary conditions would be expected to enhance the nonlinear generation of harmonics.

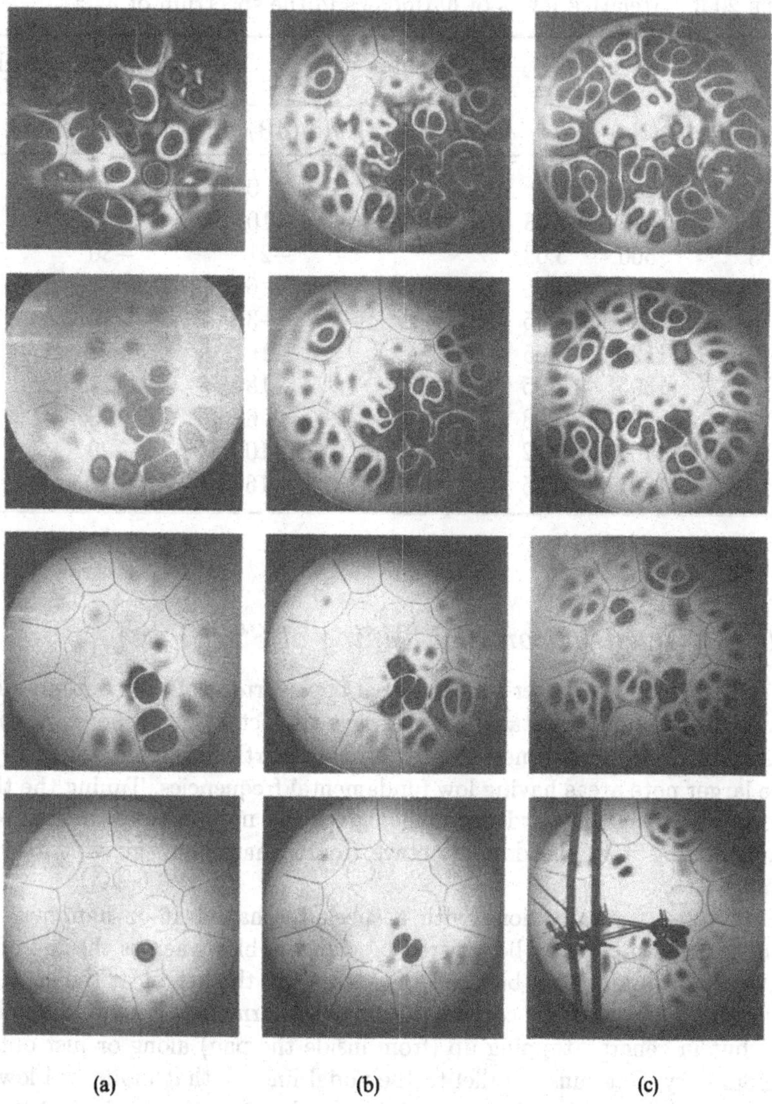

FIGURE 20.20. Holographic interferograms illustrating the motion of a double-second steel pan when the $G_5^\sharp$ note is driven at various frequencies and amplitudes: (a) driven at the fundamental frequency (815 Hz), increasing amplitude from top to bottom; (b) driven at the second resonance frequency (1605 Hz); and (c) driven at the third resonance frequency (1878 Hz) (from Rossing et al., 1986).

TABLE 20.3. Relative levels of harmonics in the spectrum of $F_3^\sharp$.[a]

			Center strike		Off-center strike	
Harmonic	f (Hz)	f/f_1	Soft (dB)	Hard (dB)	Soft (dB)	Hard (dB)
1	185	1	0	0	0	0
2	375	2.03	−8	−10	−2	−3
3	560	3.03		−21	−30	
4	750	4.05	−20	−6	−1	−1
5	935	5.05		−3	−13	+7
6	1120	6.05		−16	−27	
7	1305	7.05		−18	−27	−15
8	1495	8.08	−41	−6	−3	+1
9	1707	9.22		−10	−20	−10
10	1897	10.25		−16		

[a] Rossing et al., 1986.

20.7.3 Construction and Tuning of Steel Pans

A skilled steel pan maker tunes at least one overtone of each note to a harmonic of the fundamental (nearly always the octave). Whenever possible, a second overtone is tuned to the third or fourth harmonic, especially in the larger note areas having low fundamental frequencies. Tuning the third mode to a twelfth (third harmonic) gives the note a more mellow tone, while tuning it to the double octave (fourth harmonic) gives it a bright tone.

Tuning is typically done with a specially shaped 16-oz hammer. The fundamental pitch may be lowered by glancing blows across the top of the note area, while it may be raised by increasing the height at the center or by tapping down at the corners. Tuning the harmonics is more difficult to do, but in general, tapping up (from inside the pan) along or just outside a boundary that runs parallel to the nodal line for that mode will lower a harmonic, whereas tapping down or outward just inside this boundary will raise it (Hansen, et al., 1995).

After shaping, the pan is normally fired for several minutes to give it a partial annealing. Laboratory measurements show that the frequencies of the first three modes of the notes in the outer ring increased by 10 to 30%, suggesting that the annealing has increased the Young's modulus substantially (Rossing et al. 1996). All the notes must, of course, then be retuned. Some pan makers employ "sandwich" hardening: the surface is hardened by heating it in a nitrogen atmosphere or in a nitriding path (Rohner, 1995). A detailed guide to the making and tuning of steel pans is given by Kronman (1991).

References

Blades, J. (1970). "Percussion Instruments and Their History," Faber and Faber, London. p. 97.

Cook, R.D. (1991). Vibration of a segment of a non-circular cylindrical shell: The "musical saw" problem. *J. Sound Vibr.* **146**, 339–341.

Fletcher, N.H. (1985). Nonlinear frequency shifts in quasi-spherical cap shells: Pitch glide in Chinese gongs. *J. Acoust. Soc. Am.* **78**, 2069–2073.

Fletcher, N.H. (1993). Nonlinear dynamics and chaos in musical instruments. in "Complex Systems: From Biology to Computation" ed. D.G. Green and T. Bossomaier, IOS Press, Amsterdam.

Fletcher, N.H., Perrin, R., and Legge, K.A. (1989). Nonlinearity and chaos in acoustics. *Acoustics Australia* **18**(1), 9–13.

Hampton, D.S., Alexis, C., and Rossing, T.D. (1987). Note coupling in Caribbean steel drums. *J. Acoust. Soc. Am.* **82**, S68 (abstract).

Hansen, U.J., Rossing, T.D., Mannette, E., and George, K. (1995). The Caribbean steel pan: Tuning and made studies. *MRS Bull.* **20** (3), 44–46.

Holland, J. (1978). "Percussion," Schirmer, New York. p. 60.

Kronman, U. (1991). "Steel Pan Tuning," Musikmuseets, Stockholm.

Legge, K.A., and Fletcher, N.H. (1987). Nonlinear mode coupling in symmetrically kinked bars. *J. Sound Vibr.* **118**, 23–34.

Legge, K.A., and Fletcher, N.H. (1989). Nonlinearity, chaos, and the sound of shallow gongs. *J. Acoust. Soc. Am.* **86**, 2439–2443.

Leung, K.K., and Rossing, T.D. (1988). Sound spectra of Caribbean steel drums. Paper AE1, *American Physical Society, Crystal City, Virginia* (abstract).

Marcuse, S. (1975). "A Survey of Musical Instruments." Harper and Row, New York.

Pichary, L.(1990). "Results of the Steel Pan Survey 90." Trinidad and Tobago Bureau of Standards, Tunapuna.

Roach, K. (1992). "The Imperatives for Standardization." Trinidad and Tobago Bureau of Standards, Tunapuna.

Rohner, F. (1995). Empirical steel pan development: materials, shapes, and musical concerns. Intl. Conf. On Advanced Materials, Cancun, Mexico.

Rossing, T.D. (1982). Chladni's law for vibrating plates. *American J. Physics* **50**, 271–274.

Rossing, T.D., and Fletcher, N.H. (1982). Acoustics of a tamtam. *Bull. Australian Acoust. Soc.* **10** (1), 21–26.

Rossing, T.D., and Fletcher, N.H. (1983). Nonlinear vibrations in plates and gongs. *J. Acoust. Soc. Am.* **73**, 345–351.

Rossing, T.D., Hampton, D.S., and Boverman, J. (1986). Acoustics of Caribbean steel drums. *J. Acoust. Soc. Am.* **80**, S102 (abstract).

Rossing, T.D., Hampton, D.S., and Hansen, U.J. (1996). Music from oil drums: The acoustics of the steel pan. *Physics Today* **49**(3), 24–29.

Rossing, T.D., and Peterson, R.W. (1982). Vibrations of plates, gongs, and cymbals. *Percussive Notes* **19** (3), 31–41.

Rossing, T.D., and Shepherd, R.B. (1982). Acoustics of gamelan instruments. *Percussive Notes* **19** (3), 73–83.

Rossing, T.D., and Shepherd, R.B. (1983). Acoustics of cymbals. *Proc. 11th Intl. Congress on Acoustics (Paris)*, 329–333.

Schad, C.-R. and Frik, G. (1996). Plattenglocker. *Acustica* **82**, 158–168.

Schedin, S., Gren, P.O., and Rossing, T.D. (1998). Transient wave response of a cymbal using double-pulsed TV holography. *J. Acoust. Soc. Am.* **103**, 1217–1220.

Scott, J.F.M. and Woodhouse, J. (1992). Vibration of an elastic strip with varying curvature. *Phil. Trans. Roy. Soc. London* A**339**, 587–625.

Touzé, C., Chaigne, A., Rossing, T., and Schedin, S. (1998). Analysis of cymbal vibrations and sound using nonlinear signal processing methods. *Proc. ISMA 98*.

21

Bells

Bells are among the oldest and most cherished of all musical instruments in human history. Bells existed in the Near East before 1000 B.C., and a number of Chinese bells from the time of the Shang dynasty (1600–1100 B.C.) are found in museums throughout the world. In 1978 a set of tuned bells from the fifth century B.C. was discovered in the Chinese province of Hubei.

Bells developed as Western musical instruments in the seventeenth century when bell founders discovered how to tune their partials harmonically. The founders in the Low Countries, especially the Hemony brothers (Franois and Pieter) and Jacob van Eyck, took the lead in tuning bells, and many of their fine bells are found in carillons today.

The carillon also developed in the Low Countries. Chiming bells by pulling ropes attached to the clappers had been practiced for some time before the idea of attaching these ropes to a keyboard or handclavier occurred to bell ringers in the sixteenth century. Many mechanical improvements during the seventeenth and eighteenth centuries, including the breached wire system and the addition of foot pedals for playing the larger bells, led to the development of the modern carillon. Today, the term carillon is reserved for an instrument of 23 (two octaves) or more tuned bells played from a clavier (smaller sets are called chimes). The largest carillon in existence is the 74-bell (six-octave) carillon at Riverside Church in New York with a bourdon of more than 18,000 kg (20 tons).

Handbells also date back at least several centuries B.C., although tuned handbells of the present-day type were developed in England in the eighteenth century. One early use of handbells was to provide tower bellringers with a convenient means to practice change ringing. In more recent years, handbell choirs have become popular in schools and churches; some 2000 choirs are reported in the United States alone.

Analysis of the rich sound of a bell reveals many components or partials, each associated with a different mode of vibration of the bell. The various partials in the sound of a church bell or carillon bell are given such descriptive names as hum, prime, tierce, quint, and nominal. The most prominent partials in the sound of a tuned bell, like those of most musical instruments, are harmonics of a fundamental.

21.1 Modes of Vibration of Church Bells

When struck by its clapper, a bell vibrates in a complex way. In principle, its vibrational motion can be described in terms of a linear combination of the normal modes of vibration whose initial amplitudes are determined by the distortion of the bell when struck. In practice, such a description becomes quite complex because of the large number of normal modes of diverse character that contribute to the motion.

The acoustically important partials in a bell result from modes in which the motion is primarily normal to the bell's surface. It has been customary to classify these modes into families or groups with some common property of the nodal pattern for this component. The most important families are those that have an antinode where the clapper strikes: at the soundbow for church bells and carillon bells, and a short distance above the lip for handbells.

The first five modes of a church bell or carillon bell are shown in Fig. 21.1. Dashed lines indicate the locations of the nodes. The numbers (m, n) at the top denote the numbers of complete nodal meridians extending over the top of the bell (half the number of nodes observed along a circumference), and the numbers of nodal circles, respectively. Note that there are two modes with $m = 3$ and $n = 1$, one with a circular node at the waist and one with a node near the soundbow. Thus, we follow the suggestion of Tyzzer (1930) and others and denote the one as $(3,1^\sharp)$ in Fig. 21.1. The ratio of each modal frequency to that of the prime is given at the bottom of each diagram.

A detailed study of the vibrational modes of a church bell has compared the normal modes computed by a finite element method to the first 134 modes observed in the laboratory (Perrin et al., 1983). Modes such as (2,0), (3,1), and (4,1) are classified as "ring driven" since, in the vicinity of the soundbow, they have many of the characteristics that the thick ring at the soundbow would exhibit if it were able to vibrate in its various inextensional radial modes as an independent system. These modes are referred to as group I modes by Lehr (1965). They are excited strongly by the clapper, and they radiate most of the strong partials in the bell sound. The second important family, designated as group II, includes the $(2,1^\sharp)$, $(3,1^\sharp)$, $(4,1^\sharp)$, and higher modes. They are classified as shell-driven modes, as are other important families having $n = 2, 3, 4, \ldots$, and referred to as groups III, IV, V,.... They are characterized by a nodal circle near the mouth. Like the ring-driven modes, they are inextensional in the sense that a neutral circle in each plane normal to the bell's symmetry axis remains unstretched. This means that the radial and tangential components of the motion, u and v,

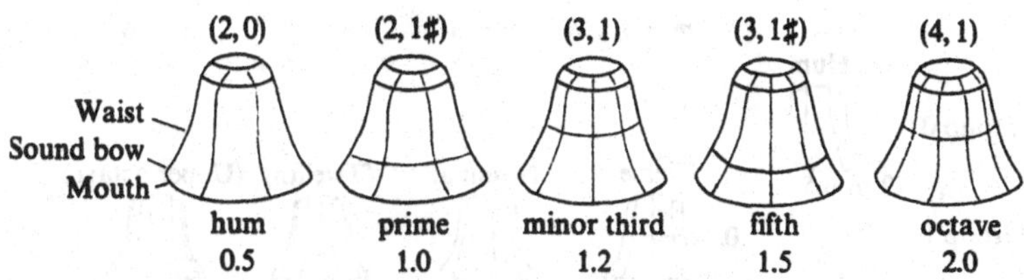

FIGURE 21.1. The first five vibrational modes of a tuned church bell or carillon bell. Dashed lines indicate the nodes. Frequencies (Hz) relative to the prime and names of the corresponding partials are given below each diagram (Rossing, 1984b).

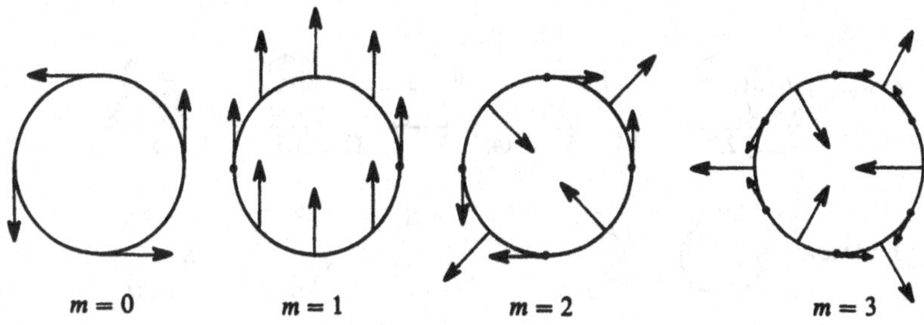

FIGURE 21.2. Motion of a bell for inextensional modes of small m. Modes with $m = 0$ and $m = 1$ require one or more nodal circles ($n > 0$).

respectively, are related by $u + \partial v/\partial\theta = 0$, where θ is the polar angle in the plane concerned (Rayleigh, 1894). Thus, we may write $u = m \sin m\theta$ and $v = \cos m\theta$. Motion of the bell for $m = 0$, 1, 2, and 3 is illustrated in Fig. 21.2. For $m = 0$, the motion is purely tangential, and we can describe these modes as "twisting" modes. Modes with $m = 1$ might be described as "swinging" modes. As m increases, these modes have radial components that become increasingly larger compared with their tangential ones.

The $m = 2$ mode is easily excited in a bell-shaped wine glass by running a moistened finger around the rim. Rayleigh (1894) pointed out that the tangential component of the motion makes this possible. In this case, the diagram in Fig. 21.2 rotates so that the point of maximum tangential motion follows the finger around the rim.

The (2,1$^\sharp$) mode deserves further discussion. Its circular node in a D_5 church bell was observed to be 16 cm above the mouth, as compared with 29 cm in the (3,1) mode (the tierce or minor third) and 10 cm in the (3,1$^\sharp$) mode (the quint or fifth). Thus, insofar as positions of nodal circles are concerned, the mode fits into group II better than group I, but in a sense, it serves both as the (2,1) and the (2,1$^\sharp$) modes in the mode classification

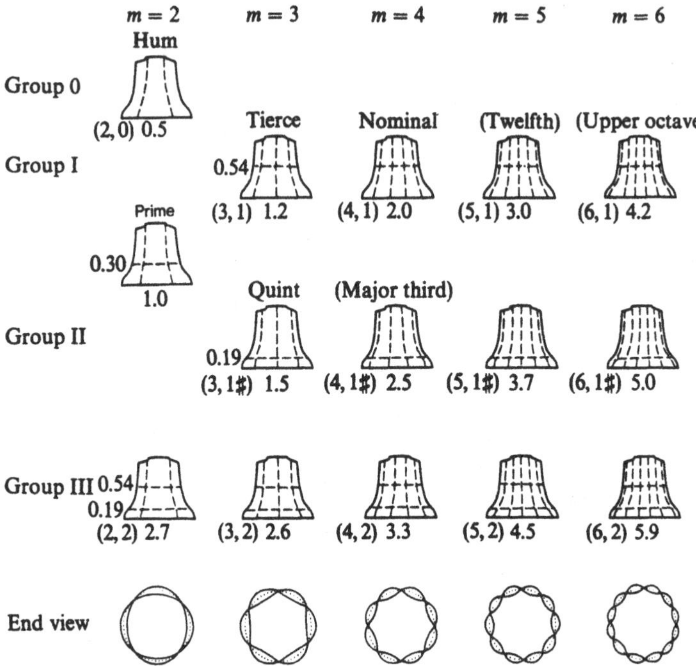

FIGURE 21.3. Periodic table of inextensional modes of vibration in a church bell. Below each drawing are the modal frequencies of a D_5 church bell relative to the prime (which has essentially the same frequency as the strike note in a bell of high quality). At lower left, (m, n) gives the number of nodal meridians $2m$ and nodal circles n (Rossing and Perrin, 1987).

scheme. Note that the (2,0) mode (hum) is the only normal mode in the modern church bell without a circular node.

Figure 21.3 is a periodic table showing some of the modes observed in a D_5 church bell. The relative modal frequencies and the locations of the nodes are indicated. To the groups previously suggested has been added a group 0 with a single member, the (2,0) mode. This classification into groups makes it easier to compare church-bell modes with handbell modes.

Vibrational frequencies of groups 0–IX in a church bell with a D_5 strike note are shown in Fig. 21.4. Arrows denote the three partials in group I that determine the strike note (see Section 21.4). Graphical displays of several modes computed by the finite element (PAFEC) method are shown in Fig. 21.5.

21.1.1 Extensional Modes

In addition to the inextensional modes we have discussed, there are a number of vibrational modes that involve stretching of the bell metal. In these extensional modes, the radial and tangential motions are related by

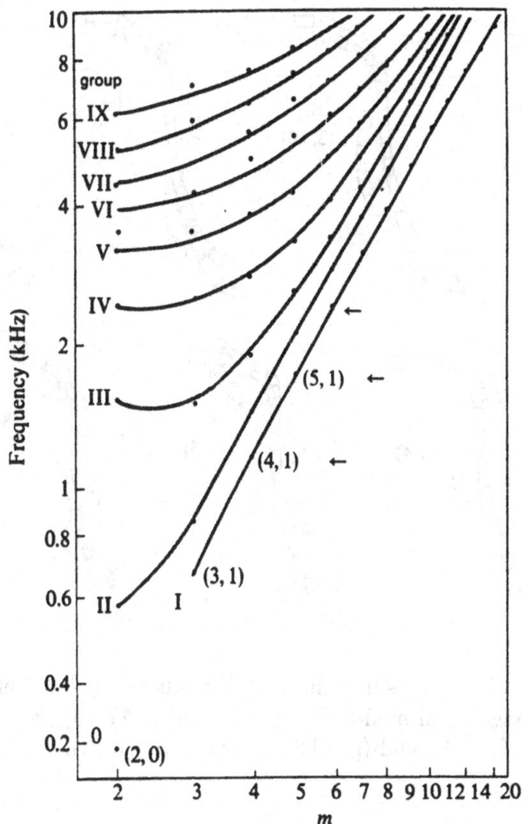

FIGURE 21.4. Vibrational frequencies of groups O–IX in a D_5 church bell (Perrin et al., 1983). Arrows denote the three partials in group I that determine the strike note (Rossing, 1984b).

$v + (\partial u/\partial \theta) = 0$, so that $u = \cos m\theta$ and $v = m \sin m\theta$. For $m = 0$, in this case, the motion is entirely radial, but with increasing m, the tangential motion increases until it takes over from the radial motion. The modes for $m = 0$, 1, and 2 are shown in Fig. 21.6. The $m = 0$ extensional mode can be described as a "breathing" mode.

Another family of ring-driven modes, not much discussed in the literature, consists of the ring axial modes in which the soundbow twists and moves in a longitudinal direction (Perrin et al., 1983). A diagram based on a finite element calculation of the $m = 3$ case for a bell is shown in Fig. 21.7. Although such modes can occur quite low in the frequency spectrum (the $m = 2$ case occurs at 2.76 times the prime in this D_5 bell), they radiate very weakly, as do the inextensional $m = 0$ "twisting" and $m = 1$ "swinging" modes and all the families of extensional modes.

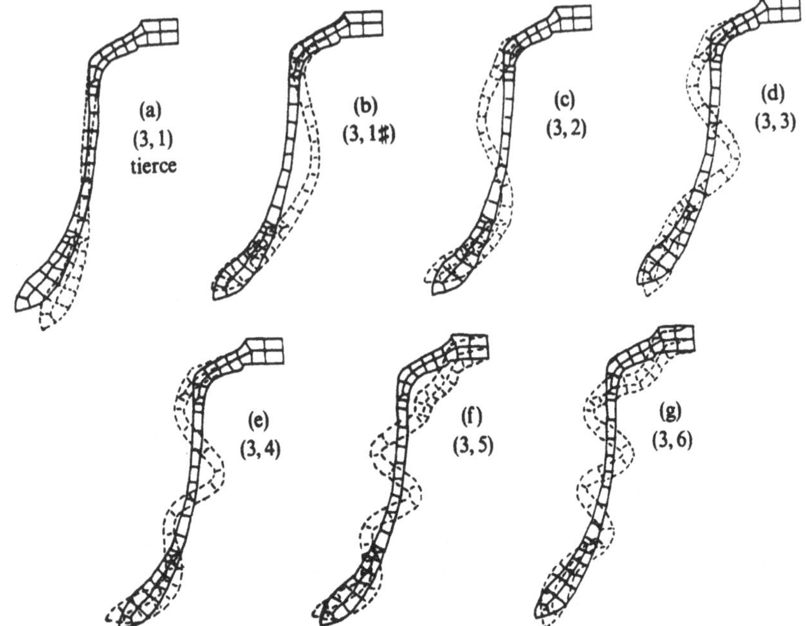

FIGURE 21.5. Modal shapes in a church bell predicted by finite element calculation for the inextensional modes with $m = 3$: (a) (3,1), (b) (3,1♯), (c) (3,2), (d) (3,3), (e) (3,4), (f) (3,5), and (g) (3,6) (Perrin et al., 1983).

21.1.2 Chladni's Law Applied to Church Bells

The modal frequencies of a flat circular plate are found to follow the empirical relationship $f_{mn} = c(m + 2n)^p$ for large values of $(m + 2n)$, a relationship sometimes called Chladni's law (see Section 3.6). Although church bells have a geometry substantially different from flat plates, it is found that the modal frequencies of a church bell can be fitted to the same

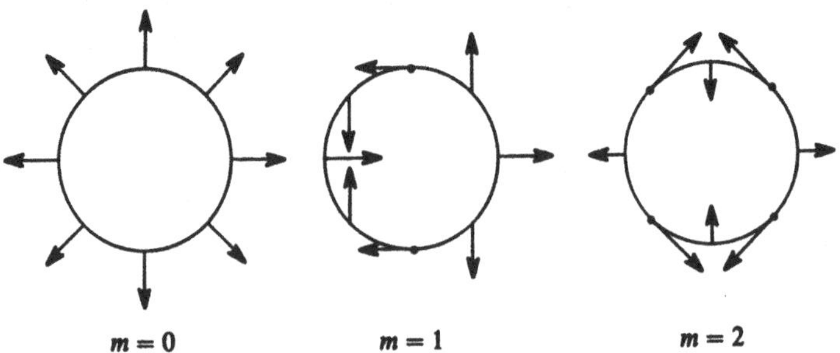

FIGURE 21.6. Motion of a bell for extensional modes of small m.

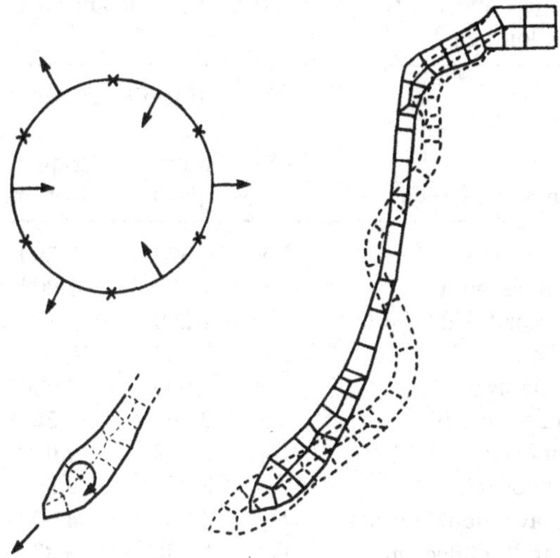

FIGURE 21.7. Motion of a bell for the $m = 3$, ring axial mode (Rossing and Perrin, 1987).

modified form of Chladni's law, $f_{mn} = c(m + bn)^p$, used for cymbals (Section 20.2) by selecting different values of c, b, and p for large and small m (Perrin et al., 1985).

21.2 Tuning and Temperament

Bell founders usually tune the lowest five modes of church bells and carillon bells so that their vibrational frequencies are in the ratios 1:2:2.4:3:4. This is done by carefully thinning the inside of the bell at selected heights while it is mounted on a bell lathe. When this tuning is done, another five or six partials take on a nearly harmonic relationship, thus giving the bell a strong sense of pitch and a very musical quality. The various names of important partials are given in Table 21.1. Also given are the relative frequencies in an ideal bell (just tuning) and a bell with partials tuned to equal temperament.

Notice that the highest four partials in Table 21.1 are raised by as much as 4% above those of the ideal bell. A similar relationship is seen in other tuned church bells and carillon bells (see Table V in Lehr, 1986). This stretching of the partial series may very well contribute a desirable quality to the bell sound (Slaymaker, 1970).

Not all church bells have harmonically tuned partials, however. A study of 363 church bells in Western Europe revealed that only 17% have the hum, the prime, and the nominal tuned in octaves (Lehr, 1987). The distribution of the various tunings is given in Table 21.2. Terhardt and Seewann (1984)

TABLE 21.1. Names and relative frequencies of important partials of a tuned church bell or carillon bell.[a]

Mode	Name of partials	Note name	Ratio to prime (or strike note) Ideal (just)	Equal temperament	Bell in Fig. 21.4
(2,0)	Hum	D_4	0.500	0.500	0.500
$(2,1^\sharp)$	Prime, fundamental	D_5	1.000	1.000	1.000
(3,1)	Tierce, minor third	F_5	1.200	1.189	1.183
$(3,1^\sharp)$	Quint, fifth	A_5	1.500	1.498	1.506
(4,1)	Nominal, octave	D_6	2.000	2.000	2.000
$(4,1^\sharp)$	Major third, deciem	$F^\sharp_6$	2.500	2.520	2.514
(2,2)	Fourth, undeciem	C_6	2.667	2.670	2.662
(5,1)	Twelfth, duodeciem	A_6	3.000	2.997	3.011
(6,1)	Upper octave, double octave	D_7	4.000	4.000	4.166
(7,1)	Upper fourth, undeciem	G_7	5.333	5.339	5.433
(8,1)	Upper sixth	B_7	6.667	6.727	6.796
(9,1)	Triple octave	D_8	8.000	8.000	8.215

[a] Rossing and Perrin, 1987.

also found a rather wide distribution in the tuning of partials in historic German bells.

21.3 The Strike Note

When a large church bell or carillon bell is struck by its metal clapper, one first hears the sharp sound of metal on metal. This atonal strike sound includes many inharmonic partials that die out quickly, giving way to a strike note or strike tone that is dominated by the prominent partials of the bell. Most observers identify the metallic strike note as having a pitch at or near the frequency of the strong second partial (prime or fundamental), but to others its pitch is an octave higher. Finally, as the sound of the bell ebbs, the slowly decaying hum tone (an octave below the prime) lingers on. (For a historical account of research on the strike note, see Rossing, 1984a, Part IV.)

The strike note is of great interest to psychoacousticians, because it is a subjective tone created by three strong nearly harmonic partials in the bell sound. The octave or nominal, the twelfth, and the upper octave normally have frequencies nearly in the ratios 2:3:4 (See Table 21.1). The ear assumes these to be partials of a missing fundamental, which it hears as the strike note, or perhaps we should say, as the primary strike note.

TABLE 21.2. Distribution of various tunings among Western European church bells.[a]

Bell type	Tonal Structure			%
	Hum	Fundamental	Nominal	
Octave bell with perfect fundamental	C_4	C_5	C_6	17.4
Octave bell with diminished fundamental	C_4	B_4	C_6	14.3
Minor-ninth bell with perfect fundamental	B_3	C_5	C_6	7.4
Minor-ninth bell with augmented fundamental	B_3	$C_5^\sharp$	C_6	7.2
Octave bell with augmented fundamental	C_4	$C_5^\sharp$	C_6	6.9
Major-seventh bell with perfect fundamental	$C_4^\sharp$	C_5	C_6	5.8
Major-seventh bell with diminished fundamental	$C_4^\sharp$	B_5	C_6	5.5
Minor-ninth bell with diminished fundamental	B_3	B_4	C_6	4.7
Major-seventh with double diminished fundamental	$C_4^\sharp$	$B_4^\flat$	C^6	4.1
Octave bell with double diminished fundamental	C_4	$B_4^\flat$	C_6	3.6
Minor-seventh bell with diminished fundamental	D_4	B_4	C_6	2.7
Minor-seventh bell with double diminished fundamental	D_4	$B_4^\flat$	C_6	2.5
Minor-seventh bell with triple diminished fundamental	D_4	A_4	C_6	2.5

[a] Lehr, 1987.

Figure 21.8 illustrates the role of each partial in determining the pitch of the strike note. The sound of a Hemony bell was recorded, and by means of a digital filter, each of the first nine partials was raised and lowered in frequency up to 10% while listeners judged the pitch of the resulting strike note in a pitch-matching experiment. The results show that partials five and six (the octave and the twelfth) are the most important, followed by partial seven (the upper octave). The other partials, including partial two (the prime, which coincides closely to the strike note in frequency), have very little effect on the pitch of the strike note, as indicated on the vertical axis (Eggen, 1986).

In very large bells, a secondary strike note may occur a musical fourth above the primary strike note and may even appear louder under some conditions (Schouten and 't Hart, 1965). This secondary strike note is a subjective tone created by four partials beginning with the upper octave. These partials are from the (6,1), (7,1), (8,1), and (9,1) modes of vibration, whose frequencies are nearly three, four, five, and six times that of the secondary strike note (see Table 21.1). In a large bell (800 kg or more), these partials lie below 3000 Hz, where the residue pitch is quite strong (Ritsma, 1967).

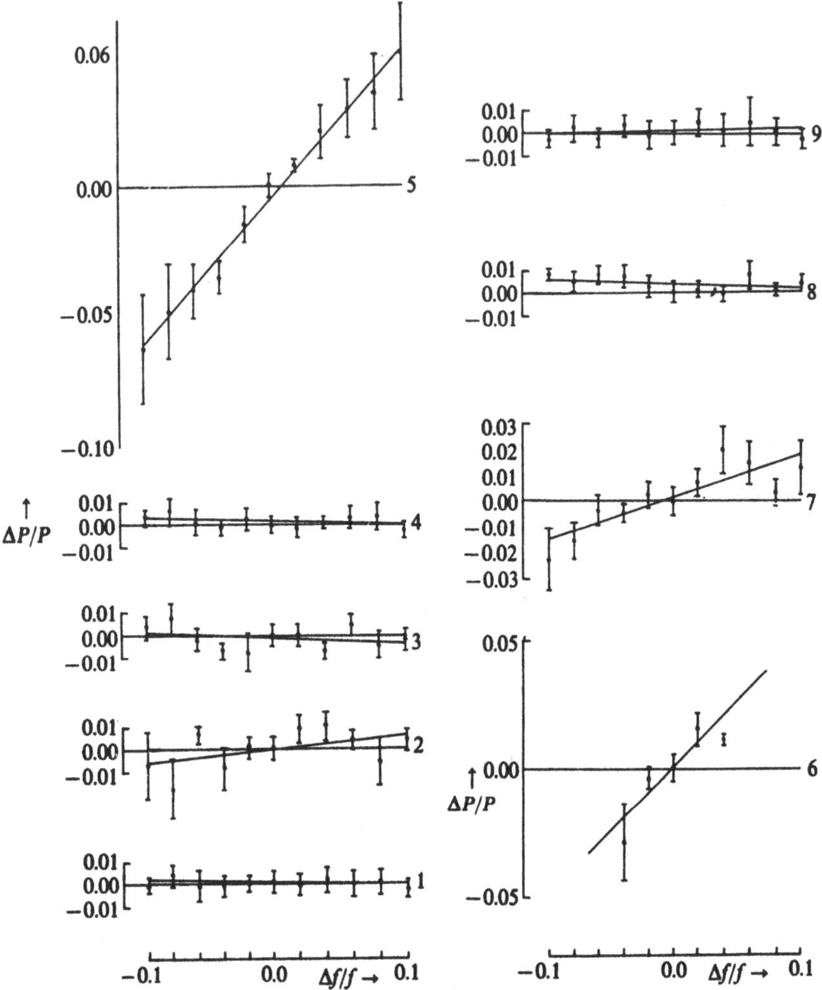

FIGURE 21.8. Effect of nine different partials in determining the strike note of a bell. $\Delta P/P$ is the relative change in pitch resulting from a change in the relative frequency $\Delta f/f$ of a partial. Note the great importance of the fifth and sixth partials (octave and twelfth) followed by the seventh partial (upper octave); other partials are relatively unimportant (Eggen, 1986).

In small bells, the higher partials lie at a very high frequency where the virtual or residue pitch is weak, and so both the primary and secondary strike notes are weak. The pitch is then determined mainly by the hum, the prime, and the nominal, which normally are tuned in octaves. It is sometimes difficult to decide in which octave the pitch of a small bell lies, especially if the frequencies of these three partials are not exactly in 1:2:4 ratio. Difference tones between these partials may be heard.

Some authorities question the existence of subjective strike notes, and suggest that the strike note is an auditory illusion resulting from the nominal, the strongest partial during the first few seconds (Milsom, 1982). Because of this, some English founders denote the pitch of the bell as the pitch of the nominal. There is some feeling that a bell rung in full circle sounds a different pitch from a bell struck somewhat more gently in a carillon (Ayres, 1983). Whether a founder tunes the nominal or the strike note makes little difference, however, because the nominal is one of the main partials that determines the tuning of the strike note.

21.4 Major-Third Bells

A new type of carillon bell has been developed at the Royal Eijsbouts Bellfoundry in The Netherlands. The new bell replaces the dominating minor-third partial with a major-third partial, thus changing the tonal character of the bell sound from minor to major. This requires an entirely new bell profile.

The idea of a bell with a major-third partial is not new. For years, some carillonneurs have felt that a composition in a major key, especially if it includes chords with many notes, might sound better if played on bells with a major character. Some authorities have even suggested a neutral third, lying between a major and minor third. Previous efforts to fabricate such bells have not been successful, however, since changing the bell profile to raise the third partial invariably changes the other harmonic partials as well.

The new bell design evolved partly from the use of a technique for structural optimization using finite element methods on a digital computer at the Technical University in Eindhoven (Schoofs et al., 1987). This technique allows a designer to make changes in the profile of an existing structure, and then to compute the resulting changes in the vibrational modes.

Based on results of the structural optimization procedure, André Lehr and his colleagues have designed two entirely different bells, both having a major-third partial, as shown in Fig. 21.9. The first bell, introduced in 1985, has more rapidly decaying partials, whereas the second bell has a longer decay, more nearly that of the traditional minor-third bell (Lehr, 1987). Frequencies of the first eight partials in the two major-third bells are compared to those of a minor-third bell in Fig. 21.10. Notice the large differences in partials seven and eight, both of which belong to group III (III-2 and III-3, respectively). Note names and relative mode frequencies are given in Table 21.3. Major-third bells are shown in Figs. 21.11 and 21.12.

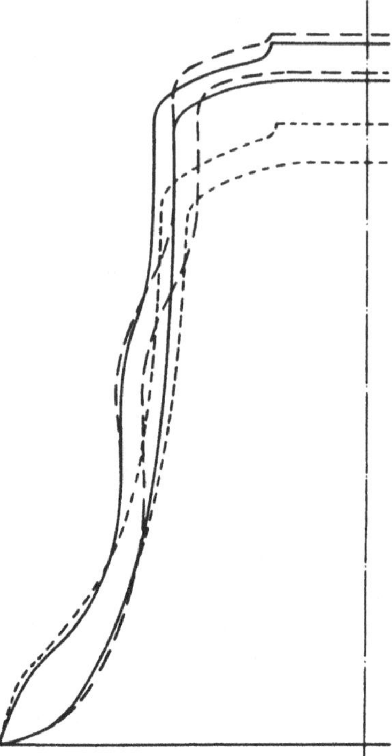

FIGURE 21.9. Profiles of major-third bells (dashed curves) compared to a minor-third bell (solid curve) (from Lehr, 1987).

21.5 Sound Decay and Warble

A vibrating bell loses energy mainly by sound radiation, although internal losses also play a role. The sound pressure level of each radiated partial decays at a constant rate (i.e., the vibrational energy decays exponentially), and thus it is customary to express the 60-dB decay time for each main partial.

The decay rates of the principal modes of vibration of the bell described in Fig. 21.4 are given in Table 21.4. Note the long decay time of the (2,0) mode (hum) and the relatively short decay times of the modes of higher frequency. This is mainly due to the greater radiation efficiency of the higher modes. Schad and Warlimont (1973) found that the damping due to internal losses was approximately the same for all the principal modes, so the large differences in decay times are indicative of different rates of radiation.

One of the prominent features in the sound of many bells is warble, which is caused by the beating together of the nearly degenerate components of a mode doublet (Perrin and Charnley, 1978). Warble is observed as a variation in amplitude of the mean frequency at a rate determined by the

TABLE 21.3. Note names and relative frequencies of partials in major- and minor-third bells.[a]

	I-2	II-2	I-3	II-3	I-4	II-4	III-2	III-3	I-5	III-4	II-5	I-6
Minor-third bell	C_4,	C_5,	$E^{\flat}_5$,	G_5,	C_6,	E_6,	$F^{\sharp}_6-$,	$F^{\sharp}_6+$,	G_6,	A_6-,	B_6,	C_7+,
	1	2.0	2.4	3.0	4.0	5.0	5.22	5.24	6.0	6.6	7.2	8.2
Major-third bell #1	C_4,	C_5,	E_5,	G_5,	C_6,	E_6,	$B^{\flat}_5-$,	C_6,				
G_5,	$A^{\flat}_6+$,	$B^{\flat}_6$,	C_7,									
	1	2	2.5	3.0	4.0	5.0	3.5	4.0	5.9	6.5	7.0	8.0
Major-third bell #2	C_4,	C_5,	E_5,	G_5,	C_6,	E_6+,	D_6,	$C^{\sharp}_6-$,	G_6,	$A^{\flat}_6$,	B_6+,	
C_7,												
	1	2.0	2.5	3.0	4.0	5.15	4.5	4.25	6.0	6.35	7.8	8.0

[a] Lehr, 1987.

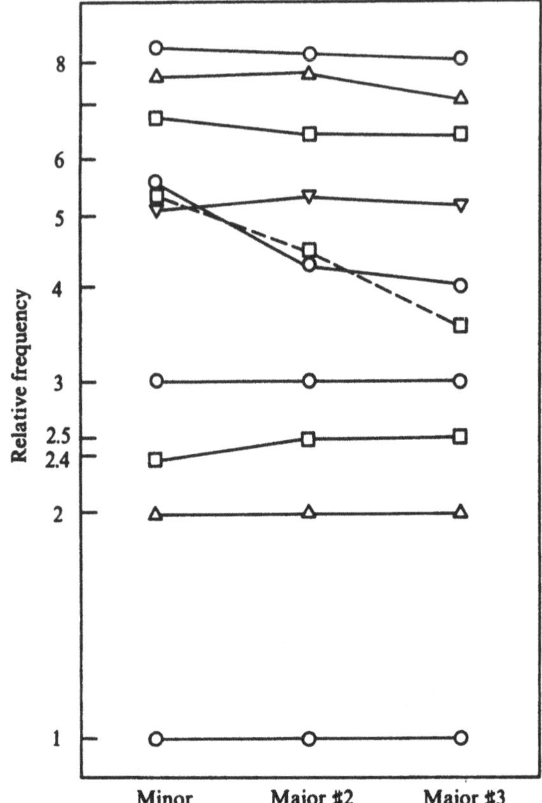

FIGURE 21.10. Relative frequencies of the first eight partials in two major-third bells compared to those in a traditional minor-third bell. Note that the hum, prime, fifth, and octave partials remain the same (adapted from Lehr, 1987).

frequency difference between the pair of modes. In theory, warble can be eliminated by selecting the strike point to lie at a node for one component and an antinode for the other. In practice, this may not be useful because doublet splitting is likely to occur in many doublet pairs and selecting the strike point to minimize warble in one pair may enhance it in another. Various methods of trying to guarantee the correct alignment of doublet pairs by suitable breaking of the axial symmetry have been suggested, including, for example, the addition of two diametrically opposite meridian ribs at whose location the clapper strikes (Perrin et al., 1982).

21.6 Scaling of Bells

The fundamental frequency of a bell has roughly the same dependence on the thickness h and diameter d as flexural vibrations in a circular plate:

FIGURE 21.11. Two new major-third bells (left) compared to conventional (minor-third) bells with the same pitches.

FIGURE 21.12. Two-octave carillon with 23 major-third bells, ranging from D_5 to D_7 (courtesy of Royal Eijsbouts Foundry).

TABLE 21.4. Decay times of the principal modes of a 70-cm church bell.[a]

Mode	Name	Decay time (s)
(2,0)	Hum	52
(2,1♯)	Fundamental or prime	16
(3,1)	Minor third	16
(3,1♯)	Fifth	6
(4,1)	Octave	3
(4,1♯)	Upper (major) third	1.4
(2,2)	Fourth	3.6
(5,1)	Upper fifth	5
(6,1)	Upper octave	4.2
(7,1)	Double deciem	3
(8,1)	Sixth	2

[a] From Perrin et al., 1983.

$f \propto h/d^2$. Thus, it is possible to scale a set of bells by making all dimensions proportional to $1/f$. This scaling approximates that found in many carillons dating from the fifteenth and sixteenth centuries (van Heuven, 1949). However, since a $1/f$ scaling causes the smaller treble bells to have a rather weak sound, later bell founders increased the sizes of their treble bells (Bigelow, 1961). The average product of frequency times diameter in several fine seventeenth century Hemony carillons has been found to increase from 100 m/s (in bells of 30 kg or larger) to more than 150 m/s in small treble bells. The measurements of the bells in three Hemony carillons are shown in Fig. 21.13. The straight, solid lines indicate a $1/f$ scale. Deviations from this scaling in the treble bells are apparent.

Studies of bells cast at the Eijsbouts bell foundry indicate that a $1/f$ scaling is used for swinging bells, but for carillon bells, the diameter of the high-frequency bells is substantially greater than predicted by this scaling (Lehr, 1952). This is probably the most notable difference between church bells and carillon bells.

The scaling of carillon bells can be conveniently described by writing the frequency of a bell as $f = k/r$ and the mass as $m = cr^3$, where r is the radius. Lehr (1952) finds that it is the factor c/k that remains more or less constant as the size of carillon bells changes from small to large.

The various modal frequencies in church bells are found to have different dependences on thickness (Lehr, 1986). In the modes of group I, which have antinodes near the soundbow, the frequency is roughly proportional to $h^{0.7}$, where h is the thickness. In the modes of group II, which have

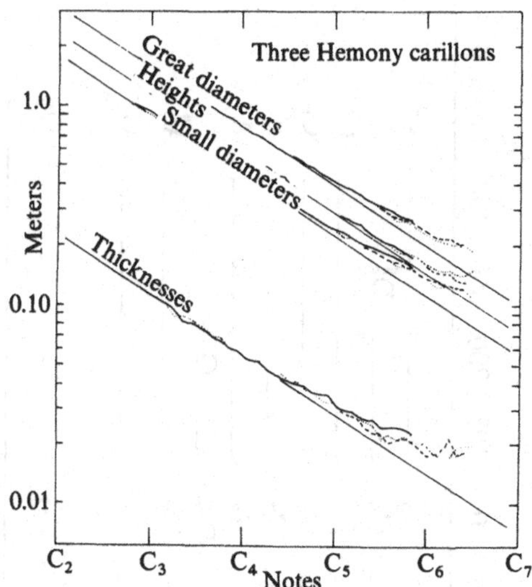

FIGURE 21.13. Measurements of the bells in three Hemony carillons: Oude Kerk, Amsterdam, François Hemony, 1658; Dom Toren, Utrecht, François and Pieter Hemony, 1663; and Heilige Sulpiciuskerk, Diest, Pieter Hemony, 1670. The solid lines represent a $1/f$ scaling law (from Bigelow, 1961).

antinodes near the waist, the frequency is roughly proportional to $h^{0.86}$, which is nearer to the $f \propto h$ behavior found in flat plates and circular cylinders.

The effect of wall thickness on the partials of a bell is illustrated in Table 21.5 by the partial frequencies of three experimental bells of the same diameter with thickness $\frac{2}{3}$ normal, normal, and $\frac{3}{2}$ normal (Lehr, 1976). The C_5 bell of normal thickness has the partials whose note names are given in the center column. To the left are shown the partials of the thin bell and to the right are the partials of the thick bell. The raising and lowering are shown on a scale of cents. Note that the harmonic relationships between partials have been altered considerably by changing the thickness. This is also the case with bells whose thickness has been altered only slightly (Lehr, 1986).

21.7 Modes of Vibration of Handbells

The vibrational frequencies of a C_5 handbell are shown in Fig. 21.14. A comparison with Fig. 21.4 shows some interesting similarities and differences. Note that each curve in Fig. 21.14 drawn for n = 1, 2, 3,. . ., shows

TABLE 21.5. Effect of wall thickness on the partials of a C_5 bell.[a]

Change in frequency (cents)	1000	900	800	700	600	500	400	300	200	100	0	100	200	300	400	500	600
Hum		E_4									C_5					F_5	
Prime					$F^\sharp_5$						C_6			$D^\sharp_6$			
Minor third		G_5									$E^\flat_6$					$G^\sharp_6$	
Fifth	A_5										G_6					$C^\sharp_7$	
Octave		$D^\sharp_6$									C_7					F_7	
Major third		$F^\sharp_6$									E_7					$A^\sharp_7$	
Fourth							$C^\sharp_7$				F_{7-}		G_7				
Fourth				A_{6+}							F_7					$A^\sharp_{7-}$	
Upper fifth		$A^\sharp_{6+}$									G_7					C_8	
Thickness				2/3							1/1					3/2	

[a] Adapted from Lehr, 1976.

a minimum in frequency at about $m = n + 2$. This is similar to the behavior of a cylinder with a fixed end cap, in which the stretching energy becomes substantial for small m (Rayleigh, 1894). Thus, the total strain due to stretching and bending in a cylindrical shell with fixed ends decreases with m until it reaches a minimum, then increases with m (see Fig. 7 in Arnold and Warburton, 1949). In the larger G_2 and G_3 handbells, the minima occur at about $m = n + 3$.

Vibrational modes of a handbell are arranged in a periodic table in Fig. 21.15. Once again the numbers at lower left give (m, n), the numbers of complete nodal meridians and nodal circles. Unlike those of a modern church bell, their order in frequency depends upon the size and shape of the handbell. The (2,0) and (3,0) modes are always the modes of lowest frequency. The next mode may be the (3,1), (4,0), or (4,1) mode, however, depending upon the size of the handbell; the (2,1) mode occurs at a considerably higher frequency (Rossing and Sathoff, 1980).

Hologram interferograms of a number of the modes are shown in Fig. 21.16. The bull's eyes locate the antinodes. Note that the upper half of the bell moves very little in the (7,1) mode; the same is true in $(m, 1)$ modes when $m > 7$. The modes of vibration calculated in a C_5 handbell using

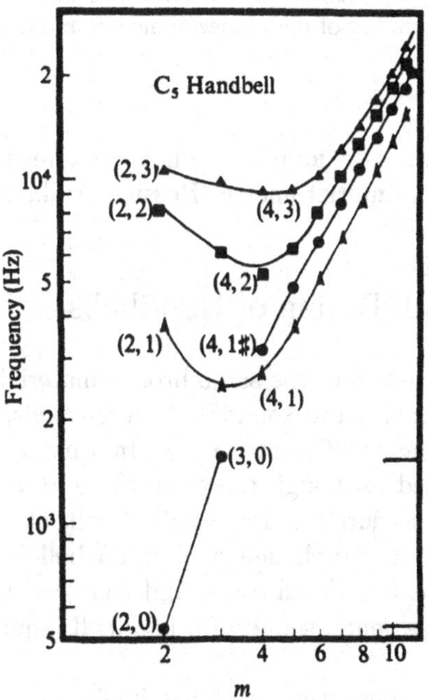

FIGURE 21.14. Vibrational frequencies of a C_5 handbell (Rossing et al., 1984).

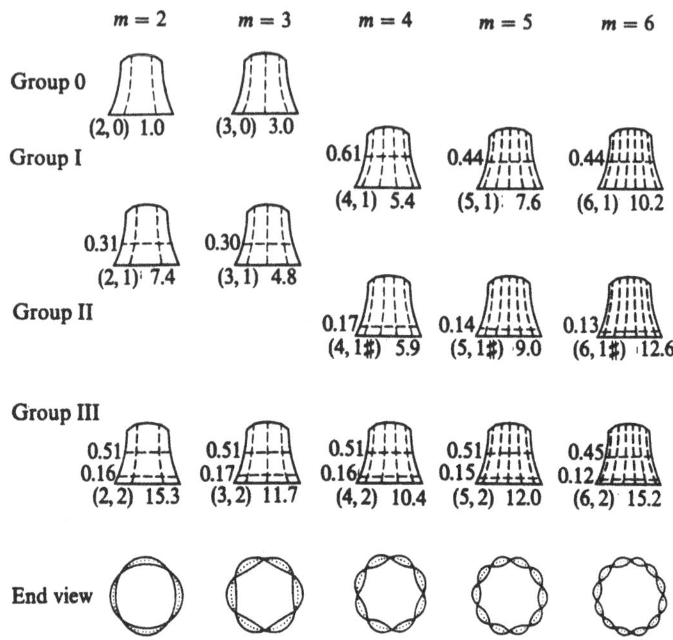

FIGURE 21.15. Periodic table of vibrational modes in a handbell. Below each drawing are the relative modal frequencies in a Malmark C_5 handbell. At lower left, (m, n) gives the number of nodal meridians $2m$ and nodal circles n (Rossing and Perrin, 1987).

finite element methods were found to be in good agreement with the modes observed by hologram interferometry (Rossing et al., 1984).

21.8 Timbre and Tuning of Handbells

Although they are cast from the same bronze material and cover roughly the same range of pitch, the sounds of church bells, carillon bells, and handbells have distinctly different timbres. In a handbell, only two modes of vibration are tuned (although there are three harmonic partials in the sound), whereas in a church bell or carillon bell, at least five modes are tuned harmonically. A church bell or carillon bell is struck by a heavy metal clapper in order to radiate a sound that can be heard at a great distance, whereas the gentle sound of a handbell requires a relatively soft clapper.

In the so-called English tuning of handbells, followed by most handbell makers in England and the United States, the (3,0) mode is tuned to three times the frequency of the (2,0) mode. The fundamental (2,0) mode radiates a rather strong second harmonic partial, however, so that

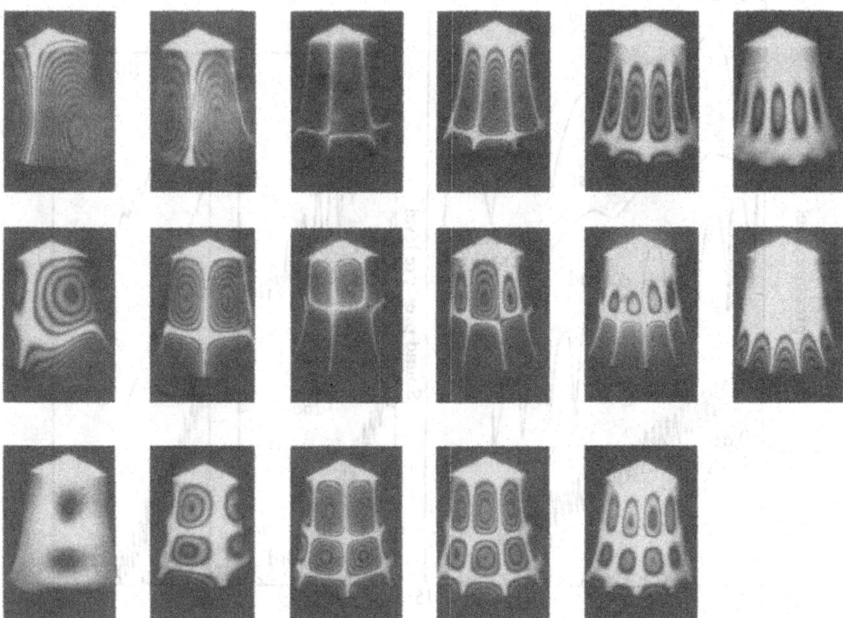

FIGURE 21.16. Time-average hologram interferograms of vibrational modes in a C_5 handbell (Rossing et al., 1984).

the sound spectrum has prominent partials at the first three harmonics (Rossing and Sathoff, 1980). Some Dutch founders aim at tuning the (3,0) mode in handbells to 2.4 times the frequency of the fundamental, giving their handbell sound a minor-third character somewhat like a church bell. Such bells are usually thicker and heavier than bells with the English-type tuning (Rossing, 1981).

A handbell, unlike a church bell, appears to sound its fundamental pitch almost from the very onset of sound. There are several reasons for the absence of a separate strike note. First of all, there is no group of harmonic partials to create a strong subjective tone. Secondly, handbells employ a relatively soft, nonmetallic clapper, so that there is no sound of metal on metal, and the partials develop a little more slowly after the clapper strikes the bell.

21.9 Sound Decay and Warble in Handbells

Sound decay curves for a large and a small handbell are shown in Fig. 21.17. Note that the small handbell decays much faster than the large one. This is partly due to the more efficient radiation of sound by the smaller bell (see Section 21.11). Note that the second harmonic decays at twice the

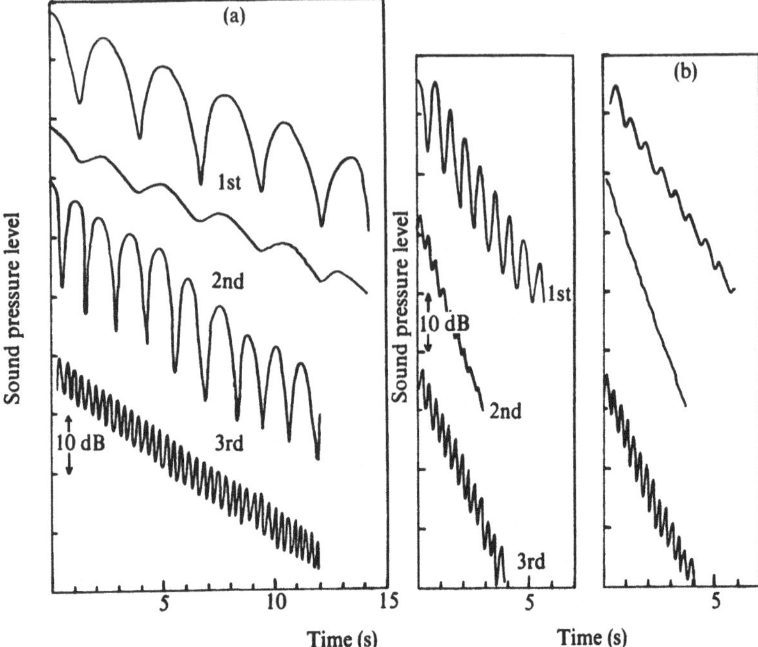

FIGURE 21.17. Sound decay from two handbells. The first and second harmonics are radiated by the (2,0) mode, the third harmonic by the (3,0) mode. (a) G_4 bell in positions that give maximum and minimum warble in the fundamental; decay times are 30, 15, and 22 s for the three harmonics. (b) C_6 bell in two different positions; decay times are 9.0, 4.4, and 7.4 s (Rossing and Sathoff, 1980).

rate of the (2,0) mode, because its intensity is proportional to the square of the amplitude of that mode (Rossing and Sathoff, 1980).

A prominent feature in Fig. 21.17 is the warble that results from the beating of the two members of the doublet mode (see Section 21.6). This warble is greatest at angles that lie midway between the maxima of the two mode members. In "voicing" a handbell, the clapper is set to strike the bell at one of these two maxima so as to minimize warble. Theoretically, this can eliminate warble in any desired partial, but in practice, the optimum strike point for the upper partials may not coincide with that of the fundamental mode.

21.10 Scaling of Handbells

Unlike church bells, which are cast nearly to their final thickness, handbells are frequently cast much thicker than the final bell, so that a considerable quantity of metal is removed on the lathe. In fact, the same size of casting

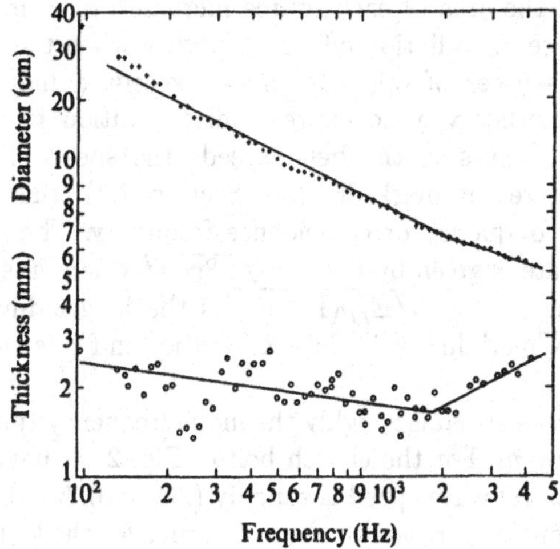

FIGURE 21.18. Scaling of a five-octave set of Malmark handbells (Sathoff and Rossing, 1983).

can be used for more than one bell by adjusting the wall thickness of the final bell. Under these conditions, the scaling curves for sets of handbells might be expected to be less regular than the scaling curves for church bells, such as the one shown in Fig. 21.8.

The general scale followed by most handbell crafters is to make the diameter inversely proportional to the square root of frequency, except for the smallest bells in which it varies inversely with the cube root of frequency instead (Sathoff and Rossing, 1983). The thickness is then adjusted, so that h/d^2 is nearly proportional to the frequency. The scaling of a five-octave set of Malmark handbells is shown in Fig. 21.18.

21.11 Sound Radiation

The most prominent partials in the spectrum of a church bell or carillon bell are radiated by the ring-driven modes belonging to group I. These modes can be considered to be due to standing flexural waves. With the exception of the (2,0) mode (hum), all modes have a nodal ring about halfway up the bell. For the purpose of understanding the general properties of the radiation field of the bell, we can model its outside surface as a collection of $4m$ sources alternating in phase [$2m = 4$ sources in the case of the (2,0) mode]. Other $4m$ sources of slightly smaller size occur on the inside surface.

The radiation efficiency of such a collection of alternating sources increases rapidly with frequency and with the size of the bell. As the bell increases in size, the area of each source increases, of course; but a more significant increase in radiation efficiency occurs when the separation between adjacent sources of opposite phase exceeds a half-wavelength of sound in air. Another way to express this condition is that when the speed of flexural waves in the bell exceeds the speed of sound, radiation efficiency increases markedly (see Section 7.3); this occurs at the so-called critical frequency or coincidence frequency. The speed of flexural waves in a plate is given by $v(f) = \sqrt{1.8c_L hf}$ where h is the thickness, f is the frequency, $c_L = \sqrt{E/\rho(1-\nu^2)}$ is the longitudinal wave velocity, E is Young's modulus, ν is Poisson's ratio, and ρ is the density (see Section 3.5).

The flexural wave speed is roughly the mode frequency times the circumference divided by m. For the church bell in Fig. 21.4, having a diameter of 70 cm, the flexural wave speed is roughly $(292.7)(0.7\pi/2) = 322$ m/s for the (2,0) mode, but it increases to about 644 m/s for the (4,1) mode and to about 1180 m/s for the (9,1) mode. Since all but the lowest mode exceed the speed of sound in air (344 m/s at 23°C), the bell radiates most of its partials quite efficiently.

In a handbell, on the other hand, the walls are much thinner, and so the flexural wave speed is considerably smaller than in a heavy church bell. In the bell described in Fig. 21.14, for example, the flexural wave speed is roughly $(523)(0.06\pi) = 100$ m/s for the fundamental (2,0) mode. Since this is considerably less than the speed of sound, the radiation is not very efficient. Air adjacent to the vibrating surface tends to flow back and forth between adjacent areas of opposite phase, creating a sort of pneumatic short circuit. The (3,0) and (4,1$^\sharp$) modes in the same handbell have flexural wave speeds of approximately 200 m/s and 300 m/s, respectively, and thus these modes tend to radiate sound a little more efficiently than the fundamental (2,0) mode. In large handbells, the radiation efficiency is low, because the flexural wave velocity for all the principal modes is less than the speed of sound in air. In a G_2 handbell, for example, the flexural wave speeds are only 41 m/s and 71 m/s for the (2,0) and (3,0) modes, respectively. This is a significant problem in handbell design.

In addition to the direct radiation of sound normal to its vibrating surfaces, a bell also radiates sound axially at twice the frequency of each vibrational mode (Rossing and Sathoff, 1980). The intensity of this axially radiated sound increases with the fourth power of the vibrational amplitude, whereas the direct radiation increases only with the square of the amplitude.

The fundamental (2,0) mode in a handbell radiates a fairly strong second harmonic partial along the axis, as well as a fundamental whose maximum intensity is perpendicular to the axis. The (3,0) mode also radiates at twice its vibrational frequency, but its partial is usually quite weak. The principal harmonic partials in the handbell sound are thus the first, second, and third harmonics (Rossing and Sathoff, 1980).

21.12 Bass Handbells

Demand for handbells of lower and lower pitches has led to the development of bass bells as low as G_0 (fundamental frequency of 24.5 Hz). Unfortunately, these large bass bells radiate inefficiently, especially the bells made of bronze, because the speed of bending waves is well below the speed of sound and therefore they operate well below the critical frequency (see Section 7.3, also compare Fig. 10.27).

In order to obtain a higher radiation efficiency and thereby enhance the sound of bass bells, the Malmark company introduced a bass handbell of aluminum rather than traditional bronze. These aluminum bells are larger in diameter, and they have lower critical frequencies, both of which lead to more efficient radiation of bass notes. In addition, they are considerably lighter in weight, thus are much more easily handled by bell ringers (Rossing, et al., 1995).

Sound spectra from large bass handbells (both of aluminum and bronze) are characterized by a strong sixth harmonic partial not readily observed in smaller handbells. This partial is radiated by the (3,0) mode in much the same manner as the second harmonic is radiated by the (2,0) mode in nearly all handbells.

21.13 Clappers

The sound of a bell is very much dependent on the size, shape, and hardness of the clapper, the point at which it strikes the bell, and the strength of the blow. Not much systematic research on clappers has been reported in the literature, however.

Church bell clappers were formerly made of wrought iron, which probably remains the preferred material, although in recent years it has been gradually replaced by cast iron. When a bell is rung full circle (as in change ringing), the clapper strikes a powerful blow, and clappers must be carefully designed to avoid breaking. Clappers in carillon bells are usually made of steel.

Handbell clappers strike the bell with a soft surface of plastic, leather, or felt. In some handbell clappers, surfaces of different hardness can be selected in order to vary the timbre of the sound. Figure 21.19 shows some of the different timbres obtainable from three handbells by changing the hardness of the clapper surface. In general, the softer clapper favors the

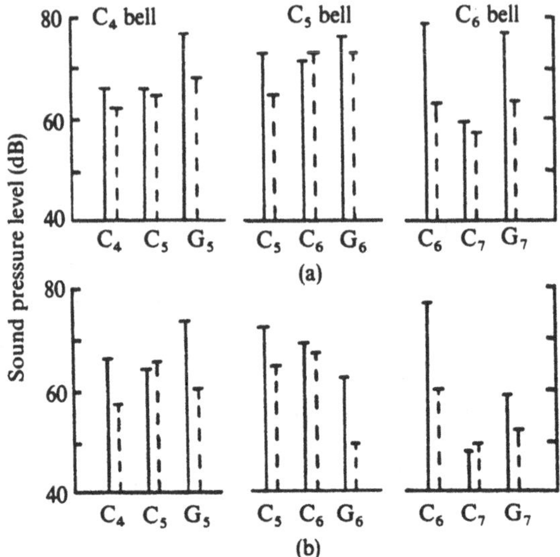

FIGURE 21.19. Comparison of the principal partials of handbell sound radiated along the bell axis (a) and at right angles to the axis (b) with hard (—) and soft (- - -) clappers, respectively. Sound pressure levels were recorded 1 m from each handbell in an anechoic room (Rossing and Sathoff, 1980).

fundamental. Note the difference between sound radiated along the bell axis and that radiated at right angles to the bell.

Bigelow (1961) reports the results of an experiment in which an F_5 carillon bell was struck by three different clappers of varying weights, each falling from three different heights. A heavy hammer was found to increase the strength of the lower partials, but to decrease those of the nominal (or octave). Since the nominal contributes in an essential way to the subjective strike note, it is implied that a heavy hammer will tend to diminish the intensity of that note. It should be noted that in a larger bell the nominal is usually found to be the most prominent partial for the first few seconds after the strike.

21.14 Ancient Chinese Two-Tone Bells

Although this book deals mainly with Western musical instruments, we would like to briefly discuss certain Oriental bells, especially ancient Chinese bells that are oval in shape and have two different strike notes. They are indicative not only of the character of ancient Chinese music but also of the advanced understanding of the art of bell casting that existed in ancient China.

In 1978, the tomb of Marquis Yi was discovered at Sui Xiang in the province of Hubei during a construction project. Among the treasures in that tomb was a set of 65 tuned bronze bells, the largest standing about 5 ft high, having a cross section that could be described as "almond-shaped" or having the shape of pointed ovals. The inscriptions on these bells revealed that each bell was designed to emit two tones, one if struck on the center of the bell face (A-tone) and one if struck about midway between the center and the lateral edge (B-tone).[1] The intended interval between these two notes in most of the bells was a minor third on the Western musical scale; in the remainder a major third was intended (Falkenhausen and Rossing, 1995).

Holographic studies of similar bells at the Shanghai Museum (Ma Chengyuan, 1980–1981) and at Northern Illinois University (Rossing et al., 1988) showed that the vibrational modes tend to occur in pairs, one with a node at the spine, or *xian*, and one with an antinode at that location. The mode with a node at the spine generally has the higher frequency. Modal frequencies in a copy of a bell from the Zhou period (1027–256 B.C.) are shown in Fig. 21.20. Note the similarities, and also the differences, between the mode families in Fig. 21.20 and those of a church bell (Fig. 21.4) and a handbell (Fig. 21.14).

Figure 21.21 compares the first six modes of vibration of the Chinese bell with the first six modes in a church bell and a small handbell. Note that the $(2,0)_a$, $(3,0)_b$, and $(4,0)_a$ modes would be efficiently excited by a blow at the center strike point (S), but the $(2,0)_b$, $(3,0)_a$, and $(3,1)_a$ modes would not. A blow at the side strike point (G), on the other hand, would excite the $(2,0)_b$, $(3,0)_a$, $(3,0)_b$, and $(3,1)_a$ modes quite strongly.

Intervals between the A-tone and B-tone fundamental pitches (determined by the $(2,0)_a$ and $(2,0)_b$ mode frequencies) in 88 two-tone bells studied by Ma Chengyuan (1980–1981) and the 64 Zeng bells (Lee, 1980) are compared in Table 21.6.

21.15 Temple Bells of China, Korea, and Japan

The rise of Buddhism in the 3rd and 4th centuries led to the casting of large and small bronze temple bells, many of which still remain in China, Korea, Japan, Burma, and India. Large iron bells appeared in some temples around the 12th century. Temple bells are generally cylindrical or barrel-shaped with a circular cross section, and there may be a slight flare at the mouth. The rims of most temple bells are plane, but some bells (especially

[1]In some papers, the A- and B-tones are referred to as the *sui* and *gu* tones, but that may be misleading, because the term gu generally refers to the entire striking area of the bell.

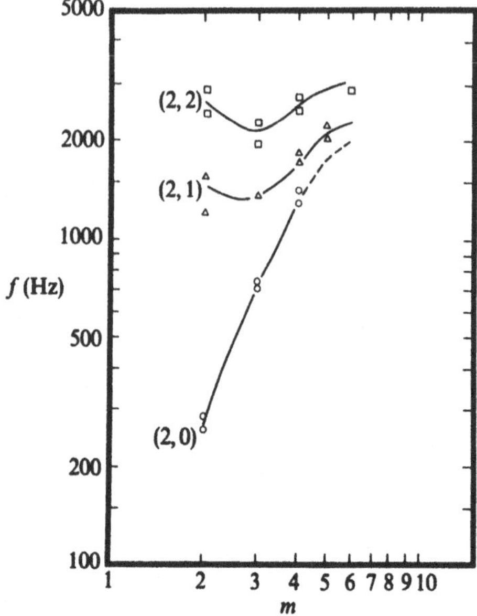

FIGURE 21.20. Modal frequencies in a Chinese two-tone bell. Note that most modes occur in pairs (Rossing et al., 1988).

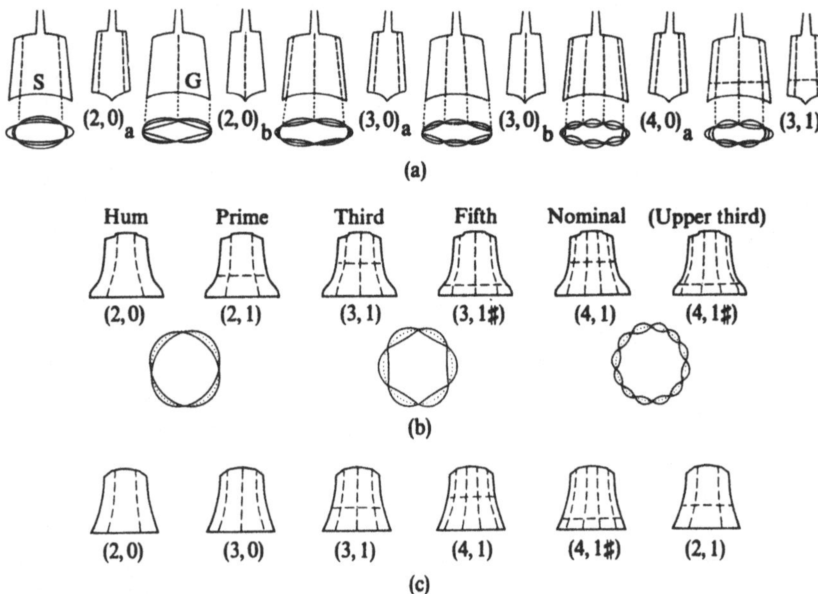

FIGURE 21.21. First six modes in (a) a Chinese two-tone bell; (b) a church bell; and (c) a handbell. Dashed lines indicate locations of nodes (Rossing et al., 1988).

TABLE 21.6. Intervals between A and B tones.[a]

	Interval						
Period	Minor second	Major second	Minor third	Major third	Perfect fourth	Augmented fourth	Minor sixth
Shang	1	4		2			1
Early and middle Western Zhou		3	2	2		1	
Late Western Zhou	1	5	14	5	1		
Spring and autumn	2	1	11	5	3		
Warring states	4	4	8	4	3	1	
Marquis Yi of Zeng		2	42	18	2		

[a] Rossing, 1989.

in China) have scalloped rims, rims with notches, or small protuberances that suggest tiny feet.

Casting of large temple bells in China reached its zenith during the Ming dynasty (1368–1620). The largest such bell was cast during the reign of the emperor Yongle (1403–1424). Its bronze body, 4.5 m high and with a maximum diameter of 3.3 m, is inscribed with over 200,000 characters (see Fig. 21.22). Its mass is estimated to be 52,000 kg. The frequency of the fundamental (2,0) mode is a very low 22 Hz, but prominent partials in its sound spectrum at 129, 164, and 218, which are near C_3, E_3, and A_3, give the bell a rather musical sound. The sound level has been measured to be 92 dB some 25 m away. Fig. 21.22(b) shows how the frequencies of seven modes of vibration depend upon the numbers of nodal meridians and nodal circles (Chen and Zheng, 1986).

Large temple bells have been cast in Korea for more than 1200 years. The most famous bell in Korea is the magnificent King Songdok ("Emille") bell cast during the Silla dynasty (771 AD). Standing 3.66 m high, it has a mass of nearly 20,000 kg. The body has a flat circular protuberance called the *dang jwa*, on which it is struck by a log suspended on ropes, and a chimney called the *eumtong* opens through the top. Many smaller temple bells are found throughout Korea.

Although Korean temple bells are round, the mass of the *dang jwa* is sufficient to create mode doublets, with one component having a node at the *dang jwa* and the other an antinode. The frequency separation of the two components is small, and this creates a slow beating or warble, which is considered to be a desirable characteristic of these bells. Modal studies of a small Korean bell are described by Rossing and Perrier (1993).

A Japanese temple bell, roughly cylindrical in shape and having a mass of 19 kg, was found to have modal frequencies of 247, 624, 850, 1100, 1380, 1640, and 2050 Hz, which are nearly in the ratios 2 : 5 : 7 : 9: 11 : 13 : 16 (Obata and Tesima, 1933-1934). The lowest three modes have $m = 2$,

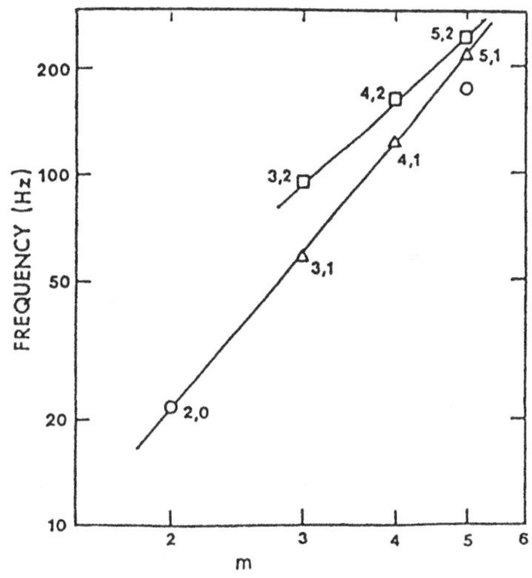

FIGURE 21.22. (a) Yongle bell in the Great Bell Temple (Beijing); (b) Frequencies of 7 modes of vibration including the fundamental (2,0) mode. Numbers of nodal meridians (m) and nodal circles are given. (Chen and Zheng, 1986).

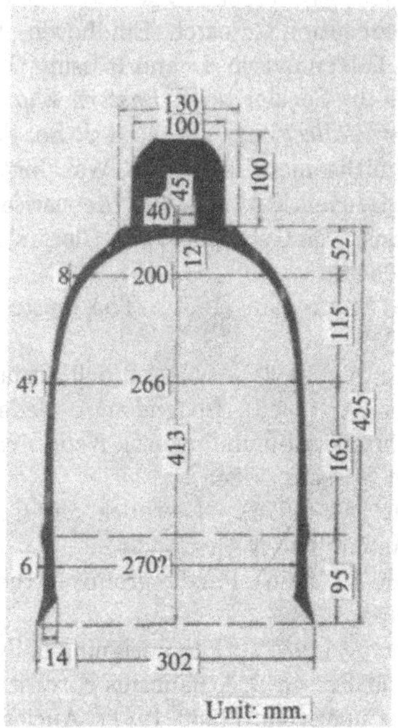

FIGURE 21.23. A Japanese temple bell 30.3 cm in diameter and a cross section showing the thickness (Obata and Tesima, 1933–1934).

3, and 4; that is they have four, six, and eight nodal meridians distributed around the circumference.

The investigators report a subjective strike note corresponding to a frequency of about 260 Hz, which is the difference between the prominent 1380-Hz and 1640-Hz tones. However, based on the experience with church bells (Section 21.4), one might be more likely to expect a strike tone around 275 Hz (the missing fundamental of 1100, 1380, and 1640 Hz) or possibly even 124 Hz (the nearly common factor in all seven frequencies above).

References

Arnold, R.N., and Warburton, G.B. (1949). Flexural vibrations of the walls of thin cylindrical shells having freely supported ends. *Proc. Roy. Soc. London* **A197**, 238–256.

Ayres, R.M. (1983). Bell tuning—modern enigma or medieval mystery? *The Ringing World*, September 23, 790.

Bigelow, A.L. (1961). "The Acoustically Balanced Carillon." School of Engineering, Princeton, New Jersey.

Chen Tong and Zheng Darui (1986). Acoustical properties of the Yongle bell. *Chinese J. Acoust.* **5**, 375–381.

Eggen, J.H. (1986). The strike note of bells. Report No. 522, Institute for Perception Research, Eindhoven, The Netherlands.

von Falkenhausen, L. and Rossing, T.D. (1995). Acoustical and musical aspects of the Sackler bells. *Eastern Zhou Ritual Bronzes from the Arthur H. Sackler Collections*, Vol 3, ed J. So. Arthur Sackler Foundation, New York and Smithsonian Institution, Washington.

van Heuven, E.W. (1949). "Acoustical Measurements on Church Bells and Carillons." De Gebroeders van Cleef, s'Gravenhage (excerpts reprinted in Rossing, 1984a).

Lee, Yuan-yuan (1980). The music of the Zenghou zhong. *Chinese Music* **3**, 3–15.

Lehr, A. (1952). A general bell formula. *Acustica* **2**, 35–38.

Lehr, A. (1965). Hedendaagse Nederlandse Klokkengietkunst (Contemporary Dutch bell-founding art). *Neth. Acoust. Soc.*, No. 7, 20–49 (English translation in Rossing, 1984a).

Lehr, A. (1976). "Leerboek der Campanologie." Nationaal Beiaardmuseum, Asten, The Netherlands.

Lehr, A. (1986). Partial groups in the bell sound. *J. Acoust. Soc. Am.* **79**, 2000–2011.

Lehr, A. (1987). "The Designing of Swinging Bells and Carillon Bells in the Past and Present." Athanasius Kircher Foundation, Asten, The Netherlands.

Ma Chengyuan (1980–1981). Ancient Chinese two-pitch bronze bells. *Chinese Music* **3**, 81–86; **4**, 18–19, 31–36.

Milsom, M.J. (1982). Tuning of bells. *The Ringing World*, September 3, 733.

Obata, J., and Tesima, T. (1933–1934). Experimental investigations on the sound and vibration of a Japanese hanging-bell. *Jpn. J. Phys.* **9**, 49–73.

Perrin, R., and Charnley, T. (1978). The suppression of warble in bells. *Musical Instr. Tech.* **3**, 109–117.

Perrin, R., Charnley, T., and Banu, H. (1982). Increasing the lifetime of warble-suppressed bells. *J. Sound Vibr.* **80**, 298–303.

Perrin, R., Charnley, T., Banu, H., and Rossing, T.D. (1985). Chladni's law and the modern English church bell. *J. Sound Vibr.* **102**, 11–19.

Perrin, R., Charnley, T., and de Pont, J. (1983). Normal modes of the modern English church bell. *J. Sound Vibr.* **90**, 29–49 (reprinted in Rossing, 1984a).

Rayleigh, Lord (1894). "The Theory of Sound," Vol. 1. Macmillan, New York. Reprinted by Dover, New York, 1945.

Ritsma, R.J. (1967). Frequencies dominant in the perception of the pitch of complex sounds. *J. Acoust. Soc. Am.* **42**, 191–198.

Rossing, T.D. (1981). Acoustics of tuned handbells. *Overtones* **27** (1), 4–10, 27.

Rossing, T.D. (1982a). "*The Science of Sound.*" Addison-Wesley, Reading, Massachusetts.

Rossing, T.D. (1982b). Chladni's law for vibrating plates. *Am. J. Phys.* **50**, 271–274.

Rossing, T.D. (1983). Tuned handbells, church bells and carillon bells. *Overtones* **29** (1), 15, 27–30; **29** (2), 27, 29–30.

Rossing, T.D. (1984a). "*Acoustics of Bells.*" Van Nostrand-Reinhold, Stroudsburg, PA.

Rossing, T.D. (1984b). Acoustics of bells. *American Scientist* **72**, 440–447.

Rossing, T.D. (1989), Acoustics of Eastern and Western bells, old and new, *J. Acoust. Soc. Jpn.* (E) **10**, 241–252.

Rossing, T.D., Gangadharan, D., Mansell, E.R., and Malta, J.H. (1995). Bass handbells of aluminum. *MRS Bulletin* **20**(3), 40–43.

Rossing, T.D., Hampton, D.S., Richardson, B.E., and Sathoff, H.J. (1988). Vibrational modes of Chinese two-tone bells. *J. Acoust. Soc. Am.* **83**, 369–373.

Rossing, T.D., and Perrin, R. (1987). Vibration of bells. *Applied Acoustics* **20**, 41–70.

Rossing, T.D. and Perrier, A. (1993). Modal analysis of a Korean bell. *J. Acoust. Soc. Am.* **94**, 2431–2433.

Rossing, T.D., Perrin, R., Sathoff, H.J., and Peterson, R.W. (1984). Vibrational modes of a tuned handbell. *J. Acoust. Soc. Am.* **76**, 1263–1267.

Rossing, T.D., and Sathoff, H.J. (1980). Modes of vibration and sound radiation from tuned handbells. *J. Acoust. Soc. Am.* **68**, 1600–1607.

Sathoff, H.J., and Rossing, T.D. (1983). Scaling of handbells. *J. Acoust. Soc. Am.* **73**, 2225–2226.

Schad, C.-R., and Warlimont, H. (1973). Akustische Untersuchungen zum Einflusz des Werkstoffs auf der Klang von Glocken. *Acustica* **29**, 1–14 (English translation in Rossing, 1984a).

Schoofs, B., van Asperen, F., Maas, P., and Lehr, A. (1985). Computation of bell profiles using structural optimization. *Music Perception* **4**, 245–254.

Schouten, J.F., and 't Hart, J. (1965). De slagtoon van klokken (The strike note of bells). *Neth. Acoust. Soc. Publ.*, No. 7, 8–19 (English translation in Rossing, 1984a).

Slaymaker, F.H. (1970). Chords from tones having stretched partials. *J. Acoust. Soc. Am.* **47**, 1569–1571.

Terhardt, E., and Seewann, M. (1984). Auditive und objective Bestimmung der Schlagtonhöhe von historischen Kirchenglocken. *Acustica* **54**, 129–144.

Tyzzer, F.G. (1930). Characteristics of bell vibrations. *J. Franklin Inst.* **210**, 55–56 (reprinted in Rossing, 1984a).

Part VI

Materials

22

Materials for Musical Instruments

As a final chapter to this book, it is instructive to look at the general principles and practical considerations that have determined the materials from which musical instruments are made. There is a great deal of tradition, and a certain amount of mystique, about this subject, and some of the conclusions we draw may be hotly contested in the instrument-making community. This is all to the good, since it can lead to discussion, experimentation, and ultimately an advance in knowledge.

The materials from which musical instruments are made have nearly always been an intrinsic part of the tradition of the instrument itself and, as such, have influenced both the visual and the tonal ideals that instrument makers have sought to achieve. In only a few cases has a significant choice of materials been possible during the evolution of an instrument; nearly always the task of construction has imposed severe limitations upon choice. In cases where the material actually affects the sound of the instrument, this has then helped to define its traditional sound, and in all cases it has helped to define its traditional appearance. Modern manufacturing techniques have now removed most of these material barriers, and sometimes new materials can be used to advantage, but usually the traditions persist for good reason.

To give structure to the discussion, we should first note that there are clearly elements of some instruments in which the material of manufacture has a large influence on the acoustic behavior. Among these are the idiophones, such as bells and cymbals, in which the whole solid material structure vibrates and radiates sound, and we could also include the primary vibrating elements of membranophones such as drums, the strings of chordophones such as guitars and violins, and the reeds of woodwind instruments. Next we should list the "sympathetic" resonant structures that are attached to these primary vibrating elements—the bodies of drums, and particularly the carefully crafted bodies of stringed instruments. It is fairly clear that material properties are also of significant importance here. Finally we come to wind instruments, in which it can be argued that the vibrating element is the enclosed air, and that the material from which the walls are made has very little influence on the sound produced.

We can also note that our discussion will embrace the relatively simple and homogeneous properties of metals and plastics, the more complex structure and properties of natural materials such as wood, and the equally complex properties of animal materials such as skin, hair and gut. We cannot hope to cover all this area in detail, but at least the general principles involved and their relation to material properties should become clear.

22.1 Mechanical Properties of Materials

Materials for musical instruments can be characterized by a large number of different properties, some of which can be measured and some only described qualitatively. The qualitative characters include such things as appearance and feel, although the former can certainly be measured objectively in terms of optical parameters, and the latter is probably a combination of measurable quantities such as surface smoothness, density, thermal conductivity, and perhaps electrochemical potential. Leaving these qualitative characteristics aside, except for a few remarks at a later stage about particular cases, let us look at those properties of materials that are directly relevant to musical instrument construction and performance. The most relevant properties are mechanical, though a few are thermal and chemical.

22.1.1 Linear Elastic Properties

For the small amplitudes associated with acoustic vibrations, the elastic properties of a homogeneous material can be specified by giving the relations between the components σ_{ij} of the stress tensor and the strain tensor ϵ_{ij}. Here the subscripts i and j refer to orthogonal directions related to axes (x, y, z) in the material. These relations, which are effectively Hooke's Law, are simply linear and can be written in two equivalent ways as

$$\epsilon_{ij} = S_{ijkl}\sigma_{kl} \qquad \text{or} \qquad \sigma_{ij} = C_{ijkl}\epsilon_{kl}. \tag{22.1}$$

The coefficients S_{ijkl} and C_{ijkl} constitute the elastic compliance and elastic stiffness matrices, respectively, for the material. (The interchange of symbols S and C from what we might expect is confusing!) The two matrices are clearly mutually inverse, so that $\mathbf{S} = \mathbf{C}^{-1}$. These 81-element matrices are fortunately reduced to manageable size by symmetry considerations, which dictate that $\sigma_{ij} = \sigma_{ji}$ and $\epsilon_{ij} = \epsilon_{ji}$, and the further requirement that the elastic body not rotate as a whole under the applied forces gives three further constraints, so that there are only six independent components for each of stress and strain, and the matrices S and C each contain 36 independent coefficients. A general notation system has been adopted that then writes S and C as 6×6 matrices S_{ij} and C_{ij} respectively, using the convention $1 \equiv xx$, $2 \equiv yy$, $3 \equiv zz$, $4 \equiv yz$, $5 \equiv zx$ and $6 \equiv xy$. A discussion of the theory can be found in standard texts on solid-state physics, such as Kittel (1966) or Pollard (1977), or in the recent monograph on wood by Bucur (1995).

Most materials have certain symmetries that simplify these relations greatly. For an isotropic material such as bulk polycrystalline metal, symmetry reduces the number of independent stiffnesses C_{ij} to only two, C_{11}, C_{12}, and these are usually written in terms of the Young's modulus E and the Poisson's ratio ν, from which the bulk modulus B and shear modulus G can be deduced. Single crystals have particular symmetries that reduce the number of independent stiffnesses as well, but these do not concern us here. Of more immediate relevance is the case of an orthotropic solid, by which we mean one that has elastic symmetry defined by three Cartesian axes (x, y, z). Solid wood is approximately orthotropic, the principal directions being respectively parallel to the trunk, radial, and tangential. For such materials, the matrix C_{ij} is defined by six independent diagonal terms, C_{11} to C_{66}, and three off-diagonal terms C_{12}, C_{13} and C_{23}. In practice, these stiffness coefficients are usually expressed in terms of three Young's moduli E_i, three shear moduli G_{ij}, and three Poisson's ratios ν_{ij}, relative to the principal axes. If two of the directions in an orthotropic material are equivalent, as for example is approximately true in plywood or other laminates with more than three plies, then the sheet has only six independent stiffness constants, while if it is actually isotropic in the symmetry plane this number is reduced to five.

22.1.2 Anelastic Behavior and Damping

All the components of the stiffness matrix are, in principle, complex. The real part then represents a normal elastic response, and the imaginary part some sort of relaxation or loss mechanism. We can easily see how this comes about. Suppose the Young's modulus E can be written $E = E_1 + jE_2$, then for a sinusoidal strain $\epsilon(t) = \epsilon e^{j\omega t}$ there is a stress $\sigma_1(t) = E_1\epsilon$ that is in-phase with the displacement, but a part $\sigma_2(t) = jE_2\epsilon$ that is in-phase with the velocity, and so dissipates energy. A complex stiffness (or Young's modulus) therefore represents a loss mechanism and damps the vibration. A measure of the loss often used is the loss-angle $\delta = \tan^{-1}(E_2/E_1)$. For any particular vibration mode, $Q \approx 1/\tan\delta$.

There is another and more physical way of looking at this, as discussed in the classic book of Zener (1948), or more recently by Pollard (1977) in a treatment oriented towards ultrasonics. Suppose that we apply a uniform strain at time $t = 0$ and examine the behavior of the stress. For a typical class of solids, the stress declines from its initial value towards some lower equilibrium value with an exponential time constant τ, as shown in an exaggerated fashion in Fig. 22.1(a). If we represent the strain by $\epsilon_0 u(t-0)$,

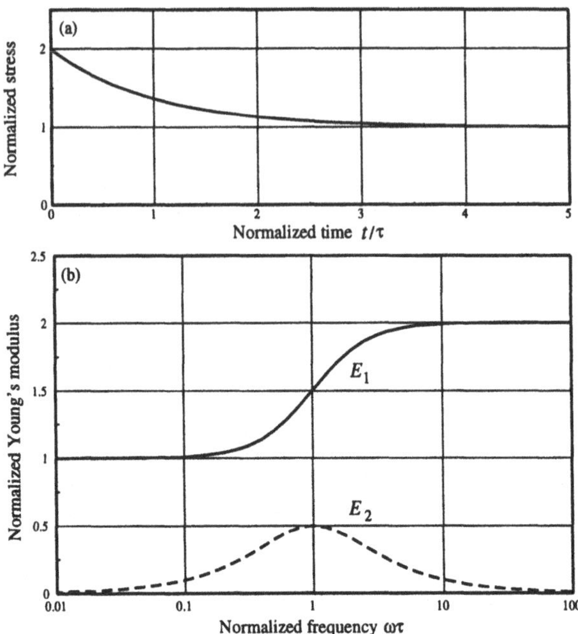

FIGURE 22.1. (a) Time behavior of elastic relaxation following application of a constant strain at $t = 0$, as caused by a single mechanism with relaxation time constant τ. The magnitude of the assumed relaxation is arbitrary. (b) Frequency variation of the real and imaginary parts of the elastic modulus $E(\omega) = E_1 + jE_2$ for this material. The extent of the relaxation (and hence the magnitude of E_2) is exaggerated by about a factor 100 compared with the behavior of a typical metal.

where u is the unit step function, then the stress behavior can be expressed as

$$\sigma(t) = [\sigma_1 + (\sigma_0 - \sigma_1)e^{-t/\tau}]u(t-0). \tag{22.2}$$

Taking the Fourier transform of this equation gives the frequency variation $\sigma(\omega)$ of the stress as

$$\sigma(\omega) = \sigma_0 + \frac{\sigma_1 - \sigma_0}{1 + j\omega\tau}. \tag{22.3}$$

Since stress and strain are related by elastic moduli, this shows why these moduli appear as complex quantities. Explicitly, for the Young's modulus E,

$$E(\omega) = E_1(\omega) + jE_2(\omega) = \left[E_\infty - \frac{E_\infty - E_0}{1 + \omega^2\tau^2}\right] + j\left[\frac{(E_\infty - E_0)\omega\tau}{1 + \omega^2\tau^2}\right], \tag{22.4}$$

where E_0 and E_∞ are the values of the real part of the Young's modulus at very low and very high frequencies respectively. As shown in Fig. 22.1(b), the real part E_1 of the Young's modulus increases from E_0 as the frequency ω goes through the value $\omega = 1/\tau$ and then levels out towards E_∞, while the imaginary part E_2, which contributes the energy loss, goes through a broad peak at this same frequency.

This anelastic behavior can arise from a number of physical processes. One common loss mechanism is thermal conduction, which is of prime importance in metals. An ordinary material with a positive thermal expansion coefficient heats when it is compressed and cools when it is stretched. A time-varying strain in the material therefore generates regions of raised and lowered temperature, the separation L between these depending upon details of the strain pattern. For simple bending of a plate, for example, L is about equal to the thickness of the sample, and for transverse vibration of a string about equal to the string diameter. For high-frequency waves in a bulk material, however, L is half the wavelength. Since heat conduction obeys a diffusion equation, the rate of heat conduction is inversely proportional to the square of the separation between hot and cold regions, and this determines the relaxation time τ. Expressed more conveniently as the frequency f^* of the loss peak, for which $\omega\tau = 1$, we have

$$f^* = A\frac{D}{L^2} = A\frac{K}{\rho C_p L^2}, \tag{22.5}$$

where D is the thermal diffusivity of the material, ρ its density, C_p its specific heat, and K its thermal conductivity. A is a constant of order unity, the exact value of which depends upon the geometry of the strain. For the transverse vibrations of a wire, $A = 2.16$ (Zener 1948). For common metals and alloys, f^* is in the range 0.1–1 Hz for a plate 10 mm in thickness, with steel near the low end and aluminum near the high end of this range. For transverse vibration of a wire 1 mm in diameter, f^* is in the range 100–1000 Hz. From Fig. 22.1(b), the loss peak is more than a decade wide in frequency, so that most of the audible range is rather similarly affected.

There is an additional loss in metals arising from the fact that they have a crystallite structure, with grains oriented in nearly random directions. Even uniform strains therefore generate different stresses in neighboring crystallites, and these in turn lead to losses through thermal conduction. The magnitude of this loss is generally much less than the macroscopic conduction loss discussed above, and its peak frequency is typically in the low kilohertz range, since grain diameters are usually of order 0.1 mm. This loss mechanism is more significant in metals with non-cubic crystal structures, since the elastic anisotropy is then greater.

Another important loss mechanism is that due to rearrangements at an atomic level. In metals this can be described by the motion of dislocations and grain boundaries, or sometimes interstitial atoms, but the characteristic relaxation times for these processes are typically greater than one second so that they are not acoustically important (except for metals such as lead). In organic materials the rearrangements usually involve hydrogen bonds between molecules, which occur much more quickly, and often have a whole spectrum of relaxation times τ_i contributed by different molecular motions. All these molecular relaxation times decrease with increasing temperature, so that the loss peaks shift to a higher frequencies, and they are usually the dominant loss mechanism in such materials at sonic frequencies.

In the case of inhomogeneous materials, there are other loss mechanisms that are peculiar to particular materials. One that is common relates to the motion of water in the material, which can contribute molecular damping, or even viscous damping if there is enough of it, as for example in living tissue such as human finger tips or lips. In pure viscous damping, the exponential decay rate of vibrations is independent of frequency, so that the loss tangent decreases at high frequencies.

In composite materials containing fibers, as for example in the gut or multi-filament nylon strings of musical instruments, an additional important loss mechanism is dry friction. Once slipping begins, the frictional force depends very little upon velocity, so that we might expect the power dissipation to be proportional to the velocity amplitude at all frequencies. If we write $\bar{v}$ for the velocity amplitude, then the stored energy is $A\bar{v}^2$, where A is a constant, and the rate of energy loss is $F\bar{v}$ where F is the frictional force. The equation of motion is thus

$$\frac{dA\bar{v}^2}{dt} = -F\bar{v} \tag{22.6}$$

from which we see that $d\bar{v}/dt = -F/2A$, which is constant. Sound decay is thus very abrupt once the velocity amplitude becomes low, and high-frequency partials of small amplitude are very quickly damped, giving the characteristic plucked-string sound on bowed instruments.

22.1.3 Nonlinear Properties

When materials are subjected to strains of more than a few percent, the distortion often becomes irreversible. In extreme cases the material fractures, and such materials are termed brittle. The metals used in musical instruments do not generally fail in this way—an exception is the cast iron often used in piano frames, though this is so thick that accidental fracture is nearly impossible. A more common result is plastic deformation, in which the material simply retains a part of its distortion when the stress is removed. In metals, this behavior can be explained in terms of grain boundary creep and the motion of dislocations past pinning sites, while in organic materials it is generally due to re-linking of molecules to different neighbors.

This nonlinear behavior is important in the forming of metal parts for musical instruments—the drawing of wires and tubes, the forging of keys, and so on—and the ease with which this can be accomplished is an important element in material choice. Pure metals are generally quite ductile, to the extent that they are usually too soft to be used. Their hardness can be increased by making additions of other metals to form an alloy, alloys with two interpenetrating phases usually being harder than those that are uniform. Most metals become increasingly hard as they suffer mechanical working, the exceptions being mostly those for which room temperature (in kelvins) exceeds about two-thirds of their absolute melting temperature. This work-hardening, which is caused by the interaction of dislocations, can be removed by annealing at a temperature greater than about two-thirds of the absolute melting point, when the dislocations become mobile and are able to untangle. Work hardening is important to the strength and rigidity of items such as piano strings, flute tubes and forged keywork.

Most materials are also subject to creep under stresses that are a good deal less than those required for plastic distortion on short time-scales. This phenomenon, which is caused by dislocation movement in the case of metals and by molecular bonding rearrangements in other materials, takes place over periods of months or years, though the rate rises rather rapidly with increasing stress and with increasing temperature. Although its effects are quite noticeable in the gradual detuning of piano strings and in the warping of soundboards and bridges, it is not generally a severe problem.

In the case of wood, the application of mechanical force tends to crush the cell structure in an undesirable way. Permanent distortion is best achieved by bending under the influence of heat and steam, both of which promote molecular rearrangements within the material. Such distortions are generally only partly reversible.

Finally there is the phenomenon of fatigue, in which the repeated application of an alternating stress leads to cracking and failure of the material. In the case of metals, this is often caused by the build-up of atomic vacancies at grain boundaries, while in organic materials it is due to the failure of molecules to re-link with their neighbors. Fatigue failure is not usually a problem in musical instruments because, although alternating stresses are common, they are generally well below the necessary magnitude.

22.2 Materials for Wind Instruments

When the solid material from which an instrument is made takes an active part in producing or radiating the sound, then it is abundantly clear that its mechanical properties will be important in determining the nature of that sound. The properties of concern will be the density, elastic moduli, and damping coefficients. Assuming that the dimensions of the instrument are fixed, these material properties determine the frequencies and widths of its mechanical resonances, and the impedance presented to whatever is the primary driving element. Most of these matters have been discussed in previous chapters in relation to particular instruments, though we will here take the opportunity to relate them more specifically to materials.

While there is no argument but that material properties are vitally important in the case of stringed and percussion instruments, the matter has been the subject of continuing controversy in the case of wind instruments. It is therefore appropriate to discuss this matter briefly before going on.

The tube walls influence the behavior of the vibrations of the air column because of the viscous and thermal losses across the boundary layer, as we discussed in Section 8.2. These losses have quite significant effect on the Q-factors of the pipe resonances, and thus on the behavior of the instrument, and vary somewhat in magnitude depending on the smoothness of the surface. Wall materials all have thermal capacity so much greater than air, however, that there is virtually no difference between them on this score. Roughness effects become significant only when the roughness itself is significant on the scale of the boundary layer thickness, or about 0.1 mm. Most claims for difference in behavior between different wall materials are based, however, upon discussions of mechanical effects.

While the mechanical virtues and aesthetic appeal of different materials are easily evaluated, the same is not true of their acoustical properties. Makers and players claim to detect clear and consistent tonal differences between otherwise similar instruments made from different materials, but physical analysis suggests that these claims may be illusory. This does not mean that wall material never has any effect, and indeed demonstrations by Miller (1909) long ago showed that the thin walls of metal pipes of square cross section can vibrate with appreciable amplitude and have a very large effect on the stability and timbre of the sound. The situation is, however, quite different for the relatively rigid walls of typical organ pipes and wind instruments.

It is easy to see why this is so. The physical quantity causing wall vibration is the acoustic pressure in the standing wave of the air column. This can couple to a vibration mode of the pipe walls only if there is reasonably close agreement between the resonance frequency of the wall mode and one of the harmonics of the air-column vibration and if the symmetry of the wall mode is such that the coupling coefficient does not vanish. It is quite easy to satisfy these conditions for a pipe of rectangular cross section, for the local "breathing" mode, in which the pipe cross section distorts successively from barrel to pincushion shape, can have a low frequency and very low impedance if the walls are thin.

The case of a pipe of circular cross section is entirely different, for the breathing mode involves an actual increase in the local radius of the tube, rather than a simple shape deformation, and therefore has a very high resonance frequency. This is true even for thin metal tubes, and the audible modes that can be excited by tapping the tube wall are in fact distortional modes in which the pipe cross section becomes elliptical. These could conceivably be excited in a thin metal tube in the vicinity of a finger hole, where the bore, and thus the pressure distribution, loses its circular symmetry, but for typical wooden instruments the excitation coupling is small and the mode frequency very high.

The discussion can be quantified for the strictly circular part of the bore by considering the relative compliances associated with expansion of the bore under pressure and with compression of the air in the tube. The ratio is about 0.001 for even a quite thin-walled tube, so that the compliance of the walls has virtually no effect upon the internal air modes and direct radiation from wall vibration is very small (Backus, 1964). Even rigid walls do, however, affect the damping of the air modes, and indeed this wall damping predominates over radiation damping except at very high frequencies. Details of wood grain and smoothness can affect the exact damping coefficient, but generally the difference between one material and another is small compared with the effects of sharp edges on finger holes, soft key pads, or even finger tips.

The outcome of this discussion is that we are led to the view that the choice of particular materials for the construction of wind instruments is governed not really by acoustics, but rather by considerations of ease of fabrication, stability, feel and appearance. We return to these matters later in relation to particular materials.

22.3 Wood

Wood is a material of great variety and individuality. Not only does each species of tree produce very different timber, but even that from individual trees, or from different parts of the same tree, varies quite greatly. An extensive survey of the acoustic properties of wood has been given by Bucur (1995), and a collection of reprints about wood for stringed instruments can be found in the volumes edited by Hutchins and Benade (1996), together with a survey article by the authors which includes references to other relevant papers.

From a chemical point of view, wood consists primarily of cellulose, with smaller fractions of hemicellulose and lignin. These materials are organized into structural elements based on tracheid cells that are greatly elongated along the trunk direction, though there are also smaller groups of cells running radially across the trunk. The dimensions of these tracheid cells, which are water-filled in the living tree, vary considerably from species to species, but are typically several millimeters in length and some tens of micrometers in diameter, so that the structure might be likened to a bundle of drinking straws. Both transverse and longitudinal sections through this microstructure are shown in Fig. 22.2.

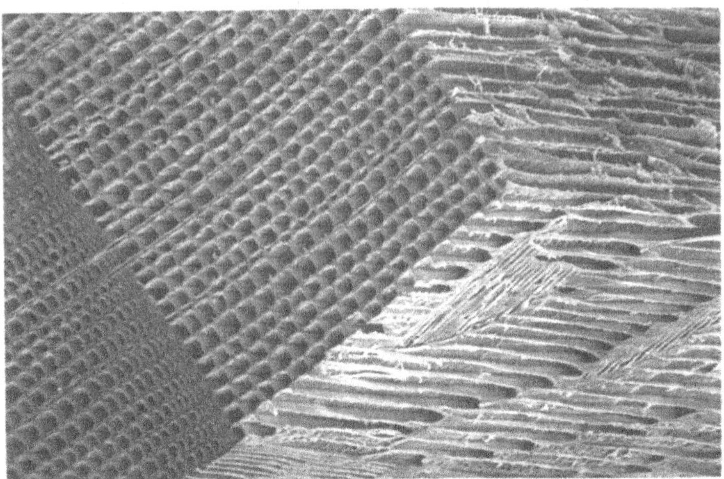

FIGURE 22.2. A scanning electron micrograph of the corner of a cube of spruce wood. The tracheid cells run from left to right along the length of the trunk, and we see a section normal to the trunk in the left-hand half of the picture (Claire Barlow).

22.3.1 Resonance Woods

The elastic properties of wood are approximately orthotropic, which means that we require 9 independent quantities to describe them. Kahle and Woodhouse (1994) have shown how these properties can be calculated from a knowledge of the cell geometry, as shown in Fig. 22.2, and the elastic properties of the cell walls, but this is scarcely a practical approach. Measurement of the elastic parameters presents difficulties and various methods have been used. Bucur (1995) gives an extensive discussion of ultrasonic methods, which do however raise the question of extrapolation to low acoustic frequencies, while Haines (1979, 1980) used a more conventional method in which the resonance frequencies of small wooden bars are studied. Both these authors give extensive tabulations of the parameters for timbers of interest to musical instrument makers. McIntyre and Woodhouse (1988) have shown how the smaller number of parameters that are of importance when the wood is in sheet form can be determined by measuring the resonance frequencies and damping of the modes of thin rectangular plates with free edges, and this is often adequate for musical instrument applications.

Schelleng (1963) has discussed the importance of the major mechanical parameters describing materials for making violins and similar instruments that rely upon the vibration of plates, and has proposed the parameter c/ρ, where ρ is the material density and c the speed of longitudinal waves, as being of prime significance. The argument goes as follows. Suppose we wish to replace a plate with density ρ_1, Young's modulus E_1 and thickness h_1 with another plate made from material described by the parameters ρ_2, c_2 and h_2 in such a way as to leave vibrational performance unaltered. The stiffness of the plate depends on the quantity Eh^3, and its inertia on the mass per unit area ρh, so we must have

$$E_1 h_1^3 = E_2 h_2^3 \qquad \text{and} \qquad \rho_1 h_1 = \rho_2 h_2. \tag{22.7}$$

If we combine these equations with the relation $c^2 = E/\rho$, then we find the requirement

$$c_1/\rho_1 = c_2/\rho_2. \tag{22.8}$$

Replacement of the material according to this rule brings with it the necessity to change the plate thickness according to either of Eqs. (22.7). Following earlier work by Barducci and Pasqualini (1948), Schelleng has classified numerous woods on this basis and shown that the resonance quality factor Q increases consistently as c/ρ increases, the resonance wood of choice for stringed instruments, Norway spruce, lying near the top of the list in both c/ρ and Q. Interestingly, materials such as magnesium or glass are quite incompatible with the sequence for woods.

Details of the properties of the many woods used in musical instrument making are too extensive to be quoted here, but a few generalizations are appropriate. An excellent survey based on his own measurements has been given by Haines (1979), and further information is given by Bucur (1995). Density ranges from about 200 kg/m^3 for balsa to about 920 kg/m^3 for pernambuco, but most lie in the range 400–700 kg/m^3. Wood is a very anisotropic material, as we might indeed expect because of its cellular structure. If we omit extreme outliers, then the general range of properties is as shown in Table 22.1. For convenience, the number of parameters given is greater than the necessary six.

The most important thing to note from Table 22.1 is the extreme anisotropy of the Young's modulus E, which is 10 to 20 times as large

TABLE 22.1. Range of elastic parameters for common woods. Elastic moduli are given in GPa, and typical values for Sitka spruce are noted in brackets.

E_{Long}	5–16	[12]	$G_{\text{Long/Rad}}$	0.7–1.6	[0.7]
E_{Rad}	0.7–2.2	[0.9]	$G_{\text{Long/Tang}}$	0.7–1.1	[0.7]
E_{Tang}	0.4–1.1	[0.5]	$G_{\text{Tang/Rad}}$	0.04–0.5	[0.04]
$\nu_{\text{Long/Tang}}$	0.4–0.5	[0.47]	$\nu_{\text{Tang/Long}}$	0.02–0.08	[0.02]
$\nu_{\text{Long/Rad}}$	0.3–0.5	[0.37]	$\nu_{\text{Rad/Long}}$	0.03–0.13	[0.03]
$\nu_{\text{Tang/Rad}}$	0.25–0.35	[0.25]	$\nu_{\text{Rad/Tang}}$	0.4–0.8	[0.43]

in the longitudinal direction as in either transverse direction. The one exception is oak, for which this ratio is only about 2.5. For spruce, which is the material of choice for the top plates of string instruments, the ratio is between 13 : 1 and 24 : 1. This same anisotropy is reflected in the shear modulus G and in the Poisson's ratio ν.

This anisotropy means that it is of vital importance that the orientation of timber from which musical instruments are to be made is carefully specified. The universal practice for the top plates of bowed-string instruments is to insist that the log be cut "on the quarter," which means that it is split into sector slices like a cheese. Two neighboring slices are then glued edge-to-edge to form a slab from which the top plate is carved. This orientation ensures that the grain runs lengthwise along the instrument body and that the growth rings in the tree are normal to the surface of the instrument. Bonamini et al. (1991) have discussed the relation between the mechanical properties of spruce wood and its anatomical features in some detail. There is not such a strict consensus about the orientation of the back plate, which is generally of less anisotropic wood, but the longitudinal direction is still along the length of the instrument. Since each wave velocity varies as the square root of the elastic modulus, this arrangement conforms generally with the approximately 3 : 1 aspect ratio of bowed-string instruments.

There are only a few studies of the detailed mechanism of damping in wood, though measurements have been reported for many species under various conditions. Fukada (1950, 1951) has studied internal damping in a variety of woods, mostly of Japanese origin, and has identified several distinct types of behavior in the frequency range 100–5000 Hz. The woods studied mostly had a loss peak of rather modest height in the upper part of this range, but also showed increased damping at very low frequencies. Studies by Dunlop (1978) showed that dry wood also has a loss peak of the visco-elastic type at a frequency that is typically in the range 50–100 kHz. The frequency of this peak changes with temperature, as expected, and its magnitude is considerably influenced by moisture—increasing the moisture content from 3% to 20% percent typically doubles the loss. Within the range of musical interest, about 100 Hz to 20 kHz, the loss factor of a "resonance wood" such as the spruce used in stringed instruments is typically nearly constant at about 0.01 for frequencies below about 2 kHz and then rises more or less proportionally to frequency, reaching a value about 0.03 at 10 kHz. Expressed another way, and from different measurements, the Q value for resonance wood typically lies in the range 30 to 100, depending upon the vibration mode studied. Indeed, the loss factor varies by as much as a factor three with strain orientation. This behavior agrees qualitatively with what we would expect from the ultrasonic measurements. Other woods have not been studied in so much detail, except to note that hardwoods such as are used in percussion instruments typically have loss factors that are smaller than this by about a factor two, while some very soft woods like balsa have significantly higher losses. The field has been reviewed by Bucur (1995).

From the discussion of bowed-string instruments in Chapter 10, it is clear that the tone quality of fine instruments is determined not only by the resonance frequencies of the body modes, but also by their damping. It is by no means desirable to minimize the material damping, however, and indeed many wood species are rejected by makers because they impart a too-bright and even scratchy quality to the tone.

Wood, along with other materials, undergoes changes when it is subjected to time-varying stresses, and of course this is what happens when an instrument is played. The stresses are generally too small to cause damage, like the fatigue cracks that can develop in repeatedly stressed metals, but a recent study by Sobue (1995) has shown that prolonged vibration at a level equivalent to that reached in playing can have an effect on the internal damping of wood. In his experiments, the vibration amplitude was about 0.1 mm at a frequency between 100 and 170 Hz, and the measurements showed a significant decrease in internal damping after a vibration period of some five hours. It is reasonable to interpret this in terms of relaxation of intermolecular bonds in the material. The observation itself is interesting, since it accords with violinist's subjective feelings about "playing in" an instrument, though this effect has also been ascribed to minor separation of the glue in the purfling.

Sometimes in the construction of instruments, wood is bent to shape, using steam to encourage plastic deformation. The effect of this process on the mechanical properties has been studied by Barlow and Woodhouse (1993), using Norway spruce as a test material. Not surprisingly, since the plastic deformation distorts and even fractures the cell walls, the effects are considerable. The Young's modulus is decreased quite significantly—by as much as 30% for some distortion directions.

Although attention has been focused in this section upon wood for stringed instruments, we must not forget the very different requirements of percussion instruments. Here the prime requirements are surface hardness, so that the instrument will not be damaged by hammer blows, and low internal damping, so that the tone will be well sustained—it is always easy to add damping felt if this is necessary. Hardwoods fill these requirements, often having internal damping less than one-half that of the softwoods used for bowed-string instruments, and a wide variety of species can be used, giving choices of color and appearance.

22.3.2 Varnishes

Because wood is a cellular material, it is most desirable to seal its surface in some way so as to preserve it, beautify it, prevent staining by dirt lodged in the cells, and prevent the ingress of moisture and other materials that might affect the mechanical properties. Techniques for accomplishing this have been known in relation to other wooden objects, and these have been adapted to musical instruments. Because the surface layer will inevitably have mechanical properties different from bulk wood, it is desirable that the coating be quite thin, and various oils and varnishes have been found suitable for the purpose. Their effect is generally to stiffen the instrument plate, so that allowance needs to be made for this.

Schelleng (1968) has carried out a detailed study of the effects of varnish on the elastic and damping properties of resonance woods. In most cases the varnish slightly increases the stiffness of the plate, and at the same time increases the damping losses. Rather similar results were obtained in a later study by Hutchins (1989) of the effects of sealers under varnish. While the mechanical changes are small with properly formulated and applied finishes, it is clear that the wrong varnish can have severe and generally detrimental effects. In complementary work, Barlow and Woodhouse (1988, 1989) have conducted an electron-microscope study of the detailed effects of various wood finishes, and have examined those used on old violins.

22.3.3 Organ Pipes

Although most organ pipes are made from metal, those of some ranks, for example stopped-flutes, are traditionally of wood. Use of wood instead of metal immediately imposes geometrical constraints upon the pipe, for it is then made with rectangular rather than circular cross-section, and its flat walls must be made quite thick in order to be adequately rigid—we discussed in Section 22.1.1 the undesirable effects that can occur if the flat walls are insufficiently stiff. The other geometrical constraints relate to the pipe mouth, for the lip must have quite a thick wedge-shaped edge, and the lower lip and sides of the mouth must be similarly thick. These geometrical constraints at the mouth are the chief determinants of the different tone of wooden pipes, quite apart from conscious design decisions such as pipe scale, lip cut-up, and blowing pressure.

When it comes to selecting wood for the construction of organ pipes, the main requirements are good working qualities, freedom from warping or splitting, and smooth straight grain. There is little discussion among organ builders about the particular merits of different timbers, and any distinction in the matter of density or elastic moduli is overshadowed by decisions on wall thickness, although relatively light wood is preferred for structural reasons.

22.3.4 Woodwind Instruments

When we come to consider suitable wood for making wind instruments, the situation is very different. The elastic and damping properties of the wood are largely irrelevant, since the acoustic impedance of the tube walls is extremely high compared with that of the air column, and its magnitude is in any case predominantly determined by the wall thickness rather than by any material properties.

The desirable properties for wind-instrument woods are rather different from those for stringed instruments. An ideal wood will have small compact cells and a close, straight grain so that it can be cleanly bored and worked, will not split or distort with time and with moisture, and takes a fine polished finish when simply oiled (Robinson, 1973). It must take sharp edges at the toneholes and embouchure hole, and must support the posts of the keywork securely and without cracking. Hardwoods from rainforest areas, such as African blackwood, ebony and grenadilla, and some fruitwoods such as pear and apple, are best suited for this purpose. They must be well seasoned, and are generally bored out well undersize and left to relax for a year or more before further working. The finish is almost always a simple polishing oil, which must be renewed regularly when the instrument is played. The smoothness of this interior finish is of considerable importance because it influences viscous losses from the air column. The choice of wood is largely a matter of availability and tradition, a black wood being preferred for modern clarinets and oboes and a light-colored fruitwood for reconstructions of Baroque period instruments.

The only real exception to these generalizations is in the case of the bassoon, which is traditionally made from maple wood. After finishing, this wood is stained and varnished to a red-brown color of the maker's choice. Because of the narrow bore and thick walls of the upper part of the instrument, cracking is a potential problem, and most high-quality instruments have this part of the bore lined with hard rubber or some plastic material such as Bakelite.

22.3.5 Reeds

The traditional material for the reeds of all woodwind instruments is some form of cane or bamboo, from the stems of which reed blanks are cut and then shaped. These plants are all giant grasses which grow in a wide variety of climates, but particular species and environments produce the most suitable materials for instrument reeds. The preferred species is *Arundo donax*, which grows wild in many parts of the world, but that most suitable for instrument reeds is found in the area near Var in the south of France. Because of pressure on the wild cane fields, however, commercial cane plantations for instrument reeds have now been established in several parts of the world where soil and climate are most suitable.

In reed manufacture, mature stems of cane are cut and dried in the sun for several months. After this seasoning, selected stems are cut to shorter lengths and sorted by diameter, tubes of larger size being used for saxophone reeds and those of smaller size for clarinet reeds. Oboe and bassoon reeds are generally made as a part-time activity by professional players, or else made from partly finished cane by individual musicians. The processes of reed making involve splitting the cane to provide appropriate blanks and then shaping these progressively on various machines or, in the case of oboe and bassoon reeds, by hand. In the case of clarinet and saxophone reeds, which are made in large quantities, the finished reeds are tested mechanically and sorted into grades (2, 3, etc.) on the basis of stiffness, particular grades being matched to particular mouthpiece curvatures according to the preferences of the player.

22.4 Plastics and Composite Materials

Recently there has been increasing interest in using new materials in musical instruments, partly because traditional timbers are more difficult to obtain and more expensive, and partly because synthetic materials offer the possibility of better manufacturing control, provided their acoustical properties are satisfactory.

The most obvious place to use plastics is in the manufacture of woodwind instruments, for here the body has a purely mechanical function and its vibratory properties are essentially irrelevant. Plastics of one kind or another have been used in the manufacture of student versions of recorders and clarinets for a long time, and have proved very satisfactory. Such criticism as there has been of these instruments has derived largely from their rather mass-produced nature and minimum of hand-finishing. Some of the plastics used have also been rather heavier than their wood counterparts. An exception has been the case of bassoons, which are expensive specialist instruments in any case, and excellent plastic-bodied instruments have been made by some major companies. Bassoon players are perhaps in any case already conditioned by knowing that top-grade instruments already have plastic liners inside nearly half of the length of the instrument! Despite these advances, there is still a traditional preference among both makers and players for wooden instruments, but this is based upon considerations other than completely acoustic ones.

When it comes to substitutes for resonance woods, the situation is rather different, since it is necessary to consider in detail the vibration behavior of the material. One early substitution was that of plywood for planks of spruce in instruments such as harpsichords. This has the advantage of inhibiting soundboard splitting, which is always a potential problem, but the vibrational properties of plywood are very different from those of solid timber with the grain direction lying in the plane of the board, and only rather cheap instruments have adopted it. On the other hand, there is no reason why plywood, with its superior mechanical strength, should not be used in other parts of large instruments such as pianos, though aesthetic considerations often dictate that solid timber is used if the component is visible.

There have been some experiments in constructing anisotropic synthetic materials to simulate wood. The approach that has been most successful is to embed unidirectional graphite fibers in a thin sheet matrix of epoxy resin and construct a sandwich plate with two such graphite-epoxy sheets surrounding a low-density core (Haines and Chang, 1975). Sheets constructed in this way provide a good match to the elastic moduli and damping properties of spruce. Their use for guitar top-plates, which are flat, is straightforward, while arched violin top-plates can be produced by hot pressing the flat sheets. Of course, subsequent adjustments to the thickness of the plates is not possible. These experiments have been broadly successful, but the material does not appear to have come into any general use.

Another anisotropic synthetic material is the synthetic cane sometimes used for clarinet and bassoon reeds. Once again it consists of strong fibers, not this time of graphite but of more glassy material, aligned in one direction and embedded in a resin matrix. Only limited scraping of the reed is possible, which restricts the acceptance of this material as a substitute for natural cane by professional players, but it does have the advantages of long life and the absence of drying out.

The other use of plastics and composites that should be mentioned is in musical instrument strings. The requirements here are rather different for plucked and bowed strings—for plucked strings the internal damping should be as low as possible to increase the sustain-time of the sound, while for bowed instruments rather high damping is required for the efficiency of the bowing process. Traditionally, strings were made from twisted dried animal intestines, which naturally have a high loss due to dry friction as well as molecular relaxation. While gut strings are still used in bowed instruments, they are often replaced by twisted multifilament nylon, which has similar properties. For plucked strings, monofilament nylon has the desired low loss properties. In both cases the string may be overwound with thin metal to increase its mass. One exception to the use of solid plucked strings that might be noted is in the case of the Japanese koto, in which the strings are made from twisted strands of silk, giving a mellow tone with a rather short decay time. Of course, in many applications, metal strings are used, and many modern plucked-string instruments also use nylon cores overspun with metal wire or tape.

A final related area in which new synthetic materials have been very successful is in drum heads. Here the traditional material is calfskin, which suffers, however, the disadvantages of nonuniformity, limited strength, and sensitivity to humidity. Modern plastic films of various compositions, particularly Mylar, have been very successful in replacing skin, since they are uniform, strong, and can be made in any desired thickness. Within broad limits, damping properties are not very important, since the membrane is thin and the major source of loss is by radiation.

22.5 Metals

Metals have had a very long historical connection with musical instruments, and the nature of that connection is related to the history of metallurgy itself. Until the last 200 years or so, the only readily available metals were those that were found native, such as gold, silver, and meteoritic iron, and those that could be easily smelted from their ores, such as lead, tin, zinc and copper. Indeed these metals have served humans for many thousands of years and in nearly all cultures, and have been used for the construction of weapons, domestic utensils, ornaments, and musical instruments.

To be useful in musical-instrument making, a metal must be able to be formed into the necessary complex shape, but must be hard enough to retain this in ordinary use, durable against corrosion, and attractive in appearance. It is surprising to what extent these requirements have dictated the metals from which instruments are made. The first requirement ruled out iron for most purposes, until the advent of modern alloys and modern technology, because it could not be completely melted with charcoal fires, and could only be worked by hammering at red heat. Its use was therefore restricted to drawn wires for plucked or hammered string instruments. Gold and silver were ideal, because they could be hammered to shape, drawn into tubes or wires, or cast into complex molds. The pure metals were rather too soft for many uses, but this could be overcome by adding small amounts of other metals to form an alloy. The main obstacle to wide use was rarity and expense. This left the extracted metals, and particularly their harder alloys, as the obvious materials of choice. Among these, bronze (lead-tin alloy), brass (lead-zinc), and pewter (tin-lead or tin-zinc) were prime choices for many purposes, being low enough in melting point to be readily cast, soft enough to work, but hard enough to be durable. A few examples will show how elementary physical properties determined the prime choice of materials in several cases.

22.5.1 Organ Pipes

While some organ pipes are made of wood, as we noted in Section 22.2.3, most are made from thin metal. The choice of an appropriate metal was dictated historically by the manner in which the pipes were made, and indeed this method is still followed by most major organ builders, and the same metal alloys are used. The first step is to melt the metal and then cast it into a thin uniform sheet, usually by flowing it onto a cloth-covered stone slab from a moving wooden box with no floor and a slight clearance gap between one side and the level of the others—an arrangement referred to as a doctor-blade. Clearly a low melting point is essential, and this is achieved with a tin-lead alloy, similar to pewter, which can have a melting point lower than that of either of its constituents. The eutectic composition, which has the lowest melting point, is 61.9% tin and melts at 183°C. This metal sheet is soft enough to be cut and rolled to form a pipe, the seam being soldered with similar alloy, and the finished pipe mouth can be cut by hand and adjusted in the voicing operation using a simple knife. The problem of softness of the alloy is not serious, since organ pipes are generally located high up out of harm's way. The preferred alloy is "spotted metal", which is on the tin-rich side of the eutectic, with a tin content of about 70%, the extra tin producing a characteristic crystalline-looking surface. High levels of lead are not favored, perhaps because it tends to oxidize in air and thus reduce the sharpness of the lip, but lead is less expensive than tin and dull-toned flute pipes are often made from thick sheet with more than the usual fraction of lead. There are generally compromises made with the larger pipes of the organ, partly because tin alloy is insufficiently strong and partly because it is rather expensive. These bass pipes are therefore generally made from a zinc alloy. Some modern organs may also use display pipes of other metals, such as nearly pure burnished tin or burnished copper, for reasons of appearance.

As discussed above, material has almost no effect upon the sound of an organ pipe, unless it somehow affects the geometry, as it does for a necessarily thick-walled wooden pipe. Despite this, it is hard to think of a material more suitable than traditional tin-lead pipe metal for its purpose, and for this reason it is still used nearly universally. Occasionally, however, organs have display pipes made from copper or brass, burnished to a fine polish. This is purely for visual effect, though the pipes form part of a normal sounding rank within the organ.

22.5.2 Flutes

The flute is one of the few instruments that has made the transition from wooden to metal construction, as a result of the work of Theobald Boehm some 150 years ago. Certainly a modern silver flute sounds very different from a reproduction of an early 19th-century flute, but the reason is that the dimensions of the instrument have been radically altered. A typical wooden flute, including Irish "folk" instruments currently being made, has a tapered bore, a cylindrical head-joint, and small finger holes closed by the finger tips. A modern metal flute, on the other hand, has a cylindrical bore, a tapered head-joint, and large tone holes closed by padded keys. Those who have played on the excellent Boehm-type wooden flutes made by Rudall Carte and Co. of London can vouch for the fact that their tone quality is quite similar to that of their modern silver counterparts.

Among the metal flutes, silver has long been the preferred material. Pure silver is far too soft, so the normal alloy is either traditional "900 fine" US coin silver (900/1000 parts of silver with the remainder mainly copper) or Sterling Silver, which is 925 fine. This silver is excellent to work; seamless tubing can be readily produced, and the tapered head-joint can be shaped by drawing a length of the tube onto a mandrel by forcing it through a hole in a lead block. Keys can also be cast or forged at a reasonable temperature and are appropriately strong. Only the axles of the keywork need to be made of steel, and 11 carat gold is used for the springs of the best models. The one disadvantage of silver is that it tarnishes to black silver sulfide, although this is not generally a severe problem.

Gold alloys possess all the advantages of silver, together with freedom from tarnish and a rich appearance. A large variety of compositions is possible, the final alloy generally being somewhere between 11 and 18 carat (24 carat is pure gold). Addition of copper gives a reddish appearance, while silver and other metals such as nickel move the color through yellow towards white. These alloys are quite hard but still ductile. Naturally, because of the cost of the gold itself, these instruments attract premium prices and, because they are generally made by the best craftsmen in the firm, are usually of very high quality. There is, however, nothing about the mechanical properties of gold that could not be duplicated by adjustment to the thickness of a silver tube, and in any case the acoustic evidence suggests that there is no detectable dependence of tone quality on wall material (Coltman 1971).

Some flutes have been made with still more exotic materials such as platinum or palladium, particularly for the head joint. Many of these are particularly fine instruments, but it has not been established that they owe any of this distinction to the metal from which they are made. In addition, metals of this particular family are not easy to work, and generally do not look as fine as gold or silver.

At the other end of the scale, student flutes are generally made from cupro-nickel alloy ("nickel silver" or "German silver") plated with silver or sometimes with chromium. These are relatively inexpensive, mass-produced instruments, and those from good makers often have excellent playing qualities. There is, however, perhaps an aesthetic objection to silver plate, and a practical objection that hand-finishing of the embouchure hole is not possible.

22.5.3 Brass Instruments

Closely related considerations lead to the selection of brass for most lip-blown instruments. Here the mechanical properties must be suitable for drawing into seamless tubes, bending to the intricate shapes of the finished instrument (a process often accomplished by first filling the tubes with water, which is then frozen to ice) and for spinning into flared bells (generally by using a rounded tool to press a sheet of metal against a pattern in a rapidly spinning lathe). The resulting instrument must also be strong enough to sustain minor bumps without bending. Brass and bronze are clear candidates from both historical and practical points of view, but brass is more easily worked and so is almost universally used. Its one minor disadvantage is that it oxidizes and suffers corrosion, so that it usually needs to be protected by a coat of lacquer or a plating of some more inert metal. More recently, various alloys, such as cupro-nickel, have become available, and are sometimes used instead of brass, but this is not standard practice. Silver and gold are obviously too expensive.

There is a strong feeling among some instrument makers that the material from which the instrument bell is made can have a significant tonal influence. Certainly, experiments by Wogram at PTB in Germany have shown that extreme material variations, such as replacing the metal bell by a soft-rubber molding of the same shape, do have a clear measurable effect on acoustic performance, but this represents a change in material properties by more than a factor 1000. If we restrict ourselves to metals, then it is fairly clear that a change in metal thickness will have a much greater effect than any change in choice of metal or alloy.

There are, of course, many alternative metals that could be used, but they are generally ruled out for practical reasons. Precious metals such as silver of gold, used in flute making, would simply make a large brass instrument too expensive, while stainless steel, an alloy of iron, nickel and chromium, although otherwise excellent, is simply too difficult to fabricate into the complex shapes required for brass instruments.

22.5.4 Bell Metals

Although small bells can be made from ceramic or glass, most bells are cast in metal. Any castable metal could be used, but bronze (nominally 80% copper and 20% tin) has always been the preferred material, because of its hardness and low internal damping. Technology for the casting of bronze articles was well established as long ago as 2000 BC. Bronze could be melted in a charcoal fire, and its pouring temperature of only 1100°C did not impose severe requirements on the mold material. There is, indeed, a close parallel between the manufacture of bells and cannon, and very similar materials and processes were historically used.

When we come to modern times, we still find these traditional bronze alloys used, and for good reason. Leaving aside fabrication criteria, the material for a bell should be hard and not too brittle, to resist clapper damage; it should have a high density and rather low elastic modulus, to give a low sound velocity; and should have low internal damping. The reason for desiring low sound velocity is that the bell can then be made rather smaller is size. This in turn reduces radiation damping, which is a major loss mechanism, and thus leads to a longer decay time for the tone.

Schad and Warlimont (1973) reported an extensive study of the effects of variations in bronze composition on these quantities. Bell bronze is normally 78–80% copper, 20–22% tin and has an impurity level of less than 2%, and the study investigated the effects of significant variation in the ratio of the principal components, together with addition of small amounts of antimony, silver, aluminum and phosphorus. Increase in percentage of tin greatly reduces the internal damping, but increases the sound velocity, while addition of lead up to 10% reduces the hardness, increases the damping, and marginally reduces the sound velocity. We conclude that the composition of standard bell bronze is close to optimal, but that close attention must be paid to reducing porosity, which can markedly increase internal damping. Interestingly, ancient bronze bells tended to have large amounts of lead in their composition, often in equal proportion to the tin fraction, while bronze for other purposes often had only 10–15% of tin, perhaps reflecting tin availability.

Because internal damping is a material property while radiation damping depends upon size and frequency, and thus only indirectly on material through the wave velocity $(E/\rho)^{1/2}$ which determines the size of the bell, it is important to assess the relative importance of these two damping mechanisms. To do this, Schad and Warlimont (1973) reduced the air pressure to zero, eliminating air damping, and then measured the change in decay rate of the bell partials as air was readmitted. The found that, for a typical large bell, the internal damping is nearly the same for all modes. When air is admitted up to atmospheric pressure, radiation damping and viscous damping assume their normal values. The damping of the sub-octave or hum tone increases by about a factor 1.5, that of the prime and tierce by about a factor 2, that of higher partials by larger factors, the air damping increasing approximately linearly with frequency. Internal damping is thus an important, but not dominant, determinant of the sound of a bell.

Cast iron and steel have also been used for large church and temple bells, because they are considerably less expensive than bronze. Since, however, the speed of sound (and consequently also of flexural waves) in iron is about 30% greater than in bronze ($c_{\text{cast iron}} \approx 4500$ m/s, $c_{\text{bronze}} \approx 3300$ m/s), this means that iron bells must be correspondingly larger in size than bronze bells if the traditional bell profile is to be retained. This in turn means more efficient sound radiation and greater radiation damping, with consequently reduced ring time. Bells made from aluminum alloys suffer from the same problem to an even greater degree ($c_{\text{aluminum}} \approx 5000$ m/s) but aluminum is now being used for bass handbells in order to enhance the radiation of low notes (Rossing et al. 1995). Aluminum is easy to cast and machine, and great hardness is not needed in handbells having relatively soft clappers.

22.6 Conclusion

This discussion has, we hope, shown the close connection that has always existed between the aesthetic ideal of a musical instrument and the materials from which it is made, together with the related connection between the necessary fabrication processes and the materials that make these possible. In some cases the choice of material has clear acoustic consequences that cannot be modified by changes in the hidden dimensions of the instrument, such as plate profile, while in other cases the choice of material has almost no acoustic consequences whatever, provided only that the dimensions and surface finishes can be maintained unchanged. Developments in materials and fabrication processes continually open up new possibilities, but these must be balanced against both tradition and aesthetic considerations.

References

Backus, J. (1964). Effect of wall material on the steady-state tone quality of woodwind instruments, *J. Acoust. Soc. Am.* **36**, 1881–1887.

Barducci, I., and Pasqualini, G. (1948) Misura dell'attrito interno e delle constanti elastiche del legno. *Nuovo Cimento* **5**, 416–466. English translation in C.M. Hutchins (ed), "Musical Acoustics, Part I," pp. 410–423. Dowden, Hutchinson and Ross, Stroudsburg, Pa. 1975.

Barlow, C.Y., and Woodhouse, J. (1988). Microsopy of wood finishes. *Catgut Acoust. Soc. J.* **1**(4), 2–9. Reprinted in Hutchins and Benade (1997).

Barlow, C.Y., and Woodhouse, J. (1989). Of old wood and varnish: Peering into the can of worms. *Catgut Acoust. Soc. J.* **1**(1). Reprinted in Hutchins and Benade (1997).

Barlow, C.Y., and Woodhouse, J. (1993). Microstructures and properties of bent spruce. *Proc. Stockholm Music Acoust. Conf. SMAC-93* Royal Swedish Academy of Music, Stockholm, No.79, pp. 346–350.

Bonamini, G., Chiesa, V., and Uzielle, L. (1991). Anatomical features and anisotropy in spruce wood with indented rings. *J. Catgut Acoust. Soc.* **1**(8), 12–16.

Bucur, V. (1995). "Acoustics of Wood." CRC Press, New York.

Coltman, J.W. (1971). Effects of material on flute tone quality. *J. Acoust. Soc. Am.* **49**, 520–523.

Dunlop, J.I. (1978). Damping loss in wood at mid kilohertz frequencies. *Wood Sci. Tech.* **12**, 49–62.

Fukada, E. (1950). The vibrational properties of wood. I. *J. Phys. Soc. Japan* **5**, 321–327. Reprinted in Hutchins (1975).

Fukada, E. (1951). The vibrational properties of wood. II. *J. Phys. Soc. Japan* **6**, 417–421. Reprinted in Hutchins (1975).

Haines, D.W. (1979). On musical instrument wood. *Catgut Acoust. Soc. Newsletter* **31**, 23–32.

Haines, D.W. (1980). On musical instrument wood. Part II. Surface finishes, plywood, light and water exposure. *Catgut Acoust. Soc. Newsletter* **33**, 19–23.

Haines, D.W., and Chang, N. (1975). Application of graphite composites in musical instruments. *Catgut Acoust. Soc. Newsletter* **23**, 13–15.

Hutchins, C.M. (editor) (1975)."Musical Acoustics, Part I: Violin Family Components." Benchmark Papers in Acoustics, vol. 5. Dowden, Hutchinson and Ross, Stroudsburg Pa.

Hutchins, C.M., and Benade, V. (editors) (1997). "Research Papers in Violin Acoustics 1975–1993," Vol. 2, pp. 765–905. Acoustical Society of America, Woodbury NY.

Hutchins, M.A. (1989) Effects on spruce test strips of four-year application on four different sealerws plus oil varnish. *Catgut Acoust. Soc. J.* **1**, 11–16. Reprinted in Hutchins and Benade (1996).

Kahl, E., and Woodhouse, J. (1994). The influence of cell geometry on the elasticity of softwood. *J. Mater. Sci.* **29**, 1250–1259.

Kittel, C. (1966). "Introduction to Solid State Physics," pp. 111–114. Wiley, New York.

McIntyre, M.E. and Woodhouse, J. (1988). On measuring the elastic and damping constants of orthotropic sheet materials. *Acta Metall.* **36**, 1397–1416.

Miller, D.C. (1909). The influence of the material of wind-instruments on the tone quality. *Science* **29**, 161–171.

Pollard, H.F. (1977). "Sound Waves in Solids." Pion, London.

Robinson, T. (1973). "The Amateur Wind Instrument Maker." John Murray, London.

Rossing, T.D. (editor) (1984). "Acoustics of Bells." Van Nostrand Reinhold, Stroudsburg, Pennsylvania.

Rossing, T.D., Gangadharan, D., Mansell, E.R. and Malta, J.H. (1995). Bass handbells of aluminum, *MRS Bulletin* **20**(3), 40–43.

Schad, C.R. and Warlimont, H. (1973). Akustische Untersuchungen zum Einfluss des Werkstoffs auf der Klang von Glocken. *Acustica* **29**, 1–14. English translation: Acoustical investigation of the influence of the material on the sound of bells, in Rossing (1984), pp.266–286.

Schelleng, J.C. (1963). The violin as a circuit. *J. Acoust, Soc. Am.* **35**, 326–338.

Schelleng, J.C. (1968). Acoustical effects of violin varnish. *J. Acoust. Soc. Am.* **44**, 1175–1183.

Sobue, N. (1995). Effect of continuous vibrations on dynamic viscoelasticity of wood. *Proc. Int. Symp. Musical Acoust.*, Dourdan. pp. 326–334. IRCAM, Paris.

Wogram, K. Private communication.

Zener, C. (1948). "Elasticity and Anelasticity of Metals." Univ. of Chicago Press, Chicago.

Name Index

Subject Index